T5-AGX-446

Developments in Precambrian Geology 9

PRECAMBRIAN GEOLOGY OF THE USSR

DEVELOPMENTS IN PRECAMBRIAN GEOLOGY
Advisory Editor B.F. Windley

Further titles in this series

1. B.F. WINDLEY and S.M. NAQVI (Editors)
   Archaean Geochemistry
2. D.R. HUNTER (Editor)
   Precambrian of the Southern Hemisphere
3. K.C. CONDIE
   Archean Greenstone Belts
4. A. KRÖNER (Editor)
   Precambrian Plate Tectonics
5. Y.P. MEL'NIK
   Precambrian Banded Iron-formations. Physicochemical Conditions of Formation
6. A.F. TRENDALL and R.C. MORRIS (Editors)
   Iron-Formation: Facts and Problems
7. B. NAGY, R. WEBER, J.C. GUERRERO and M. SCHIDLOWSKI (Editors)
   Developments and Interactions of the Precambrian Atmosphere, Lithosphere and Biosphere
8. S.M. NAQVI (Editor)
   Precambrian Continental Crust and its Economic Resources

DEVELOPMENTS IN PRECAMBRIAN GEOLOGY 9

PRECAMBRIAN GEOLOGY OF THE USSR

Edited by

D.V. RUNDQVIST

Institute of Physics of the Earth, Russian Academy of Sciences, B. Gruzinskaya 10, 123810 Moscow, Russia

and

F.P. MITROFANOV

Institute of Precambrian Geology and Geochronology, Academy of Sciences of Russia, nab. Makarova 2, Leningrad 199164, Russia

Translated by C. Gillen

The University of Edinburgh

ELSEVIER, Amsterdam — London — New York — Tokyo 1993

ELSEVIER SCIENCE PUBLISHERS B.V.
Sara Burgerhartstraat 25
P.O. Box 211, 1000 AE Amsterdam, The Netherlands

Library of Congress Cataloging-in-Publication Data

Dokembriĭskaíà geologiíà SSSR. English.
Precambrian geology of the USSR / edited by D.V. Rundqvist and F.P. Mitrofanov.
p. cm. -- (Developments in Precambrian geology ; 9)
Translation of: Dokembriĭskaíà geologiíà SSSR.
Includes bibliographical references and index.
ISBN 0-444-89380-6
1. Geology, Stratigraphic--Precambrian. 2. Geology--Soviet Union. I. Rundkvist, D. V. (Dmitriĭ Vasil'evich) II. Mitrofanov, F. P. III. Title. IV. Series.
QE653.D643513 1992
551.7'1'0947--dc20 92-8594
CIP

Precambrian Geology of the USSR was first published in Russian in 1988 as "Dokembriyskaya geologiya SSSR" by Nauka Publishers, Leningrad. ISBN 5-02-024367-1,440 pp.

ISBN: 0-444-893806

This book is printed on acid-free paper.

Printed in The Netherlands

FOREWORD

Precambrian Geology of the USSR is an attempt to draw together and generalise new geological, geochronological, petrological and geophysical material for the two fundamental continental geostructures within the Soviet Union — ancient cratons and Phanerozoic fold belts. This material reflects the results of research by a large team of geologists in the Institute of Precambrian Geology and Geochronlogy of the USSR Academy of Sciences (in Leningrad) who have been studying Precambrian regions for over 30 years. The book also takes account of the achievements of other scientific and industrial organisations in this area. It has been written by a team of authors who are the leading specialists in problems of the Precambrian and in the geology of individual Precambrian regions.

Material for the major regions — tectonic provinces or "geoblocks" — is basically presented according to a unified pattern. Since the book was written by a large team, individual chapters are to a certain extent independent essays in which the understanding of scientific questions is given from the author's point of view.

The book gives the complete Precambrian history of each region, although the early Precambrian is treated more fully than the late Precambrian, which for many regions is presented in outline only. It is not possible in a single book to describe all aspects of early Precambrian geology for each region equally comprehensively. Therefore, certain individual aspects are treated as fully as possible for one or two regions and in a more condensed form for the others. For example, the lithological composition and petrochemistry of lower Archaean rocks in high grade metamorphic complexes is given in detail for the Aldan Shield; deformation sequences and their relation to migmatites and granites are given in detail for the Belomorides and Svecofennides. The exhaustive reference lists for each region will allow the interested reader to obtain further information.

Problems in the regional geology of the Precambrian of such major regions as ancient cratons require many questions to be resolved in relation to the establishment of the time and space correlation of tectonic structures at various stages in crustal evolution; the construction of vertical and horizontal crustal sections; and an explanation of the commonest patterns in the development of the Precambrian lithosphere and in the evolution of processes taking place in the lithosphere. These questions are far from being finally resolved, and their discussion in the book reveals only the trends and directions of research in the Institute of Precambrian Geology and Geochronology. Attempts are being made in this research to combine the results of analysing recent data on Precambrian regions into a unified system. Some problems in Precambrian geology are: the evolution of sedimentation, defining the conditions existing in upper sections of the Precambrian crust in different geological periods; features of metamorphic processes of different ages at various crustal depths, reflecting the thermodynamic conditions in the Precambrian crust at various stages in its growth; the conditions pertaining to the appearance of igneous melts and the origin of magmatic bodies, which characterise P–T parameters in the crust and mantle; the definition of metallogenic cycles and ore-bearing structures — these relate to a number of topical problems which are reflected in several sections in the book and in the conclusions.

The work was designed by G.P. Pleskach and A.A. Mikhaylova. Enormous help was given in the preparation of the manuscript by G.A. Buyko and E.N. Turunova. All the tables on features of mineral deposits and ore shows were prepared by Dr. V.A. Gorelov. The authors express their sincere gratitude to them. Dr. A.N. Kazakov and Dr. G.M. Belyayev are thanked for their helpful and constructive suggestions on the manuscript.

D.V. RUNDQVIST and F.P. MITROFANOV

Authors:

V.Ya. Khiltova, A.B. Vrevsky, S.B. Lobach-Zhuchenko, K.A. Shurkin, V.A. Glebovitsky, A.K. Zapolnov, A.N. Berkovsky, V.B. Dagelaysky, D.V. Rundqvist, Yu.M. Sokolov, V.L. Dook, G.M. Drugova, R.I. Milkevich, S.I. Turchenko, N.I. Moskovchenko, L.Ye. Shustova, F.P. Mitrofanov, I.K. Kozakov, R.Z. Levkovsky, A.V. Sochava, V.A. Gorelov.

Referees:

G.M. Belyayev, A.N. Kazakov

CONTENTS

Foreword V

Introduction 1
by D.V. Rundqvist and V. Ya. Khiltova
References 7

Part I. Precambrian of the East European craton

Introduction. History of research and tectonic structure 11
by V. Ya. Khiltova and A.N. Berkovsky

Chapter 1. The eastern Baltic Shield 25
1. The Kola province (by A.B. Vrevsky) 25
2. The Karelian province (by S.B. Lobach-Zhuchenko) 70
3. The Belomoride belt (by K.A. Shurkin) 90
4. The Ladoga belt (by V.A. Glebovitsky) 105

Chapter 2. The Ukrainian Shield 125
by V.B. Dagelaysky
1. The West Ukrainian region: the Volyn-Podolsk province 133
2. Central Ukrainian province 141
3. The Azov Archaean granulite-greenstone terrain 148

Chapter 3. The Russian Platform 159
by A.K. Zapolnov
1. Crystalline basement 159
2. Precambrian of the platform cover 181

Chapter 4. Major features of Precambrian metallogeny 199
by D.V. Rundqvist and Yu.M. Sokolov

References 213

Part II. Precambrian of the Siberian Craton

Introduction. History of research and tectonic structure 235
by V. Ya. Khiltova and L.Ye. Shustova

Chapter 1. The Anabar Shield 247
by S.I. Turchenko

Chapter 2. The Aldan Shield 265
1. The Aldan granite-gneiss terrain (by V.L. Dook, R.I. Milkevich and G.M. Drugova) 269
2. The Olyokma (by G.M. Drugova, V.L. Dook and A.V. Sochava) 321
3. The Batomga granite-greenstone terrain (by G.M. Drugova, V.L. Dook and A.V. Sochava) 339

Chapter 3. Dzhugdzhur-Stanovoy province 345
by N.I. Moskovchenko

Chapter 4. Basement highs around craton margins 365
1. The Kan–Pre-Sayan terrain (by V.Ya. Khiltova) 365
2. The Baikal-Patom Highlands (by S.I. Turchenko and Yu.M. Sokolov) 388

Chapter 5. Precambrian of the cover 399
by A.K. Zapolnov

Chapter 6. Major features of Precambrian metallogeny 415
by D.V. Rundqvist and Yu.M. Sokolov

References 431

Part III. Precambrian in younger fold belts 443
by F.P. Mitrofanov and I.K. Kozakov

1. Urals-Mongolia belt 445
2. Mongolia-Okhotsk province 473
3. Pacific belt 475
4. Mediterranean belt 481
5. Crustal evolution 488

References 495

Conclusions

Crustal evolution in the Precambrian 501
by D.V. Rundqvist, V.Ya. Khiltova, A.K. Zapolnov, R.Z. Levkovsky and A.V. Sochava

Subject index 521

Introduction

The Precambrian developed in the USSR presents an ideal opportunity to study it in all its structural forms. Different vertical levels of the early Precambrian crust are exposed in the continental basement. The upper Precambrian plays an important role in the structures of the sedimentary cover. Precambrian rocks of various ages are exposed in Late Proterozoic and Phanerozoic fold belts.

Ancient cratons, in which the Precambrian is the main constituent, and Phanerozoic fold belts, in which the Precambrian is of minor importance, occupy approximately equal areas within the Soviet Union (Fig. I-1). The boundaries between them separate late Precambrian (<1650 Ma) sediments, formed on ancient continental platforms in a stable regime, from the geosynclinal sediments in younger fold belts. The basement to cratons is exposed in regions amounting to a little over 12% of the area of each craton. The East European and Siberian cratons are almost mirror images of each other in terms of the distribution of areas of exposure. In the East European craton, the northern part of the basement, the Baltic Shield, is best exposed, while the southern part, the Ukrainian Shield, is much less so. In the Siberian craton, on the other hand, areas of exposure are situated mainly in the south — in the Aldan Shield, the Stanovoy region and the marginal uplifts in the south-west of the craton, while in the north the Anabar Shield accounts for only a small area.

There are many points of view concerning the nature of the boundaries between exposed shields and cratonic areas. We take them to be historical-geological boundaries which had already been initiated in the Precambrian.

Huge areas of Precambrian rocks of the cratons are hidden beneath a sedimentary cover. Investigating them by geological and geophysical methods is of great significance in defining the space–time boundaries of the major categories of early Precambrian tectonic structures.

The basement to ancient cratons is subdivided by persistent lineaments into a system of blocks. Major crustal segments, or provinces (geoblocks), correspond to tectonic regions which developed independently and are

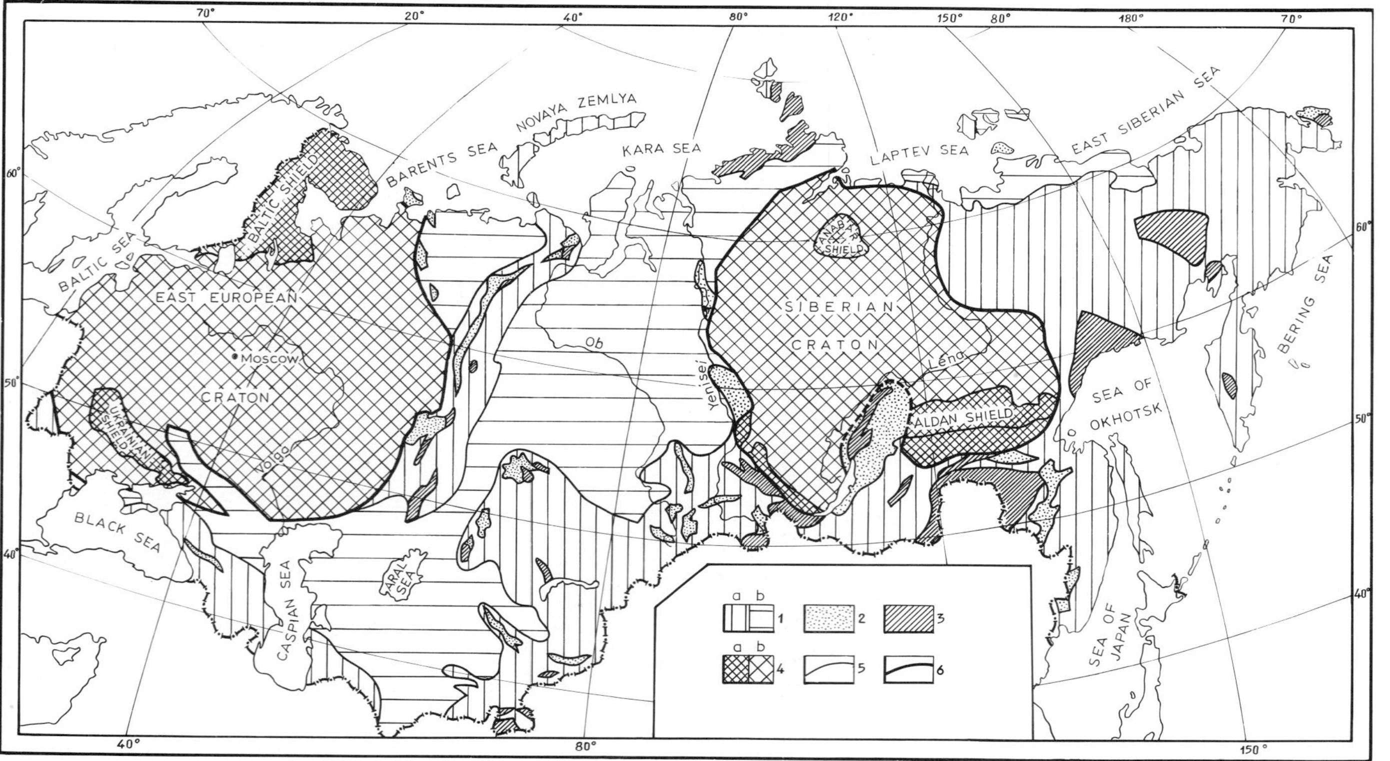

Fig. I-1. Tectonic regions of the USSR. *1* = younger fold belts: *a* = exposed, *b* = beneath platform cover; *2* = Upper Precambrian in younger fold belts; *3* = Lower (Upper in part) Precambrian in younger fold belts; *4* = Precambrian of cratons: *a* = exposed (shields), *b* = buried basement; *5* = geological boundaries; *6* = craton margins.

thereby distinguished from one another. The main difference is in the timing of the final processes: Archaean for example in the Kola and Karelian provinces of the Baltic Shield; Early Proterozoic in the West Ukrainian province of the Ukrainian Shield, etc.

The block structure of the crust, identified during the analysis of geological and geophysical data, probably reflects the earliest subdivision of the lithosphere. This is indicated by important statistical relations discovered between the composition of Archaean igneous rocks and a number of geophysical characteristics of the modern lithosphere, including the gravity potential and some parameters of the geomagnetic field (Abramovich and Klushin, 1978). Although the exposed Precambrian, especially the earlier, in younger fold belts amounts to a tiny percentage, it is nevertheless documented in belts which completed folding at different times. The Precambrian is exposed in two types of tectonic structure: median massifs and anticlinal elevations.

The study of the Precambrian in the territory of the USSR began over 100 years ago. From the 1880s to the 1940s, research was mainly carried out in crystalline shield areas. Investigations in this period by N.I. Inostrantsev, V.M. Timofeyev, A.A. Polkanov, V.A. Obruchev, M.A. Usov and others were mainly directed at solving problems in practical geology and basic methods were evolved for the successful scientific study of various aspects of Precambrian geology. The idea emerged in this period that Precambrian geology possessed certain well-expressed specific features. The reason for this was a lack of study of the late Precambrian, while the early "crystalline" Precambrian could confidently be contrasted with the Phanerozoic.

Since the 1950s, the techniques of isotope geochronology began to be applied widely by A.P. Vinogradov, E.K. Gerling, A.I. Tugarinov and N.P. Semenenko. This resulted in a qualitatively new understanding of geological processes in the Precambrian. One of the most important outcomes was the radiometric dating of major magmatic episodes, mainly granitic, and metamorphic episodes, which in reality were relatively brief intervals of intense activity, often defined as tectonometamorphic or tectonomagmatic cycles or epochs. Data from all Precambrian regions showed that they subdivided Precambrian geological history into major segments-stages of continental crustal development lasting for 800–1000 Ma (Neyelov, 1965; Kratz et al., 1981). Internal processes affected all crustal levels. Up to the present, the application of tectono-metamorphic cycles or epochs is the only useful basis for the general subdivision and interregional correlation of Precambrian rocks. They form the basis of the general geological-geochronological scale for the early Precambrian (Keller et al., 1977).

Each stage in the development of the early Precambrian crust — Early Archaean, Archaean and Early Proterozoic — is reflected in structural-lithological terrains. The geological and geochronological boundaries of these terrains are defined by periods of intense internal processes, which are dated by radiometric methods. Structural-lithological terrains include volcanogenic-sedimentary rocks and intrusive rocks, formed in the time interval between these boundaries. Since these terrains evolved over very long periods, their supracrustal components could contain formations which accumulated under varied tectonic and palaeogeographic conditions — either relatively quiet or active.

The early Precambrian terrains formed under varying P–T conditions. This book places great significance on sillimanite-andalusite and sillimanite-kyanite facies series when characterising metamorphic conditions. It is assumed that these series formed in response to different thermal gradients: high or low. High gradient metamorphism usually corresponds to processes that occurred at low P (5–6 kbar for T = 700–800°C), while low gradient metamorphism occurred at high P (7–13 kbar for T = 700–800°C).

The study of the composition of Precambrian complexes — the division into primary sedimentary and primary volcanic, and the determination of which magmatic series volcanic rocks belong to — is based on the use of various petrochemical diagrams. For sedimentary-volcanogenic rocks, the Neyelov et al. (1979), Predovsky (1979) and Golovenok (1977) diagrams were used; for igneous rock determinations, diagrams developed by the USSR Petrographic Committee were used (Andreyeva et al., 1980); and for distinguishing magmatic series, Na_2O+K_2O–Al_2O_3–MgO and others, triangular diagrams were used. The Neyelov petrochemical diagram is used throughout for analysing metamorphic rocks of the Aldan Shield; being a classification diagram for sedimentary and volcanic rocks, it provides an opportunity to compare simultaneously the compositions of the same rocks with different parageneses and to define the petrochemical features of the parageneses as a whole.

In Precambrian regions which have evolved in a cyclical fashion, the structural development of individual complexes was reflected in numerous phases of folding, designated $F_1, F_{2,\ldots n}$, during deformational cycles $D_1, D_{2,\ldots n}$.

Figure I-2 illustrates the early Precambrian structural-lithological terrains in the main regions of the USSR which are described in this book. Periods of intense internal processes are shown in Fig. I-2 as fold episodes, some of which — those that appeared over a whole region and at all crustal levels — correspond to major boundaries (solid lines), while others that appeared locally correspond to local boundaries (broken lines).

Geological-geochronological boundaries are defined using the 1978 Precambrian geochronological scale of the USSR (Keller et al., 1977; Mitrofanov, 1979), in which the Precambrian is divided into the Archaean, lower Proterozoic and upper Proterozoic (Riphean+Vendian), the main boundaries being 2600 ± 100 Ma, 1650 ± 50 Ma and 650 ± 25 Ma, and the internal

General stratigraphic scale of the Precambrian (Ma)	East European Craton: Baltic Shield (provinces): Kola	Karelian	White Sea	Svecofennian	Ukrainian Shield (Dnieper)	Siberian Craton: Aldan Shield (provinces): Olyokma	Aldan	Stanovoy	Anabar Shield	Marginal elevations
Proterozoic, Upper: V 570; R_4 650 ± 25; R_3 1000 ± 30; R_2 1350 ± 50; R_1 1650 ± 50	P L A T F O R M					S E Q U E N C E S				
				SVECOFENNIAN						
Proterozoic, Lower, Karelian: 1950 ± 50	K A R E L I A N				SAKSAGANIAN	UDOKANIAN	UNGRIAN	DZHELTULAK	KARELIAN	S A Y A N
2300	*Pechenga* *(Imandra-Varzuga)*	*Jatulian* Seletskian *Sumian*	*Jatulian (?)*	*Jatulian*	*Krivoy Rog*	*Udokanian*	*Ungrian*	*Dzheltulak* LATE STANOVOY *Gilyuy* *Odolgo*		*Urikian*
2600 ± 100	R E B O L I A N					OLONDINIAN	EARLY STANOVOY		ANABARIAN	O N O T I A N
Archaean, Upper, Lopian	*Tundra*	*Lopian*	*Tundra (?)*	*Lopian*	DNIEPROVIAN	*Olondinian*		*Early Stanovoy* *Ilikanian* *Kholodnikanian*		*Onotian*
3000 ± 100	S A A M I A N					A L D A N I A N				
Archaean, Lower, Belomorian	*Kola*	*Tonalite*	*Belomorian*		*Konka–Verkhovtsev*	*Olyokma-Chuginian*	*Timpton*	*Elgakanian* *Larbinian*		LATE SHARYZHALGAY
3500 ± 100	S A A M I A N				A U L I A N			URKANIAN	ANABARIAN	*Sharyzhalgay*
Katarchaean	*Tonalite basement*		*Keretian (?)* *(Tonalite basement)*		*Aulian*	*Kurultinian*	*Kurumkanian*	*Urkanian* *Zverevian*	*Anabarian*	*Granite-gneisses*

Fig. I-2. Correlation of structural-compositional complexes and tectonic epochs in the early Precambrian of the USSR.

boundaries 2300 Ma, 1350 ± 50 Ma and 1000 ± 50 Ma. This scale differs in certain respects from the geochronological scales of other continents. In particular, the two-fold instead of three-fold division of the Proterozoic, and the lesser signficance compared to elsewhere of the 1000 ± 100 Ma boundary. These differences relate significantly to the fact that the 1978 geochronological scale of the USSR was established on the basis of material from the Precambrian of ancient cratons, in the geological history of which the 1700 ± 100 Ma date represents the boundary between the latest events in the mobile development of these regions and their aulacogen-pericratonic stage, which in Vendian to Palaeozoic times merged with the stage of formation of the platform cover succession. Geological and geochronological material for the early Precambrian, so far accumulated, has assisted the authors in a number of cases to subdivide major geochronological units into smaller ones. For example, three subdivisions are recognised, as in the Archaean of the Siberian craton: $AR_1^1 > 3400$ Ma, $AR_1^2 > 3000$ Ma and $AR_2^1 > 2600 \pm 100$ Ma.

In determining the age of structural-lithological terrains, the timing of metamorphism and the age of igneous rocks, account has been taken of all dates using different radiometric methods: U–Pb and Rb–Sr isochrons, Pb–Pb zircon ages, K–Ar and Sm–Nd, the latter at present used only to a very small extent. Preference was given to U–Pb, Rb–Sr and Sm–Nd methods. Results from the Pb–Pb thermo-emission method (Zr evaporation technique) were regarded as reflecting the minimum age of an event, although in many cases the figures obtained by this technique were close to or even coincided with U–Pb zircon isochron dates. Dates obtained from the K–Ar method were taken into consideration when evaluating the time of final crust-forming processes, including vertical uplift of blocks, and the age of rocks formed during platform-type development.

It has now become evident that intense internal processes of the same type which mark the completion of one stage in crustal evolution can be of different ages in different regions. In the East European craton for example, this difference is of the order of 300–400 Ma (Kratz et al., 1981; Khiltova et al., 1986), rejuvenation having occurred from south to north.

Each stage in crustal evolution, embracing an enormous slice of time, is characterised by the dominant development of a distinctive type of mobile belt with general features of supracrustal and igneous rocks. It is now known that mobile belts of the same type, for example Archaean greenstone belts, did not develop simultaneously within one geoblock, but successively (the Karelian granite-greenstone belt, for example), with the difference in time between the final processes in the extreme members of this sequence being not less than 250 Ma, thus each stage in crustal evolution is reflected not in the history of one belt, but in the history of an entire tectonic domain.

The early Precambrian basement of platforms is not uniform and has been regionally subdivided differently by various authors. The regional tectonic scheme used in this book is set out below. Here we simply emphasise that for well-studied exposed regions, the crustal blocks identified by geological and geophysical means differ in the time and mechanism of formation of crustal structures — tectonic or mobile domains, which include mobile belts (e.g. greenstone belts). The latter structures are smaller in scale than domains.

Early Precambrian internal processes formed various types of Precambrian mobile belts, for which Kratz et al. (1981) proposed a two-fold subdivision, into those with or without geosynclinal precursors. In the first category, belts with early basin phases, periods of intense internal processes immediately followed the formation of "geosynclinal-type" volcanogenic-sedimentary rock sequences in troughs. Belts without geosynclinal phases developed from stable structures, and internal processes transformed their pre-existing "proto-platform" (sub-platform) cover and basement complex, which formed long before the latter processes.

Apart from the two best expressed types of mobile belt, others can be recognised. In particular, belts originated on stabilized regions, where rift-type volcanogenic-sedimentary assemblages ("geosynclinal" in composition) formed, distinct from belts without geosynclinal precursor stages. Granitic magmatism in these belts was weakly expressed or entirely absent, while basic magmatism often has an inverse trend, from tholeiites to late komatiites (the Imandra-Varzuga belt on the Kola Peninsula, for example).

In studying the structural and evolutionary features of Precambrian mobile belts, fundamental importance has been attached to explaining the most general and characteristic features in each type. However, it has been established for belts of the same type and even the same age, e.g.

greenstone belts in the Baltic and Aldan Shields, in which internal processes were complete by 2600 ± 100 Ma ago, that there are differences in the composition of supracrustal complexes, and in particular features of their metamorphism and tectonism. In order to understand the reasons for these differences, it was necessary to study not only the greenstone belts proper (the suprastructure) but also the infrastructure, the basement complexes on which they formed. Studying the composition, tectonic complexity and internal processes in the suprastructure and infrastructure allowed us to identify granite-greenstone and granulite-greenstone terrains for the Archaean and granite-schist and granulite-schist terrains for the Early Proterozoic (Khiltova et al., 1986). Tectonic regions form part of larger structures, such as tectonic domains. The boundaries of tectonic domains observed at the present time only occasionally correspond to the initial, earliest palaeotectonic boundaries. In the majority of cases, various types of mobile belts developed along these boundaries or suture zones.

A large body of factual material on Precambrian geology will encourage researchers to construct models of the structure and evolution of the crust in different regions in a lateral and vertical sense. Such models have been made possible thanks to radiometric dating of the main geological boundaries and the erection of structural and metamorphic scales for different structural-lithological zones at different crustal levels in the various regions. Petrochemical and geochemical data for basic and ultrabasic rocks make it possible to determine the composition of mantle material and the conditions under which melts are generated. These achievements in the study of Precambrian geology have allowed us to undertake the first attempts at explaining a mechanism for the formation and time of appearance of a layered lithosphere and to identify general patterns of change in crustal structure and evolutionary processes.

The widespread acceptance of plate tectonics ideas in Phanerozoic geology presents Precambrian researchers with the question of the possibility of using these ideas in interpreting geological facts from Precambrian regions. There is no doubt that individual points of this hypothesis are quite applicable in explaining geological relationships. However, in the early Precambrian plate tectonics most likely operated in different ways and at different scales compared to the Phanerozoic (Musatov et al., 1983).

REFERENCES

(In Russian unless otherwise stated)

Abramovich, I.I. and Klushin, I.G., 1978. The petrochemistry and deep structure of the Earth. Leningrad, Nedra, 375 pp.

Andreyeva, E.D., Bogatokova, O.A. and Borodayevskaya, M.B., 1980. Classification and nomeclature of igneous rocks. Moscow, Nedra, 160 p.

Golovenok, V.K., 1977. High-alumina formations in the Precambrian. Leningrad, Nedra, 268 p.

Keller, B.M., Kratz, K.O., Mitrofanov, F.P., Semikhatov, M.A., Sokolov, B.S., Sokolov, V.A. and Shurkin, K.A., 1977. Achievements in developing a general stratigraphic scale for the Precambrian of the USSR. Izvestiya Akad. Nauk SSSR, Ser. Geol., 11: 16–21.

Khiltova, V.Ya., Lobach-Zhuchenko, S.B., Mitrofanov, F.P., Dook, V.L. and Abramovich, I.I., 1986. Structure and evolution of the Archaean lithosphere. In: Evolution of the Precambrian lithosphere. Leningrad, Nauka.

Kratz, K.O. et al., 1981. States and types of evolution of the Precambrian crust of ancient shields. Leningrad, Nauka.

Mitrofanov, F.P. (Editor), 1979. Subdivision of the Precambrian in the USSR. Proc. 5th session of the Scientific Council on Precambrian geology, Ufa, 1977, Interdepart. Strat. Committee, USSR Acad. Sci. Leningrad, Nauka, 164 pp.

Mitrofanov, F.P., Khiltova, V.Ya. and Vrevsky, A.B., 1986. Composition, structure and processes in the Archaean lithosphere. In: Early Precambrian tectonics and metallogeny. Moscow, Nauka, pp. 135–144.

Musatov, D.I., Fedorovsky, V.S. and Mezhelovsky, N.V., 1983. Archaean tectonic regimes and geodynamics. Review of the All-Union Inst. for Economic Minerals Res. and Geol. Surveying. General and regional geology and geological mapping. 42 pp.

Neyelov, A.N., 1965. Precambrian palaeotectonics of the Siberian Platform and evolutionary processes in Precambrian mobile regions. In: Precambrian Geology. 23rd session Int. Geol. Congress, Leningrad, pp. 41–51.

Neyelov, A.N. and Milkevich, R.I., 1979. Petrochemistry of metamorphic complexes in the south of East Siberia. Leningrad, Nauka, 311 pp.

Predovsky, A.A., 1979. Geochemical reconstruction of the primary composition of Precambrian metamorphosed volcano-sedimentary formations. Apatity, 116 pp.

Yanshin, A.L. (Editor), 1964. Tectonic map of Eurasia, 1 : 5,000,000 scale. Moscow, Geodesic Commission.

Part I. Precambrian of the East European craton

Introduction

HISTORY OF RESEARCH AND TECTONIC STRUCTURE

The East European craton embraces almost the entire territory of the European part of the USSR. The broad outlines of the craton in general terms had been formed by the beginning of the Riphean and are defined by younger fold belts which occur around the margins (Nalivkin, 1974).

The largest major structural units of the East European craton are the Baltic Shield, the eastern part of which crops out in the USSR, the Ukrainian Shield and the Russian Platform. Precambrian formations have been studied in the shield areas and in many parts of the platform area where they have been found by drilling beneath sedimentary cover formations (Fig. I-3). The general picture of the structure of the crystalline basement of the platform has been put together mainly on the basis of geological and geophysical data.

* * *

Scientific views on the structure of the basement of the East European craton have undergone substantial evolution during a century of investigation. Karpinsky (1887) was the first to show that the basement of the platform was composed entirely of Precambrian rocks. Arkhangelsky et al. (1937) developed these ideas on the basis of the first geophysical research; he worked out a method for interpreting magnetic and gravity anomalies and went on to produce a model for the internal structure of the basement. Later, Shatsky (1946), using magnetic maps and geological field data from the shields, devised a scheme for the basement divided into areas of Archaean, Karelian and Baikalian folding.

By the mid 1950s, regional land-based and gravity surveys had been completed in the European part of the USSR and many surveys had been carried out using seismic and electrical methods. Combined interpretation of geophysical data together with the results of stratigraphic test drilling allowed Fotiadi (1958) to compile the first basement relief map and a model of the internal structure. Slightly later, thanks to a greatly expanded drilling programme, an intensive study of the petrography of the basement rocks of the Russian Platform was begun. The first summaries appeared for the basement of the Volga-Urals region, the Kursk-Voronezh region, Byelorussia, the Baltic republics and the southern edge of the Baltic Shield.

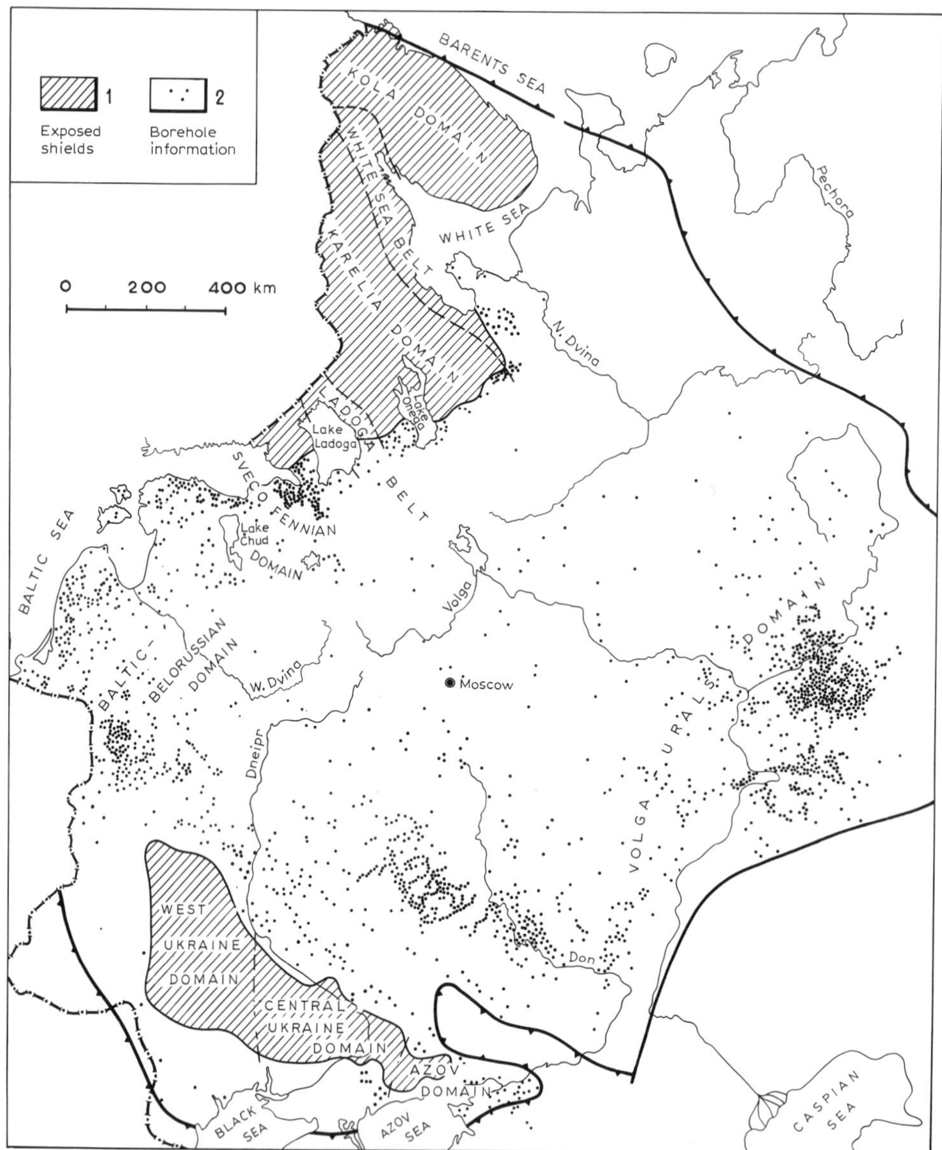

Fig. I-3. Level of study of the crystalline basement of the East European craton and its main structural units.

From this information, Vardanyants (1964) was able to compile a geological map of the crystalline basement of the Platform.

Since the end of the 1950s, aeromagnetic surveying has played an especially important role in basement studies. As a result of regional surveys, a complete picture was built up of the magnetic field which was a combination of zones of linear stripe anomalies and fields of a mosaic character. The linear zones, radiating out from an east–west focal point in the centre of the platform, form a type of vortex, between which mosaic anomaly fields are situated. The general structure of the magnetic field was interpreted by Gafarov (1963), who proposed a scheme of tectonic subdivisions consisting of Archaean massifs (mosaic fields) and Svecofennide-Karelide fold systems surrounding them (linear zones). According to this model, the fold systems are superimposed on an Archaean foundation.

Later achievements in studying the basement are related to improvements in the scale of geological surveying and interpretation. The work of Zander et al. (1967) was mainly based on aeromagnetic survey data, while that of Nevolin et al. (1971) involved interpreting the gravity field. This research played an important role in studying the relief and internal structure of the basement. The geological map (Zander et al., 1967) was the first attempt to explain the geophysical fields of the Russian platform on the basis of the geophysical characteristics of structural-petrographic provinces of the Baltic Shield and well-drilled regions.

The internal structure of the basement was reflected in the Basement map of the USSR (Nalivkin, 1974) and the Metamorphic belts map of the USSR (Kratz and Glebovitsky, 1975). Work on the latter map opened up new possibilities which were developed in the Metamorphic map of the basement of the Russian platform, compiled by Bondarenko et al. (1978). Problems of the structure and evolution of the platform basement are also viewed from the standpoint of geoblocks, or crustal segments, which differ in their age of consolidation and particular features of their crustal development (Dedeyev and Shustova, 1976), and from the point of view of horizontal movements (Valeyev, 1978). At the start of the 1970s, the structural model of the basement, which is a combination of ancient massifs and younger linear zones (Gafarov, 1963; Zander et al., 1967; Nalivkin, 1974), began to contradict the petrological data. It was established that within the platform, the linear stripe anomaly zones were caused mainly by Early Archaean supracrustal rocks. It turned out that upper Archaean and more especially lower Proterozoic rocks were of limited extent in the Russian Platform and were restricted to areas which from geophysical regional mapping project as massifs.

During the 1970s under the leadership of K.O. Kratz, new investigations were undertaken at the Institute of Precambrian Geology and Geochronology into the basement using interdisciplinary methods of interpreting geological and geophysical data. In carrying out this work, use was made of material from all detailed and medium-scale aeromagnetic and gravity surveys, data from over 3000 bore holes, and geological field observations on the Baltic and Ukrainian Shields. The work was supported by the simultaneous analysis of the magnetic and gravity fields and petrological description of cores. For the first time it was possible on the scale of the Russian Platform to organise research in such a way that models based on geophysics were backed up by direct examination of core material and an analysis of the conditions and sequences of metamorphic and igneous events. This significantly widened the possibility of correlating borehole data with type complexes on the shields and in the end introduced some certainty into the geological interpretation of geophysical fields. The resulting publications and associated magnetic, geological, metamorphic and tectonic maps represent the most up to date fundamental summary of the structure of the crystalline basement of the Russian Platform (Kratz et al., 1979).

The geological map of the basement which resulted from this work is shown in simplified form in Fig. I-4. The map covers the entire platform within the USSR, except for the Cis-Caspian depression where the basement is deeply buried and borehole data are absent, hence the geophysical possibilities are extremely limited.

The methodology of compiling the geological map is now described briefly. Geological models were constructed using regional geophysical mapping of the basement surface, by combining magnetic and gravity fields. The possibility of such regional mapping is determined by the particular features of the anomaly sources in both fields and the interrelation between them.

The sources of magnetic and gravity anomalies in the overwhelming majority of cases result from the internal structure of the basement and appear at its surface. Rare deep anomalies are usually fairly broad and constitute the background on which anomalies associated with inhomogeneities of the basement surface appear without any significant deviation. Anomalies caused by local objects within the sedimentary cover on the other hand show up against a background of anomalies associated with basement inhomogeneities, causing characteristic changes in the trend of some contours. In the majority of cases, both deep anomalies and those from the sedimentary cover, can be identified by qualitative and quantitative analysis and are excluded in regional mapping of the basement surface.

The mapping out of regions on the basis of combined magnetic and gravity fields is possible because they are in agreement everywhere. In very rare cases this consists in the simple coincidence of individual magnetic and gravity anomalies. Agreement between fields is usually expressed by the coincidence of regions identified in the gravity field with zones which are characterised by definite distribution of magnetic bodies. In particular, linear regions in the gravity field correspond to zones with linear magnetic bodies elongated in the strike direction of the regions. Isometric regions correspond to zones containing magnetic bodies of various shapes, which strike parallel to the boundaries of the regions.

Regional geophysical mapping consisted of delineating homogeneous zones with particular characteristics in the anomaly field caused by inho-

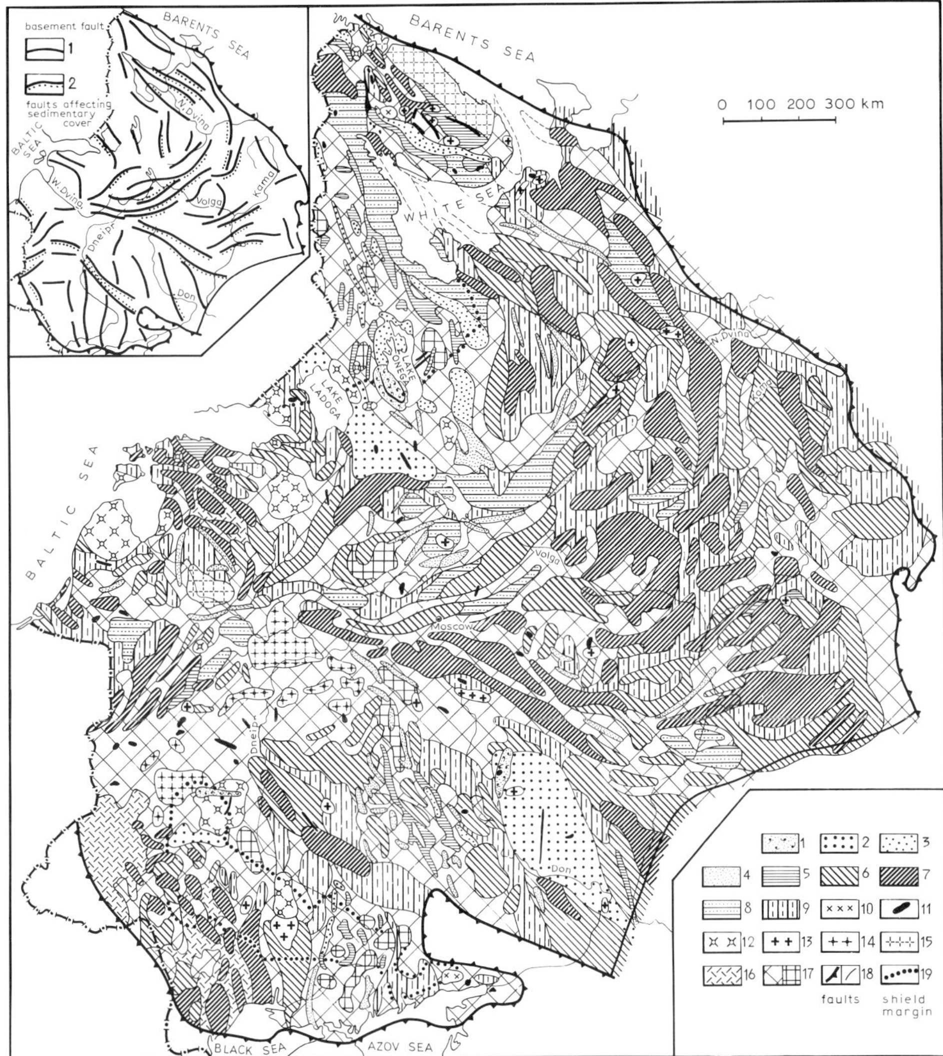

Fig. I-4. Geological map of the basement of the East European craton (compiled by A.N. Berkovsky and L.P. Bondarenko). See Table I-1 for legend *1–17*; *18a* = craton margins, *18b* = boundary fault of Cis-Caspian depression, *19* = shield margins. *Inset*: Major fault system in basement.

mogeneities within the near-surface parts of the basement. Subdivision into regions was carried out according to the nature of gravity anomalies and the mutual disposition, shape and magnetization of magnetic formations. Boundaries of geophysically homogeneous zones were initially outlined from lines of maximum gradient in the gravity field and subsequently refined from the magnetic map. Apart from broad zones, local objects — probably intrusions — were identified on the basis of individual magnetic and gravity anomalies, especially in cases where their position was discordant relative to the surrounding situation.

A comparison between the results of geophysically subdividing the basement into regions and geological and geophysical maps of the Baltic and Ukrainian Shields convinces us that geophysically homogeneous zones on the whole correspond to areas within which rocks of particular structural-

TABLE I-1

Main geophysical characteristics used to delineate various early Precambrian complexes in the East European

		Data from legend to geological map (Fig. I-4)	Type complexes from the Baltic Shield (BS), Ukrainian Shield (US) and Russian Platform (RP)	Shape of zones on regional geophysical maps	Intensity of gravity field	Characteristics on magnetic map: saturation with bodies; magnetization of bodies; background	Characteristics on magnetic map: outline of bodies; mutual arrangement of bodies
SUPRACRUSTAL ASSEMBLAGE	L. PROTEROZOIC	1	BS — Vepsian US — Pugachev Gp. Novgrad-Volyn member of Klesov Gp RP — Grazin, Baygorov & Adazhi Fms, Hoglandian	elongate	high	weak; weak	small, elongate; ordered
		2	BS — Ladoga Group US — Teterev Group (?) RP — Vorontsov Gp, Oskol & Uniy members	elongate broad	high	weak to average weak to average	small, elongate; ordered and disordered
		3	BS — Jatulian, Suisarian, Pechenga, Imadra-Varzuga Gps US — Krivoy Rog Group RP — Kursk Group	elongate usually narrow	high or absent	average to strong; moderate and strong	elongate, sinuous steady
		4	Kursk & Mikhaylov Gps (undifferentiated) and geophysical analogues	elongate, narrow linear	high or absent	strong; strong & average	long non-linear; steady
	ARCHAEAN	5	BS — Lopian, Keyvy Group US — Konka-Verkhovtsev Group RP — Mikhyalov, Okolov, Sarman Gps, Yagala member	elongate	high and low	weak; weak &moderate	small, sometimes long; steady
		6	BS — Volshpak, Chudzyavr (part), Pinkelyavr Fms, Kola Gp US — Ingulo-Ingulets Gp, Kamenno-Kostovat & Roshchakhov Fms of Bug Group* RP — Bolshecheremshan Gp, Besedin member, Alutaguz complex	elongate, less commonly isometric	high	weak; moderate	long, linear and sinuous; ordered
		7	BS — Chudzyavr, Pinkelyavr, Volshpak (part) Fms, Kola Group US — Dniestr-Bug Gp; Koshar-Alexandrov Fm, Bug Goup RP — Ivyev & Otradnin Groups	elongate, linear, less commonly isometric	extremely high	strong or average to strong; background	elongate, linear and sinuous; constantly ordered
		8	BS — Belomorian Group	elongate, linear, maybe isometric	low and high	weak weak	elongate; disordered
		9	undifferentiated gneissose granitoids	elongate, narrow, sometimes broad, various outlines	high or absent	average, sometimes strong	elongate, ordered
PLUTONIC ASSOCIATIONS	L. PROTEROZOIC	10	nepheline syenites: BS — Kola Peninsula (PZ) US — Sea of Azov region (PR_1)	rounded	low	weak weak	elongate;
		11	mafic & ultramafic intrusions	narrow linear minor rounded	high or absent	strong; strong	isometric; ordered
		12	rapakivi & similar granites. Intrusions: BS — Vyborg, Salmi, Ulyaleg, Umba; Litsa-Araguba US — Korosten, Korsun-Novomirgorod; Riga	isometric broad	low and extremely low	weak; weak	varied; disordered
		13	Proterozoic granites. Intrusions: BS — alkaline granites of Kola Peninsula US — Novoukrainka, Uman	isometric and irregular shapes	low and extremely low	weak; weak	varied; disordered
		14	granitoids: US — Osnits complex	isometric	high	strong; medium & strong; background	varied; ordered
		15	granitoids: BS — Murmansk complex	isometric broad	high and low	average, inhomogeneous; moderate; background	varied; ordered and disordered
	ARCHAEAN	16	enderbites: US — Podolsk block	elongate	low	strong, sometimes average; strong and average; background	elongate sinuous sometimes isometric ordered
		17a	undifferentiated granite-gneisses: BS — Kola Peninsula, Karelia US — Dnieper & Western Ukraine RP — Oboyan Group	elongate and isometric, broad	low and extremely low	weak; inhomogeneous background	minor, elongate; ordered
		17b	Basic rocks within granitoids of 17a: BS — Vodlozero, Tulos, Voknavolok regions US — Slavgorod region	isometric	high	weak and average background	minor elongate; disordered, less commonly ordered

* These formations in the Ukrainian Shield show up as a single complex in the geophysical field. The compilers of the basement map consider them to be Archaean, intruded by early Proterozoic granitoids.

petrographic terrains are developed. Such a comparison in well-studied regions of the Russian Platform shows that each of the zones delineated is characterised by quite distinct associations of metamorphic and igneous rocks belonging to a single terrain, which are different from those of neighbouring zones. This is also confirmed by drilling. Such agreement has allowed us to identify geological complexes in well-studied regions of the platform with standard geophysical characteristics. These, together with drill hole data used for control, then enable us to move from a regional geophysical map to a geological map of the basement. Table I-1 shows the main geophysical characteristics of the various terrains in the platform basement.

It follows from the method of construction that terrains as shown on the map are in their main distribution zones, with averaged boundaries. Such averaging is greater with increasing depth to the basement, with correspondingly less resolution possible for geophysical methods. On the map this is expressed as a coarser general picture for more deeply buried basement.

The structure of the shields (for convenience in comparing with the territories of platforms) is represented in a more generalised fashion than on medium-scale geological maps.

Terrain boundaries delineated on the map satisfy the criterion of fault tracing, used in geophysics. In most cases they separate terrains that differ in structure and history of development. Figure I-4 (inset) shows the longest and most clearly expressed ancient deep lineaments. Many of them express themselves as flexures in the sedimentary cover and are detectable in satellite photographs. These must be considered as long-lived faults.

* * *

The generalized geological and geophysical material on the geological map helps to uncover the main structural features of the basement of the East European craton and to draw its regional tectonic boundaries (Fig. I-5).

The major crustal segments or geoblocks which stand out in the basement of the East European craton represent different tectonic structures. From the period 1600 Ma ago they developed in a single tectonic regime. Earlier (up to 1700 Ma ago), many segments were different tectonic structures: in the Early Proterozoic, cratons or mobile regions; in the Archaean, mobile regions which differ not only in the time of stabilization (which is usually used in defining Precambrian tectonic domains), but in their internal structure and the evolution of their particular types of crustal processes.

We may surmise that beginning as early as the Katarchaean (>3500 Ma ago), the Precambrian crust of the platform basement was differentiated in composition, structure and internal processes, judging in particular by the composition of the oldest supracrustal assemblages. From the period

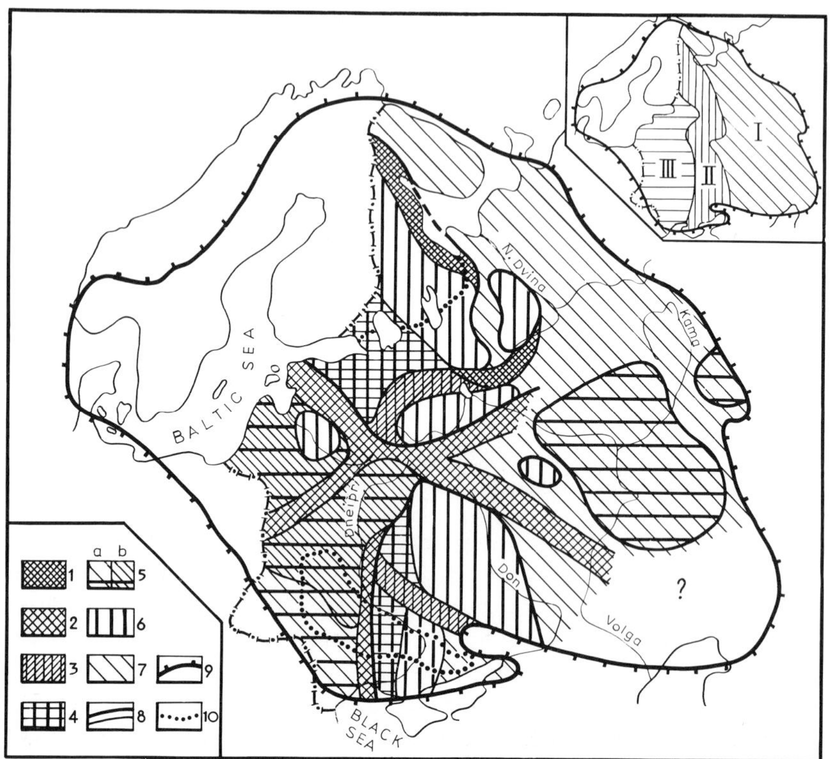

Fig. I-5. Sketch map showing tectonic regions of the Precambrian in the East European craton. Early Proterozoic mobile belts without geosynclinal precursors, types: *1* = Belomoride, *2* = Odessa-Belotserkov, *3* = Krestetsk-Cherepovets. Early Proterozoic mobile regions (cratons), types: *4* = Svecofennian, *5* = Volynye-Podolsk (*a* = lower sections, *b* = upper sections). Archaean mobile regions (epi-Archaean cratons): *6* = granite-greenstone, *7* = granulite-greenstone, *8* = major structural boundaries, *9* = craton margins, *10* = shield margins. *Inset*: provinces (*I* = Eastern, *II* = Central, *III* = Western).

3200 Ma ago, such differentiation was established at two levels: (1) at the suprastructural level, where different types of mobile belts formed, and (2) for the crust as a whole, including the suprastructure and the infrastructure. It is this second level that enables the various types of mobile regions to be identified within the basement which formed at a stage in the development of the continental crust.

Territories are assigned to mobile regions of the same age, in which similar crust-forming processes operated that formed the regional structure, were developed pervasively and are revealed at different crustal levels (in the suprastructure and infrastructure). Most of the characteristic features in mobile regions, such as the dominant strike direction of tectonic movements, metamorphic regime, sequence of magmatic processes, etc., which usually provide a basis for classifying tectonic structures, are established mainly by studying exposed territories. In the Baltic and Ukrainian Shields, mobile regions belong to two age groups: where final processes are Late Archaean (1) or Early Proterozoic (2). No mobile belts of Katarchaean age have been found in the East European craton. The only terrains known to be of this age form part of younger mobile regions.

(1) *Archaean mobile belts* are regions in which the intense processes responsible for the formation of crustal structures are older than 2500 Ma. For the platform as a whole, the time of stabilization of Archaean mobile regions with the same type of structural evolution (granite-greenstone, for example), has been determined to lie within the interval 3000–2600 Ma. The oldest stabilization age characterises the south of the platform, i.e. the Ukrainian Shield, and the youngest for the Baltic Shield.

The Kola and Karelian provinces in the Baltic Shield are Archaean mobile regions. Each displays individual evolutionary features, identifiable in the time interval 3200–2600 Ma. These areas of the Baltic Shield represent the two apparently commonest types of Archaean mobile regions: granite-greenstone (Karelia) and granulite-greenstone (Kola). The term granulite-greenstone terrain is introduced in order to distinguish those parts of granulite-gneiss terrains in which Late Archaean supracrustal complexes have been recognised, and represented as greenstone and/or paragneiss belts. Like the Late Archaean granite-greenstone terrains, they have a two-stage structure. These regions differ in geological structure (see Fig. I-4) and a number of other features, such as: (1) composition of volcanic rocks in the upper Archaean supracrustal assemblage and associated terrigenous rocks (see Sections 1 and 2); (2) composition of the basal complex (granite-gneiss in granite-greenstones and granulite-gneiss with relatively dense basic granulites in granulite-greenstones); (3) style of tectonic fabric (concentric in granite-greenstones and linear in granulite-greenstone terrains); (4) the type of final metamorphism (steep gradient in the first type, moderate to low in the second); and (5) direction of material transport in the crust (vertical at all crustal levels in granite-greenstone terrains, horizontal in the lower levels and vertical in upper levels of granulite-greenstone terrains) (Khiltova et al., 1986).

In the Ukrainian Shield, all workers classify the Dnieper block (the eastern part of the Central Ukraine province) as a granite-greenstone terrain. The Azov province can be considered as granulite-greenstone type and the Volyn-Podolsk block (West Ukraine) probably also developed similarly in the Archaean, but was reworked in the Early Proterozoic and only partially preserved earlier features.

(2) *Early Proterozoic terrains* (2600–1600 Ma). Within the East European craton, as in all other ancient cratons, the first stable regions, the epi-Archaean cratons, appeared at this time, as well as mobile regions.

We can distinguish at least two categories of Early Proterozoic tectonic structures, which are represented in turn by probably independent types.

The first category are mobile regions, within which Early Proterozoic supracrustal assemblages accumulated in sedimentary basins in cases where the belts had an initial depositional precursor phase. This category can provisionally be subdivided into a granite-schist type, the type structure being the Svecofennian Domain of the Baltic Shield, and a schist-granulite type, the type structure for this being the Volyn-Podolsk Domain of the Ukrainian Shield.

In the *Svecofennian type* of mobile region, mobile belts of the Ladoga type developed, containing supracrustal rocks of volcanic-terrigenous composition. The infrastructure is dominated by granite-gneiss with relicts of Archaean greenstone belts (see Section 4, "Ladoga belt"). This suggests that in the Archaean, Svecofennian-type terrains developed in a similar manner to Archaean granite-greenstone terrains.

The *Volyn-Podolsk type* of mobile region contains Early Proterozoic belts with terrigenous supracrustal rocks in its suprastructure. It differs from the previous type in that basic granulites are an important constituent of the infrastructure. Dome-shaped structures mapped in the basement have mafic granulite and amphibolite facies metamorphic rocks in their cores, and are surrounded mainly by granitic rocks. The Archaean prehistory of Volyn-Podolsk type terrains exhibits features that are typical of Archaean granulite-greenstone terrains. Within the Archaean granulites, two metamorphic episodes can be identified, the later one being at high pressures (Kratz et al., 1979). Amphibolite facies complexes in these regions can probably be correlated with the greenstone belts of Archaean granite-greenstone terrains, but they differ in often having been affected by uniform metamorphism in the sillimanite-kyanite facies series. In the Volyn-Podolsk type of terrain, as distinct from unreworked Archaean granulite-greenstone terrains, upper crustal sections were composed mainly of granulites by the Early Proterozoic, probably as a result of thrusting or some other mechanism.

The second category of early Proterozoic mobile structures are mobile belts which lacked any early sedimentary phase. They occur in the junction zones between different types of Archaean mobile regions (granite-greenstone and granulite-greenstone), for example the Belomorides and the adjacent Lapland belt in the Kola province of the Baltic Shield, or the Odessa-Belotserkov province (the Golovanev suture zone and the Belotserkov block) of the Ukrainian Shield.

Internal crustal processes developed in these belts in lower and upper Archaean complexes and possibly also in the Early Proterozoic sedimentary cover of the platform. The belts are often structurally and compositionally heterogeneous, on three accounts. Firstly, they developed throughout extremely varied types of craton, in which the structural-petrographic terrains differ in composition and structure; secondly, the degree of reworking for this time interval is usually unequal in the belts; finally, vertical movements of constituent blocks differ in scale, but are on the whole substantial. It is mainly in this type of belt that the deepest levels of Archaean crust are exposed at the surface.

Various types of Archaean and Proterozoic mobile regions, which possess individual internal structural features, as reflected in their separate development, constitute the basement to the East European craton and are its main structural elements.

Terrains which evolved through various types of mobile regions possess their own particular and individual distinctive patterns in their geophysical field characteristics, which allow the distribution of the provinces to be determined beneath the platform cover. The whole basement can be subdivided into three provinces, based on the distribution of the different types of terrain: Eastern, Central and Western (see Fig. I-5, inset).

The *Eastern province* is composed mainly of granulite-greenstone terrains and deep sections of these, in which basic granulites and amphibolites are widespread, while gneiss and schist complexes are fewer. The other constituent of this province are Volyn-Podolsk type Early Proterozoic mobile regions.

From geophysical data, the structural pattern of terrains in the Eastern province is two-fold, consisting of linear belts and isotropic regions. Terrains with linear structure (a typical region being the North-West province) geologically, geophysically and petrologically are entirely comparable with Archaean granulite-greenstone mobile belts in exposed regions, possibly only with deeper erosion levels. Terrains with isotropic structure (of which the Volga-Urals province is typical) are characterised by widespread basic granulites (Bogdanova et al., 1982). Two metamorphic episodes have been identified in these granulites, as shown by petrological studies on drill cores: an earlier low-pressure event (Kratz et al., 1979) and a later high-pressure event. Amphibolite complexes in these terrains can be correlated with late Archaean greenstone belts in exposed terrains. They have been metamorphosed under kyanite-sillimanite facies series. Mafic volcanics belonging to the basalt-komatiite series were altered to garnet amphibolites. Among the Late Archaean supracrustal rocks are frequent occurrences of clastic terrigenous sediments in association with acid volcanics. The dominant basic intrusive rocks of this age are ultrabasic-gabbro-anorthosite complexes. This suggests that the second high-pressure granulite facies metamorphism, with which is associated the formation of eclogitic rocks in the basement complex, was synchronous with the metamorphism of the late Archaean supracrustal rocks belonging to the kyanite-sillimanite series.

The *Volga-Urals* province bears some resemblance to the Early Proterozoic Volyn-Podolsk mobile region, but differs from it in having a thinner crystalline basement and a smaller volume of Early Proterozoic supracrustal rocks.

The *Central province* consists mainly of granite-greenstone terrains, including the Karelia province in the Baltic Shield, the Dnieper block in Central Ukraine, the Central Russian block in the central part of the Russian Platform, and the Voronezh block in the southern part (Fig. I-4). These terrains have a similar general geological and tectonic structure and show up clearly in the geophysical fields.

Although individual granite-greenstone terrains display many structural and compositional features in common, there are some differences in geophysical characteristics, for example in P-wave velocity at the base of the crust (Dedeyev and Shustova, 1976). Velocities are highest in the Voronezh massif (Kursk-Voronezh region) and lowest in Karelia. The Voronezh massif, which is a granite-greenstone terrain, has a low average crustal density, resulting from significant decompaction of rocks in the middle and lower (10–40 km) crust. The Dnieper block in the Central Ukraine province has a higher density compared to Karelia. These differences reflect particular features of the regional geology, mainly that of Archaean times.

The *Western province* has a large-scale mosaic structural pattern due to the interplay of terrains with different geological and tectonic structures. Of the three tectonic provinces in the basement, the Western is the most inhomogeneous as a result of the presence of terrains belonging to different regional types and different crustal sections through them. The northern part consists of the Early Proterozoic Svecofennian granite-schist terrain which typically lacks any significant development of low-pressure granulites. In the southern part of the province is a schist-granulite terrain (the Volyn-Podolsk), which is seen at various crustal levels. Early Proterozoic granitoids are widespread in the upper levels. Being characteristic of upper and middle crustal levels in terms of crystallisation conditions, the granitoids evolved from rocks in deeper sections — basic granulites, intruded at higher crustal levels by the Early Proterozoic (Kratz et al., 1979).

The south-western part of the Ukrainian Shield is a deep section through a Volyn-Podolsk type of mobile region, in which Early Proterozoic crustal processes typical of deep (occasionally mid-crustal) levels are represented. Metamorphism took place at amphibolite or granulite facies.

To summarise, each province has its own characteristic association of mobile regions. In the Eastern, Archaean granulite-greenstone terrains are associated with Early Proterozoic granulite-schist terrains of the Volyn-Podolsk type. Archaean granite-greenstone terrains occupy an insignificant area. The Central tectonic province consists almost exclusively of Archaean granite-greenstone terrains. The Western province includes two types of Early Proterozoic region, the Svecofennian and Volyn-Podolsk, the latter in upper and lower sections. Only tiny areas of Archaean granite-greenstone terrains are preserved.

The crystalline basement of the East European craton as a whole therefore consists of tectonic domains of different ages and types, which are separated by mobile belts. The belts differ depending on the type of basement upon which they formed, either granite-greenstone or granulite-greenstone.

In the Baltic Shield for example, the Lapland belt which formed along the southern edge of the Kola granulite-greenstone terrain, has a pronounced linear structure to which are confined the deepest granulites and Early Proterozoic basic intrusions. Similar characteristics are found in the Odessa-Belotserkov belt (the Golovanev suture zone and the Belotserkov block) restricted to the eastern edge of the West Ukrainian province of the Ukrainian Shield, which also evolved in the Archaean on the pattern of granulite-greenstone type terrains. Similar belts are probably buried beneath the sedimentary cover to the south of the Central Russian block and to the north of the Kursk-Voronezh region. In both cases, they separate granite-greenstone terrains from granulite-greenstone terrains.

The Kresttsy-Cherepovets belt is compositionally and structurally different. It separates the Central Russian crustal segment from the Karelian province and the Svecofennian Domain (Fig. I-4), which are mobile regions of different ages, although in the Archaean they evolved from the same granite-greenstone type of terrain. Since Riphean aulacogens developed from these same suture zones (see Part I, Chapter 3), we may conclude that the belts separating the provinces formed in the basement and on the shields before the upper Proterozoic.

This suggests that the crystalline basement of the platform and the shields evolved according to one tectonic model during the early Precambrian (>1700 Ma). Differences between them began to appear from the end of the Early Proterozoic. During the interval 1700–1500 Ma, these differences began to become more marked, and determined the different evolutionary trends of the shields and the platform.

The model considered for the tectonic regional zonation of the basement of the East European craton (Fig. I-5) reflects its structure at the end of the Early Proterozoic. For the Late Archaean (3.0–2.6 Ga), we suggest that the basement was tectonically more uniform. In essence, it was made up of granulite-greenstone terrains, which comprised its eastern and western parts. Granite-greenstone terrains occupied its central part, which in turn was divided into two independently evolving provinces (Fig. I-6).

Granulite-greenstone terrains were affected over a wide area by Early Proterozoic crustal processes. Granite-greenstone terrains, on the other hand, were subjected to high-grade processes only over small areas in the early Precambrian.

We may go so far as to suggest that the crustal differentiation observed

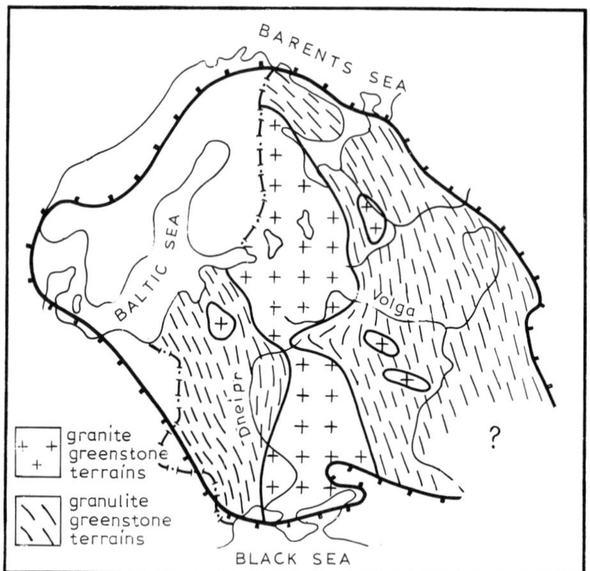

Fig. I-6. Distribution of Archaean mobile belts in the basement of the East European craton.

for the Archaean also occurred in the Early Archaean (Katarchaean), since supracrustal complexes of this age in granulite-greenstone terrains have a more mafic composition compared to granite-greenstone terrains.

Chapter 1

THE EASTERN BALTIC SHIELD

The modern tectonic structure of the Earth's crust in the eastern part of the Baltic Shield is defined by a system of geoblocks or tectonic provinces, each with a different internal structure and history of geological evolution. Within the USSR are the Kola and Karelian provinces and the Belomorian (White Sea) and Ladoga belts.

1. THE KOLA PROVINCE

Modern studies of the Precambrian of the Kola Peninsula date from the 1930s and to a great extent are associated with the name of Academician A.A. Polkanov, who laid the foundations for the Precambrian stratigraphy of the region and established a framework for its geological structure. The results of subsequent work are set out in a number of fundamental publications on stratigraphy (Gilyarova, 1974; Zak et al., 1975; Makiyevsky, 1973; Perevozchikova, 1974; Kharitonov, 1966), evolution of metamorphic and igneous processes (Kratz et al., 1978; Batiyeva et al., 1985; Petrov et al., 1986; Belyayev et al., 1977) and major structures and Precambrian intrusive complexes (Belkov, 1963; Gilyarova, 1972, 1974; Goryainov, 1976; Zak, 1980). The metallogeny of Precambrian formations in the Kola Peninsula is discussed in a number of works concerned with individual ore deposits (Bogachev et al., 1964; Goryainov, 1976; Kozlov, 1973; Bilibina, 1980; Gorbunov et al., 1981, 1985).

The Precambrian rocks of the Kola Peninsula were the first rocks to be dated by Polkanov and Gerling (1960), who developed the K–Ar absolute age dating method. Further successful geochronological research was carried out at the Institute of Precambrian Geology and Geochronology, the Geological Institute of the Kola Branch of the Academy of Sciences and the Academy of Sciences' Geochemical Institute. By using a number of different dating techniques, this work formed the basis for the erection of a geochronological scale for the Baltic Shield, including the Kola Peninsula (Kratz, 1972).

The Kola province occupies the northern part of the eastern Baltic Shield. Its southern margin is taken here to be the southern edge of the

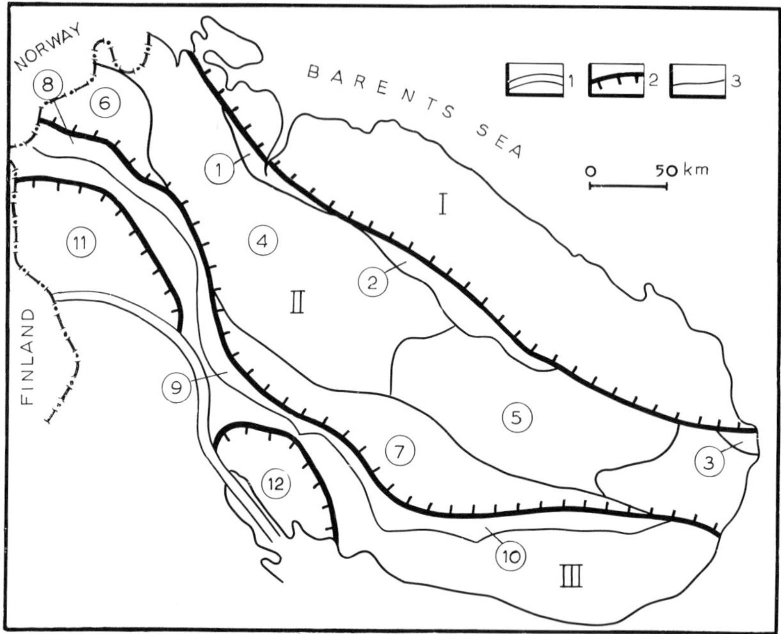

Fig. I-7. Tectonics of the Kola province. *1* = province boundaries; *2* = segment boundaries; *3* = boundaries of zones and structural belts. Segments: *I* = Murmansk, *II* = Central Kola, *III* = Tersk. Numbers in circles: *1–3*: Kolmozero-Voronya greenstone belt (*1* = Uraguba, *2* = Polmos = Porosozero, *3* = Ust-Ponoy belts); *4* = Chudzyavr zone; *5* = Keyvy belt; *6* = Pechenga belt; *7* = Imandra-Varzuga belt; *8–10*: Tersk-Allarechka greenstone belt (*8* = Kasmak-Allarechka, *9* = Cis-Imandra, *10* = Tersk belts); *11* = Tuadesh-Salnotundra zone; *12* = Kandalaksha-Kolvitsy zone.

Tuadesh-Salnotundra and Kandalaksha-Kolvitsy zones or structural belts (Fig. I-7).

At the present erosion level, the tectonic structure of the Kola province is defined by three segments (Murmansk, Central Kola and Tersk) which differ in geophysical characteristics and geological structure (Kratz et al., 1978; Fig. I-7). The block structure of the Kola province is thought to result from the exposure at the surface of geological formations belonging to different depth levels, or of different crustal layers (Zhdanov and Malkova, 1974).

Using a combination of all the available data, we can subdivide the Kola province into at least six complexes (Fig. I-8):

– the Saamian complex (apparent age > 3.5 Ga) of ancient tonalite gneisses in places including intensely reworked supracrustal rocks;

– the Kola complex (3.5–3.0 Ga) of various gneisses and other supracrustal rocks belonging to the Kola, Barents Sea and Varzuga Groups;

– the Lopian sedimentary-volcanic complex (3.0–2.6 Ga);

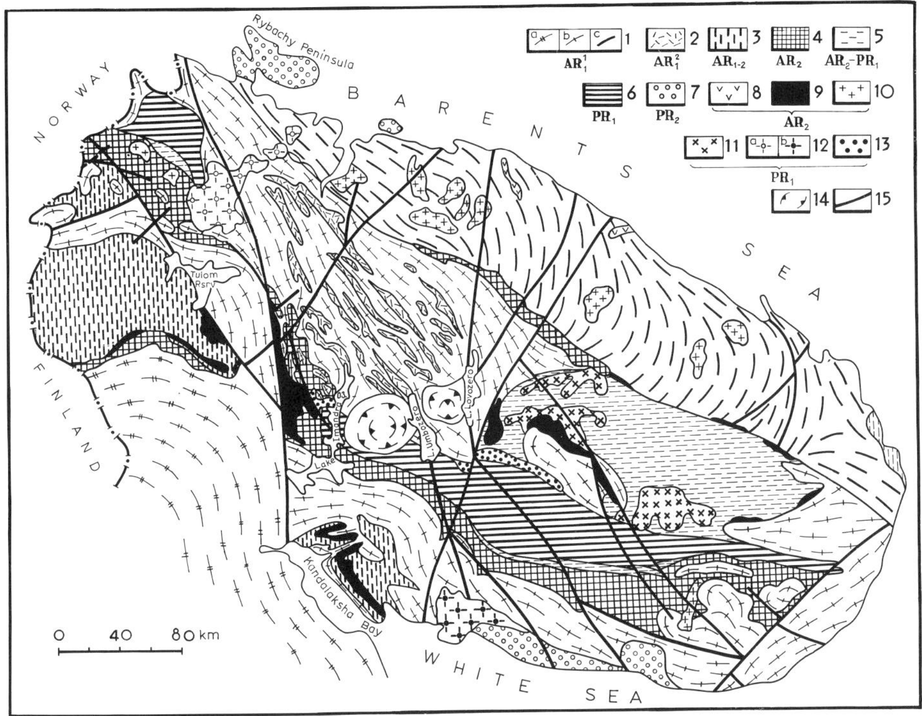

Fig. I-8. Geological sketch map of the Kola Peninsula (based on Zagorodny and Radchenko, 1983; Ivanov, 1986; Gorbunov et al., 1981) *1* = Saamian in various segments (*a* = Belomoride, *b* = Central Kola and Tersk, *c* = Murmansk); *2* = Kola Group; *3* = Kola Group and Lopian complex (undiff.); *4* = Lopian complex; *5* = Lopian-Sumian complex (Kevyv and Strelnin Groups and undiff. formations); *6* = Karelian complex (Pechenga and Imandra-Varzuga Groups); *7* = Riphean-Vendian; *8–14*: intrusions (*8* = gabbro-amphibolite, *9* = gabbro-labradorite, *10* = microcline and plagiomicrocline granite, *11* = alkali granite, *12a* = granodiorite and leucogranite, *12b* = charnockite-granite, *13* = ultrabasic-gabbro-norite, *14* = nepheline syenite); *15* = faults.

- the Sumian complex (2.6–2.4 Ga);
- the Karelian complex (2.4–1.8 Ga); and
- the upper Proterozoic (Riphean and Vendian) complex (<1.65 Ga).

The lower Archaean Saamian and Kola complexes

In the *Central Kola segment*, these complexes consist of polymetamorphosed gneiss, schist and orthogneiss of various compositions. These lithological groups of Archaean rocks with a radiometric age definitely older than 2.7–2.8.Ga (Table I-2) are not easily subdivided, since they often possess a uniform structural pattern with NW-trending folds. However, many workers consider the tonalite gneisses to be the oldest rocks of this segment (Table I-3), being relicts of a primary sialic crust (the group I granites of

TABLE I-2

Radiometric age of lower Archaean formations in the Kola Peninsula

No.	Locality, dated material	Sample	Method	Age (Ma)	Reference
	Murmansk segment				
1.	Barents Sea Fm — two-pyroxene schist	whole rock	Pb-Pb	2700±160	Mints et al. 1982
2.	Kachalov assemblage amphibolite	whole rock	Pb-Pb	2500 ±160	Mints et al. 1982
3.	Plagioclase-microcline granite (Teribersk type)	whole rock	Pb-Pb	2820±160	Mints et al. 1982
4.	Ditto (Geokan type)	whole rock	Pb-Pb	2370±130	Mints et al. 1982
5.	Ditto (Chiliyavr type)	whole rock	Pb-Pb	2420±160	Mints et al. 1982
6.	Charnockite	whole rock	Pb-Pb	2600	Tugarinov & Bibikova 1980
	Central Kola segment				
1.	Kola gneiss	whole rock	Pb-Pb	3150±150	Sobotovich et al. 1963
2.	Ditto	whole rock	Rb-Sr (isochron)	2700	Kratz 1972
3.	Kola high-Al gneiss	whole rock	Rb-Sr (isochron)	2660±40	Gorokhov et al. 1976
4.	Enderbite	zircon	U-Pb (isochron)	2820±15	Pushkarev et al. 1978 Pushkarev et al. 1979
5.	Tonalite	zircon	Pb-Pb U-Pb (isochron)	2790±100 2710±60	Pushkarev et al. 1979 Pushkarev et al. 1979

TABLE 1-3

Chemical composition of the oldest granitoids of the Kola Peninsula (Vetrin, 1984)

Oxide	1 (n = 6)	2 (n = 16)	3 (n = 45)	Oxide	1 (n = 6)	2 (n = 16)	3 (n = 45)
SiO_2	61.47	66.42	70.45	Na_2O	4.50	4.48	4.93
TiO_2	0.75	0.49	0.31	K_2O	1.28	1.49	1.42
Al_2O_3	17.48	46.18	15.18	H_2O^-	0.10	0.10	0.10
Fe_2O_3	1.93	1.54	0.97	H_2O^+	0.31	0.10	0.08
MnO	0.06	0.09	0.05	loss on ign.	0.67	0.72	0.57
MgO	2.48	1.35	0.81	P_2O_5	0.14	0.12	0.08
CaO	5.56	3.91	3.15	CO_2	0.07	0.13	0.10

1 = quartz diorite, 2 = tonalite, 3 = trondhjemite. ("plagiogranite" in Russian)

Polkanov), overlain by Kola Group supracrustal rocks. In some cases, the tonalite gneiss strikes NE and E–W, contrary to the regional trend, with broad antiformal and domal structures (Batiyeva and Belkov, 1968; Vetrin, 1984).

This two-fold relationship between plagiogneiss-granite (i.e. tonalite-trondhjemitic rocks) and Kola Group supracrustal rocks has long been known (Bondarenko and Dagelaysky, 1968; Polkanov, 1937). On the one hand, the Kola Group is geologically younger than the granite-gneiss, but on the other hand the tonalites (Polkanov's group II tonalites) have affected Kola Group rocks.

In order to explain this, a number of workers (Batiyeva and Belkov, 1968; Vetrin, 1984) have suggested that the tonalites and some of the gneisses (the extrusive equivalents of granites) of the Kola Group belong to a single Early Archaean volcanic-plutonic association which evolved over a long period. However, from detailed structural and metamorphic field observations, these relationships can be explained by remobilization of plagiogneiss-granite during high-temperature metamorphism of the overlying younger Kola Group.

The high-grade rocks of the Kola Group form narrow linear N–W-trending synforms and are divided into two different supracrustal sequences on the basis of composition and stratigraphic position in the section.

The *lower unit* consists of strongly migmatized biotite and hornblende-biotite melanocratic gneisses which have been affected by microcline metasomatism. The base of this unit is unknown. It forms the cores of antiforms; along the margins of lithological and structural rock varieties are alaskite, pegmatite and orthoamphibolite bodies.

The *upper unit* is subdivided into three formations. The lowest is the *Pinkelyavr Formation* of garnet-biotite gneiss, interleaved with mafic schist and magnetite schist. Its estimated thickness is 600 m. The *Chudzyavr Formation*, 350–500 m thick, consists of mafic schist and amphibolite with units of rhythmically-banded[1] schist and basic gneiss. The topmost *Volshpakh Formation* consists of rhythmically-banded high-alumina sillimanite-garnet-biotite and cordierite-sillimanite-garnet-biotite gneiss and eulysite.[2]

Results from the drilling of the Kola superdeep borehole in the NW part of the Central Kola segment have significantly widened and refined our understanding of the nature and thickness of the Kola Group in general, and its lower formations in particular (Kozlovsky, 1987). High-grade metamorphic rocks of the Kola Group around the borehole occur from a depth of 6842 m and continue to 11662 m (the depth reached in 1984). Borehole results generally confirm the two-fold structure of the Kola Group (Fig. I-9). Taking all the geological, petrographic, mineralogical and geophysical data from the superdeep project together, three genetic groups of rocks emerge:

(1) metavolcanics (basalts, andesite-basalts, dacites);
(2) metaterrigenous (alumino-siliceous formation); and
(3) metavolcanic-terrigenous.

The average chemical compositions of the most important rock types in the Kola Group are shown in Table I-4. Land-based seismic survey data in

---

[1] Russian original has "rhythmically-bedded".

[2] High-grade banded iron formation.

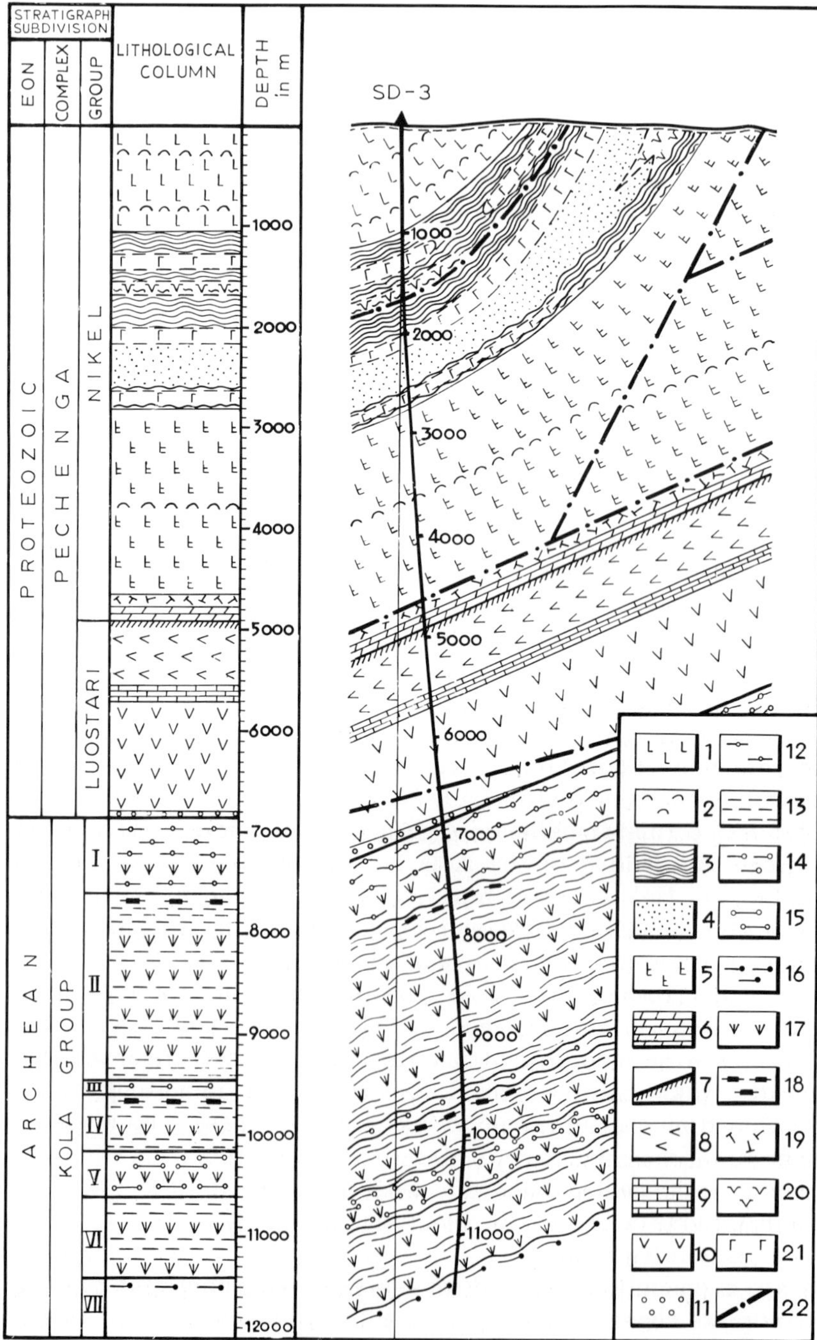

Fig. I-9. Geological section of the Kola superdeep borehole (SD-3) (simplified from Kozlovsky, 1987). *1* = augite diabase with pyroxene porphyrite; *2* = mafic tuff; *3* = phyllite and siltstone with tuff; *4* = rhythmically-bedded sandstone and minor siltstone; *5* = actinolitized diabase; *6* = dolomite and arkose; *7* = sericite schist; *8* = meta-diabase; *9* = dolomite and polymict sandstone; *10* = schistose porphyritic diabase; *11* = polymict sandstone; *12* = bi-plag gneiss with high-Mg minerals; *13* = migmatized bi-plag gneiss; *14*: as *13*, with silli+gt; *15*: as *13*, with ky; *16* = gt-bi-plag-silli gneiss; *17* = amphibolite; *18* = hb-mag schist; *19–21*: intrusions (*19* = andesitic porphyrite, *20* = wehrlite, *21* = gabbro-diabase); *22* = faults.

TABLE 1-4

Average chemical composition of major rock types in the Kola Group

Oxide	1	2	3	4	5	6
SiO_2	64.36	70.48	63.7	61.0	50.01	50.8
TiO_2	0.49	0.34	0.7	0.8	1.40	1.1
Al_2O_3	16.92	15.68	16.0	17.8	14.20	13.9
Fe_2O_3	1.02	0.51	1.5	1.7	3.10	3.9
FeO	3.47	2.18	6.4	6.7	10.31	9.8
MnO	0.035	0.03	0.1	0.1	0.22	0.1
MgO	1.84	0.88	3.0	3.7	6.00	6.3
CaO	4.47	2.09	2.3	2.1	9.95	10.4
Na_2O	4.61	5.54	2.9	2.5	2.31	2.1
K_2O	1.35	1.48	2.3	2.5	0.63	0.7
n	17	130	18	29	68	23

1 = hornblende-biotite gneiss (Kozlovsky, 1987), 2 = biotite gneiss (Kozlovsky, 1987), 3 = garnet-biotite gneiss (Kryukov, 1978), 4 = garnet-sillimanite gneiss (Kryukov, 1978), 5 = amphibolite (Kozlovsky, 1987), 6 = pyroxene-amphibole schist (Kryukov, 1978).

the vicinity of the Kola borehole allow seismic boundaries to be fixed at depths of 12.5–13.5 km, which may correspond to the lower boundary of the Kola gneisses. In this case, the gneisses are estimated to be 6 km thick at the present erosion level.

The *Murmansk segment* (Fig. I-10) in the northern part of the Kola province occupies the entire north-eastern shoreline of the peninsula as well as the shelf zone of the Barents Sea. The SW margin of the Murmansk segment is the Kolmozero-Voronya greenstone belt and a steeply-dipping deep fault, along which the Murmansk segment has been moved (at an average dip angle of 50–60°) over the Central Kola segment and the supracrustal rocks of the Kolmozero-Voronya belt (Platunova, 1975). In the NE part, the Murmansk segment is overlain by non-metamorphosed and undeformed deposits of the Riphean sedimentary cover.

The Murmansk segment differs significantly in terms of geophysical characteristics and geological structure from other segments of the Kola province, the main feature being the dominant oligoclase granite-gneiss and migmatite, enderbite and charnockite, and pervasive Late Archaean microclinization and granitization. Some of the oligoclase granites are thought to be Katarchaean primary sialic crust (Kratz, 1978).

Supracrustal rocks which have not undergone substantial granitization and migmatization have been identified in the Barents Sea Group (Makiyevsky, 1973). The lower part of the group consists of two-pyroxene-plagioclase schist (granulite), while the upper part, which is mainly exposed along the shore line and in the SE part of the segment, consists of amphibolite, hornblende-biotite and biotite gneiss. The upper part of the group has been considered by some workers (Mints et al., 1980) to be an independent stratigraphic subdivision (the Kachalov and Orlov Formations)

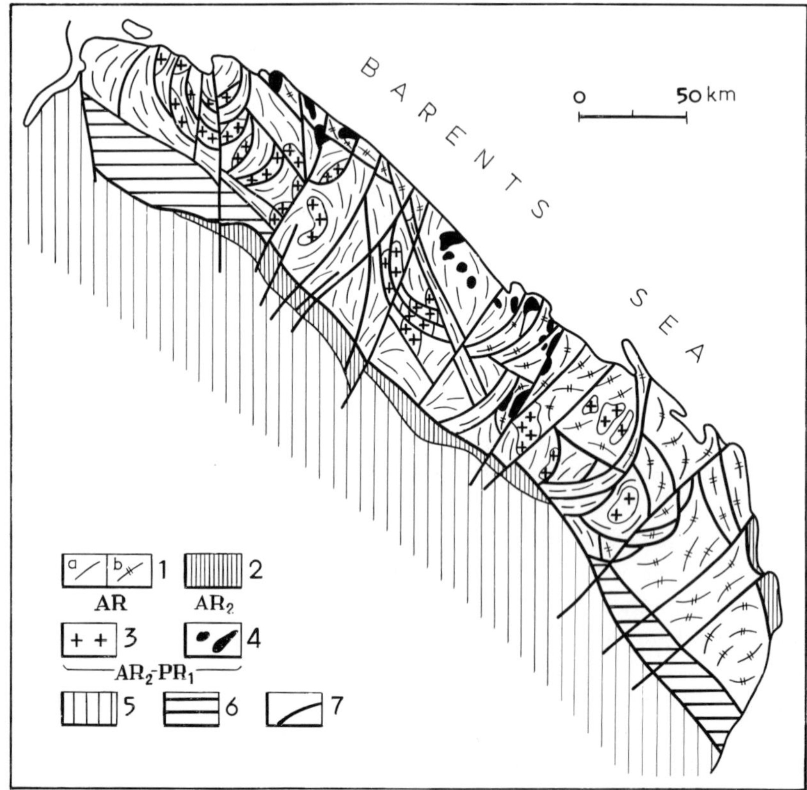

Fig. I-10. Tectonics of the Murmansk segment (simplified from Mints, 1980). *1* = Lower Archaean: *a* = enderbite, tonalite, migmatite, *b* = two-pyroxene schist, hb-bi gneiss (Barents Sea Group); *2* = Upper Archaean (Lopian) sedimentary-volcanic complex; *3* = plag-microcline granite; *4* = mafic dykes and intrusions; *5* = Central Kola segment; *6* = undiff. formations of Murmansk and Central Kola segments; *7* = faults.

which, from structural data, overlie the high-grade metamorphic rocks of the Barents Sea Group. The Kachalov and Orlov Formations may possibly correlate with upper Archaean supracrustal rocks of the Polmos-Porosozero structural belt.

The internal tectonic structure of the Murmansk segment from geophysical data and satellite imagery (Mints et al., 1980) is defined by a combination of arcuate and semicircular faults (Fig. I-10) along which stacking of thrust sheets occurred. Intrusions of plagioclase-microcline granite are found at the base of the nappe pile, which in turn are bounded by the arcuate faults. Several of the mapped ring structures may be eroded tonalite-migmatite domes.

Modern geochronological data (Table I-2) do not allow us to delimit different ages of metamorphic and igneous complexes sufficiently definitively within the lower structural stage of the Murmansk and Central Kola segments. On the whole, during their Early Archaean (Katarchaean) stage of evolution, these segments had much in common with respect to the sequence of tectonometamorphic processes. This allows us to associate structural differences (more widespread lower Archaean metamorphosed supracrustal assemblages in the Central Kola segment) to the elevation and greater degree of erosion of the Murmansk segment during the Proterozoic. This is also confirmed by differences in crustal thickness in the Murmansk and Central Kola segments, 37–38 km and 40–42 km respectively (Kratz et al., 1978).

The *Tersk segment* of the Kola province occupies the territory on the south-western and southern shores of the Kola Peninsula. To the north, it is bounded by the Early Proterozoic Imandra-Varzuga zone of the Karelides, while to the south it is bounded by a deep lineament which passes through the White Sea and Kandalaksha Bay (Fig. I-7). A number of workers (Shurkin et al., 1980; Sudovikov, 1939) consider the Tersk segment to be part of the Belomorian province with its component rocks therefore belonging to the Belomorian Group. However, the geophysical characteristics of this segment as well as certain features of its geological structure and evolution allow it to be treated separately from the Belomorian province. This difference is defined by the absence in the Tersk block of polymetamorphic processes, widespread Proterozoic granite formation and high-alumina gneisses. Furthermore, the Tersk segment is separated from the Belomorian province proper by the Kandalaksha-Kolvitsy granulite zone or the Lapland suture (Kratz et al., 1978). To the south, in the coastal part, early Precambrian rocks of the segment are unconformably overlain by Middle Riphean platform sediments, the Tersk Formation.

Early Archaean rocks of the Tersk segment form the Varzuga Group (Belyayev et al., 1977) or the Kola-Belomoride complex (Ivanov et al., 1983; Kozlov, 1983), but the presence of magnetite schist among the gneisses allows them to be referred to the Kola Group (Kharitonov, 1966). Subdivision of Archaean rocks in the Tersk segment is provisional, since there are no radiometric dates at present and the structural and metamorphic history of the region are extremely poorly known.

The tectonic structure of the lower Archaean complex of the Tersk segment is defined by the aggregation of dome-shaped anticlinal blocks of varying dimensions (the Varzuga, Chvang, Pyalitsa, Ingozero and Berezov) imbricated along nappes towards the Imandra-Varzuga zone (Kozlov, 1979, 1983). The rocks which constitute the Early Archaean complex of the Tersk segment are biotite, two-mica, garnet-biotite, hornblende-biotite and amphibole gneiss and schist, often intensely migmatized and granitized. Despite the complex interleaving of these rock types, it is possible to identify three major units (Ivanov, 1986; Ivanov et al., 1983) which differ in

TABLE 1-5

Average chemical composition of lower Archaean rocks in the Tersk Segment

Oxide	1	2	3	4	5	6
SiO_2	48.77	54.42	58.18	64.67	60.84	71.61
TiO_2	1.01	0.95	0.65	0.42	0.66	0.37
Al_2O_3	16.59	16.72	17.54	15.20	17.75	13.03
Fe_2O_3	3.55	2.89	2.46	1.16	1.91	0.96
FeO	7.49	6.13	4.97	2.79	5.82	3.14
MnO	0.21	0.17	0.12	0.06	0.06	0.06
MgO	6.10	4.10	2.91	1.33	3.28	1.57
CaO	9.43	7.19	5.74	3.66	1.09	1.46
Na_2O	3.31	4.11	3.99	4.74	1.69	3.10
K_2O	1.29	1.66	1.98	1.84	3.89	2.88
P_2O_5	0.29	0.33	0.28	0.15	0.12	0.08
CO_2	0.16	0.19	0.03	0.03	0.11	0.03
H_2O^-	0.08	0.10	0.16	0.13	—	—
H_2O^+	1.25	0.78	0.71	0.56	—	—
n	22	16	6	19	6	8

1 = metabasalt (feldspar and garnet amphibolite), 2 = meta-andesite-basalt (amphibole gneiss), 3 = meta-andesite (amphibole-mica gneiss), 4 = meta-dacite (mica gneiss), 5 = metapelite (sillimanite-garnet-biotite gneiss), 6 = meta-subgreywacke (hornblende-biotite-plagioclase gneiss). *Sources*: Ivanov and Bolotov (1984), Ivanov et al. (1983).

the degree of migmatization and the volume of metavolcanics present. The lowest unit contains every chemical variety of metavolcanic rock from mafic to felsic (Table I-5).

Terrigenous rocks make up the middle unit, represented by biotite, garnet-biotite and garnet-two-mica gneiss with sillimanite, kyanite and staurolite. Chemically the unit corresponds to metamorphosed pelite, greywacke and sub-greywacke (Table I-5).

Early Archaean intrusive and ultrametamorphic rocks. The Early Archaean intrusive complex of granitoids or group II granites embraces a wide spectrum of rocks — tonalite, diorite, enderbite and charnockite. Establishing the relationships between these rock types and between the rocks of the Kola Group in most cases is extremely difficult due to their spatial separation, regional Late Archaean microclinization and superimposed amphibolite and granulite facies metamorphic events.

The relative age of the granitoids in this group has been determined from their effects on the gneisses and schists of the Kola Group and their occurrence as pebbles in conglomerates from Late Archaean greenstone belts.

Two suites can be distinguished in these Early Archaean granitoids, based on primary origin and, or subsequent metamorphism: tonalite (trondhjemite) and enderbite-charnockite.

Rocks of the enderbite-charnockite suite occur as separate intrusions or as parts of intrusions, mainly in the SW and SE parts of the Murmansk segment (the Kanentyavr intrusion, Vinogradov and Vinogradova, 1981),

TABLE I-6

Average composition of lower Archaean tonalite complex

n	SiO_2	TiO_2	Al_2O_3	Fe_2O_3	FeO	MnO	MgO
67	68.69	0.39	15.63	1.19	2.02	0.06	1.09
	CaO	Na_2O	K_2O	P_2O_5	CO_2	H_2O^-	H_2O^+
	3.54	4.87	1.42	0.10	0.10	0.12	0.11

Source: Vetrin, (1983).

the middle of the Central Kola segment (the Chudzyavr zone, Batiyeva et al., 1985) and in the junction zone between these segments (Vezhetundry, Avakyan et al., 1984). The three main chemical types are tholeiite, andesite and crustal anatectic series (Shurkin et al., 1980). For all the Early Archaean granitoids Na is typically > K, Fe is low and Mg high (Table I-6).

The upper Archaean Lopian complex

There is no consensus at present among research workers in the Kola Peninsula concerning the extent and boundaries of lower Archaean and upper Archaean (Lopian) complexes, due to the effects of Late Archaean–Early Proterozoic high-grade metamorphism, granite formation and intense folding in the region. Nevertheless, new geochronological, petrological and lithoformational data allow us to identify in some structures of the region an independent upper Archaean (Lopian) structural-lithological terrain. This

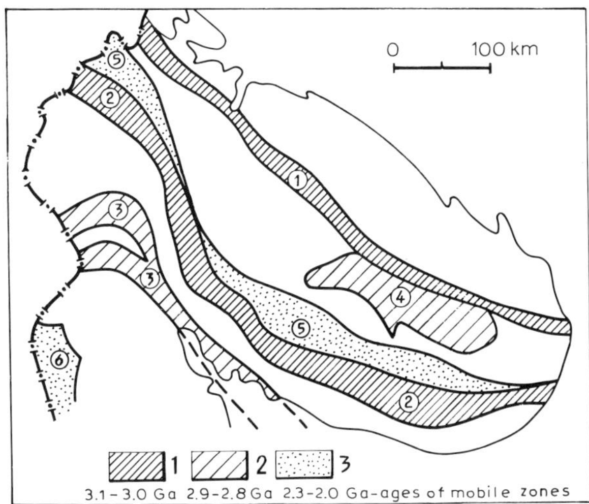

Fig. I-11. Sketch showing palaeotectonic reconstruction of Late Archaean and Proterozoic mobile zones of the Kola Peninsula. Numbers in circles: *1* = Kolmozero-Voronya and *2* = Tersk-Allarechka greenstone belts; *3* = Korva-Kolvi and *4* = Keyvy zones; *5* = Pechenga-Imandra-Varzuga and *6* = Panajärvi zones.

terrain incorporates supracrustal volcanogenic-sedimentary and associated intrusive rocks, the majority of which are considered to be analogous to the rocks of greenstone belts (Vrevsky, 1985; Vrevsky and Kolychev, 1986).

At the present erosion level, the greenstone belts of the Kola province are relicts of extensive zones (Fig. I-11) which have been preserved after

Fig I-12. Rhythmically-bedded unit of polymict conglomerates, Lake Litsa, Polmos-Porosozero belt.

the Late Archaean granite formation, metamorphism and folding. The geological position of the greenstone belts in the Kola Peninsula is defined on the one hand by the fact that they are overlain by lower Proterozoic (Karelian) formations and are intruded by plagioclase-microcline granites with an age of 2.6–2.7 Ga, and on the other hand they are unaffected by the Early Archaean deformation, granulite metamorphism, migmatization and plagioclase granitization which are of regional extent in the basement to the greenstone belts. In a number of regions, the supracrustal complexes of the greenstone belts overlie the Early Archaean rocks of the basement above an erosional unconformity, expressed as a basal gneissose conglomerate horizon (Figs. I-12–I-14; Table I-7).

The overall structural plan of the greenstone belts is linear in form, but in a number of cases the tectonic pattern of greenstone structures (Tersk, Allarechka and Kaskam) is influenced to a significant extent by Early Archaean basement blocks, which may have been partly remobilized later in the Archaean, and also by major diapiric intrusions of plagioclase-microcline granite (Fig. I-11). The internal structure of the greenstone belts is most commonly tight synforms and monoclines, sometimes asymmetrical. The stratigraphy of individual upper Archaean volcanogenic-sedimentary rocks has been dealt with in a number of publications (Belkov, 1963; Belolipetsky et al., 1980; Gilyarova, 1967; Makiyevsky and Zagorodny,

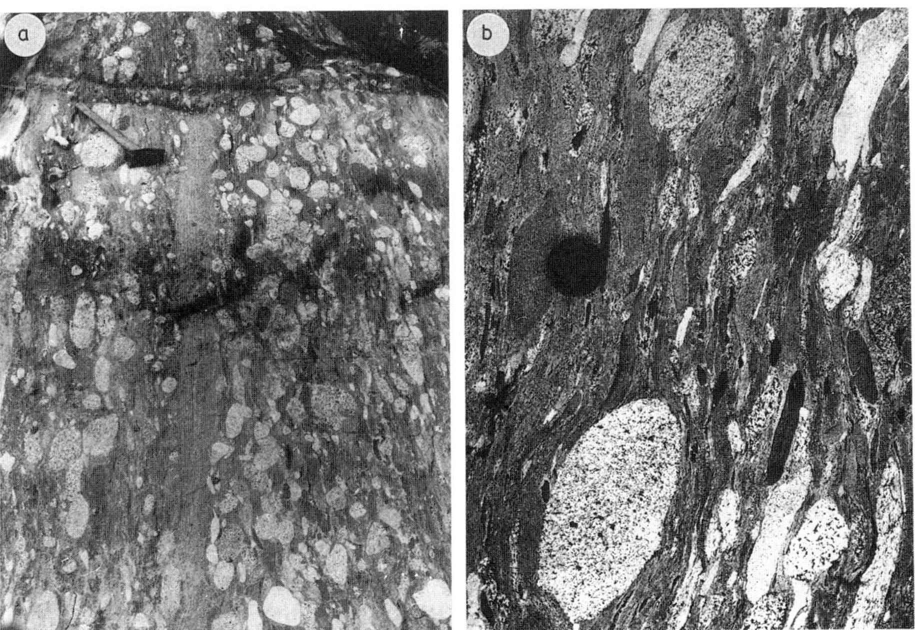

Fig. I-13. Polymict conglomerates, Lake Voche-Lambi, Cis-Imandra belt.

TABLE I-7

Correlation of Late Archaean supracrustal assemblages in the Kola Peninsula

Age, Ga	Kolmozero-Voronya belt	Tersk-Allarechka belt					Korva-Kolvitsy zone			Keyvy zone
	Polmos-Porosozero structure	Kaskam structure	Allarechka structure	Cis-Imandra structure	Trans-Imandra structure	Tersk structure	Kolvitsy structure	Korva structure	Terma-Kerka structure	Keyvy structure
Karelian = 2.60±0.1	2730 (G—A)			(G—A)	(G—A)		2700 (G-A)	(G—A)	(G—A)	2750 (G—A)
	Chervurt Fm (Al)	Talyin Formation		Arvarench Fm (A—D) 2450	Volcheozero Formation (Al)	Bezymyannaya (Sergozero) Fm	Jaurijoki Fm	Korva Fm (Al)		Keyvy Fm (Al)
	Voronyetundra (B—A—D) (Fe) 2740 2930	(B—A—D)		Viteguba Fm (B—A—D)	Olyenye (Trans-Imandra) Fm (B—A—D) (Fe) 2560	(Babozero member) (B—A—D)	Poreguba Fm (D–R)			Lebyazhi Fm (D—R) 2780
Lopian							Ploskotundra Fm (AB-D)			Ponoy (Patcher-tundra) Fm (AB—D)
		Kaskam				Pyaloch	Kandalaksha Fm (B-A) (K) 2700	Podas Fm (B—A)	Kerka (B—A) (Fe)	Kolovay member (K)
	Polmostundra Fm	Formation (K—T)	Allarechka (Annam) Fm (K—T) (Fe)	Kisloguba (K—T)		Formation (Purnach member) (K-T)				
3.1±0.1 Saamian	Lyavozero Fm (K) 2790	Vyrnim Fm	Koposov Fm	Vochelambi Fm (K) 3080 2530	Rovkvun Fm	Peschano-ozero (Pulon) Fm (K)				

Symbols: Isotopic age: [box] = Rb–Sr isotopic, [double box] = Pb–Pb isochron, [box with dashed sides] = U–Pb isochron, and [dashed box] = ^{207}Pb–^{206}Pb Zr evaporation technique on zircons; (K) = basal gneissose conglomerate horizon; (Fe) = ferruginous quartzite (banded ironstone); (G–A) = gabbro-anorthosites; (Al) = high-alumina sediments. Volcanic formations: (K–T) = komatiite-tholeiite, (B–A–D) = basalt-andesite-dacite, and members (AB–D) = andesite-basalt-dacite, (B–A) = basalt-andesite, (D–R) = dacite-rhyolite.

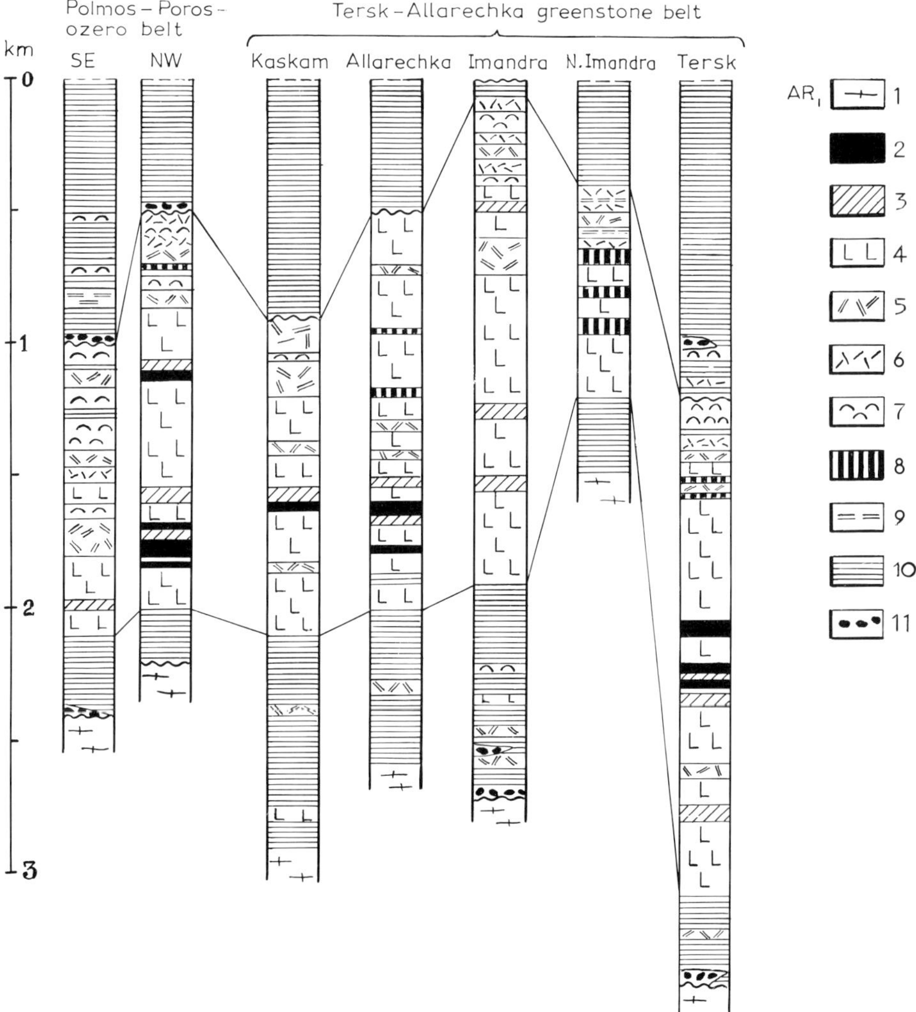

Fig. I-14. Correlation chart of summary stratigraphic sections of Late Archaean greenstone belts of the Kola Peninsula. *1* = basement granite-gneiss (AR_1); *2* = peridotitic komatiite; *3* = pyroxenitic and basaltic komatiite; *4* = basalt; *5* = andesite; *6* = dacite; *7* = rhyolite; *8* = banded ironstone; *9* = chert; *10* = arkose, greywacke, pelite; *11* = conglomerate.

1981; Mirskaya, 1978; Kharitonov, 1966), some of whose data have been used in compiling Table I-7. The Kolmozero-Voronya greenstone belt has been selected as a tectonotype and stratotype, since it is geologically and geochronologically the best studied.

The Kolmozero-Vornya greenstone belt

The volcanogenic-sedimentary units of this belt constitute the major Polmos-Porosozero structure (Fig. I-15) as well as a number of minor synclines and monoclines in its N–W extension. The most detailed geological and geochronological work has been done on the *Polmos-Porosozero structural belt* (Vrevsky, 1985; Vrevsky and Kolychev, 1986; Belolipetsky et al., 1980). The Polmos-Poros complex consists of four volcano-sedimentary formations, the Lyavozero, Polmostundra, Voronyetundra and Chervurt (from bottom to top).

The *Lyavozero*, or lower terrigenous formation, consists of garnet-biotite and biotite schist, forming narrow outcrops which can be traced along the entire SW flank of the structure, while on the NE flank they are observed only sporadically. In the NW part of the belt, the formation is 100–150 m thick, and 300–500 m thick in the SE (Fig. I-14).

The *Polmostundra Formation*, or komatiite-tholeiite series, is the thickest of a number of amphibolite units, most of which are metamorphosed tholeiitic basalts with thin interbedded volcano-sedimentary rocks (Table I-8, Fig. I-16). Peridotitic komatiite, interleaved with pyroxenitic and basaltic komatiite have also been discovered here (Vrevsky, 1980, 1985). Peridotitic komatiites form several flows, totalling over 250 m in thickness, each of which has an upper zone displaying spinifex textures (Fig. I-16). Individual

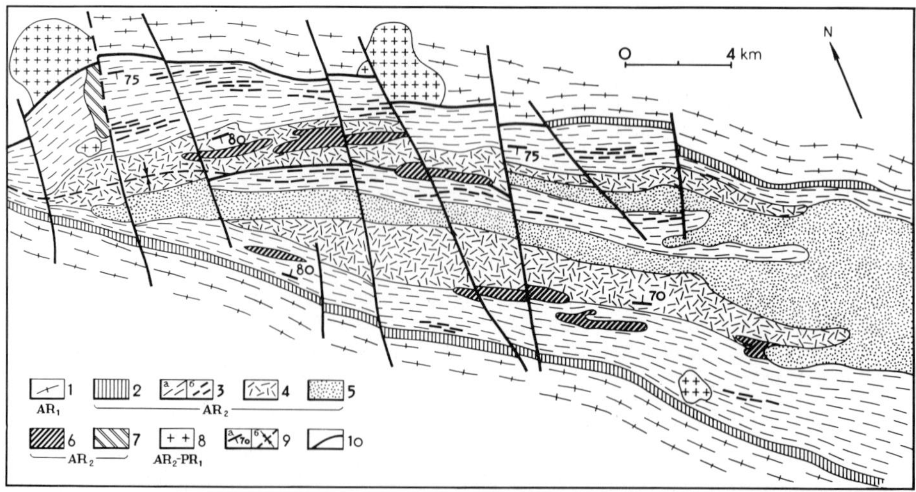

Fig. I-15. Geological sketch map of the NW part of the Polmos-Porosozero belt. *1* = tonalite gneiss, plag-microcline migmatite, granite, granite-gneiss; *2* = Lyavozero Formation (gt-bi schist); *3* = Polmostundra Formation (*a* = metatholeiite, *b* = komatiite); *4* = Voronyetundra Formation (metabasalt, andesite, dacite); *5* = Chervurt Formation (aluminous gneiss and schist); *6* = gabbro-amphibolite; *7* = metaperidotite; *8* = microcline granite; *9a* = dip and strike, *9b* = synclinal axes; *10* = faults.

TABLE I-8

Average chemical compositions of volcanics in the Polmos-Poros complex

No.	*n*	SiO_2	TiO_2	Al_2O_3	Fe_2O_3	FeO	MnO	MgO	CaO	Na_2O	K_2O	Ni	Cr	Rb	Sr	Co
1	10	39.71	0.23	3.43	7.65	7.25	0.19	28.85	3.42	0.08	0.04	1896	4265	—	18	147
2	32	39.09	0.19	3.72	7.28	6.98	0.20	28.80	3.02	0.08	0.05	1845	4039	—	41	151
3	22	46.58	0.48	9.25	2.10	9.46	0.19	17.12	9.72	0.48	0.15	430	2952	—	49	86
4	13	48.89	0.72	14.96	1.86	8.93	0.18	9.43	10.79	1.72	0.10	254	—	—	146	85
5	19	49.21	0.88	15.06	2.31	9.25	0.19	7.30	9.84	2.44	0.25	110	—	—	155	64
6	12	48.85	0.77	15.24	2.15	9.19	0.19	8.68	10.34	1.56	0.17	73	n.d.	n.d.	73	38
7	15	57.58	0.80	14.81	2.03	6.87	0.15	5.09	7.46	2.89	0.57	n.d.	n.d.	n.d.	43	n.d.
8	24	70.30	0.39	14.07	0.78	2.23	0.09	1.66	3.62	3.50	1.55	n.d.	n.d.	n.d.	n.d.	n.d.

No.	*n*	La	Ce	Sm	Eu	Gd	Tb	Yb	L	Y	Zr
1	1	0.155	0.450	0.084	0.040	—	0.020	0.070	0.012	5	11
2	1	0.173	0.695	0.159	0.078	—	0.061	0.220	0.038	5	14
3	1	0.366	1.100	0.263	0.088	—	0.065	0.260	0.031	9	23
3a	1	1.300	3.730	0.995	0.407	1.520	0.283	1.090	0.180	—	—

Notes: Ni, Cr, Rb, Sr, Co, REE, Y, Zr in ppm; n.d. = element not determined; *n* = number of analyses. 1–5—komatiite-tholeiite series: 1 = spinifex textures, 2 = cumulate peridotitic komatiite, 3 = pyroxene komatiite, 3a = basaltic komatiite, 4 = Olenye Range metabasalt, 5 = ditto, from Polmostundra Formation; 6–8—basalt-andesite-dacite series: 6 = metabasalt, 7 = meta-andesite, 8 = acid metavolcanics. *Source*: Vresky (1985).

Fig. I-16. Peridotitic komatiite with spinifex texture, Mt Leshaya, Polmos-Porosozero belt.

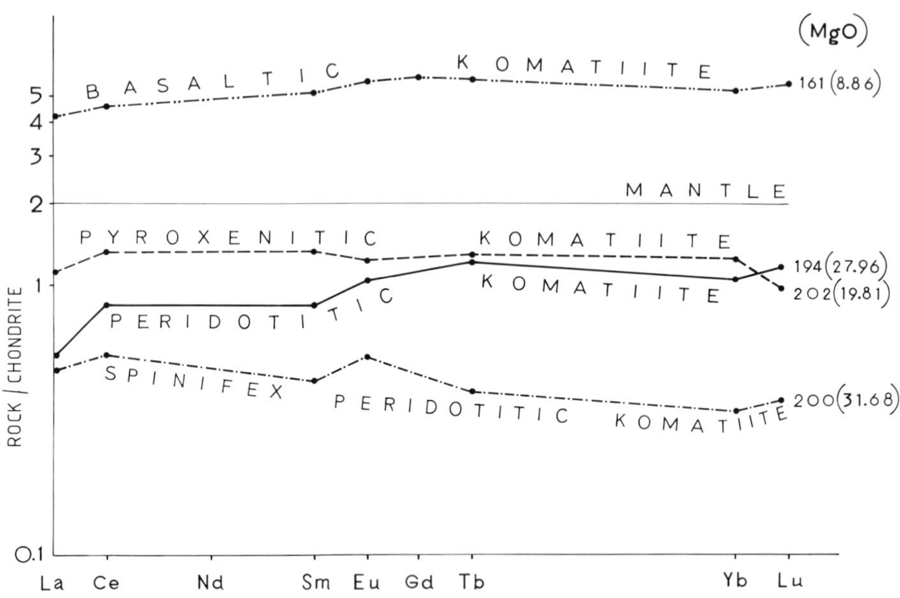

Fig. I-17. Distribution of rare-earth elements, chondrite-normalised, in komatiites from the Polmos-Porosozero belt.

flows can be traced for 1500–1800 m along strike. Basaltic komatiite is found in close spatial association with pyroxenitic and peridotitic komatiite and in a number of cases they form independent flows among tholeiitic metabasalts. In terms of petrochemistry and REE distribution (Table I-8, Fig. I-17), the komatiites are comparable with similar rock types in many greenstone belts around the world.

The *Voronyetundra Formation* or basalt-andesite-dacite series (Table I-8), is an 800 m thick unit, the constituent rocks of which have the most varied composition. These are metamorphosed mafic volcanics, interleaved with intermediate and felsic metavolcanics which increase in volume upwards, eventually dominating the top part of the succession (Fig. I-14).

The *Chervurt Formation*, or upper terrigenous unit, consists of alumina-rich gneiss and schist with andalusite in the NW and kyanite in the SE parts of the belt. Polymict conglomerate has been found at the base, in which the pebbles contain oligoclase granite together with practically all varieties of the underlying rocks, including komatiite.

The supracrustal rocks of the Polmos-Porosozero belt have been affected by five episodes of folding and were metamorphosed at amphibolite facies, the thermodynamic conditions of which corresponded to the transition in baric type from andalusite-sillimanite to kyanite-sillimanite facies series (Vrevsky, 1985). From geochronological data (Table I-7), the major features of the structure were established 3.0 Ga ago, while metamorphism took place in the interval 2.7–2.8 Ga. The widespread K–Ar dates of the order of 2.0–1.9 Ga from the SE part of the belt are associated with the opening of the isotopic system as a result of block faulting and the intrusion of alkali granites.

The Tersk-Allarechka greenstone belt

This belt is the largest Late Archaean structure in the Kola Peninsula. It extends from SE to NW for over 40 km and is 60 to 20 km wide, and in a sense frames the Pechenga and Imandra structures (Fig. I-18).

The greenstone belt consists of a number of smaller synclines and monoclines (Tersk, Cis-Imandra, Trans-Imandra iron-ore belt, Allarechka and Kaskam), which gradually pass one into another or are separated by major regional faults.

The *Tersk structural belt* is situated on the southern margin of the Imandra-Varzuga zone within the Tersk segment and extends from the mouth of the White Sea to the middle reaches of the river Umba. It is separated from the Imandra-Varzuga structure by a tectonic zone. The volcanic and sedimentary rocks of the Tersk belt form several small synclines and monoclines around the margins of dome-shaped protrusions of the granite-gneiss basement (Fig. I-18; Ivanov, 1986; Ivanov and Bolotov, 1984; Ivanov et al., 1983), which are clearly delineated on the basis of their low magnetic and gravity characteristics.

The supracrustal rocks of the Tersk belt are subdivided into three

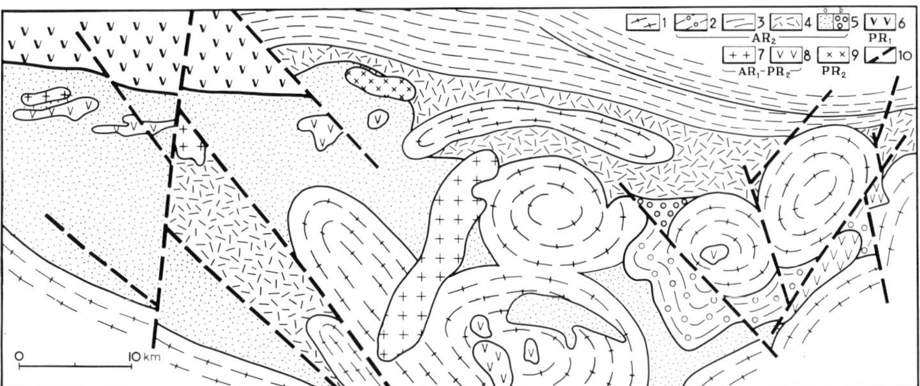

Fig. I-18. Geological sketch map of the eastern part of the Tersk belt (after Ivanov and Bolotov, 1984; Kozlov, 1979; Petrov and Voloshina, 1978; Kratz et al., 1979). *1* = tonalite, plag-microcline gneiss, granite, migmatite; *2* = Peschano-ozero (Pulonga) Formation (bi-ga paraschists); *3* = Pyaloch Formation (metabasalts, partly komatiites); *4* = Babozero Formation (metadacite and meta-andesite); *5* = Bezymannaya (Sergozero) Formation (*a* = bi, bi-musc, bi-hb paragneiss, *b* = conglomerate); *6* = Karelian complex, Imandra-Varzuga Group; *7* = microcline granite; *8* = gabbro-amphibolite; *9* = alkaline granite; *10* = faults.

TABLE 1-9

Average chemical composition of major rock types in the Tersk-Allarechka greenstone belt

Oxide	1	2	3	4	5	6	7	8	9	10	11	12
No. anal.	18	10	1	3	21	17	20	11	7	5	11	18
SiO_2	48.84	55.00	58.56	67.24	50.66	58.89	70.69	49.50	47.16	44.48	38.39	40.14
TiO_2	1.34	1.06	0.66	0.71	1.11	0.84	0.33	1.42	0.56	1.17	0.20	0.48
Al_2O_3	14.21	16.60	13.64	11.94	14.64	17.09	15.90	13.79	17.60	9.45	1.63	1.90
Fe_2O_3	3.42	2.68	1.83	2.12	3.04	2.12	0.74	2.76	3.16	2.52	3.25	3.89
FeO	9.97	8.15	6.75	4.41	9.58	6.05	1.73	10.22	9.82	9.94	11.23	11.85
MnO	0.26	0.16	0.15	0.11	0.18	0.12	0.03	0.20	0.23	0.24	0.20	0.24
MgO	6.51	4.32	4.60	1.35	7.13	3.94	0.75	6.95	6.64	18.14	39.11	36.00
CaO	10.32	7.35	4.37	3.52	10.67	6.20	2.33	9.58	11.65	8.91	2.08	1.80
Na_2O	2.23	3.46	4.42	3.85	2.39	3.69	5.28	2.53	1.26	0.99	0.23	0.30
K_2O	0.66	0.93	2.17	2.40	0.58	1.25	2.20	0.84	0.30	0.27	0.12	0.39
Structures	Tersk				Trans-Imandra			Kaskam–Allarechka				

1–4: Pyaloch Formation (Ivanov and Bolotov, 1984): 1 = tholeiitic basalt, 2 = andesite-basalt, 3 = andesite, 4 = dacite; 5–7: Trans-Imandra Formation (Kratz et al., 1978): 5 = basalt, 6 = andesite, 7 = dacite; 8 = basalt, Annam Formation; 9 = basalt, Kaskam Formation (Belyayev, 1978); 10 = picrite, Kaskam Formation (Bolotov, 1983); 11 = olivinite, 12 = harzburgite (Zak, 1980).

formations, the general succession being as follows, from bottom to top:

The *Peschano-ozero* or *Pulonga Formation* consists of biotite and biotite-garnet schist with bands of amphibolite and hornblende-biotite schist. A polymict conglomerate member lies at the top of the formation (Fedorova and Shustova, 1980).

The *Pyaloch Formation* is a thick unit of metavolcanics which is subdivided into two. The lower *Purnach* member consists of amphibolite and hornblende schist showing a variety of textures and compositions. The commonest types are garnet and feldspar varieties, as well as monomineralic amphibolite which correspond in composition to tholeiitic basalt and in part to komatiite (Table I-9). The upper *Babozero* member consists of up to 600 m of various feldspar amphibolite, amphibole, hornblende-biotite and amphibole-two-mica schists, which are mainly metamorphosed acid volcanics, intermediate and basic volcanics being less common (Table I-9). Individual horizons and bands of muscovite and ferruginous quartzites and biotite-muscovite gneiss appear at the top of the formation.

The *Bezymyannaya* or *Sergozero Formation* occurs mainly in the NW part of the structure (Fig. I-18). The rocks constituting this formation can be divided into two main groups: (a) biotite, biotite-muscovite and hornblende-biotite schist and gneiss, often containing garnet, staurolite and kyanite, account for about 80% of the volume of the formation; and (b) amphibolite, amphibole and biotite-amphibole schist, which are metamorphosed basic to intermediate volcanics. The rocks of the Bezymyannaya Formation unconformably overlie both the granite-gneiss of the basement and the rocks of the Pyaloch Formation.

All the rocks of the Tersk belt were metamorphosed at low temperature amphibolite facies conditions, initially in the andalusite-sillimanite and later the kyanite-sillimanite facies series (Petrov et al., 1986; Belyayev et al., 1977).

The *Cis-Imandra belt* is the north-western continuation and closure of the Tersk greenstone belt on the west shore of Lake Imandra. The NW and W flanks of the structure are bounded by the Glavny Range gabbro-labradorite intrusions (Fig. I-19).

The supracrustal assemblage of the Cis-Imandra belt has been studied fairly well in cross-section both geologically and geochronologically (Gilyarova, 1972; Latyshev, 1984) and serves as a reference for the entire Tersk-Allarechka greenstone belt, although a number of workers correlate these rocks with the Karelian complex.

At the base of the supracrustal assemblage lies a gneissose conglomeratic unit, the *Vochelambi Formation*, up to 800 m thick, consisting of banded biotite, garnet-biotite and biotite-muscovite gneiss and schist (meta-

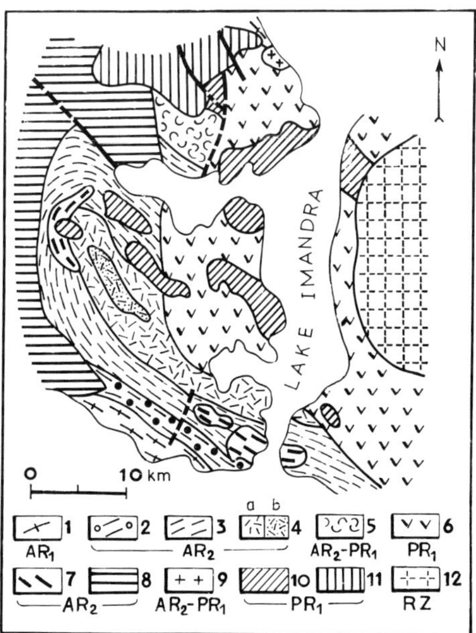

Fig. I-19. Geological sketch map of the Cis-Imandra belt (modified from Zagorodny et al., 1982; Kharitonov, 1966). *1* = tonalite, plag-microcline gneiss, granite, migmatite; *2* = Vochelambi Formation (paragneiss, conglomerate, metavolcanics); *3* = Kisloguba Formation (komatiite, metabasalt, meta-andesite-basalt); *4* = Viteguba Formation (*a* = metabasalt, meta-andesite, *b* = aluminous gneiss and schist); *5* = Arvarench Formation (meta-andesite, dacite, rhyolite); *6* = Karelian complex (Imandra-Varzuga Group). Intrusions: *7* = gabbro-amphibolite, *8* = gabbro-labradorite of the Glavny Range; *9* = granodiorite, *10* = gabbro-norite-diorite, *11* = Monchegorsk pluton, ultrabasic-gabbro-norite, *12* = nepheline syenite.

morphosed arkose, greywacke and tuff) interbanded with meta-andesite, metadacite (biotite-quartz-plagioclase schist) and metabasalt, which increase regularly up through the succession. In the middle of the formation are several lenticular horizons of polymict conglomerate (Fig. I-13a, b). Rocks of the Vochelambi Formation are separated from underlying plagioclase gneiss and granite-gneiss by a tectonic zone of blastomylonite (Latyshev, 1984).

The overlying *Kisloguba Formation* consists of interbanded amphibolite, amphibole and biotite-amphibole schist with thin bands of biotite and garnet-biotite gneiss and schist, which gradually give way upwards to a monotonous unit of schistose amphibolite, the *Viteguba Formation*. In the upper part of the succession, the Viteguba Formation contains thin bands of highly aluminous schist with staurolite, kyanite, andalusite and, less commonly, cordierite.

Around the northern shores of Lake Imandra near Ar-Varench is an isolated outcrop of volcanic rocks, referred to as the *Arvarench Formation*, the rocks of which are intruded by gabbro-labradorite of the Glavny Range. At the base are metamorphosed amygdaloidal basalt, pyroxenite and basaltic komatiite, which in many respects are similar to the basic metavolcanics of the Viteguba Formation. The sequence extends to a 500 m thick series of interbanded acid and intermediate volcanics. Palaeovolcanic reconstructions (Kozlov et al., 1974) suggest a fissure-type subvolcanic structure.

The stratigraphic position of the Arvarench Formation is open to debate. A number of workers (Gilyarova, 1972; Kharitonov, 1966) consider that it belongs to the pre-Karelian (Lopian) complex, while others (Kozlov et al., 1974; Petrov et al., 1986) refer it to the topmost Early Proterozoic division.

Data obtained by the Leningrad Institute of Precambrian Geology and Geochronology (Table I-8) for supracrustal rocks of the Cis-Imandra structure suggest that it was initiated around 3.0 Ga ago and that its development was completed by 2.5 Ga, this being the time of formation of the plagioclase-microcline granites which intrude rocks of the Vochelambi Formation. Acid metavolcanics of the Arvarench Formation have a similar age.

The Trans-Imandra (Olenegorsk) structure. Supracrustal assemblages in this belt form a series of narrow synclinal and monoclinal structures, elongate in a near N–S orientation, surrounding oval and lensoid dome-like basement blocks (Fig. I-20). These blocks consist of Early Archaean tonalite-migmatite, granodiorite and diorite, which were partly remobilized and subjected to microclinization in the Late Archaean.

The following is a summary section of the supracrustal rocks of the Trans-Imandra structure, from the base upwards (Goryainov, 1976; Makiyevsky and Zagorodny, 1981).

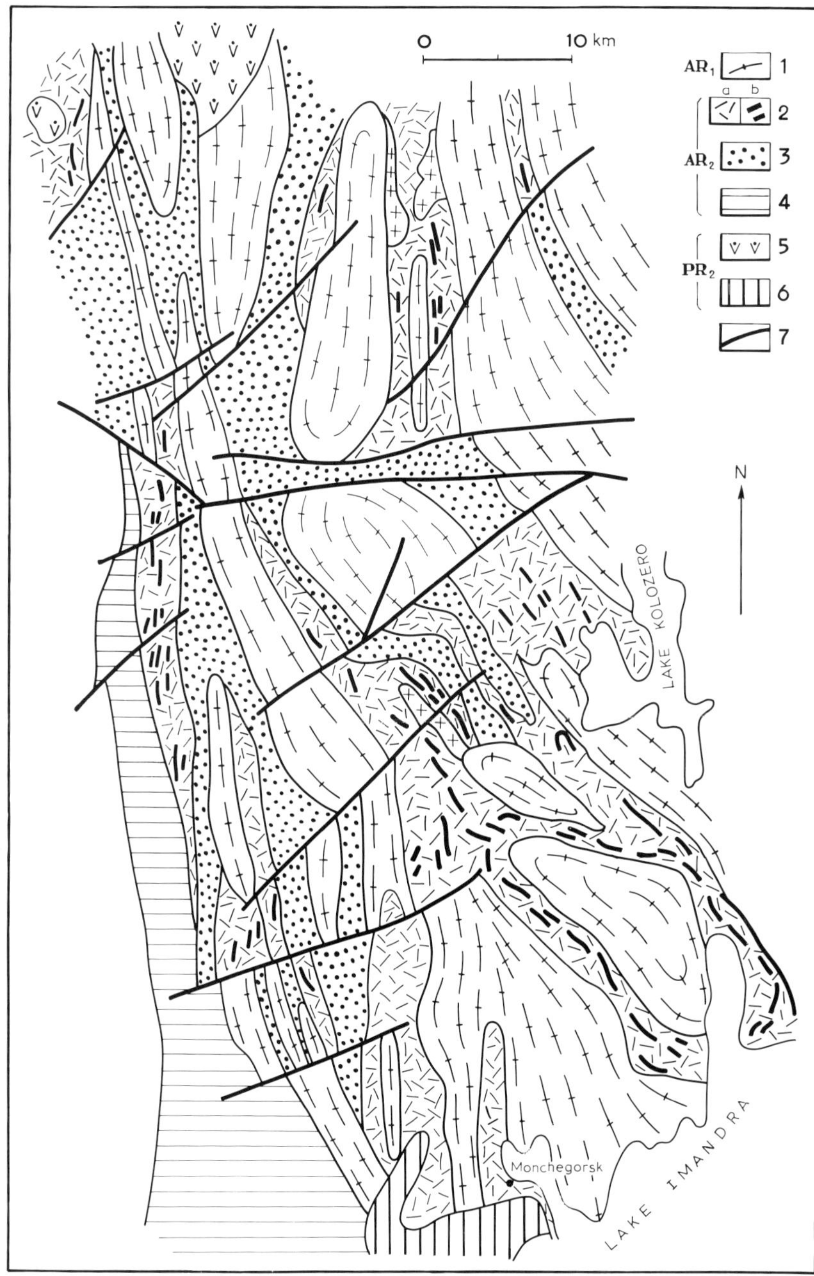

Fig. I-20. Geological sketch map of the Trans-Imandra iron ore belt (simplified from Goryainov, 1976). *1* = tonalites, plag-microcline gneiss, granite, migmatite; *2* = Imandra (Olyenye) Formation (*a* = metabasalt, andesite, dacite, rhyolite, *b* = ferruginous quartzite); *3* = Volcheozero Formation (aluminous gneiss and schist); *4* = Glavnyy Range gabbro-labradorite; *5* = alkaline gabbroid; *6* = ultrabasic-gabbro-norite, Monchegorsk pluton; *7* = faults.

(1) Biotite and garnet-biotite gneiss with sillimanite (the *Rovkvun Formation*), 100–500 m thick, corresponding in composition to arkose and greywacke with intercalated tuff.

(2) The *Trans-Imandra (Olenye) Formation*: amphibolite, hornblende and biotite gneiss with thin bands of metavolcanics and chert. Productive horizons up to 200 m thick and major banded ironstone deposits are restricted to the middle part of the succession, and are overlain by a unit of amphibole and biotite gneiss and metamorphosed acid volcanics. The formation varies in thickness from 200 to 600 m. The metavolcanics which comprise the greater part of the section form a complete series from metabasalt to metarhyolite (Table I-9).

(3) The section concludes with the 400 m-thick *Volcheozero Formation* of aluminous gneiss.

At the present time, the age and structural position of the supracrustal rocks of the Trans-Imandra iron ore belt remain a matter of dispute. Several workers (Bondarenko and Dagelaysky, 1968) consider them to be a constituent of the Early Archaean Kola Group, while others (Goryainov, 1976) assign them to the Late Archaean.

We consider that the Trans-Imandra iron ore belt belongs to the Tersk-Allarechka greenstone belt (or the upper Archaean Lopian complex) on the following basis:

(a) The similarity between the sedimentary-volcanic complex and the stratotype section of the Lopian complex in the Polmos-Porosozero belt.

(b) The structural and metamorphic evolution of the rocks in the Trans-Imandra structure encompassed five stages of deformation (Kazakov, 1979) and two episodes of amphibolite facies regional metamorphism, the first of which belongs to the andalusite-sillimanite and the second to the kyanite-sillimanite facies series (Yevdokimov et al., 1978). This coincides exactly with information available from the Polmos-Porosozero belt (Drugova et al., 1982) and does not correlate with the Kola Group, in which 12 deformation stages have been identified (Dobrzhinetskaya, 1978).

(c) The age of the first stage of metamorphism in the rocks of the Trans-Imandra belt has been determined by the Rb–Sr isochron method at 2560 ± 60 Ma (Table I-8), while the K–Ar dates on biotites and amphiboles in the interval 1750–1580 Ma probably correspond to the time of the second stage (Gorokhov et al., 1981).

The *Kaskam-Allarechka belt* extends in an approximately east–west direction for over 150 km (Fig. I-21) from the USSR–Finland–Norway state boundary to the Verkhnetulom reservoir in the SE. Its southern margin extends along a regional fault which separates it from the Lotta-Salnotundra block of the Kola Peninsula Granulite belt, while its northern margin coincides with the Poritash fault zone which separates it from the Pechenga belt. Tectonically the Kaskam-Allarechka structure is subdivided into two blocks, the Allarechka in the east and the Khihnajärvi (or Kaskam) in the west, separated by the Veshyaur crush zone.

Broad dome-like anticlinal granite-gneiss and granodiorite structures of the lower Archaean basement complex are widespread in both blocks, containing relict gneisses and schists of the Kola Group and major plagioclase-microcline granite intrusions (Belyayev and Zagorodny, 1974; Fig. I-21).

The Late Archaean supracrustal complexes form, on the one hand, large-scale synclinal structures (the Kaskam block), in which the succession is most fully represented and contains the thickest volcanic sequences, while on the other hand they form minor synclines and monoclines between blocks (the Allarechka block), in which the succession is characterised by a more widespread development of continental clastic sequences.

The *Kaskam (Khihnajärvi) block* was previously considered by many

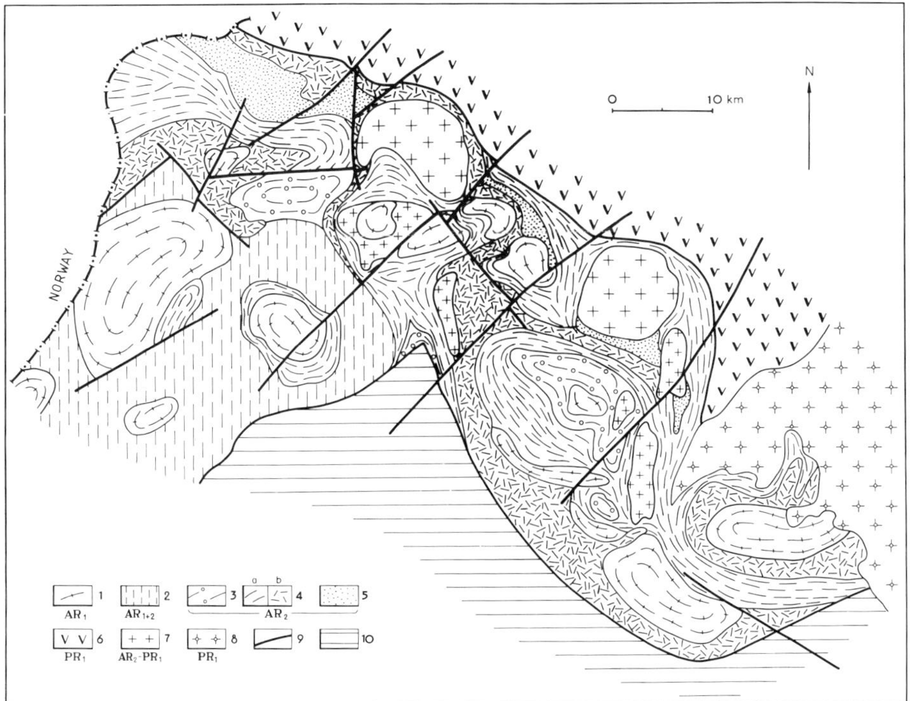

Fig. I-21. Geological sketch map of the Kaskam-Allarechka belt (from Belyayev, 1978; Belyayev and Zagorodny, 1974; Gilyarova, 1972; Kozlovsky, 1987). *1* = tonalite, plag-microcline gneiss, granite, migmatite; *2* = undiff. gneiss and schist, Kola Group and Lopian complex; *3* = Koposov and Vyrnim Formations (metagreywacke); *4* = Kaskam and Allarechka Formations (*a* = komatiite, picrite, metabasalt, *b* = meta-andesite, bi-gt paragneiss and paraschist); *5* = Talya Formation (two-mica, bi-gt paragneiss and paraschist); *6* = Karelian complex (Pechenga Group); *7* = granodiorite and plag-microcline granite; *8* = granite, granodiorite, granosyenite; *9* = faults; *10* = Lotta-Salnotundra block.

workers (Zagorodny et al., 1964) to be the southern eroded limb of the Pechenga belt. However, Gilyarova (1967) and later many other researchers (Belyayev, 1978; Kremenetsky, 1979; Kozlovsky, 1987) confirmed that the supracrustal rocks of the Kaskam block are older than those of the Pechenga structure. The upper Archaean volcanosedimentary complex lies with angular unconformity (near Mt Kuroaivi) on a granite-gneiss basement.

The complex is divided into the following three formations, from bottom to top: the *Vyrnim Formation*, 1000–1200 m thick, consisting of biotite and garnet-biotite gneiss and schist of greywacke composition, sometimes strongly migmatized and granitized (Bolotov et al., 1981); the overlying *Kaskam Formation* consists of a number of structural varieties of amphibolite and hornblende-biotite schist, corresponding compositionally to tholeiitic basalt (Table I-9). Biotite and garnet-biotite schist occur in the upper part. Among the amphibolites are highly magnesian rocks, corresponding to picrite and pyroxene komatiite (Table I-9). The supracrustal complex concludes with a unit of interleaved two-mica and biotite-garnet gneiss and schist (the *Talya Formation*), formed from clay-rich continental sediments (Kremenetsky, 1979; Suslova, 1971). The formation is estimated to be 700–900 m thick. Rocks of the Talya Formation occupy small synclines, situated in the N and SE of the Kaskam-Allarechka structural belt.

In the *Allarechka block* the supracrustal formations form narrow synclines and monoclines, bordering dome-like basement blocks (the Koposov, Allarechka, Annam, etc.; Fig. I-21) and are subdivided into two formations. The lower *Koposov Formation* consists of various paragneisses and schists of unknown thickness. The overlying *Allarechka* (or *Annam*) *Formation* consists of feldspathic amphibolite, amphibole, hornblende-biotite and biotite schist, sometimes with structural evidence of volcanics, which correspond in composition to picrite, tholeiite, andesitic basalt and andesite (Table I-9) with tuff bands. Among the biotite and biotite-amphibole schist are patches of interleaved manganese rocks (garnet-cummingtonite-pyrrhotite schist), sulphide-graphite rocks and ferruginous quartzite (Bolotov et al., 1981). Sheet-like and lenticular ultrabasic bodies, some 200 in total, occur within the feldspathic amphibolite units and were folded and cleaved with them (Bogachev et al., 1964; Zak, 1980; Gorbunov et al., 1985). A number of ultrabasic bodies occur in the granite-gneisses of the basement. Two Kola Peninsula copper-nickel deposits, the Allarechka and Vostok, are associated with the meta-ultrabasics (olivinite, harzburgite, pyroxenite, Table I-9). Most workers consider these meta-ultrabasics to be intrusions, although there is no general consensus about their age, which is considered to be from Early Archaean to Early Proterozoic. However, in a number of cases these "bodies" are really interleaved sheets (flows?) of ultrabasic rocks with amphibolite, amphibole and biotite schist (Zak, 1980), suggesting that the copper-nickel deposits of the region are associated with a komatiitic volcano-plutonic complex, typical for many Archaean greenstone belts around the world.

The metamorphism of the supracrustal rocks of the Kaskam-Allarechka belt corresponds on the whole to high-temperature amphibolite facies conditions. There are signs of polymetamorphism, such that in the Allarechka block, possibly because of deeper erosion, plagioclase-microcline migmatization of the supracrustal rocks is significantly greater than in the Kaskam block. Earlier (Archaean) metamorphic assemblages correspond to low-pressure conditions, while superimposed metamorphic changes which proceeded under higher pressure conditions of the kyanite facies series are related to the Proterozoic prograde metamorphism of the Pechenga structure (Belyayev and Petrov, 1974; Kremenetsky, 1979).

Most of the age dates for the upper Archaean Kaskam-Allarechka supracrustal complex have been obtained by the K–Ar method and lie within the range 1.7–1.9 Ga (Bogachev et al., 1964; Tugarinov and Bibikova, 1980). However, two age determinations of the order of 2.2–2.6 Ga (Table I-7) were obtained by the U–Pb isochron method, suggesting that these rocks belong to the Late Archaean.

Late Archaean intrusions

During the Late Archaean stage in the evolution of the crust in the Kola province, some 12 plutonic suites have been identified (Batiyeva et al., 1985), the largest and most important being gabbro-labradorite, diorite-tonalite and anatectic-palingenic (metasomatic) granite.

Gabbro-labradorite intrusions are of widespread occurrence in the Kola Peninsula (Fig. I-22). Practically all the bodies are spatially and genetically associated with Late Archaean volcanogenic complexes of the Kolmozero-Voronya (the Patchemvara intrusion) and the Tersk-Allarechka (Glavny Range intrusion) greenstone belts, as well as the Keyvy (Tsagin, Achin and other intrusions) and Korva-Kolvi (Kolvi, Kandalaksha and Salnotundra intrusions) zones.

The distribution and emplacement of the intrusions is controlled by deep fault zones between Early Archaean granite-gneiss complexes and upper Archaean supracrustal assemblages, with which they were simultaneously metamorphosed and deformed. Contacts between the intrusions and the country rocks are usually tectonic, but sometimes, as in the case of the Kolvi and Glavny Range intrusions, it is possible to identify hornfelsed contact zones in which the country rocks have granulite facies mineral assemblages. The intrusions can be divided into two types on the basis of their internal structure. The first typically has a well-expressed rhythmic layering (the Tsagin and Yelskozero bodies) with the following sequence: olivinite → clinopyroxenite → gabbro-norite → gabbro-labradorite → titanomagnetite gabbro and ultrabasics → late magmatic titanomagnetite ores. The Tsagin

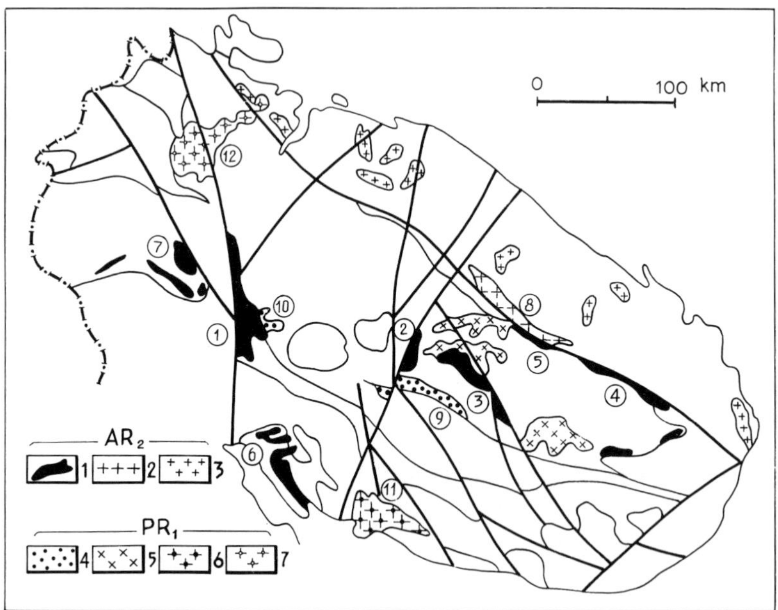

Fig. I-22. Distribution of major plutonic complexes in the Precambrian of the Kola Peninsula (Batiyeva et al., 1985; Yudin, 1980). Intrusions: *1* = gabbro-labradorite, *2* = diorite-tonalite, *3* = partial melt granites, *4* = ultrabasic-gabbro-norite, *5* = alkali granite, *6* = charnockite-granite, *7* = granodiorite-granite. Numbers in circles are named intrusions: *1* = Glavny (Main) Range, *2* = Tsagin, *3* = Yelskozero, *4* = Achin, *5* = Patchemvara, *6* = Kolvitsy, *7* = Salnotundra, *8* = Kolmozero, *9* = Pana-Fyodorova-tundra, *10* = Monchegorsk, *11* = Umba, *12* = Litsa-Araguba.

intrusion occupies an area of some 180 km^2 and is a flat-lying sheet-like body approximately 1.5 km thick, with an eroded apical zone (Radchenko, 1984; Yudin, 1980). The average chemical composition of the Tsagin intrusion corresponds to subalkaline basalt, while its thermodynamic conditions of crystallization (6–7 kbar) indicate moderate depths (Sharkov, 1984).

Intrusions of the second type (the Glavny Range, Kolvi, Salnotundra and Keyvy massifs) are characterised by the following more or less complete layered series (from the contact to the centre): gabbro and gabbro-norite → coarse-grained gabbro → gabbro-labradorite → labradorite and megacrystic labradorite → gabbro-labradorite → gabbro and gabbro-norite.

Tectonically, the Kolvi, Salnotundra and Keyvy masses are sill-like intrusions, while the Glavny Range body is a fault intrusion, associated with a suture or deep fault (tectonic zone).

The Glavny ("Main") Range intrusion is the largest Precambrian plutonic body in the Kola Peninsula and occupies an area of around 440 km^2. The geological position of the mass is defined by the fact that it cuts upper Archaean supracrustal rocks belonging to the Tersk-Allarechka greenstone

TABLE 1-10

Average chemical composition of gabbro-labradorite suites in the Kola Peninsula

Oxide	1	2	3	4	5	6
SiO_2	50.32	48.04	50.71	50.11	52.31	51.92
TiO_2	2.13	0.90	0.45	0.55	0.20	0.86
Al_2O_3	15.67	16.09	22.94	21.51	27.19	22.33
Fe_2O_3	4.48	2.81	1.06	1.45	0.64	2.18
FeO	11.08	9.66	4.86	5.52	1.74	4.70
MnO	0.22	0.15	0.09	0.10	0.04	0.10
MgO	6.10	4.10	2.91	1.33	3.28	1.57
CaO	6.36	12.04	10.40	12.26	12.50	9.98
Na_2O	3.28	2.30	2.95	2.85	3.64	3.79
K_2O	0.96	0.48	0.98	0.24	0.25	1.06
P_2O_5	0.02	—	—	—	—	—
H_2O^+	0.67	1.33	—	—	—	—
Σ	99.98	99.91	100.00	94.85		

1 = average composition of the Tsagin intrusion (Sharkov, 1984a); 2 = average composition of the Achin intrusion (Sharkov, 1984a); 3 = average composition of the Kolvi-Salnotundra complex (Sharkov, 1984b); 4 = average (from 13 analyses) composition of the Glavny Range gabbro intrusion (Sharkov, 1984b, Yudin, 1980); 5 = average (from 6 analyses) composition of labradorite from ditto; 6 = average (from 9 analyses) composition of gabbro-labradorite from the Yelskozero intrusion (Yudin, 1980).

TABLE I-11

Intrusion	Average chemical composition	Crystallisation P, kbar
Patchemvara	High-Al olivine tholeiite	5-7
Achin	Tholeiite	
Kolvi, Salnotundra	High-Al andesite-basalt, calc-alkaline series	6-7

Source: Sharkov (1984).

belt. The average composition of the Glavny Range intrusion (Yudin, 1980; Sharkov, 1984) corresponds to tholeiitic basalt (Table I-10), which was differentiated at a pressure of 8 ± 1 kbar.

The Keyvy, Salnotundra and Kolvi gabbro-labradorite massifs are conformable sheet-like intrusions with a similar internal composition and were to a large extent subjected to superimposed metamorphism at amphibolite and granulite facies. The average chemical composition and conditions of crystallisation of intrusions in this group differ slightly from one another (Table I-11; Fig. I-23).

Diorite-tonalite intrusions occur in western (Litsa and Kaskelyavr complexes), central (Kolmozero complex) and eastern (Ustponoy complex) parts of the Kola Peninsula (Batiyeva et al., 1985; Darkshevich et al., 1984) and are spatially associated with the Kolmozero-Voronya greenstone belt (Fig. I-22). The majority of these bodies underwent deformation and metamor-

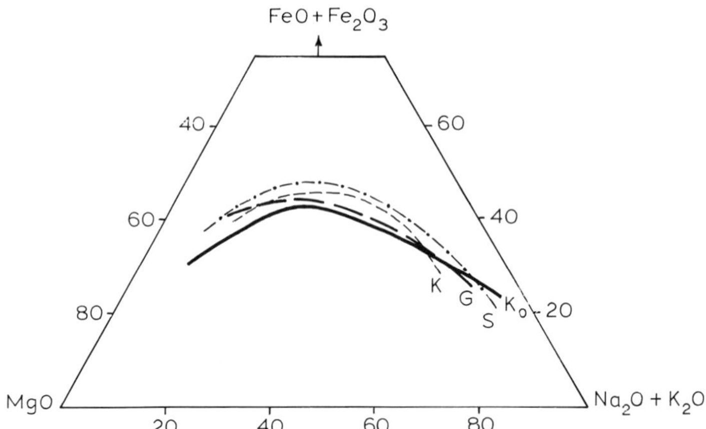

Fig. I-23. Petrochemical diagram showing composition of rocks in the gabbro-labradorite association in the Kola Peninsula, in wt% (after Batiyeva et al., 1985). Intrusions: *G* = Glavny Range, *K* = Keyvy, *Ko* = Kolvitsy, *S* = Salnotundra

phism at the same time as the supracrustal assemblages of this greenstone belt. In some of the masses, the following pattern of rock type transitions from the centre to the margins has been observed: tonalite → biotite granodiorite → hornblende-biotite granodiorite → hornblende granodiorite. Chemically the rocks form a series from gabbro-diorite to plagioclase-microcline granite. The age relationships of the diorites-tonalites have been determined from the fact that they intrude upper Archaean volcano-sedimentary complexes. Individual isotope dates (Table I-12) for these rocks confirm a Late Archaean age.

Anatectic and palingenic-metasomatic granites are widely developed in the Murmansk segment but are less widespread in the Central Kola and Tersk segments (Batiyeva et al., 1985). They are subdivided into anatectic, metasomatic, palingenic-metasomatic, porphyroblastic, intrusive and migmatitic granites, corresponding to different depth levels. The most chemically inhomogeneous types are the porphyroblastic and migmatitic granites, which vary from granodiorite to normal K–Na granite.

The geological and tectonic setting of these granitoids is determined from intrusive and other effects on already metamorphosed Late Archaean greenstone belt assemblages, while radiometric ages obtained for them (Table I-12) define an upper age limit for the formation of these structures.

The upper Archaean–lower Proterozoic Sumian complex

At the present time, the identification in the Kola Peninsula of an independent pre-Karelian, or Keyvy-Strelnin complex, is becoming more

TABLE I-12

Radiometric age of upper Archaean formations of the Kola Peninsula

No.	Locality, rock dated	Material dated	Method	Age, Ma	Laboratory Source reference
	I. Polmos-Porosozero belt				
1	Plagioclase-microcline granites, cutting the supracomplex	Whole rock	Rb-Sr Pb-Pb U-Pb	2730 2805 2735	Kola [1]
2	—ditto—	Zircon	U-Pb	2735	
3	Meta-andesite of the Voronyetundra Formation	—ditto—	Pb-Pb*	2930±90	
4	Gneiss, Chervurt Fm (Lake Litsa-Lake Poros area)	—ditto—	U-Pb	2790±45	PC [2]
5	—ditto—	Whole rock	Rb-Sr	2650±60	
6	—ditto—, L. Okhmylk area		Rb-Sr	2700±40	
	II. Cis-Imandra belt				
7	Meta-andesite, Vochelambi Formation	Zircon I Zircon II Zircon	Pb-Pb* —do— U-Pb	3080±65 2300±50 2860±40	PC
8	Garnetiferous amphibolite same locality	Amphibole Biotite	K-Ar —do—	2530±20 1980±20	
9	Plagioclase-microcline granite, cutting rocks of Vochelambi Formation	Zircon	Pb-Pb*	2455±30	PC
10	Meta-andesite-dacite of Arvarench Formation	Zircon I Zircon II	—do— —do—	2420±40 2350±40	
	III. North Imandra iron ore belt				
11	Hornblende-biotite gneiss	Whole rock	Rb-Sr	2560±60	PC [3]
12	—ditto—	Biotite	K-Ar	2060	
13	—ditto—	—ditto—	—do—	1940	
14	—ditto—	Amphibole	—do—	2470	
15	Basement tonalite	Whole rock	Pb-Pb	2790±40	
16	—ditto—	—ditto—	U-Pb	2800	Kola [1]
17	—ditto—	—ditto—	—do—	2690	
	IV. Kaskam-Allarechka belt				
18	Allarechka gneiss	Whole rock	U-Pb	2660	GC [4]
19	Migmatite, from rocks of the Allarechka Fm	—ditto—	—do—	2210	
	V. Keyvy belt				
20	Gneissose acid meta-volcanics, Lebyazhi Fm	Whole rock	Pb-Pb	2780±100	Kola [1]
21	Pegmatites cutting gabbro-anorthosite intrusions of Patchemvarek (younger than Lebyazhi Fm gneisses)	—ditto—	Pb-Pb	2730±15	[5]
22	Kyanite schists, Keyvy Fm	—ditto—	Pb-Pb	1600±120	[5, 6]
23	Alkali granosyenite, cutting Lebyazhi Fm gneisses	Zircon	Pb-Pb	1810±200	[1]
24	Kolovay porphyroblastic granite, supposedly after Lebyazhi Fm gneisses	Zircon	U-Pb	2620±70	[1]
	VI. Intrusive complexes				
25	Ust-Ponoy granodiorite complex	Whole rock	U-Pb	2730±70	Kola [7]
26	Gabbro-anorthosites of Glavny (Main) Range	—ditto—	Rb-Sr	2020±150	[8, 9]
27	—ditto—	—ditto—	Pb-Pb	2885±380	[4]
28	Coarse-grained leuco-cratic gabbro of Glavny Range	Zircon I Zircon II	Pb-Pb* —do—	2320±35 2390±30	PC

* Pb–Pb Zr evaporation technique.

References: [1] Pushkarev et al. (1979); [2] Ovchinnikova et al. (1985); [3] Gorokhov et al. (1981); [4] Tugarinova and Bibikova (1980); [5] Pushkarev et al. (1978); [6] Petrov (1979); [7] Vetrin et al. (1977); [8] Yudin (1980); [9] Birck and Allègre (1973). PC = Institute of Precambrian Geology and Geochronology, GC = Institute of Geochemistry, Kola = Kola Branch, all USSR Academy of Sciences.

of a topical question in terms of palaeotectonic reconstructions of the evolution of the region. This complex has unique structural-tectonic and lithoformational features (products of deep weathering and redeposition and substantial andesitic-basaltic vulcanism), resulting from a transition in crustal conditions from mobile Archaean to a more stable Proterozoic regime (Predovsky, 1980).

It is also possible that during this period of evolution of the Earth's crust, several gabbro-labradorites were intruded and eventually exposed at the surface.

In the present work we assign the supracrustal rocks of the Late Archaean–Early Proterozoic Keyvy structure to the upper Archaean–lower Proterozoic structural-lithological terrain, while the assemblages of the Strelin Group and the Korva-Kolvitsy zone, which we correlate with the Sumian of the Karelian province, but with certain reservations, are traditionally regarded as being lower Proterozoic formations.

The Keyvy structural belt

This is one of the biggest structures in the Kola Peninsula and occupies almost the entire SE half of the Central Kola segment (Figs. I-7, I-24). Its NE boundary is the north Keyvy system of faults, separating it from the Murmansk segment. The western boundary of the structure is a zone of deep faults, with which are associated the Panatundra and Fyodorovtundra major ultrabasic and gabbro-labradorite intrusions. Numerous alkaline granites occur on the SW margin of the Keyvy structure, south of which extends a tectonic zone separating the Keyvy structure from the Imandra-Varzuga zone (Fig. I-7). Alkali granites occupy the greater part of the area of the Keyvy structural belt. They intrude kyanite schists of the Keyvy Formation, which have been reworked by alkaline metasomatism over a wide area (Fig. I-24). There are a number of views concerning the internal structure of the Keyvy belt. It is considered variously as a graben (Mirskaya, 1979), a synclinorium (Belyayev et al., 1977) or a recumbent syncline (Shurkin et al., 1980).

The age of the supracrustal assemblage of the Keyvy belt, from field relations (basal conglomerate with pebbles of plagioclase-microcline granite and intrusion by microcline granite) is younger than the gneisses of the Early Archaean Kola Group. The stratigraphic relationship between rocks of the Keyvy belt and the upper Archaean assemblages of the Polmos-Porosozero greenstone belt has not yet been established. Differences in chemical composition between rocks of these structures do not permit direct correlation (Belolipetsky et al., 1980).

The stratigraphy of the supracrustal assemblages in the Keyvy belt has been well studied by a number of authors (Antonyuk, 1976; Belolipetsky et al., 1980; Belkov, 1963; Mirskaya, 1978, 1979; Kharitonov, 1966). The following four stratigraphic units have been identified.

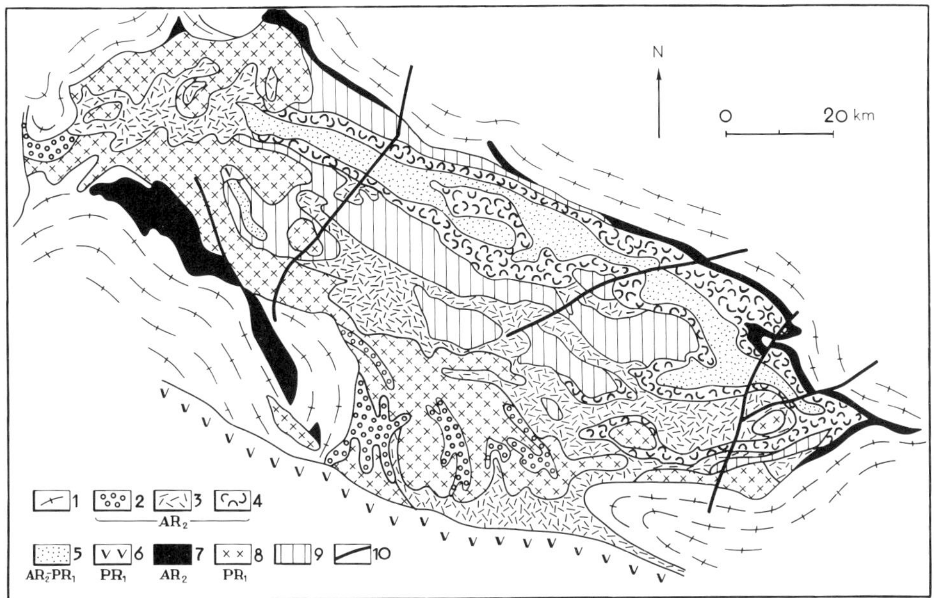

Fig. I-24. Geological sketch map of the Keyvy belt (simplified from Belolipetsky et al., 1980). *1* = plag-microcline gneiss, granite, migmatite; *2* = Kolovay assemblage (paragneiss and schist); *3* = Ponoy Formation (meta-andesite-basalt, meta-andesite); *4* = Lebyazhi Formation (meta-andesite, meta-rhyolite); *6* = Karelian complex (Imandra-Varzuga Group); *7* = gabbro-labradorite; *8* = alkali granite; *9* = zones of alkali metasomatism; *10* = faults.

The *Kolovay* member crops out around the margins of all the Early Archaean basement inliers bounding the Keyvy belt (Fig. I-24). The dominant rock types in this belt are gneiss and schist with biotite, garnet-biotite, hornblende-biotite or two micas; from their initial composition these rocks (Table I-13) are considered to be polymict sandstone and tuff with acid and intermediate material (Belolipetsky et al., 1980). Lenses of polymict conglomerate containing tonalite pebbles from the basement are found in the lower part of the Kolovay member, around Lake Yefimozero (Batiyeva and Belkov, 1958). The total thickness of rocks in the member varies from 500 to 800 m.

The *Ponoy* (or *Patchervtundra) Formation* consists of an assemblage of metavolcanic rocks with an apparent thickness of 500–1000 m (Mirskaya, 1978, 1979). The members of the assemblage are amphibolite, hornblende-biotite and biotite gneiss and schist. Compositionally, the overwhelming majority are andesite-basalt and andesite (Table I-13; 4), there being a regular decrease in basicity of the metavolcanics up through the succession until acid metavolcanics predominate, which are distinct from the overlying *Lebyazhi Formation*. This formation includes a unit of biotite-quartz-feldspar gneiss with volcanic rocks showing well-preserved evidence of explosive,

TABLE I-13

Average chemical composition of major rock types in the Keyvy belt

Oxide	1	2	3	4	5	6	7	8
SiO_2	70.18	55.17	56.38	67.20	70.35	69.56	75.96	63.30
TiO_2	0.45	1.11	1.14	0.70	0.43	0.63	0.42	1.26
Al_2O_3	13.80	13.97	13.88	13.14	12.62	14.05	10.75	31.29
Fe_2O_3	0.85	2.83	4.42	2.23	1.67	1.64	1.86	0.75
FeO	3.06	9.05	8.44	5.27	4.78	4.87	3.46	0.56
MnO	0.06	0.20	0.19	0.12	0.12	4.92	3.99	0.93
MgO	0.83	4.14	2.78	0.97	0.41	0.09	0.05	0.03
CaO	2.44	7.98	6.69	2.54	1.58	0.52	0.34	0.26
Na_2O	0.97	3.18	3.28	3.45	3.92	1.85	0.67	0.28
K_2O	2.74	0.63	1.58	3.03	2.93	2.13	2.08	0.45
n	5	7	8	12	26	7	6	28

1 = biotite gneiss, Kolovay member; 2 = plagioclase amphibolite (andesite-basalt), Patchervtundra Formation; 3 = ditto, Ponoy Formation; 4 = meta-dacite, Lebyazhi Formation; 5 = meta-rhyodacite, ditto; 6 = garnet-mica and two-mica-plagioclase schists (meta-arkose), Keyvy Formation; 7 = garnet-biotite gneiss (metatuff and siltstone), Keyvy Formation; 8 = kyanite schist (kaolin and hydromica clay). *Source*: Belolipetsky et al. (1980).

effusive, vent and subvolcanic facies (Mirskaya, 1979). Compositionally, around 75% of these gneisses correspond totally to acid metavolcanics — dacite and rhyolite (Table I-13) — and only 15% are metasediments (the two-mica schist), which are interpreted as greywacke, arkose and tuff (Table I-13). The Lebyazhi Formation is 1000–1500 m thick.

The metavolcanics are unconformably overlain by the *Keyvy Formation*, which has been studied in most detail by Belkov (1963). The high-Al schists of the Keyvy Formation, which crop out mostly in the area of the Keyvy Highlands, are subdivided into four units: (A) staurolite-garnet, (B) kyanite, (C) quartzite, and (D) kyanite-staurolite, with a total thickness of up to 600 m. The entire highly differentiated schist sequence is characterised by lateral lithological continuity of these units and a rhythmic alternation within the succession of essentially siliceous and aluminous rocks. The Keyvy schists are chemically equivalent to polymict and arkosic sandstone, greywacke, kaolinitic and hydromicaceous clays (Belolipetsky et al., 1980; Table I-13).

The supracrustal assemblage of the Keyvy belt was metamorphosed in the kyanite-sillimanite facies series of the amphibolite facies (Petrov, 1979). However, particular features of the mineral parageneses of the rocks (the presence of kyanite pseudomorphs after andalusite) suggest that they have undergone at least two metamorphic episodes. There are two opinions concerning the tectono-metamorphic history of the formation of the Keyvy belt. According to some (Kratz et al., 1978; Mirskaya, 1978, 1979), the Keyvy aluminous schists are younger than the rest of the succession and represent a Proterozoic (1.9–2.0 Ga) stage of development of the Keyvy structure with its kyanite metamorphism, synchronous with the metamorphism of the Imandra-Varzuga complex. Others (Belyayev et al., 1977), basing their conclusions on the presence of kyanite-bearing assemblages in Lebyazhi gneisses, consider that all magmatic and metamorphic processes in the Keyvy belt were completed in the Late Archaean. In this case, the available K–Ar isotopic age dates of the order of 1600 Ma for kyanite schists in the Keyvy Group can be explained by the opening of the isotopic system under the influence of alkali granites and associated metasomatism.

The Korva-Kolvitsy zone

One of the most complex zones in the geological and tectonic structure of the regions of the Kola province is where it meets the White Sea province (Figs. I-6–I-8), which is known in the literature as the Lapland or Main White Sea suture (Kratz et al., 1978; Priyatkina and Sharkov, 1979), or the Granulite belt (Vinogradov et al., 1980), or the Korva-Kolvitsy zone (Vrevsky and Kolychev, 1986). The structural-lithological rock complex constituting this zone consists mainly of amphibolite, schist and anorthosite. These high-grade metamorphic rocks extend from Kandalaksha Bay in the White Sea north-westwards for 500–600 km through the Nyavka, Salnyye and Korva tundras to the region of Lake Inari in Finland (Fig. I-7).

The "granulite complex" concept has traditionally been used to describe the stratigraphic extent and lateral distribution of the rocks in the Korva-Kolvitsy zone (Kratz et al., 1978; Priyatkina and Sharkov, 1979; Suslova, 1984), although in view of the zonal nature of the metamorphism (Vinogradov et al., 1980) the products of amphibolite facies metamorphism should also be considered as stratigraphic units (e.g. the Korvatundra amphibolites and the Inari-Tana schists). The tectonic structure of the Korva-Kolvitsy zone is determined by the existence of two major blocks (the Kandalaksha-Kolvitsy and the Lapland or Tuadesh-Salnotundra), which are separated by a roughly N–S-striking thrust of Karelian age along the mountain massif of the Glavny Range. The geology, structure and geochronology of these two blocks have not been studied to the same extent and their main features are therefore outlined below separately.

The *Kandalaksha-Kolvitsy block* is situated in the SE of the Kola Peninsula in the present-day junction zone between the White Sea province and the Tersk segment. The overall structure is thought to be a block-synclinorium (Belyayev, 1971) or a graben with a complex internal fold structure (Bogdanova and Yefimov, 1983).

The intensely deformed high-grade rocks of the Kandalaksha-Kolvitsy zone essentially form a volcanogenic complex which on the western limb of the structure overlies a granite-gneiss complex (Belomoride) with angular and stratigraphic unconformity, basal polymict conglomerate being present (Bogdanova and Yefimov, 1983, 1986). There are four supracrustal formations: Kandalaksha gneiss and amphibolite, Ploskotundra garnet-pyroxene schist, Poryeguba garnet-pyroxene-plagioclase schist, and acid granulite. Some authors separate the Yauriok Formation of essentially continental metasediments from the Poryeguba Formation (Belyayev, 1971).

The volcanic formations are chemically equivalent to three series (Table I-7): basalt, andesite-basalt and dacite-rhyolite. The percentage of volcanogenic rocks gradually decreases upwards as the percentage of continental (pelite and psammite) and volcanosedimentary formations (tuffite, greywacke, etc.) increases (Bogdanova and Yefimov, 1982, 1986).

The *Lapland (Tuadesh-Salnotundra) block* is a major synclinorium(?), bounded by deep faults, and with an overall N–W strike which swings to near E–W in the region of the supposed closure of the structure, around the Great Salnyye tundras (Suslova, 1984).

The polyphase deformed high-grade metamorphic rocks of the Lapland block have a broad two-fold structure. The lower part of the succession consists of various amphibolites, the Kandalaksha Formation, above which lie mafic schist, granulite and garnet-biotite gneiss, the Beloguba, Vuim, Ploskotundra and Yauriok Formations (Belyayev, 1971; Suslova, 1984).

From petrological and geochemical considerations, the rocks of the Lapland block form a volcanogenic-sedimentary complex, the lower part of which (the Kandalaksha Formation) is essentially volcanogenic in nature (tholeiitic basalt with thin bands of komatiitic and picritic varieties, andesite, dacite and related tuffs, which alternate with high-alumina basalt; Kozlov, 1983; Suslova, 1984). Rocks of the upper part of the section are of mixed genesis, the "pelite series" (Ivlev, 1979). Any estimate of the thickness of the Lapland complex rocks can be only very approximate (1000–2000 m), due to the fact that there has been no sufficiently detailed interpretation of the internal fold structure of the block.

The long and complex history of tectono-metamorphic transformation of rocks in the Korva-Kolvitsy zone is interpreted in various ways. Up to four major fold episodes and two to three metamorphic episodes (Bogdanova and Yefimov, 1983; Vinogradov et al., 1980) have been identified, creating certain difficulties in tying three groups of isotopic dates to particular geological events (2.7–2.8, 2.15–2.0, 1.9–1.8 Ga) using U–Pb isochron (Tugarinov and Bibikova, 1980; Bubnoff, 1952) and Pb–Pb thermochron (Bogdanova and Yefimov, 1986; Suslova, 1984) techniques.

The lower Proterozoic Karelian complex

Volcanogenic-sedimentary assemblages in the Karelian complex crop out over large areas of the Kola Peninsula and occur mainly in two major structures — the Pechenga and Imandra-Varzuga zones, as well as in a number of smaller remnants and fragments of synclinal and monoclinal structures, which previously had constituted a single Pechenga-Varzuga belt in the Karelides (Figs. I-7, I-11), over 500 km long. At the present time, this belt is broken into two parts (Pechenga and Imandra-Varzuga), by an apparently late Karelian north–south thrust along the Glavny mountain range, with a horizontal translation of the order of 100 km (Zagorodny et al., 1983).

The Pechenga structural belt

This structure, which occupies an area in excess of 2300 km^2, extends for 70 km in a north-westerly direction. From its internal structure it is an asymmetric graben-syncline with a broad NE limb (Fig. I-25) and is greatly complicated by cross-faults and limb-parallel faults at the margins.

The core of the Pechenga synform has its own independent structure consisting of two saucer-shaped second-order short broad synclines, separated by a narrow anticline. The south-western boundary of the structure is the Poritash shear belt — a zone of parallel, narrowly-spaced deep faults, 60 km long and 6–8 km wide. Within the Poritash "crush zone" the rocks consist of complexly interleaved and highly schistose phyllite, siltstone, sandstone and basic and acidic tuffs, the stratigraphic position of which is still a matter of dispute.

The entire northern part of the Pechenga graben-syncline is filled with a stratified rhythmically-alternating sequence of volcanogenic and continental rocks of the Pechenga complex. Four complete macrorhythms (formations) have been defined, each of which commences with a sedimentary unit and concludes with a volcanic unit (Gilyarova, 1967; Zagorodny et al., 1964; Polyak, 1968; Predovsky et al., 1974). At the base of the sequence of complexes (from 6840 to 6842 m in the Kola SD-3 superdeep borehole) a mature weathering crust is present (the first sedimentary unit) with conglomerate overlying lower Archaean Kola Group gneisses (Fig. I-9). The volcanosedimentary assemblages of the Pechenga complex form two groups, the Luostari with up to three sedimentary and one or two volcanic formations and the Nikel with four sedimentary and three or four volcanic formations (Fig. I-9). The boundary of these groups is a metamorphosed weathering crust, the Zapolyarny Formation[1] (at a depth of 4884 m in the Kola borehole, Kozlovsky, 1987). These two parts of the section of the Pechenga complex correspond to two Early Proterozoic volcanic megarhythms, the lower being trachyandesitic to basaltic and the upper picritic to basaltic.

Of the intrusive bodies associated with the Pechenga complex, special attention has been devoted to the layered intrusions in tuffs and sediments: serpentinised peridotite, olivinite, pyroxenite and gabbro, with which are associated all the known copper-nickel deposits of Pechenga.

The rock formations constituting the Pechenga structure were subjected to zonal metamorphism under conditions of the kyanite-sillimanite series ($\sim$30°/km, P_{tot} = 3–6 kbar; Dook, 1977), the metamorphic zones being

[1] "Transpolar Suite" in Pharaoh et al (1987).

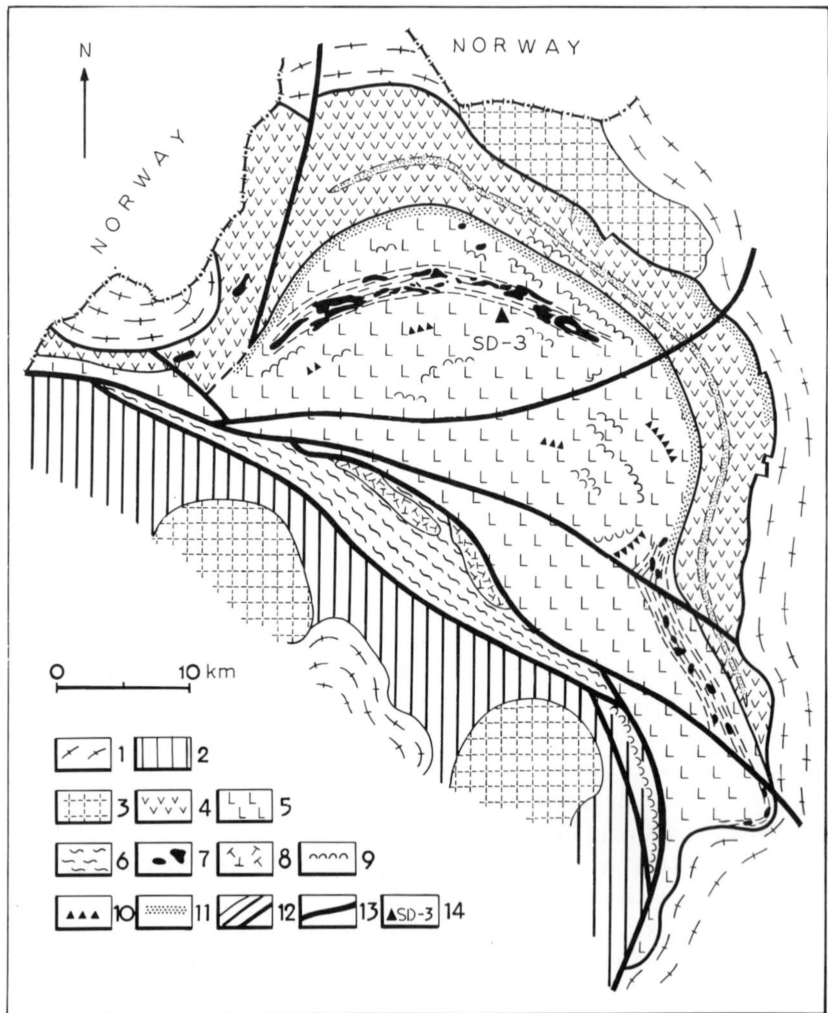

Fig. I-25. Geological sketch map of the Pechenga belt (simplified after Kozlovsky, 1987). *1* = plag-microcline gneiss, granite, migmatite; *2* = Kaskam-Allarechka greenstone belt; *3* = plagio-microcline granite intrusions; *4–11*: Pechenga complex (*4* = Luostari Group (trachyandesite, trachybasalt, andesite-basalt), *5* = Nikel Group (tholeiitic basalt, graphitic phyllite, tuffite, siltstone, sandstone), *6* = Poryitash "crush zone", *7* = layered basic-UB intrusions, *8* = diorite, *9* = tuffs and related schistose rocks, *10* = UB and high-Mg volcanics (picrites), *11* = metasandstone, metaconglomerate, quartzose sandstone, dolomite); *12* = geological boundaries; *13* = major tectonic breaks; *14* = site of Kola superdeep borehole (SD-3).

oriented close to lithostratigraphic boundaries. From SD-3 cores, the rocks of the Nikel Group were metamorphosed from prehnite-pumpellyite to epidote-amphibolite facies, while the assemblages in the Luostari Formation were metamorphosed from epidote-amphibolite (at 4880–6000 m depth level in SD-3) to amphibolite facies (6000–6842 m). Moreover, using chemical differences in metamorphic minerals, Dook (1977) has proven the existence of an earlier phase of zonal metamorphism in the Pechenga complex, in the andalusite-sillimanite facies series (~40°/km, P_{tot} = 2–4 kbar)

From geological and geochronological studies (Kratz et al., 1980; Dook, 1977; Kozlovsky, 1987) it is possible to identify two basic structural-metamorphic stages in the evolution of the Pechenga synform. Of the isotopic age determinations obtained by various techniques, only the oldest ages within the interval 2150–1750 Ma (around 1950 Ma) reflect the first stage in the development of the structure — the accumulation of sedimentary and volcanic assemblages and their metamorphism under andalusite-sillimanite series conditions. The second group of values (1750–1500 Ma) reflects the second stage of metamorphism, kyanite-sillimanite type with a peak at 1650 ± 100 Ma.

The Imandra-Varzuga belt

This structure, consisting of Early Proterozoic sedimentary-volcanic complexes, is one of the biggest structures in the Karelides of the Baltic Shield and has the most complete sequence of supracrustal rocks.

The Imandra-Varzuga belt extends for 330 km in a north-westerly direction and is 40–50 km wide in the central part (Fig. I-26). The northern boundary of the structure is the long-lived Pana-Babozero fault, while the southern boundary, which is less sharply defined, lies along the junction with the upper Archaean assemblages of the Tersk greenstone belt. The total thickness of the Imandra-Varzuga supracrustal complex is around 13 km, over 10 km of which consists of volcanogenic formations (Zagorodny et al., 1982; Fedotov, 1985). In a regional sense, the structure is an asymmetric graben-syncline (Zagorodny and Radchenko, 1983; Petrov and Voloshina, 1978), complicated by a network of faults of various orders (Fig. I-26). The lower horizons of the complex have a monoclinal structure, with beds dipping SW at 10–60°, while the upper horizons form a narrow graben-syncline. The inhomogenous nature of the tectonic structure of the eastern, central and western parts of the Imandra-Varzuga belt results in major differences in the lithology and thickness of partial sequences of the volcano-sedimentary complex. Nevertheless, the total stratigraphic sequence of the complex, compiled on the basis of a formational subdivision of units (Zagorodny and Radchenko, 1983), is divided into three groups, the Strelnin Group volcano-sedimentary formations, missing from the Pechenga structure, being present here, owing to the earlier origin of the Imandra-Varzuga structure.

Each of the component groups in the Imandra-Varzuga complex is separated from the others by stratigraphic breaks and unconformities, while the formations as a rule comprise a lower sedimentary and an

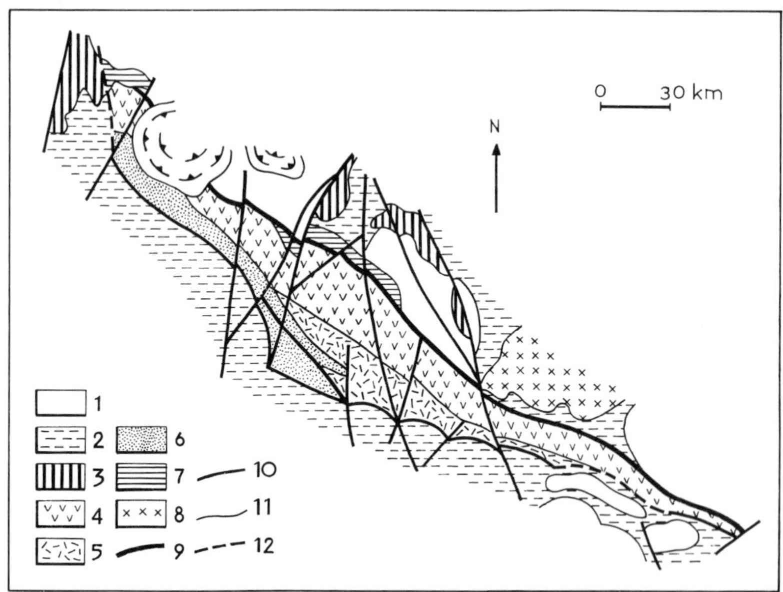

Fig. I-26. Diagrammatic sketch of the Imandra-Varzuga structural belt (simplified from Zagorodny and Radchenko, 1983). *1* = gneiss and granite-gneiss (AR_1); *2* = volcano-sedimentary assemblage in Tersk and Imandra greenstone belts (AR_2); *3* = gabbro-labradorite (AR_2); *4–6*: Imandra-Varzuga complex (PR_1) (*4* = Strelnin Group, *5* = Varzuga Group, *6* = Toming Group); *7* = UB-gabbro-norite; *8* = alkali granite; *9*, *10* = faults; *11* = geological boundaries; *12* = tentative boundaries.

upper volcanic member. Differences in thickness and to a certain extent in the lithology of the supracrustal complex in different parts of the structure are explained by deposition having taken place on basement blocks of differing stability (Fedotov, 1985). The formation of the supracrustal complex to a large extent determined the intensity of volcanic activity, the total thickness of volcanic products being around 10 km. There are six volcanic units (members) identifiable in the complex, with a uniform trend in composition in each volcanic cycle. The evolution of the composition of volcanics in the complex as a whole is much more complicated. Vulcanism in the Strelnin Group is characterised by a uniform trend in evolution from tholeiitic basalt through andesite-basalt to dacite and rhyolite, with a preponderance of intermediate-basic varieties.

During the middle stage of development of the complex (Varzuga Formation) vulcanism was also uniform in general, from picrite-basalt series to sub-alkaline (Ilmozero Formation) and trachybasalt series (Umba Formation), allowing these formations to be correlated with the Pechenga complex. The concluding volcanic cycle (Toming Formation) typically has

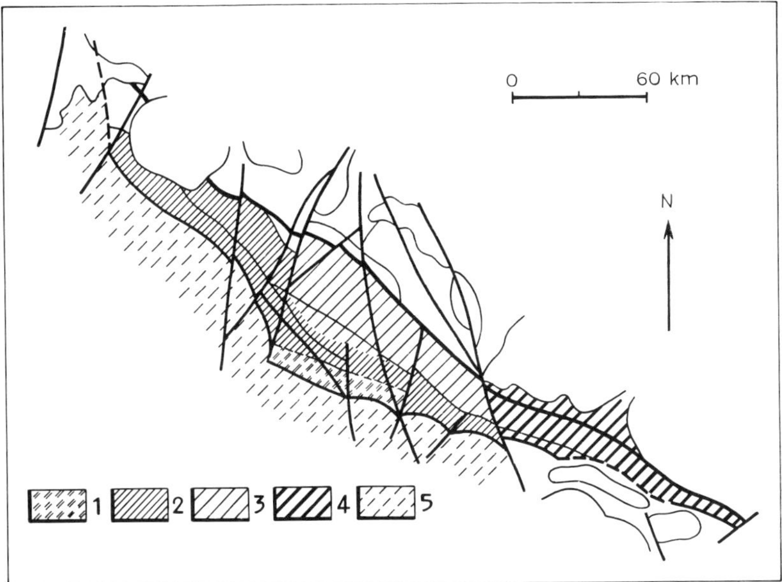

Fig. I-27. Sketch of metamorphic zonation within the Imandra-Varzuga belt (simplified from Zagorodny et al., 1982). *1–4*: kyanite-sillimanite type metamorphic facies (*1* = Chl-Epi-Act zone, *2* = Bi-Chl-Act zone, *3* = Bi-Epi-Act zone, *4* = epidote-amphibolite and amphibolite facies); *5* = greenschist retrogression of kyanite-sillimanite series following low-temperature amphibolite facies andalusite-sillimanite series.

a wide spectrum of volcanics (picrito-basalt, tholeiitic basalt and dacite) of the subalkaline series (Fedotov, 1985).

All the rocks of the Imandra-Varzuga structure are zonally metamorphosed, from greenschist to amphibolite facies, within the kyanite-sillimanite facies series (~30°/km; Zagorodny et al., 1982; Sergeyev et al., 1985). Metamorphic zoning within the belt has an irregular, tectonically disrupted pattern (Fig. I-27). The structure is divided into two parts by a system of cross-faults: a western part, within which are developed greenschist facies formations, and an eastern part, with outcrops of rocks displaying epidote-amphibolite and amphibolite facies assemblages (Zagorodny et al., 1982; Petrov et al., 1986). There is on the one hand a regular increase in metamorphic grade down the sequence, and on the other an oblique pattern of metamorphic facies boundaries relative to lithostratigraphic boundaries as a result of tectonic disruption of the structure.

At present there are practically no isotopic dates for rocks of the Imandra-Varzuga complex. Available figures for the age of alkali granites (1770 ± 15 Ma, Pushkarev et al., 1978) may be considered as the upper age limit for the Imandra-Varzuga complex. It is for this reason that most authors, ourselves included, hold the view that the formation of the lower

Strelnin Group must be considered pre-Karelian (the Keyvy-Strelnin or Keyvy-Sumian complex, Predovsky, 1980).

The lithology and geochemistry of Strelnin Group rocks are similar in many respects to assemblages in the Ponoy and Lebyazhi Formations of the Keyvy belt. This, together with recently published data, including geochronology, allows us to argue with more certainty for Sumian-type formations in the Kola Peninsula, and hence to define and evaluate more precisely the significance of the boundary of these with the Lopian complex, i.e. the Archaean-Proterozoic boundary.

Upper Proterozoic Riphean and Vendian supergroups

Upper Proterozoic formations of the Kola Peninsula comprise sedimentary and volcanic assemblages of the Riphean and Vendian Supergroups, with minor igneous intrusions. This rock complex crops out around the southern periphery of the peninsula in a few inland depressions and does not play a major role in the geological structure of the Kola province (Fig. I-8).

Supposed Lower Riphean formations are found only in the south of the Kola Peninsula (Turin Formation, Sergeyeva et al., 1971), where they consist of quartzose and quartz-feldspar sandstone with quartzose conglomerate at the base. Middle Riphean formations include a flysch sequence (the Rybachy Group on Rybachy Peninsula; Dobrokhotov et al., 1981; Kalyayev et al., 1980), together with continental red beds (Tersk Formation along the shores of the White Sea). Upper Riphean rocks consist of a sequence of rhythmically-bedded sandstone, mudstone and siltstone (the Kildin and Volokov Groups, Lyubtsov, 1980).

Proterozoic intrusions. Some 13 plutonic sequences of Karelian age and one suite of Riphean continental-type dolerites have been identified in the Proterozoic (Batiyeva et al., 1985). The largest and most widely distributed Karelian plutonic sequences are as follows: (1) ultrabasic-gabbro-norite; (2) charnockite-granite; (3) granodiorite-granite; (4) alkaline granite and granosyenite; and (5) alkali gabbro. They belong to formations of the early inversion stage of protogeosynclinal structures (2), deep faults (1, 4, 5) and sub-platform stages (3) (Shurkin et al., 1980). The *ultrabasic-gabbro-norite association* includes such major layered intrusions as Monchegorsk (65 km^2), Fyodorov Tundra (30 km^2) and Pana Tundra (200 km^2) and a number of smaller masses (Fig. I-22). All the masses have the same structure — lensoid or funnel-shaped, each with an independent internal structure. They are formed of a series of rocks: dunite, harzburgite, bronzite, norite, gabbro-norite, gabbro-norite-anorthosite, titanomagnetite gabbro, which sometimes form major megarhythms. Intrusions vary greatly in thickness, from 200–500 m to

3000 m. Judging by the chemistry of chilled margins to the intrusions, the original composition of the melt which underwent magma-chamber differentiation was a tholeiitic basalt (Sharkov, 1984) or an undersaturated tholeiite (Batiyeva et al., 1985). All the bodies are broken into blocks, with vertical uplift of up to 1.5 km. There are no reliable isotopic age dates for these intrusions.

Within the Kola Peninsula the so-called Umba complex (Fig. I-22) belongs to the *charnockite-granite association* and is located at the intersection of three long-lived deep fault zones (Shurkin et al., 1980). The complex contains three granitoid series or intrusive phases: (1) diorite and quartz diorite; quartz monzodiorite and tonalite of the enderbite series; (2) quartz diorite, granodiorite and tonalite of the charnockite series; and (3) adamellite and plagioclase-microcline leucogranite. The granitoids are overlain by upper Proterozoic quartzose sandstone of the Turin Formation and have yielded isotopic age dates of 2150 and 1950 Ma by the U–Pb and Rb–Sr isochron methods respectively (Pushkarev et al., 1978).

Granodiorite-granite plutons occur within the bounds of the Yuovoay-Uraguba tectonic zone, which extends for over 10 km in the north-west of the Kola Peninsula (Fig. I-23; Batiyeva et al., 1985). The association forms the Litsa-Araguba complex, consisting of seven bodies with a total area of some 900 km^2. There are five intrusive phases in the complex (Shurkin et al., 1980): (1) quartz diorite and monzonite with a dyke facies of dioritic porphyrite and lamprophyre; (2) porphyritic plagioclase-microcline granite and granodiorite; (3) fine-grained porphyritic granite and granodiorite; (4) aplite, granite-porphyry and pegmatite; and (5) granosyenite and quartz monzonite. According to Levkovsky (1976), the Litsa-Araguba complex can be considered analogous to rapakivi granites.

The age of the Litsa-Araguba complex is defined by intrusive episodes, which coincide in time with the final stage of development of the Pechenga belt. This is confirmed by their isotopic age of 1755 $\pm$ 25 and 1840 $\pm$ 50 Ma by the Rb–Sr and U–Pb isochron methods, respectively.

The suite of *alkaline granites and granosyenites* is widely distributed in the central part of the Kola Peninsula and has a maximum development of over 1200 km^2 within and along the southern edge of the Keyvy belt (Figs. I-22, I-24).

Depending on the level of erosion of the intrusions, there is a progressive change in composition from syenite through nordmarkite to alkaline microcline-albite granite. Wide zones of metasomatically-altered country rock are usually associated with alkaline granite intrusions.

The geological-stratigraphic position of the massifs can be deduced from their effects on rocks of the Strelnin and Lebyazhi Groups, the Pana Tundra intrusion, and their removal by erosion in pre-Polisari time (Batiyeva et al., 1985), which suggests the existence of two age groups of alkaline granites. Also, isotopic age data for these formations are contradictory

TABLE I-14

No.	Deposit or show; useful mineral	Tectonic setting	Type of mineralization (ore formation and genetic type)	Geological and host rock or mineralized rock	mineralogical features: shape of deposit, morphology of ore body	Ore type	Mineral composition, secondary minerals	Age of ore mineralization
				1. Saamian and Kola complexes				
1.	Bazarnaya Guba; zinc, lead	Central Kola segment	Pb-Zn and barite-calcite veins; hydrothermal	Biotite and biotite-garnet gneisses	Veins, vein zones	Massive, fissure	Galena, sphalerite (chalcopyrite)	PR
2.	Strelnin; mica, microcline	Tersk segment	Muscovite; pegmatite	Gneisses	Veins, lenses	Clusters, blocks	Muscovite, microcline (quartz, plagioclase)	AR
				2. Lopian complex				
3.	Olenogorsk; iron	Trans-Imandra greenstone belt	Iron ore in quartzites; magnetite, metamorphic	Amphibole and biotite gneisses; amphibolites, leptites	Concordant sheets & lenticular deposits	Dense clusters, massive	Magnetite, hematite (pyrrhotite)	Late AR
4.	Allarechka; Ni, Cu, Co, etc.	Kaskam-Allarechka greenstone belt	Cu-Ni sulphides; magmatic-metamorphic	Harzburgites, pyroxene olivinites (komatiites)	Complex sheet and lenticular deposits	Massive, disseminated breccia type	Pyrrhotite, pentlandite, chalcopyrite, magnetite (pyrite, violarite, bornite)	Late AR–Early PR
5.	Kolmozero–Voronya; rare metal	Kolmozero–Voronya greenstone belt	Rare metal; pegmatitic	Pegmatites of alkaline granites	Concordant and discordant veins	Clusters and disseminated	Polucite, columbite (magnetite, lepidolite, etc.)	Late AR
6.	Kolmozero–Voronya zone (Pelaphakhk); Mo, Cu, etc.	Kolmozero–Voronya greenstone belt	Cu-Mo (porphyry?) hydrothermal–metasomatic	Biotite, hornblende–biotite schists and gneisses	Lenticular deposit	Disseminated	Molybdenite, chalcopyrite (pyrite, pyrrhotite, etc.)	Late AR
7.	Tsagin; Ti, Fe, V	W. margin of Keyvy belt	Ilmenite-titanomagnetite; magmatic segregation	Gabbro–anorthosite, norite	Sheet-like deposits of disseminated–vein ores, lenses of solid ores	Massive, disseminated	Titanomagnetite, ilmenite, magnetite (hematite, chalcopyrite, etc.)	Late AR–Early PR
				3. Sumian complex				
8.	Chervurt (Keyvy group of deposits); aluminous	Keyvy belt	High alumina sillimanite–kyanite in schists; metamorphic	Gneisses and schists	Layered bodies	Massive	Sillimanite, kyanite (ilmenite, pyrrhotite, etc.)	Late AR
9.	Lovnozero; Ni, Cu, Co, etc.	Lapland block, granulite belt	Cu-Ni sulphides; magmatic–metamorphic	Norites	Concordant lens-shaped deposits	Disseminated, massive	Pyrrhotite, pentlandite, chalcopyrite, pyrite (magnetite, etc.)	Early PR
10.	Jaurijoki (Yugas); Mo, fluorite	Lapland block, granulite belt	Mo in quartz veins and greisens; hydrothermal	Granites	Quartz veins with disseminated ore mineralization	Disseminated, vein–type disseminations	Molybdenite, fluorite (euxenite, scheelite)	Early PR
				4. Karelian complex				
11.	Kaula (Pechenga group of deposits); Ni, Cu, Co	Pechenga–Varzuga zone	Cu-Ni sulphides; magmatic liquation	Wehrlites	Concordant sheets and lenticular bodies, rarely cross–cutting bodies	Breccia–type, massive, disseminated	Pentlandite, chalcopyrite, pyrrhotite, magnetite (pyrite, violarite, etc.)	Early PR
12.	NKT (Monchegorsk group of deposits); Cu, Ni, Co, etc.	Central part of Pechenga–Varzuga zone	Cu-Ni sulphides; magmatic liquation	Peridotite, pyroxenite, norite	Veins and sheet–like deposits	Massive, disseminated	Pyrrhotite, pentlandite, chalcopyrite, magnetite (pyrite, violarite, etc.)	Early PR
13.	Gremyakha–Vyrmes; Ti, Fe, V, P	Central Kola segment	Apatite–ilmenite–titano-magnetite with zircon; magmatic	Olivine gabbro–pyroxenite, troctolite	Concordant sheets and lenticular deposits	Disseminated, disseminated–vein type	Ilmenite, titanomagnetite, apatite (zircon, magnetite, pyrrhotite, chalcopyrite, etc.)	Early PR

Source: Gorbunov et al. (1981).

(Pb–Pb isochron method on zircons yields 2060 Ma, 1770 ± 15 Ma and 2440 ± 50 Ma; Pushkarev et al., 1978) and the question has no simple answer.

Precambrian ore deposits of the Kola Peninsula

The Kola Peninsula is one of the most remarkable Precambrian regions of the USSR in terms of the number and variety of useful minerals. Table I-14 presents a brief summary of the most characteristic deposits and ore shows in the Kola Peninsula.

Early Archaean. The Early Archaean tonalite-trondhjemite complex, despite its wide areal distribution (c. 40%) in the Kola Peninsula has an extremely poor ore content and is of no practical value, save for ore shows in ferruginous quartzites in the Kola Group supracrustal rocks (Goryainov, 1980).

Late Archaean. In broad terms, the metallogeny of Late Archaean greenstone belts in the Kola Peninsula is primarily determined by ore formations in volcanogenic series, to the extent that the development of a particular volcanic formation has a major influence on the specific metallogeny of each particular structural belt.

The greatest variety of ore formations occurs in the Kolmozero-Voronya and Tersk-Allarechka greenstone belts. Associated with komatiite-tholeiite series intrusive and extrusive igneous rocks we have sulphide-copper-nickel deposits at various scales and a number of other ore formations of igneous and metamorphic genetic types; metamorphic-type porphyry copper ore mineralisation is associated with volcanics of the basalt-andesite-dacite series; while iron and manganese ore formations are associated with volcanoclastic and volcanosedimentary facies (Table I-14, Nos. 3, 4).

Late-stage processes of granite formation and medium-gradient amphibolite facies metamorphism in these belts produced deposits of rare earth metals, ceramic and muscovite pegmatites (Table I-14, No. 5). The Keyvy and Korva-Kolvitsy zones have a much narrower spectrum of ore formations, the most significant of which are high-alumina metasedimentary types (Table I-14, No. 8). The gabbro-anorthosites associated with these zones contain iron-titanium-vanadium ores (Table I-14, No. 7; Yudin, 1980).

The Lapland granulite belt proper contains deposits of copper-nickel and molybdenum ores (Table I-14, Nos. 9, 10).

Proterozoic. The most productive rocks in the Kola Peninsula are the lower Proterozoic formations with associated commercial deposits of sulphide-copper-nickel in the Pechenga and Monchegorsk orefields, as well as numerous ore shows of copper and nickel in the ultrabasic-gabbro-norite intrusions of Pana Tundra, Fyodorov Tundra, General, etc. (Batiyeva et al., 1985).

The apatite-ilmenite-titanomagnetite ore deposits in the Gremyakha-Vyrmes alkaline gabbros (Table I-14, No. 13) occupy a special place in the Precambrian complexes of the Kola Peninsula. This magmatism, restricted to the sub-platform development stage of the Kola province, appeared as a sort of precursor to the much later Palaeozoic alkaline magmatism (the Khibiny and Lovozero nepheline syenite intrusions and bodies of alkaline-ultrabasic rocks), with which are associated major apatite deposits.

2. THE KARELIAN PROVINCE

Introduction

The Karelian province occupies the south-eastern part of the Baltic Shield. It possesses its own specific geological structure, allowing this territory to be identified as an independent tectonic structure — the zone of the Karelides (Kratz, 1963), a stable massif (Kharitonov, 1966), and the Karelian granite-greenstone terrain (Gorlov, 1975).

The provinces identified in the Baltic Shield by geophysical means correspond to geochronological zones (Kratz and Lobach-Zhuchenko, 1972). The Karelian province is situated in the oldest geochronological zone, characterised by the least intense activity of late (Proterozoic) internal processes.

The Karelian province (Fig. I-28) stretches in a north-westerly direction; it occupies an area, including the western part of the granite-greenstone terrain situated in Finland, of some 125,000 km^2. It borders with the White Sea province in the north-east. Geological field relations testify to the complex nature of this boundary: it separates zones with different P–T regimes, although more and more data are appearing on the similarity between the terrains in the Belomoride zones and the Karelian province. The tectonic nature of the boundary in the Late Archaean to Early Proterozoic is obvious — blastomylonite zones formed here, there are charnockite intrusions (Shurkin et al., 1974), the Vetreny riftogene belt originated along the boundary, and we observe differences in the baric conditions of metamorphism.

In the SE, the boundary of the province coincides with the margin of the shield; it is expressed as the Ladoga-Mezen fault zone (Kratz et al., 1978). The SW and western boundaries follow the eastern edge of the Ladoga-Kalevi zone of the Karelides of Karelia and Finland.

The most significant contributions to the study of the geology and petrology of Karelia have been by V.M. Timofeyev, A.A. Polkanov, N.G. Sudovikov, L.Ya. Kharitonov, K.O. Kratz, M.A. Gilyarova, V.A. Perevozchikova and E.K. Gerling. Investigations of the deep structure of the region have been carried out by G.A. Porotova, I.V. Litvinenko and R.V. Bylinsky. The regional metallogeny of Karelia has been studied by K.D. Belyayev, T.V. Bilibina, A.V. Sidorenko and A.I. Bogachev.

The Karelian province is a craton, the basement of which consists of Early Archaean (Saamian), Late Archaean (Lopian) and Early Proterozoic (Sumian-Sariolian) terrains (Fig. I-28). The composition and structure of complexes in the basement indicate that it is a granite-greenstone terrain. Around 80% of the craton basement consists of gneiss and granite and around 20% is greenstone rocks of the Lopian complex and Sumian-Sariolian rocks. In the Proterozoic (Jatulian), this region formed as a craton with a protoplatform sedimentary cover, having an age of 2.1 ± 0.1 Ga. The concluding phase of stabilization of the territory occurred during the Riphean. During the Proterozoic, the Karelian craton bordered with mobile

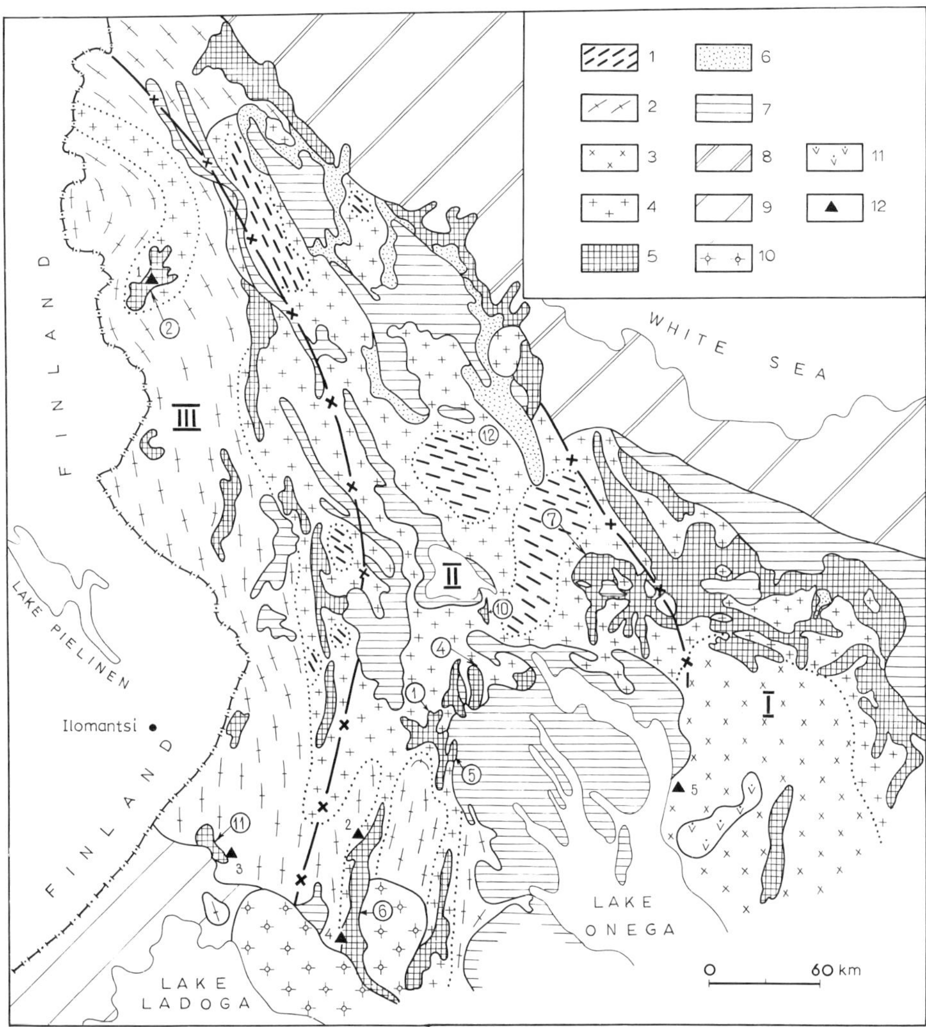

Fig. I-28. Sketch map showing geological structure of Karelia (compiled by S.B. Lobach-Zhuchenko and V.P. Chekulayev, using maps by A.N. Berkovsky and S.I. Rybakov). Granite-gneiss terrains AR_1+AR_2: *1* = Ondozero, *2* = West Karelia, *3* = Vodlozero types; AR_2: *4* = Segozero type of granitoid; *5* = greenstone belts; PR_1: *6* = Sumian-Sariolian; *7* = Jatulian; *8* = Belomoride zone (AR–PR); *9* = Svecofennide zone (PR_1); *10* = rapakivi granite; *11* = gabbro-norite-peridotite (Burakov intrusion); *12*: ore deposits (*1* = Kostomuksha, *2* = Hautavaara, *3* = Yalonvara, *4* = Vodlozero, *5* = Pudozhgora); *13*: boundaries between zones (*I* = East Karelian, *II* = Central Karelian, *III* = West Karelian). Numbers in circles are structural zones: *1* = Semcha, *2* = Kostomuksha, *3* = Kamennozero, *4* = Palalambi, *5* = Koykara, *6* = Hautavaara, *7* = Shilos, *8* = Kozhozero, *9* = Toksha, *10* = Oster, *11* = Jalonvaara, *12* = Parandov-Nadvoitsk.

belts in the NE and SW, during the development of which there was partial activation of the territory in the marginal parts of the craton, expressed mainly as magmatism.

The lower Archaean Saamian complex

Extensive granite formation, synchronous with Lopian greenstone complexes, complicates the reliable identification of older rocks over most of Karelia. Nevertheless, around Lake Tulos and Verkhneye Kuito in western Karelia, relicts of gneiss, schist, tonalite and migmatite are referred to the Saamian complex (Sviridenko, 1974). The main evidence in favour of these rocks being identified as belonging to a basement complex is the presence of granulite facies minerals and corresponding areas of gravity maxima. There are as yet no convincing data (structural or isotopic) on the old age of the granulite metamorphism, but it is quite probable that the material corresponds to the basement complex (Kozhevnikov, 1982). More reliable basement rocks are found in SE Karelia, in the Vodlozero block, where they are preserved in small areas among younger granitoids. They consist of hornblende-biotite gneiss with rare amphibolite bands (Table I-15; Lobach-Zhuchenko et al., 1984). The complex has been migmatized by leucocratic tonalite. The gneiss typically has higher Ti, P, Zr and Y contents, and moderately fractionated REE. The amphibolites are characterised by a calc-alkaline differentiation trend and REE fractionation, which makes them substantially different from younger (Lopian) metabasalts. The complex underwent several stages of deformation, the earliest of which took place under high-temperature amphibolite facies conditions. Isotopic age determinations on zircons from gneiss and amphibolite, carried out by Sergeyev et al. (1985), are older than 3.15 Ga (Table I-16). But even for the Vodlozero complex there is no absolute certainty that the observed fold structures and mineral associations in the gneisses originated during the Early Archaean rather than the Late Archaean.

The upper Archaean Lopian complex

Rocks of the Lopian complex predominate within the territory of Karelia. Supracrustal rocks in greenstone belts, as well as the majority of granites and granite-gneisses separating the greenstone belts, belong to the Lopian (Fig. I-28). At the present erosion surface, the Lopian supracrustal rocks occupy less than 15% of the area covered by $AR_1 + AR_2$ rocks. Palaeogeographic and palaeotectonic reconstructions (Svetova, 1979) suggest that late Lopian sedimentary basins were much more extensive.

Along the contacts between greenstone belts and basement granite-gneiss are frequent instances of younger granites, which are clearly delineated on magnetic maps (Krylov et al., 1984; Lobach-Zhuchenko et al., 1985).

TABLE 1-15

Average chemical composition of Archaean igneous rocks

Oxide	1	2	3	4	5	6	7	8	9
SiO_2	61.20	68.62	52.12	49.35	62.32	43.25	47.54	48.83	50.11
TiO_2	0.84	0.31	1.50	0.98	0.67	0.32	0.48	0.96	0.83
Al_2O_3	15.67	15.99	17.36	15.47	14.99	6.76	10.45	14.51	14.79
Fe_2O_3	3.37	0.98	4.32	3.78	4.14	4.42	2.46	3.10	2.34
FeO	3.42	1.65	6.26	7.43	4.14	6.87	8.72	9.40	8.61
MnO	0.12	0.04	0.13	0.18	2.17	0.18	0.21	0.21	0.19
MgO	2.17	1.15	4.08	6.59	1.79	23.97	15.29	7.48	7.31
CaO	5.51	3.98	7.41	11.85	5.51	7.28	9.53	10.58	10.58
Na_2O	4.16	4.72	4.43	2.83	3.87	0.11	1.10	2.32	2.12
K_2O	1.26	1.22	1.14	0.68	1.41	0.04	0.43	0.61	0.32
Rb	43	45	23	—	—	—	12	17	11
Sr	461	500	438	—	—	—	36	114	119
Zr	245	188	258	—	—	—	—	48	63
Y	17	8	26	—	—	—	—	22	26
No.	3	7	5	6	3	19	22	29	65

Oxide	10	11	12	13	14	15	16	17	18
SiO_2	50.36	48.76	49.74	61.27	69.43	67.20	70.68	68.95	74.86
TiO_2	0.88	0.87	0.91	0.71	0.22	0.40	0.38	0.37	0.20
Al_2O_3	15.17	14.81	14.56	15.81	15.99	16.61	14.51	15.51	13.17
Fe_2O_3	2.38	2.89	2.61	1.83	0.72	1.29	0.87	1.11	0.59
FeO	9.51	8.74	8.80	3.90	1.51	1.95	1.82	2.04	1.50
MnO	0.24	0.20	0.19	0.10	0.13	0.05	0.08	0.06	0.05
MgO	6.94	7.73	7.65	3.45	0.72	1.23	1.08	1.22	0.52
CaO	10.54	10.18	10.36	4.73	3.80	3.11	2.48	3.58	0.64
Na_2O	1.95	2.43	2.17	3.62	5.04	4.58	5.06	4.50	3.08
K_2O	0.28	0.24	0.36	1.63	0.77	1.93	1.56	1.27	4.87
Rb	5	8	—	480	24	73	58	45	274
Sr	115	131	—	380	441	474	253	385	37
Zr	50	48	—	120	211	133	151	151	121
Y	21	19	—	10	12	8	6	11	80
No.	26	77	139	136	14	14	4	142	6

Notes: 1–3: Vodlozero complex, AR (1 = mesocratic gneisses, 2 = leucocratic gneisses, 3 = amphibolites); 4–5: Tulosozero complex, AR (4 = two-pyroxene schists, 5 = hornblende gneisses); 6–7: volcanics in AR greenstone belts (6 = peridotitic komatiites, 7 = komatiitic basalts); 8–12: basalts (8 = Palalambi, 9 = Semch, 10 = Kostomuksha, 11 = Shilos belts, 12 = Karelia as a whole); 13 = andesites, Hautavaara belt, 14 = rhyodacites, Kostomuksha belt; 15–18: AR granitoids (tonalites) (15 = Central Karelia, 16 = Western Karelia, 17 = Karelia as a whole, 18 = post-folding granite, Kartashi intrusion, Central Karelia).

Subsequent ultrametamorphic processes, for example in SW (Kratz, 1969) and W Karelia (Sviridenko, 1984) led to the replacement of Lopian supracrustal rocks by migmatite. These factors are also largely responsible for the absence of direct observations of the relationships between Lopian supracrustals and older rocks. Nevertheless, at the base of Konto Group basalts in the Kostomuksha belt, there are muscovite quartzite and gneiss which are considered to be basal formations (Gorkovets and Rayevskaya, 1983). Our geochemical studies of microcline gneiss at the base of the basalts in this same structural belt have also enabled us to demonstrate that they are metasediments which accumulated during the weathering of mainly tonalites. In the Palalambi belt in central Karelia, the likely

TABLE I-16

Isotopic age of rocks in the Karelian province (U–Pb method on zircons)

No.	Rock type; structural belt, location	Age, Ma	Lab., refs
1	Tonalite gneisses and amphibolites; Vodlozero complex, SE Karelia	3150	PC 1, 2
2	Tonalite (migmatite substrate); basement to Palalambi belt, Central Karelia	3100	PC 3, 2
3	Granite (vein material of migmatite); as ditto	2920	PC 3
4	Meta-andesite; Palalambi and Oster greenstone belts, Central Karelia	3020	PC 3, 2
5	Metadacite; Kojkara greenstone belt, Central Karelia	2935	GC 3, 2
6	Gabbro-diorite; cuts the Semch greenstone belt, Central Karelia	2890	PC 3, 2
7	Gabbro; cuts Palalambi greenstone belt, Central Karelia	2840	PC 3, 4
8	Granite-porphyry; dyke, cutting upper terrigenous succession of the Oster belt, Central Karelia	2830	PC 3, 2
9 & 10	Granite, post-folding; Suna river region, Central Karelia	2740 2820	GC, PC 5, 2
11	Granite, post-folding; Pala Lambi region, Central Karelia	2810	PC 3, 2
12	Granites and granodiorites of E. Karelia, near lakes Viksa, Nizhneye, Mashozero and Nadvoitsy station	2740	GC 5, 2
13	Granites and migmatites; Lake Suojärvi region, SW Karelia	2760	PC 3, 2
14	Enderbite; Lake Tulos region, Western Karelia	2740	GC 5, 2
15	Tonalites and granites cutting upper terrigenous succession of Kostomuksha greenstone belt, Western Karelia	2740	GC 5, 2
16	Granulites; Lake Tulos region, Western Karelia and near Shalsky village, SE Karelia	2650	PC 3, 2
17	Sumian quartz porphyry, Lehti belt, E. Karelia	2760	PC 3, 2
18	Dolerite; dyke, cutting Middle Jatulian	2150	PC 2

Note: PC = Institute of Precambrian Geology and Geochronology, Leningrad; GC = Institute of Geochemistry, USSR Academy of Sciences, Moscow. *References*: 1 = Sergeyev et al. (1985); 2 = Anon (1986); 3 = Lobach-Zhuchenko et al. (1986); 4 = Sergeyev et al. (1983); 5 = Tugarinov and Bibikova (1980).

contact between basalt-komatiite series and older tonalites is also exposed. The tonalites and metabasalts were metamorphosed and deformed together. Isotopic data (Table I-16, No. 2) confirm the view that the tonalites are older than the metavolcanics of the Palalambi structure.

The isotopic age of rocks in the Lopian complex is shown in Table I-16. From this information we can demonstrate that the age of the volcanic rocks in central Karelia is 3020–2935 Ma or older, while in western Karelia the dates are younger: 2800 Ma. The age of basic and granitic intrusions which cut rocks of the greenstone belts varies from 2890 to 2680 Ma. Some of the figures are for granite-gneiss (Table I-16, Nos. 12, 13) and granite (Table I-16, Nos. 9–11) located amongst granite-gneiss far from greenstone belts, as well as for rocks formed during the Late Archaean granulite facies metamorphism.

The Lopian supracrustal assemblage includes metamorphosed volcanic rocks: peridotitic, pyroxenitic and basaltic komatiite, basalt, andesite, andesite-dacite, dacite and rhyolite and sedimentary rocks of volcano-terrigenous, continental and chemogenic origin: tuff, tuffite, greywacke,

Fig. I-29. Variolitic textures in pyroxene komatiite in the Koikari belt, Central Karelia (photo by A.I. Svetova).

arkose, micaceous, graphitic and pyritic schist, ferruginous quartzite, quartzite, chert and carbonate rocks. Table I-15 shows the chemical composition of the volcanics. The volcanic rocks are grouped into suites: komatiite-basalt, andesite-dacite and dacite-rhyolite (Rybakov and Svetova, 1981).

No primary minerals are preserved in the deformed and metamorphosed volcanics, but primary structures and textures are well seen, such as pillows, amygdales and varioles (Fig. I-29), breccia, spinifex, etc., making it possible to determine the primary volcanic nature of the rocks. Mafic and ultramafic volcanics are typically lavas, while intermediate and acid volcanics are pyroclastics.

Supracrustal rocks form isolated greenstone belts, varying in length from 2–3 km to 100–150 km. As can be seen from Fig. I-30, the structures differ considerably from one another in thickness, in the composition and sequence of volcanics and in the composition of sediments. There are a number of ideas on the reasons for the observed differences in the sequences. The extreme points of view are:

(1) the differences reflect local and/or regional differences in volcanic and sedimentary environments, and

(2) the successions were identical; as a result of tectonic and metamorphic processes we are thus dealing with various relict parts of a single succession.

Nevertheless, general features have been established for practically all the successions; these are also typical for greenstone belts in other shield areas of the world: (a) a predominance of volcanic rocks in the lower part of a

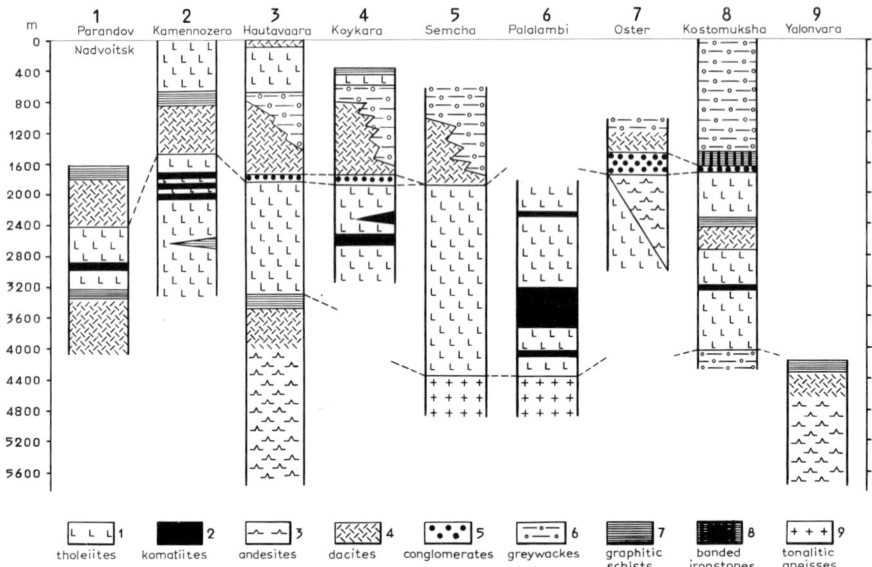

Fig. I-30. Stratigraphic sections of Lopian sequences forming greenstone belts in Karelia (after Rybakov and Svetova, 1981; Kratz, 1978).

sequence and sediments in the upper part, and (b) a predominance of basalts among the volcanic rocks (Fig. I-30). In a number of belts, upper and lower parts of a sequence are separated from each other by conglomerate (Fig. I-30). Using the Oster belt as an example, where polymict conglomerates are very thick, pebble composition is most varied and where the pebble composition has been studied in detail (Kratz et al., 1977; Kratz, 1978), it has been shown that the conglomerates are intraformational. Their formation preceded the deformation and metamorphism of the lower volcanic part of the sequence.

A comparative analysis of the sections and their spatial distribution has allowed us to join the greenstone structures together as belts (Rybakov and Svetova, 1981) and to identify three zones (Krylov et al., 1984), the greenstone structures in which differ in composition and in the age of the volcanics. The east Karelian zone (Fig. I-28) includes the Sumozero-Kenozero and Vygozero belts, the first of which is the biggest and longest. Typical for this zone is bimodal vulcanism with an abundance of mafic and ultramafic rocks (Kulikov and Kulikova, 1979; Kulikova and Kulikov, 1981). In the central Karelian zone we have the Segozero-Vodlozero belt. Basalt, dacite, andesite and andesite-dacite are present, as well as komatiite.

Special mention will be made of the Hautavaara belt, the exposed sequence of which begins with andesite; andesite and andesite-dacite dominate the section (Fig. I-30), which sharply distinguishes it from other

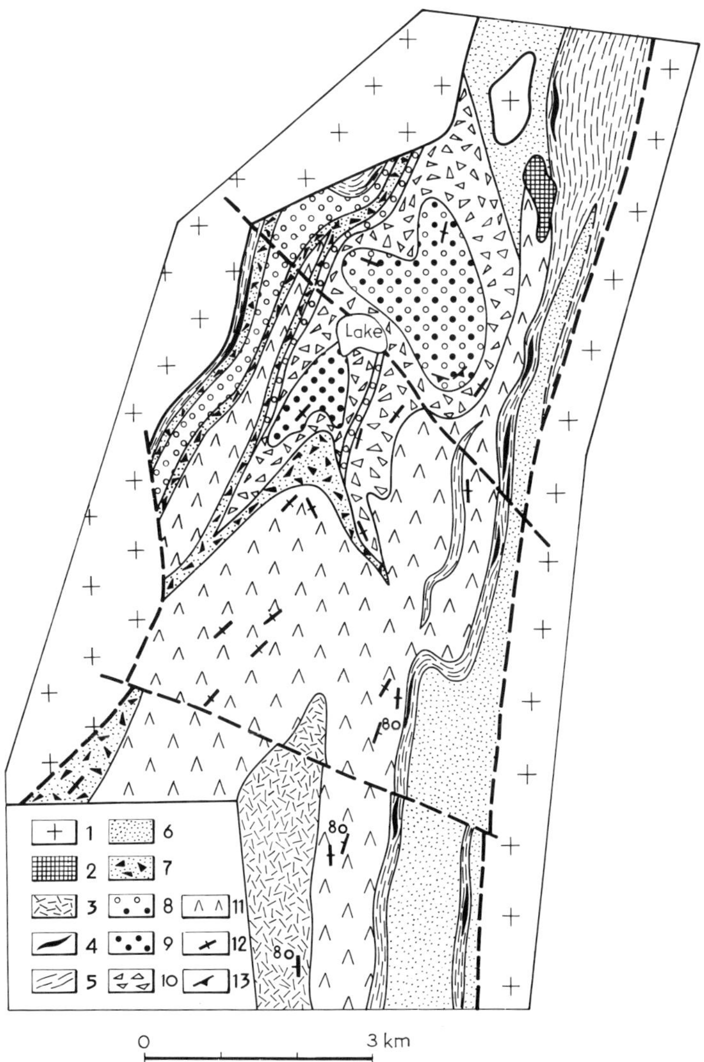

Fig. I-31. Geological sketch map of the Chalki palaeovolcanic edifice (Rybakov and Svetova, 1981). *1* = granitoid; *2* = gabbro-diabase; *3* = dacite tuff; *4* = pyrite deposits; *5* = quartz-sericite-graphite schist; *6* = andesite-dacite crystalloclastic tuff; *7* = andesitic agglomerate-lappili tuff; *8* = dacite and andesite-dacite lavas; *9* = subvolcanic andesite; *10* = andesitic boulder-agglomerate tuff; *11* = andesite lavas. Orientations: *12* = schistosity; *13* = dip and strike of volcanic flows.

structural belts. By means of palaeofacies analysis (Robonen et al., 1975; Svetova, 1979), palaeovolcanic reconstructions have been mapped out in the Hautavaara and several other belts in central Karelia. One of these, the Chalkin (Fig. I-31) is situated in the northern part of the Hautavaara belt. A zone of necks near a crater consists of pyroclastic andesite, which gives way westwards and southwards to a lava facies; the lavas in the crater zone were erupted into shallow water in a near-shore basin. Around the periphery of the edifice are outcrops of psammitic and crystalloclastic tuff, tuffite, tuffaceous sandstone, siliceous tuffite and quartzite. A lava field with an area of 300–400 km^2 formed as a result of activity in the Chalkin palaeovolcano. The nature of the distribution of volcanic facies in space and the age relations of subvolcanic dykes have allowed the authors to distinguish explosive-effusive and effusive phases of activity in this Archaean volcano, and to establish a uniform basic to acid nature of the evolutionary trend of the volcanism.

In many belts (Fig. I-30) the section commences with basalt, with associated komatiite in a number of cases. The greatest thickness of komatiite is attained in the Palalambi belt (Kratz, 1978; Rybakov, 1981), where pyroxenitic komatiite predominates; peridotitic komatiite plays a minor role. In the west Karelian zone are the Kostomuksha, Khedozero-Bolsheozero, Gimola and other structures, which together make up the Gimola-Kostomuksha greenstone belt (Sokolov et al., 1978). The largest and best studied structure is the Kostomuksha (Gorkovets et al., 1981) which characteristically has bimodal volcanism. A volcanogenic unit occurs in the basal formations, consisting of basalt, dacite-rhyodacite and komatiite, with thin magnetite quartzite. They are separated from overlying sedimentary units by a conglomerate member. The upper unit consists of rhythmically-layered flysch-type quartz-biotite, quartz-garnet-biotite and graphitic schist; ferruginous quartzite is restricted to the lower part of this unit (the Kostomuksha iron ore deposit).

The rocks of the Lopian complex were metamorphosed at amphibolite to greenschist facies. On the whole, the metamorphism corresponds to low-pressure conditions (Drugova, 1979). At the same time, a detailed structural and metamorphic study of individual structures (Kratz et al., 1978; Chekulayev, 1981; Chekulayev and Baykova, 1984) has revealed a more complex picture: different trends in the evolution of the metamorphic processes — from low- to medium-pressure and vice versa. Variations in metamorphic grade are frequently observed within the same structure (metamorphic zonation). The origin of zonation is related to domal-block movements of granitoids in the basement complex (Kratz et al., 1978; Chekulayev, 1981). Supracrustal rocks are cut by ultramafic, gabbro-diorite, tonalite and granite intrusions. Taken together they constitute gabbro-tonalite, ultramafic, migmatite-granite and granite associations (Sviridenko, 1985). Of these, gabbro and diorite are most closely associated with greenstone belts, while the ultramafics (Zak et al., 1975) and granitoids are encountered in both greenstone belts and in granite-gneiss terrains separating them. Rocks of gabbro-diorite composition are divided geologically into two groups. The earlier gabbro and diorite intrusions occur in strong spatial association with volcanics and they were metamorphosed and deformed together. They form dykes, sills and minor intrusions and are probably comagmatic with the volcanics. The second group consists of major compound intrusions of gabbro-diorite composition. Intrusion of these bodies took place after early deformation and metamorphism of metavolcanics along fault zones, which were long-lived zones of eruptive breccias (Kratz et al., 1978; Arestova and Pugin, 1985). The formation of the largest intrusion, the Semch, took place under relatively stable tectonic conditions, which assisted melting during the formation of the rocks (Arestova and Pugin, 1985). Differences in composition and the degree of REE differentiation in the metatholeiites of the greenstone structures and of the Semch gabbro-diorites are evidence that these magmas formed at different depths and by different mechanisms.

Ultrabasic rocks (Sviridenko, 1985) form small bodies or occasionally larger intrusions of peridotite, which are almost completely altered and consist of serpentine, amphibole, chlorite, talc, ore minerals and carbonate. The primary assemblages were olivine + magnetite, or olivine, magnetite, chromite + enstatite (Sviridenko, 1985). The ultrabasic bodies occur among greenstone rocks and granitoids (Zak et al., 1975; Kratz et al., 1978; Lavrov, 1979). They were intruded later than the gabbroids; in the Oster structure they cut the upper terrigenous part of the sequence, in which the basal conglomerates contain pebbles of gabbro and diorite (Kratz et al., 1978). A distinguishing feature of the ultrabasics is their uniformly weak differentiation and lack of any relationship with gabbroids (Sviridenko, 1985) and komatiites.

Late Archaean tonalites crop out all over Karelia and are typical of other granite-greenstone terrains. They occur among supracrustal rocks and within areas of granite-gneiss; they vary in genesis and mode and time of intrusion. Early tonalites are situated in marginal parts of greenstone belts; they underwent all the same structural-metamorphic transformations as the volcanics (Chekulayev, 1981). Later minor intrusions of tonalite and dykes of tonalite porphyry were emplaced after the early deformation and metamorphism of the volcanic layer, but before the deposition of continental beds containing conglomerate; finally, hypabbysal dykes of tonalite porphyry cut the conglomerate and metaquartzite. The chemical composition of tonalites is shown in Table I-15.

In areas of granite-gneiss, rocks of tonalitic composition are also widespread. Among these, we can distinguish plagioclase-rich partial-melt migmatite, intrusive tonalite and enderbite. Tonalites have been noted which originated by metasomatism of supracrustal rocks or tonalites (Sviridenko, 1974; Lobach-Zhuchenko, 1984). In the latter case, the tonalites arose as a result of Si-metasomatism under granulite facies conditions. The majority of intrusive tonalites of this phase display a number of geochemical characteristics (e.g. Figure I-32) which sharply differentiate them from tonalites of Phanerozoic gabbro-granite associations, island arc

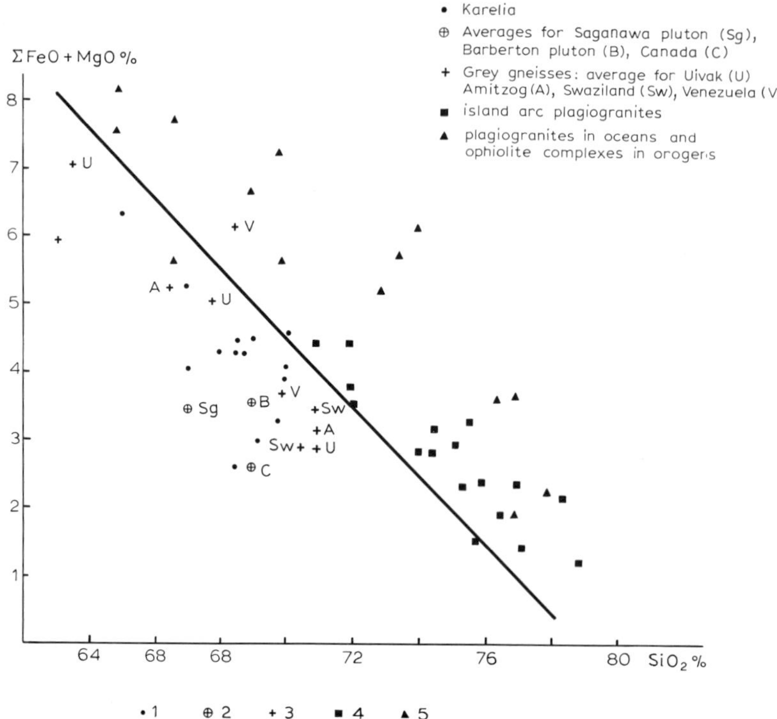

Fig. I-32. Relationship between SiO_2 and mafic oxides in plagiogranite and tonalite from granite-greenstone terrains.

trondhjemites and tonalites in ophiolite complexes (Kratz, 1978; Lobach-Zhuchenko, 1979, 1984, 1985).

Magmatism of the Lopian stage concluded with the intrusion of late- and post-folding granites (Table I-17). These minor intrusions occur at both contacts with greenstone belts (e.g. the Kartash mass, with an age of 2810 Ma (No. 11 in Table I-16), and among granite-gneisses of the reworked basement. The chemical composition is shown in Table I-15.

Useful minerals of Lopian time occur in structural-metallogenic zones: the Gimola-Kostomuksha, Jalonvaara, Hautavaara-Vygozero and Sumozero-Kenozero, as expressions of the tectonic and lithologic heterogeneity of the Karelian granite-greenstone region. In the Gimola-Kostomuksha zone, volcano-sedimentary and sedimentary-type iron ores are of major importance: the Kostomuksha and other deposits (Bogachev and Zak, 1980; Gorkovets et al., 1981). In addition, metasomatic ores are also developed in this zone: iron ore, pyrite, sulphide and magmatic ores — copper-nickel in ultrabasic intrusions, rare-earth and rare-metal deposits in pegmatites, iron-titanium ores in gabbroids, and rare-earth–rare-metal deposits in micaceous picrites.

TABLE I-17

Correlation of supracrustal and plutonic complexes and internal processes in the Karelian province

Proto-platform	Karelian	Vepsian volcanogenic complex; alkali-gabbroic intrusions Suisarian volcano-sedimentary and intrusive complex North Onega volcano-sedimentary complex Jatulian volcano-sedimentary complex (2-2.1 Ga)		
Orogenic	Karelian	Sumian-Sariolian volcano-sedimentary formations, layered basic-ultrabasic intrusions		
Mobile	2600 ±100	West Karelian zone	Central Karelian zone	East Karelian zone
Mobile	Lopian	Local low-P granulite metamorphism, 2.65 Ga Metamorphism and partial melting, 2.76 Ga Intrusive granites, 2.75-2.80 Ga Kostomuksha greenstone belt (bimodal volcanism) 2.8 Ga	Rhyolites, Lehti belt 2.75 Ga; granites, 2.74 and 2.82 Ga; migmatites Gabbro-diorites, 2.89 Ga Zonal metamorphism and migmatites, 2.9 Ga	Local low-P granulite metamorphism, 2.65 Ga Metamorphism, Sumozero-Kenozero greenstone belt (2.9 Ga)
Mobile	Lopian		Segozero-Vedlozero greenstone belt (polymodal volcanism), 2.9-3.0 Ga	
Mobile	Saamian	Gneisses, schists (Voknavolok and Tulos complexes)	Gneissose diorites (substrate of migmatites) 3.1 Ga	Gneisses, amphibolites (Vodlozero complex), 3.15 Ga

In the Hautavaara-Vygozero zone, several ore formations of igneous origin have been identified (sulphide-copper-nickel in pyroxenites, iron-titanium-vanadium in gabbro-amphibolites, chromite in serpentinites), a group of pyrite formations of complex genesis, and various ore shows of metasomatic and hydrothermal origin. Pyrite ores (sulphide-copper-nickel formation) are the main useful mineral component of the Lopian complex in the Hautavaara-Vedlozero and Jalonvaara zones (the Hautavaara, Nyalmozero, Vedlozero and Jalonvaara deposits). Stratiform volcano-sedimentary mineralization was of primary pyritic origin, which subsequently underwent metamorphic regeneration (Rybakov, 1980, 1982).

Sulphide-copper-nickel, polymetallic sulphide and high-vanadium iron-ore formations are also associated with metasomatic processes (Slyusarev and Kulikov, 1973; Slyusarev et al., 1975).

In the Sumozero-Kenozero zone there are stratiform volcano-sedimentary and sedimentary pyrite and ironstone formations. Associated with the greenstone belts there are also chromite, sulphide, copper-nickel and iron-titanium formations of igneous origin, related to basic and ultrabasic rocks. A brief summary of the most characteristic ore deposits in the Lopian complex is shown in Table I-18.

TABLE I-18

Typical mineral deposits and ore shows

Deposit or ore show; minerals	Tectonic setting	Type of mineralization: ore-forming and genetic	Geological and mineralogical features				Age of host rocks and mineralization
			Host rocks and mineralized rocks	Shape of deposit, ore body morphology	Ore types	Mineral composition of ores (secondary minerals)	
I. Lopian complex							
Kostomuksha; iron	Archaean greenstone belt	Iron ore, in quartzites—magnetite; metasedimentary	Hornblende and biotite gneisses	Conformable layered and lenticular deposits pyrrhotine,	Dense clusters and massive	Magnetite (hematite, pyrrhotine, pyrite)	Late Archaean
Parandovo; sulphur, iron	Segozero greenstone belt	Pyrrhotine-pyrite; sedimentary exhalative, hydrothermal metamorphic	Schists after volcanics of the keratophyre-spilite formation	Layered, lenticular	Massive and disseminated	Pyrite, pyrrhotine (sphalerite, magnetite, galena, pentlandite) propyllites	Early PTZ
Hautavaara; sulphur, iron	Suojärvi block-anticlinorium	Pyrrhotine-pyrite; metasomatic hydrothermal-metamorphic after volcano-sedimentary rocks of keratophyre-spilite formation	Quartz-sericite schists, siliceous tuffs	Conformable layered, lenticular deposits, veins	Massive and disseminated	Pyrite, pyrrhotine (chalcopyrite, sphalerite, magnetite, etc.)	Early PTZ
Lebyazhi (Kamennozero group of ore shows); nickel, copper, cobalt	Archaean greenstone belt	Copper-nickel sulphides, magmatic-liquation up to metamorphogenic related to ultramafic bodies	Ultramafics, olivinites to wehrlites	Concordant lenticular deposits	Disseminated	Pyrrhotine, pentlandite hislewoodite, chalcopyrite (magnetite, millerite, pyrite)	Early PTZ
II. Karelian complex							
Lukkulajsvaara; nickel, copper, cobalt	Panajärvi-Tikshozero belt	Copper-nickel sulphides, magmatic-liquation in layered peridotite-pyroxeneite-norite intrusions	Peridotites, norites	Conformable lenticular deposits	Disseminated, clusers in thin veins	Pyrrhotine, chalcopyrite pentlandite (pyrite, magnetite, ilmenite, vallerite)	Early PTZ
Kuzoranda; copper, cobalt	Onega synclinorium	Chalcopyrite-pyrite with cobalt in sandstones and shales; volcano-sedimentary	Siltstones, sandstones	Conformable layered	Layered-disseminated	Pyrite, chalcopyrite, (chalcosite, pyrrhotine); carbonitization and silicification	Early PTZ
Pudozhgora; titanium, iron, vanadium	Onega synclinorium	Titanomagnetite; magmatic segregation in gabbro-diabases	Gabbro-diabase	Conformable sheetlike deposits	Dense clusters	Titanomagnetite, ilmenite (magnetite, hematite, pyrite, chalcopyrite)	Early PTZ
Majmjärvi; copper	Segozero graben-synclinal zone	Chalcopyrite-chalcosite; volcano-sedimentary	Quartzose sandstones	Conformable layered and lenticular deposits	Laminated-disseminated	Chalcopyrite, chalcosite bornite (pyrite, magnetite, molybdenite, sphalerite, etc.)	Early PTZ
Päävara; molybdenum	Päozero-Vodlozero block	Molybdenum; hydrothermal	Granites	Stockworks	Disseminated, clusters in thin veins	Molybdenite (chalcopyrite, pyrite, etc.)	Early PTZ
Zazhogin; shungite	Onega basin	Shungite; metamorphic	Metadiabases, tuffs, siltstones	Conformable multi-layered beds	Massive	Shungite, quartz (pyrite, chlorite)	Proterozoic

The lower Proterozoic Karelian complex

Rocks of the Karelian complex crop out in small isolated structures which are found throughout almost the entire territory of the Karelian province (Fig. I-33). In terms of age and conditions of formation, Proterozoic supracrustal rocks are divided into a number of superhorizons (Sokolov, 1984; Table I-19). Igneous rocks formed contemporaneously with the supracrustals, and these are grouped into complexes and associations. They consist of volcanics, which form a natural constituent of the supracrustal assemblages, and intrusions, which are sometimes separated from the synchronous supracrustal rocks.

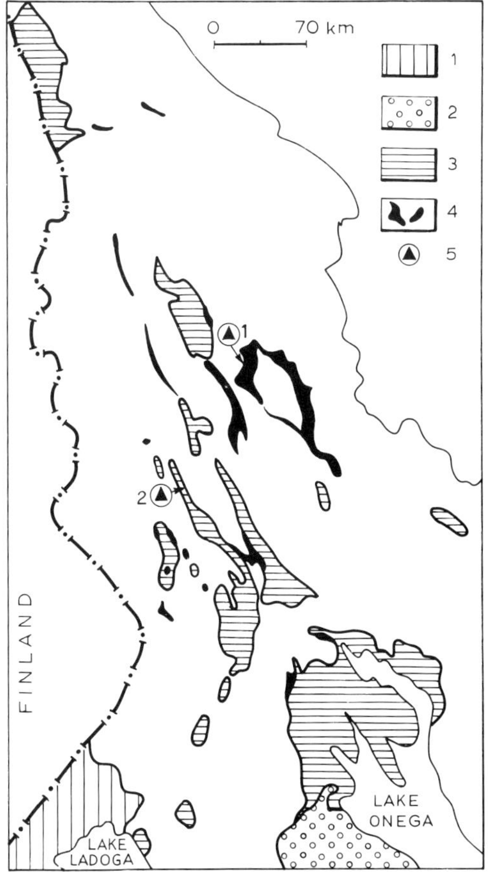

Fig. I-33. Sketch map showing distribution of lower Proterozoic supra-crustal assemblages in Karelia. Supergroups: *1* = Vepsian, *2* = Livvian, *3* = Jatulian, *4* = Sariolian and Sumian; *5*: isotope dating localities (*1* = Lehti Sumian structure, *2* = Yangozero Jatulian structure).

TABLE I-19

Schematic stratigraphic subdivision of supracrustal rocks [a] in Proterozoic igneous complexes of Karelia and their formational classification [b]

Supergroup	Igneous complex	Formations	
		effusive	intrusive
Vepsian	Vepsian	Tholeiite-basaltic	Alkaline-gabbroic
Livvian			
Lyudikovian	Suisaari North Onega	Picrite-basaltic Tholeiite-basaltic	Gabbro-peridotitic
Jatulian	Jatulian	Tholeiite-basaltic	
Sariolian	Sariolian	Andesite-basaltic	
Sumian	Sumian	Dacite-rhyolitic	Peridotite, gabbro-norite

[a] After Sokolov (1984); [b] after Sviridenko (1985).

Sumian and Sariolian. Sumian and Sariolian formations have been investigated by L.Ya. Kharitonov, M.A. Gilyarova, V.A. Perevozchikova, V.I. Robonen, M.M. Stenar, T.F. Negrutsa, V.Z. Negrutsa, K.I. Heiskanen and others. Sumian and Sariolian rocks are andesite-basalt and polymict conglomerate which overlie granitoids and upper Archaean rocks of the Lopian complex with a structural break (Kratz, 1963). Heiskanen has identified three main types of sequence (Fig. I-34a). The Selet type mainly includes continental and volcano-sedimentary rocks. The Kumsi type shows a more complete succession due to the presence of andesite-basalt layers at the base of the section. The third type, the Shuyezero, reflects the presence of patches of K-rhyolites and tholeiites, which occur beneath the andesite-basalt (Fig. I-34a). There are two viewpoints concerning the position of the boundary between the Sumian and the Sariolian, as shown in Fig. I-34a. Most geologists (Sokolov, 1984) support the view that the polymict conglomerate and andesite-basalt lava (the Kumsi sequence) belong to the Sariolian, while the Sumian Supergroup consists only of the metabasalt and rhyolite of the Shuyezero sequence. According to others, the andesite-basalts also belong to the Sumian. Sumian rhyolites of the Lehti structure have been dated by the U–Pb method and yield an age of 2760 Ma (Krylov et al., 1984; Table I-18, No. 17; Fig. I-33).

Rocks of the Sumian and Sariolian Supergroups are weakly deformed, and form gently inclined troughs. Metamorphism varies from greenschist to epidote-amphibolite facies. Rocks of dacite-rhyolite composition are represented by hypabbysal, lava, tuff and mixed volcano-sedimentary facies. A characteristic petrochemical feature of the volcanics is the strong predominance of K over Na (Table I-20). Andesite-basalts occupy a significant area of outcrop. In some structural zones, the andesite-basalts contain high-Mg varieties, varying according to SiO_2 content from basaltic komatiite to

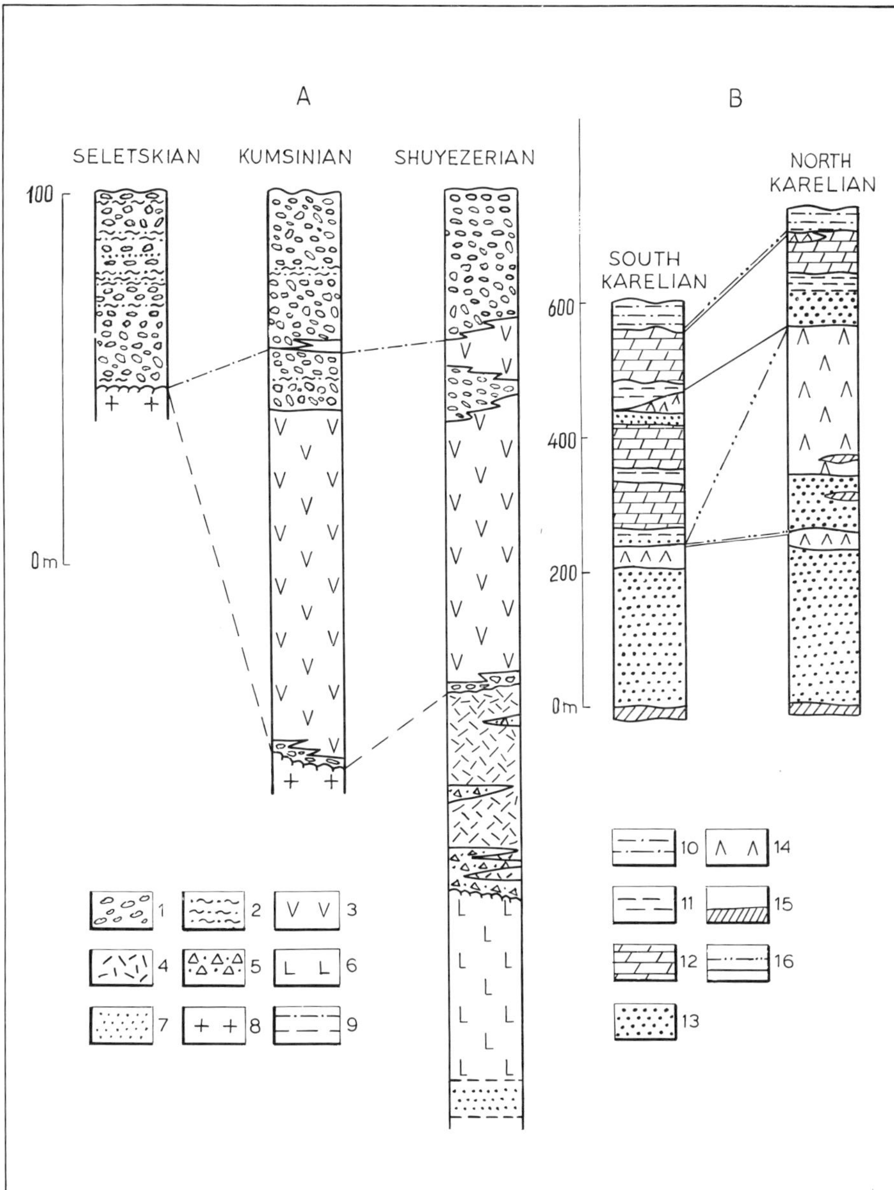

Fig. I-34. Types of Sumian-Sariolian (A) and Jatulian (B) successions and correlation variants (after Sokolov, 1984). A. *1* = acid and basic clastic rocks; *2* = sandstone (tuffite), shale and siltstone (tuff); *3* = andesite-basalt; *4* = rhyolite; *5* = sedimentary-pyroclastic rocks; *6* = basalt; *7* = quarzitic rocks; *8* = granitoids; *9* = Sariolian/Sumian boundary (two variants). B. *10* = Lyudikovian sediments; *11* = carbonate and terrigenous-carbonate; *12* = meta-argillaceous; *13* = quartzose grit and conglomerate; *14* = basalt; *15* = chemical weathering crust; *16* = Lower, Middle and Upper Jatulian boundaries (two variants).

TABLE I-20

Average chemical composition of Proterozoic igneous rocks

Oxide	1	2	3	4	5	6	7	8	9
SiO_2	77.43	77.80	55.62	53.05	49.98	48.74	48.14	47.60	49.28
TiO_2	0.26	0.22	0.92	0.67	1.61	1.29	2.41	2.71	1.31
Al_2O_3	11.62	11.60	13.84	11.87	13.22	14.77	13.32	13.72	14.21
Fe_2O_3	1.55	0.95	2.16	1.88	6.79	5.44	9.14	6.67	3.00
FeO	n.d.	n.d.	7.92	7.81	7.39	7.22	6.49	11.28	9.82
MnO	0.02	0.01	0.16	0.25	0.19	0.17	0.14	0.21	0.20
MgO	0.30	0.17	5.97	9.51	5.80	7.62	6.71	4.15	6.78
CaO	0.22	0.17	6.00	9.14	6.57	7.64	6.37	7.80	9.26
Na_2O	0.60	0.58	3.78	2.27	3.46	3.43	4.81	3.50	2.56
K_2O	7.30	7.40	1.18	0.91	0.59	0.34	0.35	0.63	0.49
No. of analyses	13	35	253	10	68	159	24	49	5

Oxide	10	11	12	13	14	15	16	17	18
SiO_2	49.26	51.80	48.39	43.54	42.63	52.10	49.17	44.60	40.37
TiO_2	1.37	1.49	1.86	1.42	1.66	0.76	0.65	0.43	0.33
Al_2O_3	14.04	14.61	12.69	8.97	9.10	13.45	11.60	8.00	4.51
Fe_2O_3	2.62	3.06	2.10	1.74	2.14	1.78	1.57	2.94	4.69
FeO	10.30	6.96	9.44	10.58	10.24	8.60	9.20	8.51	7.41
MnO	0.21	0.12	0.18	0.21	0.23	0.18	0.19	0.21	0.18
MgO	6.73	8.17	9.67	18.22	19.42	9.19	13.70	23.37	30.08
CaO	9.24	6.00	8.87	8.59	7.82	10.19	9.11	6.83	2.97
Na_2O	2.33	3.35	1.87	0.49	0.28	1.93	1.50	0.58	0.22
K_2O	0.97	0.82	0.83	0.15	0.03	0.59	0.38	0.17	0.12
No. of analyses	4	16	47	11	10	24	130	24	138

Note: Sumian-Sariolian: 1 = lava, 2 = subvolcanic facies of Sumian rhyolites. Lehti belt: 3 = Sumian-Sariolian andesite-basalts (1—3 after Heiskanen et al., 1977), 4 = high-Mg Sumian-Sariolian andesite-basalts, Bolsheozero belt. Jatulian assemblage: 5 = Lower Jatulian basalts, 6 = Middle Jatulian basalts, 7 = Upper Jatulian basalts, 8 = gabbro-dolerites of Givras ore-bearing sill (Jatulian). North Onega complex: 9 = North Onega basalts, 10 = gabbro-dolerites. Suisari complex: 11 = plagioclase basalts of Suisari volcanic zone, 12 = basalts dykes in Suisari volcanic zone (5–12 after Gobulev and Svetov, 1983), 13 = picritic basalts of Rovnozero volcanic zone, 14 = picritic basalt dykes of Suisari volcanic belt. Vetreny belt: 15 = tholeiitic basalts, 16 = olivine basalts, 17 = picritic basalts, 18 = peridotites (13—18 after Slyusarev and Kulikov, 1973; Sviridenko, 1985).

boninite (Table I-20). Intrusive igneous activity of this period is represented by a peridotite-gabbro-norite association (Sviridenko, 1985). It includes a number of layered intrusions: Kivakka, Tsipringa, Lukkulajsvaara, Kivach and Petusjärvi and the largest intrusion, Burakov, with an area of some 700 km^2. The lower age boundary of the intrusions is quite well defined, since they cut Late Archaean granites. The north Karelian intrusions have been studied in detail (Lavrov, 1979). In the Tsipringa and Lukkulajsvaara intrusions, the lower horizons consist of gabbro-norite, olivenite, peridotite and pyroxenite, the upper horizons being mesocratic and leucocratic gabbro-norites. The Kivakka body has the most complex finely-layered structure.

The Burak intrusion (Sviridenko, 1985) is a lopolith, divided into three blocks by faults. It consists of a complexly differentiated series of rocks from ultrabasic to basic composition. The north-eastern Aganozero block possesses its own independent structure. Here there is a zone of ultrabasic rocks, alternating peridotite-pyroxenite-plagioclasite, and a zone of gabbro-norite. The lower parts of the Burakov and Shalozero blocks consist of gabbro-norite, while the upper parts are gabbro-diorite. The intrusions contain varying mineral compositions: plagioclase (An_{61-31}), orthopyroxene and clinopyroxene. Sariolian clastic rocks in the Selet type of sequence overlie eluvial breccia, but overlie andesite-basalt in other sections and show different relations: either there is an erosional unconformable contact, or volcanic flows wedge out within the clastic rock units. Sariolian clastic rocks consist of rhythmically-layered psephite and psammite, tuff (mainly andesite-basalt composition), tuffo-conglomerate and conglomerate, agglomerate and xeno-agglomerate. The conglomerates contain pebbles of gneiss and granite, schist, greenstone and andesite-basalt volcanics. This is the first stratigraphic level of conglomerates with pebbles of microcline-bearing granite.

Jatulian. Small outliers of Jatulian rocks occur in Karelia (Fig. I-33) and eastern Finland. At the base there is a coarse chemical weathering crust. The Jatulian varies in thickness from 300 to 2000 m. The composition of the sedimentary rocks of the Jatulian Supergroup indicates the important role of chemical weathering during their formation (Sokolov, 1984). Four depositional regimes have been identified, in which four types of sequence formed. The south Karelian type of sequence typically has thick carbonates, while in the other three, continental red beds predominate (Sokolov, 1972, 1984). Three rhythms have been identified in Jatulian sequences, each with sediments followed by volcanics: lower, middle and upper Jatulian (Sokolov et al., 1970). The three-fold division of the Jatulian and variations in the correlation of these parts in the different types of section are shown in Figure I-34. The isotopic age of zircon extracted from an albitic gabbro-diabase dyke cutting a lower Jatulian quartzose sandstone is 2150 Ma (Table I-18, No. 18).

Jatulian rocks were metamorphosed under greenschist facies conditions. They form discordant basins and superimposed synclines (Kratz and Lazarev, 1961). In plan, discordant basins are isometric, gently-inclined synclinal structures, basically synsedimentary (e.g. the Segozero, Suojärvi and Tulomozero basins). Besides basins, we also find pronounced linear belts (superimposed synclines or graben-synclines) of synclinal structure, complicated by marginal and axial faults (e.g. the Lubosalmi graben-syncline).

A detailed study of Jatulian volcanism was carried out by Svetov (1979). The volcanics, which account for a significant proportion of the succession, are mainly basalt lavas (Table I-20), with minor amounts of subvolcanic and pyroclastic facies, and chemogenic volcano-sedimentary rocks. Thick sills of gabbro-diabase, for example the Kojkara ore-bearing intrusion, are characteristic of the upper Jatulian. The lava facies consists of flows and sheets of submarine and continental eruptions. Massive, pillow, brecciated and pahoehoe basalts and basaltic hyaloclasites have been mapped. Flows vary in thickness from a few cm to 10 m. The basaltic eruptions formed a lava field up to 2500–3500 km^2.

Lyudikovian. This stratigraphic subdivision has relatively recently been separated from the upper Jatulian (Sokolov, 1980). It also contains the Suisaari complex and the Besovets Group.

Basal horizons of the Lyudikovian rest unconformably on pre-Karelian granitoids and upper Jatulian rocks (Sokolov, 1984). The lower part of the supergroup (the Onega complex) consists of carbonate-clay sediments and shungitic volcano-sedimentary rocks. The assemblage is some 600–650 m thick.

Onega vulcanism is associated with the opening of the Onega, Pana-Kuolajärvi, North Ladoga and several other volcano-tectonic depressions in the marginal zones of the Karelian province (Sokolov, 1984). The volcanics — pyroxene basalt and tuff — attain their maximum thickness of 900 m in the Pana-Kuolajärvi belt. This stage of magmatism is characterised by the widespread intrusion of sills and dykes, comagmatic with the lavas.

The upper part of the supergroup consists of the Suisaari volcanic complex and the Besovets Group, which is essentially sedimentary. In the stratotype region, on Suisaari island, the volcanics are 620–650 m thick. Suisaari volcanism was controlled by essentially the same structures as Onega volcanism. Igneous rocks of the Suisaari complex (Sviridenko, 1985; Svetov, 1979) belong to the picrite-basalt association. Volcanic series of the Vetreny belt also belong to this time interval, forming a linear structure (palaeorift) at the boundary between the Belomorian and Karelian provinces. Differences in tectonic regimes were responsible for differences in the successions. For example, thick (650 m) basalts (with a transition to andesite-basalt) and rare flows of picrite-basalt formed in the Pana-Kuolajärvi depression. In the riftogene zone of the Vetreny belt, the volcanic series has a thickness of 2 km and consists of basalt and komatiite, together with numerous peridotite bodies (wehrlite-lherzolite association). The chemical composition of the major rock types in the Suisaari complex is shown in Table I-20.

In the Onega belt, the sedimentary and volcano-sedimentary rocks (shungitic schist, tuff, chert, siltstone and mica schist) of the lower formation of the Besovets Group belonging to this supergroup, were formed slightly later than the Suisaari volcanics.

Livvian and Vepsian. The Livvian Supergroup (Sokolov, 1984) contains sediments of the Onega belt which had previously been regarded as belonging to the two upper formations (Vashezero and Pados) of the Besovets Group (Kayryak, 1973). The formations contain volcanomict sandstone, tuffaceous sandstone with conglomerate lenses, silicified silty argillite, chert, limestone, sandstone, arkose and greywacke, 110–200 m thick.

The Vepsian Supergroup (Sokolov, 1984) consists of the Petrozavodsk and Shokshin Formations, previously referred to the Jotnian. Vepsian rocks crop out along the western shores of Lake Onega, where they form a flat-lying syncline striking north–west, with a gently inclined (10–20°) eastern limb and a steep (up to 70°), faulted western limb. The 400–500 m thick Petrozavodsk Formation consists of siltstone, feldspathic and polymict sandstone and breccio-conglomerate and has an erosive base. The Shokshin Formation consists of 1000–2000 m of pink, red and purple quartzite, quartzose sandstone, siltstone and conglomerate.

Vepsian volcanism is manifest in the Onega belt in the form of individual tholeiitic basalt flows and gabbro-dolerite dykes and sills.

In the junction zone between the Belomorian and Karelian provinces, the oldest alkaline intrusions formed at this time — the Yeletozero and Tikshozero masses. The Yeletozero mass formed in two stages (Kukharenko et al., 1969; Bogachev and Zak, 1980). During the first stage there were sequential intrusions of magnetite peridotite[1], layered gabbro-peridotite with titano-magnetite ore mineralization, orthoclase gabbro and micaceous peridotite. During the second stage, nepheline syenite and pegmatite formed. The Tikshozero mass has a very similar composition (Slyusarev et al., 1975), but it also contains metasomatites and carbonatites which do not belong to this association (Safronova, 1982).

Mineral resources. Useful mineral deposits in the Karelian complex formed during two metallogenic epochs: the Sumian-Jatulian and Svecofennian (Bogachev and Zak, 1980). Brief details of characteristic ore deposits and ore shows associated with the rocks of this complex are shown in Table I-18.

The Sumian dacite-rhyolite association has potential for Cu, Ti and V exploration and for ceramic raw materials. Basic to ultrabasic intrusions (Tsipringa, Kivakka, Lukkulajsvaara and Burakovka) have Ni, Cu, Co, Cr, Ti and Fe potential, while the Gaykol, Gomoselg and Kolozero sills are potential sources of Cr and Ni ores. Granite masses — Nuorunen, Hizhjärvi and Medvezhozero — contain Sn, Ta, Nb, Th and Mo prospects. Jatulian tholeiite-basalt sills have ore shows of copper and titanium (the Koykari and Mednyye Gory ("Copper Hills") sills).

Ore shows within the epi-Archaean Karelian craton, belonging to the

---

1 "Sideronitic" peridotite in Russian, i.e. a matrix of skeletal crystals with silicates filling the embayments.

Svecofennian epoch, are mainly associated with zones of secondary activation (Sviridenko, 1985). Mention must be made in the first place of the Yeletozero and Tikshozero intrusions with their Ti, Nb, Ta, Tl and Pt ore shows. The Onega-Suisaari complex has a black schist-carbonate-volcanic formation with Pb, Zn, Cu and As prospects. There are several localities in older rocks along the NE boundary of the Karelian craton where polymetallic mineralization has been noted. This formed in the Svecofennian epoch, as shown by geological and isotopic data (Vinogradov et al., 1959).

Conclusions

A characteristic feature of the structural evolution of the Karelian province which distinguishes it from the other provinces of the Baltic Shield is its development in the Proterozoic, when internal processes were active mainly along the margins of the province or geoblock. Beginning with the Jatulian, most of this was a protoplatform type of development (Sokolov, 1982; Krylov et al., 1984), which led to the formation of the epi-Archaean Karelian craton. The most active endogenic processes in the Karelian province took place during the Late Archaean (Table I-18). It was at this time that the overwhelming majority of the rocks were formed which crop out at the present erosion level. Among these rocks, plagioclase-microcline migmatite and granite gneiss are noticeably dominant. This fact distinguishes the Karelian granite-greenstone terrain from other similar structures in the basement of the East European craton in which Late Archaean processes of metamorphism, K-metasomatism and partial melting are less clearly expressed. Another characteristic feature in the development of the Karelian province during the Archaean is its heterogeneity. This is expressed (Table I-17) in the shape of volcanism of different compositions in greenstone belts and a regular spatial pattern of greenstone belts with bimodal vulcanism, metamorphic and partial melt processes of different ages, trends and intensities in different zones, in the late local appearance of high-gradient granulite metamorphism and its association in time with basic magmatism (Krylov et al., 1984; Sakko, 1971).

3. THE BELOMORIDE BELT

The White Sea province is a major fault-bounded deformed lens-shaped Archaean massif. It extends for 1500 km in a north-westerly direction and has a maximum width of around 200 km. The central part of the province is hidden by the waters of the White Sea and the southern half is beneath a cover of Phanerozoic continental platform deposits and is exposed only in isolated boreholes. The area of the White Sea province accessible to direct geological study within the Karelia-Kola region is about 25,000 km^2.

In the geological literature, this territory is usually called the north-west White Sea or simply White Sea region, while as a geotectonic structure is it referred to as the Belomoride fold belt or the Belomorides[1].

The first ideas of the geological evolution of the Belomorides emerged in the late 1930s, from the work of V.M. Timofeyev, N.G. Sudovikov, A.A. Polkanov, G.H. Buntin, P.K. Grigoryev, P.A. Borisov and others. The early notion was that the Belomorides formed during two Archaean tectonomagmatic cycles — the Svionian and post-Bothnian, separated by a period of stabilization and the intrusion of basic rocks called "drusites". Intrusive granite, migmatite and pegmatite were associated with each cycle, of which commercial pegmatites belong to the post-Bothnian cycle. These ideas are set out in publications by Grigoryev (1937), Sudovikov (1939a, b) and Polkanov (1939).

Mapping and thematic work in the 1950s and '60s led to the stratigraphic subdivision of the Belomorian Group, but during this period there arose different interpretations of the number of formations — from three (Amelandov, Misharev) to five (Gorlov, Chuikina) or eight (Dook, Shurkin). The reason for this difference in opinion was primarily different interpretations of the structure of the Belomorides in the western White Sea region. In one alternative, it was seen as a monoclinal structure with a sequence of five gneissic units from bottom to top (west to east), with a generally shallow (20–30°) dip to the east. According to another opinion, the Belomorides represent a system of anticlinal and synclinal folds, the axial surfaces of which are overturned to the west. This alternative point of view is examined in more detail below.

In the last few years an alternative notion has emerged to explain the evolution of the Belomorides. In contradistinction to the original idea of two tectonomagmatic cycles — Early and Late Archaean — a new concept has been advanced that the Belomorides represent one cycle, formed during one extensive Archaean phase — the Belomoride tectonomagmatic epoch, including sequential complexes of igneous and partial melt rocks pre-, early-, late- and post-folding (Shurkin et al., 1962). The considerable number of radiometric investigations undertaken by Shurkin, Polkanov and Gerling on granitoids, migmatites and pegmatites have been of no assistance in deciding on this alternative. Numerous K–Ar dates on micas fall into a broad range, from 1720 to 2670 Ma. Uranium–lead age determinations on zircons from the same rocks, including isochrons for mica and ceramic pegmatites, yielded 1900 ± 100 Ma (Vinogradov and Tugarinov, 1964), from which it was concluded that they formed during an epoch of tectonomagmatic activity, which in the framework of the modern geochronological scale of the Karelia-Kola region coincides with the Svecofennian cycle.

As a result of research carried out in the 1970s–'80s on many facets

[1] "Beloye Morye" is Russian for "White Sea".

of these rocks, we now have at our disposal material which allows us to present fundamentally new interpretations of the history of formation of the Belomorides.

The Belomorides now appear as a structural unit which began its development in the Early Archaean and subsequently, in the Late Archaean and Proterozoic, underwent tectonomagmatic activation.

The most complicated part, as yet reconstructed only in the broadest outlines, is the early Precambrian history of formation of the Belomorides, when internal processes of various types and complexity were accompanied by profound structural and compositional reworking of the White Sea province. Table I-21 outlines the main sequence of early Precambrian geological events.

The Belomorian Group constituting the base of the White Sea province is a polymetamorphic rock complex. Judging from the rare relicts of rhythmic banding in some varieties of gneiss and textures peculiar to effusive rocks (interseptal, amygdaloidal, porphyritic) in ortho-amphibolites, the Belomorian Group represents an original thick volcano-sedimentary assemblage. It was deposited on an unknown basement in the Early Archaean, but the time of formation of its constituent primary sedimentary and volcanic rocks has not yet been successfully dated by isotopic methods. We can say only that this event took place more than three billion years ago, since Lopian deposits of that age overlie metamorphic rocks of the Belomorian Group (Kratz, 1963; Balagansky et al., 1986).

The stratigraphic succession of the group remains debatable, since it relies only on petrographic criteria, and an evaluation of the boundaries and extent of the formations and members described by various workers has no strict lithostratigraphic definition. Here we adhere to the scheme proposed by Shurkin et al. (1962), which in the case of the lower formations was confirmed by later work (Stenar, 1973; Systra, 1978). In accordance with this scheme, the Belomorian Group is divided into eight formations (Fig. I-35), the distribution of which is shown in Fig. I-36. These formations are as follows, from the bottom upwards: Keret, Hetolambi, Loukhi, Chupa, Knyazheguba, Kaytatundra, Kanda and Yona. Some geologists (e.g. N.V. Gorlov) consider the Keret granite-gneiss to be a remobilized Belomoride basement, and others are of the opinion that the Chupa and overlying formations are Lopian age sediments. However, neither variant on the extent of the Belomorian Group can be considered proven.

The composition of the rocks in all eight formations of the Belomorian Group is shown in the description accompanying Figure I-35.

The mineral composition of Belomorian metamorphics is determined on the one hand by the chemistry of the initial rocks and the P–T parameters of regional metamorphic processes, which repeatedly affected the White Sea province, and on the other hand by the appearance of partial melt mineral growth under conditions of regional "granitization"

TABLE I-21

Schematic sequence of Early Precambrian geological events in the White Sea province

Early Precambrian tectonic cycles of the Baltic Shield; isotopic age, Ma	Main geological events	Deformational events. Dominant structural types	Regional metamorphism	Magmatism and partial melting (ultrametamorphism); granitoids
		Consolidation		
Early Proterozoic (PR_1) IV. Svecofennian 1900±100 Ma	Local tectonism and partial melting, partly inherited from cycle II and III structures	Discontinuous N-S zones of schistosity and fracture (IV*D*) Open recumbent folds, N-S and NE strike (IV*F*) Faults	Stage of acid leaching (400°C, P ≈ 5.5 kbar) Almandine-amphibolite facies (B_2 facies series). Subfacies of Ky-Gr-Bi-Mi and Ky-Gr-Bi-Musc gneisses (570-630°C; P ≈ 6-8 kbar)	Dykes of diabase and and plagioclase-microcline granite (γ_4'') Muscovite and ceramic pegmatites, orthotectites, quartz-muscovite metasomatites. Local magma chambers of anatectic granite (plag-microcline, two-mica, hb-bi, γ_4'); migmatite (IV mig). Group III drusites (olivine gabbro-norite)
		Consolidation		
Early Proterozoic (PR_1) III. Šeletskian 2400±100 Ma	Tectonometamorphic activity, local, in part inherited	Near E-W and NE fault zones (III*D*). Isolated moderately open folds in NW and WNW striking granitized zones (III*F*) Faults	Almandine-amphibolite facies (AB, B_2 facies series) Ky-Gr-Bi-Mi gneisses subfacies (640°C; P = 6.5-7.5 kbar	Plagioclase-microcline anatectite-aplite (γ_3) Alkali metasomatism Porphyroblastic leucosome of plag-qz and plag-mic composition in separate portions (III mig) Group II drusite (gabbro, gabbro-norite)
		Consolidation		
Late Archaean (AR_2) II. Rebolian 2700±100 Ma	 Main phase of superimposed folding, affecting the entire Belomoride belt Formation of Lopian volcanosedimentary assemblages in local tectonic depressions	Shear zones and faults (II*D*) Moderately open and isoclinal folds, mainly ENE and NE striking ("cross-folds") (II*F*) Discordances — cross-cutting and interlayer Graben-like tectonic depressions	Regressive stage of acid leaching (450-550°C) Almandine amphibolite facies (B_2 facies series) Ky-Ga-Bi-Orth subfacies (660-700°C; P ≈ 9 kbar)	Pegmatites (uneconomic) Plag-mic granite (γ_2'', γ_2''') Plag and plag-mic leucosome (regional migmatization and granitization; II mig_1, II mig_2) Granodiorite, tonalite (γ_2') Group I drusite (wehrlite, plagiolherzolite, olivine gabbro-norite) Basaltic volcanics in superimposed tectonic depressions and associated with central-type volcanoes
		Consolidation		
Early Archaean (AR_1) I. Saamian (Belomorian) >3000 Ma	Formation of polyphase deformed granulite gneisses of Belomorian complex Sedimentation of thick Belomorian volcano-sedimentary assemblage (5-8 km thick)	Shear zones and zones of blastomylonitization (I*D*) Formation of general NW striking Belomoride folds (330°). Yona-Loukhi synclinorium, Western and Maritime-Kandalaksha anticlinoria (I*F*)	Regressive stage: Ky-Gr-Bi-Orth subfacies (B_2 facies series) 500-650°C; P ≈ 9.5-10 kbar Granulite facies two-pyroxene gneisses and eclogitic rocks (700°C; P ≈ 12-14 kbar)	Plagioclase pegmatite Plagioleucosome (non-anatectic composition) in granite migmatite (I mig_2) and charnockite-migmatite (I mig_1) Intrusive tonalitic rocks (γ_1) Andesite- and tholeiite-basaltic volcanics, mafic and ultramafic intrusions

Note: Table based on material by the Geological Institute, Karelian Branch, USSR Academy of Sciences, Petrozavodsk and the Institute of Precambrian Geology and Geochronology, USSR Academy of Sciences, Leningrad.

with the redistribution of material within the Belomorian Group and the introduction of new elements. In other words, the mineral associations of rocks in the Belomorian Group are genetically varied: as well as those that are thermodynamically equilibrated, there are just as many examples of non-equilibrated rocks resulting from silica-alkali metasomatism and retrogression.

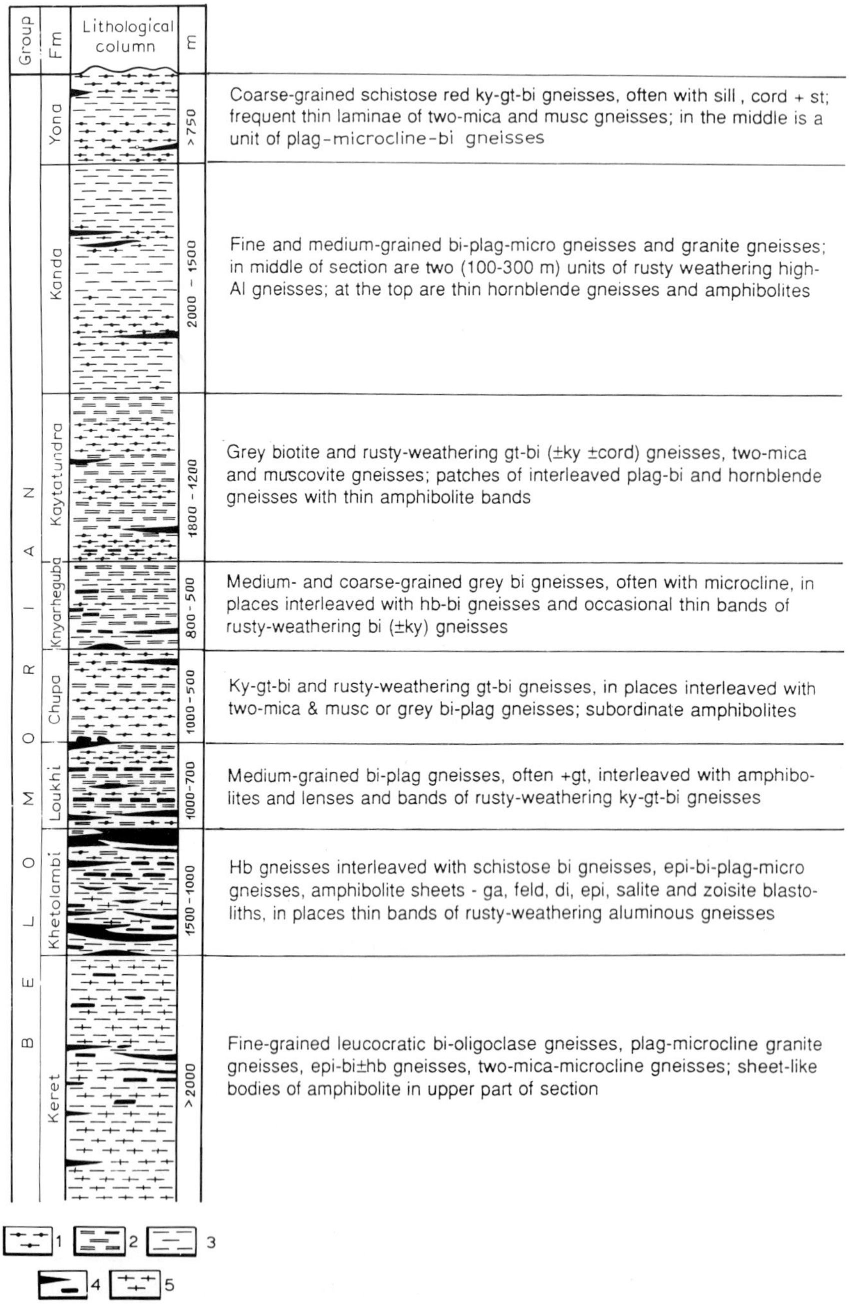

Group	Fm	Lithological column	m	
BELOMORIAN	Yona		>750	Coarse-grained schistose red ky-gt-bi gneisses, often with sill, cord + st; frequent thin laminae of two-mica and musc gneisses; in the middle is a unit of plag-microcline-bi gneisses
	Kanda		2000 - 1500	Fine and medium-grained bi-plag-micro gneisses and granite gneisses; in middle of section are two (100-300 m) units of rusty weathering high-Al gneisses; at the top are thin hornblende gneisses and amphibolites
	Kaytatundra		1800 - 1200	Grey biotite and rusty-weathering gt-bi (±ky ±cord) gneisses, two-mica and muscovite gneisses; patches of interleaved plag-bi and hornblende gneisses with thin amphibolite bands
	Knyarheguba		800 - 500	Medium- and coarse-grained grey bi gneisses, often with microcline, in places interleaved with hb-bi gneisses and occasional thin bands of rusty-weathering bi (±ky) gneisses
	Chupa		1000 - 500	Ky-gt-bi and rusty-weathering gt-bi gneisses, in places interleaved with two-mica & musc or grey bi-plag gneisses; subordinate amphibolites
	Loukhi		1000-700	Medium-grained bi-plag gneisses, often +gt, interleaved with amphibolites and lenses and bands of rusty-weathering ky-gt-bi gneisses
	Khetolambi		1500 - 1000	Hb gneisses interleaved with schistose bi gneisses, epi-bi-plag-micro gneisses, amphibolite sheets - ga, feld, di, epi, salite and zoisite blastoliths, in places thin bands of rusty-weathering aluminous gneisses
	Keret		>2000	Fine-grained leucocratic bi-oligoclase gneisses, plag-microcline granite gneisses, epi-bi±hb gneisses, two-mica-microcline gneisses; sheet-like bodies of amphibolite in upper part of section

1 2 3

4 5

Fig. I-35. Reference stratigraphic column of the Belomorian supracrustal assemblage (after Shurkin, 1962, 1968). Composition of formations is shown schematically. *1* = rusty-weathering aluminous bi-gt-ky gneiss; *2* = grey gt-bi plagiogneiss; *3* = finely foliated biotite and epidote-biotite gneiss; *4* = amphibolite; *5* = granite gneiss and nebulous granitic migmatite.

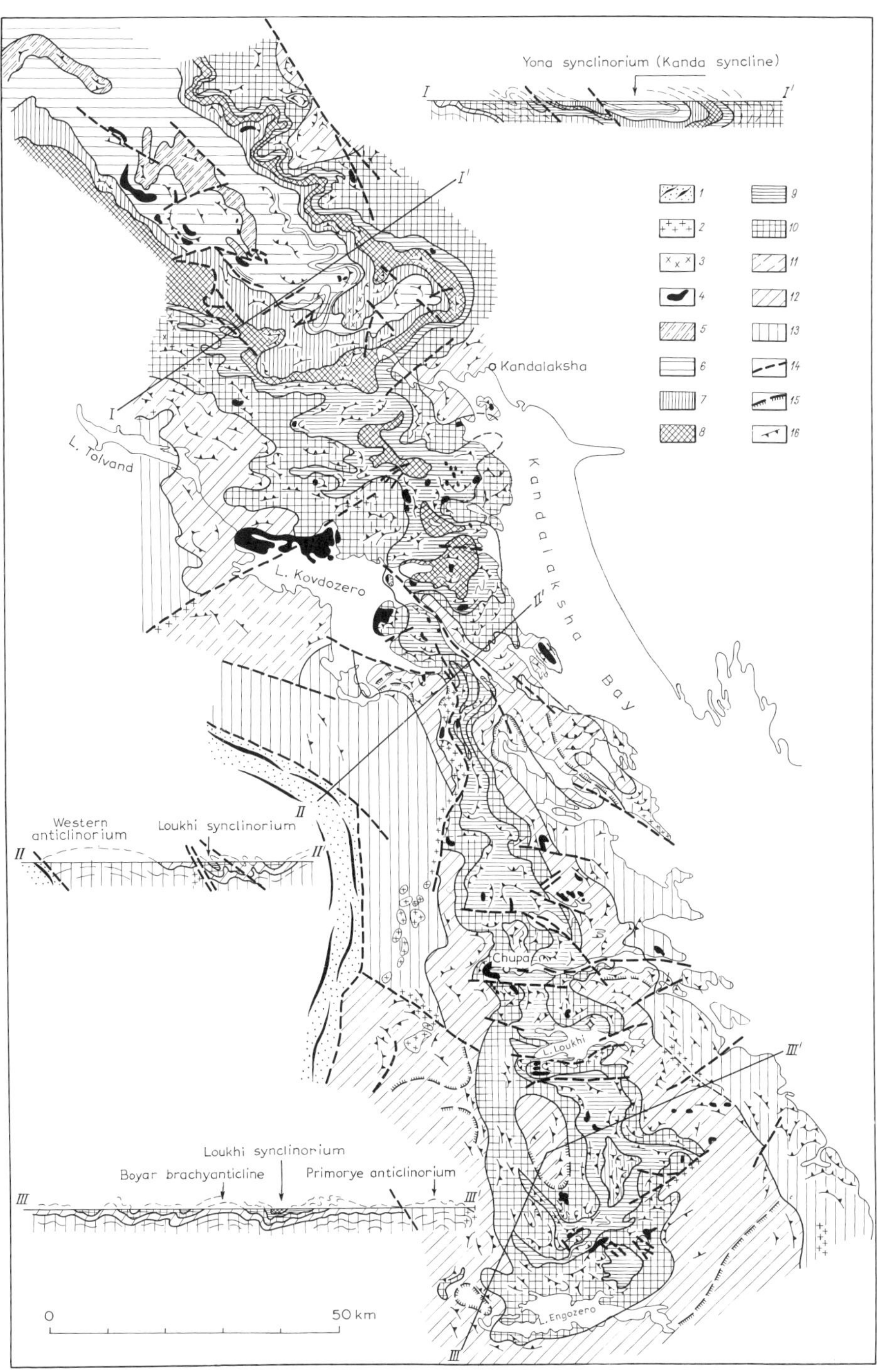

Fig. I-36. Sketch map showing geological structure of the NW White Sea province. *1* = Belomoride/Karelide junction zone; *2* = anatectic granite; *3* = hypersthene diorite; *4* = basic rocks (drusites). Formations in Belomorian Group: *5* = Yona, *6* = Kanda, *7* = Kaytatundra, *8* = Knyazhe-guba, *9* = Chupa, *10* = Loukhi, *11* = poorly exposed, undifferentiated, *12* = Hetolambi, *13* = Keret; *14* = major faults; *15* = boundary of strongly granitized gneisses (symbol in direction of increase of granitization); *16* = generalized structural trend-lines of gneisses.

TABLE I-22

Average chemical composition of rocks in the Belomorian Group

No.	Rocks	n	SiO_2	TiO_2	Al_2O_3	Fe_2O_3	FeO	MnO	MgO	CaO	Na_2O	K_2O	SO_3	loss on ignition	Total
	Group I. Brown biotite-plagioclase gneisses														
1.	Medium-grained biotitic	7	65.06	0.50	15.01	0.85	5.06	—	3.52	2.72	2.38	3.00	0.26	1.30	99.31
2.	Medium-grained garnet-biotite	22	67.06	0.52	15.25	1.41	4.59	0.06	3.00	3.08	2.86	1.57	0.19	0.78	100.37
3.	Medium-grained kyanite-garnet-biotite	20	61.31	0.58	18.01	1.31	6.49	0.09	3.46	2.51	2.52	2.50	0.17	0.91	99.86
4.	Schistose, coarsely crenulated kyanite-garnet-biotite	36	67.59	0.51	15.33	1.03	3.62	0.13	2.19	3.52	3.19	1.44	0.26	1.25	100.06
5.	—Ditto—, kyanite-garnet-muscovite-biotite (±sillimanite-cordierite)	5	58.44	0.62	20.04	1.51	5.07	0.10	3.59	2.73	2.84	3.73	0.37	0.96	100.00
	Group II. Grey biotite-plagioclase gneisses														
6.	Medium- to coarse-grained plagioclase-biotite	9	69.70	0.23	16.39	0.87	1.52	0.04	1.05	3.42	4.68	1.77	0.10	0.43	100.20
7.	—Ditto—, garnet-biotite	7	69.83	0.35	14.17	1.31	2.82	0.05	2.21	3.09	3.43	1.39	0.30	0.41	99.36
8.	—Ditto—, epidote-biotite	4	70.22	0.28	15.71	1.65	1.84	0.02	0.95	3.36	3.15	2.16	0.16	0.51	100.19
	Group III. Leucocratic finely crenulated biotite-plagioclase gneisses														
9.	Light grey biotite gneisses	4	70.10	0.21	15.99	0.74	2.01	0.02	1.18	2.41	4.20	1.65	0.80	0.57	100.03
10.	Pink microcline-biotite with muscovite	6	70.90	0.22	15.28	0.88	2.09	0.02	1.03	3.19	4.47	1.31	0.13	0.45	99.97
	Group IV. Amphibolitic gneisses														
11.	Biotite-amphibole	6	63.12	0.10	16.77	2.01	3.84	0.07	3.08	3.98	4.16	1.70	0.17	0.77	99.77
12.	Garnet-amphibole	3	57.07	0.88	15.84	2.53	6.52	0.04	4.17	8.18	3.47	0.68	0.07	0.68	100.13
	Group V. Amphibolites														
13.	Feldspathic orthoamphibolites	6	51.55	1.01	13.41	2.18	8.61	0.35	9.75	9.92	1.62	0.61	0.10	0.84	99.95
14.	Garnetiferous orthoamphibolites	11	48.43	1.69	16.48	3.14	9.10	0.82	5.87	10.78	1.93	0.89	0.07	0.81	100.02
15.	Diopside (salite) para-amphibolites	2	45.46	0.75	13.14	3.90	10.01	0.29	9.48	13.95	1.30	0.30	0.12	1.27	99.97

Note: Analyses taken from Shurkin et al. (1962), Volodichev (1975), Glebovitsky et al. (1985), Misharev et al. (1960).

Chemical compositions of the commonest gneisses and amphibolites are shown in Table I-22. As well as those referred to in the table, the Belomorian Group also contains sporadic occurrences of rare streaks and lenses of calciphyre, dolomitized marble, diopside, zoisite and kyanite-garnet-gedrite rocks with blastic texture and eclogite-like diopside-garnet rocks.

Tremolite-actinolite amphibolites frequently containing talc, serpentine, breunerite and dolomite as rock-forming minerals, constitute a special group of rare rocks in the series. They occur sporadically as small (10–20 m) sheet-like bodies among garnet and feldspar ortho-amphibolites of the Hetolambi Formation and the possibility cannot be excluded that they represent intensely metamorphosed komatiites.

Textural varieties of rocks are linked on the one hand to processes of metamorphic differentiation (banded, gneissose, schistose textures) and on the other to patchily-developed regional processes of migmatization, anatectic granite formation and metasomatic feldspathization, leading to the formation of migmatites with highly varied textural morphologies where the leucosome is banded, vein-like, diffuse or metablastic (Figs. I-38, I-39).

The overall structure of the Belomorides (Figs. I-36, I-37), which strike

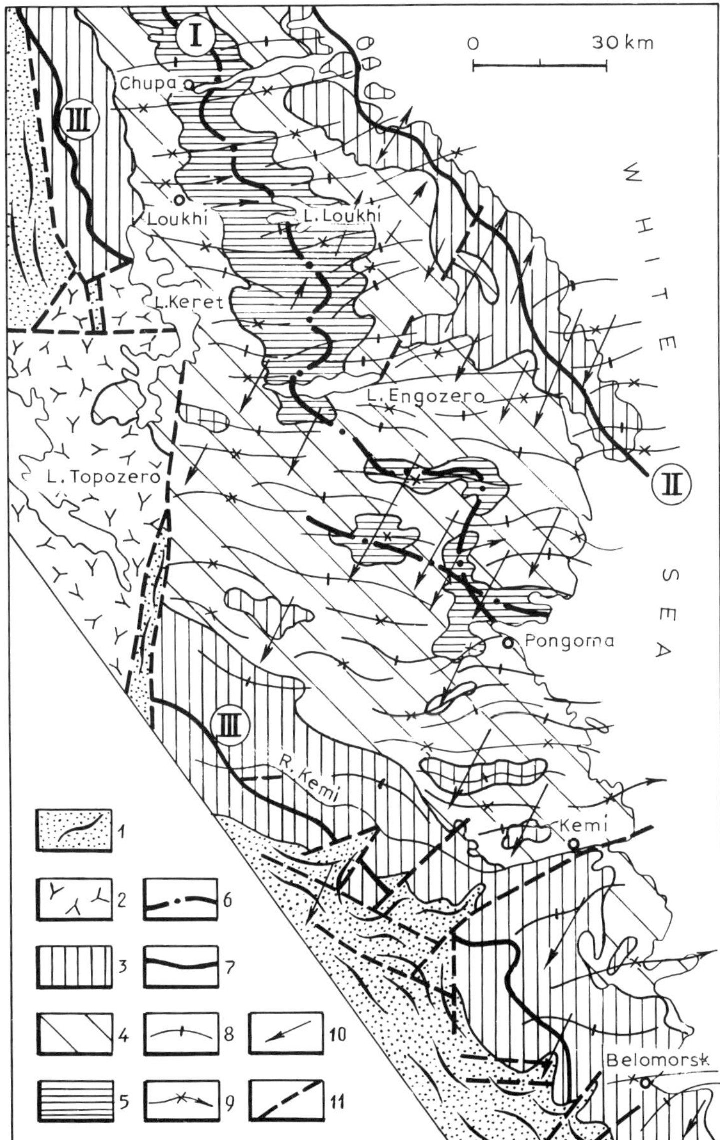

Fig. I-37. Sketch map showing tectonic structure of the Belomorides of the NW White Sea province (compiled by Stenar, 1973). *1* = Lower Proterozoic; *2* = undifferentiated Archaean and Proterozoic; *3–5*: Archaean [*3* = Keret Formation (bottom), *4* = Hetolambi (incl. Loukhi member), *5* = Chupa Formation (top)]; *6*, *7*: major structures (*I* = Loukhi synclinorium, *II* = Maritime (Primorsky) anticlinorium, *III* = Western (Zapadny) anticlinorium). Axial traces of Rebolian folding (arrows show plunge direction): *8* = anticlinal, *9* = synclinal; *10* = axial traces and plunge directions of superimposed Proterozoic folding; *11* = major faults.

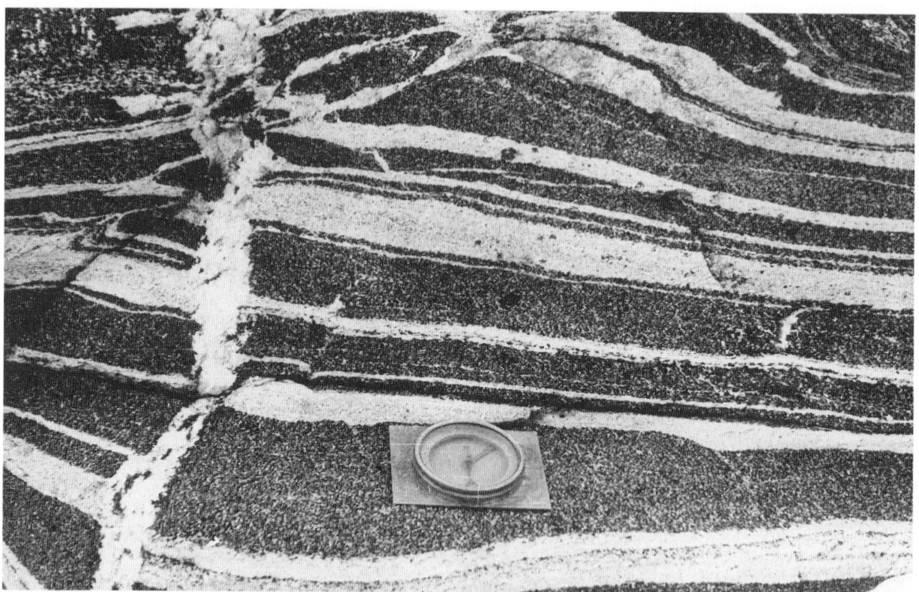

Fig. I-38. Coarsely-banded migmatites. Substrate is garnet amphibolite, metatect is plagioaplite. Hetolambi Formation, White Sea region.

north–west and north–south, is of the Yona synclinorium in the north, south of which is the Loukhi synclinorium surrounded by the Primorye (Maritime) and Western anticlinoria. The axial surfaces of all these structures, as mentioned above, are overturned to the west and are deformed by second-order cross folds. It is these particular folds which control the outcrop pattern of the formations in the Belomorian Group. The western boundary of the Belomorides in Karelia is a zone of deep faults, following the junction zone with the Karelian province. Associated solitary faults and charnockite intrusions cut the adjacent limb of the Western anticlinorium (Shurkin et al., 1974).

The formation of the above-named regional fold structures is related to the Early Archaean Saamian (Belomorian) epoch of tectonomagmatic activity, the first phase in the history of the Belomorides (Table I-21). No morphological features of minor structural forms of this phase have been preserved. Judging from isolated relicts and fragments of fold limbs and closures, the folds were isoclinal and formed at the same time as a regional metamorphic event at high pressure granulite facies conditions. According to Volodichev (1975), mineral growth took place within the interval T = 650–700°C and P_{tot} = 12–14 kbar.

In the non-migmatized Chupa kyanite-garnet-biotite gneiss, garnet contains up to 31–34% pyrope and the biotite is low in Fe (f = 22–28 mol%). In central parts of the western White Sea region, the Hetolambi amphibole

Fig. I-39. Gneissose-nebulous migmatite. In inhomogeneous non-uniform grain-size metatect, granitized remnants of plagioclase-biotite gneiss. Keret Formation, White Sea region.

gneiss is known to contain relicts of "eclogitic" type rocks, of hyp-di-gt-pl-qz composition.

Episodes of migmatization, synchronous with regional metamorphism, are also found throughout the outcrop area of the oldest metamorphics (Table I-21). The migmatites (Im_1) that are associated with the main phase of Belomorian folding mostly have a banded fine-grained quartz-feldspar leucosome of low-alkali granitoid composition, which is the product of isofacies metamorphic differentiation of the gneissose substrate. During the waning stages of partial-melt granite formation, late fold migmatites formed (Im_2), the leucosome of which shows signs of having originated by differential anatexis of early leucosome with movement of melt along fractures and faults.

Magma chambers of anatectic granite seem to have formed at the same time as Im_2 migmatites, and the magmas migrated along fault and fracture zones, forming higher level small fault intrusions of leucocratic granite (γ_1) of anchi-cotectonic composition. Homogenization temperatures of fluid inclusions in these granites indicate that they crystallized in a narrow temperature interval no lower than 760°C, i.e. barely above the temperature of regional metamorphism of the country rocks intruded by these granites.

The first episode of metamorphism concluded with the formation of zones of schistosity and blastomylonitization during a retrogressive stage.

New mineral parageneses belong to the same facies of kyanite-orthoclase gneiss, but with somewhat lower temperatures (655–550°C) and pressures (P_{tot} = 9.5 kbar; Salye et al., 1985).

It has not yet been possible to establish the isotopic age of the oldest complex of metamorphic rocks and granites, but the upper boundary is clearly delimited both geologically and radiometrically relative to the tectonomagmatic and metamorphic events of the subsequent *Rebolian* cycle, with an age of 2700 Ma (Tugarinov and Bibikova, 1980). The initial phases of the Rebolian tectonic activity of the Belomorides (Table I-21) consist of numerous minor intrusions (10s to a few 100s metres, rarely larger) of peridotite, gabbro-norite, gabbro and leucogabbro-labradorite with corona textures (hypersthene-diopside-garnet-amphibole rims at olivine and plagioclase boundaries) and widespread formation of the well-known "coronites" or "drusites".

Immediately after the basic igneous rocks, early syn-folding intrusions of granodiorite-tonalite (γ_2') formed, the source melts of which may belong to late differentiates of the same epigenetic series. They were usually altered later than the superimposed processes of schistosity and blastomylonitization in gneissose biotite granodiorite and plagiogneiss-granite, which often contain garnet.

Fold structures of the Rebolian cycle (II*F*) are predominantly north–east and east–west striking; they deform axial surfaces and hinges of (I*F*) folds of the Belomorian cycle (Fig. I-37).

Structural analysis indicates that Rebolian tectonism involved from two to four deformational episodes, including the formation of isoclinal folds, followed by open and close folds at various scales ("cross folds") and in the final stage — thrust-nappe structures, soft faults and shear zones.

Simultaneously with the folding, rocks of the Belomoride complex were affected by regional metamorphism of kyanite-microcline gneiss type in the same high-pressure kyanite-sillimanite facies series. Parameters of the Rebolian stage of prograde metamorphism were 600–650°C and 9–10 kbar (Volodichev, 1975).

New mineral associations in aluminous, plagioclase-biotite and amphibole gneiss and amphibolite in all formations of the Belomorian Group inherit associations of early (pre-Rebolian) metamorphics, but from paragenetic analysis, the mineral phases themselves already have a different composition. Biotite becomes more ferruginous (f = 31–38 mol%, compared to 22–28 mol%) and the pyrope end-member content in garnet decreases to 23.8–28.2% (as against 31–34% in the granulite facies).

Indistinct signs of partial-melt granite formation, expressed as quartz-feldspar neosome segregation (IIm_1), are associated with early folding, but the culminating stage of migmatization in the Rebolian cycle is accompanied by the regional formation of migmatites (IIm_2). In this regard, it has been shown that the leucosome composition in the Chupa aluminous gneiss does not correspond to anatectites. More likely, it originated by segregation during metamorphic differentiation with subsequent partial metasomatic reworking. Anatectites seem to have arisen only as a result of selective melting of the migmatite substrate (IIm_2) with a thinly-banded and diffuse neosome of plagioclase-microcline composition, distributed among thinly-banded Keret plagioclase-biotite and plagioclase-microcline-biotite gneiss. They usually form aureoles around the indistinct margins of autochthonous syndeformational plagioclase-microcline migmatite-granites (γ_2'').

Rare occurrences of fluid inclusions in the leucosome of migmatites (IIm_2) with homogenization temperatures of 700–710°C suggest a relationship between the syntectonic granitoids and migmatites with melts of anatectic origin, enriched in volatiles, ensuring the postmagmatic reworking of both.

Products of the stage of granite formation in the Rebolian cycle are late-kinematic leucocratic biotite and garnet-biotite-microcline-plagioclase granite (γ_2'''). These include both subautochthonous granitoids with a "ghost stratigraphy" of the substrate and skialith rocks, as well as intrusive granite bodies. They are associated with deep-level magma chambers containing water-saturated melts with a temperature not less than 760–780°C, significantly greater than the temperature of the country rocks at the level of formation of the intrusions. Solidification of the bodies released fluids and hydrothermal solutions, which had a significant, essentially potassic, metasomatic effect on the rocks in shear zones and soft faults. The Rebolian cycle of tectonometamorphic activity of the Belomorides concluded with the emplacement of numerous thin (0.5–2 m) pegmatites, of no economic significance. U–Pb dates on Rebolian migmatites from the Keret, Hetolambi and Chupa Formations are all 2700 ± 50 Ma (Tugarinov and Bibikova, 1980), corresponding to the geochronological boundary between the Archaean and Proterozoic.

The subsequent *Seletskian* cycle of tectonomagmatic activity in the Belomorides belongs to the Early Proterozoic. Geological events of this time within the Belomorides are expressed far less extensively than the Rebolian. The most intense deformations are localised in linear zones, and it is only here that there is any new mineral growth.

The initial phase consists of the imposition of a system of faults with a north–west strike (320–345°), controlling the emplacement of group II "late orogenic" coronites (Shurkin et al., 1962). Most of the intrusions are in the form of dykes, less commonly stocks, cutting migmatites and granitoids of earlier cycles, as well as group I coronite intrusions, amphibolized and deformed during the Rebolian cycle. These intrusions are most often gabbro-norite and peridotite, but there are less common occurrences of pyroxene and plagioclase porphyrite, gabbro-diabase and microgabbro. Inner "chilled margin" contacts are characteristically present, and in isolated cases very small volumes of exo-contact palingenic granophyric granitoids, from selective melting of the leucosomes of host migmatites.

All the rocks in these intrusions display corona textures, as do the early coronites, but not so well developed. In relation to the superimposition of deformation and metamorphism of both Seletskian and later Svecofennian periods, these coronites were subjected to folding, faulting, amphibolization, and they are cut by granite veins and pegmatites.

Salye (1983) has identified three phases of folding and north–south faulting in the structural evolution of the Seletskian cycle of activity in the Belomorides, which partly follow the direction of Rebolian structures. Morphologically, the Seletskian folds (III*F*) consist of packets (a few hundreds of metres to 2–3 m wide) of tight isoclinal folds, rarer isolated semi-open folds and flexural warps in gneiss along faults. Wide (330–500 m) zones of faulting and fracturing conclude the tectogenesis.

Regional metamorphism, which took place simultaneously with III*F* folding, was at kyanite-microcline subfacies, and mineral parageneses in the Chupa aluminous gneiss correspond to temperatures of 600–650°C and a pressure of 7.5–8.5 kbar. According to Salye et al. (1985), these rocks preserve an earlier mineral association: ky-ga-bi-pl-qz, but the Fe-content of the biotite is higher, up to 36.3–44.5 mol%, as against 32–37% in Rebolian metamorphics, while the pyrope component in garnet is lower, at 20.1–23.6 mol%.

During the retrograde stage of Seletskian metamorphism, which took place at 500–550°C, retrograde rocks developed in linear schist belts, with mineral parageneses of the kyanite-muscovite facies (Volodichev, 1975), represented by two-mica ($\pm$ kyanite) gneiss, garnet-biotite gneiss and muscovite gneiss.

Processes of migmatization and granite formation took place locally, only in zones of activation. The leucosome of IIIm migmatites, found in grey plagioclase-biotite, amphibole and other gneisses of coarse- to medium-grained granitoids, consists of vein-like orthotectite with large flakes of biotite in aluminous gneiss. There are many occurrences of migmatites with porphyroblastic leucosome, which in places grade into autochthonous porphyro-nebulitic migmatite-granites (γ_3'). Late orogenic intrusive granitoids (γ_3'') comprise concordant and subconcordant bodies of biotite-microcline-plagioclase composition and aplitic appearance.

The time of formation of metamorphic rocks in the Seletskian cycle was 2370 $\pm$ 45 Ma, and 2320 $\pm$ 110 Ma for granitoids, by the Rb–Sr method. In the regional chronostratigraphic scale for the Precambrian of Karelia, the 2400 $\pm$ 100 Ma boundary is taken as the boundary between the Sumian-Sariolian and the Jatulian.

The last, *Svecofennian*, cycle of tectonomagmatic activity begins with the formation of the group III coronite ("drusite") dyke swarm. Corona textures are poorly developed — only incomplete garnet and hornblende rims are observed around clinopyroxene (augite) and plagioclase (An_{36-40}).

Svecofennian deformational events are expressed as open folds at various scales (IV*F*), mostly striking north–east and nearly north–south, and linked thrust-nappe structures, zones of superimposed (repeated) schistosity and zones of fracturing, shearing and extension, controlling the distribution and sites of emplacement of clusters of micaceous and ceramic pegmatitic veins. Kotov (1985) describes three phases of deformation in the Svecofennian cycle, while Salye (1983) identifies four. They refer the NW, N–S and NE steep shear zones, controlling the emplacement of economic pegmatite deposits of muscovite and ceramic raw materials, to the Svecofennian. A brief outline of some of these deposits is presented in Table I-23.

The final phase of folding was accompanied by regional metamorphism at kyanite-muscovite gneiss facies conditions, evidence for this being the paragenesis which formed at 570–620°C and 6.5–7.5 kbar (Volodichev, 1975). The metamorphic rocks are coarse-grained biotite and two-mica (± kyanite, ± garnet) gneisses with large mica flakes; biotite has a high iron content (up to 44–48 mol.%) and garnet is low-pyrope (13.4–17.8%).

The formation of migmatite and granite took place locally during the Svecofennian — only within restricted areas and zones of activity, most of these being inherited from earlier Rebolian and Seletskian structures. These repeated processes of ultrametamorphism played a vital role in the formation of the leucosome of syndeformational (IVm_1) migmatite and associated autochthonous migmatite-granite (γ_4'), then late-kinematic porphyroblastic migmatite (IVm_2) and discordant sheets of subautochthonous aplitic granite (γ_4'') and the muscovite and ceramic pegmatites referred to above. Granite formation in the Svecofennian concluded with the intrusion of a suite of plagioclase-microcline dykes (γ_4''').

Newly formed leucosome (IVm_1) consists of granitic material which is usually coarse-grained and not uncommonly shows orthotectitic and pegmatitic textures. There are both autochthonous and suballochthonous distributions of leucosome (IVm_1). We frequently observe direct mutual transitions, with transport of mobilized granitic material from its original source to fracture cavities and fault zones between boudins.

In areas of very intense migmatization with diffuse neosome in a granitized substrate, there is a field of more or less uniform autochthonous plagioclase-microcline migmatite-granites (γ_4'). At the highest stratigraphic levels, for example within the Chupa Formation, it is possible to map out fields of leucocratic granite-aplite (γ_4''), forming concordant stratiform sheets a few metres to tens of metres thick. They show signs of having formed both autochthonously and suballochthonously. These granites (γ_4' and γ_4'') have cotectic compositions and their genesis is consistent with an anatectic model of granite formation. They are characteristically highly saturated in gas–water–salt fluid, which arises from an abundance of gas-fluid and melt (salt, glass) inclusions in quartz and feldspar. The dominant fluid phases are H_2O, H_2 and CO_2. Homogenization temperatures of melt inclusions have been determined to lie in the interval 720–750°, which

TABLE I-23

Characteristic ore deposits and ore shows in the White Sea Province

Ore deposit or ore show; useful minerals	Tectonic setting	Type of mineralization, ore-forming and genetic	Geological & mineralogical features				Geological age (host rock and mineralization)
			host or mineralized rock	shape of deposit, ore body morphology	ore type	ore mineral composition (secondary minerals) and hydrothermally altered rocks	
Yona; muscovite, microcline	Chupa-Loukhi block. Pegmatite field	Muscovite, pegmatite	Aluminous gneisses	Pegmatite veins	Clusters, block	Muscovite (microcline, quartz, plagioclase)	Archaean
Tedin	—ditto—	—ditto—	—ditto—	—ditto—	—ditto—	—ditto—	—ditto—
Kuruvaara; ceramic pegmatites	—ditto—	Quartz-feldspar, pegmatite	Gneisses, amphibolites	Pegmatite veins	Block	Microcline (albite, quartz, biotite)	—ditto—
Kivguba; cobalt	Lapland-Kandalaksha polymetamorphic zone	Cobalt-pyrrhotine (fahlbands); metamorphic	Garnet-quartz amphibolites	Conformable lenticular deposits	Vein-like dissemin-ations, dis-seminated	Pyrrhotine, pyrite, magnetite (chalco-pyrite, pentlandite, ilmenite, rutile)	Early Proterozoic (?)
Kovdor; iron, phosphorus, zircon, phlogopite, and others	Kovdor alkaline ultramafic massif (central type) in Archaean Yona block	Apatite-magnetite and phlogopite; magmatic and metasomatic associated with Kovdor intrusion	Pyroxenites, ijolites, carbonatites	Steeply-dipping tubular deposits, lenticular bodies	Massive, vein-like dissemin-ations	Magnetite, apatite (baddeleyite, zircon pyrrhotine)	Early Palaeozoic
Kovdor; phlogopite, vermiculite	—ditto—	Phlogopite; magmatic, metasomatic with vermiculite weathering crust associated with Kovdor intrusion	Olivine pyroxenites, peridotites	Steeply-dipping branched tubular deposits	Massive, megacryst	Phlogopite, vermiculite (apatite, magnetite, pyroxene, calcite)	Early Palaeozoic

suggests that the granite magma had a source in deep crustal magma chambers.

During retrograde metamorphism, which occurred at the same time as the formation of the shear zones, there was intense development of metasomatic processes of acid leaching under thermodynamic conditions of $T = 400°C$ and $P = 5.5$ kbar. These processes in aluminous gneisses produced a quartz-muscovite mineral association and led to the formation of economic deposits of muscovite (Salye and Glebovitsky, 1976).

Svecofennian granite formation concluded with granite dykes (γ_4'''), 0.2–10 m thick, which are found sporadically in the western White Sea region. They were not subjected to metasomatic alteration and seem to have been intruded later than the pegmatites. The dykes consist of tonalite, adamellite and essentially potassic (microcline) granite. The isotopic age of the paragenetic association of partial-melt granitoid-migmatite-pegmatite (micaceous and ceramic) is 1860 Ma (U–Pb isochron method, with numerous determinations on zircon, uraninite and monazite).

The early Precambrian history of development of the Belomorides concludes with the formation of Svecofennian granites. From this point onwards, the lithosphere of the White Sea province was not subject to internal structural and chemical reworking on a regional scale. During the Late Proterozoic within the confines of the White Sea province, the wide, trough-shaped, fault-bounded Kandalaksha-Tersk synform developed, and was filled by sediments of the Middle Riphean Tersk Formation of quartzose sandstone, arkose and conglomerate. No igneous rocks of this age have been found in the White Sea region.

During the Palaeozoic, an extensive zone of faulting and fracturing developed, with a NW strike, running from Kandalaksha Bay through the Onega peninsula and stretching as far as the Russian Platform in the south. Associated with this zone are early Caledonian alkali basalt dykes (420–480 Ma), late Caledonian fourchite (olivine-free monchiquite) and monchiquite dykes and explosion breccias (345–375 Ma), nephelinite, ålnöite and other alkaline basaltoids (270–300 Ma), numerous central-type alkaline-ultrabasic intrusive rocks (Kovdor, Vuorijärvi, Afrikanda, etc.) with rare-metal–phosphorus–iron ore, titanomagnetite, phlogopite and vermiculite mineralization, forming Kovdor-type economic deposits (Table I-23).

In addition to the above-mentioned vein rocks, the Kandalaksha fracture zone also has many localities where fahlband-type sulphide-quartz metasomatites have been found. They are structurally localized in the form of linear (tens to a few hundreds of metres long), thin (a few metres to over 10 m in rare cases), solitary or more commonly *en échelon* zones, which are found within Hetolambi and Keret biotite-amphibole gneiss and amphibolite. Sulphide mineralization consists in the main of disseminated pyrite and pyrrhotite, with chalcopyrite present as an admixture. Sulphide concentration is extremely irregular, from dispersed to dense, with rare instances of small concentrations, forming monolithic ore bodies. Ferromagnesian silicate minerals of the host rocks are preserved in these zones — garnet, hornblende and pyroxene — but feldspars are replaced by quartz. Belomoride fahlbands are of no economic interest, since the volume of the ore bodies in each ore show is extremely small, and the content of any of the valuable metals is quite insignificant. The isotopic age of these metasomatic rocks has not been determined. As an example, Table I-23 outlines the brief characteristics of the Kivguba ore show.

4. THE LADOGA BELT

The Ladoga belt is situated in the NW part of the Svecofennian province. Its NE margin is a series of faults, discordant to the near N–S structural belts of the Karelian granite-greenstone terrain. These faults are developed in a 50 km wide transition zone, within which are found both Proterozoic suprastructure and infrastructure, as well as Archaean genetically varied granitoids and, to a much lesser extent, volcano-sedimentary rocks.

Within the USSR, Early Proterozoic structural-lithological complexes are well exposed only in the Karelian isthmus and the North Ladoga region.

The Archaean complex

Archaean formations are mainly granitoids belonging to the Rebolian (Late Archaean) cycle, with an age in excess of 2800 Ma, from U–Pb determinations on zircons (Kouvo, 1970). Lopian supracrustals (Gimola Group, Ilomantsi assemblage) are exposed within troughs close to the NE boundary of the Svecofennian Domain, where they are cut by shear zones and superimposed folds.

The lower Proterozoic Ladoga complex

The Early Proterozoic Ladoga complex includes the Sortavala and Ladoga supracrustal assemblages and various associations of partial melt and igneous rocks. In the lower parts of the supracrustal sequence, there are Jatulian quartz-carbonate units, grading into volcanogenic-carbonate-schist units in inner regions ("marine" Jatulian), while higher up there are the flyschoidal formations of the Ladoga Group (Kalevian). Leucodiabase sills within the Jatulian have been dated at 2150 Ma by the U–Pb method on syngenetic zircon.

Granitoid magmatism of this complex occurred within the interval 1880–1750 Ma, from U–Pb studies on zircons (Kouvo, 1970). Rapakivi granite formation is associated with a period of cratonization at 1650 Ma ago.

Stratigraphy. As the result of work by many researchers in the North Ladoga region, we have now developed a scheme for the stratigraphic subdivision of the lower Proterozoic supracrustal assemblage of this region. This is in fact reflected in earlier publications by Kratz and Lobach-Zhuchenko (1972) and Kharitonov (1966) and remains essentially unaltered. The scheme envisages two groups, Sortavala and Ladoga.

The Sortavala Group crops out in the immediate vicinity of inliers (domes) of granite-gneiss basement. The main rock type is hornblende schist with thin bands of carbonate rocks, quartzite and aluminous schist. The hornblende schists include both ortho-amphibolite and para-amphibolite. Demonstrably extrusive rocks crop out in the northern part of the territory, around Ruskeala and Ruttyjärvi (Kratz and Lobach-Zhuchenko, 1972). Volcanics are represented by metatholeiite and metatuff of the same composition, with rare occurrences of more acid rocks. Just as characteristic of this region are calcite and dolomite marbles and associated amphibole para-schist. These rocks occur at the base of the section of this assemblage and in its upper parts. The total thickness of the Sortavala Group is 1600 m.

As a whole the Sortavala Group correlates well with the marine Jatulian of south-west Finland, where these sediments rest directly on a granite-gneiss basement, with patches of basal conglomerate locally. Judging from available geophysical data, the Sortavala Group and analogous rocks occur only in the zone of granite-gneiss basement uplift. It follows, that during sedimentation this zone was a chain of highs or islands.

The Ladoga Group occurs in synforms separating upwarps. It consists mainly of schists containing biotite, biotite-staurolite, andalusite, garnet and two micas, which nearly everywhere show good graded bedding. As long ago as 1928, E. Wegman referred this series to the flysch of the Karelides.

The relationship between the Ladoga Group and the underlying Sortavala Group has been studied by many researchers. According to some, the assemblages are conformable with each other (Predovsky et al., 1967), although in the northern part of the region at the boundary with the Karelian massif there was intraformational erosion, one consequence of which was the formation of the conglomerates at Partonen farm and quartzites on Kontiosaari island. Coarsely fragmental facies are typical only for the marginal part of the sedimentary basin. In adjacent territories in Finland, a coarse boulder conglomerate has been found at the base of the Ladoga Group (Kalevian), with pebbles containing quartzite from the continental Jatulian, an indication not only of the relatively young age of the Ladoga Group but also of the source being in the north where relevant Jatulian rocks are found. The coarse clastic facies is replaced southwards by finer-grained formations.

Complex folding and high-grade metamorphism, the practically total absence of marker horizons and poor knowledge of cyclicity within each succession have prevented any detailed internal subdivision of the Ladoga Group. We can ascertain only that towards the top of the group there is a quantitative increase in aluminous schists, which usually form the upper parts of rhythms.

Tectonics. There are a number of zones in the North Ladoga region which differ in fold style and in several specific features concerning the composition of sedimentary and volcano-sedimentary units (Fig. I-40). These are: (1) the Jänisjärvi synclinorium; (2) the Sortavala-Pitkäranta zone of granite-gneiss basement domes (anticlinorium); (3) the Putsaari synclinorium; and (4) a zone of linear folds within which anticlinal structures alternate with synclines at the same scale.

The Jänisjärvi synclinorium is a broad negative belt with a highly asymmetric structure. On its NE limb, Ladoga Group formations normally cover Jatulian quartzite-dolomite units lying directly on the SW margin of the Karelian massif. On the SW limb is a system of very tight, often isoclinal, folds, overturned towards the core of the synclinorium; these are complicated on the limbs by thrusts and nappes.

As noted previously, the Sortavala-Pitkäranta zone was a chain of upwarps (volcanic islands) even during the period of sedimentation and constituted an additional source of eroded material during the period of

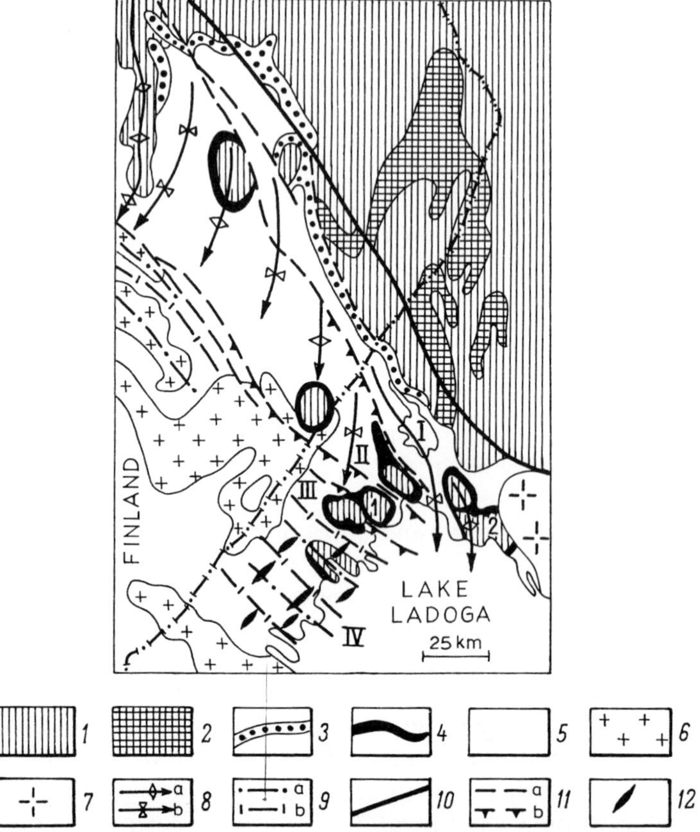

Fig. I-40. Tectonic sketch map of the NE part of the Svecofennian province. *1* = granitoids of pre-Proterozoic basement; *2* = supracrustal assemblages in Archaean greenstone belts. Proterozoic supracrustals: *3* = Jatulian, *4* = Sortavala, *5* = Ladoga; *6* = autochthonous and parautochthonous Proterozoic granitoids; *7* = rapakivi granite; *8* = major folds, episode I (*a* = anticline, *b* = syncline); *9* = major folds, episode II (*a* = anticline, *b* = syncline); *10* = zone of boundary faults; *11* = other dislocations (*a* = reverse faults, *b* = thrusts); *12* = orientation of faults and folds, episode III.

accumulation of Ladoga Group terrigenous units (Predovsky et al., 1967). It remained as a positive structure during early deformational episodes, when isolated positive structures, isometric in plan, emerged, further evolution of which led to the formation of the Sortavala and Pitkäranta group of basement domes (Fig. I-40).

On the southern limb of the Sortavala-Pitkäranta anticlinorium is a zone of nappes with an overall north–west strike; consequently, they intersect the north–south folds that are present between the domes. Sudovikov et al. (1970) have demonstrated that in the northerly-directed nappe fronts there are folds which had flattened earlier structures.

Within the Putsaari synclinorium the dominant folds are NW-striking isoclines which Sudovikov et al. (1970) suggested were more or less synchronous with the nappes.

An examination of the relationships between major and minor structural forms leads to the conclusion that there were three main deformational episodes (Fig. I-41).

During the first episode, anticlinal elevations of the pre-Karelian basement formed, surrounded by NW-striking linear fold belts and a system of nappes, formed under conditions of lateral compression of the entire mobile zone during the second episode. The third deformational episode was the result of axial compression of the major folds.

Each of these episodes includes the formation of several generations of folds. A more detailed account follows of the deformation sequence in Sortavala and Ladoga Group rocks and the basement complex, using material published by Glebovitsky et al. (1981).

The deformational history of the first episode is well interpreted for the Pitkäranta group of domes and for the zone separating it from the Sortavala group. The major mapped structures of this region are linear folds with an approximately north–south strike, usually strongly flattened to isoclinal in the east and more open in the west. From the work of Kazakov (1977a, b) and others (Sudovikov et al., 1970), the N–S folds are at least third generation. F_1 folds are also early, belonging to the first episode, and occur in the Ladoga Group "from changes in the orientation of rhythmic layering across its strike from normal to overturned" (Kazakov, 1977b). Naturally, such folds are better exposed in zones free of migmatization, and in which rhythmically-bedded flysch deposits are found.

F_1 folds as a rule are tight to isoclinal, with wavelengths up to 1 km and amplitude over 5 km. The present orientation of F_1 axial surfaces is defined by the nature of superimposed deformation. There is reason to believe that in the zone of basement domes, the axial surfaces and axes of these folds were approximately horizontal, which is reflected in the radial compression of Ladoga Group formations during the growth of the domes.

Second generation folds (F_2) are mainly established on the basis of analysing the attitude of F_3 folds, which are the major folds of this region. In the south they plunge gently southwards, but going northwards they become steeper and in places they are overturned. In a similar fashion, in the south of the Impilahti dome the F_3 fold axes plunge to the south at an angle of 25–30°.

This indicates that prior to the appearance of the third deformational episode, the Ladoga complex and the basement were folded into approximately E–W striking open folds, upright or slightly overturned to the north.

Definite F_2 folds are observed in the west of the region (Kazakov, 1977b) where F_3 folds are open. F_2 folds deform F_1 axial surfaces and the F_1 axial planar schistosity. The E–W and NW orientation of axial surfaces observed

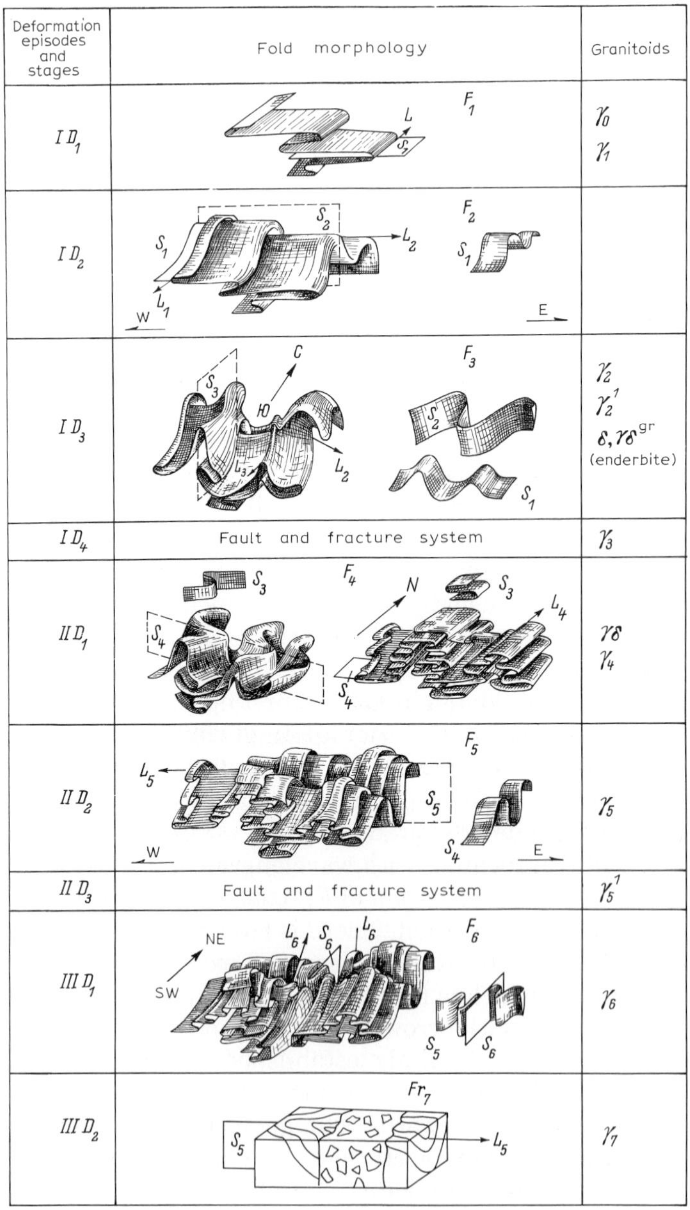

Deformation episodes and stages	Fold morphology	Granitoids
ID_1		γ_0 γ_1
ID_2		
ID_3		γ_2 γ_2^1 $\delta, \gamma\delta^{gr}$ (enderbite)
ID_4	Fault and fracture system	γ_3
IID_1		$\gamma\delta$ γ_4
IID_2		γ_5
IID_3	Fault and fracture system	γ_5^1
$IIID_1$		γ_6
$IIID_2$		γ_7

Fig. I-41. Evolution of structures in the Ladoga complex.

here suggests that they were formed by lateral compression in a N–S or NE–SW direction.

Thus, during the first stage of deformation, at least three generations of folds were formed (Fig. I-41). The first of these (phase ID_1) were caused by compression in a zone of basement upwarps. Lateral variation in the intensity of compression related to irregularities in the elevation of the basement caused the formation of fairly major folds during horizontal flow. The second folds (phase ID_2) were caused by lateral compression oriented approximately north–south, while the third folds (phase ID_3) formed in east–west compression. Fold interference patterns explain the isometric plan of folds in the Ladoga Group (Kazakov, 1977b) and the surrounding dome-like shapes of the Pitkäranta group, the morphology of which was not substantially altered by subsequent deformations.

Relationships between structures of the first and second episodes of deformation are observed in the northern part of the Putsaari synclinorium, where there is a system of nappes involving not only rocks of the Ladoga complex, but also the granite-gneiss basement. Sudovikov et al. (1970) have shown that the Sortavala dome and similar adjacent structures to the north and west were detached from their root zones and translated northwards along low-angle thrust-nappes. During these movements, minor folds developed at the nappe fronts, with east–west or north–west striking axial surfaces. These folds deform F_3 axial surfaces and therefore are fourth generation structures or possibly later.

At the same time, on the basis of observations on structures in the zone under discussion, we may assert that during the period of nappe formation, a normal compressive regime prevailed. At least they can be used to give a satisfactory explanation of the occurrence here of a system of recumbent isoclinal folds with axes plunging constantly at low angles to the south, which may be related to these folds being superimposed on F_3, since in this region they have a constant north–south strike. Perpendicular compression was probably irregularly distributed over the area, and there was significant horizontal movement of material from the axial zone of a thermal structure, where processes of granitization and melting at the base of the crust are most strongly expressed, to the periphery, which also led to the production of nappes. In such a case, the vertical compression of F_3 folds and the growth of recumbent F_4 folds with axes plunging gently southwards (phase IID_4) are not strictly synchronous with the lateral compression of F_3 folds in nappe fronts (phase IID_5).

Nappe surfaces are crenulated by north–west striking folds, which become more close towards the south-west. In the axial zone of the Putsaari synclinorium and south of that in the entire zone of linear folding they become the dominant folds and they belong to at least fifth generation structures (F_5, phase IID_5).

Later deformations, of the third episode, as a rule resulted in the formation of open folds or *en échelon* shear zones striking north–east. They are quite typical of the entire zone of linear folding and of the axial part of the Putsaari synclinorium. Parallel to their axial surfaces are faults which controlled the emplacement of partial-melt granitoids. Judging from the polyphase nature of the intrusion of these granitoids, there are several phases of deformation belonging to the third episode. However, there are few data concerning their mutual relationships and in the present work it was important not so much to present the complete sequence of deformations as to explain the overall trend of the structural evolution and to emphasise that the IID_5–$IIID_6$ boundary represents a change from folding due to general compression to *en échelon* brittle fractures in a rigid medium.

Metamorphic zoning. The North Ladoga area is a classic region for the development of andalusite-sillimanite type metamorphic zoning (Fig. I-42). The degree of metamorphism varies from greenschist to granulite facies.

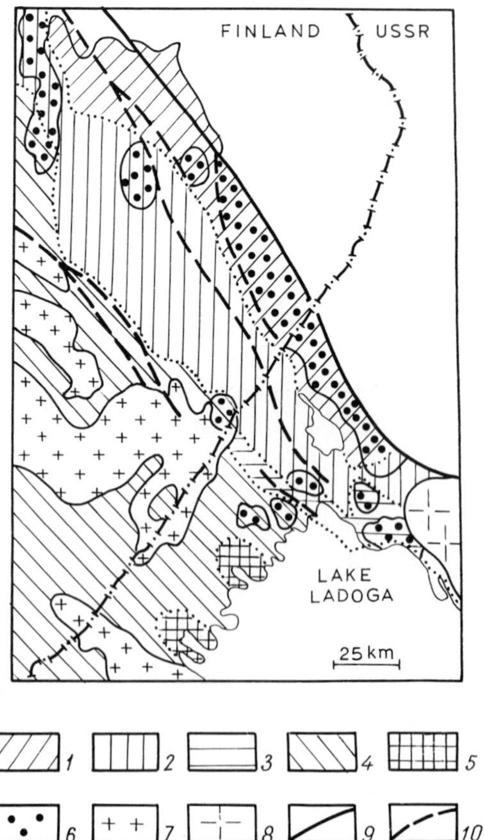

Fig. I-42. Sketch map of metamorphic zonation in the East Finnish zone of the Karelides. *1* = greenschist facies; *2* = staurolite subfacies; *3* = gt-silli-bi-musc subfacies; *4* = gt-cord-bi-orth subfacies of the cummingtonite-amphibolite facies; *5* = granulite facies; *6* = amphibolite facies, Archaean; *7* = autochthonous and parautochthonous granitoids in partial-melt zones; *8* = rapakivi granites; *9* = zone of boundary faults; *10* = other faults.

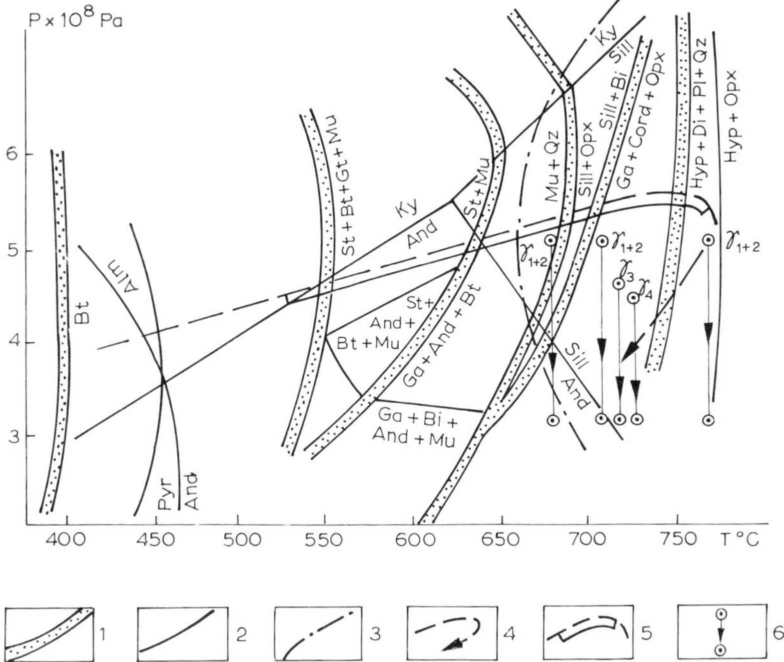

Fig. I-43. Thermodynamic conditions of metamorphism and ultra-metamorphism. *1* = mineral reactions-facies boundaries; *2* = other mineral reactions; *3* = granite melting minimum; *4* = *P–T* evolutionary trend during metamorphism; *5* = temperature interval used for estimating *P–T* parameters in garnet-biotite equilibrium; *6* = interval of migmatization conditions determined by homogenizing melt inclusions and cryometry of CO_2 inclusions.

Several features of the mineral composition of the metamorphic rocks allow us to evaluate the thermodynamic conditions of metamorphism on the basis of a facies scheme for metapelites (Fig. I-43). In the greenschist facies zone there are pyrophyllitic quartzites in which kyanite and andalusite appear at higher temperatures, which is probable at a pressure of around 4–4.5 kbar and *T* around 450–480°. Almandine garnet and staurolite in metapelites appear at similar temperatures, which is possible at a pressure of about 4 kbar. At higher values of *P*, the garnet becomes noticeably lower temperature than staurolite. Wherever andalusite-two-mica schists are sufficiently common, cordierite with staurolite and garnet are extremely rare in potassic metapelites, and even then it occurs as a late mineral, replacing staurolite. This means that pressure was not substantially lower than 4 kbar. Evidence for this is the dehydration of staurolite in the sillimanite stability field and the almost complete absence of gneisses with the assemblage sillimanite + biotite + garnet + potash feldspar. In such rocks there is always an abundance of cordierite, while the garnet-sillimanite-biotite-orthoclase subfacies forms only a narrow field (Fig. I-43).

The first migmatites in the North Ladoga region appear in the biotite-garnet-sillimanite-muscovite subfacies field (I), while the transition to the biotite-garnet-cordierite-orthoclase subfacies of the amphibolite facies (field IIa) takes place in the region of partial melting, which also embraces the granulite facies field (IIb). By homogenization of melt inclusions in anatectites belonging to the earliest stages of migmatization, temperature determinations in the zone where it just begins are 680°C on average, which agrees well with estimates made by mineralogical thermometers. Table I-24 presents some data on the conditions of formation of the rocks of the Ladoga Group close to the boundary of the zone of partial melting (ultrametamorphism). It emphasises the fact that muscovite assemblages are very high temperature. If we take the average (around 680°C) as an actual temperature, then we must conclude that H_2O fugacity in the fluid was unusually high, around 3.0–3.5 kbar for a pressure of 4.0–4.5 kbar (Glebovitsky et al., 1985). This temperature also reflects the conditions of the initial, essentially anatectic, migmatization (γ_0, γ_1 and γ_2). Cordierite assemblages are in fact higher temperature than muscovite assemblages. The temperature variations of 680 to 775°C obtained are totally realistic, since the garnet-cordierite-orthoclase subfacies field replaces the biotite-sillimanite-muscovite subfacies field over a very narrow transition zone, while definite cordierite gneisses sometimes coexist with hypersthene-bearing rocks, which appear at a temperature of 750°C.

Pressure estimates using cordierite gneiss are less reliable than those using muscovite gneiss. This seems to be related to an increase in Mg redistribution between garnet and biotite at the time migmatites form and in post-migmatite stages. At the very least, a direct comparison of pressures estimated from the country rock and vein material almost always indicate a lower value in vein rocks.

From the freezing point of CO_2 inclusions, it has been shown that as migmatization progressed and subsequent crystallization of anatectic magmas took place, there was decompression from 4.5–5 kbar at the metamorphic peak to approximately 3 kbar at the time of solidification of the final melts (Glebovitsky et al., 1985).

Temperatures, determined for highly migmatized rocks, never exceed 700°C. This may be explained by the fact that at the time of formation of significant quantities of melt, the temperature stabilizes in large volumes of rock.

Thus, we may conclude that migmatization in the North Ladoga region commenced and proceeded at a temperature of around 680°C and a total pressure of around 4.5 kbar which later dropped to 3 kbar, and an initial water fugacity of around 3.0 kbar.

Partial melting and intrusive magmatism. The earliest intrusive rocks belong to a gabbro-peridotite suite, forming small bodies in the northern

TABLE I-24

Metamorphic conditions for rocks of the Ladoga Group in the transition zone to partial melting

Sample No.	X_{Mg}^{Gr}	X_{Mg}^{Bt}	K_{Mg}^{Bt-Gr}	Fe^{Gr}	Mn^{Gr}	Ca^{Gr}	Fe^{Bt}
139	0.157	0.505	0.182	2.22	0.25	0.10	1.07
55	0.187	0.513	0.218	2.25	0.13	0.21	1.12
56	0.168	0.493	0.196	2.31	0.11	0.09	1.13
60	0.185	0.520	0.210	2.26	0.12	0.10	1.07
197 b	0.161	0.538	0.165	2.19	0.20	0.16	1.04
273	0.240	0.542	0.267	1.94	0.05	0.14	1.23
225	0.225	0.530	0.257	2.04	0.04	0.13	1.08
134	0.198	0.462	0.287	2.26	0.04	0.12	1.23
105 b	0.228	0.475	0.326	2.02	0.04	0.14	1.14
818	0.242	0.555	0.255	2.16	0.06	0.08	1.06
65/77	0.245	0.519	0.301	2.06	0.03	0.10	1.05
122c	0.232	0.572	0.226	2.02	0.07	0.26	1.04
122d	0.238	0.522	0.258	2.07	0.10	0.14	1.14
140c	0.227	0.542	0.248	2.17	0.04	0.10	1.02
140e	0.201	0.502	0.250	2.14	0.07	0.10	1.13
142k	0.213	0.511	0.259	2.19	0.05	0.10	1.09
142l	0.180	0.465	0.252	2.24	0.10	0.11	1.22
59a	0.157	0.505	0.182	2.33	0.05	0.10	1.12
59d	0.172	0.441	0.263	2.26	0.07	0.08	1.24

Sample No.	ΔK	K_{isp}	T^{I}, °C	T^{II}, °C	p, kbar	Paragenesis
139	+0.054	0.236	680		4.2	Bt + Mu + Sill + Gr + Qu
55	+0.033	0.251	699		4.0	—ditto—
56	+0.034	0.290	672		4.0	»
60	+0.040	0.250	698		4.5	»
197 b	+0.056	0.221	661		4.4	»
273	−0.053	0.209		683	4.8	Bt + Gr + Cord + KFsp
225	−0.070	0.187		695	4.8	—ditto—
134	−0.037	0.250		726	4.2	»
105 b	−0.100	0.226		775	3.6	»
818	−0.018	0.237		694	4.1	»
65/77	−0.067	0.234		749	3.6	»
122 c	−0.031	0.195	638		4.5	Bt + Gr + Pl + Qu + Cord
122 d	−0.040	0.218	658		3.9	Bt + Gr + Pl + Qu
140 c	−0.001	0.247	694		4.3	Bt + Gr + Pl + KFsp + Qu
140 e	−0.037	0.213	651		3.8	—ditto—
142 k	−0.011	0.248	695		3.8	Bt + Gr + Pl + Qu
142 l	−0.027	0.225	666		2.9	—ditto—
59a	+0.057	0.239	684		3.2	
59 d	−0.041	0.222	662		3.4	

Note: All T and P determinations refer to gneisses and schists, except determinations for four granite samples (γ_4);

X_{Mg}^{Gr}, X_{Mg}^{Bt} are molar % Mg in Gr and Bt, $K_{Mg}^{Gr+Bt} = \dfrac{X_{Mg}^{Gr}(1 - X_{Mg}^{Bt})}{(1 - X_{Mg}^{Bt} X_{Mg}^{Bt})}$;

Fe^{Gr}, Mn^{Gr}, Ca^{Gr} and Fe^{Bt} are quantities of Fe, Mn, Ca in Gt and Bt in formula units;

$\Delta K = 0.43(Fe^{Gr} - 2.19) + 0.25(Mn^{Gr} - 0.10) + 0.23(Ca^{Gr} - 0.19) - 0.38(Fe^{Bt} - 1.06)$;

T' is temperature calculated from the formula $T' = 1250 K_{Mgisp}^{Gr-Bt} + 385(0°C)$;

T'' is temperature calculated from the formula $T'' = -0.334 \ln K_{Mg}^{Gr-Bt} - 0.154 Mg^{Gr} + 0.613$ (Mg^{Gr} – molar % pyrope in garnet).

part of the region. Since they are clearly deformed by ID_3 structures, being folded by north–south striking linear folds (Sudovikov et al., 1970), their intrusion was preceded by the most widespread early migmatites (γ_2). Compositionally they are amphibolites with a variable plagioclase content.

Rocks of the gabbro-tonalite association are also early intrusions. In south-west Finland they have been shown to be comagmatic (Arth et al., 1978). There, gabbro, diorite and trondhjemite occupy significant areas. Although they underwent intense high-grade metamorphism, relicts of igneous textures and minerals can often still be observed. The age of such rocks from Rb–Sr isochron determinations is 1900 Ma (Arth et al., 1978). In the North Ladoga region, rocks of this association form the huge Valaam pluton and a number of smaller Välimäki-type intrusions. The least altered rocks are norite and gabbro-norite. At the contacts of the intrusions, strong metamorphic alteration effects are observed, and there is also sometimes evidence of granitization of basic rocks. Chemically the basic rocks correspond to high alumina basalt, or less usually olivine basalt. The Impiniemi and similar masses are sometimes regarded as the tonalitic part of this association, ahead of a migmatization front. However, from their geological setting they correlate with a complex of high-grade (incomplete) anatectic tonalite and granodiorite (see below).

Detailed studies have shown that partial melting processes are clearly related to the three main deformational episodes that have been identified (Fig. I-41).

During the first deformational episode, at least four generations of migmatitic vein material formed, γ_0, γ_1, γ_2, γ_3, which is established from their mutual intersections. γ_0 is developed insignificantly and is evident only if in the same outcrop γ_0, γ_1 and γ_2 are present, usually oriented along the S_1 penetrative schistosity, and S_0 coincides with primary layering (although there are cases where this is not so); then γ_0 cuts the layering. γ_1 cuts γ_0, restricted to F_1 fold axial surfaces. γ_2 uses the S_3 penetrative schistosity and is located along F_3 fold axial surfaces, which in turn fold γ_0 and γ_1. During the development of F_3 under conditions of prolonged compression, γ_2 are folded into a system of minor flattened folds of migmatitic character. In the majority of cases, γ_0, γ_1 and γ_2 form a series of subparallel veins, similar in composition and structure, hence when studying the geochemistry they are treated together owing to the impossibility of separating them. Such an assumption is perfectly justifiable, since these granitoids are the products of fractional crystallization of anatectic melt formed at an earlier stage of ultrametamorphism during the sequence of deformations of the first episode.

All generations of veins are cut by γ_3 granites, which are linked in time with a series of subparallel or mutually intersecting fault zones at the end of the first episode. As they get closer to these zones, the migmatites lose their banding and become more homogeneous, with an averaging out of the melanosome and leucosome compositions, while compositionally the rock approaches granodiorite or tonalite, in which it is possible to make out nebulous portions of the host migmatites. They disappear towards the centre of the zone, and the rock becomes increasingly homogeneous with a patchy occurrence of leucocratic aplitic or pegmatitic-like segregations.

In the granulite facies field, γ_{1+2} migmatites are cut by dykes and small concordant or discordant bodies of basic rocks and enderbites (β and $\gamma\delta^{Gt}$) with sharp contacts. They form part of an assemblage of intrusive granitoids of very wide distribution in the Svecofennian Domain. The largest enderbite intrusion with an area of 45–50 km^2 is the Kurkijoki massif (Saranchina, 1972). They are comagmatic with gabbroids, forming numerous minor intrusions in the axial part of the fold belt in question (Glebovitsky et al., 1985). Compositionally the basic members of the association correspond to high alumina basalt, while enderbites correspond to quartz andesite and sodic dacite (Table I-25), and they are all chemically close to an island arc calc-alkaline suite. These bodies may possibly be synchronous with intrusions of the gabbro-tonalite association and should be incorporated into a single formation. But even if this is the case, it is essential to take account of the specific nature of the thermodynamic and fluid regimes under which the acid members of the igneous series originated, leading to the formation of high temperature and "dry" enderbites.

Small enderbite and gabbroid bodies were boudinaged during IID_4 or IID_5 deformational episodes. Larger bodies take the shape of F_4 and F_5 folds, in which case they often have a marginal schistosity (see below). In the well-exposed eastern contact of the Kurkijoki massif, a number of satellite sheets all agree with the attitude of the S_4 schistosity. These slightly discordant sheets are part of the same generation of folds that affect the massif, the main folds in this zone, which developed during the deformation of the intrusion and the country rocks (Glebovitsky et al., 1985). In other words, the intrusion of melts which formed the enderbite bodies was controlled by pre-IID_4 structures. The upright isoclinal folds in the linear shear belt formed after the rocks of the intrusions had completely crystallized. Consequently, all the metamorphic and partial melt processes from phase IID_4 onwards are superimposed.

Enderbites in turn are cut by minor, irregular-shaped veins of tonalite which are morphologically and structurally analogous to γ_3. These relationships between generations of veins of different age, their association with particular fold structures of the first deformational episode, and the gradual increase in metamorphic grade towards the south-west suggest parallel development of vein material generation in granulite and amphibolite facies migmatites.

Thus, the rock complex under discussion is on the whole synchronous with metamorphism, separating early anatectites from later migmatites. Therefore we consider these formations (Glebovitsky et al., 1985) to be

TABLE I-25

Average chemical composition of substrate gneisses, migmatite vein material of various generations and intrusive rocks in the North Ladoga region

Oxide	gn	δ^{Gt}_{1+2}	γ^{Gt}_{1+2}	γ^{a}_{1+2}	γ_{1+2}	γ_3	γ_4	γ_5
SiO_2	63.34	55.05	70.36	72.81	72.07	71.94	69.52	74.09
TiO_2	0.82	1.13	0.35	0.36	0.36	0.43	0.38	0.39
Al_2O_3	15.93	17.41	14.18	13.78	13.90	13.96	15.31	12.93
Fe_2O_3	1.65	2.49	1.01	0.59	0.72	0.78	1.08	0.45
FeO	4.93	5.76	2.93	2.04	2.31	2.34	2.23	1.89
MnO	0.09	0.12	0.09	0.06	0.07	0.05	0.06	0.06
MgO	3.03	3.50	1.49	1.17	1.27	1.33	1.04	0.88
CaO	2.57	6.68	2.70	2.21	2.36	1.97	2.73	1.55
Na_2O	2.66	3.72	2.78	3.12	3.02	3.11	3.42	2.60
K_2O	2.98	1.90	2.48	2.43	2.43	2.94	2.99	4.03
P_2O_5	0.16	0.73	0.12	0.14	0.14	0.11	0.21	0.09
n	65	10	10	23	33	21	40	4

Oxide	γ_6	β	$\gamma\delta^{Gt}$	$\gamma\delta$	$\gamma\delta_P$	$\gamma\delta_L$	$\gamma\delta_I$	γ_7
SiO_2	63.39	53.41	62.44	64.41	63.45	64.02	70.39	71.46
TiO_2	0.78	0.95	0.75	0.76	0.82	0.78	0.36	0.33
Al_2O_3	15.18	17.52	17.52	15.87	16.49	15.40	14.65	14.26
Fe_2O_3	0.62	2.58	1.02	1.06	1.22	0.98	0.55	0.77
FeO	1.55	5.60	4.07	3.93	3.84	4.43	2.54	2.03
MnO	0.02	0.14	0.07	0.08	0.09	0.08	0.05	0.05
MgO	0.81	3.60	2.13	2.27	1.97	2.98	1.23	0.75
CaO	1.99	7.21	4.67	3.48	3.41	3.97	2.11	1.50
Na_2O	2.83	3.64	3.96	3.53	3.62	3.28	3.96	3.14
K_2O	5.66	1.98	1.82	2.55	2.95	1.87	3.00	4.66
P_2O_5	0.24	0.88	0.28	0.34	0.47	0.22	0.13	0.13
n	10	6	30	56	29	21	6	40

Note: gn = Ladoga gneisses; δ^{Gt}_{1+2} = early vein material in basic rocks of the granulite facies; γ^{a}_{1+2} = early vein material in acid rocks (gneisses) of the granulite facies; γ_{1+2} = early vein material in amphibolite facies gneisses; γ_2, γ_1, γ_5, γ_6, γ_7 = subsequent generations of vein material in migmatites; β = gabbroids comagmatic with enderbites; $\gamma\delta^{Gt}$ = enderbites; $\gamma\delta$ = granodiorite complex in total, including $\gamma\delta_P$ (from minor intrusions in the Putsari region), $\gamma\delta_L$ (from the Lauvatsari intrusion), $\gamma\delta_I$ (from the Impiniemi intrusion).

differentiates of a major mantle asthenolith, whose high-level position immediately beneath the M-boundary caused rapid heating of crustal sediments and the appearance of steep thermal gradient metamorphism.

A compositionally inhomogeneous group of granodiorites, tonalites, quartz diorites and syenodiorites formed during the second deformational episode; these are named below according to the dominant variety of rocks of the granodiorite series (Glebovitsky et al., 1985). Rocks of this group usually belong to early fold formations (Glebovitsky et al., 1985) or to a complex of moderately-acid granitoids of the first formational type (Saranchina, 1972). In the muscovite-biotite-garnet-sillimanite subfacies field and sometimes in the low-temperature part of the garnet-cordierite-biotite-orthoclase subfacies they form a series of major bodies, attaining

50–100 km^2. On the whole these bodies, especially the large ones, are concordant with the structures of the host rocks, although in a number of cases they form a series of cross-cutting bodies from which emerge apophyses that exploit early fold structures. In the higher part of the garnet-cordierite-biotite-orthoclase subfacies, that is when hypersthene appears in mafic schists, and also often in the low-temperature part, the nature of the mutual relations between granodiorites and host rocks changes in some cases. Together with intrusive contacts, homogenization of migmatites is observed in individual zones and there is a gradual transition through a zone of patchy granodiorite to homogeneous types which finally form independent bodies and become mobile. Such relationships may be regarded as the *in situ* inception of granodiorite, tonalite and diorite by melting of migmatite of the Ladoga complex.

Granodiorites which display clear signs of mobility and emplacement contain xenoliths of more basic rocks — diorites and quartz diorites. It is thought that this group of rocks did not form in a single event. The first melt fractions were more basic, with segregation of their initial magmas having taken place somewhat deeper than their emplacement level where the relevant temperatures obtained for the formation of these magmas. Granodiorites cut γ_{1+2} and γ_3 and were subjected to the second migmatization episode (γ_4 and γ_5).

The geological position of the granodiorites is defined according to their relations with folds. It has been established that many minor bodies were boudinaged during the formation of F_5 folds and at the same time they cut F_4 folds. They therefore fall in the IID_4–IID_5 time interval. Coarsely-banded and nebulous migmatites formed from granodiorites, basic rocks and early migmatites, the vein material of which is represented by two generations of granitoids γ_4 and γ_5 (Fig. I-42). They appear in the low temperature part of the garnet-biotite-cordierite-orthoclase subfacies field and in previously migmatized metasediments they form a series of subparallel veins from a few cm thick to 1–2 m, cutting earlier structures. Overall, they are oriented parallel to axial surfaces of recumbent isoclinal folds in nappes and are folded by north–west to roughly east–west trending F_5 folds along whose axial surfaces γ_5 developed, often with pegmatoid or porphyroblastic textures. In the granodiorites and basic rocks γ_4 and γ_5 form thicker bodies, often without evidence of deformation. Sometimes between the granodiorites it is possible to observe gradual transitions, suggesting that granitization was widespread during this stage of development of the region, which is a reason for the formation of the major granite-gneiss bodies that are typical of the entire Svecofennian Domain.

The intrusion of a large swarm of basic dykes is associated with the final stages of the second episode of deformation, or the beginning of the third episode (Sudovikov et al., 1970).

The formation of group γ_6 veins is associated with the third deformational episode (Fig. I-41), and includes varieties that developed slightly successively. They are oriented along F_6 axial surfaces, with N–S and NE strikes. Associated with these is the formation of coarsely-banded migmatites and agmatites in fault zones a few tens to hundreds of metres wide. Among these granitoids are abundant veins of pink leucocratic granites, sometimes with pegmatoid segregations, and zoned veins in which biotite-rich bands alternate with aplitic and pegmatitic varieties, oriented parallel to contact surfaces. The veins of almost monomineralic microcline rocks and pegmatites cut these zoned veins.

In the final stages of the third deformational episode, γ_7 microcline granites formed, which make up the Tervuss, Kuznechnoye and other intrusions, known as group II granites (Saranchina, 1972) or post-folding granites (Glebovitsky et al., 1985). They are similar in age to the essentially potassic granites in the southern part of the Svecofennian province (Simonen, 1960). Judging from morphological features, early migmatite vein material (γ_0, γ_1, γ_2) had an anatectic origin. However, this was not simple selective melting in a closed system. The absence of all indications of incongruent melting of micas, the unidirectional trend in feldspar composition from melanosome to leucosome and several other features of the migmatites is evidence that anatexis proceeded under conditions of perfectly mobile behaviour of alkalis. One striking feature of Table I-25 is the low total K + Na and the strong basics for granites with a normal silica content. Melting apparently proceeded in the presence of a highly acid fluid which reacted with the melt, stabilizing the micas in the solid state and preventing the widespread development of granites with potash feldspar, despite the fact that anatexis occurred in gneisses rich in potassium.

The anatectic nature of early migmatites is confirmed by finds of inclusions of crystallized melt. By homogenizing these, the average temperature of anatexis was determined to be 680°C in the garnet-biotite-sillimanite-muscovite subfacies field (I), 710°C in the garnet-biotite-cordierite-orthoclase subfacies field (IIa) and 770°C in the granulite facies field (IIb). Such a steep temperature gradient in the melting field can exist only if it occurs to an insignificant extent, evidence for this being the small volume of early generation granite veins in the metasediments.

Granite veins of the third generation, judging by their morphology, signify the beginning of high-grade, incomplete anatexis, when during the melting process, only relatively acid rocks were involved. The composition therefore differs little from γ_{1+2} (Table I-25). Homogenization temperatures of melt inclusions also show little variation, from 730° to 745°C. The appearance of rocks in the granodiorite group is associated with the massive development of high-grade incomplete anatexis. Their average chemical composition is close to that of the migmatites (Table I-25). Three varieties can be distinguished. The first, tonalites, form major bodies (the Lauvatsaari intrusion), which subsequently underwent only minor changes ($\gamma\delta_L$). Distinct from these are minor bodies of granodiorite which were subjected to marked feldspathization during the second episode, and are more potassic. The third variety, those that formed ahead of a migmatization front (e.g. Impiniemi), are acid differentiates of the same tonalitic magmas, judging from their composition. Following the crystallization of the high-grade anatectic magmas, there was a renewed burst of anatectic migmatization, leading to the formation of γ_4 and γ_5 leucosomes and granitization. The slightly more basic composition of these granitoids compared to γ_{1+2} may be related to the large volumes of melt material or the redeposition in the veins themselves of strong basics leached from the magma as aggregates of garnet grains which in fact are frequently encountered in γ_4. The melting process itself is demonstrated by finds of inclusions of crystallized melts, whose homogenization temperature changes little in zonation and is 715°C on average, which agrees well with garnet-cordierite-orthoclase subfacies conditions, where γ_4 also first appear. Also corresponding to this subfacies is the level of thermostasis of the entire volume subjected to ultrametamorphism, including the field which at the first stage of evolution was situated in granulite facies conditions.

The third stage of evolution is characterised by local residual migmatization, mostly injection type. Typical are granites (γ_6, Table I-25) which are essentially potassic and less siliceous than earlier generations. The third stage concluded with the intrusion of potassic granites from magma chambers considerably deeper than the present erosion surface. This fairly deep level of magma chambers, coupled with anhydrous melts that were being generated, were the reasons for the high potash content of the granites.

The rapakivi granite complex

In the exposed part of the Svecofennian province which we have examined, there are three intrusions of rapakivi granite: Vyborg, Salmi and Ulyaleg. In these three masses there are six textural varieties of granite: (1) vyborgite (biotite-hornblende rapakivi in which ovoids have oligoclase rims), (2) even-grained biotite granite, (3) biotite (lepidomelane) rapakivi with perthitic microcline ovoids which lack plagioclase rims, (4) inequant-grained granite, (5) porphyritic granite with large ovoid clusters, and (6) porphyritic rapakivi with trachytoidal texture. In the majority of cases these varieties represent phases of magma intrusion. A general feature of the petrographic composition of all the rapakivi granites is a sharp dominance of potash feldspar over plagioclase, while potassium exceeds sodium, their total being high, and strong basics have low concentrations.

TABLE I-26

Characteristic ore deposits and ore shows

		Geological and mineralogical features				
Deposit or ore show, useful minerals	Type of mineralization, ore-forming and genetic	host rocks and mineralized rocks	shape of deposit, ore body morphology	ore type	mineral composition of ores (secondary minerals)	Geological age (host rocks & ores)
Pienijänisjärvi* (Jalonvara) molybdenum	Chalcopyrite-molybdenite in granites; hydrothermal	Granites	Stockworks	Disseminated vein-type disseminations	Molybdenite (chalco-pyrite, arsenopyrite, sphalerite	Early Proterozoic
Pitkäranta; tin, copper, zinc	Chalcopyrite-sphalerite/ cassiterite related to skarns	Skarns in limestone/ granite contact zones	Bedded, conformable and cross-cutting deposits	Disseminated, massive	Cassiterite, sphalerite, chalco-pyrite (magnetite)	Early Proterozoic
Latvasyra wolfram	Scheelite; in skarns	Skarns with carbonate rocks and hornblende schists	Bedded and lenticular bodies	Cluster-disseminated	Scheelite (pyrite, magnetite, sphalerite wolframite, molybd-enite, chalcopyrite)	Late Proterozoic
Jalonvara; sulphur, iron, copper	Chalcopyrite-pyrrhotine-pyrite; sedimentary-exhalative, hydrothermal-metamorphic	Schists derived from keratophyre-spilitic formation	Bedded, lenticular and vein bodies	Massive, disseminated	Pyrite, pyrrhotine, chalcopyrite (sphalerite, magnetite galena, pentlandite); propyllites	Early Proterozoic

The gabbro-diabase complex

Among post-Svecofennian intrusions, apart from a suite of diabase and dolerite dykes, there are Upper Riphean gabbro-diabase intrusions, the largest of which make up the Valaam archipelago in the northern part of Lake Ladoga. This intrusion is an almost horizontal layered pluton, in which the lower parts consist of medium-grained gabbro-diabase and the upper parts consist of coarse-grained syenodiorites, with a gradual transition between the two (Glebovitsky et al., 1981). Subhorizontal and steeply dipping dykes of syenodiorite and aplite-granophyre occur. Early generations belong to a weakly expressed sodic type, while later ones are potassic.

Bilibina et al. (1980) note that the Ladoga belt is characterised by deep vault-like upwarps of lower levels of the Earth's crust. The reworking of deep structures occurred during a period of protoactivation, with which is associated the formation of major intrusions of gabbro-anorthosite and rapakivi granite with a tin, molybdenum and tungsten signature, giving the Ladoga belt a lithophile metallogenic character. Table I-26 presents a brief summary of some mineral deposits typical of this belt.

Conclusions

(1) Proterozoic supracrustal complexes accumulated on a sialic Archaean basement, to a large degree stabilized, as shown by the continental facies of the Jatulian, underlying the Ladoga Group flysch. The duration of the Svecofennian cycle was 500 Ma.

(2) The tectonic cycle includes three characteristic episodes of formation of folds and faults of varying morphology and significance. This sequence

may be used as a local time scale for studying the evolution of internal processes. The morphology of structures reflects two geodynamic trends: a general lateral compression of the entire fold system, and vertical compression with the formation of recumbent folds. The latter correlate with bursts of melting and may be explained by density inversion phenomena in gravitationally unstable systems.

(3) The region under investigation is characterised by the appearance of classic andalusite-sillimanite-type metamorphic zoning, reflecting the existence of a highly intense positive thermal anomaly. The maximum temperature of 750–770°C was attained at a depth of the order of 15 km (P_{tot} = 4.5 kbar), which means an average geothermal gradient of 45°/km. The appearance of this anomaly may be associated with an uprise of the mantle asthenolith, differentiates of which appear as igneous complexes comprising gabbroids and enderbites, broadly synchronous with metamorphism and partial melting.

(4) Partial melt (ultrametamorphic) processes appear in the sequence: anatectic migmatization — high-grade, incomplete anatexis and formation of tonalite-granodiorite magmas — renewed burst of anatectic migmatization and granitization with the formation of granite-gneiss massifs — arrival of a melting front and formation of late injection migmatites — intrusion of essentially potassic granites.

Chapter 2

THE UKRAINIAN SHIELD

The southern outcrop of the Precambrian basement, the Ukrainian Shield, is some 900 km long by 60–150 km wide. The shield and its flanks are treated as a single structure, separated by a system of marginal faults from the Pripyat basin and the Dnepropetrovsk-Donetsk depression in the north, the Volyn-Podolsk and Moldavian plates in the west and the Black Sea depression in the south (Fig. I-44). The boundary fault system, with a total vertical displacement of from 300–400 m to 2–5 km, originated in the Riphean (Klochkov et al., 1983). The deepest burial of the pre-Riphean basement is 20–22 km beneath the Dnieper-Donetsk aulacogen (Sollogub et al., 1980). The thickness of the sedimentary cover on the shield varies from 0 to 100 m.

The Precambrian of the Ukraine has been studied since the 19th century. By the 1940s the first stratigraphic schemes for the Precambrian had been compiled (Bezborodko, 1935; Luchitsky, 1939; Polovinkina, 1940; Sobolev, 1936). Structural and geological maps of the Ukrainian crystalline massif were prepared and a stratigraphy was worked out in the 1950s (Antropov, 1958; Polovinkina, 1953; Semenenko, 1958).

During the period 1960–1970, the accumulated data allowed the compilation of new geological and tectonic maps for the shield as a whole and for a number of subregions (Kalyayev, 1965). Archaean crystalline rocks were subdivided into a number of groups on these maps, and several cycles of Precambrian structural evolution of the shield were identified, allowing the shield to be divided into protogeosynclinal and protoplatform blocks, separated by deep faults.

More intense study of Precambrian formations in the shield, including discussions and debates on the problem of Precambrian stratigraphy (Belevtsev, 1981; Dobrokhotov et al., 1981), the collection of field data, geochemistry, litho-formational analysis and isotope geochronology provided the basis for erecting a number of schemes for the tectonic structure of the Ukrainian Shield, in which the early Precambrian was approached from the standpoint of the geosynclinal concept or the concept of the outward growth of previously consolidated crust (Kalyayev, 1965; Ryabenko, 1970). In recent years, schemes for the tectonic structure and evolution of the Ukrainian Shield have been developed on the basis of plate tectonic ideas (Kalyayev, 1980; Kalyayev et al., 1984; Peyve et al., 1976; Stupka, 1980).

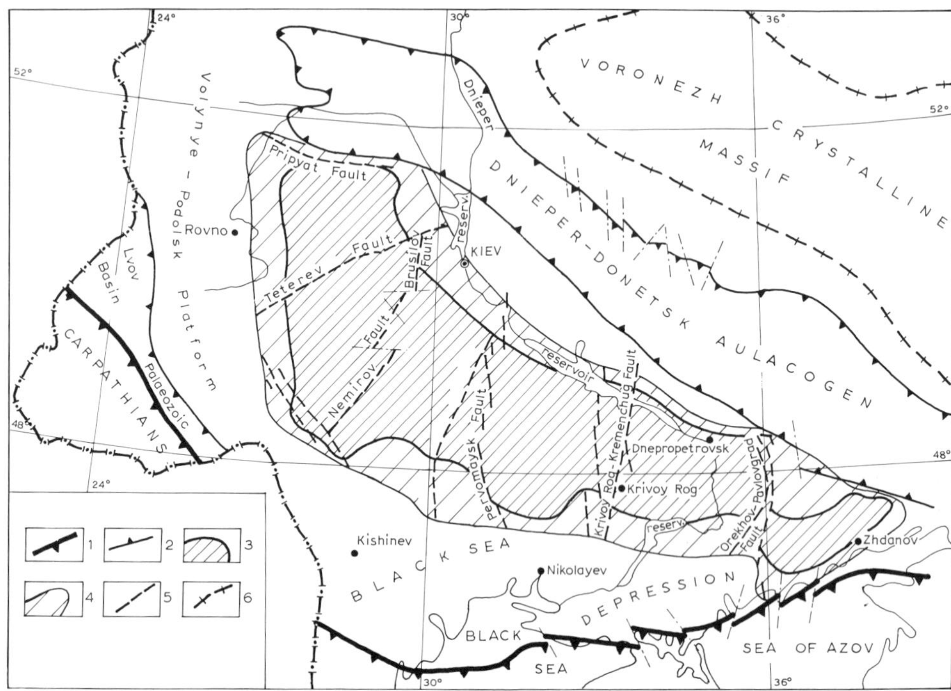

Fig. I-44. Position of the Ukrainian Shield within the East European craton (simplified from Krutikhovskaya et al., 1982). *1* = boundaries of East European craton; *2* = boundaries of negative platform structures; *3* = boundary of exposed part of shield; *4* = conventional boundary of shield from 300 m depth contour to Precambrian basement; *5* = deep faults of shield; *6* = conventional boundary of Voronezh crystalline massif from 300 m depth contour to basement.

Much use has been made of formational analysis of supracrustal assemblages in the Ukrainian Shield (Shcherbakov et al., 1984; Semenenko et al., 1978a, b, 1979, 1982; Lazko, 1982; Lazko et al., 1975; Sivoronov et al., 1981; Polovinkina, 1954). Intensive work has been done over the last 20 years in developing isotopic geochronological methods for subdividing the Precambrian of the Ukrainian Shield (Semenenko et al., 1964, 1965; Shcherbak, 1975, 1978, 1980; Shcherbak et al., 1981, 1984).

Of great significance in understanding the geology and metallogeny of the Ukrainian Shield has been the study of iron ores associated with particular geological formations and restricted to at least three levels of iron accumulation (Belevtsev and Galetsky, 1981; Semenenko et al., 1978a, b; Kalyayev et al., 1984; Yaroshchuk, 1983). Stratigraphic correlation of iron ore levels has shown that in the Ukrainian Shield, a leading role was played by an Early Proterozoic epoch of iron ore accumulation in primary sedimentary and, to a lesser extent, volcano-sedimentary formations (Belevtsev et al., 1981). Summaries of the metamorphism, metasomatism, ultrametamorphism, petrography, mineralogy and petrochemistry of the crystalline rocks of the Ukrainian Shield are in a number of works (Bespalko et al., 1976; R. Belevtsev, 1982; R. Belevtsev et al., 1983; Usenko, 1982, 1960; Usenko et al., 1980; Nalivkina, 1964, 1978; Venediktov et al., 1979; Strygin, 1978; Shcherbakov, 1975). Intensive development of geophysical research on the deep structure of the Earth's crust in the Ukrainian Shield has yielded data on crustal thickness and the nature of crustal layering within different provinces ("geoblocks") (Sollogub and Chekunov, 1975; Sollogub et al., 1980; Chekunov, 1972; Shcherbak et al., 1981). A magnetic model (Krutikhovskaya et al., 1982) shows lateral inhomogeneity of the crust in the Ukrainian Shield and the existence of blocks of predominantly femic and sialic composition, as well as a relationship between deep and surface structures.

Information about various aspects of ore formation, including metamorphic types, has enabled the compilation of metallogenic maps and general theoretical conclusions to be drawn about epochs of ore formation in the Precambrian of the Ukrainian Shield (Belevtsev and Galetsky, 1984; Usenko et al., 1976; Belevtsev et al., 1975; Belevtsev, 1974) and permitted the working out of a classification of ore-bearing structures and useful mineral deposits of metamorphic genesis (Belevtsev, 1977, 1981a, b; Bilibina et al., 1984).

The combined analysis of data obtained from geology, gravity, magnetics and drilling forms a basis for delineating several major provinces within the shield, which are further subdivided into blocks, differing to a greater or lesser extent in the composition of Precambrian rock formations, deep structure and tectonic domains, as well as specific features of magmatism, metamorphism and metallogeny (Fig. I-45).

The Ukrainian Shield can be divided tectonically into three parts: Azov, Central Ukrainian (Dnieper and Kirovograd blocks) and West Ukrainian. The Azov part, from the composition of the component Archaean rocks and particular features of the internal processes that gave rise to their development, and the presence of Early Proterozoic sub-platform type normal and alkali granite intrusions, may be regarded, with certain reservations, as an analogue of granulite-greenstone terrains. The Dnieper block is a typical Archaean granite-greenstone terrain, while the Kirovograd block is its reworked part, identified by analogy with similar structures in the Baltic Shield. Along the boundary of the first two parts lies the Orekhov-Pavlograd megasuture.

A large area of the shield is occupied by the West Ukrainian region, consisting of a number of blocks of various origin which formed during an intense period of activity in the Early Proterozoic. The structure of some of the blocks probably reflects the primary inhomogeneity of the crust on which the blocks developed, while in others sharply different erosion levels mean that structural-lithological complexes belonging to the upper

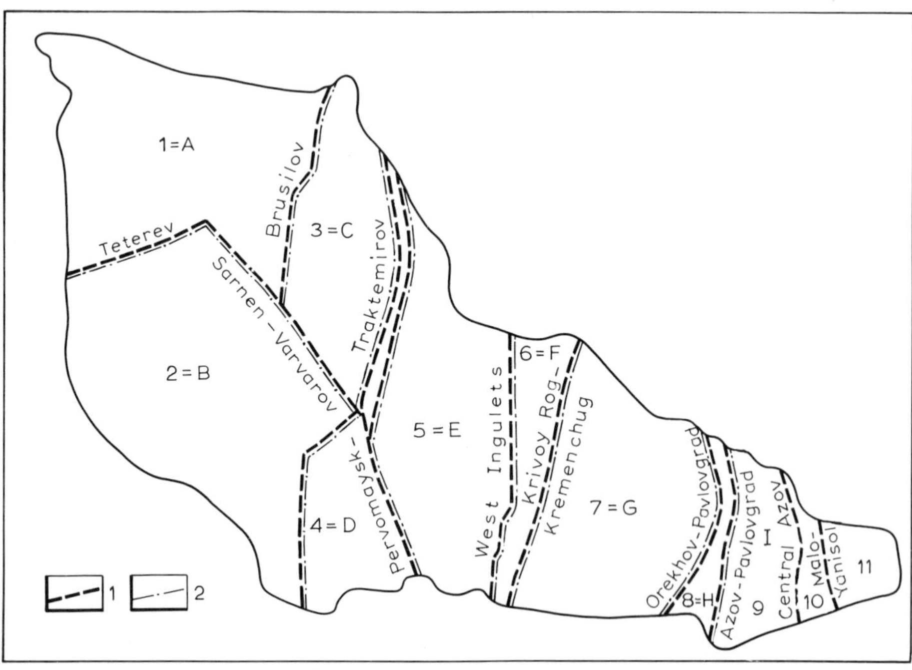

Fig. I-45. Block structure of the Ukrainian Shield and regions in the Ukrainian metallogenic province. Block structure of the Ukrainian Shield: 1 = first order deep faults, bounding first order blocks. Volyn-Podolsk province: *1* = Volyn block, *2* = Podolsk block, *3* = Belotserkov block. *4* = Golovanevsk suture. Central Ukrainian province: *5* = Kirovograd block, *6* = West Ingulets suture, *7* = Dnieper block. *8* = Orekhov-Pavlovgrad suture. Azov province: *9* = West Azov block, *10* = Central Azov suture, *11* = East Azov block. Regions in the Ukrainian metallogenic province: 2 = boundaries between metallogenic provinces and zones. Numbers in brackets correspond to tectonic blocks and zones. Provinces: *A* = Volyn (= *1*), *B* = Podolsk (= *2*), *C* = Belotserkov (= *3*), *E* = Kirovograd (= *5*), *G* = Dneprovsk (= *7*), *I* = Azov (= *9+10+11*). Zones: *D* = Golovanev (= *4*), *F* = Krivoy Rog–Ingulets (= *6*), *H* = Orekhov-Pavlovgrad (= *8*). (After Storchak, 1984.)

structural stage may be preserved in some blocks. In all cases, prolonged polyphase internal processes are of great importance. In this regard, the northern Volyn block is the least eroded part and includes not only Early Proterozoic supracrustal rocks but also younger, Upper Proterozoic, formations such as the Ovruch Group.

Of particular importance is the suture zone between the Dnieper and Kirovograd blocks, which coincides with the long-lived West Ingulets zone that developed on the site of an earlier structure and which is here regarded as an Early Proterozoic mobile belt.

The delineation of these three tectonic regions in the general structural scheme of the Ukrainian Shield is essential for correlating with other parts of the East European craton. Each of the provinces and blocks in the Ukrainian Shield is characterised by particular features of fold styles and fault patterns, due to the prolonged history of tectonic development (Ryabenko, 1983).

The northern part of the West Ukrainian region, the Volyn block, consists of a basement of Archaean granulites and metamorphosed lower Proterozoic groups, with widespread development of platform sediments (the Pugachev and Ovruch Groups) and granitoids (e.g. Osnitsk and Korosten). The gneisses (Teterev Group) typically show brachyform folds and dome structures consisting of migmatites and granites. In the southern part of the region (the Podolsk block), impersistent brachyanticlinal upwarps are observed, with narrow, often isoclinal, synclinal structures. Here there are few domes, of small dimensions. The West Ukrainian region on the whole is characterised by the wide distribution of granitoids and migmatites which weld together blocks of lower Archaean supracrustal rocks (the Dnestr-Bug Group).

In the Central Ukrainian region, the Kirovograd block is sometimes regarded as a Proterozoic protoplatform (Ryabenko, 1970). Intense granite formation is developed here in separate parts of minor fold structures — cores of anticlines and synclines with volcanosedimentary formations (the Bug and Ingul-Ingulets Groups). In the Dnieper block, a granite-greenstone terrain, there are major synclinoria consisting of metamorphosed Archaean volcanics and sediments (mainly the Konka-Verkhovtsevo Group), separated by granite-gneiss anticlinoria containing abundant dome-like structures.

The Azov region consists of lower and upper Archaean and possibly lower Proterozoic formations, containing respectively two cycles of sedimentation and volcanic activity, with the formation of the West Azov and Central Azov Groups and also the Azov and East Azov granitoid complexes. The metamorphic rocks form a series of synclines and anticlines, complicated by minor folds. Rocks of the upper structural stage (upper Archaean or lower Proterozoic) occupy narrow synclines, separating blocks of predominantly lower Archaean rocks.

The provinces (geoblocks), first order blocks and sutures separating them typically have different deep structures and component rock compositions and as a consequence different geophysical characteristics. These characteristics are already reflected in the book in a general sense for the whole East European craton (Table I-1, Fig. I-4).

The geophysical pattern of the Ukrainian Shield shows broad regions in which a mosaic field pattern is developed, with narrow zones of linear anomalies dividing such regions. The mosaic field reflects structural features of the different blocks. The extensive linear anomalies correspond to suture zones.

The Azov block is characterised on the whole by high magnetic and gravity fields. The complex pattern of local magnetic anomalies is due mainly to the structures in the gneisses that make up the blocks.

The Dnieper block in a regional sense has low-intensity fields. Against this background a number of intense gravity maxima stand out, caused by greenstone structures, and less intense but broad maxima in regions where rocks of approximately dioritic composition crop out. Individual intense elongate magnetic anomalies are caused by iron ore formations and ultrabasic rocks of greenstone belts.

For the Kirovograd block we have a typically low overall magnetic field with rare weak intensity anomalies. In the gravity field, the block in the south is characterised by higher gravity values, while in the central part and in the north there are intense gravity minima, due to the Novoukrainka granites and rapakivi granites of the Korsun-Novomirgorod complex.

The Belotserkov block is marked by having low values for both fields. Individual small local magnetic anomalies are explained by the presence of formations belonging to the Rosin-Tikich Group.

The Volyn block has a different magnetic field from the Belotserkov block, being rather high in general. There are some weak local magnetic anomalies, but these are very rare. Deep gravity lows in the north-east correspond to the Korosten rapakivi granites, and the intense gravity and magnetic fields in the north-west are due to the Osnitsk granitoids.

The Podolsk block stands out on account of the coarsely mosaic or massive structure of its high magnetic and gravity fields, reflecting the variety of granulite facies rocks exposed here.

Volcanogenic-sedimentary rocks account for 15–20% of the area of the shield, the remaining part being mainly granitoids.

For the Ukrainian Shield as a whole, it is possible to identify several episodes of metamorphism, differing in scale and conditions of development. The distribution and sequence of metamorphic facies as developed in the Ukrainian Shield are shown in Figure I-46. Early Archaean regional metamorphism is reflected in granulite and amphibolite facies conditions (R. Belevtsev et al., 1985); in the Late Archaean there was metamorphism at greenschist and epidote-amphibolite facies in the Dnieper granite-greenstone terrain. In the Early Proterozoic there was regional metamorphism and ultra-metamorphism from granulite to greenschist facies and local retrogression of Archaean granulite complexes. The metamorphic rocks of the Ukrainian Shield correspond to moderate and low pressure facies series. In areas of granulite facies metamorphism in the Azov region and in the south of the West Ukrainian region, evidence of eclogitization and other indications of high pressure have been found.

Granulite facies rocks are most widely distributed in the SW part of the shield in the Podolsk block and in the Golovanevsk suture zone. In the Volyn and Dnieper blocks, granulite facies rocks occupy limited areas, while in the Belotserkov block intensely retrogressed granulites occur. In the Kirovograd block and the West Ingulets zone, granulite facies rocks account

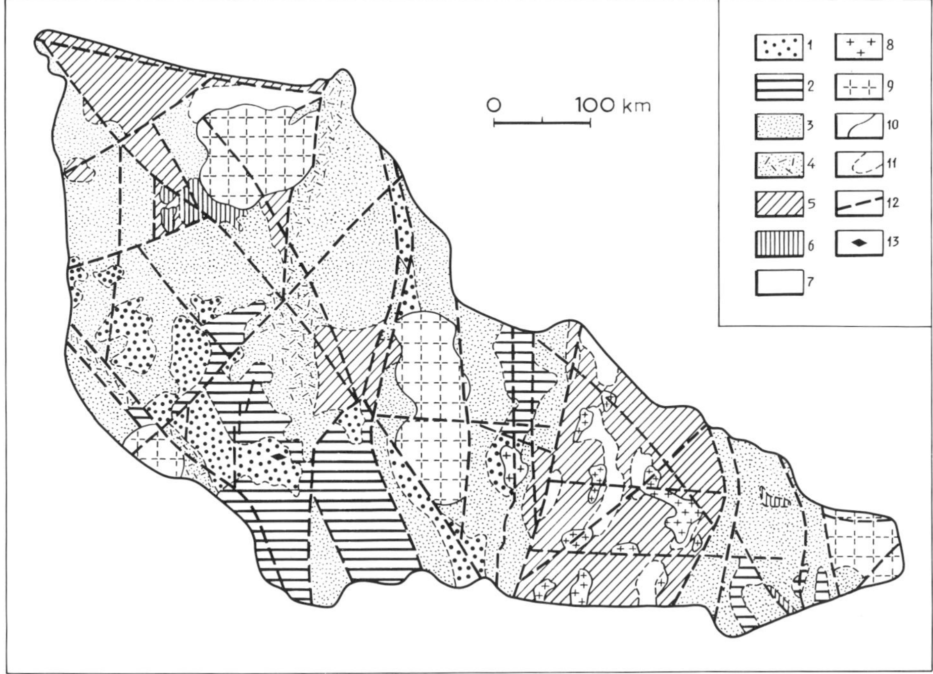

Fig. I-46. Sketch map showing distribution of regional metamorphic facies in the Precambrian of the Ukrainian Shield (compiled from Storchak, 1984). *1* = granulite facies; *2* = amphibolite facies superimposed on granulite; *3* = amphibolite facies; *4* = low-*T* amphibolite facies superimposed on high-*T*; *5* = epidote-amphibolite facies superimposed on amphibolite; *6* = epidote-amphibolite facies; *7* = greenschist facies; *8* = Archaean partial melt granitoids; *9* = post-metamorphic Proterozoic granitoids; *10* = boundaries of geological bodies; *11* = boundaries of metamorphic fields; *12* = deep faults and crustal faults; *13* = eclogitization.

for discontinuous strips, while in the Azov province they occur as inliers in areas of migmatites and granites. For the amphibolite facies, which occurs in all the provinces in the shield, and for both epidote-amphibolite and greenschist facies, two branches of metamorphism are established, prograde and retrograde.

The age relationships of the structural-lithological complexes of many tectonic regions of the Ukrainian Shield are debatable. In this work they are described following Shcherbak (1983) and Storchak (1984), and the metamorphic patterns follow Storchak (1984) and R. Belevtsev et al. (1985). Figure I-47 shows the rock complexes of the Ukrainian Shield.

It is essential to note that for the Ukrainian Shield the geological position and age of a number of sedimentary-volcanic units remain disputed at present. This concerns, firstly, the volume and age of the Bug and Central Azov Groups, the volume of the Konka-Verkhovtsev Group as a whole and the age of the Belozero Formation, the age of the Osipenko Formation

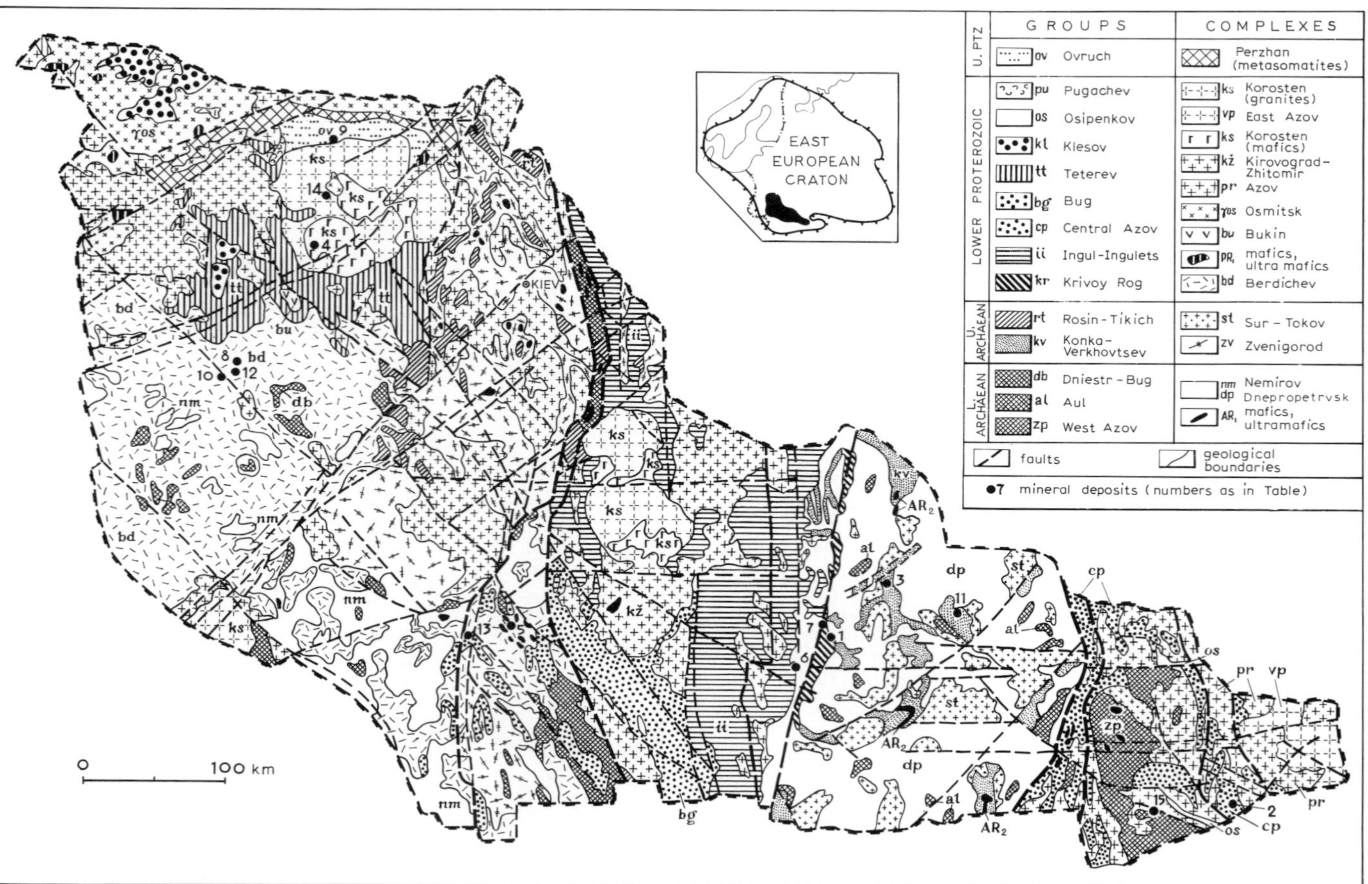

Fig. I-47. Geological sketch map of Precambrian formations in the Ukrainian Shield (simplified from Storchak, 1984).

and the volume of the Ovruch Group (Grechishnikov, 1983; Kiselev, 1977; Berzenin et al., 1982; Kushinov, 1981, 1985; Polunovsky et al., 1983; Shulga et al., 1982; Piyar et al., 1983).

1. THE WEST UKRAINIAN REGION — THE VOLYN-PODOLSK PROVINCE

This region includes the Volyn, Podolsk and Belotserkov blocks and the Golovanevsk suture zone, in which three structural-lithological complexes are distinguished.

Lower Archaean

Supracrustal rocks of this complex form the *Dnestr-Bug Group*. It is widely distributed in the Podolsk block and less so in the Belotserkov and the southern part of the Golovanevsk zone (in the Upper and Middle Bug region and the region around the Dnestr). It is observed as small areas, strips and remnants among granitoids in the Archaean Nemirov and Proterozoic Berdichev complexes. There are two units in the undifferentiated formations of the group. The lower consists of two-pyroxene and hypersthene-plagioclase gneiss, amphibole-pyroxene schist and gneiss, pyroxene and amphibole-magnetite quartzite and calciphyre; the upper consists of biotite and garnet-biotite gneiss, sometimes with cordierite and biotite-graphite schist, 800–900 m to 1600 m thick in individual sections. Metamorphism of the rocks corresponds to granulite and high temperature amphibolite facies. An assemblage of basic and ultrabasic rocks is associated with lower Archaean supracrustal rocks. They are intimately associated with and grade into the mafic schists of these groups. Nalivkina (1977) used this fact to identify an Early Archaean ophiolite association. Small fragments and large ophiolite massifs occur within granitoids or as xenoliths in Archaean enderbites and tonalites.

Representatives of the *Nemirov granitoid complex* occur widely within the Podolsk block and the Golovanevsk suture zone as partial melt rocks — enderbite and small bodies of tonalite and plagio-migmatite, often containing garnet and hypersthene, which may have formed from the lower and upper units of the Dnestr-Bug Group.

Upper Archaean

The upper Archaean complex includes the *Rosin-Tikich Group* which crops out in the Podolsk and Belotserkov blocks, where it forms isoclinal and monoclinal structures striking north–west and approximately north–south, as well as inliers and remnants among Late Archaean (the Zvenigorod complex) and Early Proterozoic granitoids. Age relations with older rocks (the Dnestr-Bug Group) have not been established geologically. The rocks of the Dnestr-Bug Group in contact with the Rosin-Tikich Group are retrogressed. Undifferentiated formations of this group consist of amphibolite (porphyritic metadolerite) and amphibole, hornblende-biotite and biotite schists. In individual sections there are associations of gneiss and biotite, amphibole-biotite and pyroxene schists, together with magnetite quartzite. These rocks are grouped into andesite-basalt and ferruginous quartzite assemblages. The degree of metamorphism is amphibolite facies.

The *Zvenigorod complex* includes granitoids of partial melt derivation. They are widespread in the Podolsk and Belotserkov blocks, less so in the Golovanevsk suture zone. The granitoids are intimately associated with rocks of the Rosin-Tikich Group, the Dnestrovsk-Bug Group and the Nemirov granitoid complex. There is a sharp predominance of migmatites of dioritic and granodioritic composition, tonalites and biotite-hornblende-plagioclase migmatites. There is a limited development of aplitic and pegmatitic granites in veins and dykes. The Zvenigorod complex concludes the Archaean stage of evolution of the Ukrainian Shield. Its radiometric age is 2700–2600 Ma, determined by the U–Pb and K–Ar methods (Shcherbak, 1975, 1978).

The Archaean or Early Proterozoic complex

The extent and stratigraphic subdivision of this complex as a whole and its constituent groups still remains the subject of dispute. The age of the complex has been evaluated in several ways by different researchers, which is also reflected in its name. Several workers place it in the Archaean, mainly on the basis of radiometric dating.

The *Bug Group* crops out widely within the Podolsk block and the Golovanevsk suture zone. The group comprises the following formations, from bottom to top: Kamenno-Kostovat, Roshchakhov, Koshar-Aleksandrov, Khashchevat-Zavalevsk and Sinitsevsk. In the Bratsk synclinorium (Golovanevsk zone), the first two formations form isoclinal to monoclinal fold belts striking NW and dipping to the NE, or they are present as remnants in the later Kirovograd-Zhitomir Proterozoic complex. Rocks of the Koshar-Aleksandrov, Khashchevat-Zavalevsk and Sinitsevsk Formations occur in narrow linear synclinal and monoclinal structures also striking NW or as remnants in the same Kirovograd-Zhitomir granitoids. The total thickness of the Bug Group is over 6000 m.

The *Kamenno-Kostovat Formation* is divided into up to six horizons, each around 200–400 m thick, consisting of a variety of hypersthene- and garnet-bearing gneisses and hypersthene and two-pyroxene schists. There are also a few rare occurrences of garnet-biotite gneiss and schist with cordierite and hypersthene, sometimes containing magnetite. The total thickness of the formation is around 2000 m; it was metamorphosed under granulite facies conditions.

The 2000 m thick *Roshchakhov Formation* is conformable on the Kamenno-Kostovat Formation. It consists of garnet-biotite and cordierite gneiss and schist, (± sillimanite); thin bands and boudins of pyroxene gneiss and schist are also present. In the middle part of the section is fine-grained granular leucocratic gneiss with biotite and less commonly sillimanite and garnet — the so-called "leptites". Graphite has been found in the aluminous gneiss and schist of the upper part of the formation. The degree of metamorphism is from granulite to high-temperature amphibolite facies.

The *Koshar-Aleksandrov Formation* overlies enderbites of the Nemirov complex, the relationships with which have not been established. At the base of the formation is a unit of biotite and garnet gneiss. Above this are rhythmically interbanded quartzite, aluminous gneiss and mafic schist (cordierite-biotite-sillimanite, biotite-graphite, biotite-pyroxene, amphibole-pyroxene-biotite-garnet). The total thickness of the formation exceeds 2500 m. Metamorphism took place under granulite and high temperature amphibolite facies. The radiometric age of rounded zircons from quartzite of the Koshar-Aleksandrov Formation is 2800 ± 100 Ma, and for zircons from schist is 2650 ± 100 Ma (Storchak, 1984).

The *Khashchevat-Zavalevsk Formation* conformably overlies the Koshar-Aleksandrov Formation and forms the cores of common synclinal structures. The base of the formation is dominated by two-pyroxene schist interleaved with hypersthene-bearing garnet-biotite gneiss, amphibolite and pyroxene-magnetite quartzite. In this part of the section, mafic schists (komatiitic in composition; Yaroshchuk et al., 1982; Fomin and Pastukhov, 1981) are associated with layered ultrabasic bodies. In the middle part there is a predominance of garnet-biotite, sillimanite-garnet-biotite and biotite-graphite gneiss interleaved with marble and calciphyre. Graphite is ubiquitous. In the upper part of the formation, calciphyre is abundant while marble is less common. The thickness of the formation exceeds 1400 m, and metamorphism took place under hornblende-granulite subfacies conditions of the granulite facies. A lead isochron date on marble from the Khashchevat-Zavelevsk Formation gave a value of 3300–3600 Ma (Storchak, 1984).

The *Sintsevo Formation* is of limited distribution. Its position in the section of the Bug Group has not been reliably established. It consists of biotite and garnet-biotite gneiss with thin bands of pyroxene-amphibole schist, amphibolite and quartzite with sillimanite and garnet. The thickness is over 2000 m and the degree of metamorphism is amphibolite facies.

Lower Proterozoic complexes

The *Teterev Group* occurs within the Volyn and the northern part of the Podolsk blocks. The group contains the following formations, from the base upwards: Vilensk, Kocherov and Gorod.

Vilensk Formation. Relationships with underlying Archaean formations have not been established. The formation consists of biotite and garnet-biotite gneiss and schist with sillimanite, cordierite and graphite, as well as amphibole-biotite varieties; less commonly there are amphibolite, calciphyre and amphibole-pyroxene schist, mostly in the form of individual bands in the upper part of the 2000 m thick formation. Metamorphism took place at amphibolite facies.

The *Kocherov Formation* conformably overlies the Vilensk Formation. Two types of succession pass into each other laterally. In the first type, biotite and amphibole schist and orthogneiss are present, together with ortho-amphibolite with subordinate calciphyre bands. The thickness of this unit is 1800 m. The second type consists of marble and calciphyre, biotite-amphibole, amphibole-pyroxene and pyroxene paragneiss and paraschist, and para- and ortho-amphibolite. It is several thousand metres thick. Volcanics belong to the andesite-basalt series. Metamorphic conditions were at amphibolite and epidote-amphibolite facies.

The *Gorod Formation* is conformable on the Kocherov Formation. At the base of the section are bands of metaconglomerates. The formation is around 2000 m thick, and consists of biotite, two-mica and sillimanite-biotite-garnet paragneiss and schist which sometimes contain graphite. The rocks were initially flyschoidal silty sandstones. Metamorphism took place at amphibolite and epidote-amphibolite facies. The age of metamorphism of the Teterev Group rocks is 230–2100 Ma (Shcherbak, 1975).

The *Klesov Group* occurs in the north-western part of the Volyn block, where it forms small areas or xenoliths in granitoids of the Osnitsk complex. The Klesov Group itself consists of quartzofeldspathic granulite and gneiss, diabase and porphyry (apodacitic, apoandesitic and apodiabasic[1]); acid effusives sharply predominate, and there are rare occurrences of albitophyre, keratophyre and fine-grained quartz-feldspar gneiss. Volcanics on the whole form a basalt-andesite-liparite series, with an estimated thickness of 4000–5000 m. Epidote-amphibolite facies metamorphism has been superimposed on amphibolite facies.

In the stratigraphic scheme for the Ukrainian Shield under discussion, the Klesov Group includes the Novograd-Volyn assemblage, which crops out only within a volcanic structure of the same name in the southern part of the Volyn block. Rocks of this assemblage unconformably overlie rocks of the Teterev Group. The formation consists of diabase and andesite por-

[1] Apo- = metosomatically altered rock, without destruction of original texture.

phyry with thin intercalations of tuffaceous schist and quartzofeldspathic granulite at the base (1500 m thick), while the upper part consists of a unit of rhythmically interleaved metasiltstone and metasandstone (100 m thick). The volcanics of the formation belong to the basalt-andesite series. The degree of metamorphism is epidote-amphibolite facies. The time of formation of acid effusives of the Klesov Group has been evaluated at 2100 Ma on zircons by the U–Pb method (Shcherbak et al., 1978).

The *Pugachev Group* is developed locally within the Volyn block, where it forms a synclinal structure on the Proterozoic Kirovograd-Zhitomir granites and also occurs as xenoliths in rocks of the Korosten pluton. The group has two formations, the Belokorovich (below) and the Ozeryansk.

The *Belokorovich Formation* lies with angular stratigraphic unconformity on rocks of the Teterev Group and on granites of the Kirovograd-Zhitomir complex. It is 400 m thick and consists of two members, the lower of which has rhythmically alternating sandstones, quartz-sericite schists, conglomerates, shales and siltstones; there are also small quantities of tuffaceous material and diabase flows. The upper member includes quartzose and polymict sandstones with thin gravelly conglomerates. Metamorphism occurred at greenschist facies.

The *Ozeryansk Formation* is conformable on the Belokorovich Formation in a synclinal structure. In the lower part there are thin flows of diabase porphyry and higher up are siltstone, shale and quartz-sericite schist. The formation is up to 700 m thick, and metamorphism was at greenschist facies.

Within the boundaries of the entire Ukrainian Shield, there are several Early Proterozoic rock associations — intrusive and ultrametamorphic. The oldest is a complex of basic and ultrabasic rocks and within the Volyn-Podolsk province there is a basic to intermediate intrusive complex and the Berdichev ultrametamorphic granitoid complex.

Basic and ultrabasic rocks form small bodies and intrusive massifs, folded together with lower Proterozoic rocks. They are lensoid and sheet-like bodies of gabbro-diabase, gabbro-norite and gabbro and undifferentiated ultrabasic rocks.

The *Buki intrusive complex* cuts rocks of the Teterev Group in the extreme north of the Podolsk block and forms isometric, lenticular and irregular zoned bodies. It consists of a single genetic series of gabbro-monzonite-diorite-quartz diorite-granodiorite containing pyroxene-, pyroxene-amphibole- and biotite-bearing varieties. The rocks formed during the interval 2350–2100 Ma according to radiometric data (Shcherbak, 1978).

The *Berdichev complex* is widely distributed within the Volyn and Podolsk blocks, also in the Golovanevsk suture zone, and formed by partial melting ("ultrametamorphism") during the granitization of mainly Archaean Nemirov complex rocks and the inliers of rocks belonging to the Dnestr-Bug Group preserved in the Nemirov complex. The dominant rock types are the Chudno-Berdichev garnet- and biotite-bearing granite and migmatite, and granite and migmatite containing garnet, biotite and hypersthene — garnetiferous migmatite and charnockite. Younger members of the complex include leucocratic granite with blue quartz, granite and migmatite with apatite and magnetite, aplitic and aplo-pegmatitic granite. The isotopic age of the Chudno-Berdichev granites and charnockites (U–Th–Pb method on monazites and zircons) lies in the interval 2380–2150 Ma (Shcherbak, 1978, 1975). The latest members of the complex — aplo-pegmatitic granites — give an isotopic age of 2200–2000 Ma.

The *Kirovograd-Zhitomir granitoid complex* occurs most widely in the Volyn and Belotserkov blocks. It formed as a result of granitization of lower Proterozoic assemblages and by reworking of Archaean formations. Autochthonous and parautochthonous varieties predominate, these being granodiorite, diorite and monzonite; and tonalite and plagio-migmatite of granodioritic and dioritic composition. Relatively younger rocks include a varied group of granitoids and migmatites which were previously called by local names (Zhitomir, Kirovograd, Umansk, Novoukrainsk, etc.). They are biotite and two-mica granites, porphyroblastic with biotite and amphibole, and contain plagioclase + microcline or microcline. Allochthonous members are represented by microcline and plagioclase-microcline granites, forming small intrusions. Other varieties include subalkaline albite-microcline granites in small zoned intrusions — coarsely porphyritic at the centre and fine- to medium-grained at the edges. Metasomatic formations are associated with the Kirovograd-Zhitomir complex, localized along tectonic zones and represented by microclinite, albitite and albite-microcline rocks.

The formation of the plagio-microcline and essentially microcline granites (of the Kirovograd-Zhitomir complex) has been dated at 2100–1800 Ma by K–Ar and U–Pb methods; the most intense ultrametamorphic granitization occurred in the interval 2000–1900 Ma (Shcherbak, 1975).

The *Osnitsk complex* is exposed solely in the extreme north-west of the shield, in the Volyn block. It formed as the result of granitization of a complex of basic and ultrabasic rocks (amphibole and biotite-amphibole diorite and granodiorite, giving intrusions of irregular shape) and granitization of quartzofeldspathic granulites of the lower Proterozoic Klesov Group (biotite and biotite-amphibole prophyritic granite and biotite granite). U–Pb dates on zircons give an age of 2000–1800 Ma for the formation of the Osnitsk complex (Fomin and Pastukhov, 1981).

Dyke complex. Dykes are common throughout the entire shield, forming east–west, north–south and north–west swarms and include quartz porphyry, diabase, gabbro-diabase, gabbro, gabbro-norite, pyroxenite and peridotite and they cut the Krivoy Rog and Konka-Verkhovtsev Groups and the Dnepropetrovsk and Sursk-Tokov granitoid complexes.

The youngest Early Proterozoic association consists of the Korosten granitoid complex and a dyke swarm. The rocks of the complexes are younger than formations of the Klesov and Pugachev Groups. The *Korosten complex* is found in the Volyn and Podolsk blocks (the Korosten pluton), as well as the Kirovograd block (the Korsun-Novomirgorod pluton). There are active intrusive contacts with the country rocks. In the outer contact zone there are skarns and hornfelses and in the case of the Korosten granite there is also evidence of alkali-silica metasomatism. Three groups of rocks make up the complex. The oldest contains gabbro and gabbro-norite, anorthosite and gabbro-anorthosite. The second group crops out at the contact between basic rocks and granites, and consists of hybrid rocks — gabbro-monzonite, monzonite, quartz monzonite and gabbro-syenite. The youngest group consists of rapakivi granite and similar rocks, subalkaline granite, granite porphyry, pegmatite and aplite. The Korosten pluton has a circular shape and complex structure. The basic rocks form several layered intrusions. Rapakivi and similar granitoids separate and intrude basic intrusions. The Korsun-Novomirgorod pluton has an elongate shape, with predominant rapakivi granite. Basic rocks form masses at granite margins, but the age of the basic rocks in the complex has not been established. Age determinations for the granites, including rapakivi, using a variety of methods are in agreement at 1750 ± 100 Ma (Shcherbak, 1978; Shcherbak et al., 1981).

Dyke complex. The history of formation of the intrusions of the Korosten and East Azov complexes concludes with the emplacement of a swarm of lamprophyre, quartz porphyry, diabase, gabbro-diabase and diabase and andesite porphyrite dykes.

Late Proterozoic complexes

Ovruch Group. This is present in the Volyn block, in one structure, the so-called Ovruch graben-syncline. There are two formations, the Zbrankov and Tolkachev. The *Zbrankov Formation* occurs at the base on an eroded surface of Korosten granite. Towards the base there are amygdaloidal diabase and quartz porphyry with thin conglomerate layers, quartzose sandstone and phyllitic schist. Higher up there are trachyandesite, porphyrite and dolerite lava flows with thin fine-grained conglomerate and quartzose sandstone containing volcanogenic intercalations. The thickness is less than 300 m. Metamorphism was at greenschist facies. The association is trachybasalt-trachyandesite-trachyliparite. The age of the effusives of the Zbrankov Formation of the Ovruch Group lies in the interval 11500–1000 Ma (K–Ar and U–Pb methods; Shcherbak et al., 1981; Shcherbak, 1978).

The *Tolkachev assemblage* is unconformable on the Zbrankov Formation and consists of a monotonous unit of red quartzose and polymict sandstone with rare conglomerate horizons and pyrophyllite schist at the base. The thickness is up to 900 m. Metamorphism was at greenschist facies.

Perzhan metasomatites. These crop out in the Volyn block, mostly in the Sushchan-Perzhan tectonic zone. The biotite and two-mica granite, grano-syenite, syenite, quartzite and quartz-sericite schist form small separate masses or trails of lensoid bodies. Siliceous-alkaline metasomatism played an important role in the formation of these rocks. Metasomatic rocks of this type are described in particular for quartz porphyries of the Zbrankov Formation (Ovruch Group). Metasomatic processes in the Sushchan-Perzhan tectonic zone have been dated at 1300–1200 Ma by the U–Pb method on zircons (Shcherbak, 1978).

Dyke complex. This includes rocks developed at the edges of the shield in the Volyn block. The diabase, gabbro-diabase, trachyandesite, quartz porphyry and granite porphyry dykes cut the Korosten complex, the Ovruch Group and the Perzhan metasomatites.

Metamorphism

The West Ukrainian region is characterised by the presence of rocks which reflect wide variations in *P–T* metamorphic conditions. In the SW part of the region, in the outcrop area of granulite facies rocks, the rocks are polymetamorphic in nature, based on coexisting minerals in two-pyroxene and hypersthene schists which in places are amphibolized.

In the southern part (the Podolsk block) and the Golovanevsk suture zone, there are high-temperature granulite facies rocks, showing indications of eclogitization. Metabasic rocks have the assemblage garnet-two-pyroxene; in association with sillimanite the garnet has a low iron content of c. 55% (780°C, 7–10 kbar). As a result of eclogitization of metabasites, clinopyroxene + plagioclase symplectite coronas have formed; this process took place at 900°C and a total pressure greater than 6–7 kbar (Shcherbakov, 1975). In the southern part of the Golovanevsk zone, similar granulite facies conditions prevailed.

In the Belotserkov block, granulite facies rocks are intensely granitized and retrogressed at amphibolite facies. A characteristic feature in the rocks of the Belotserkov block is the inequilibrium mineral assemblage: garnet–biotite–sillimanite–andalusite–cordierite–staurolite–muscovite–plagioclase–quartz, defining prograde metamorphism from epidote-amphibolite to amphibolite facies (staurolite is included in cordierite and garnet). Corresponding temperatures are 640°C from the biotite-garnet thermometer and 600°C from the staurolite-garnet. In the north of the region (the Volyn block) the lower Proterozoic rocks of the Teterev Group belong to the low temperature part of the amphibolite facies (T = 610–660°C, total pressure = 3–5 kbar from assemblages containing biotite, garnet, cordierite and sillimanite in metapelites, from the almandine to pyrope ratio in garnets, using various geothermometers). The upper part of this group was metamorphosed at epidote-amphibolite facies conditions (550–620°C in metapelites containing biotite-muscovite-quartz assemblages with relicts of metasiltstone and metapsammite textures). In the outcrop area of the Chudno-Berdichev granitoids and migmatites, metamorphism corresponds to the high temperature part of the amphibolite facies (in granites we have found a paragenesis consisting of high-Fe garnet, f = 82%, cordierite and sillimanite). The epidote-amphibolite facies — which has the association biotite + muscovite in metapelites, and relicts of primary textures in metavolcanics — is restricted to the outcrop area of the Novograd-Volyn assemblage.

In the rocks of the Ovruch Group, we have observed the parageneses chlorite + sericite and chlorite + sericite + albite (the low-temperature part of the greenschist facies); rocks of the Pugachev Group have been metamorphosed in a similar fashion. Rocks of the Klesov Group were metamorphosed and granitized at high temperature amphibolite facies conditions (640–670°C, pressure = 3.5–4.0 kbar).

2. CENTRAL UKRAINIAN PROVINCE

This region includes the Kirovgrad and Dnieper blocks and the West Ingulets junction zone.

Lower Archaean complexes

Aul Group. This crops out mainly in the Dnieper block, in the form of small inliers among Archaean ultrametamorphic granitoids of the Dnepropetrovsk and Sursk-Tokov complexes. Two rock units emerge on examination of partial sections. An older consists of pyroxene and amphibole schist, amphibolite and minor bands of calciphyre and quartzo-feldspathic granulite; the upper unit is made of biotite, garnet-biotite and amphibole gneiss, schist and amphibolite and magnetite-amphibole-garnet quartzite. The lower unit contains layered bodies of metamorphosed gabbro and ultrabasics. The total thickness of the group is 600 m. Metamorphism was at granulite and high temperature amphibolite facies. In the basic and ultrabasic rocks associated with the Aul Group (in the contact zone between the Dnieper and West Azov blocks), accessory zircons from ultrabasics in the Novo-Pavlovsk area have been studied, which are morphologically and geochemically close to zircons in tonalites that include ultrabasic relicts. Isotopic dates obtained for these zircons (Shcherbak, 1984; Shcherbak et al., 1984) are 3700 ± 200 Ma, and an older U–Pb isochron age is 3810 Ma, which defines the minimum age of the oldest ultrabasic-tonalite complex in the shield (Fig. I-48).

Dnepropetrovsk complex. This crops out over the major part of the Dnieper block, where there are dome-shaped structures, restricted to

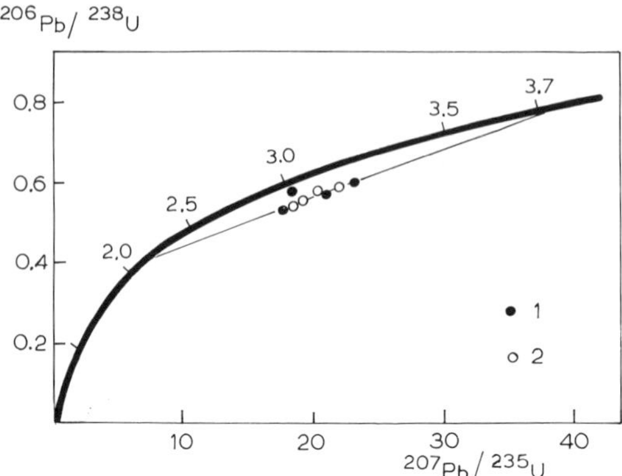

Fig. I-48. Concordia diagram for accessory zircons from meta-ultramafics in the West Azov region (Novo-Pavlovsk sector) (after Shcherbak and Bibikova, 1984). *1, 2* = sample numbers.

the area where Aul Group rocks occur, and also found in the West Ingulets junction zone. The complex consists of several varieties of granite. Associated with the lower parts of the Aul Group are diorite and quartz diorite, granodiorite, migmatite and biotite-hornblende tonalite, sometimes with pyroxene; there is a characteristic relationship with rocks of basic and ultrabasic composition. In the main, tonalite and plagio-migmatite occur in the upper unit of the Aul Group. Less commonly there are porphyroblastic tonalites, forming small intrusions in the western part of the Dnieper block. The radiometric age of the tonalite-granodioritic rocks of the Dnepropetrovsk complex has been determined at 2970 ± 20 Ma on accessory zircons by the U–Pb method (Shcherbak, 1984) and 3090–3160 Ma (Shcherbak, 1983).

Upper Archaean complex

Supracrustal rocks of this complex make up the Konka-Verkhovtsev Group. It comprises structures, regarded as greenstone belts, which are mainly distributed within the Dnieper block. Here the group takes the shape of major synclines and monoclines, and it occurs as inliers among granitoids in the West Ingulets zone. The group is subdivided into three formations: Konka (bottom), Belozero and Teplov (top).

The *Konka Formation* occurs at the base of the section in all synclinal structures in the Dnieper region. In several more complete sections the formation is divided into three members. There are instances where Konka amphibolites occur on a metamorphosed weathering crust of Aul Group gneisses; more often, though, there is a tectonic contact between the two. The lowest member consists of amphibolite with less common metadiabase and metaspilite and some insignificant ferruginous quartzite and schist;

the thickness is of the order of 3000 m. The middle member contains garnet-biotite, biotite and garnet-amphibole schist as well as bands of amphibolite and ferruginous quartzite, with a thickness not exceeding 700 m. In the top member, metamorphosed intermediate and acid effusive volcanic rocks predominate — metaporphyrite and metakeratophyre, in places amphibolite and metadiabase. Quartz-sericite and biotite-chlorite schist together with amphibole-magnetite quartzite account for a small percentage of the approximately 3000 m of this member. Some workers recognise komatiites and tholeiitic basalts as being members of the Konka Formation (Sivoronov et al., 1981; Fomin and Pastukhov, 1981). Rocks of the Konka Formation were metamorphosed at greenschist and epidote amphibolite facies. The rock associations are komatiite-tholeiite, jaspilite-tholeiite and dacite-andesite-tholeiite.

The *Belozero Formation* overlies various horizons of the Konka Formation, with an unconformity and structural break. It is divided into two members. The lower, terrigenous, unit consists of up to 2000 m of metamorphosed sandstone, siltstone, conglomerate, micaceous quartzite and quartz-chlorite-sericite schist with bands of metakeratophyre and metadiabase. In places, metavolcanic effusive rocks play an important part. The member is up to 2000 m thick. The upper terrigenous-chemogenic member consists of 900 m of ferruginous quartzite, quartz-biotite and quartz-chlorite schist and metasandstone. In a number of places there are alternations of schist and metavolcanics of various composition. The total thickness of the formation is up to 2.5 km. Metamorphism was almost everywhere at greenschist facies. Rock associations are conglomerate-schist, jaspilite-chert-schist and andesite-dacite-schist.

The *Teplov Formation* is recognised in axial regions of major synclines. It consists essentially of volcanogenic rocks — metamorphosed andesitic and doleritic porphyrite, dolerite, keratophyre, quartz-amphibole and quartz-chlorite-sericite schist and ferruginous quartzite. The formation is up to 800 m thick. Metamorphism was at greenschist and epidote-amphibolite facies conditions. The rock associations are jaspilite-chert schist and andesite-basalt.

Primary igneous zircon from metamorphosed volcanics (andesites) of the upper part of the Konka Formation (Konka-Verkhovtsev Group) has an age of 3250 ± 120 Ma from different zircon fractions (U–Pb method), which is confirmed by a Sm–Nd model age of 3300–3240 Ma (on minerals from these volcanics; Shcherbak et al., 1981, 1982; Shcherbak, 1984, 1978).

The *Sursk-Tokov granitoid complex* crops out mainly in the Dnieper block. This autochthonous group of granitoids forms a wide area of dome-like structures, surrounded by rocks of the Konka-Verkhovtsev Group. They consist of biotite and hornblende-biotite tonalite and plagio-migmatite. Allochthonous granitoids form individual intrusions, which forcefully intruded rocks of the Konka-Verkhovtsev Group, in the outer contact of which there is evidence of microclinization and greisenization. Allochthonous granitoids are represented by biotite and hornblende-biotite tonalite, microcline and microcline-plagioclase granite. There are limited occurrences of granite porphyry, in the form of small bodies and stocks. Aplite-pegmatite granites are common as dykes and minor intrusions.

Biotite and amphibole tonalite of the Sursk-Tokov complex have been dated by the uranium–lead method on zircons, and yield a value of 3090–2830 Ma (Shcherbak et al., 1983).

Dyke complex. This is well-developed in the Dnieper block, where it consists of diabase, amphibolized diabase, gabbro-diabase, gabbro and gabbro-norite.

Lower Proterozoic complex

Ingul-Ingulets Group. This occurs in the border zones around domes within the Kirovograd block (around the Novoukrainsk and Korsun-Novomirgorod granitoid intrusions) and the West Ingulets junction zone. The Proterozoic age of this group is determined from the fact that it overlies a metamorphosed weathering crust of Archaean granitoids. The group is divided into five formations. From bottom to top these are:

Zelenorech Formation. A unit of barren quartzite lies at the base of the section immediately above Archaean granitoids. Above there are lenses and sheets of biotite, two-mica and amphibole-biotite gneiss, sometimes with sillimanite, and ortho-amphibolite. The average thickness is 700–800 m. Metamorphism was at amphibolite facies. The initial rock association was siliceous-schist.

The *Artemov Formation* forms broad structures with the Zelenorech Formation, on which it is conformable. It consists of biotite-sillimanite, garnet-biotite and amphibole gneiss and schist, magnetite quartzite and skarnoids. The thickness is up to 200 m. Metamorphism was at amphibolite facies. The primary rock formations were ferruginous chert and carbonate shale.

The *Rodionov Formation* is conformable on the Artemov Formation, or else it lies with angular stratigraphic discordance on a metamorphosed weathering crust of Archaean granitoids (undifferentiated Dnepropetrovsk and Sursk-Tokov complexes). It is more widely distributed than the previous two formations. At the base occur quartzites and metamorphosed sandstone with thin layers of graphite-mica schist. Higher are graphite-biotite schist, sometimes with actinolite, calciphyre and metasandstone. The maximum thickness is 2000 m. The degree of metamorphism is epidote-amphibolite facies; in the northern part of the region, where it was amphibolite facies, we find biotite gneiss with graphite and quartzite with minor bands of marble and calciphyre. The primary sedimentary association was carbonate-terrigenous.

The *Spasov Formation* is widely developed in the West Ingulets junction zone and the Kirovograd block, where it forms the cores of minor anticlinal folds surrounding the Korsun-Novomirgorod granitoid pluton. It is conformable on the Rodionov Formation, sometimes with gradual transitions. The formation consists of two-pyroxene, hypersthene-biotite (± magnetite) and pyroxene gneiss and schist and quartzite; in the southern part of the region there are biotite gneiss and schist, sometimes with garnet and graphite, and actinolite-biotite schist. The thickness reaches 3000 m. The degree of metamorphism is granulite and amphibolite facies.

The *Chechelev Formation* occurs widely within the margins of the Kirovograd block. It is conformable on the Spasov Formation, and consists of a monotonous assemblage of biotite and garnet-biotite (± sillimanite) aluminous gneiss, cordierite-bearing at the base of the section, and frequently containing thin bands and lenses of clinopyroxene and amphibole-clinopyroxene rocks. The thickness of the formation is 3500–5000 m. The degree of metamorphism is amphibolite facies.

The age of the polyphase metamorphism of the Ingul-Ingulets Group (granulite, amphibolite and epidote-amphibolite facies) has been determined to lie in the interval 2400–2100 Ma, while ultrametamorphic granitization with the formation of granites of the Kirovograd-Zhitomir type, is 2100–1800 Ma, by the U–Pb method (Shcherbak et al., 1983).

The *Krivoy Rog Group* is the stratotype for lower Proterozoic formations of the Ukrainian Shield. It crops out in the western part of the Dnieper block. An Early Proterozoic age for the group has been determined from the fact that it lies unconformably on a metamorphosed weathering crust of Archaean granitoids. In the discontinuous outcrop area (Krivoy Rog–Kremenchug), rocks of the group form a series of synclines and monoclines. It is divided into five formations (from the base upwards): Novokrivoyrog, Skelevat, Saksagan, Gdantsev and Gleyevat; some of these are subdivided into members, and the Saksagan Formation is also divided into horizons.

The *Novokrivoyrog Formation* lies at the base of the Krivoy Rog Group on a metamorphosed crust of weathered Archaean granitoids. It is divided into two members. Volcanics predominate in the lower member — amphibolite, metadiabase, amphibole and biotite schist; there are also thin layers of conglomerate, metasandstone and white quartzite. The upper member contains interbanded amphibolite with chlorite, biotite-chlorite and amphibole-chlorite tuffaceous schist, biotite schist and metamorphosed porphyritic dolerite. The thickness of the most complete sequence is up to 1200 m. Metamorphism was at greenschist and epidote-amphibolite facies. The primary rock formation was andesite-basalt.

The *Skelevat Formation* is unconformable on the Novokrivoyrog Formation or on Archaean granitoids. The lower member consists of quartzose and arkosic metasandstone, metaconglomerate and bands of garnet-biotite schist. The middle phyllite member consists of sericite, chlorite-sericite and two-mica (± staurolite) schist, in places with significant amounts of graphite, and quartzite. The upper member consists of amphibole-chlorite-serpentine-talc and carbonate-talc schist (altered ultrabasic lava), metasandstone and metaconglomerate. The thickness of each member varies from 20 m to 160–180 m. Metamorphism was at greenschist facies. The primary rock associations were siliceous shale and conglomerate with sandstone.

The *Saksagan Formation* (1200–1500 m thick) is conformable on the Skelevat Formation. It is the main ore-bearing assemblage in the Krivoy Rog–Kremenchug iron ore basin. The formation contains up to seven horizons of ferruginous quartzite, separated by schist, in three members. The schist horizons consist of biotite- and sericite-chlorite, sericite, amphibole-chlorite, graphite-sericite and carbonate schist. The ferruginous horizons contain goethite-hematite-martite, magnetite-martite-carbonate and amphibole-magnetite rocks. Metamorphism was at greenschist facies. The primary rock association was chert-jaspilite.

The 1600 m thick *Gdantsev Formation* lies with a structural and stratigraphic break on rocks of the Saksagan Formation, on which a metamorphosed weathering crust has been found at a number of localities. Typical of the formation are lithological variations and a sharp change in facies along strike. In the lower part of the section there are breccia, metasandstone, quartzite, chlorite and magnetite-chlorite schist, carbonate and iron ores; in the upper part are carbonate-biotite and sericite-graphite schist and dolomitic marble. Metamorphism was at greenschist facies. The primary association was chert-carbonate-sandstone.

The *Gleyevat Formation* (3500 m thick) crops out in the axial region of the Krivoy Rog structure. It lies unconformably on Gdantsev Formation rocks. In the lower member are weakly metamorphosed conglomerate with clastic material from underlying formations, sandstone, siltstone, garnet-biotite and hornblende-biotite schist. Metamorphism was at epidote-amphibolite facies. The primary association was conglomerate + sandstone.

The age of clastic zircons and monazites from metasandstones and metaconglomerates of the Skelevat Formation of the Krivoy Rog Group is 2800 ± 100 Ma (Shcherbak, 1984); material was derived from eroded granites of the Dnepropetrovsk complex. The upper age boundary of the Krivoy Rog Group is defined by the age of cross-cutting granites of the Kirovograd-Zhitomir complex (1980 ± 40 Ma) and the age of superimposed soda metasomatism, 1850 ± 50 Ma (by the uranium–lead method). An age of 2500 ± 100 Ma was determined for the lower part of the section of the Krivoy Rog Group, by the U–Pb isochron method on authigenic concentrations of uranium (Tugarinov et al., 1963).

Metamorphism

In the Kirovograd block, there are outcrops of granulite facies rocks within the Ingul-Ingulets Group with a temperature of formation of 680–710°C and high temperature amphibolite facies rocks (630–650°). Farther north, amphibolite facies mineral assemblages occur in metapelites, with garnet, biotite, cordierite and sillimanite, and minor amphibolites. In the most northerly part of the block, near the Golovanevsk zone, ferruginous garnet in metapelite is characterised by a high Mn content (7–10% on average, sometimes up to 29% spessartine), which may indicate low pressure conditions or some specific rock composition. In the central and eastern parts of the block, there are assemblages with granulite facies metamorphism containing hypersthene-orthoclase and two-pyroxene parageneses in association with garnet, cordierite and biotite; the temperature of formation corresponds to the low temperature part of the granulite facies (650–680°). In the West Ingulets junction zone we can differentiate between high-temperature amphibolite facies (metamorphism of the Archaean basement) and low-temperature amphibolite and epidote-amphibolite facies (lower Proterozoic unit). In the eastern part of the Kirovograd block there is lateral metamorphic zonation (R. Belevtsev, 1978) from andalusite-staurolite to sillimanite-muscovite subfacies, and also from low-temperature and high-temperature parts of the amphibolite facies and further to the granulite facies (in the upper part of the lower Proterozoic Ingul-Ingulets Group). The total pressure during the metamorphism of the rocks in the Kirovograd block and the West Ingulets zone was estimated at 5–6 kbar (Usenko et al., 1980). In the Dnieper block, Early Archaean assemblages were metamorphosed and granitized under granulite and high-temperature amphibolite facies conditions (pyroxene schist, tonalite, biotite and hornblende-biotite gneiss and amphibolite). There was widespread retrogression of these rocks at epidote-amphibolite facies conditions (actinolite + epidote assemblages in altered basic rocks).

In the outcrop areas of the upper Archaean Konka-Verkhovtsev Group, the majority of synclines have rocks metamorphosed at greenschist facies — biotite and almandine-chlorite subfacies. At individual exposures we encounter rocks at epidote-amphibolite facies metamorphism — the staurolite subfacies (blue-green hornblende in metabasics, ferruginous quartzite with epidote, actinolite and cummingtonite).

In the Krivoy Rog–Kremenchug structural belt, Early Proterozoic zonal metamorphism (Usenko et al., 1982) in the central part corresponds to the almandine-chlorite subfacies of the greenschist facies (with chlorite, chloritoid, biotite and almandine). Along strike, from the central part of the structure, the greenschist facies develops progressively, and westwards across strike there is an increase in metamorphic grade, with a transition to the epidote-amphibolite facies (muscovite-almandine-staurolite subfacies, and — in the western and southern parts of the structure — sillimanite-muscovite subfacies). Zoning is isobaric, with a total pressure of 4–5 kbar. Metamorphic temperatures, determined from comparing experimental data with geothermometry using the composition of coexisting minerals, are in the following order: garnet zone 450–570°C, staurolite zone 480–580°C and sillimanite-muscovite zone 530–630°C (Storchak, 1984).

3. THE AZOV ARCHAEAN GRANULITE–GREENSTONE TERRAIN

The region includes the West and East Azov blocks and the Central Azov junction zone, and also includes the Orekhov-Pavlovgrad suture zone which separates the Azov and Dnieper regions. Such a division is supported by geophysical data. From magnetic characteristics, the Azov region is subdivided into two parts. In the West Azov part there is a regional positive magnetic anomaly, related to ancient granulites, which can be traced into the Orekhov-Pavlovgrad zone. For the East Azov part, there is a characteristic negative magnetic anomaly (the Azov and East Azov granitoid and alkali complexes). The entire Azov region as a whole and the Orekhov-Pavlovgrad suture zone display fairly high magnetic values, due to the presence of Archaean metamorphosed volcanosedimentary rocks.

According to the latest geochronological data, it is mainly Archaean (lower and upper) terrains which are exposed in the Azov province. The status of the Central Azov Group and the Osipenko and Gulyaypol Formations remains debatable, as yet unresolved by field data.

Lower Archaean terrain

West Azov Group. This crops out in the West Azov block as broad regions or small inliers among granitoids. The composition and succession of the group have not been well studied. In the most complete sections, the lower part consists of 3500 m of two-pyroxene and pyroxene schist, garnet-biotite gneiss and amphibolite; the upper 2500 m thick unit is composed of biotite, garnet-biotite and amphibole gneiss, amphibolite and ferruginous quartzite. Metamorphism was at granulite and high-temperature amphibolite facies.

Upper Archaean–lower Proterozoic terrain

Central Azov Group. This occurs widely in the West Azov block and the Central Azov junction zone. There are also rarer undifferentiated exposures in the Orekhov-Pavlovgrad suture zone and the East Azov block. The group is divided into three formations. From below upwards these are:

The *Temryuk Formation* lies with angular unconformity on rocks of the West Azov Group. At the base of the formation there are amphibole-biotite,

biotite and garnet-biotite gneiss with sillimanite, sometimes also corundum and spinel, graphitic gneiss, garnetiferous and micaceous quartzite. Above we find interbanded biotite and garnet-biotite gneiss with marble and calciphyre, which are replaced along strike by amphibole and pyroxene gneiss and schist and amphibolite. The formation is 1700–2100 m thick and metamorphism was at granulite and high temperature amphibolite facies.

The *Sachkin Formation* overlies rocks of the Temryuk Formation above a stratigraphic break in short, broad synclines which contain the upper parts of the succession. The main iron ore deposits of the Azov region belong to this formation. Four rock units are recognised (from the bottom upwards): (1) biotite and garnet-biotite gneiss, sometimes with sillimanite and graphite; calciphyre and barren quartzite are present as thin bands towards the top; (2) amphibole and pyroxene gneiss and schist, pyroxene-magnetite quartzite and solitary bands of aluminous gneiss; (3) pyroxene-amphibole and amphibole gneiss and schist with subordinate biotite and garnet-biotite gneiss with graphite and pyroxene-magnetite quartzite; and (4) biotite and garnet-biotite gneiss, sometimes with graphite, marble and calciphyre and thin bands of amphibole-biotite gneiss. There are gradual transitions between these rock units. Metamorphism was at amphibolite facies.

The *Karatysh Formation*, which is of limited distribution, occurs in minor short, broad anticlines and as inliers amongst granitoids. It consists of 4500 m of alternating biotite, amphibole-biotite and amphibole gneiss; also present are patches of garnetiferous and graphitic gneiss, amphibole-pyroxene gneiss and amphibolite, with sillimanite gneiss observed at the top of the section. Metamorphism was at amphibolite facies.

The *Gulyaypol Formation* crops out in a basin structure of the same name in the West Azov block. It lies above a metamorphosed weathering crust of Early Archaean tonalites. At the base of the group there is a patch of metamorphosed sandstone and conglomerate and sericitic and andalusite-muscovite schist. Higher up there are silicate-magnetite quartzite with thin bands of garnet-amphibole and amphibole-biotite schist and barren quartzite. At the very top of the section there are biotite and sericite-biotite schist, sometimes with garnet, containing carbonaceous material. Throughout the 1000 m thick formation, there are metamorphosed volcanics and volcanoclastics of acid and intermediate composition (fine-grained quartzo-feldspathic granulite, so-called "hälleflintas" and porphyroids). The volcanics belong to the andesite-leptite series. Metamorphism was at epidote-amphibolite and amphibolite facies.

The *Osipenko Formation* crops out in two grabens in the West Azov block. Three units are present, aluminous, carbonate and graphitic (from bottom to top). In the lowest unit there are two-mica, garnet-, staurolite- and sillimanite-biotite schist with rare thin bands of muscovite quartzite at the base; higher up we have hornblende schist and amphibolite with thin garnet-biotite schist and hornblende-magnetite quartzite and at the top two-mica, garnet- and sillimanite-biotite schist with ubiquitous graphite and thin solitary bands of micaceous quartzite and calciphyre. In the middle unit there are interbanded marble and calciphyre with thin hornblende and biotite schist. The transition between the lower and middle units is gradual. The top unit contains interbanded mica-, sillimanite- and garnet-biotite and hornblende-biotite schist; graphite is present in all varieties of aluminous rocks. A break has been found between the middle and upper units. The formation is 1000 m thick on average. Metamorphism was at epidote-amphibolite and amphibolite facies. We have obtained data indicating an Archaean age for the Osipenko Formation. Granodiorite in cross-cutting contact with these rocks gave a radiometric age of 2790 Ma, while clastic(?) zircon from the biotite gneiss yielded 3260 $\pm$ 10 Ma (U–Pb isochron method on zircons; Artemenko et al., 1986).

Lower Proterozoic terrain

The *Azov granitoid complex* is compared with the Kirovograd-Zhitomir complex, although there are no radiometric age data. The complex is widely distributed within the entire Azov province. It contains formations recognised as autochthonous and parautochthonous, derived from rocks of the West Azov and Central Azov Groups. These are diorite, granodiorite, tonalite and migmatite, containing biotite, amphibole + biotite and pyroxene. Minor intrusions contain aplo-pegmatitic granite, pegmatite, aplite, biotite-orthite and sphene granite (Saltychan) and leucocratic monazite-bearing granite (Anadol).

Chernigov alkaline ultrabasic complex. This is found in the West Azov block, in a fault zone of the same name, among high grade metamorphic rocks of the Archaean West Azov Group. Component rocks are pyroxenite, phlogopitic pyroxenite, micaceous peridotite; nepheline syenite, alkali syenite; fennite and carbonatite. Retrogressive effects are widespread: albitization, biotitization, amphibolization, etc.

The only available radiometric age for rocks of the Chernigov complex was made by the K–Ar method, which yielded a figure within the interval 2000–1850 Ma (Shcherbak, 1978; Shcherbak et al., 1981).

The *East Azov syeno-granite complex* is developed mainly in the East Azov block and the Central Azov junction zone as a number of minor intrusions. Three age groups of rocks have been identified. The first are olivine gabbro-peridotite, olivinite, pyroxenite and isolated blocks at the margins of the Oktyabr (October) nepheline syenite intrusion. The second group is of hybrid rocks — monzonite and gabbro-monzonite. The third and dominant group comprises nepheline syenite, foyaite, mariupolite (albite-nepheline syenite), syeno-granite and subalkaline granite. The most typical intrusion in the Oktyabr massif has a zoned structure; the central part consists of alkaline and nepheline syenite, with syenite and syenogranite at the margins among which are blocks of basic and ultrabasic rocks. Intrusions in the East Azov complex are fault intrusions, discordant to the country rocks. According to radiometric data, rocks of only the first two stages of formation of the East Azov complex (syenite and subalkaline granite; alkaline syenite and syenite-pegmatite) have an age of 1950–1740 Ma (Yeliseyev et al., 1965; Shcherbak et al., 1981), comparable to the age of granites in the Korosten complex.

The *Kamennomogila complex* includes intrusive granitoids, forming a suite of isometric stocks in the Central Azov junction zone which occur at the intersections of variously oriented tectonic zones. The stocks consist of biotite and two-mica granite which have been affected to varying extents by greisenization and albitization. The presence of silica-alkali metasomatism in the concluding phase of formation of the intrusions makes these bodies similar to the Perzhan metasomatites.

Metamorphism

In the Orekhovo-Pavlograd suture zone, the westernmost part of the Azov region, the great majority of rocks are at amphibolite facies, with isolated relicts of granulite facies rocks. In the northern part of the zone, zonal metamorphism has been mapped out, with an increase in metamorphic grade from west to east, from sillimanite-biotite-orthoclase to sillimanite-garnet-cordierite-orthoclase subfacies of the granulite facies. The garnet-biotite equilibration temperature is 630–670°C, and pressure is 4.8–5.2 kbar. At higher temperatures, cordierite and sillimanite develop at the expense of biotite. For iron-rich rocks in association with metapelites at amphibolite facies, temperatures of 750°C (two-pyroxene geothermometer) and 565°C (garnet-orthopyroxene) are obtained. In some places, observations have been made of a superimposed greenschist facies metamorphic event on amphibolite facies rocks, as well as amphibolization of two-pyroxene ferruginous rocks, linked with granitization.

Metamorphic rocks in the West Azov block, where they form inliers among granitoids and migmatites, display features of amphibolite facies and less commonly granulite facies. The metamorphic assemblage seen in aluminous rocks is garnet (31% pyrope end-member), orthopyroxene (Al_2O_3 = 5.89%) and gedrite, characterizing high-pressure granulite facies conditions. Most of the pelites have been metamorphosed at amphibolite facies conditions, in the sillimanite-muscovite subfacies, with T = 600–640°C and P = 4.6–4.8 kbar (using the garnet-biotite geothermometer). In the eastern part of the block, metamorphism at hornblende-granulite subfacies proceeded at a temperature of 650–670°C (from mineral equilibria in two-pyroxene schists with hornblende). Granitization of these rocks at granulite facies conditions led to the formation of charnockites and enderbites. In metapelites of this same facies with garnet, sillimanite, cordierite and biotite, the pyrope content of the garnet is 28–30%. Metamorphic conditions were 720°C and 6–8 kbar, from the garnet-biotite pair. Quartz-free sapphire-cordierite-biotite schists, also containing sillimanite and orthopyroxene, are regarded as granulite facies relicts (Siroshtan et al., 1982; Rusakov et al., 1979). Metamorphic conditions for parts of the amphibolite facies corresponded to 655°C and 6.1 kbar, using the garnet-biotite pair.

In the Central Azov junction zone, metamorphic conditions were similar to those already described: amphibolite and granulite facies rocks are present. It is interesting to note the presence among ferruginous rocks of granulite facies eulysites, characterized by the assemblage olivine (fayalite), high-Fe orthopyroxene, clinopyroxene and quartz, similar to those described by Bondarenko and Dagelaysky (1968). Conditions of formation of these rocks were 780–800°C and 5–6 kbar (Storchak, 1984).

The East Azov block contains only remnants of schists among granitoids in the western part; these were metamorphosed in the hornblende-granulite subfacies.

Ore formation in Precambrian complexes of the Ukrainian Shield

Investigations into the metallogeny of the Ukrainian Shield, specific features of ore-forming processes, their relations with particular geological complexes and formations have been particularly intense over the last 30 years. In the 1960s, metallogenic and ore-deposit forecasting maps of the shield and its separate regions were compiled under the direction of Y.N. Belevtsev. Summaries were published of work that looked at various aspects of ore deposits, including principles and methods of metallogenic analysis of Precambrian formations in the Ukrainian Shield (Belevtsev, 1974, 1981; Belevtsev and Galetsky, 1984; Usenko et al., 1976; Belevtsev et al., 1975). By the end of the '70s and during the '80s, broad generalizations were compiled on the problem of metamorphic ore formation, geological, formational and age categorization of iron ore complexes, and regional metallogenic mapping of the whole territory of the shield. The following summary is based on work by Shcherbak (1983), Storchak (1984) and Belevtsev (1974).

Regional metallogenic mapping in the Precambrian of the Ukrainian Shield was undertaken as follows. The whole shield, as a fold-block province, corresponds to the Ukrainian metallogenic province. Blocks identified as being first order in general correspond to metallogenic regions, which sometimes include suture zones and junction zones between blocks. For example, the Belotserkov metallogenic region embraces the Golovanevsk suture zone, while the Orekhov-Pavlovgrad suture zone is included in the Dnieper metallogenic region.

The metallogenic regions of the Ukrainian Shield have the following characteristic features (Belevtsev, 1974):

Volyn region — deposits and ore shows of titanium, tin, kyanite and pegmatites; molybdenum, nickel, lead and zinc, tungsten and graphite shows.

Podolsk region — ore deposits and shows in pegmatites; iron, titanium, copper, molybdenum, nickel ore shows, zircon, apatite, graphite and fluorite.

Belotserkov region — deposits and ore shows of nickel and cobalt, iron, graphite, sillimanite and pegmatites; ore shows of titanium, bismuth, molybdenum, lead and zinc, tungsten, apatite and corundum.

Kirovograd region — pegmatite deposits; ore shows of titanium, bismuth, copper, molybdenum, nickel, lead, tungsten, zirconium and silver.

West Ingulets–Krivoy Rog region — deposits and ore shows of iron, graphite and talc; ore shows of titanium, copper, molybdenum, nickel, gold, silver, apatite and asbestos.

Dnieper region — deposits and ore shows of iron, aluminium, nickel and cobalt, talc and pegmatites; ore shows of titanium, copper, molybdenum, lead, tungsten and gold.

Azov region — ore deposits and shows of iron, titanium, molybdenum, graphite, fluorite and pegmatites; ore shows of copper, nickel, tin, lead and zinc, apatite and asbestos.

Metallogenic epochs in the history of the Ukrainian Shield that have been identified are: Archaean, Early Proterozoic and Late Proterozoic, each of which is characterised by the accumulation of particular useful minerals, linked in a regular manner to identifiable formational types. During the Archaean epoch, iron accumulated in volcano-sedimentary assemblages and carbonaceous and highly aluminous assemblages, while a number of elements are found in basic–ultrabasic rocks. The Archaean epoch is manifest in the Podolsk, Belotserkov, Dneprovsk and Azov metallogenic regions. In the Early Proterozoic epoch, iron accumulation in sedimentary assemblages formed rich ores, as did the formation of deposits of non-ferrous metals, graphite, ceramic raw materials and muscovite. In the final stage of the Early Proterozoic epoch, specific formations and associated deposits of titanium and apatite, fluorite, etc. originated. This epoch is widely developed in the Volyn, Kirovograd and Azov metallogenic regions. In the Late Proterozoic epoch, the formation of useful minerals associated with metasomatic rocks in regions of tectonic activity took place. This epoch is represented in the Volyn and Azov metallogenic regions.

Whereas in the Archaean epoch the accumulation of useful material mostly occurred in the so-called protogeosynclinal ("pre-geosynclinal") stage, in the Early Proterozoic ore-forming epoch, the characteristic conditions were protoplatform, early and late orogenic and activation stages.

The characteristics of some features of metallogeny in the Ukrainian Shield are shown in Table I-27, while Table I-28 lists the major features of the deposits mentioned in Table I-27. The locations of the deposits are shown on Fig. I-49.

TABLE I-27

Some characteristic metallogenic features of the Ukrainian Shield

Metallogenic province; main metallogenic epochs	Geological groups and complexes	Main ore-bearing formations	Ore formations (deposits & ore shows)
Volyn; Early Proterozoic and Late Proterozoic	Ovruch Group	Trachyandesites + sandstones	Quartz-pyrophyllite (Nagoryan, Zbrankov)
	Teterev Group	Schists + carbonates	Metacarbonate
		Graphitic schists	Graphitic gneisses
		Ferruginous schists	Metamorphic iron ores
	Korosten complex	Leucocratic granites;	Ceramic and micaceous pegmatites;
		rapakivi granites	cavity pegmatites (Korosten pegmatites)
		Gabbro-anorthosites	Titanium (Volodar-Volyn and Chepovich intrusions)
	Perzhan metasomatites	Alkaline granites	Tin-tungsten; fluorite
Podolsk; Archaean and Early Proterozoic	Dnestr-Bug Group	Kinzigite	Graphite-kinzigite (Makharinets); high-alumina garnetiferous (Slobodskoye)
	Berdichev complex	Leucocratic granites	Tungsten-copper-molybdenum (Lyubar, Ostropolsk)
	Complex of alkaline ultramafic rocks and carbonatites	Alkaline ultramafics and carbonatites	Apatite-carbonatite
Belotserkov; Archaean, partly Early Proterozoic	Bug Group	Kondolite	Graphite-kondolite (Zavalev)
		Ferruginous chert	Volcanosedimentary iron ores
	Mafic-ultramafic rock complex	Dunite-peridotite; dunite-clinopyroxenite-gabbro	Chromite (Kapitanov); copper-nickel sulphides
		Komatiite-tholeiite	—ditto—
Kirovograd; Early Proterozoic	Bug Group	Flyschoid	Graphitic gneisses
	Kirovograd-Zhitomir complex	Subalkaline granitoids	Copper-molybdenum; tin-tungsten skarns
	Korosten complex	Rapakivi granites	Quartz; tungsten-molybdenum
		Gabbro-anorthosites	Phosphoro-titanium; high-Al feldspathic
West Ingulets-Krivoy Rog zone; Early Proterozoic	Ingul-Ingulets Group	Chert-schist-carbonate	Graphitic gneisses
		Ferruginous chert	Volcanosedimentary & sedimentary iron ores
		Meta-andesite-metabasalt, ferruginous chert	Copper-nickel sulphides (Karachunov-Lozovat, Ingulets)
	Krivoy Rog Group	Chert-jaspilite	Metamorphic and metasomatic iron ores; sedimentary iron ores (Krivoy Rog-Kremenchug iron ore basin)
Dneprovsk; Archaean, partly Early Proterozoic	Aul Group	Jaspilite–meta-tholeiite	Volcanogenic iron ores
	Konka-Verkhovtsev Group	Dunite-peridotite	Talc-magnesite (Pravdin)
		Dunite-pyroxenite-gabbro	Copper-nickel sulphides
		Ferruginous chert	Volcanogenic iron ores (Verkhovtsev, Chertomlyk)
	Krivoy Rog Group	Minor intrusions of aplo-pegmatitic granites	Tungsten-molybdenum (Annovo)
Azov; Archaean, Early and Late Proterozoic	East Azov complex	Alkali nepheline syenites	Phosphorus
	West Azov Group	Ferruginous chert	Volcanosedimentary iron ores (Mariupolsk)
	Gulyaypol and Osipenkov Fms	Ferruginous chert	Chemogenic-sedimentary iron ores
	Azoc complex	Metasomatites; granodiorites	Pegmatites (Yeliseyev pegmatite field)
	East Azov complex	Syenite-granosyenite alkali nepheline syenite	Nepheline-feldspar-apatite
	Kamennomogila complex	Alkali granitoid	Fluorite
	Chernigov complex	Alkali ultramafic	Apatite-carbonatite

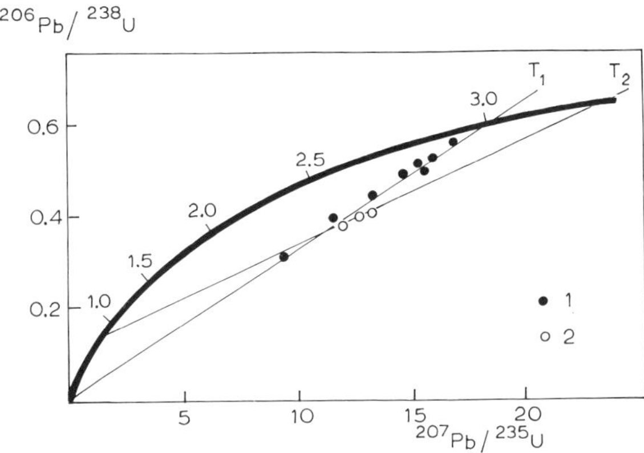

Fig. I-49. Concordia diagram for zircons from metavolcanics in the Konka-Verkhovtsev Group and synorogenic granites. *1* = zircons from granites; *2* = zircons from metavolcanics.

Great attention has been paid to the study of iron ores while researching metal ore deposits in the Ukrainian Shield. Within the shield, iron ore formations are encountered at various stratigraphic levels, and their formation mainly occurred during two epochs, the Archaean and Early Proterozoic. The banded iron formations of the region are divided into three types (Belevtsev et al., 1981), shown in Table I-29. The metallogenic regions and zones identified within the confines of the Ukrainian Shield iron ore province are characterized by definite formations of iron ores. The Odessa-Belotserkov metallogenic zone has ores of the iron-chert volcano-sedimentary formation; the Krivoy Rog–Kremenchug zone has an iron-chert sedimentary formation; the Dnieper metallogenic region has ores of the volcano-sedimentary formation; the Belozero-Orekhovo zone has volcanic and sedimentary formations; the Azov metallogenic region has ores of the iron-chert sedimentary and volcano-sedimentary formations.

The stratigraphic position of iron ore formations (ferruginous quartzite and schist, silicate- and carbonate-bearing ores) is determined for the above-named regions and zones on the basis of the following data. In the Odessa-Belotserkov metallogenic zone, the ores are present in the Rosin-Tikich Group of questionable age (Archaean or lower Proterozoic). In the Krivoy Rog–Kremenchug zone, including the unique Krivoy Rog iron ore basin, the ores occur in the lower Proterozoic Krivoy Rog Group and the stratigraphically equivalent Ingul-Ingulets Group. In the Dnieper metallogenic region, iron ore formations are associated with rocks of the Archaean Konka-Verkhovtsev Group. The Belozero-Orekhovo zone is subdivided into two subzones. In the Konka-Belozero subzone, ores occur in the Konka Formation of the Archaean Konka-Verkhovtsev Group and in the Belozero Formation, which some workers compare with the Krivoy

TABLE I-28

Characteristics of a number of ore deposits and ore shows of the Ukrainian Shield

Useful mineral, deposit or ore show	Tectonic setting; blocks and zones after [3]	Mineralization type, ore-forming and genetic	Geological and mineralogical features			Ore mineral composition (secondary minerals) and hydrothermally altered rocks	Age of host rocks; age of ore mineralization	Refs
			Host or mineralized rocks	Shape of deposit, morphology of ore bodies	Ore type			
Iron Ingulets, Novokrivoy Rog and others (Krivoy Rog-Kremenchug basin)	West Ingulets interblock zone; sutural interblock basin with terrignous infill [8]	Ferruginous quartzites of Krivoy Rog type; metamorphosed sedimentary formation	Terrigenous-chemogenic rocks of the Krivoy Rog (Saksagan Fm) and Ingul-Ingulets Groups (shales, schists, amphibolites)	Conformable beds	Banded	Magnetite, quartz (hematite martite; chlorite, carbonate; amphibole, pyroxene, garnet)	Early Proterozoic	[1]
Mariupol and others	Azov geoblock; granulite-gneiss terrain	Ferruginous quartzites of Mariupol type metamorphosed	Terrigenous-volcanogenic rocks of the Central Azov Group (Sachkin Fm, gneisses, schists, marbles, calciphyres)	—ditto—	Banded and massive	Magnetite, quartz (grünerite, clino- and orthopyroxene, olivine, garnet)	Early Proterozoic (?)	[1]
Verkhovtsev, Chertomlyk and others	Dnieper block; ancient granite-greenstone terrain	Ferruginous quartzites of Verkhovtsev type; metamorphosed volcanogenic formation	Volcanogenic, occasionally terrigenous rocks of the Konka-Verkhovtsev Group (Konka Fm, hornblende schists and gneisses, amphibolites, paraschists)	Conformable beds and lenses	Banded	Magnetite, quartz, carbonate (biotite, amphibole, chlorite)	Late Archaean	[1]
Titanium Volgograd-Volyn, Chepovich	Volynye block; mafic rocks forming part of a polyphase platform-type intrusion	Titanium; magmatic	Gabbros, gabbro-norites and gabbro-anorthosites of the Korosten complex	Disseminations	Disseminated	Ilmenite (ilmenite-magnetite, titanomagnetite, apatite)	Late Proterozoic	[2]
Chromium Kapitanov, Skholnoye, etc.	Belotserkov block; early orogenic ultramafic intrusions in ancient fault zones	Chromite; magmatic	Ultramafics (apodunites, apoharzburgites, serpentinites)	Lenticular and pipe-like?	Disseminated and massive	Chrome spinels	Early Archaean	[3]
Copper, nickel Karachunov-Lozovat, Ingulets	West Igulets zone; areas of hydrothermal alteration along shear zones and crush zones	Copper-nickel sulphides, vein-type hydrothermal-metasomatic	Basic rocks of meta-andesite-metabasalt formation and ferruginous quartzites of the Ingul-Ingulets Group; migmatites and granites	Veins and vein-like disseminations	Vein and vein-type disseminated	Cubanite, pyrrhotite, chalcopyrite, pentlandite	Early Proterozoic	[2]
Tungsten, molybdenum Annov	Dnieper block; contact zone between granitoids and metamorphic rocks	Tungsten-molybdenum; hydrothermal-metasomatic	Amphibolites at the base of the Krivoy Rog Group, near contact with aplo-pegmatitic granites; zones of potash metasomatism	Parallel zones of disseminations	Finely disseminated	Molybdenite (& scheelite)	Early Proterozoic	[2]

Molybdenum Lubarsk, Ostropol	Podolsk block; tectonic knots and zones in partial-melt granitoids	Tungsten-copper-molybdenum;hydro-thermal-metasomatic	Granites, charnockites, pyroxene granites of the Berdichev complex, metasomatically altered	Stocks and pipe-like accumulations	Disseminated, aggregate	Molybdenite (pyrrhotite, pyrite, chalcopyrite, magnetite, ilmenite) potash feldspar, albite, quartz, biotite, chlorite	Early Proterozoic	[2]
Pyrophyllite Nagoryansk, Zbrankov	Volyn block; epi-cratonic basins of platform stage	Quartz-pyrophyllite; metamorphic	Sandstones of the Ovruch Group	Sheet-like and lenticular	Massive	Pyrophyllite	Late Proterozoic	[3]
Garnet Slobodsk	Podolsk block; areas of older rocks affected by migmatiz-ation and metasomatism	High-alumina garnetiferous; metamorphic	Leucocratic granites of the Berdichev complex	Lenses and deposits	Disseminated, massive	Garnet	Early Archaean; Early Proterozoic	[2, 3]
Talc-magnesite Pravdin, Veselyansk	Dnieper block; orogenic ultramafic intrusions in ancient granite-greenstone belt	Talc, magnesite; metasomatic ($CaCO_3$ metasomatism)	Ultramafics (dunites, peridotites)	Thin bands and lenses	Massive	Talc, magnesite	Late Archaean	[2, 3]
Graphite Makharinets	Podolsk block; mafic granulite basement highs	Graphite, kinzigite; metamorphic	High-Al gneisses and schists of Dnestr-Bug Group	Lenticular and pocket-like disseminations	Disseminated, massive	Graphite (sillimanite)	Early Archaean	[4, 5]
Zavalevsk	Belotserkov block; terrigenous-carbonate sediments in palaeo-lagoonal zones	Graphite-kondolite; metamorphic	High-Al gneisses and schists and carbonates of the Bug Group (Khashchevat-Zavalevsk Formation)	—ditto—	—ditto—	—ditto—	Early (?) Proterozoic	[5, 6]
Cavity pegmatites Korosten pegmatite field	Volyn block; poly-phase intrusion of basic rocks and granitoids	Cavity pegmatites; magmatic pneumato-lytic	Endocontact zones between granitoids and basic rocks	Pegmatite bodies & groups of bodies	Filled chambers (cavities)	Quartz, fluorite, precious (topaz, morion, etc.)	End Early Proterozoic	[7]
Pegmatites Yeliseyev pegmatite	Azov block; associated with ancient fault	Rare-earth-rare-metal pegmatites	Metamorphic rocks of the West Azov Group (schists, gneisses, amphibolites)	Pegmatite bodies & groups of bodies	Dispersed mineralization	Columbite, tantalite, zircon (wolframite)	Early Archaean; Early Proterozoic	[2]

References: [1] Belevtsev et al. (1981); [2] Belevtsev (1974); [3] Storchak (1984); [4] Bukharev and Polyanski (1977); [5] Kalyayev et al. (1981); [6] Valeyev (1978); [7] Lazarenko et al. (1973); [8] Bogachev and Gorelov (1968).

TABLE I-29

Classification of Precambrian banded iron formations in the Ukrainian Shield

Precambrian iron formation	Conditions of formation	Mineral composition of rocks	Shape of body	Examples of deposits and ferruginous quartzite areas
Sedimentary (mio-geosynclinal), Krivoy Rog or Superior type	Sedimentation, diagenesis and greenschist facies metamorphism	Quartz-magnetite, chlorite-magnetite, carbonate-chlorite-magnetite	Beds 10's-100's m thick, 10's km long	Ingulets, Skelevat—magnetite, Novokrivoyrog, Gorishneplavnin, Gulyaypol, and others
	Sedimentation, diagenesis and amphibolite or granulite facies metamorphism	Quartz-magnetite, amphibole-magnetite, pyroxene-magnetite (± garnet)	Beds usually a few 10's m thick and several km strike	Pervomaysk, Annov, Artemov, Petrovsk, North Tersyan, Vasinovsk
Volcano-sedimentary (mio-eugeosynclinal) Mariupolsk type	Sedimentation and minor submarine volcanism, diagenesis and greenschist, amphibolite and granulite facies metamorphism	Quartz-magnetite, carbonate-magnetite (± iron silicates) amphibole-magnetite, pyroxene-magnetite	Beds up to a few 10's m thick and several km strike	Mariupolsk, Konka, Kuksungur
Volcanogenic (eu-geosynclinal) Verkhovtsev or Algoma type	Submarine volcanism, sedimentation, diagenesis, greenschist and amphibolite facies metamorphism	Carbonate-magnetite, silicate-carbonate-magnetite	Beds and lenses, a few m to a few 10's m thick and 10's-a few 100's km along strike	Verkhovtsev, Chertomlyk, Sursk, West Belozersk

Source: Belevtsev et al. (1981).

Rog Group. In the Orekhovo-Pavlovgrad subzone, iron ores also occur in these two assemblages. In the Azov metallogenic region, ores are present in the Archaean West Azov Group. However, the main iron ore deposits in the Azov region are spatially and genetically associated with rocks of the Sachkin Formation in the (?)lower Proterozoic Central Azov Group, which is compared with the Krivoy Rog Group. In this region we also have the Gulyaypol Formation of lower Proterozoic age, which is found only within the confines of the ore deposit of the same name.

Chapter 3

THE RUSSIAN PLATFORM

1. CRYSTALLINE BASEMENT

Several schemes have been published which show the basement subdivided into major lithospheric blocks or geoblocks. On the whole, these schemes are all quite similar, although while constructing them the authors proceeded from an analysis of different geophysical and geological data (Simonenko and Tolstikhina, 1968; Dedeyev and Shustova, 1976; Kratz et al., 1979). At the level of the present day surface of the crystalline basement, the lithospheric blocks correspond to tectonic domains which differ in the unifying nature of their geological structure. This is defined by the association of supracrustal assemblages, the nature and sequence of igneous activity and regional metamorphic effects, the time of stabilization and the type of infrastructure formed.

The data presented below on the geological structure of the basement are grouped according to the division into blocks which we have proposed earlier (Kratz et al., 1979, pp. 69–72, 107–108). For the sake of brevity, we have included in the discussion of tectonic regions, corresponding to lithospheric blocks, an examination of the structure of adjacent regions which correspond to junction zones with a long history of development between blocks (Fig. I-50). In this section we have retained the provisional name of "block" to refer to these unified regions.

The Estonian block

The Estonian block includes Estonia and parts of the Leningrad, Pskov and Novgorod regions. The block has sharp boundaries. Along the southwestern and south-eastern border zones, geophysical field maps show intense positive strip anomalies, which appear to correlate with linear zones of high grade metamorphic rocks in the basement. The comparatively shallow (0.5–0.6 km) depth of burial of the basement has been a helpful factor in the fairly good coverage of the block by drill cores (especially in northern Estonia and the north of the Leningrad region). On the whole, drilling results confirm the division into geological regions, carried out on the basis of geophysical survey data.

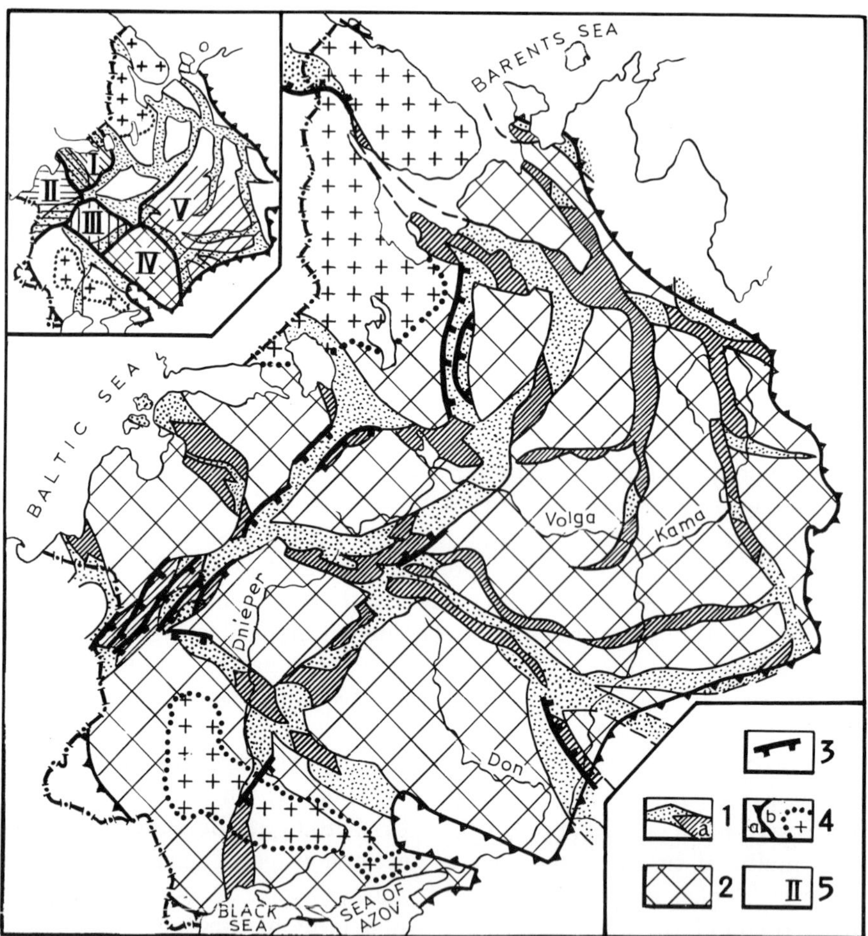

Fig. I-50. Sketch map showing location of tectonic provinces (blocks) in the crystalline basement of the Russian Platform (after Zapolnov and Neyelov, 1984; Kratz et al., 1979; Zapolnov et al., 1980). *1* = interblock zones (*a* = linear zones of mafic granulites); *2* = blocks; *3* = conjectural reverse faults and steep thrusts; *4a* = platform boundaries, *4b* = exposed basement; *5* = tectonic provinces (*I* = Estonian, *II* = Belorussian-Baltic, *III* = Smolensk, *IV* = Kursk-Voronezh, *V* = Volga-Urals).

Lower Archaean complexes have been studied in Estonia and the Karelian isthmus (Leningrad region). In southern Estonia there is a NW-trending belt, up to 100–150 km wide, of granulite facies schists — two-pyroxene, two-pyroxene-amphibole, biotite-hypersthene, and others (Table I-30). From their chemical composition, they belong to a group of mafic (and less commonly intermediate) Ca-rich rocks. Hornblende-biotite gneisses are encountered much less commonly. Individual bore holes show amphibolite, quartzite and carbonate rocks (calc-silicate marbles) to a limited extent. In

TABLE I-30

Correlation scheme of basic supracrustal assemblages of the crystalline basement to the Russian Platform

Age, Ma	Estonian block	Belorussian-Baltic block	Smolensk block	Kursk-Voronezh block West	Kursk-Voronezh block East	Volga-Urals block
1650±50	Hoglandian	Adazh assemblage	Quartz porphyries, Bobruysk region	Glazunov Fm	Baygorov Fm	
1800±100	High geothermal gradient metamorphism, from greenschist to amphibolite facies					
	Zonal?				Zonal	
	Uljaste assemblage			Oskol Gp Kursk Gp	Vorontsov Gp Losev Group	Uniy assemblage
2600±100	Low geothermal gradient metamorphism, from greenschist to amphibolite facies					
	Yagala complex	Okolovo Group, Inchukalns assembl.	Zhitkovichi Group	Mikhaylov Group		Sarman Group
3000±100	High geothermal gradient metamorphism, mainly granulite facies					
	Aluminous gneisses of NW Estonia and Leningrad region		Kulanzhi Group	Besedino assemblage		Bolshecheremshan Group
	Crystalline schists of S. Estonia	Ivyev Group	Bryansk Group			Otradnin Group

northern Estonia, the schist complex has been found by drilling around the town of Tapa.

Effects of charnockitization are widespread and intense in the rocks of this complex. Massive charnockite bodies display relicts of two-pyroxene and biotite-hypersthene schist. The charnockites themselves in fact usually occur as venite migmatite veins. It is possible that some migmatites of granitic composition also formed at granulite facies conditions; the K-spar in these rocks is orthoclase (Puura et al., 1983, pp. 29, 67–69).

In western Estonia, in the continuation of the South Estonia anomaly zone, a rock complex of similar chemical composition is developed, but the degree of metamorphism does not reach granulite facies. The Kohila borehole, south of Tallinn, penetrates 340 m of hornblende-biotite and biotite gneiss and amphibolite of this assemblage.

Another type of lower Archaean granulite facies section has been drilled in NE Estonia. Along the shores of the Gulf of Finland, around the town of Iihvi, there is an aluminous gneiss complex (garnet-cordierite-sillimanite), interleaved with magnetite quartzite and ferruginous chert. Around Lake Chudskoye, aluminous gneiss predominates, with rare thin bands of biotite gneiss and hypersthene schist with granite veins containing garnet and cordierite.

There is no general agreement about the position of the aluminous gneisses in the overall section. Some workers, interpreting them as an original assemblage of flysch-type rocks, refer the complex to Svecofennian supracrustal assemblages and correlate them with the Ladoga Group (Puura et al., 1983). Other geologists, who point to the undoubted overall similarity of granulite complexes of Estonia with complexes in neighbouring regions, assign them to the lower Archaean (Kratz et al., 1979).

Indeed, in the Karelian isthmus at a depth of 530–700 m beneath the basement surface, thinly-banded migmatized garnet-biotite gneiss with thin bands of hypersthene and hypersthene-amphibole schist are exposed in individual boreholes. This association is assumed to have formed from thinly laminated pelites and greywackes. The rocks are penetrated at a shallow depth.

Above is an assemblage of mafic schists with anorthosite and charnockite — the probable equivalent of the South Estonia granulite complex — which in turn is succeeded by an assemblage of aluminous gneisses.

The aluminous gneisses have been penetrated by several boreholes in the Karelian isthmus at depths of up to 600 m. They consist of garnet-cordierite-sillimanite-biotite gneiss with large quantities of garnetiferous granitoid vein material. In terms of composition and metamorphic grade, these rocks are equivalent to the aluminous gneisses of NE Estonia. As far as the rock association near Iihvi is concerned, it shows undoubted similarities with the section through the lower part of the Volshpakh Formation of the Kola Group. In the region of Volshpakh-tundra, a similar rock association conformably overlies a rhythmically-banded assemblage of mafic schist and this in turn is overlain by a thick unit of aluminous gneiss (Kratz et al., 1979).

The schist and gneiss complexes under consideration here were regionally metamorphosed at granulite facies conditions. Metamorphism was essentially uniform, at moderate pressure (5–6 kbar) and temperature (720–780°C). In a number of boreholes, such rocks as eulysite (feldspar-free two-pyroxene-garnet granulite) and magnetite-garnet-hypersthene schist have been encountered; these may be referred to high-pressure granulites. In southern Estonia as well as in the Kola Group of the Kola Peninsula, both types of granulite parageneses occur together. In such cases, medium-pressure mineral assemblages are earlier. The medium-pressure granulite facies assemblages have developed in turn at the expense of higher pressure granulite facies assemblages, related to granitization in the concluding stage of granulite metamorphism. For amphibolite facies rocks in western Estonia, P–T metamorphic conditions are 670–680°C and 4–5 kbar (Puura et al., 1983).

An Early Archaean age for the complexes being considered is taken on the basis of uniform high-grade metamorphism, typical for all the oldest supracrustal assemblages, and the overall similarity in sections from southern Estonia, the Leningrad region and the Kola Group on the Kola Peninsula. The majority of K–Ar mineral determinations fall in the interval 1700–1850 Ma, suggesting substantial heating during the Svecofennian cycle. A Rb–Sr isochron age, obtained on granulite facies rocks in cores from six boreholes in southern Estonia, is also 1740 Ma.

The upper Archaean *Jagala complex* (also known as the *Tallinn-Loksa complex*) is exposed in bore holes in NW Estonia. The complex consists of finely banded quartz-feldspar gneiss ("leptite"), alternating with hornblende-biotite and biotite tonalite and amphibolite, together with aluminous gneiss (cordierite-biotite-sillimanite) and amphibolite dykes. The gneisses and amphibolites of Hogland Island also appear to belong to this same complex.

The complex is polymetamorphic (Bondarenko et al., 1978; Kratz et al., 1979). The earliest metamorphism took place at almandine to epidote-amphibolite facies conditions. Detailed petrographic research on these garnet-cordierite-biotite-sillimanite gneisses indicates an earlier episode of cordierite growth and acid leaching, with the formation of the mineral assemblage sillimanite-muscovite-quartz. Locally, along blastomylonite zones, the effects of subsequent reworking at staurolite subfacies conditions of the andalusite facies series have been observed. This particular type of metamorphism (staurolite or greenschist grade) shows the effects of retrograde reworking of the granulite complexes of southern Estonia and the Karelian peninsula. This last metamorphic episode broadly correlates with the Early Proterozoic Ladoga-type prograde zonal metamorphism.

However, there is another point of view concerning the structural position and age of the Jagala complex. According to Puura and his Estonian co-workers, the rocks of this complex are equivalent to the Lower Proterozoic Sortavala and Ladoga Group metasediments. As evidence they point to the reconstructed primary composition; sedimentary flyschoid sequences (alternation of aluminous and biotite gneiss) and acid or intermediate lavas (some quartz-feldspar and hornblende-biotite gneiss with blastoporphyritic textures). Weakly expressed metamorphic zoning is noticed in the region where Jagala complex rocks are developed, right up to the appearance of mineral associations with hypersthene, indicating a transition to granulite facies conditions. The discovery of the polymetamorphic nature of the complex with an early kyanite facies series metamorphism prevents us from adhering to this viewpoint.

Lower Proterozoic rocks are the *Uljaste assemblage*, found in NE Estonia around Uljaste and Haljalaa. These are garnet-cordierite-biotite gneiss, amphibolite, amphibole-diopside schist, calc-silicate marble and quartzite. The rocks were metamorphosed in the staurolite zone of the cummingtonite-amphibolite facies. Aluminous gneisses often contain large quantities of graphite (up to 9 or even 25%, the so-called "black gneisses") and sulphides (pyrrhotite, pyrite less commonly, sphalerite and galena). An analogous rock assemblage can be traced from geophysical data as sinuous bands in both NW Estonia and most of the Gulf of Finland. Overall, we are confident that the rocks of the Uljaste assemblage correlate with the Pitkäranta Formation of the Ladoga Group.

The youngest supracrustal formations in the lower Proterozoic are represented by the *Hoglandian*. On Hogland Island in the Gulf of Finland, the Hoglandian rests on an eroded surface of metamorphosed formations which correlate with the Jagala complex. Hoglandian rocks dip gently eastwards at 5–30°. The sequence commences with quartzite and quartz conglomerate, succeeded by porphyrite, volcanic agglomerate and quartz porphyry, to a total thickness of up to 130 m.

Graben-type structures, consisting of Hoglandian or its analogues, are spatially related throughout the Baltic Shield with anorogenic intrusions of rapakivi and alkali granites of the same age. A Rb–Sr isochron age for sub-Jotnian porphyries in Central Sweden (Upper Dala — a probable equivalent of the Hoglandian) is 1670 ± 15 Ma.

Intrusive complexes (PR_1). Within the block there are widespread post-tectonic (early platform) complexes. Rapakivi granites and associated rocks form the major Vyborg and Luga intrusions.

The Vyborg rapakivi intrusion is situated on the northern shores of and partly beneath the Gulf of Finland. It consists of various porphyritic granites which differ in the size and percentages of the K-spar megacrysts (ovoids), granodiorite, quartz syeno-diorite and syeno-diorite, labradorite and olivine gabbro-norite. According to Velikoslavinsky et al. (1978), the Vyborg massif is a sheet-like polyphase undifferentiated intrusion. The early granitoid phase consists of quartz syeno-diorite, followed by ovoidal rapakivi (vyborgite or wiborgite). Among the vyborgite are small bodies of late trachytic rapakivi and porphyritic rapakivi with a fine-grained groundmass. Levkovsky (1975) provides evidence of an even more complex picture for the origin of the massif, with up to 15 phases.

The other massif of rapakivi and gabbroid rocks, the Luga intrusion, is situated in the eastern part of the Estonian block. It has been identified from geophysical survey results.

The age of the rapakivi granites is usually taken as 1650 ± 50 Ma, based on numerous determinations using a variety of methods. Some workers present evidence for the time boundary of the initial appearance of the complex as around 1900 Ma (Levkovsky, 1975, p. 182–183). The age of the porphyritic potassic granites of Estonia, from K–Ar, Rb–Sr and lead-isochron methods is put at 1680–1710 Ma.

Mineral resources associated with the crystalline basement of the Estonian block are not yet sufficiently well studied (Anishchenkova et al., 1979). Most interest in the granulite complex rocks has been devoted to magnetite mineralization in the Iihvi zone, connected with ferruginous quartzite. Around the town of Tapa, there are ore shows of Cu–Ni–Co mineralization, associated with ultrabasic and basic rocks. In addition to the graphite-pyrrhotite mineralization of the "black gneisses" in the gneiss complexes of northern Estonia already mentioned, there are also characteristic vein-type pyrite-pyrrhotine (also developed in the magnetite quartzite), titanium-magnetite and rare-metal mineralization.

An Early Proterozoic date has been obtained for telethermal copper-polymetallic and polymetallic ore shows, restricted to zones of tectonic breaks in the basement (Puura and Sudov, 1975). In many cases there are telescopically superimposed ore mineralization episodes of various ages, localized both in the basement and in the sedimentary cover. Against this background, the type of mineralization can also vary with time. For example, in Uljaste, Hercynian polymetallic mineralization is superimposed on Svecofennian pyrite mineralization.

The rapakivi granites may be the source of many types of useful minerals. Piezoquartz and fluorite occur in pegmatite veins in the granites. In skarns, greisens and quartz veins there are concentrations of polymetallic mineralization with high concentrations of Sn, W, Mo, Cu, Ag, Au and REE (Levkovsky, 1975).

The Belorussian-Baltic block

The Belorussian-Baltic block is located SW of the Estonian block and includes Latvia, Lithuania and the Kaliningrad region, together with the NW part of Belorussia (approximately along a line from Minsk to Polotsk). The block has not been studied to the same uniform level. Much more is known from drill core information about the region adjacent to the river Neman in the south of Lithuania and NW Belorussia, as well as the coastal regions in the southern part of the Baltic republics.

Lower Archaean complexes have been studied mostly in the border regions between Belorussia and Lithuania and in SW Latvia.

The Belorussian-Lithuanian zone has a linear character and consists of a series of narrow (up to 20 km) elongate (up to 100–250 km) imbricate blocks, made of rock complexes which are broadly analogous to the granulite rocks of Estonia, and an amphibolite-gneiss complex. The structure of the zone is defined as stacked thrust sheets, as indicated by the conformable arcuate contours of the granulite blocks and the wedging-out of individual blocks at depth, as shown by geophysical data. This structure was previously thought to be a block-fold type, with uplift of relatively elevated blocks of older plagiogneiss and the development of relatively depressed blocks of biotite-amphibole gneiss, amphibolite and two-pyroxene schist. All these rocks belong to the Shchuchin Group (Pap, 1977). Recently (Aksamentova et al., 1982), the sequence of these formations has been treated differently. The oldest rocks belong to the *Ivyev Group* of amphibole-two-pyroxene and two-pyroxene schist, feldspar and pyroxene amphibolite. There are very rare occurrences of garnet-clinopyroxene eclogitic rocks. Chemically the mafic schists are regarded as the products of highly primitive basaltic magmas.

Associated with these rocks are charnockites, genetically related to partial melting under granulite facies conditions. Immediately adjacent to thrust emplaced schists, the rocks display an essentially plagioclase composition with a small amount of hypersthene. Larger bodies of hypersthene-plagioclase rocks (enderbite) may have an igneous origin.

Similar granulite complexes, although on a smaller scale, are found in Latvia and Lithuania. Two-pyroxene and biotite-two-pyroxene schists occur in SW Latvia. In the south of Lithuania, two-pyroxene schists alternate with partial-melt plagioclasites. Superimposed lower temperature metamorphism and amphibolization affect all the rocks, and metasomatic magnetite formed at the expense of pyroxene. The formation of late plagioclase-microcline granitoids and migmatites appears to be an Early Proterozoic event.

Bore hole sections in western Lithuania are dominated by aluminous gneisses: garnet-sillimanite-biotite-cordierite, sillimanite-cordierite, garnet-cordierite, etc. Charnockitized rocks of basic composition are rarely encountered. Biotite-orthoclase-plagioclase migmatites are developed in the aluminous and acid gneisses. The rocks are noticeably retrogressed at amphibolite facies conditions.

Rocks of the Ivyev Group and its correlatives were regionally metamorphosed at medium-pressure granulite facies conditions. Temperature conditions have been calculated to be 700–800°C (Pap, 1977). Subsequent changes in the rocks are related to granitization during a retrogressive stage of the granulite metamorphism. Practically everywhere there are zones of blastomylonitization with the development of complex and nonequilibrium mineral associations, corresponding to two superimposed metamorphic episodes. The first of these took place in the high-temperature subfacies of the almandine amphibolite facies. The second episode did not exceed low pressure staurolite grade. This episode is usually referred to the Svecofennian cycle (Ladoga-type metamorphism).

There is even less information available on other complexes which are also conventionally regarded as lower Archaean.

In western Belorussia is the Ozery amphibolite-gneiss complex which consists predominantly of biotite, hornblende-biotite and plagioclase gneiss with thin bands of hornblende gneiss and amphibolite near the base of the section (Aksamentova et al., 1982). Recently, this complex has been considered to occupy a higher stratigraphic level than the Ivyev Group, but we cannot say that this position is definitive. Another point of view is possible: that these complexes are the same age but tectonically juxtaposed.

Upper Archaean supracrustal complexes include the Okolovo Group of Belorussia and its analogues. The upper unit of the Okolovo Group consists of garnet-biotite-staurolite schist and garnet-biotite and sillimanite gneiss. The following rock associations are seen in sequence going down through the lower unit: (1) interbanded cordierite-garnet-gedrite-chlorite-quartz schist and garnet-amphibole-magnetite quartzite; (2) alternating plagioclase-gedrite-staurolite-garnet-quartz schist and magnetite quartzite; and (3) alternating amphibole-plagioclase, plagioclase-biotite-staurolite-quartz and magnetite-amphibole schist. Magnetite quartzite bands are from 1 to 32 m thick. Ore-bearing formations show the following rock sequence: cummingtonite-magnetite quartzite → amphibole-magnetite quartzite → magnetite quartzite. Evidently, the magnetite-bearing schist and ferruginous quartzite were formed by metasomatic processes during a retrograde stage of regional metamorphism in a primary sedimentary rock association (Kratz et al., 1979, p. 38–39).

The thermodynamic conditions of metamorphism of Okolovo Group correspond to low grade almandine-amphibolite facies (cordierite-staurolite-gedrite subfacies).

The metamorphic grade is higher in rocks of the Inchukalns assemblage, central Latvia (high temperature subfacies of ultrametamorphism zones). Here, east of the Gulf of Riga, there are biotite-plagioclase and hornblende-biotite schist, together with plagioclase-anthophyllite gneiss. Retrogression in these rocks is usually marked by the growth of quartz and microcline in zones of cataclasis and mylonitization.

Archaean intrusive complexes. The earliest igneous rocks, associated with the Ivyev Group, are concordant and discordant bodies of peridotite, as well as rare small (up to a few kilometres) gabbro and gabbro-norite bodies (Pap, 1977).

Associated with the amphibolite-gneiss complex are folded granites and migmatites. The granites are intensely gneissified, they have concordant boundaries with the country rock gneiss and amphibolite and contain xenoliths of the gneiss. Granodiorite and tonalite bodies have also been mapped.

Post-tectonic Archaean granites crop out in the Minsk region (the Vygonov complex), where they form large intrusions. The granites are medium- to coarse-grained leucocratic rocks with equal amounts of plagioclase and microcline. They are probably the product of synorogenic granitization of gneisses (Pap, 1977).

In SE Latvia, a very few bore holes penetrate gneissose plagioclase and microcline-plagioclase granites of the broad Latgal intrusion. This massif is inhomogeneous, being composed of both metasomatic and rheomorphic granitoids of various ages. The main body of the intrusion appears to be related to Archaean amphibolite facies metamorphism.

In NW and central Lithuania, much of the basement is made of gneissic granitoids and granite gneiss, including aluminous gneiss. These rocks are usually granitized by younger (Early Proterozoic?) microcline granites.

Lower Proterozoic. Parallel to the Hoglandian of the Estonian block is the weakly metamorphosed Adazh assemblage, covering the formations of the Inchukalns assemblage. The Adazh assemblage consists of carbonate-sericite-chlorite, biotite-hornblende and biotite-hornblende-epidote schist and andesite-basalt volcanics. In addition, quartz porphyry occurs along the shores of western Latvia, at Pavilosta.

Intrusions (PR_1) are represented by anorthosite and rapakivi. The Riga rapakivi granite intrusion occupies the Kurzeme peninsula and extends under the Gulf of Riga and the Baltic Sea. The massif has been found in numerous boreholes, the cores of which contain in addition to rapakivi and similar rocks (without orthoclase ovoids), syenite and quartz syenite, gabbro, gabbro-norite and labradorite. On the basis of gravity data, the thickness of the intrusion is calculated at not less than 8–10 km.

Mineral resources. Prospects of finding ore deposits in lower Archaean complexes are mainly limited to gabbroid bodies. The earliest gabbroids (the Berezov complex) have potential for Cu–Ni ore prospects (Makhnach et al., 1979). Ilmenite-magnetite mineralization with vanadium is associated with post-tectonic intrusions (the Korelichi complex). Data from the Novoselkov deposit (Pap, 1977) show that the formation of ore minerals is related to the last stage of metamorphism. The ore minerals are represented by magnetite and ilmenite, with admixtures of spinel and sulphides. The Okolovo iron deposit is related to the deposition of the upper Archaean Okolovo Group. Here, the ore mineral is almost exclusively magnetite.

The Smolensk block

The Smolensk block is located SW of the Belorussian-Baltic block and occupies the territory of the southeastern part of Belorussia, as well as the Smolensk, Bryansk and Kaluga regions of the Russian Federation. The boundaries of the block are sharp and can be traced as zones of linear positive magnetic and gravity anomalies, which are associated with linear zones in the crystalline basement, which consists predominantly of mafic schists.

Lower Archaean. The oldest rocks belong to the *Bryansk Group*, drilled in the Bryansk-Kaluga strip. These are granulite-facies metamorphic and ultrametamorphic assemblages: pyroxene and biotite-pyroxene schist, garnet-biotite-plagioclase gneiss, often with pyroxene, charnockitic and enderbitic gneiss. Amphibolite facies retrograde metamorphic effects have been observed. The vertical thickness of the complex is estimated to be 5–6 km (Krestin and Berdnikov, 1982; Krestin, 1984).

Farther west, in Belorussia, is the major massif of the Kulanzhi Group aluminous paragneiss (garnet-biotite, hypersthene-garnet-biotite and sillimanite-garnet-biotite) with thin bands of pyroxene and feldspar amphibolite in the lowest levels of the section. As in the Estonian block, it is assumed that the aluminous gneisses comprise the upper part of the section of the lower Archaean granulite complex, although not exclusively, so that they are at least partial facies analogues of the metabasic section (Aksamentova et al., 1982).

The greater part of the block consists of a lower Archaean granite-gneiss complex within which are major intrusions, with prominent granitoids of various ages. The granite-gneiss complex is dominated by biotite, hornblende-biotite and plagioclase gneiss, containing thin bands and patches of hornblende gneiss and amphibolite. Rocks of the complex are intensely metamorphosed and granitized. Gneissose granite and microcline granite may form quite broad outcrop areas amongst the granite gneisses.

Mineral parageneses of the gneisses correspond to almandine-amphibolite facies in the partial melt field. There are no reliable indications of the age of the gneiss complex. It seems to be made mainly of lower Archaean supracrustals, stratigraphic analogues of granulite complexes. Retrograde effects, restricted mainly to blastomylonite zones, are widespread and were at epidote amphibolite facies.

In the south of Belorussia is the locally developed *Zhitkovichi Group*, which must be referred to the *upper Archaean*. It consists of the Lyudenevichi Formation of metasediments — quartz-sericite and chlorite schist and quartzite, and the Kozhanovichi Formation — porphyritic andesite to rhyolite effusives. Pap (1977) considers the Zhitkovichi Group to be lower Proterozoic and claims it is the equivalent of the leptite formation in the Svecofennides of the Baltic Shield. However, Bondarenko (in Kratz et al., 1979) views the group as an analogue of the Tallinn-Loksa (Jagala complex) metamorphic complex and on this basis refers it to the upper Archaean.

The quartz-rich rocks of the Zhitkovichi Group are metasomatic rocks which formed during two phases of acid leaching. Kyanite quartzites were formed in an early quartz-kyanite alkaline episode, related to metamorphism at staurolite subfacies conditions of the almandine-amphibolite facies. Muscovite-quartz schists formed in a subsequent quartz-muscovite late alkaline episode, associated with a later greenschist facies retrogressive metamorphic episode (Kratz et al., 1979).

Intrusive complexes. Among the undifferentiated Archaean to Early Proterozoic igneous complexes are microcline granites and the Orsha and Osnitsa (Mikashevichi and Zhitkovichi) complexes of inhomogeneous composition.

In southern Belorussia, around Smolensk and Kaluga, intrusions of microcline granitoids and gneissose granites are common. They are structurally concordant with the surrounding granite-gneiss complexes. Relicts of Archaean schists and charnockite series rocks, as well as amphibolites and magnetite-bearing formations, occur in the Kaluga gneissose granitoids. Plagioclase-microcline-muscovite granitoids, often with high quartz and epidote contents, predominate in this complex. According to L.P. Bondarenko, they formed as a result of superimposed metasomatic processes of acid leaching facies, related to Early Proterozoic regional metamorphism. The complex as a whole is therefore the metasomatic and partial melt product of reworking in a basically lower Archaean but also partly lower Proterozoic substrate.

In the northern part of the block inhomogeneous intrusions are widespread, from diorite to granite (the Orsha complex). Gneissose granite, granodiorite and diorite have been penetrated in drill holes, as well as intensely blastomylonitized hornblende-biotite and biotite gneiss which have undergone microcline-quartz metasomatism. This granitoid complex is believed to have formed from lower Archaean gneisses, as well as from formations similar to the Okolovo Group.

Rocks of the Osnitsa complex in southern Belorussia form intrusions of the gabbro-diorite-granite series, cutting both Archaean and lower Proterozoic assemblages. The most complete expression of this complex is found in the region of the Mikashevichi-Zhitkovichi basement outcrop, which includes hornblende gabbro, diorite (with gneissose textures), granodiorite, tonalite and granite. All the rocks show signs of hybrid origin. Field evidence proves that the Zhitkovichi porphyroblastic biotite-hornblende granites (zircon age c. 1700 Ma) cut the Zhitkovichi Group. The age of zircons from plagioclase-microcline granites in the Mikashevichi massif is 1865–1935 Ma.

Lower Proterozoic sediments within the block are developed to an extremely limited extent. Around the Bobruysk region, quartz porphyries have been drilled which are probably equivalent to the Hoglandian.

Metamorphic complexes in the Smolensk block have practically not been studied from the point of view of their ore mineral content. There are indications of copper-nickel, rare-metal–tin and rare-earth metallogenic types in the Orsha complex. Microcline granites and gneissose granites contain REE and W, Sn, Mo. Rare-earth mineralization may be associated with alkaline metasomatic rocks (Makhnach et al., 1979).

The Kursk-Voronezh block

The Kursk-Voronezh block is located SE of the Smolensk block, and incorporates a number of regions in the south European part of the Russian Federation, from the Kursk and Orel regions in the NW to the Volgograd region in the SE. The territory of this block is distinct in terms of possessing a unique and highly complex geological structure.

The *Oboyan Group* is regarded by all workers as belonging to the *lower Archaean*. Within the Kursk Magnetic Anomaly (KMA), Oboyan granite-gneisses occupy the so-called inter-anomaly zone, interpreted as anticlinorial zones between synclinoria filled with stratified Archaean formations.

The Oboyan Group consists of biotite and biotite-hornblende-plagioclase gneiss, strongly reworked by partial melting and regional silica-soda metasomatism and altered to migmatite and gneissose tonalite. Mineral parageneses suggest regional metamorphism under high-temperature almandine-amphibolite facies conditions. There is also reason to believe that there was an earlier metamorphism at medium-pressure granulite facies conditions and therefore the amphibolite facies metamorphic event could be considered as retrograde (Zaytsev et al., 1978).

The structure of the Kursk-Besedino and Maloarkhangelsk anomaly zones is a more complex matter. Zaytsev and others have grouped the metabasites and associated iron-ore formations found here under the name of the *Kursk-Besedino assemblage*, which is considered to be part of either the Oboyan Group or older formations. Recently, metavolcanics (the Pokrov Group) and minor gabbro-norite and pyroxenite intrusions (the Besedino complex) have been regarded as belonging to the final stage of the Early Archaean cycle (Chernyshov et al., 1982; Krestin and Berdnikov, 1982).

However, this does not exhaust the range of geological possibilities regarding the content of the anomaly zones. Bondarenko (in Kratz et al., 1979) has described high alumina gneisses (typical metapelites) from the top of the section in a number of drill cores, also patches of hypersthene-magnetite and garnet-two-pyroxene-magnetite quartzite and eulysite, interleaved with two-pyroxene and garnet-two-pyroxene schist. We may possibly be dealing here with analogues of lower Archaean complexes described above from other blocks. Polishchuk and Polishchuk (1978) previously considered the Besedino unit to be the age equivalent of the upper Archaean Mikhaylov Group.

The *upper Archaean* consists of *Mikhaylov Group* sediments, which form a number of narrow elongate greenstone belts, drilled and exposed in quarries within the KMA. The group consists mainly of basic rocks: actinolite-hornblende and hornblende-feldspar amphibolite, amphibole-biotite gneiss, amphibole-magnetite schist, and rarer garnet-biotite and biotite-amphibole gneiss. Thin beds of silicate-magnetite and magnetite quartzite have been found (1–40 m or more thick), also ore-free quartzite.

Krestin (1978) showed that metamorphosed lavas at the base of the section are komatiitic basalts in the main, alternating with flows of pyroxenitic and peridotitic komatiite. Compositionally the basalts are continental tholeiites (plateau basalts, Krestin and Berdnikov, 1982). Ultrabasic intrusions, everywhere containing sulphide mineralization, are strongly associated with the basic–ultrabasic part of the section. The overlying part consists of andesite and rhyolite (probably lavas and tuffs).

The sediments were metamorphosed at greenschist facies conditions, but in places the metamorphic grade can reach almandine amphibolite facies. Metamorphism was accompanied by the intrusion of major granodiorite-tonalite bodies, the Saltykov complex. A Rb–Sr date on muscovite from late pegmatite dykes is 2950 ± 40 Ma.

Over the greater part of the block the *lower Proterozoic* consists of iron ore schists of the Kursk Group and the volcano-sedimentary Oskol Group. These are restricted to relatively narrow (up to 50–70 km) zones which can be traced along strike for several hundreds of kilometres. The zones are complex graben-synclines, bounded by steep faults, sometimes dissected into chains of strongly flattened basins lying above faults, or small isolated troughs. There is a marked change in succession along and across strike, and intraformational erosion surfaces have been observed. Recently (Krestin and Berdnikov, 1982) these structures have been regarded as "protoaulacogens", i.e. essentially ancient platform structures.

The Kursk Group consists of the Stoylen and Korobkov (upper) Formations. The former consists of rhythmically layered phyllitic, often carbonaceous, shale, metasiltstone, quartzose sandstone and dolomite. In the most complete sequences, where the thickness can attain 1.0–1.5 km, three transgressive–regressive rhythms can be recognised. At the base of the section there is a basal gravel conglomerate unit with pebbles of tonalite that cuts the Mikhaylov Group. The Korobkov Formation consists of ferruginous quartzite and quartz-mica, graphite-mica and other schists, sometimes with garnet and amphibole. The ferruginous quartzites form two horizons, the lower one containing abundant magnetite, silicate and magnetite quartzite, and the upper iron-mica-magnetite quartzite. The thickness of the series of ore bodies of ferruginous quartzite reaches 500 m and more.

Oskol Group rocks unconformably overlie Kursk Group rocks, in places with an angular unconformity. The development of a pre-Oskol weathering crust has been noted. The succession is dominated by frequently carbonaceous quartz-sericite and quartz-chlorite-sericite schist, as well as limestone and dolomite. There are specular iron fragments in the schists. The lower part of the succession is coarsely clastic. Conglomerate lenses are up to

100 m thick. Ferruginous quartzite predominates as clastic material. Discontinuous lenticular beds of silicate- and carbonate-magnetite quartzite are up to 150 m thick. Individual structures have thin bands of altered volcanogenic rocks (potassic rhyolite and tholeiitic basalt). Generally speaking, the metamorphism of lower Proterozoic formations is inhomogeneous, varying from greenschist to lower amphibolite facies in low pressure conditions.

A totally different lower Proterozoic sequence occurs in the eastern part of the block. West of the Losevo-Mamon regional fault there are metavolcanics of a basalt-andesite-dacite suite (the Losevo Group or the Voronezh Formation, according to Yu.S. Zaytsev) totalling at least 1000 m in thickness. Eruption of the volcanics was accompanied by deposition of very small quantities of coarse clastic terrigenous rocks and the intrusion of gabbro-diabases. A number of workers (Zaytsev et al., 1978; Krestin and Berdnikov, 1982) believe that the formation of the volcano-sedimentary Losevo Group was completed prior to the deposition of Vorontsov Group sediments.

East of the Losevo-Mamon fault is a wide blanket of Vorontsov Group terrigenous sediments, consisting of over 1500 m of phyllitic quartz-plagioclase-mica, graphite, andalusite, staurolite and garnet-staurolite schists, deformed into major open folds.

Vorontsov Group sediments were metamorphosed mainly at greenschist facies conditions. There is a clearly marked zonation with an increase in metamorphic grade to epidote-amphibolite facies and the low-temperature field of the cummingtonite-amphibolite facies. Overall, the metamorphic grade increases towards the south and south-west. In isolated localities, the metamorphic grade reaches the high temperature subfacies of the amphibolite facies, with the development of garnet-biotite and garnet-cordierite gneisses and the appearance of anatectic granitization. Andalusite-type zonal metamorphism has been reliably established (Zaytsev et al., 1978; Kratz et al., 1979).

The youngest lower Proterozoic sediments are a terrigenous-volcanic complex, which crops out in relatively minor basins over the entire block. In the north of the Michurinsk-Volgograd zone (where the Vorontsov Group crops out) is the Baygorov Formation, over 400 m of tuffaceous schist and tuffaceous sandstone, chloritic schist and quartzose sandstone belonging to this complex. In the lower part of the succession are 10–20 m thick bands of conglomerate. In the Kursk Magnetic Anomaly region, the complex consists mainly of basaltic rocks (the Glazunov Formation), a few hundred metres thick. The rocks are either subhorizontal or they form monoclines with moderate dip angles. Greenschist facies metamorphic effects are weakly expressed. The most likely correlation of these sediments is with the Hoglandian.

Intrusive complexes PR_1. The Kursk-Oskol evolutionary cycle concluded with the intrusion of a small number of minor syenites (2220 ± 50 Ma) and gabbro-diabases (2170 ± 50 Ma).

In the east of the block, activation of deep faults at this time led to the intrusion of ultramafic–mafic associations (the Mamon and Peskovat complexes) and a gabbro-diorite-granodiorite suite of intrusions (the Stoylo-Nikolayev complex). In northern regions of the block we have the Maloarkhangelsk gabbro-norite complex. The widespread distribution throughout the entire block of palingenic plagioclase-microcline granites, such as the Liskin and Bobrov intrusions, is probably related to deep fault zones. The youngest members of this group appear to be the ovoidal (rapakivi) granites of the Olym complex.

Associated mineral deposits. The outcrop area of the Kursk Magnetic Anomaly belongs to a number of ore regions with a developed extractive mining industry. The main target in iron ore extraction is the ferruginous quartzite in the Kursk Group. The Kursk iron ore unit is regarded as a schistose ferruginous-chert banded iron formation (Golivkin et al., 1982; Bondarenko et al., 1977) which makes up the cores or limbs of synclines within several ore zones. The thickness of the series of ore bodies exceeds 500 m, and is 5–100 km or more along strike (the Lebedinsk, Korobkov, Stoylensk and other deposits).

The ore content of other complexes in the Kursk-Voronezh block is at the investigation stage. Magnetite ore mineralization is associated with the lower Archaean Besedino unit in the area of the Kursk-Besedino and Maloarkhangelsk anomalies. Ferruginous-silicate rocks account for up to 20% of the volume of this unit. The ferruginous quartzites have high TiO_2 values (Polishchuk and Polishchuk, 1978).

Metabasites of the upper Archaean Mikhaylov Group have associated amphibole-magnetite and magnetite quartzite, but they do not form economic deposits. Ultrabasic intrusions, and probably also in part metavolcanics, have associated nickel-bearing sulphide and silicate ores and chromites. Late-stage mineralization related to a cycle of acid igneous activity concluding the Late Archaean, is represented by hydrothermal polymetallic pyrite and gold-sulphide mineralization (Chernyshov et al., 1982; Krestin and Berdnikov, 1982).

In the lower Proterozoic Kursk Group, as well as the iron-ore horizons, basal metasandstone horizons with lenses and beds of oligomict quartz-pebble conglomerate are of interest also. Gold ore shows are restricted to this horizon; the gold appears to be both hydrothermal and sedimentary in origin (Kononov, 1975).

A ferruginous-chert carbonate schist formation is associated with the Oskol Group in the cores of major synclines, such as the Belgorod and others. Here there is a high graphitic schist content, usually with pyrite and pyrrhotite mineralization. Associated minerals are sphalerite (disseminated and in veins), galena, etc. The Tim-Yastrebovka syncline has a graphitic schist-carbonate formation with associated phosphorite and manganese shows. Rich iron ore deposits are restricted to a weathering crust on a ferruginous and carbonate schist formation, while there are also bauxite deposits in the Belgorod ore zone.

There are a number of ore shows related to Early Proterozoic intrusive complexes. The Mamon gabbro and peridotite complex has Pechenga-type liquation Cu–Ni mineralization, while in the Peskovat peridotite, pyroxenite and gabbro-norite complex the Cu–Ni mineralization is Monchegorsk type. Rocks belonging to the Stoylo-Nikolayev complex have associated hydrothermal pyrite-polymetallic, galena-sphalerite and gold-sulphide ore mineralization (Krestin and Berdnikov, 1982). Finally, hydrothermal Mo, Pb, Zn, Sn, Bi and REE ore shows are known from the latest microcline-plagioclase and other granites belonging to the Bobrov, Liskin etc. complexes.

The Volga-Urals block

The Volga-Urals block or province encompasses a broad area in the south and south-east of the Russian Platform, with a sharp wedge jutting in towards the centre, in the direction of Ryazan and Moscow. In the east, the block lies beneath the Urals fold belt, while to the SE it is cut off by the superimposed Caspian depression with a highly altered and basified basement. The NW and SW margins of the block are sharp and geophysically well-expressed, although there is very little information concerning their geological structure. This results from sudden and very deep burial of the crystalline basement in the marginal zones of the block and the development above the margins of a thick Riphean sedimentary assemblage.

The structure of internal regions of the block has been studied very unevenly, in spite of the large number (over 3000) of bore holes which have reached the crystalline basement. Two holes (Tuymazy-2000 and Minnibayevo-20000) have penetrated the basement to 2.3 km and 3.2 km, respectively.

Lower Archaean. In general terms, the rock sequence of the lower Archaean granulite complex corresponds to that found in western regions of the platform. The lower group within isometric granulite massifs (the *Otradnin Group*) consists of two-pyroxene schist, hypersthene-plagioclase gneiss and minor amounts of aluminous plagiogneiss (the Nurlat complex, according to S.V. Bogdanova). The vertical thickness encountered reaches 1 km (the Ulyanov field). Outside the granulite massifs, the group comprises wide-spread ultrametamorphic enderbitic gneiss, charnockite and alaskaite

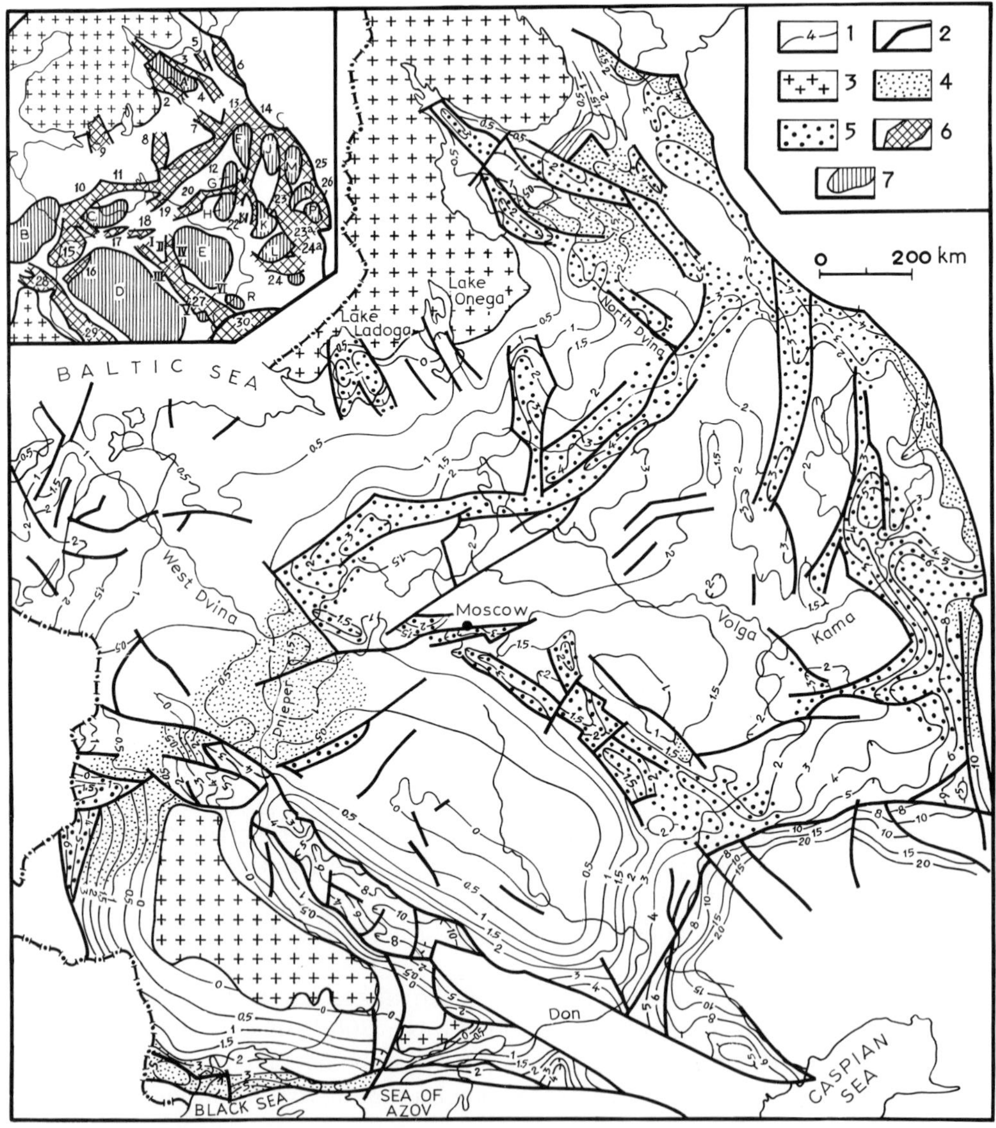

Fig. I-51. Sketch map showing distribution of Middle–Upper Riphean sediments on the Russian Platform. *1* = contours and depth to basement (in km); *2* = major faults; *3* = exposed basement; *4* = Riphean in pericratonic and intracratonic basins; *5* = Riphean in aulacogens. Contour intervals: 0.5 km (<2 km), 1 km (to 6 km), 2 km (to 10 km), 5 km (>10 km). *Inset*: basic structural elements of platform cover (regions drawn for top of basement). *6* = negative structures; *7* = positive structures. Basins (numbers): *1* = Kandalaksha, *2* = North Dvina, *3* = Kildin, *4* = Pinezh, *5* = Leshukon, *6* = Safonov (Safonov and Omsk basins), *7* = Toymen, *8* = Vozhe-Lacha, *9* = Ladoga (W Ladoga and Pasha grabens), *10* = Krestets, *11* = Rybinsk, *12* = Sukhona, *13* = Yarensk, *14* = Vychegda, *15* = Orsha, *16* = Klintsov, *17* = Gzhatsk, *18* = Moscow, *19* = Yaroslavl, *20* = Puchezh graben system, *21* = Kazhim, *22* = Kazan-Vyatsk graben system, *23* = Kama-Belsk (*23a* = Osa basin), *24* = Sernovodsk-Abdulin (*24a* = Bavlin

granite (the Kinel complex). Using accessory zircons from charnockite, an isochron age of 2790 ± 10 Ma has been published for this complex (Bibikova et al., 1984). Chemically, the mafic schists correspond to differentiated tholeiitic basalts. Hypersthene-plagioclase gneisses on average correspond to dacite (Bogdanova et al., 1982), and the averaged chemical composition of the group is close to andesite. The Nurlat complex has a number of geochemical features similar to "grey gneisses".

The succeeding *Bolshecheremshan Group* has been exposed in drill holes over most of the Almetyev dome in the Tatar arch. The position of this basement arch and its component domes is shown in Fig. I-51. A borehole in Minnibayevo has penetrated some 3 km of this group. From geophysical data, the total vertical thickness has been estimated to be 6–8 km. The Bolshecheremshan Group consists predominantly of high-alumina gneiss, more leucocratic biotite-garnet and biotite-garnet-sillimanite gneiss and migmatite. Major intrusions of tonalite and microperthitic garnet-cordierite granitoids are frequently associated with the gneisses. In the lower part of the sequence are thin bands of mafic schist. The primary composition of the high-alumina rocks was pelitic, while the aluminous plagiogneisses were greywackes and pyroclastics (Bogdanova et al., 1982).

The Otradnin and Bolshecheremshan Groups were metamorphosed under medium-pressure granulite facies conditions. According to data obtained by Sitdikov (1982), the origin of the enderbites in the Otradnin Group was related to an initial stage of partial melting during a granulite episode, which was accompanied by soda metasomatism. In a subsequent episode, potash-series charnockites formed. The P–T conditions of enderbite and charnockite formation have been determined to be 700–730°C and 6–7 kbar. On the southern flank of the Almetyev dome (the Nurlat region), there are garnet-clinopyroxene eclogitic rocks, garnet-two-pyroxene granulite and eclogitized gabbro in the granulite complex. According to a few isolated determinations, the mineral assemblages of the garnet-pyroxene rocks formed at T = 910–920°C and P = 8.0–10.3 kbar.

The high-pressure granulites probably also include the eulysites described from the Kuybyshev area in the Volga region (Nesmeyanov field). Rocks close to eulysite — garnet-ferrohypersthene schist — occur in a 120 m thick horizon in the Minnibayevo borehole. Granitization of these rocks was accompanied by magnetite segregation.

---

graben), *25* = Solikamsk, *26* = Yuryuzan, *27* = Ryazan-Saratov (grabens: *I* = Ryazan, *II* = Nepeytsin, *III* = Kaverin, *IV* = Sasov, *V* = Pachelma, *VI* = Petrovsky), *28* = Pripyat, *29* = Dnieper-Don, *30* = Cis-Caspian syncline. Positive forms (letters): *A* = Archangelsk upwarp, *B* = Belorussian-Lithuanian anteclise, *C* = Nelidov arch, *D* = Voronezh anticline, *E* = Tokmov arch, *F* = Sysola arch, *G, H* = Kotelnik upwarp (N and S closures), *J* = Komi-Permyatsky arch, *K, L* = Tatar arch (N = Kukmor, S = Almetyev closures), *M* = Kama upwarp, *N* = Perm arch, *P* = Kungur upwarp, *R* = Pugachev arch.

Epidote-amphibolite to kyanite facies series (amphibolite facies) retrogressive metamorphism took place in separate narrow zones or isometric massifs. In a number of cases metamorphism seems to have been accompanied by tectonic reworking, with the formation of major linear zones (Kratz et al., 1979, pp. 32, 78). In some places there is evidence for later metamorphic events at lower pressure.

Also belonging to the lower Archaean is an intensely granitized gneiss complex which is most extensively developed in the east of the block in the basement to the Kama-Belsk depression. In the eastern regions of the Tatar basement high, Sitdikov (1982, Sitdikov and Polyanin, 1976) has grouped plagiogneiss, granite gneiss and migmatite into the *Ik Group*. The Tuymazy-2000 borehole has reached more than 1000 m into a complex of orthogneiss-granodiorite, charnockite, labradorite, etc., beneath the Ik Group granite-gneiss (1200 m thick in the drill core). All the rocks are intensely granitized and retrogressed at metamorphic conditions of lesser depths. The gneiss-granitoid-charnockite complex probably forms a Katarchaean(?) basement complex. Petrographically, the complex resembles the ultrametamorphic complex in the central part of the Kola Peninsula (Kratz et al., 1979; Sitdikov, 1982).

The most typical *Early Archaean intrusive complexes* are the Tuymazy and Chubov (Bogdanova, 1982; Bogdanova et al., 1982). The Tuymazy metagabbro-norite-anorthosite complex consists of major sheet-like bodies up to 50–70 km thick. The rocks are often altered to two-pyroxene schist, plagiogneiss, etc. Their primary igneous nature has been established from relict igneous textures and minerals. Spinel peridotite, pyroxenite and melanogabbro-norite of the Chubov complex form minor (up to 3 km) isolated bodies.

Upper Archaean complexes have been found on various occasions while drilling the Tatar dome. In various regions these formations are locally developed, they are conditionally correlated and have different names, but frequently they are grouped under the term Sarman Group (Sitdikov, 1982; Bogdanova, 1982; Bogolepov and Votakh, 1977). Rocks of this group are restricted either to narrow zones of lower Archaean basement reworking (the Kukmor assemblage), or to broad areas of reworking of granulite massifs (the Uratmin assemblage).

The *Kukmor assemblage* consists of high-alumina kyanite-garnet-two-mica schist and fine-grained biotite gneiss with staurolite. Anthophyllite-tremolite-chlorite, actinolite-biotite and amphibole-plagioclase schist of this assemblage, which have been found in a few boreholes in the Kukmor region (the Northern Domes field), are termed the Privyatsk complex by Lapinskaya and Bogdanova (1982). The particular chemistry and banded nature of the succession led them to refer the hornblende schist and basaltic and peridotitic komatiite-type meta-effusives to Archaean greenstone belts.

The *Uratmin assemblage* consists predominantly of biotite-hornblende and hornblende-diopside schist, fine-grained hornblende-biotite gneiss and biotite-gedrite schist. A patch of gedrite-bearing schists, discovered in the Uratmin area, according to Bondarenko (in Kratz et al., 1979, p. 40) is identical to the upper barren member of the Okolovo Group in Belorussia. Here there is also a mineral assemblage of quartz or amphibole with magnetite which formed during an episode of retrograde metamorphism.

The degree of metamorphism of all the assemblages referred to the Sarman Group never reaches the level of migmatization and granitization. Almandine amphibolite (staurolite zone) facies metamorphism is typical (or garnet-kyanite-biotite-muscovite subfacies for the kyanite-bearing schists in the Kukmor assemblage). Taking into account the P–T regime, metamorphic grade and the nature of the rock distribution, these formations could be considered to be the age equivalents of the upper Archaean Okolovo and other groups.

The *Bakal (Sviyazh) granitoid complex* is probably Late Archaean. Tonalite predominates, with granodiorite and quartz diorite forming intrusions of varied shapes and dimensions.

The *lower Proterozoic* complex (*Uniy assemblage*) is noticeably different in the nature of the metamorphism from Sarman Group formations. The Uniy assemblage is locally developed within the Chernokholunitsky (upper reaches of the river Vyatka) and Almetyev basement blocks. The rocks are finely-crystalline andalusite-biotite knotted schist with limestone concretions, biotite, quartz-biotite-chlorite-tourmaline and actinolite schist, metamorphosed at staurolite zone conditions (cummingtonite-amphibolite facies). From their lithology and metamorphic regime, Uniy assemblage sediments can be correlated with the lower Proterozoic Vorontsov and Ladoga Groups.

The metallogenic signature of early Precambrian complexes in the Volga-Urals block has practically not been studied.

As far as the geological structure of central and northern regions of the Russian Platform is concerned, all we have at our disposal are regional geophysical data. As shown on the map (Fig. I-50), we assume that Archaean and lower Proterozoic rocks are developed here, which on the whole are similar in composition and metamorphic history to corresponding complexes in western and southern regions of the platform.

TABLE I-31

Riphean to Lower Vendian assemblages of the Russian Platform

Phyteme	Horizon	Kama-Belsk aulacogen		Bavlin-Baltayev graben		Sergiyevsk–Abdulin aulacogen			Ryazan–Saratov aulacogen	Moscow aulacogen		Orsha basin		Volyn basin	
Vendian	Lower Vendian							Drevlyan Gp	Zubovopolyansk *vt* Partsin *tg*				Sviloch *vt* Vilchan *tg*	Volyn Group	Berestovets *tv* Gorbashev *t* Vilchan *tg*
Kudash	Pachelm		660±20 Ma Shtandin *ct* Gozhan *tr* 700±25 Ma	Serafimov Gp	Leonidov *t* Olkhov *tr* Tukayev *tr*	Mishkin Gp	Leonidov *t* Shtandin *ct* Gozhan *tr*	Pachelma Gp	Sosedsk *tr* Krasno-ozersk *t* Voronya *tr* Vedenyapin		Nogin Group *t* Pavlovo-Posad *t*		Blonsk *tr*		
Upper Riphean	Tangaur	Kyrpin Group	Kaltasy *c* Arlan *ct* 920±30 Ma	Kudash Group	Malokamysh *c* Mizgirev *ct*			Peresypkin (Serdob) Gp	Sekretarkin *tc* Belynsky *c* Irgiz *t*	Loginov Gp	Zhukov *tc* Dulev *tc* Ignatev *t* Ulitin *t*		Lapich (Osipovichi) *tc*		
Upper Riphean	Kipach	Kyrpin Group	Tyuryushevo *tr* 1050±50 Ma	Kudash Group	Troitsky *tr*		Borovsky *tr*	Somovo Gp	Tsnin *tr* Rtishchev *tr*	Ramensky Group	Monin *tr* Runov *tr*	Belorussian (Polyesye) Group	Orsha *t* Rudnya *t* Rogachev *tr*		Polyese *tr*
Middle (?) Riphean								Kaverino Gp	Inkash *tr* Tyrnitsk *tr*						

2. PRECAMBRIAN OF THE PLATFORM COVER

The Precambrian complexes of the platform cover are considered here as belonging to the upper Proterozoic. Riphean to Lower Vendian sediments almost exclusively fill relatively narrow elongate rifts, usually called aulacogens, in the crystalline basement (Fig. I-51). Upper Vendian (Valday Group) sediments are more widely distributed. They form part of the cratonization structural stage.

Despite the fact that Riphean platform sediments have been actively studied for over two decades, many questions concerning their correlation and stratigraphy remain disputed and authors differ in their opinions for a whole number of reasons. So far, the thickest and probably the most complete Riphean sections have not been exposed by drilling. Riphean biostratigraphic subdivisions have had to be worked out not on the basis of fossil evolution but mainly from trace fossil evidence. Nor do radiometric data provide sufficiently reliable results in any quantity. Up to the present time, no satisfactory correlation between K–Ar and Rb–Sr glauconite dates has been obtained, which is an indication of the poorly understood evolution of isotopic systems in glauconite and other clay group authigenic minerals.

In accordance with the decision of the Inter-Departmental Stratigraphic Committee in 1978 on the unified regional stratigraphic scale for the Precambrian of the European part of the USSR, the Riphean is divided into Lower, Middle and Upper.

The overlying Kudash sediments are often included in the Riphean. The taxonomic rank of the Kudash cannot yet be considered to be completely defined: different authors consider it variously as either a phytem (a major Precambrian biostratigraphic unit defined on the basis of stromatolites), i.e. a subdivision, adequate for the whole of the Upper Riphean, or as a Regional Stage, following on from the Kipchak and Tangaur stages of the Upper Riphean.

Various authors at times adhere to the most divergent views concerning the age of individual formations and groups. In this summary, the interpretation of the data for key regions in the Volga-Urals province is based mainly on the work of Lagutenkova and Chepikova (1982), who have made a special study of the palaeogeographic facies conditions of Riphean sedimentation in the region. Sections for central and western regions of the Russian Platform are compared with the Kama-Belsk depression (Table I-31).

Riphean–Lower Vendian assemblages

The *Kama-Belsk (Kaltasy)* aulacogen has a north–west strike, swinging to near north–south in latitudes north of the town of Izhevsk. It separates the Tatar and Komi-Permyatsk arches in the west and the Kama, Perm and Kungur arches in the east (Fig. I-51). The width of the northernmost part of the aulacogen does not exceed 30–40 km, but in the south in widens markedly in the direction of the Urals. The length of the entire structure is around 700 km, and the depth of the deepest part probably reaches 10–12 km, of which less than 5 km has been penetrated by even the deepest boreholes. The most tightly folded SE part of the structure is named the Osa depression. This consists of two parallel troughs, the Sarapul in the west and the Bolsheusinsk in the east. The upper surface of Riphean and especially of Vendian sediments is highly smoothed.

The oldest exposed sedimentary assemblage is the Tyuryushev Formation (Table I-31, Fig. I-52), the exposed thickness of which reaches 690 m in the Kokar graben. Seismic survey data show that sediments of the Tyuryushev Formation or its analogues occur throughout the entire Osa depression. The formation consists of two units of red quartz-feldspar sandstone, separated by a horizon of coarser-grained rocks — even-grained sandstone and fine-grained conglomerate with a patch of breccio-conglomerate at the base. As a rule, all the sandstones are poorly sorted, especially those in the lower unit.

In the western and central parts of the Osa depression (the Arlan and Oryebash areas, for example) there are carbonate-terrigenous sediments, the Arlan Formation. The quartz-feldspar sandstones are red or pink, while the finer-grained fractions are mostly dark grey or sometimes greenish. Dark grey to black shales constitute the upper part of the section. Among the carbonate layers, there is a predominance of dolomites and dolomitic marls. The lower boundary of the formation — the contact with the Tyuryushev Formation — is not exposed. The supposed relationships between the Arlan and Tyuryushev Formations is confirmed by relations of their analogues, the Mizgirev and Troitsky Formations, in the Bavlin-Baltayev graben. The exposed thickness of the Arlan Formation reaches 980 m.

Above lie the sediments of the Kaltasy Formation, widespread in the Osa depression. These sediments are more widely developed than the Arlan Formation and around the periphery of the aulacogen they may lie immediately on the crystalline basement, as in the region of Severokamsk.

In the majority of sections the formation consists of dolomite (pelitomorphic chemogenic, biogenic stromatolitic, fragmental), grey or sometimes reddish-brown, and frequently silicified. Generally speaking, the facies composition of the formation is quite varied. For example, in the central part of the Osa depression (Kushkul), argillaceous siltstones and shales play a significant role in the succession. Terrigenous material also occurs in north-eastern and western sections together with gypsum and anhydrite inclusions. The sediments are cut by gabbro-diabase intrusions, 57 m thick in cores. The total thickness of the Kaltasy Formation reaches 1980 m at Oryebash.

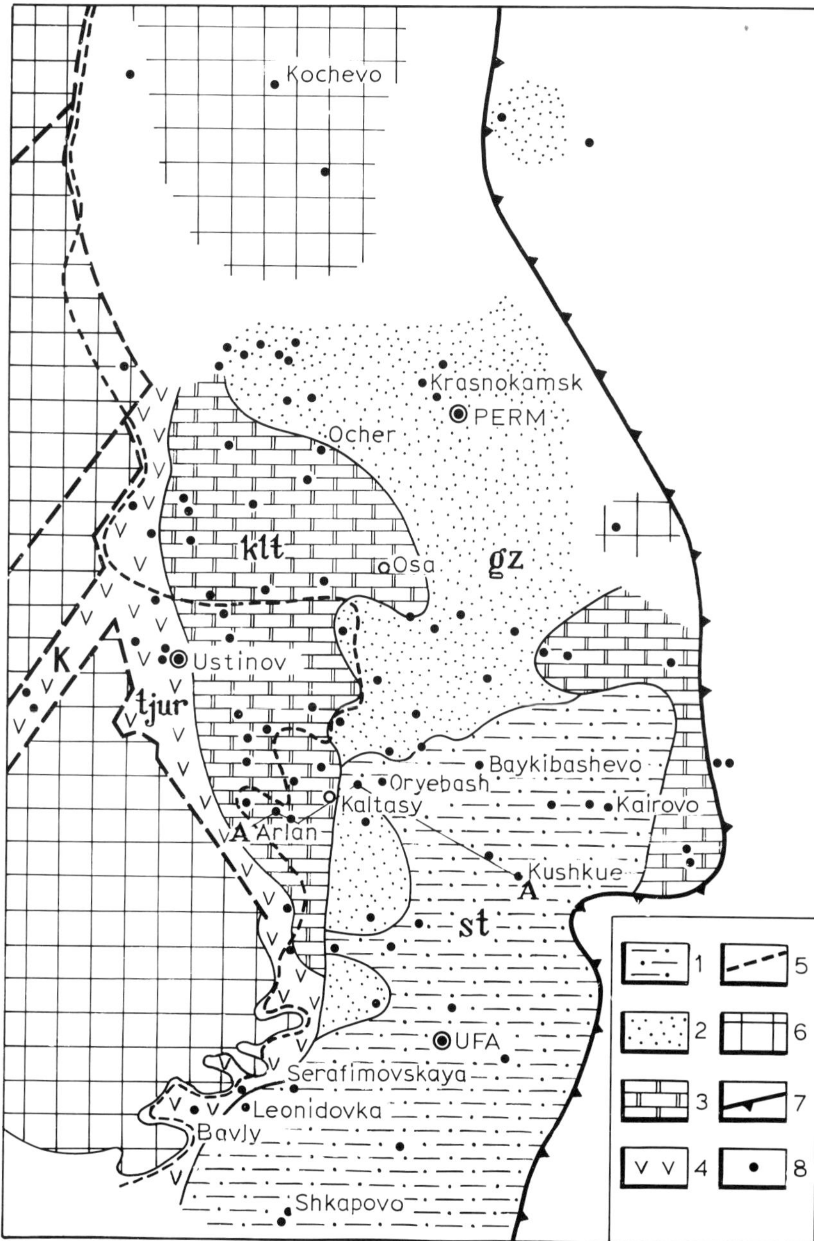

Fig. I-52. Sketch map showing distribution of upper Precambrian in the Kama-Belsk aulacogen. Vendian and younger sediments removed. (Modified from Lagutenkova and Chepikova, 1982.) Formations: *1* = Shtandin (*st*), *2* = Gozhan (*gz*), *3* = Kaltasy (*klt*), *4* = Tyuryushev (*tjur*); *5* = W boundary of Upper Vendian; *6* = sediments absent; *7* = W boundary of Urals; *8* = boreholes. *A–A* = seismic profile line (Fig. I-53). *K* = Koykar graben.

The age of the formations discussed above (they are sometimes grouped together into the Kyrpin Group) has been and is still treated variously by different research workers. A comparison is also made between the Kaltasy carbonate Formation and one of the carbonate formations in the Southern Urals stratotype section respectively (with the Satkin R_1, Avzyan R_2 or Minyar R_3, Tangaur). K–Ar isotopic determinations on glauconite from the Arlan Formation and from gabbro-diabases are highly contradictory and cannot at present be used as the basis for determining the stratigraphic position of the formations. Arising from this is an uncertainty in evaluating the age of formations belonging to the Kyrpin Group. A number of authors, including M.A. Semikhatov, Ye.M. Aksenov and K.E. Yakobson, refer the entire Kyrpin Group to the Lower Riphean. I.Ye. Postnikova and E.A. Revenko refer only the Tyuryushev and Arlan Formations to the Lower Riphean, and they consider that the Kaltasin Formation, which lies above a break, to belong to the Upper Riphean. A.A. Klevtsova refers the Kyrpin Group to the Middle Riphean. A characteristic feature of the majority of these schemes is a definite hiatus in the scale — the absence of sedimentation throughout the entire Middle Riphean, i.e. around 400 million years.

The least contradictory idea at present is that all the exposed Kyrpin Group sediments are Late Riphean in age, a view held most consistently by Lagutenkova and Chepikova (1982). This viewpoint is based on the following considerations.

Within the Osa basin, immediately adjacent to the Riphean stratotype region which includes the Lower Riphean, the deepest boreholes, 4 km deep, have not passed through Riphean sediments. Considering that the Lower Riphean has already been reached in such boreholes, then its total thickness would exceed 6–7 km; this in itself is unlikely and contrasts sharply with the thicknesses of all the overlying subdivisions of the Riphean. Furthermore, from an analysis of seismic sections and geological profiles (Frolovich, 1970) it follows that beneath the Arlan and Tyuryushev clastic formations there is a regionally distributed assemblage of very dense rocks with an average layer velocity of 6.5–7.5 km/s, underlain by a rock complex with lower velocities of 5.5–6.0 km/s. For comparison, we may quote the average layer velocity of the Arlan-Tyuryushev sediments of 4.7 km/s and the Kaltasin sediments of 6.5 km/s. The total thickness of pre-Tyuryushev sediments in the central part of the Osa basin is not less than 3 km (Fig. I-53), while in the eastern part of the basin it probably reaches 5–6 km.

Much of the Kaltasy Formation contains microphytoliths (determined by, among others, E.A. Revenko, Z.A. Zhuravleva and I.K. Chepikova), characteristic of Riphean assemblages III and even IV. This allows us to refer sediments containing these fossils confidently to the Upper Riphean (Tangaur) equating the Kaltasy Formation with the Minyar Formation of the stratotype section. Detailed lithofacies work carried out by Lagutenkova

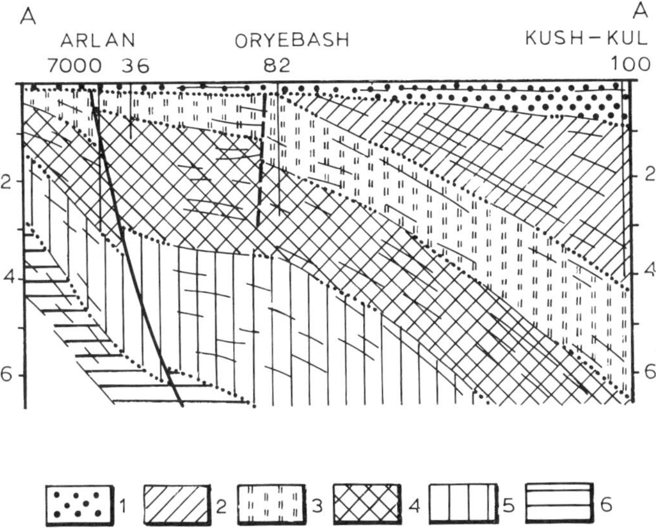

Fig. I-53. Seismic profile along line *A–A* (modified from Frolovich, 1980, and incorporating borehole survey data). *1* = terrigenous Devonian and Vendian. Formations: *2* = Shtandin and Gozhan, *3* = Kaltasy, *4* = Arlan and Tyuryushev, *5* = sequence of very dense rocks beneath Tyuryushev Formation, with average layer velocity of 6.5–7.5 km/s; *6* = rock unit with average layer velocity of 5.5–6.0 km/s.

and Chepikova (1982) support earlier data (Ivanova, 1970) concerning a gradual transition between the Kaltasy and underlying Arlan Formations. A possibility emerges from this that the Kaltasy and Arlan Formations may be correlated with the Peresypkin (Serdob) Group in the Pachelma aulacogen. Accordingly the Tyuryushev Formation may be equivalent to the Somovo Group in central parts of the platform and therefore belongs to the lower part of the Upper Riphean (the Kipchak).

The following is also pertinent. Kyrpin Group sediments together with overlying sediments of the Mishkin Group form a complete major sedimentary cycle, concluding with uplift of the territory and pre-Late Vendian erosion (Fig. I-54). Restricting the start of this cycle to the beginning of the Late Riphean also underlines the significance of the Middle-Late Riphean boundary (1050 ± 50 Ma) for eastern regions of the Russian Platform.

We propose with a fair measure of certainty that the Avzyan phase of folding, manifest on the western slopes of the Urals, is also present in the pre-Urals, significantly curtailing the area of the Riphean Kama-Urals pericratonic basin and marking the beginning of the evolution of the Kama-Belsk aulacogen.

Sediments lying immediately above the Kaltasy Formation, both on the basis of microphytolith (assemblage IV) and acritarch determinations, as well as lithological similarity, can confidently be correlated on the one

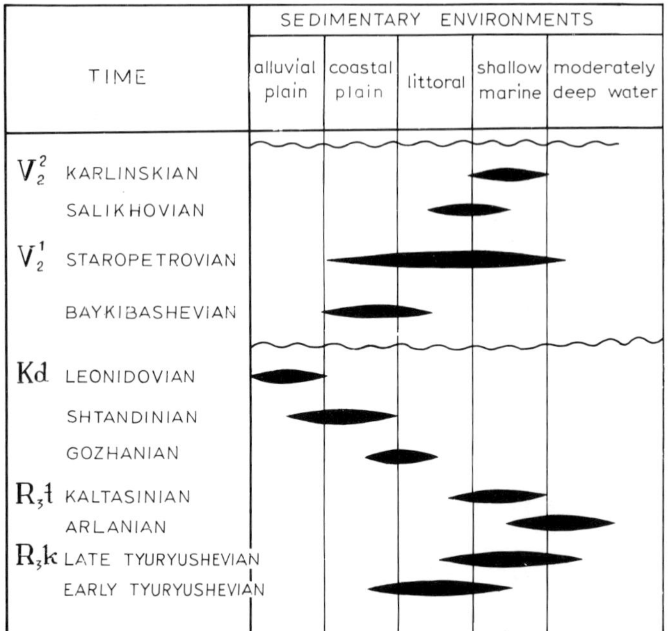

Fig. I-54. Change in facies conditions of sedimentary environments during the Riphean to Vendian in southern regions of the Osa basin (compiled from material in Lagutenkova and Chepikova, 1982).

hand with sediments of the Pachelma Group, and on the other hand with sediments of the Uk Formation of the Urals (Kudash). These sediments are sometimes grouped together as the Mishkin Group.

The Gozhan Formation consists predominantly of fine-grained sandstone with thin siltstone and shale. Clastic material is mostly quartz (60–95%) and potash feldspar. The rocks are pink and reddish-brown. Within the Osa basin the formation does not exceed 200 m in thickness, while in the west and north of the basin rocks of this formation are completely absent. The thickness increases sharply on the SE margin of the basin, where it reaches 800 m and over (at Kushkul). In the NE the sediments extend beyond the margins of the basin into the Perm arch and the Solikamsk foredeep. In the Perm-Berezniki region there is a major palaeobasin in which the formation may attain a thickness of 1000 m. The sediments are cut by gabbro-diabase intrusions, up to 140 m thick.

The Shtandin Formation is conformable on the Gozhan. It comprises reddish-brown and green siltstone, shale and dolomitic marl. The content of dolomitic marls (and rarer dolomites and dolomitic limestones) increases from west to east. The formation varies greatly in thickness, due to erosion in pre-Valday time. The maximum thickness, up to 400 m, is seen in the southern part of the Osa basin. The sediments wedge out northwards from the Oryebash-Baykibashevo line.

The *Sergiyev-Abdulin (Sernovodsk-Abdulin)* aulacogen is situated to the south of the Kama-Belsk aulacogen. In the west it is divided into the Almetyev arch and its Tuymazin tectonic salient, but farther east the boundary between the two aulacogens is tentative. The Sergiyev-Abdulin aulacogen can be traced for 500 km in an east–west direction. It does not exceed 50 km wide in the west, but this increases to 150–200 km in the east. In cross-sectional profile the aulacogen is highly asymmetric, with the main trough following the southern margin, where depths to basement reach 5–6 km. In the NE part of the structure, parallel to the main trough, is the separate narrow but deep Bavlin-Baltayev (Serafimovsk) graben. From the pre-Devonian surface, we observe a monocline plunging southwards in step-like fashion, in the region of the Sergiyev-Abdulin aulacogen and adjacent areas to the south.

The Riphean succession is very similar to that in the Kama-Belsk aulacogen. The oldest exposed rocks are coarse red sandstones of the Borov Formation, and in the Bavlin-Baltayev graben, red sandstones of the Troitsky (Tyuryushev) Formation. The overlying two formations are known only from the Bavlin-Baltayev graben. These are greenish-grey and grey-brown quartz-feldspar siltstones with thin shales, sandstones and dolomites of the Mizgirev Formation (up to 150 m) and pink dolomites of the Malokamysh Formation (up to 105 m). All these formations correlate well with sections in the Osa basin.

The overlying sediments in the Sergiyev-Abdulin aulacogen are usually grouped together as the Mishkin Group (Postnikova, 1977). Different workers correlate borehole sections slightly differently, though most usually identify the Gozhan and Shtandin Formations, as in the Osa basin. The exposed thickness of the Gozhan Formation reaches 1340 m (at Shkapovo), and the Shtandin 400 m (at Shkapovo and Kipchak). In the Bavlin-Baltayev graben, the equivalent to these formations are Serafimov Group sediments.

Above comes the locally distributed Leonidov Formation (in some schemes, see for example Semikhatov et al. (1979), Frolovich (1980), the post-Kaltasy part of the sequence is composed of the Leonidov and Priyutov-Shikhan Formations). The lower contact of the formation has not been studied in boreholes. The formation consists entirely of grey, pink and brick-red quartzose sandstone and occasional poorly sorted fine-grained quartz-feldspar conglomerate and minor siltstone and shale. The formation varies in thickness, the exposed thickness in the Bavlin-Baltayev graben being 760 m.

The *Kazan-Kazhim (Vyatka)* aulacogen is located west of the northern N–S striking part of the Kama-Belsky aulacogen. The structure separates a system of elevations in the Sysola arch and the Kotelnik upwarp in the west (Fig. I-51) and the Tatar arch (northern dome) and the Komi-Permyatsk arch in the east. The aulacogen is over 400 km long and 40–70 km wide. The depth to the crystalline basement surface is 3–3.5 km in the northern and central parts of the depression.

Initiation of this aulacogen probably occurred at the same time as that of the Kama-Belsk aulacogen. Riphean sediments consist of red quartz-feldspar sandstones of varied grain size with small clasts and thin conglomerate horizons. These Kazhim Formation sediments are equivalent to the Tyuryushev Formation in eastern regions of the Russian Platform. The thickness of Riphean sediments is estimated to be 500–1500 m, but the total thickness is not exposed. In the north of the aulacogen, Riphean sediments are covered by Upper Vendian sediments (up to 30 m) and elsewhere they are immediately overlain by the Middle Devonian.

The *Ryazan-Saratov (Pachelma)* aulacogen is situated between the Voronezh anteclise and the Tokmov arch. It is 550 km long by 130–170 m wide. From geophysical data, the depth to basement is 3.5–4.0 km. A second trough, the Petrov-Sasov zone, occurs along the Tokmov arch and is confirmed by drilling results in the region of the Nepeytsin and Sasov grabens. Between the trough zones there is a zone of basement highs, where the Riphean succession begins with the Peresypkin Group (Upper Riphean, Tangaur). The amplitude of these highs relative to the troughs is 0.5–1.5 km. The basic structural elements of the aulacogen are reflected in very broad general terms in the pre-Devonian surface. The surface of terrigenous Devonian sediments in areas of negative structures describes a uniform slope from the Voronezh anteclise in the NE towards the Tokmov arch. The aulacogen was initiated in the Middle Riphean according to Yakobson (in Bogolepov and Votakh, 1977), Klevtsova (1979), Semikhatov et al. (1979) and Lagutenkova and Chepikova (1982), although Postnikova (1977) maintains that it is Early Riphean. The uncertainty arises because of a lack of information and the monotonous nature of the lower redbed part of the succession.

The most complete section has been exposed in the Somovo borehole, 3703 m deep. Here the lower part of the succession consists of conglomerate and quartzose sandstone of the Tyrnitsk Formation and poorly sorted quartz-feldspar sandstone and conglomerate of the Inkash Formation (the Kaverino Group, total thickness 915 m). Above come pink quartzitic sandstone of the Rtishchev Formation and coarse-grained unsorted red quartz-feldspar sandstone and fine conglomerate of the Tsnin Formation (Somovo Group, total thickness over 1100 m). Pink quartzitic sandstones comparable with the Rtishchev Formation are known from the Saratov region in the Volga lowlands (Tatishchev Formation), where they have a total thickness of up to 740 m. Probable analogues of the Somovo Group occur in other boreholes in the Kaverino and Ryazan grabens. However, they cannot reliably be correlated with the Somovo reference section.

A Kipchak age for the Somovo Group is based on analogy with the

Tyuryushev Formation and is strengthened by several determinations of an acritarch assemblage (Postnikova, 1977, p. 34) from the overlying Irgiz Formation, similar to a microfossil assemblage from the Inzer Formation in the South Urals stratotype section. The situation regarding the Kaverino Group is much less certain. These sediments have not been characterised by microfossils. The contact with the overlying Somovo Group sediments has not been studied in drill cores, and a break is assumed on the basis of changes in commercial geophysical characteristics. Lagutenkova and Chepikova (1982) and Semikhatov et al. (1979) provisionally refer the Kaverino sediments to the Middle Riphean. If we consider that the initiation of the Kama-Belsk aulacogen, a genetically similar structure, occurred at the Middle-Late Riphean boundary, then we cannot exclude the probability that the Kaverino Formation could be younger, i.e. Late Riphean.

Overlying sediments lie above an erosional break on various Riphean horizons and partly overstep the margins of the aulacogen itself and lie immediately on the crystalline basement. There is a simultaneous sharp drop in the thickness gradient. Sections of post-Kipchak sediments differ in NW and SE parts of the aulacogen. In the south, in the Pachelma and Pugachev zones, only the lower part of the section is developed — Peresypkin (Serdobsk) and Pachelma Groups. The Peresypkin reference section consists of a lower grey quartzose sandstone, the Irgiz Formation (50–750 m), followed by mainly organic dolomite and limestone of the Belyn Formation (up to 110 m), which correlate with the Malokamysh and Kaltasy Formations (Tangaur) in eastern regions of the Russian Platform. Belyn Formation carbonates occur only in the Pachelma and Pugachev zones. The Peresypkin Group concludes with dolomitic marl and limestone with mottled sandstone and siltstone of the 60–130 m thick Sekretarkin Formation (southern regions of the Pachelma zone — the Serdobsk-2 borehole and the Pugachev zone), which more often tend to be referred to the Kudash from the typical microphytolith assemblage.

The overlying Pachelma Group sediments consist of grey sandstone at the base, followed by up to 200 m of grey and greenish shale, siltstone and marl of the Vedenyapin Formation, then up to 340 m of red sandstone, siltstone and shale, the Voron Formation, and locally in the north of the Pachelma and the south of the Kaverino graben (the Morsovo borehole), dark grey shale and greenish siltstone of the Krasnoozero Formation, up to 190 m thick. Additionally, in the centre of the Pachelma graben, there are well-sorted pink and red quartzose sandstones of the Sosedsk Formation (up to 240 m).

In north-western regions of the aulacogen, the exposed part of the succession consists of the Pachelma and Drevlyan Groups, previously referred to the Lower Vendian. The Drevlyan Group consists of a lower unit of tillite, glacio-lacustrine sediments and cross-bedded fluvioglacial sandstones, with various local names, up to 100 m thick and an overlying transgressive unit of light grey poorly-sorted quartz-feldspar sandstone and fine conglomerate which alternate towards the top with tuffaceous sandstone and siltstone (up to 115 m).

Yakobson (in Bogolepov and Votakh, 1977, p. 150) and Salop (1982, p. 150), using results from the Kaverino borehole, from where Krasnoozersk Formation sediments are missing, reject the existence of an erosion surface between the tillite horizon and the Voron Formation of the Pachelma Group. Instead, they correlate the Lower Vendian (Partsin Formation) tillite-bearing horizon with the Krasnoozersk Formation of more southerly regions.

The *Moscow* and *Gzhatsk* aulacogens occur *en échelon* south of the Nelidov arch. The structures are distinguished by an enormous thickness for relatively small dimensions, up to 250 km long by 25–30 km wide. East of Moscow, around Pavlovsky Posad, the most complete sequence of upper Proterozoic sediments in the central part of the platform is penetrated by a deep borehole. The hole has reached a depth of 4783 m, with upper Precambrian sediments first encountered at 1279 m. The overlying rocks are Middle Devonian.

Some 1900 m of predominantly red sandstones and siltstones are referred to the Riphean and Kudash. Overall, the succession has many features in common with that in the Kaverino graben. The stratigraphic position of red quartzose and oligomict sandstones, the Ramensk Group (depth interval 4783–3681 m), in the reference section is debatable. Different authors at different times have considered it to be Lower, Middle or Upper Riphean (Postnikova, 1977; Yakobson in Bogolepov and Votakh, 1977; Valeyev, 1978). This group is probably best compared with the Somovo Group in the Pachelma aulacogen.

An interbedded unit of sandstone (quartzose at the base to polymict at the top), siltstone and shale in the 3681–3206 m interval is usually separated out from the Ramenskaya Group and correlated with the basal part of the Peresypkin Group. Loginov Group sediments (3206–2891 m interval) are correlated with the carbonate part of the Peresypkin Group succession. These are predominantly grey to black shale, sandstone, limestone and dolomite. Overlying sediments of the Pavlovo-Posad (Table I-31) and Nogin Groups are correlated with the Pachelma Group on the basis of lithological composition and the presence of pronounced rhythmic layering, but they are much thicker (2891–1760 m interval). In both groups, the rocks are mainly coloured grey. Only the lower beds of the Pavlovo-Posad Group are red and brown. The middle (Dreznin) formation of the Nogin Group is noteworthy for being mottled. Lava and ash fragments have been found in the poorly sorted sandstones of this formation. Lower Vendian sediments appear to be missing from the Moscow aulacogen.

The *Orsha basin* follows on immediately south of the graben system of the Krestetsky basin, although there are pronounced morphological differences, both in amplitude and in the overall indistinct isometric shape. The most deformed central regions of the basin are between Smolensk and Vitebsk (the Rudnya region) and south of Orsha. In addition, there are a number of small narrow grabens (e.g. the Vitebsk) on the western flank of the basin. There was undoubtedly a connection, palaeogeographically, between the Orsha basin and the Volyn (Lvov) basin, via the Polesye saddle.

The Riphean succession in the Orsha basin is represented by a thick (up to 1200 m) sequence of red and mottled quartzose sandstones, the Belorussian (Polesye) Group. The stratigraphic relationships of these sandstones has not been determined precisely. Overall, they most probably correlate with the Somovo Group (Semikhatov et al., 1979), i.e. they belong to the Upper Riphean Kipchak horizon. As far back as the 1974 Kishinev meeting on upper Precambrian stratigraphy, the Belorussian Group in the Orsha basin and the Polesye Group in the Volyn basin were referred to the Upper Riphean. Relying on K–Ar age determinations on sills and cross-cutting sheets of gabbro-diabase in the Polesye Group (1040–1175 Ma), a number of workers continue to regard them as Middle Riphean (Makhnach et al., 1979).

In the Orsha basin, sediments of the lower, Rogachev, Formation (up to 340 m) are restricted to the axial zone of the depression. These are red sandstones, unsorted greywacke types below and fine- to medium-grained pink and orange quartzose varieties towards the top.

The overlying formations of the Belorussian Group are unconformable on the lower, and occur throughout the basin. They consist of quartzose and arkosic sandstones of the Rudnya Formation (up to 310 m) and mottled lilac, pink, orange and brick-red monomineralic quartzose sandstones of the Orsha Formation (up to 600 m). The surface of the Orsha Formation forms a buried erosional relief with vertical amplitude up to 350 m.

Situated south of the Orsha basin, in Volyn and Polesye, are the Polesye Group redbeds. These are widespread and are correlated with the Belorussian Group. The group is everywhere represented by quartzose sandstone and minor beds of siltstone and clay together with dykes and sills of gabbro-diabase have been found. The known thickness of the group reaches 725 m (Gorokhov borehole, SW of Lutsk).

Younger Riphean sediments in the region under discussion are developed to a limited extent. In the Orsha basin within hollows in the erosional relief are thin (20–80 m) terrigenous-carbonate sediments of the Lapich Formation. This consists of rhythmically alternating mottled conglomerate and sandstone with a dolomitic cement and algal dolomite. The dolomites contain a mixed III and IV assemblage of microphytoliths. The Lapich Formation would appear to correlate with sediments of the Sekretarkin Formation (Semikhatov et al., 1979, p. 38; Makhnach et al., 1979, p. 112), the age of which has been put at Tangaur or Pachelma by different authors.

Even more locally, only in the centre of the Orsha basin, is red feldspar-quartz sandstone of the Blonsk Formation (up to 115 m), which is usually correlated with the Pachelma Group in central regions of the Russian Platform [or it is considered by some to belong to the Lower Vendian (Yakobson, in Bogolepov and Votakh, 1977; Semikhatov et al., 1979)]. Possible analogues of the formation occur in the Pripyat depression: mottled sandstone with thin dolomite bands, the Luchkov Formation (Postnikova, 1977).

Lower Vendian sediments are strongly unconformable on various Riphean horizons, or they overstep directly onto the crystalline basement. The Lower Vendian consists of the Vilchan and Volyn sedimentary groups. The glacial sedimentary assemblage of the Vilchan Group (or Formation) is most fully developed within the Orsha basin, where it levels out the buried pre-Vilchan erosional relief and reaches 400 m thick. Tillites may account for the entire succession (e.g. around the town of Osipovichi), but more often tillite patches grade into fluvioglacial deposits. Analogues of the Vilchan sediments in Volyn are tilloids found in the Kremenets and Brody boreholes south of Lutsk.

The stratotype locality for the Volyn Group is the Volyn basin. Here the group consists of basal gravel conglomerate and sandstone of the Gorbashev Formation (up to 50 m) and overlying volcanics and tuffs of the Berestovets Formation. The major components of the formation are basalt or occasionally dolerite. Amygdaloidal basalts are common in lower horizons. Available K–Ar age dates do not usually exceed 630 Ma, although individual determinations go down as far as 1040–1120 Ma. The known thickness of the formation is 430 m.

In the Brest basin, analogues of the Volyn lavas are known as the Rotaychits Formation. In the middle of the succession are andesitic and andesite-dacitic porphyrite and dacitic porphyry. Intermediate tuffs also occur. K–Ar age determinations lie in the 600–640 Ma interval.

Within the Orsha basin, Volyn effusives are replaced by the Svisloch Formation, up to 150 m of tuffaceous sediments (Semikhatov et al., 1979, p. 39; Sokolov, 1979, p. 54). It consists typically of lilac and brown volcano-sedimentary and volcanoclastic rocks at the bottom and mottled siltstone and clay at the top.

The structure of other Riphean basins has been studied to a much lesser extent. A chain of basins, from the Krestets to the Yarensk, stretching across the entire platform, is referred to as the *Central Russian* aulacogen. The depth to basement in individual grabens is as much as 4 km or more. In the eastern half of the aulacogen, Vendian sediments describe major inversion arches with an amplitude from the top surface of the Vendian of 0.1 km (the Yarensk arch) and 0.3–0.5 km (the Sukhonsky arch). The lower part of the Riphean succession is exposed in the west, in the Krestets basin. At the base are red sandstones, giving way to poorly sorted mottled feldspar-quartz sandstones, which are cut by gabbro-diabase intrusions. At the top of the succession, the sandstones alternate with basic tuffs and tuffo-breccias. K–Ar age determinations of gabbro-diabases lie in the interval 1080–355 Ma (Yakobson, in Bogolepov and Votakh, 1977). The unit reaches 670 m in thickness. Semikhatov et al. (1979) provisionally correlate these sediments with the Middle Riphean(?) Kaverino Group. According to Postnikova (1977), the sandstone and tuffaceous sandstone of the Krestets basin are analogues of the Rogachev and Rudnya Formations of the Belorussian Group. Since the overlying sediments are correlated with the Pachelma Group, the latter point of view is preferred.

An assemblage of finely laminated micaceous shale, siltstone and sandstone with a Pachelma acritarch assemblage at the base is assigned to the Pachelma. The maximum thickness is 2700 m, which occurs in the Roslyatino borehole (on the south-eastern marginal zone of the Sukona basin). The lower grey part of the succession is usually referred to the Vologod Group and the upper to the Bologoye Group. In the Rybinsk basin, analogues of these formations lie directly on the basement and have a thickness of around 1200 m (the Bologoye borehole). An even more attenuated succession (85 m) has been found by Yakobson and Klevtsova in the Krestets basin (Bogolepov and Votakh, 1977). The oldest sediments in a system of basins in the North-East Russian Platform appear to be those of the Nyonoksa Formation, which occur in the north-western termination of the North Dvina basin. They are represented by yellow and orange fine-grained quartzose sandstones in the main, with thin beds of mottled coarser-grained varieties. This formation lies directly on the basement and is overlain by the Upper Vendian. The thickness of the formation is 335 m at Nyonoksa. In the lower part of the succession, cut by the Solozero borehole, there are basalt lava flows and dolerite dykes with a K–Ar age of 1300 Ma. Regionally developed is a sequence of grey siltstones and shales, the Safonov Group, which is cut by a number of boreholes and has a proven thickness of up to 1800 m. It contains assemblage IV microphytoliths and is correlated with the Pachelma Group.

The aulacogen-type *Ladoga basin* coincides with the area of Lake Ladoga. The lower, Priozersk, formation occurs in boreholes in the SE and west Ladoga regions (Kayryak, 1972). At the base of the formation there is usually a patch (1–44 m) of coarse clastic rocks. Above come mottled feldspar-quartz and quartzose sandstones. The thickness of the formation in the SE Ladoga region reaches 430 m. In the Kondratyevo borehole (in the river Pasha estuary), the section of the Priozersk Formation is capped by andesite-basalt porphyrite (47.7 m).

Lying above an erosion surface are poorly-sorted arkosic sandstones with thin conglomerates, the Salmi Formation. Along the east shore of Lake Ladoga these sediments lie directly above Ladoga Group schists or on rapakivi granite. In such cases the granite weathering crust is up to 4 m thick, but it can exceed 40 m for the schists. The sandstones are usually less than 40 m thick, but in several boreholes in the river Pasha valley they reach 160 m. In places the terrigenous part of the Salmi Formation succession is overlain by alternating basaltic and andesite-basaltic porphyrites and tuffs of the same composition.

The widely held view that the Salmi Formation is Middle Riphean in age (Semikhatov et al., 1979) is based exclusively on single K–Ar age determinations: 1350 Ma for volcanics of the Salmi Formation and 1220 Ma for Priozersk Formation volcanics. On this point, Kayryak (1979, p. 125) has drawn the interesting conclusion that there is a possible correlation between the Priozersk Formation and the Polesye Group. If this is so, then the late Precambrian succession in the Ladoga basin also turns out to be Upper Riphean. Makhnach et al. (1979, p. 112) hold a similar view. They correlate the Priozersk Formation with the Polesye Group and the dolomitized limestones at the very top of the Priozersk Formation with the Lapich Formation.

The Late Proterozoic basins (or aulacogens) in the Russian Platform are thus filled with mainly Upper Riphean to Lower Vendian sediments. Sediments which may be referred to the Middle Riphean are encountered in very few boreholes. It may be true that some of the sediments in more deeply folded zones (the Moscow and Pachelma aulacogens), presently beyond the reach of boreholes, belong to the Middle Riphean. We can probably expect Lower to Middle Riphean sediments to crop out only in pericratonic basins (Kama-Urals, Timan and, possibly, Lvov), whose development within the platform preceded the initiation of the aulacogen structural stage.

The above does not exclude the possibility of finding Lower to Middle Riphean sediments in local intracratonic basins. Such sediments could be the quartzite in the Krestets basin (Bologoye-I borehole), the poorly sorted grey quartzose sandstone in the Vozhe-Lacha basin (Konosha borehole) and the grey and brown shale with thin quartzose sandstone in the Yulovo-Ishim borehole in the south of the Tokmov arch. The last named dip at 15–20° and have a K–Ar age of 1700 Ma. All these sedimentary assemblages can be compared with the Middle Riphean Jotnian of the Baltic Shield, based on similarities in the degree of epigenetic changes and the distribution pattern in isolated basins.

The majority of sediments forming the successions in the aulacogens are continental, near-shore marine and shallow-water marine. Continental, mainly fragmentary redbed associations, predominate in the Upper Riphean Kipchak Regional Stage and in the Kudash. In the Tangaur Regional Stage, finer-grained grey and mottled fragmentary and clay-carbonate rocks are very common, together with a carbonate rocks with stromatoliths. The Lower Vendian forms an independent structural stage which is not developed in all aulacogens, but where it is present, it forms quite wide basins which partly overstep the margins of the aulacogens themselves.

Characteristic of this stratigraphic level are tillites, fluvioglacial sediments and abundant volcanics.

The aulacogens were not initiated randomly at a particular stage in the evolution of the platform, but were controlled by the earlier growth of the region in a pre-platform stage (Kratz et al., 1979). During the creation of the basin network, the block structure of the Earth's crust seems to have been revealed, having formed in the very earliest stages of evolution. Aulacogens are spatially related to the very active inter-block zones (Fig. I-50), which also underwent subsequent activity, including during the neotectonic stage. This explains the reflection of the aulacogens in the topography of the present-day relief.

Sediments in Riphean aulacogens have prospect potential for mineral deposits formed by both internal and external processes, in particular stratiform deposits of non-ferrous metals. In this regard, the widely-held views on the metallogeny are supported by the fact that the aulacogen network undoubtedly also continued to act as a tectonic factor in the localization of useful minerals in sediments belonging to the overlying platform structural stage. We are here referring not only to the lead-zinc, copper and mercury mineralization in Vendian to Palaeozoic carbonates, sandstones and claystones, but also to the distribution of several types of non-metallic ores, including fluorite, sulphur, borates and halides. Valeyev (1978) for example has shown that boron mineralization in the sedimentary cover is restricted to junction zones between aulacogens, such as the zone where the Kama-Belsk and Sergiyevsk-Abdulin aulacogens meet. In the Volyn basin (around the town of Rovno), copper ore shows — mainly in native form — are associated with basalts in the Berestovets Formation of the Volyn Group.

Upper Precambrian basins in the Russian Platform contain major deposits of underground water, both fresh and mineralized. The entire upper Precambrian sedimentary succession in eastern regions of the platform appears to be of regional significance in oil and gas content, although oil flows have so far been obtained only from wells in the Upper Vendian (Siva and Sokolovka). Oil and bitumen shows are also known in aulacogens in central regions of the platform.

Upper Vendian assemblages

Upper Vendian (Valday Group and its correlates) sediments form a blanket-like cover over the entire northern half of the Russian Platform (Fig. I-55), unlike Riphean to Lower Vendian sediments which occupy isolated basins. The Upper Vendian consists mainly of terrigenous and partly volcanomict rocks, a few hundreds of metres thick as a rule. Only in the NE of the platform does the thickness of the Upper Vendian probably increase to thousands of metres. The Valday type section is the Moscow syncline,

Fig. I-55. Sketch map showing distribution and thickness (in m) of Upper Vendian sediments. (Yakobson, in Bogolepov and Votakh, 1977).

where the group has three constituent formations: Redkin, Lyubim and Reshmin.

The Redkin Formation consists of a sandstone member, a brown clay member — a marker horizon — with thin ash beds, and an argillaceous member, rich in organic matter. The richest finds of the ediacarian faunal assemblage belong to the Redkin level (the Ustpinezh Formation on the Onega peninsula, the Yaryshev Formation in the Dnestr region, etc.). The Lyubim Formation consists of alternating greenish sandstone and silty-clayey members. Lyubim sediments are preceded by a break, and in western regions of the platform they rest directly on the basement. The Reshmin Formation consists predominantly of fine-grained sandstone and siltstone. Outcrops are restricted to the north and east of the platform. Its correlates may also be present in sections in Moldavia (Bogolepov and Votakh, 1977; Semikhatov et al., 1979). Upper Vendian sediments are overlain by mudstones of the Lower Cambrian Baltic Formation.

Valday to Baltic and younger lower and middle Palaeozoic sediments as a whole constitute the concluding cycle of sedimentation, corresponding to the lower structure stage of the plate cover of the platform. Sedimentation ceased at the end of this evolutionary stage. Lower Devonian continental red beds accumulated only around the margins of the plate. The overall form of the lower structural stage is simple, with beds dipping at less than one degree in the centre of the plate. On the buried slope of the Baltic Shield, dip angles reach 10–14° in places.

Chapter 4

MAJOR FEATURES OF PRECAMBRIAN METALLOGENY

In this chapter we examine the commonest space–time distribution patterns of useful mineral deposits. The detailed metallogeny of the East European craton has been elucidated by T.V. Bilibina, K.D. Belyayev and V.Y. Popov for the Soviet part of the Baltic Shield and by V.S. Domarev, N.K. Lazarenko, Ya.N. Belevtsev, Ye.M. Lazko, N.M. Chernyshov, L.S. Galetsky and others for the Ukrainian Shield and the Voronezh crystalline massif.

General metallogenic features of the Precambrian of the East European craton

The major metallogenic features of the Precambrian of the craton are determined by the pre-Riphean history of its geological evolution. The metallogenic signature is characterised by the presence of unique concentrations of ores in ferruginous quartzites of various ages and genesis, copper-nickel deposits, rare-metal mineralization in pegmatites, quartz veins, greisens, skarns and albitites, titanium magnetite mineralization and among non-metals muscovite, high-alumina raw materials (kyanite and sillimanite schists), rock crystal, graphite, shungite and pyrophyllite.

Also typical is the development of major pyrite deposits in Archaean greenstone belts, accompanied by weak occurrences of copper, polymetallic and gold–pyrite mineralization. It is interesting to note the overall insignificant development compared to other Precambrian regions (in particular the Siberian craton) of deposits of talc, magnesite, asbestos, metamorphic apatite, magnetite skarns and phlogopite ores, i.e. types of ores that appear under the specific conditions of metamorphism and granitization of terrigenous-carbonate assemblages.

As examples of other characteristic features of metallogeny in the East European craton, we can cite the formation during Palaeozoic tectono-magmatic episodes (mainly Devonian in age) of unique concentrations of apatite, phlogopite, vermiculite, magnetite and rare earths associated with alkaline nepheline syenites, alkaline ultrabasic rocks and carbonatites.

Main ore-bearing structures

Noteworthy among ore-bearing Precambrian structures in the East European craton are firstly the granite-greenstone terrains with major iron deposits. These are the Kostomuksha deposit in Karelia, Olenegorsk in the Kola Peninsula, Verkhovtsev and Chertomlyk in the Dnieper block and ore shows of ferruginous quartzites in the Mikhaylov Formation in the Voronezh massif (Kursk-Voronezh region).

Typical of greenstone belts are copper-nickel shows, such as the well-known Kamennozero and Allarechka, as well as pyrite mineralization of the type seen in the Parandovo and Hautavaara deposits and several other rarer types of raw material (gold, platinum, asbestos).

Less significant is rare-metal mineralization: porphyry-type copper-nickel, disseminated-stockwork molybdenum among quartz-sericite rocks, rare-metal greisen and pegmatite-types associated with manifestations of granitoid magmatism in greenstone belt "envelope" structures.

Granite-gneisses and granulites within Archaean consolidation structures — epi-Archaean cratons — are practically barren. The main productive structures in the Precambrian of the East European craton are suture zones associated with deep faults on various scales, which were rejuvenated several times during their evolution.

Of prime importance are Early Proterozoic rift-type sutures. One such structure is the Pechenga-Imandra-Varzuga zone in the Kola province with copper-nickel mineralization and pyrite and copper-epidote ore shows among basic volcanics. A related but on the whole distinct rift-type structure with lesser quantities of basic volcanics but widespread minor ultrabasic intrusions containing copper-nickel mineralization is typical for the eastern part of the Voronezh massif. We may also refer the Krivoy Rog–Kremenchug zone in the Ukrainian Shield to this type of ore structure. It bears ferruginous quartzites, mainly restricted to PR_1 terrigenous assemblages.

The examples quoted above are three different types of Proterozoic taphrogenic structures: volcanogenic, volcano-sedimentary with numerous ore-bearing intrusives, and mainly sedimentary shales with minor volcanics.

Specific ore-bearing structures are rift zones consisting of gabbro-diabase dyke systems and gabbro-pyroxenite layered intrusions. Two main age intervals for the development of these emerge — at the Archaean–Proterozoic boundary and in the Late Proterozoic. Associated with them are titano-magnetite ores in gabbro-diabases (Pudozhgora), and chromite and copper-nickel ore shows in layered intrusions (the Burakov and Monchegorsk intrusions). A characteristic Proterozoic structural type is suture zones on the sites of deep-level basement faults. Within the Karelian granite-greenstone terrain, volcanogenic-terrigenous basins are of this type, transitional between protogeosynclinal and protocratonic (e.g. Panajärvi and Lehti). Their constituent sedimentary and volcanic rocks form pronounced synclinal folds, often "superimposed" on greenstone basins. A number of structures contain copper-cobalt mineralization in sandstones, native copper in basic volcanics, and precious metals in conglomerates.

Another variant of sutures are those structural belts separating blocks which were stabilized earlier, in the Archaean. This is demonstrated to the fullest extent in the junction zone between the Belomorian (White Sea) and Kola provinces, known as the Lapland granulite belt. It is typified by titano-magnetite mineralization in gabbro-anorthosites and copper-nickel mineralization among meta-ultramafics (Lovno-ozero), which is related to Early Proterozoic igneous activity and metasomatism.

A different type of ore-bearing Proterozoic structure is illustrated by the Svecofennides in the Lake Ladoga region. This structure, a continuation of the Main Sulphide Belt from Finland into the USSR, is usually regarded as a typical protogeosyncline. The metallogeny of the Svecofennides around Ladoga is determined by several factors — the primary ore-bearing nature of volcano-sedimentary rocks and superimposed ultrametamorphism and granitization, as well as the presence of rapakivi granites and gabbro-anorthosites in the concluding phase of igneous activity. Volcano-sedimentary rocks display a normal rhythmic structure, typical for geosynclinal regions with a change from eugeosynclinal types of successions to miogeosynclinal. In this type of structure are polyformational deposits of different ages: tin-polymetallic, skarn, rare-metal greisens and albitite-types, stratiform tungsten deposits and others. A very important general pattern in their distribution is the close association with structures enclosing granite-gneiss domes and localization of ores in a chemically contrasting compositional unit of lower Proterozoic rocks — ortho-amphibolites, marbles, skarns, etc. In Finland, it is in just this type of unit, where this zone controls the location of pyrite-cobalt-copper-polymetallic deposits (e.g. Outokumpu and Vihanti), that graphitic quartzites and talc-serpentine schists are present.

It is a characteristic of the Soviet part of the East European craton that the orogenic volcanic-plutonic belts at the end of the Early Proterozoic are atypical, since in adjacent countries — Sweden and Finland — they have an important metallogenic significance. Within the Svecofennides of the Ladoga region, including the Karelian isthmus, as well as the western part of the Ukrainian Shield, major orogenic associations of this age include gabbro-anorthosite-rapakivi granite intrusions and associated rare-metal skarn and greisen mineralization. Volcanics of this age have a very restricted distribution and are represented by the barren Salmi plateau basalts. A unique analogue of the orogenic associations that concluded the preceding Archaean stage are volcano-sedimentary formations belonging to the Tungud-Nadvoitsy complex and Keyvy-type high-alumina kyanite and sillimanite schists which are usually considered to be metamorphosed weathering crusts and/or Late Archaean acid volcanics.

Metallogenic zoning

In a very broad sense, metallogenic zoning in the East European craton, considered for the pre-Vendian denudation surface, is determined by the development of pre-Riphean useful mineral associations (ferruginous quartzites, copper-nickel-sulphide and copper pyrite ore mineralization and micaceous, rare metal, ceramic, rock crystal etc. pegmatites) originating due to internal and external processes. They are localized in basement highs in the NW and SW of the Baltic and Ukrainian Shields and the Voronezh massif. Information about useful minerals in the pre-Riphean basement which lies beneath a thick cover of platform sediments in central and eastern regions of the craton (the Volga-Urals block; see Chapter 3, Section 1) is practically non-existent, in spite of the large number of boreholes that have reached the basement.

The most general scheme of tectonic and metallogenic zoning in the pre-Riphean basement of the East European craton emphasises the presence of two fragmentary belts: a western external belt, richly ore-bearing, which includes the large elevations of the Baltic and Ukrainian Shields and the Voronezh massif, and an eastern internal belt in the form of a depression with a deeply buried basement consisting of more basic granulites that are practically barren of ores (the Volga-Urals region and adjacent territories). This scheme agrees well with data presented previously by Levkovsky (1976). The general first-order metallogenic zonation for the Precambrian of the East European craton shown in outline in Fig. I-56 may reflect primary crustal inhomogeneity during an older "lunar" period (>3.8 Ga) with widespread isometric "basaltic seas" and anorthosite highlands around the periphery (Sharkov, 1984).

In considering the second-order metallogenic zoning for the exposed Baltic and Ukrainian Shields which have been studied in detail, and the shallowly buried Voronezh massif, we may emphasise an overall close sequence of changes in the ore-bearing structural belts, geological formations and useful mineral associations within these widely separated crustal segments. The metallogenic zoning shown schematically in Fig. I-56 may reflect a general tectonic zoning in the craton basement with zones replacing one another in sequence from east to west, from blocks which had been completely consolidated earlier, to blocks with a more prolonged evolution (zones A, B, C on Fig. I-56). Within the Soviet part of the Baltic Shield this takes the form of the replacement in a north-east to south-west direction of the granulite-greenstone terrain in the Kola province by the granite-greenstone terrain in the Karelian province

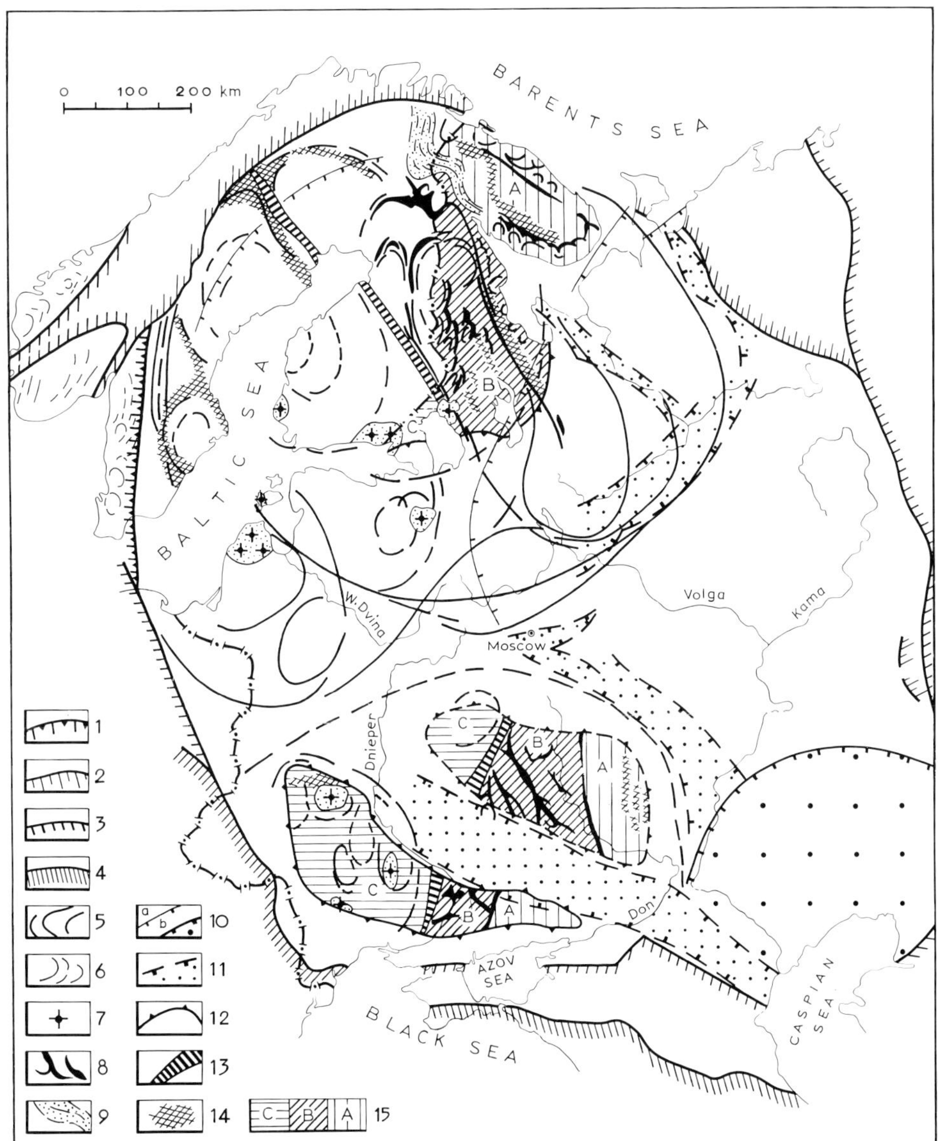

Fig. I-56. Main types of metallogenic provinces in the overall structure of the basement to the East European craton. *1–4*: craton boundaries, related to Dalslandian (*1*), Baikalian and Caledonian (*2*), Hercynian (*3*) and Alpine (*4*) fold episodes; *5* = early Precambrian arches and domes; *6* = Dalslandian domes; *7* = rapakivi granites; *8* = greenstone belts; *9* = Lapland granulite belt; *10* = boundaries of depressed regions; *11* = aulacogen boundaries; *12* = shield margins (Voronezh massif shown with broken line); *13* = major ore-bearing sutures; *14* = other major ore-bearing zones. *15*: *A*, *B*, *C* = metallogenic provinces of various types (see text).

and finally the Proterozoic granite-schist mobile belt in the Svecofennian province (the Lake Ladoga region). The same sort of alternating sequence of block structures is also seen in the Ukrainian Shield. Here from east to west we have the change from a granulite-greenstone terrain (the Azov block) to a granite-greenstone terrain (the Dnieper block) and then to a granulite-(granite)-schist terrain which developed during the Svecofennian (Volyn, Podolsk, Belotserkov and Kirovograd blocks). The differences noted between the constituent structures can be explained by differences in their depths of erosion. In particular, we consider that the wide distribution of granulites in the Podolsk and Belotserkov blocks is the result of a deeper erosion level in the Svecofennides. Within the Voronezh massif, two main zones in the sequence are expressed: A and B, with zone C occurring only partly in the extreme west.

Metallogeny in type A zones is in granulite-greenstone terrains (the Kola province, the eastern part of the Voronezh massif and the Azov province). It is defined in the first place by ore deposits in linear greenstone belts (Kolmozero-Voronya and Allarechka type) with copper-nickel and copper-molybdenum ore mineralization and rare-metal pegmatites of AR_2 age and Keyvy-type paraschist basins (AR_2) with deposits of high alumina raw materials and rare-metal pegmatites. It is interesting to emphasise the close association between precisely these rift-type basin terrains (PR_1) and copper-nickel, iron-titanium, copper pyrite and copper-epidote ore mineralization, as well as alkaline complexes of PR_{1-2}–PZ age, with which are associated rare earth, apatite, phlogopite, magnetite, ilmenite, vermiculite and several rare-metal deposits (the Mariupol deposit in the Azov block, and the Khibiny, Kovdor and Gremyakha-Vyrmes deposits in the Kola province).

Zone B metallogeny is found in granite-greenstone terrains (Karelia province, Voronezh massif and the Dnieper block) with their typical copper sulphide ore mineralization, ferruginous quartzites and copper-nickel shows. Granulites are insignificant here and crop out in the cores of deeply eroded structures, where they are practically devoid of ore minerals. The subsequent Proterozoic history of development of these structures differs from that of granulite-greenstone terrains. Instead of Early Proterozoic rift-type suture zones, here we have broad Lehti-type sub-platform depressions with sparse disseminated copper mineralization in basic volcanics, copper-cobalt ores in sandstones and traces of gold in conglomerates. Alkaline complexes of Proterozoic and Palaeozoic age are quite rare here and consequently the mineralization associated with them is not characteristic.

Zone C metallogeny belongs to the Lake Ladoga region Svecofennian terrain and the whole western part of the Ukrainian Shield, characterised by the development of protogeosynclinal structures with rare-metal mineral deposits of skarn, greisen and albite types with widespread rapakivi granite complexes and gabbro-anorthosites containing iron-titanium-vanadium and apatite mineralization. In addition to rare-metal skarns, greisens and albitites, vug-type pegmatites containing rock crystal are associated with the rapakivi granites.

Thus in addition to the provisionally outlined "relict" concentric zonation of the East European craton, second order metallogenic zoning determines the internal zoning of major positive crustal segments. A similar change in zones A, B, C in them reflects a polyphase sequence of crustal reactivation which reached a maximum in the extreme western and south-western regions of the craton.

This second structural plan of metallogenic and tectonic zoning is complicated by a system of domes and inter-dome zones at various scales, as well as by a system of linear suture zones of varying tectonic nature, whose important ore-bearing significance has been emphasised above. This is third order metallogenic zoning and in it we can point out two main types with concentric and linear symmetry. The first is particularly clearly seen in zones where granite-gneiss domes and ovoids crop out. For this type, "crustal metallogeny" is characteristic, with a regular pattern of zones from the centre of a dome to its margins: rare metals, tin, tungsten in the internal contact and the immediate external contact of the domes; copper, polymetals, iron and titanium in the parts nearest the periphery. The second type of metallogenic zoning is characteristic of suture zones (rifts, troughs and aulacogens), the zonation vector in this case being oriented across the strike of such structures; ore mineralization is predominantly of mantle origin from copper-nickel and iron-titanium at the centre to rare-metal at the periphery.

The first type of zoning is widespread in the Ladoga region and the Volyn block of the Ukrainian Shield. An example of the second type may be suture zones of the Imandra-Varzuga and Vetreny belt rifts. Suture zones are often restricted to inter-dome basins and in this case both types of zoning coincide.

A characteristic feature of both domes and sutures is their long history of development, involving superimposition of mineralization on previously active structures. For example, Early Archaean domes are transformed in younger zones into Late Archaean and Early Proterozoic domes, as a result of which there is telescoped mineralization, related to granitization and metasomatism at various stages. In particular, this may be illustrated by the example of the White Sea zone, where pegmatites formed several times, at 2600, 2400–220(?) and 1800 Ma, although productive muscovite-bearing pegmatites are the youngest (1800 Ma).

Because the terrains evolved over a long period, epochs of mineralization belonging to different time intervals coincide in them. A typical feature of such terrains, in which younger structures have inherited some features from older structures, is the preservation of the main features of metallogenic signatures belonging to different epochs. A clear example of this is the Krivoy Rog–Kremenchug basin. Uneconomic ferruginous quartzite deposits are present here in formations as old as Late Archaean. However, commercial concentrations appear only in Proterozoic (2400–2200 Ma) generations of rift-type depressions. Polyphase evolution with repeated similar volcanic and sedimentary formations is also typical for the Pechenga synclinorium, although economic ores appeared in it only at the boundary between the third and fourth rhythms. Even more significant in this respect is the Imandra-Varzuga zone and the Vetreny belt with coincident rhythms of basic volcanism in the Lopian (AR_2) and Karelian (PR_2) complexes.

Another example of inheritance in metallogeny is provided by the Lehti structure in the Karelian province. Here, economic ore mineralization has so far not been discovered. However, the metallogenic signature of Proterozoic volcano-sedimentary complexes (copper, cobalt, gold, etc.) to a significant extent is inherited from older greenstone belts.

From the general patterns, we should make special mention of the following: sutures with the richest ore deposits are those which separate blocks of different age and type of development. Respectively, in the Baltic Shield these are the Vihanti-Outokumpu-Ladoga zone which separates the Svecofennides from Archaean granite-greenstone terrains; in the Voronezh massif it is the Kursk-Belgorod belt, the zone in which ferruginous quartzites are developed; and in the Ukrainian Shield it is the Krivoy Rog–Kremenchug zone of ferruginous quartzites, occupying a similar tectonic position and separating the Svecofennides from Archaean granite-greenstone terrains. A similar position is the case of ore-bearing zones in the Vetreny belt in Karelia and the Orekhovo-Pavlograd belt in the Ukraine. However, these differ from the preceding types in that they separate Archaean terrains of different evolutionary types.

In any analysis of the metallogenic zoning of the East European craton, we must take account of the important significance of the different ages and genetic types of granulites. Some occur in dome cores and are represented by granulites, retrogressed to varying degrees, belonging to the earliest Archaean evolutionary epochs. Others developed along suture zones at dome margins and in the spaces between domes. These are zonal granulites, frequently high-pressure types, of Archaean to Proterozoic age. The first type has practically no economic mineralization, except for places where they have been retrogressed or migmatized. In such cases, they contain graphite-kinzingite and graphite-kondolite mineralization, together with high-alumina (sillimanite) schists and gneisses. Suture zone granulites contain major concentrations of titanomagnetite, apatite and aluminous raw materials due to the presence of associated intrusive and metamorphic-metasomatic gabbro-anorthosites.

As a general conclusion to this section, we would emphasise that in the tectonic belts of the East European craton, the zonation which we see in ore mineral emplacement takes on a progressively more complex structure with time. Central type zoning (first order zoning) gives way to suture zone type linear zoning and in sum total a complex mosaic picture arises of blocks with different metallogenic signatures.

Metallogenic epochs

The concepts "metallogenic epoch", "metallogenic pulse" and "metallogenic boundary" are used with different meanings by various authors. Following V.I. Smirnov and G.A. Tvalchrelidze, we take "metallogenic epoch" to mean a relatively long time interval, during which a complete series of useful mineral deposits produced by internal and external processes in a single tectonomagmatic cycle develops. The term "metallogenic pulse" is used widely in a number of works (Sokolov and Kratz, 1984). It is taken to mean a relatively short time interval (part of an epoch), characterized by the maximum development of useful mineral deposits.

The appearance of metallogenic epochs and pulses is particularly complex for the Precambrian. This is due to the protracted way in which mineral deposits formed, generally bearing signs of polyphase reworking and remobilizing of material. Primary disseminated concentrations and stratiform deposits assumed features of metamorphic or newly-formed hydrothermal-metasomatic deposits due to regional and contact metamorphic effects. The formation of mineral deposits in this way occurred frequently and was stretched out throughout an entire epoch and in a number of cases through two or even three epochs. In particular, ferruginous quartzite deposits, metamorphic and contact metasomatic apatite, magnetite, phlogopite, etc. deposits are of this type. However, in this regard we should emphasise that during the formation of mineral deposits that spanned several epochs, the ore forming processes themselves, as indicated by field studies and the results of geochronological research, were short-lived and corresponded to separate pulses; pulses of reworking were mainly associated with metamorphic peaks and late granites in succeeding epochs.

Targets studied in detail by geochronological methods (by E.K. Gerling, I.M. Gorokhov, L.A. Neymark and others), such as the Monchegorsk and Pechenga deposits, the mica and rare-metal pegmatites in the White Sea region and the Ukraine, and the banded ironstones provide a complete picture of the long history of formation of mineral deposits and individual pulses from the closure times of different isotopic systems (K–Ar, Rb–Sr, Pb–Pb, etc.).

The following metallogenic epochs have broadly been established for the East European craton: AR_1 — Belomorian, Saamian, AR_2 — Lopian, Rebolian; PR_1 — Svecofennian, Late Karelian and less clearly PR_1 — Early Karelian (Seletskian) and in part R_{2-3} — Dalslandian or Grenvillian; Phanerozoic–Caledonian-Hercynian PZ_1–PZ_2.

We present below a brief outline of the characteristics of metallogenic epochs in the Precambrian of the East European craton.

1. *AR_1 ($\geq$3.0 Ga)*. Iron ore mineralization is typical for this epoch: olivine-quartz-hypersthene-magnetite (-eulysite); olivine-quartz-hypersthene-amphibole-two-pyroxene-magnetite; and garnet-hypersthene-magnetite. In some structural belts, accumulations of primary disseminated syngenetic sulphide concentrations have been found (the Kola and Karelia provinces, the Ukrainian Shield and the Volga-Urals province). Finds of quartz-feldspar pegmatites within the Kola province with a K–Ar isotopic age of 3.6 Ga prove that pegmatite-forming processes were already occurring right at the start of this epoch.

It must be emphasised that this period of geological history has still not been fully studied as far as its metallogenic aspect is concerned.

2. *AR_2 (3.0–2.6 Ga)*. In this epoch the East European craton was host to major deposits of Fe, Ti, S and ore shows of Cr, Ni, Cu and other useful minerals. Several different types of mineralization have been identified within the Karelia-Kola region: (1) Fe–Ti–Cr — associated with basic layered intrusions and dykes (the Pudozhgora deposit and the Burakov intrusion); (2) a sulphur-chalcopyrite association with minor expressions of copper-molybdenum and noble metal mineralization (Hautavaara, Parandovo, Jalonvaara, Kolmozero-Voronya); (3) iron-chert, accompanied by subsidiary noble metals (Kostomuksha and Olenegorsk); (4) rare-metal–rare-earth pegmatites (the Kolmozero-Voronya zone); (5) high-alumina kyanite schists (the Keyvy Group); and (6) copper-nickel (Allarechka).

In greenstone belts of the Voronezh crystalline massif at this time, Ni and Cr ore shows formed in association with komatiites and harzburgite intrusions, and pyrite shows are ubiquitous. In the Ukrainian Shield there was pyrite mineralization.

Also in the Ukraine, ironstone-chert associations form major iron ore deposits in belts where carbonate-metabasic associations are developed. During amphibolite facies retrogression of granulites, iron ore deposits in the Belotserkov-Odessa zone arose, in association with W, Mn, Sn, Pb, Zn, Cu and graphite ore shows. Similar ore shows are also known within the Azov metallogenic zone (Mariupol and Konsungur), and in the Voronezh massif. The rare-metal pegmatites which were emplaced in the Ukrainian Shield at 2.7–2.8 Ga ago formed a culminating pulse of metallogenic activity.

3. *PR_1 (2.6–1.8, 1.65 Ga)*. The Early Proterozoic era breaks into two metallogenic epochs. During the Early Karelian or Seletskian epoch (2.6–2.2 Ga), jaspilites and iron ores formed in Krivoy Rog, Kursk Magnetic Anomaly, Byelorussia and the Volga-Urals province, pyrite ores, also high-alumina and graphitic schist ore-bearing assemblages. Taphrogenic basins

originated in the period 2.4–2.2 Ga, with picrite-komatiite-basalt volcanism, accompanied by sulphur-, copper-nickel and copper-zeolite ore shows. The Pechenga and Vetreny Belt ores plus copper-nickel mineralization in layered intrusions (the Monchegorsk pluton) are associated with metamorphosed ultramafics belonging to the concluding stages of formation of the rift-type structures. In the Voronezh massif we know of norite intrusions containing copper-nickel mineralization, and similar ore shows have also been noted in ultramafics in the Ukrainian Shield.

The *Svecofennian epoch (2.2–1.8, 1.65 Ga)* is of fundamental importance in terms of the metallogenic signature of major structures in the craton basement. Reworking of the structural pattern and a change in internal *P–T* regimes in the basement occurred at 1.9–1.8 Ga within the Baltic and Ukrainian Shields, as a result of which new types of ore deposits arose. Muscovite pegmatites formed during kyanite-sillimanite-type metamorphism in the White Sea belt, in which we find the unique muscovite deposits of the White Sea muscovite province. In the Lake Ladoga region, the Kola Peninsula and the Ukrainian Shield, rare-metal association pegmatites and metasomatites formed during andalusite-sillimanite-type metamorphism. Regional metamorphism during the Svecofennian cycle influenced the appearance of hydrothermal-metasomatic sulphide ores in the Karelian province, the formation of asbestos deposits in the Karelian province and the Ukrainian Shield, and a talc formation. Also associated with this period of activity in the Baltic and Ukrainian Shields and the Byelorussian massif are metasomatic phenomena, characterised by varying alkali-metal content and responsible for varied rare-metal mineralization.

A substantial change in ore quality and epigenetic ore genesis during a retrogressive amphibolite facies metamorphic episode in the Svecofennian epoch occurred in primary sedimentary and volcano-sedimentary formations during acid leaching stages, when rich accumulations of kyanite and iron and abrasive garnet deposits with economic reserves formed in the Ukraine, Karelia and the Kola Peninsula.

During the transition to the craton evolutionary stage (c. 1.65 Ga), the unique rapakivi-anorthosite association formed, with its associated variety of skarn and greisen deposits of Mo, Sn, W, Pb–Zn in the Ladoga region, and vug-type pegmatites containing rock crystal in the Ukrainian Shield.

PR_2 (1.8, 1.65–0.6 Ga). During the Riphean, the Dalslandian epoch was the most active (1.4–1.0 Ga). This epoch is mainly documented outside the USSR. In Norway and Sweden, ceramic pegmatites, copper-molybdenum, copper-silver and gold mineralization are associated with granite magmatism of this age, in addition to ore shows in cupriferous sandstones. Activity is only weakly expressed in the eastern Baltic Shield,

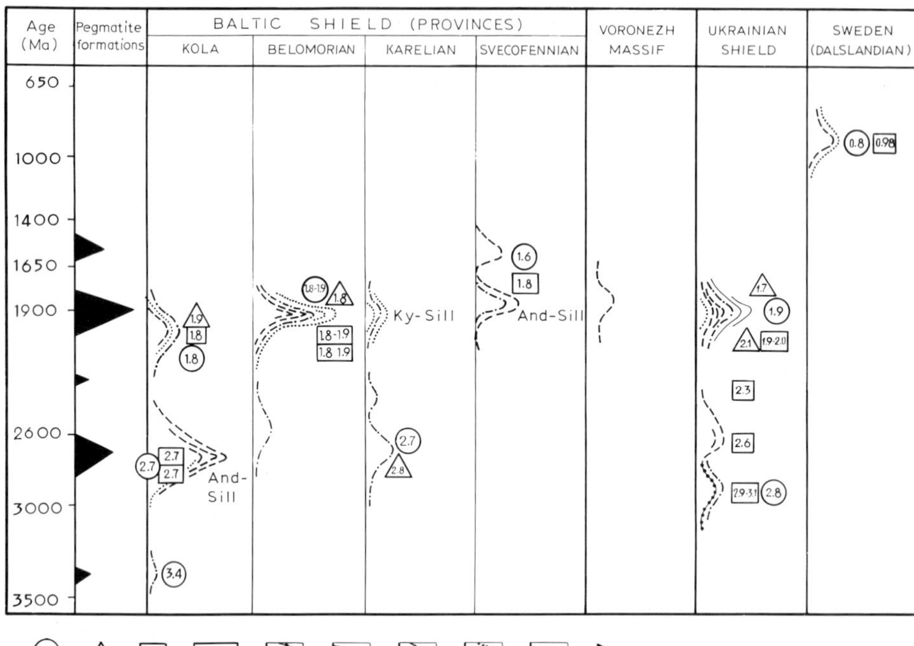

Fig. I-57. Evolution of pegmatite formations in the Precambrian of the East European Platform. *1–4* = isotope dating methods. Pegmatite formations: *5* = rare-earth, *6* = quartz-feldspar (ceramic), *7* = rare-metal, *8* = muscovite, *9* = vug-type pegmatites, *10* = distribution maxima of pegmatite formations, *11, 12* = metamorphic maxima.

as distinct from the Devonian (the Khibiny alkaline intrusions and the Kovdor carbonatites, for example).

The evolution of internal ore forming processes in the geological history of the East European craton can be illustrated using pegmatite formations as an example (Fig. I-57). They are divided into quartz-feldspar (ceramic), rare-earth (rare-earth-ceramic), muscovite, rare-metal-muscovite, rare-metal and rock crystal types (Rundqvist and Sokolov, 1986). The types are widely distributed throughout the entire territory in question, allowing us to identify the major pegmatite-forming pulses in the metallogenic cycles and to trace their genetic links to various *P–T* regimes and determine the migration of pegmatite types with time. From an analysis of pegmatite-forming processes in the East European craton, we are able to establish that each metallogenic epoch has its own particular geochemical suite of pegmatites. For example, in complexes belonging to the Early Archaean epoch, for which the characteristic metamorphism is granulite facies, quartz-feldspar (ceramic) and rare-earth pegmatites formed. In zones where andalusite-sillimanite type metamorphism occurs, rare-metal pegmatites began to appear, beginning with the Late Archaean epoch (in the Baltic and Ukrainian Shields and the Voronezh massif). During the Svecofennian, contrasting types of metamorphism led to the formation of rare-metal pegmatites in the andalusite-sillimanite regime in the Svecofennian province and muscovite pegmatites in kyanite-sillimanite metamorphism in the White Sea province.

The following metallogenic pulses have been established for the pegmatites: 2.7, 2.4, 1.8 and 1.6 Ga. The most important of these are 2.7 and 1.8 Ga. For the 2.7 Ga pulse, the typical pegmatite-forming geochemical trend is rare-earth–rare-metal. The 1.8 Ga pulse differs in that during this time, pegmatites belonging to all formations formed, while rock crystal pegmatites were generated at 1.6 Ga in the Volyn block. This, then, is the geochemical evolution of pegmatites in time, with an apparently global trend which is most typical for pegmatite suites in the Baltic and Ukrainian Shields.

In conclusion, we emphasise the fact that the major ore-forming epochs identified within the boundaries of the East European craton generally correlate well with major tectonomagmatic, tectonometamorphic and tectonometasomatic epochs established in the Precambrian of other ancient cratons. The constancy of epochs of endogenic activity has important practical consequences, since the biggest and most valuable useful mineral deposits reveal a well-defined distribution pattern on a geological time scale.

REFERENCES

(All references are in Russian unless otherwise stated)

Anon., 1973. Geochronology of the USSR, Vol. 1, Precambrian. Leningrad, Nedra, 347 pp.

Anon., 1976. Geology, petrology and metallogeny of crystalline formations in the East European Platform. Vol. 1, Mapping of the buried basement. Moscow, Nedra, 224 pp., Vol. 2, Geology, petrology and metallogeny of igneous and metasedimentary complexes. Moscow, Nedra, 263 pp.

Anon., 1986. Methods of isotope geology and the geochronological scale. Moscow, Nedra, 236 pp.

Aksamentova, N.V., Naydenkov, I.V. and Arkhipova, A.A., 1982. Stages in the formation and structure of the basement to ancient platforms — Belorussia as an example. Geotectonics, No. 5, pp. 24–31.

Anishchenkova, O.N., Berkovsky, A.N. and Garbar, D.I., 1979. Geological structure and metallogeny of the basement on the southern and south-eastern slopes of the Baltic Shield. In: Metallogenic features of the Precambrian in the USSR. Leningrad, pp. 9–18.

Antonyuk, Y.S., 1976. Composition and primary nature of metamorphic rocks in the Lebyazhi gneiss-schist complex. In: Geochemical evolution of Precambrian metamorphic complexes in the Kola Peninsula. Apatity, Kola Branch, USSR Acad. Sci., pp. 51–62.

Antropov, P.Y. (Editor), 1958. Geology of the USSR, Vol. 5, part 12, Ukraine and Moldavia, Gosgeoltekhizdat, 1000 pp.

Arestova, N.A. and Pugin, V.A., 1985. Distribution of Sr, Ba, Cu, Cr, V, Ni and Co in rocks of the basalt-rhyolite series as an indicator of their liquation nature. Geokhimiya, No. 7, pp. 946–952.

Arkhangelsky, A.D., Kolyubakin, V.V. and Roze, T.N., 1937. Precambrian tectonics of the basement to the East European Platform using data from the general magnetic survey of the USSR. Izvestiya Akad. Nauk SSSR, Geophysics and Geography Series, No. 2.

Artemenko, G.V., Zhukov, G.V. and Klets, V.A., 1986. A lower age limit for the Osipenkov Formation (Azov block). Doklady Akad. Nauk UkrSSR, Series B, No. 3, pp. 3–5.

Arth, J.G., Barker, F., Peterman, Z.E. and Friedman, I., 1978. Geochemistry of the gabbro-trondhjemite suite of south-west Finland and its implications for the origin of tonalitic and trondhjemitic magmas. J. Petrology, Vol. 19, No. 2, p. 289 [in English].

Avakyan, K.K., Bogdanova, S.V. and Dobrzhinetskaya, L.F., 1984. Structural status of charnockites in the central Kola Peninsula. In: S.B. Lobach-Zhuchenko (Editor), Natural associations of Archaean grey gneisses — geology and petrology. Leningrad, Nauka, pp. 123–133.

Balagansky, V.V., Bogdanova, M.N. and Kozlova, N.Y., 1986. Structural and metamorphic evolution of the north-west White Sea Region. Apatity, 100 pp.

Batiyeva, I.D. and Belkov, I.V., 1958. Basal conglomerates in the Keyvy metasedimentary series in the West Keyvy region. Izvestiya Karel. and Kola Branch, USSR Acad. Sci., No. 4, pp. 48–54.

Batiyeva, I.D. and Belkov, I.V., 1968. Granitoid formations in the Kola Peninsula. In: Essays on petrology, mineralogy and metallogeny of granites in the Kola Peninsula. Leningrad, Nauka.

Batiyeva, I.D., Belkov, I.V. and Vetrin, V.R., 1985. Precambrian igneous associations in the north-eastern Baltic Shield. Leningrad, Nauka, 176 pp.

Batiyeva, I.D., Belolipetsky, A.P. and Belkov, I.V., 1980. Early Precambrian volcanics in the Kola Peninsula. Leningrad, Nauka, 160 pp.

Belevtsev, R.Y., Yakovlev, V.G. and Shcherbakova, T.G., 1985. The granulite facies in the Ukrainian Shield. Kiev, Naukova Dumka, 220 pp.

Belevtsev, Y.N. (Editor), 1974. Metallogeny of the Ukraine and Moldavia. Kiev, Naukova Dumka, 507 pp.

Belevtsev, Y.N., 1977. Geological and physico-chemical conditions of metamorphic ore formation. In: Metamorphic ore formation. Moscow, Nauka, pp. 5–24.

Belevtsev, Y.N., 1978. Metamorphic zoning in the Ukrainian Shield. In: Metamorphism of the Ukrainian Shield and its margins. Proc. 1st Republic Conf. Kiev, pp. 79–81.

Belevtsev, Y.N., 1981a. On the stratigraphy of the Ukrainian Shield. Geol. Zhurnal, Vol. 41, No. 4, pp. 1–5.

Belevtsev, Y.N., 1981b. Major concepts in the modern theory of metamorphic ore formation. Kiev, Preprint IGFM-81, 53 pp.

Belevtsev, Y.N., 1981c. Development of the theory of ore formation and Precambrian metallogeny in the Ukrainian Academy of Sciences. In: Ore formation and metallogeny. Kiev, Naukova Dumka, pp. 5–36.

Belevtsev, Y.N., 1982. Progressive zonal metamorphic regime in the Precambrian of the Ukrainian Shield. Kiev, Naukova Dumka, 151 pp.

Belevtsev, Y.N., Borisenko, S.T., Galitsky, L.S. and Kalyayev, G.I., 1975. Metallogeny of the Ukrainian Shield. In: Precambrian metallogeny. Leningrad, pp. 95–97.

Belevtsev, Y.N. and Galetsky, L.S., 1984. Metallogeny of the Ukrainian Shield. In: Geology and useful mineral resources of ancient cratons. Proc. Indo-Soviet Earth Sciences Symp., Moscow, Sept., 1981. Moscow, pp. 60–64.

Belevtsev, Y.N., Koval, V.B., Domarev, V.S. and Kulish, Y.A., 1984. Fundamentals of the theory of metamorphic ore formation. Geol. Zhurnal, Vol. 44, No. 3, pp. 1–11 [in English].

Belevtsev, Y.N. and Shcherbakov, I.B., 1983. Metamorphism and stratigraphy of supracrustal assemblages in the Ukrainian Shield. Abstracts of the 1st regional stratigraphic conference, Dnepropetrovsk. Kiev, pp. 33–35.

Belevtsev, Y.N., Yepatko, Y.M. and Verigin, M.I., 1981. Iron ore deposits in the Precambrian of the Ukraine and an evaluation of their prospects. Kiev, Naukova Dumka, 230 pp.

Belkov, I.V., 1963. Kyanite schists in the Keyvy Formation. Leningrad, Nauka, 320 pp.

Belolipetsky, A.P., Gaskelberg, V.G. and Gaskelberg, L.A., 1980. Geology and geochemistry of Early Precambrian metamorphic complexes in the Kola Peninsula. Leningrad, Nauka, 240 pp.

Belyayev, K.D., 1971. New data on the structure, geology and metallogeny of a granulite assemblage. In: Magmatism in the Baltic Shield. Leningrad, Nauka, pp. 218–225.

Belyayev, O.A., 1978a. Composition and genesis of amphibolites in the Kola-Belomoride complex. In: Basic-ultrabasic magmatism in the Kola Peninsula. Apatity, Kola Branch, Acad. Sci. USSR, pp. 20–31.

Belyayev, O.A., 1978b. Pre-Karelian successions in the NW Kola Peninsula (southern border of the Pechenga anticlinorium). In: Stratigraphic subdivisions of the Precambrian in the Kola Peninsula and their correlation. Apatity, Kola Branch, Acad. Sci. USSR, pp. 19–25.

Belyayev, O.A., 1980. The oldest basement of the Tersk structural belt. In: Geological evolution and structure of Precambrian structural belts in the Kola Peninsula. Apatity, Kola Branch, Acad. Sci. USSR, pp. 5–14.

Belyayev, O.A. and Petrov, V.P., 1974. Regional metamorphism of crystalline rocks in the Pechenga-Allarechka region. In: Regional geology, metallogeny and geophysics. Apatity, Kola Branch, Acad. Sci. USSR, pp. 9–15.

Belyayev, O.A. and Zagorodny, V.G., 1974. Structural and stratigraphic features of the Annama-Hihnajärvi zone. In: Regional geology, metallogeny and geophysics. Apatity, Kola Branch, Acad. Sci. USSR, pp. 16–27.

Belyayev, O.A., Zagorodny, V.G., Petrov, V.P. and Voloshina, Z.M., 1977. Facies of regional metamorphism in the Kola Peninsula. Leningrad, Nauka, 88 pp.

Berzenin, B.Z., Bylinskaya, Y.P. and Bryansky, V.P., 1982. Refining the stratigraphic correlation scheme for Precambrian formations in the Ukrainian Shield. Geol. Zhurnal, Vol. 42, No. 6, pp. 43–53.

Bespalko, N.A., Donskoy, A.N. and Yeliseyeva, G.D., 1976. Accessory minerals of the Ukrainian Shield. Kiev, Naukova Dumka, 260 pp.

Bezborodko, N.I., 1935. Petrogenesis and petrogenetic map of the Ukrainian crystalline region. Kiev, Akad. Nauk UkrSSR, 389 pp.

Bibikova, E.V., Bogdanova, S.V., Kirnozova, T.I., Popova, Z.P., 1984. U–Pb age of charnockitoids in the Volga-Urals region. Doklady Akad. Nauk SSSR, Vol. 276, No. 4, pp. 916–919.

Bibikova, E.V. and Krylov, I.N., 1983. Isotopic age of Archaean volcanics in Karelia. Doklady Akad. Nauk SSSR, Vol. 268, No. 5, pp. 1231–1234.

Bilibina, T.V. (Editor), 1976. Geology and metallogeny of ancient platform shields in the USSR. Leningrad, Nedra, 338 pp.

Bilibina, T.V. (Editor), 1980. Metallogeny of the eastern Baltic Shield. Leningrad, Nedra, 246 pp.

Bilibina, T.V., Kazansky, V.I. and Laverov, N.P., 1984. Main types of ore-bearing structures in the Precambrian. In: Metallogeny of the Early Precambrian in the USSR. Leningrad, Nauka, pp. 14–32.

Birck, J.L. and Allègre, C.I., 1979. Rb–Sr systematics of the Montsche Tundra pluton (Kola Peninsula, USSR). Earth and Planet. Sci. Letters, Vol. 20, No. 2, pp. 266–273 [in English].

Bogachev, A.I. and Gorelov, V.A., 1968. Some aspects of the structure and ore mineralization of the Allarechka copper-nickel sulphide deposit. Geologiya Rudnykh Mestorozhdeniy, No. 2, pp. 74–78.

Bogachev, A.I., Gorelov, V.A. and Konchev-Pervukhov, V.I., 1964. Main features of the structure and sulphide ore mineralization of the region between the rivers Pechenga and Lotti. In: Precambrian geology and geochronology. Moscow, Nauka, pp. 306–311.

Bogachev, A.I., Grib, V.P. and Grishin, A.S., 1982. Main metallogenic features of Karelia. In: The metallogeny of Karelia. Petrozavodsk, Karelian Branch, Acad. Sci. USSR, pp. 5–40.

Bogachev, A.I. and Zak, S.I., 1980. An alkaline-gabbroid association. In: Early Precambrian magmatic facies in the USSR, Vol. 3, Moscow, pp. 144–165.

Bogdanova, M.N. and Yefimov, M.I., 1983. Geological structure of the Kandalaksha-Kolvitsy structural-lithological zone. In: Precambrian geology of the Kola Peninsula. Apatity, Kola Branch Acad. Sci. USSR, pp. 19–30.

Bogdanova, M.N. and Yefimov, M.I., 1982. Metabasites in the Kandalaksha-Kolvitsy structural-formational zone (composition and conditions of formation). In: Volcanism and sediment genesis in the Precambrian of the Kola Peninsula. Apatity, Kola Branch Acad. Sci. USSR, pp. 89–100.

Bogdanova, M.N. and Yefimov, M.I., 1986. On the combination of mobile belts and cratogenes in the north-eastern Baltic Shield: the granulite belt in the Kola Peninsula. In: Problems in the evolution of the Precambrian lithosphere. Leningrad, Nauka, pp. 174–181.

Bogdanova, S.V., 1982. Tectonic regional mapping of the basement in the Volga-Urals oil and gas province. In: The basement and intervening complex of ancient and young platforms in the USSR (Trudy MINKH i GP, Vol. 161). Moscow, pp. 15–25

Bogdanova, S.V., Veselovskaya, M.M. and Lapinskaya, T.A., 1982. Early Precambrian igneous complexes in eastern and central regions of the Russian Platform and their formational correlation. In: Geology, petrology and correlation of crystalline complexes in the European USSR. Leningrad, Nedra, pp. 61–71.

Bogolepov, K.V. and Votakh, O.A. (Editors), 1977. Ancient platforms of Eurasia. Novosibirsk, Nauka, 312 pp.

Bolotov, V.I., 1983. Picrite metavolcanics in Archaean gneiss and amphibolite units, Kola Peninsula. In: Precambrian igneous complexes in the north-eastern Baltic Shield. Apatity, Kola Branch, Acad. Sci. USSR, pp. 109–113.

Bolotov, V.I., Balabonin, N.L. and Ivanov, A.A., 1981. Emplacement and conditions of formation of manganese ore shows in Archaean metamorphic rocks of the Allarechka region. Priroda i khozyaystvo Severa (Natural environment and economy of the North), Vol. 9, pp. 24–27.

Bondarenko, L.P. and Dagelaysky, V.B., 1968. Geology and metamorphism of Archaean rocks in the central Kola Peninsula. Leningrad, Nauka, 168 pp.

Bondarenko, L.P., Dagelaysky, V.B. and Berkovsky, A.N., 1977. Iron-ore formations in the Early Precambrian of the basement to the Russian Platform. In: Problems in Precambrian geology. Leningrad, Nauka, pp. 117–129.

Bondarenko, L.P., Dagelaysky, V.B. and Berkovsky, A.N., 1978. Evolution of metamorphic processes in the basement to the Russian Platform. In: Metamorphic complexes in the basement to the Russian Platform. Leningrad, Nauka, pp. 199–215.

Bubnoff, S., 1952. Fennosarmatia. Acad. Verlag, Berlin, 450 p. [in German].

Bukharev, V.P. and Polyansky, V.D., 1977. The role of metamorphism in the formation of graphite deposits in the Ukrainian Shield. In: Metamorphic ore deposits. Moscow, pp. 247–262.

Chekulayev, V.P., 1981. Structural evolution of the oldest granites in the Sunski complex, Central Karelia. In: The oldest granitoids of the USSR. Leningrad, Nauka, pp. 58–66.

Chekulayev, V.P. and Baykova, V.S., 1984. Granulite associations of grey gneisses in Western Karelia. In: S.B. Lobach-Zhuchenko (Editor), Natural associations of Archaean grey gneisses Leningrad, Nauka, pp. 141–150.

Chekunov, A.V., 1972. Structure of the Earth's crust and tectonics of the southern European USSR. Kiev, Naukova Dumka, 176 pp.

Chernyshov, N.M., Bocharov, V.L. and Berdnikov, M.D., 1982. Magmatism and endogenic metallogeny of the Voronezh crystalline massif. In: Geology, petrology and correlation of crystalline complexes in European USSR. Leningrad, Nedra, pp. 41–45.

Dagelaysky, V.B. and Sokolov, Y.M. (Editors), 1977. Problems in Early Precambrian geology. Leningrad, Nauka, 243 pp.

Darkshevich, O.Ya., Schleifstein, B.A. and Antonyuk, Y.S., 1984. New data on Late Archaean magmatism in suture zones in the Kola Peninsula. Apatity, Kola Branch, Acad. Sci. USSR, pp. 40–57.

Dedeyev, V.A. and Shustova, L.E., 1976. Geoblocks in the European USSR. "Scientific papers" Preprint Series, No. 25, Syktyvkar, 52 pp.

Dobrokhotov, M.N., Berzenin, B.Z. and Boyko, V.L., 1981. A stratigraphic correlation scheme for Precambrian formations in the Ukrainian Shield. Geol. Zhurnal, Vol. 41, No. 4, pp. 6–13.

Dobrzhinetskaya, L.F., 1978. Structural and metamorphic evolution of the Kola series. Leningrad, Nauka.

Drugova, G.M., 1979. Geothermal regime in early stages of metamorphism in Precambrian polymetamorphic complexes. The White Sea region and the South-west Pamirs. In: Processes of deep-level petrogenesis and mineral genesis in the Precambrian of the USSR. Leningrad, pp. 52–71.

Drugova, G.M., Glebovitsky, V.A. and Dook, V.L., 1982. High gradient metamorphic regimes in the evolution of the Earth's crust. Leningrad, Nauka, 229 pp.

Dubrovsky, M.I. and Miklyayev, A.N., 1984. Anatexis and problems of granulite facies metamorphism. In: Early Precambrian metamorphism and metamorphic ore mineral genesis. Apatity, Kola branch, Acad. Sci. USSR, pp. 85–96.

Dook, G.G., 1977. Structural and metamorphic evolution of the Pechenga complex. Leningrad, Nauka, 106 pp.

Dook, V.L., 1976. Brief sketch of the tectonic evolution of the Eastern Baltic Shield. In: Metallogenic signatures of pegmatites in the Eastern Baltic Shield. Leningrad, Nauka, pp. 6–17.

Fedorov, E.E., 1974. Major structural features of the central Kola Peninsula. In: Regional geology, metallogeny and geophysics. Apatity, Kola Branch, Acad. Sci. USSR, pp. 31–44.

Fedorova, M.E. and Shustova, T.N., 1980. Chatom conglomerates. In: Geological structure and evolution of Precambrian structural zones in the Kola Peninsula. Apatity, Kola Branch, Acad. Sci. USSR, pp. 20–26.

Fedotov, Ye.Ye. and Migdisov, S.A., 1974. Evolution of Proterozoic structures in the Kola Peninsula. Bull. Moscow Nat. Sci. Soc., Geology Section, No. 4, pp. 16–17.

Fedotov, Zh.A., 1985. Evolution of Proterozoic volcanism in the eastern part of the Pechenga-Varzuga belt; petrochemistry. Apatity, Kola Branch, Acad. Sci. USSR, 119 pp.

Fomin, A.B. and Pastukhov, V.G., 1981. Evolution of Early Precambrian basic and ultrabasic magmatism in the Ukrainian Shield. In: Petrology and ore content of the lithosphere. Abstracts, 6th All-Union Meeting on Petrography, Leningrad, pp. 59–60.

Fotiadi, E.E., 1958. Geological structure of the Russian Platform from regional geophysical data and test drilling. Trudy VNII Geofiziki (Proc. All-Union Geophys. Res. Inst.), No. 4.

Frolovich, G.M., 1980. Comparison of Precambrian sedimentary successions in the Kama-Belsk basin. Izvestiya Akad. Nauk SSSR, ser. geol., No. 4, pp. 75–85.

Gafarov, R.A., 1963. Structure of the folded basement to the East European Platform from geophysical data. Izvestiya Akad. Nauk SSSR, Ser. Geol., No. 8, pp. 56–67

Gilyarova, M.A., 1967. Stratigraphy and structure of Pechenga. Leningrad, Leningrad State University, 95 pp.

Gilyarova, M.A., 1972. Stratigraphy and structure of the Precambrian in Karelia and the Kola Peninsula. Leningrad, Leningrad State University, 217 pp.

Gilyarova, M.A., 1974. Stratigraphy, structure and magmatism of the eastern Baltic Shield. Leningrad, Nedra, 323 pp.

Glebovitsky, V.A., Golubev, A.I. and Kazakov, A.N., 1981. Field guide to geological excursions in southern Karelia. Leningrad, 112 pp.

Glebovitsky, V.A., Drugova, G.M. and Moskovchenko, N.I., 1971. Metamorphic complexes and belts in the eastern Baltic Shield. In: Metamorphic belts of the USSR. Leningrad, Nauka, pp. 5–23.

Glebovitsky, V.A., Zinger, T.F. and Kozakov, I.K., 1985. Migmatization and granite formation in various thermodynamic regimes (Editor F.P. Mitrofanov). Leningrad, Nauka, 311 pp.

Golivkin, N.I., Leonenko, I.N. and Belykh, V.I., 1982. Early Precambrian metamorphic formations of the Voronezh crystalline massif and their ore content. In: Geology, petrology and correlation of crystalline complexes in European USSR. Leningrad, Nauka, pp. 129–138.

Golubev, A.I. and Svetov, A.P., 1983. Geochemistry of basalts in platform volcanism of Karelia. Petrozavodsk, 192 pp.

Golubev, A.I. and Svetov, A.P., 1975. Morphology and chemistry of lava flows in the Sumian-Sariolian volcano-sedimentary complex, north-west Onega region. In: V.A. Sokolov (Editor), Geology and mineral resources of Karelia. Petrozavodsk, Karelian branch, Acad. Sci. USSR, pp. 32–37.

Gorbunov, G.I., Belkov, I.V. and Makiyevsky, S.I. (Editors), 1981. Mineral deposits of the Kola Peninsula. Leningrad, Nauka, 271 pp.

Gorbunov, G.I., Zagorodny, V.G. and Robonen, V.I., 1985. Copper-nickel deposits of the Baltic Shield. Leningrad, Nauka, 329 pp.

Gorkovets, G.Y., Rayevsky, M.B., Belousov, Y.F. and Inina, K.A., 1981. Geology and metallogeny of the region around the Kostomuksha iron-ore deposit. Petrozavodsk, 143 pp.

Gorkovets, G.Y. and Rayevskaya, M.B., 1983. First find of an Archaean chemical weathering crust in Karelia. Doklady Akad. Nauk SSSR, Vol. 272, No. 6, pp. 1425–1428.

Gorlov, N.V., 1975. The oldest structural domains of the continental crust. Izvestiya Akad. Nauk SSSR, Ser. Geol., No. 2, pp. 13–27.

Gorokhov, I.M., Dagelaysky, V.B. and Morozova, I.M., 1981. Age of the Olenegorsk iron ore deposit, Kola Peninsula, using Rb–Sr and K–Ar data. Geologiya Rudnykh Mestorozhdeniy, Vol. 23, No. 3, pp. 67–79.

Gorokhov, I.M., Krylov, I.N. and Baykova, V.S., 1976. A geochronological study of the Kola Group polymetamorphic rocks. In: Development and use of methods in nuclear geochronology. Leningrad, Nauka, pp. 177–192.

Gorokhov, I.M., Kutyavin, E.P. and Varshavskaya, E.S., 1971. Rb–Sr age of phyllites in the Pechenga Group (Kola Peninsula). In: Abstracts of 22nd session of the Commission for determining the absolute age of geological formations. Moscow, pp. 72–73.

Goryainov, P.M., 1976. Geology and genesis of banded ironstone-chert formations in the Kola Peninsula. Leningrad, Nauka, 147 pp.

Goryainov, P.M., 1980. The Kola-Norwegian megablock — the oldest craton in the Precambrian of the Kola Peninsula. In: A.K. Zapolnov and K.O. Kratz (Editors), Regional tectonics of the Early Precambrian of the USSR. Leningrad, Nauka, pp. 88–10.

Grechishnikov, N.P., 1983. Correlating the stratigraphic scheme for Precambrian formations in the Ukrainian Shield. Geol. Zhurnal, Vol. 43, No. 1, pp. 28–33.

Green, A.H. and Naldrett, A.J., 1981. The Langmuir volcanic peridotite associated nickel deposits: Canadian equivalents of the western Australian occurrences. Econ. Geol., Vol. 76, No. 6, pp. 1503–1523 [in English].

Grigoryev, P.K., 1937. Mica pegmatites in Northern Karelia. In: Micas in the USSR, Leningrad and Moscow, ONTI NKTP SSSR, pp. 159–191.

Grigoryeva, L.V., 1979. Precambrian crustal activation. Geotectonics, No. 2, pp. 49–59.

Heiskanen, K.I., Golubev, A.I. and Bondar, L.R., 1977. Orogenic vulcanism in Karelia. Leningrad, 216 pp.

Hörmann, P.K., Raith, M., Raase, P., Ackermand, D. and Seifert, F., 1980. The granulite complex of Finnish Lapland: petrology and metamorphic conditions in the Ivalojoki-Inarijärvi area. Geol. Surv. Finland Bull., Vol. 308, 95 pp. [in English].

Ivanov, A.A., 1986. On the existence of two Archaean supracrustal complexes in the Tersk block of the Kola Peninsula. Doklady Akad. Nauk SSSR, Vol. 287, No. 4, pp. 928–930.

Ivanov, A.A. and Bolotov, V.I., 1984. Petrochemistry of Archaean metavolcanics in the eastern Kola Peninsula. In: Precambrian igneous complexes in the NE Baltic Shield. Apatity, Kola Branch, Acad. Sci. USSR, pp. 114–124.

Ivanov, A.A., Martynov, E.V. and Bolotov, V.I., 1983. Chemistry of metapelites in the Tersk block. In: Sedimentary basins and volcanic zones in the Precambrian of the Kola Peninsula. Apatity, Kola Branch, Acad. Sci. USSR, pp. 61–68.

Ivanova, T.V., 1970. Some questions concerning sediment genesis in Lower Riphean deposits in North-east Bashkiria. Trudy Geol. Inst., Vol. 26, Kazan, pp. 7–14.

Ivlev, A.I., 1979. The relationship between stratigraphic and zonal structural-metamorphic complexes and Proterozoic mobile zones in the north-eastern Baltic Shield. In: Problems in Precambrian metamorphism. Apatity, Kola Branch, Acad. Sci. USSR, pp. 52–65.

Kalyayev, G.I., 1965. Precambrian tectonics of the Ukrainian iron ore province. Kiev, Naukova Dumka, 192 pp.

Kalyayev, G.I., 1980. The Early Precambrian and plate tectonics. In: Problems of tectonics in the Early Precambrian. Leningrad, Nauka, pp. 169–179.

Kalyayev, G.I., Glevassky, E.B. and Dmitrov, G.K., 1984. Palaeotectonics and structure of the Earth's crust in the Precambrian iron ore province of the Ukraine. Kiev, Naukova Dumka, 240 pp.

Kalyayev, G.I., Krutikhovskaya, Z.A. and Ryabenko, V.A., 1980. Early Precambrian tectonics of the Ukrainian Shield. In: K.O. Kratz and A.K. Zapolnov (Editors), Early Precambrian regional tectonics of the USSR, Leningrad, Nauka, pp. 18–32.

Kalyayev, G.I., Verbitsky, V.N., Gorlitsky and Snezhko, A.M., 1981. Early Precambrian graphitic sediments in the Ukrainian Shield. In: A.V. Sidorenko (Editor), Problems in Precambrian sedimentary geology. Graphitic sediments and their ore content. Vol. 7, pt 1, Moscow, Nauka, pp. 80–85.

Karpinsky, A.P. 1887. A summary of physico-geographic conditions of European Russia in past geological periods. Acad. Sci. Notes, Vol. 60, No. 8.

Kayryak, S.I., 1973. The Besovets Group in the Onega structure. Leningrad, 176 pp.

Kayryak, A.T., 1979. Justification for placing the boundary between the Karelian and the Riphean in the south of the Baltic Shield. In: Stratigraphy of the Upper Proterozoic in the USSR (Riphean and Vendian). Leningrad, Nauka, pp. 121–125.

Kazakov, A.N., 1976. Deformation and superimposed folding in metamorphic complexes. Leningrad, Nauka, 237 pp.

Kazakov, A.N., 1977a. The Ladoga Group of the Baltic Shield. In: A.N. Kazakov et al. (Editors), Structural evolution of metamorphic complexes. Leningrad, Nauka, pp. 79–97.

Kazakov, A.N., 1977b. Synchronous structural and structural-metamoprhic paragenesis. In: A.N. Kazakov et al. (Editors), Structural evolution of metamorphic complexes. Leningrad, Nauka, pp. 5–16.

Kazakov, A.N., 1979. Structure of the Olenegorsk iron ore deposit, relationships between rock deformation and the crystallization of minerals. In: Processes of petrogenesis and mineral genesis in the Precambrian of the USSR. Leningrad, Nauka, pp. 253–265.

Kazansky, V.I. (Editor), 1978. Endogenic ore mineralization in ancient shields. Moscow, Nauka, 196 pp.

Khain, V.E. and Leonov, Y.G. (Editors), 1979. Tectonic map of Europe and adjacent regions, scale 1 : 10,000,000. Moscow, GUGK.

Kharitonov, L.Y., 1966. Structure and stratigraphy of the Karelides in the eastern Baltic Shield. Leningrad, Nauka, 359 pp.

Khiltova, V.Y., Lobach-Zhuchenko, S.B. and Mitrofanov, F.P., 1986. Structure and evolution of the Archaean lithosphere. In: B.S. Sokolov and F.P. Mitrofanov (Editors), Evolution of the Precambrian lithosphere. Leningrad, Nauka, pp. 53–63.

Kiselev, A.S., 1977. New data on the stratigraphic relations of the Krivoy Rog and Ingulets Groups. Geokhimiya i rudoobrazovaniye (Geochemistry and ore genesis), No. 6, pp. 46–54.

Klevtsova, A.A., 1979. Subdivisions and correlation of the Riphean in the Russian Platform on the basis of identifying natural stages in its evolution. In: Upper Proterozoic stratigraphy of the USSR (Riphean and Vendian), Leningrad, Nauka, pp. 139–143.

Klochkov, V.M., Solovitsky, V.N. and Pastukhov, V.G., 1983. Additional data on the boundaries of the Ukrainian Shield and its slopes. Geol. Zhurnal, Vol. 43, No. 1, pp. 80–85.

Kononov, N.D., 1975. Main patterns of gold localization in coarse-grained fragmentary rocks from the basal horizon in the Proterozoic of the Kursk Magnetic Anomaly. In: Precambrian metallogeny, Leningrad, pp. 129–131.

Konopleva, N.G., 1974. On the age of sedimentary units on Rybachy Peninsula. Doklady Akad. Nauk SSSR, Vol. 214, pp. 410–413

Kouvo, O., 1970. Kallioperamme ikäsuhtesta (The relative age of basement rocks). Geologi, 73 pp. [in Finnish].

Kozhevnikov, V.N., 1982. Conditions of formation of structural-metamorphic parageneses in Precambrian complexes. Leningrad, Nauka, 288 pp.

Kozlov, Y.K., 1973. Natural rock series in nickeliferous intrusions and their metallogeny — the Kola Peninsula as an example. Leningrad, Nauka, 288 pp.

Kozlov, M.T., 1979. Fault tectonics of the north-east Baltic Shield. Leningrad, Nauka, 140 pp.

Kozlov, M.T., Latyshev, L.N. and Markitakhina, T.M., 1974. Palaeovolcanic structures on mount Arvarench. In: Regional geology, metallogeny and geophysics. Apatity, Kola Branch, Acad. Sci. USSR, pp. 78–81.

Kozlov, N.E., 1983. A new variant of the stratigraphic interpretation of the Kolvitsy zone of the granulite belt. In: Sedimentary basins and volcanic zones in the Precambrian of the Kola region. Apatity, Kola Branch, Acad. Sci. USSR, pp. 69–81.

Kozlovsky, E.A. (Editor), 1987. The Kola superdeep borehole. Berlin, Springer Verlag, 490 pp. [in English].

Kratz, K.O., 1963. Geology of the Karelides of Karelia. Proc. LAGED Akad. Nauk SSSR (Laboratory of Precambrian Geology), Vol. 16, 208 pp.

Kratz, K.O. (Editor), 1969. Geology and petrology of granite-gneiss terrains in south-west Karelia. Leningrad, Nauka, 226 pp.

Kratz, K.O. (Editor), 1972. Geochronological boundaries and geological evolution of the Baltic Shield. Leningrad, Nauka, 192 pp.

Kratz, K.O. (Editor), 1978. Geology and petrology of an Archaean granite-greenstone complex in Central Karelia. Leningrad, Nauka, 263 pp.

Kratz, K.O. (Editor), 1984. Fundamentals of metallogeny of Precambrian metamorphic belts. Leningrad, Nauka, 339 pp.

Kratz, K.O., Berkovsky, A.N., Bondarenko, L.P. et al., 1979. Basic problems in the geological structure of the Russian Platform. Leningrad, Nauka, 120 pp.

Kratz, K.O., Gerling, E.K. and Dook, G.G., 1980. Geochronological studies of metamorphic processes: the Pechenga complex as an example. In: Problems in Precambrian metamorphism. Apatity, Kola Branch, Acad. Sci. USSR, pp. 7–14.

Kratz, K.O. and Glebovitsky, V.A. (Editors), 1975. Map of metamorphic belts of the USSR, 1 : 5,000,000 scale. Leningrad.

Kratz, K.O., Glebovitsky, V.A. and Bylinsky, R.V., 1978. The Earth's crust in the Eastern Baltic Shield. Leningrad, Nauka, 232 pp.

Kratz, K.O. and Lazarev, Y.I., 1961. Main features of the tectonic structure of the Jatulian in Karelia. In: Problems in the geology of Karelia and the Kola Peninsula. Murmansk, pp. 43–57.

Kratz, K.O., Levchenkov, O.A., Shuleshko, I.K., Yakovleva, S.Z., Makeyev, A.F. and Komarov, A.N., 1976. Age boundaries of the Jatulian complex, Karelia. Doklady Akad. Nauk SSSR, Vol. 231, No. 5, pp. 1191–1194.

Kratz, K.O. and Lobach-Zhuchenko, S.B., 1972. Geochronological boundaries and geological evolution of the Baltic Shield. In: Geochronological boundaries and geological evolution of the Baltic Shield. Leningrad, Nauka, pp. 162–177.

Kratz, K.O., Lobach-Zhuchenko, S.B. and Shuleshko, I.K., 1977. Discovery of ancient conglomerates in central Karelia and their significance for Early Precambrian stratigraphy. Doklady Akad. Nauk SSSR, Ser. Geol., Vol. 234, No. 1, pp. 142–144.

Kratz, K.O., Shurkin, K.A., Lobach-Zhuchenko, S.V. Maslennikov, V.A., 1971. A regional scheme for the stratigraphy of isotope geochronology of the Eastern Baltic Shield. Leningrad, pp. 120–129.

Kremenetsky, A.A., 1979. Metamorphism of Precambrian basic rocks and the genesis of amphibolites. Moscow, Nauka, 112 pp.

Krestin, E.M., 1978. A first find of komatiites in the USSR. Doklady Akad. Nauk SSSR, Vol. 242, pp. 412–415.

Krestin, E.M., 1984. Natural associations of Early Archaean supracrustal assemblages in the Kursk-Voronezh crystalline massif: an interpretation of their geology and genesis. In: S.B. Lobach-Zhuchenko (Editor), Natural associations of Archaean grey gneisses. Leningrad, Nauka, pp. 168–175.

Krestin, E.M. and Berdnikov, M.D., 1982. Relationship between and evolution of tectonics, magmatism, metamorphism and ore formation in the Precambrian of the Kursk-Voronezh crystalline massif. In: Geology, petrology and correlation of crystalline complexes in European USSR. Leningrad, Nedra, pp. 202–213.

Krutikovskaya, Z.A., Pashkevich, I.K. and Silina, I.M., 1982. A magnetic model and the structure of the Earth's crust in the Ukrainian Shield. Kiev, Naukova Dumka, 215 pp.

Krylov, I.N., Levchenkov, O.A., Lobach-Zhuchenko, S.B. and Chekulayev, V.P., 1984. Heterogeneity in the structure and evolution of the Archaean lithosphere in the Karelian granite-greenstone terrain. In: Proc. 27th sess. Int. Geol. Congr., Vol. 5, Moscow, Nauka, pp. 100–106.

Kukharenko, A.A., Orlova, M.P. and Bagdasarov, E.A., 1969. Alkaline gabbroids in Karelia. Leningrad, 183 pp.

Kulikov, V.S. and Kulikova, V.V., 1979. Identifying the boundaries of the Sumozero-Kenozero Archaean greenstone belt on the eastern margin of the Baltic Shield. In: The Early Precambrian geology of Karelia. Petrozavodsk, pp. 70–76.

Kulikova, V.V. and Kulikov, V.S., 1981. New data on Archaean peridotitic komatiites in Eastern Karelia. Doklady Akad. Nauk SSSR, Vol. 259, No.3, pp. 693–697.

Kushinov, N.V., 1981. Geological structure of the Belozero iron ore region. Geol. Zhurnal, Vol. 41, No. 4, pp. 14–19.

Kushinov, N.V., 1985. The Teplov Formation in the Belozero synclinorium. Geol. Zhurnal, Vol. 45, No. 4, pp. 71–74.

Lagutenkova, N.S. and Chepikova, I.K., 1982. Upper Precambrian sediments in the Volga-Urals region and their oil and gas prospects. Moscow, Nauka, 111 pp.

Lapinskaya, T.A., 1982. Early Precambrian stratigraphy of the Volga-Urals oil and gas province. In: The basement and intervening complex of ancient and young platforms of the USSR. Trudy MNKH i GP, Vol. 161. Moscow, pp. 25–36.

Latyshev, L.N., 1984. Geological structure of pre-Karelian formations in the Lake Vochelambi–Kislaya Guba region. In: Geology and history of formation of Precambrian structures in the Kola Peninsula. Apatity, Kola Branch, Acad. Sci. USSR, pp. 20–27.

Lavrov, M.M., 1979. Ultrabasic rocks and peridotite-gabbro-norite layered intrusions in the Precambrian of Northern Karelia. Leningrad, 136 pp.

Lazarenko, E.K., Pavlishin, V.I. and Latysh, V.T., 1973. Mineralogy and genesis of cavity pegmatites in Volynye. Lvov, Lvov Univ. Press, 360 pp.

Lazko, E.M., 1982. The construction of a rational stratigraphic scheme for the Precambrian of the Ukrainian Shield. Geol. Zhurnal, Vol. 42, No. 3, pp. 77–78.

Lazko, E.M., Kirilyuk, V.P., Sivoronov, A.A. and Yatsenko, G.M., 1975. The Lower Precambrian of the western Ukrainian Shield (age complexes and formations). Lvov, Lvov Univ. Press, 240 pp.

Levkovsky, R.Z., 1975. Rapakivi. Leningrad, Nedra, 223 pp.

Levkovsky, R.Z., 1976. Subplatform granitoid complexes in the north-western Kola Peninsula. Leningrad, Nauka, 156 pp.

Lobach-Zhuchenko, S.B., 1979. Early Precambrian tonalites — genetic relationships and formational types. Trudy Inst. geol. i geokhim. UNTs AN SSSR (Proc. Inst. Geol. Geochem., Urals Scientific Centre, Acad. Sci. USSR), Vol. 155, pp. 140–161.

Lobach-Zhuchenko, S.B., 1984. The tonalite series in the Archaean of Karelia: geological types and petrogenesis. In: Proc. 27th Int. Geol. Congress, Moscow, Aug. 1984, Vol. 9, Petrology. Moscow, Nauka, pp. 141–149.

Lobach-Zhuchenko, S.B., 1985. Particular features of Archaean tonalite-trondhjemite complexes and problems of their genesis. In: tectonics and some questions of Precambrian metallogeny. Moscow, Nauka.

Lobach-Zhuchenko, S.B., Chekulayev, V.P. and Baykova, V.S., 1974. Epochs and types of granite formation in the Precambrian of the Baltic Shield. Leningrad, 208 pp.

Lobach-Zhuchenko, S.B., Chekulayev, V.P. and Berkovsky, A.N., 1985. Gneiss-granite terrains in Karelia. In: Problems in the evolution of the Precambrian lithosphere. Leningrad.

Lobach-Zhuchenko, S.B., Dook, V.L. and Krylov, I.N., 1984. Geological and geochemical types of tonalite-trondhjemite series associations in the Archaean. Leningrad, Nauka, pp. 17–52.

Lobach-Zhuchenko, S.B. and Levchenkov, O.A., 1985. New data on the geochronology of Karelia. In: Isotope methods and problems in the Precambrian geology of Karelia. Petrozavodsk, pp. 5–26.

Lobach-Zhuchenko, S.B., Levchenkov, O.A. and Chekulayev, V.P., 1986. Geological evolution of the Karelian granite-greenstone terrain. Precambrian Research, Vol. 33, pp. 45–65 [in English].

Lobikov, A.F., 1982. On the age of early Karelian metavolcanics using data from the lead-isochron method. In: Problems in the isotopic dating of volcanic and sedimentary processes. Abstracts. Kiev, pp. 90–91.

Luchitsky, V.I., 1939. Precambrian stratigraphy of the Ukrainian crystalline massif. In: Stratigraphy of the USSR, Vol. 1. Moscow, Acad. Sci. USSR.

Lyubtsov, V.V., 1980. Riphean and Vendian lithostratigraphy and microphytofossils in the NW Kola Peninsula. In: Geochemistry and conditions of formation of sedimentary units in the

Precambrian of the Kola Peninsula. Apatity, Kola Branch, Acad. Sci. USSR, pp. 97–113.

Makhnach, A.S., Davydov, M.N. and Dominikovsky, G.G., 1979. Episodes of tectono-magmatic activation and metallogenic features of the crystalline basement in Belorussia. In: Special metallogenic features of the Precambrian in the USSR. Leningrad, pp. 35–39.

Makhnach, A.S., Veretennikov, N.V. and Shkuratov, V.I., 1979. Basic problems in Upper Precambrian stratigraphy of the USSR-materials from the western subregion of the Russian Platform. In: Upper Proterozoic stratigraphy of the USSR (Riphean and Vendian). Leningrad, Nauka, pp. 111–114.

Makiyevsky, S.I. and Zagorodny, V.G., 1981. Distribution of meta-sedimentary iron-ore formations in the Kola Peninsula. In: Geology of ore deposits in the Kola Peninsula. Apatity, Kola Branch, Acad. Sci. USSR, pp. 59–64. Makiyevsky, S.N., 1973. Geology of metamorphic rocks in the NW Kola Peninsula. Leningrad, Nauka, 152 pp.

Meriläinen, K., 1976. The granulite complex and adjacent rocks in Lapland, northern Finland. Geol. Surv. Finland Bull., Vol. 281, 129 pp. [in English].

Mints, M.V., Kolpakov, N.I. and Puzanov, V.I., 1980. Tectonic structure of the Murmansk block in the Baltic Shield. In: K.O. Kratz and A.K. Zapolnov (Editors), Early Precambrian regional tectonics of the USSR. Leningrad, Nauka, pp. 133–145.

Mints, M.V., Sobotovich, E.V. and Tsyon, O.V., 1982. Lead isotopic dating of rocks in the Murmansk block and its surroundings, Kola Peninsula. Izvestiya Akad. Nauk SSSR, Ser. Geol., No. 10, pp. 5–17.

Mirskaya, D.D., 1978. Stratigraphy and correlation of formations in the Keyvy and Kolmozero-Voronya zones. In: Stratigraphic subdivisions of the Precambrian of the Kola Peninsula and their correlation. Apatity, Kola Branch, Acad. Sci. USSR, pp. 4–17.

Mirskaya, D.D., 1979. An Early Precambrian stratotype section in the Kola Peninsula — the Keyvy zone. In: Archaean and Early Proterozoic stratigraphy of the USSR. Leningrad, Nauka, pp. 42–46.

Misharev, D.T., Amelandov, A.S., Zakharchenko, A.I. and Smirnova, V.S., 1960. Stratigraphy, tectonics and pegmatites in the NW White Sea region. Trudy VSEGEI, New Series, Vol. 31, 110 pp.

Mitrofanov, F.P. (Editor), 1979. General questions concerning the subdivision of the Precambrian in the USSR. Proc. 5th sess. scientific council on Precambrian geology. Conference Proc., Ufa, 1977, 164 pp., Acad. Sci. USSR, Interdepart. Stratig. Committee. Leningrad, Nauka, pp. 127–146.

Musatov, D.I., Fedorovsky, V.S. and Mezhelovsky, N.V., 1983. Archaean tectonic regimes and geodynamics; regional and model aspects. VIEMS Review, Moscow, 42 pp.

Nalivkin, D.V. (Editor), 1974a. Tectonic map of the USSR basement, 1 : 5,000,000 scale. Leningrad.

Nalivkin, D.V. (Editor), 1974b. Basement structure of platform regions in the USSR. Explanatory note to the 1 : 5,000,000 scale tectonic map of the basement in the USSR. Leningrad, Nauka, 400 pp.

Nalivkina, E.B., 1964. Charnockites in the south-western part of the Ukrainian crystalline massif and their genesis. Moscow, Nedra, 123 pp.

Nalivkina, E.B., 1977. Early Precambrian ophiolite associations. Moscow, Nedra, 183 pp.

Nalivkina, E.B., 1978. The role of metamorphism in the formation of the Earth's crust in the Ukrainian Shield. Proc. 1st Republican conf. Kiev, Naukova Dumka, pp. 85–87.

Negrutsa, V.Z., Zagorodsky, V.G. and Stenar, M.M., 1980. Early Precambrian tectonics of the eastern Baltic Shield. In: K.O. Kratz and A.K. Zapolnov (Editors), Early Precambrian regional tectonics of the USSR. Leningrad, Nauka, pp. 5–17.

Nevolin, N.V., Belyayeva, E.V. and Berezina, G.A., 1971. A study of the geological structure of the East European craton by geophysical methods. Moscow, Nedra, 120 pp.

Neyelov, A.N., 1977. Evolution of the Earth's crust and geological periodicity. In: Correlation of internal processes. Leningrad, pp. 378–399.

Nikitin, V.D., 1952. Processes of recrystallization and metasomatism in micaceous and ceramic pegmatites. Zap. Len. Gorn. In-ta (Proc. Leningrad Mining Inst.), Vol. 27, No. 2, pp. 107–158.

Nikitin, V.D., 1955. Towards a theory of pegmatite genesis. Zap. Len. Gorn. In-ta (Proc. Leningrad Mining Inst.), Vol. 30, No. 2, pp. 44–117.

Ovchinnikova, G.V., Yakovleva, S.Z. and Kutyavin, E.P., 1985. U–Pb systems in gneisses around Lake Litsa (the Kolmozero-Voronya zone, Kola Peninsula). In: Recent data on isotope geochemistry and cosmochemistry. Leningrad, pp. 78–81.

Pap, A.M., 1977. The crystalline basement of Belorussia. Moscow, Nedra, 125 pp.

Perevozchikova, V.A. (Editor), 1974. Tectonics of the eastern Baltic Shield. Leningrad, Nedra, 288 pp.

Petrov, V.P., 1979. On the age and nature of metamorphism of the Keyvy-Lebyazhi supracrustal complex. In: Problems of Precambrian metamorphism. Apatity, Kola Branch, Acad. Sci. USSR, pp. 66–80.

Petrov, V.P., Belyayev, O.A. and Voloshina, Z.M., 1986. Metamorphism of Early Precambrian supracrustal complexes in the north-eastern Baltic Shield. Leningrad, Nauka, 272 pp.

Petrov, V.P. and Voloshina, Z.M., 1978. On the stratigraphy of volcanosedimentary formations on the eastern flank of the Imandra-Varzuga zone. In: Stratigraphic subdivisions of the Precambrian of the Kola Peninsula and their correlation. Apatity, Kola Branch, Acad. Sci. USSR, pp. 44–61.

Petrov, V.P., Voloshina, Z.M., Latyshev, L.N. and Rezhenov, S.A., 1984. New data on the metamorphism of supracrustal rocks in the Lake Imandra region and their geological significance. In: Precambrian geology of the Kola Peninsula. Apatity, Kola Branch, Acad. Sci. USSR, pp. 58–78.

Peyve, A.V., Yanshin, A.L. and Zonenshayn, L.P., 1976. Origin of the continental crust in Northern Eurasia. Geotectonics, No. 5, pp. 6–24.

Pharaoh, T.C., Warren, A. and Walsh, N.J., 1987. Early Proterozoic metavolcanic suites of the northernmost part of the Baltic Shield. In: T.C. Pharaoh, R.D. Beckinsale and D.T. Rickard (Editors), Geochemistry and mineralization of Proterozoic volcanic suites. London, Geological Society, Spec. Publ. No. 33 (IGCP Project 217, Proterozoic Geochemistry), pp. 41–58 [in English].

Piyyar, Y.K., Klochkov, V.M. and Pastukhov, V.G., 1983. Some questions about the stratigraphy of Precambrian metamorphic formations in the Podolsk block and the Golovanev suture zone. Geol. Zhurnal, Vol. 43, No. 5, pp. 39–44.

Plaksenko, N.A. (Editor), 1976. Geology and metallogeny of the Precambrian in the Voronezh crystalline massif. Voronezh, Voronezh Univ. Press, 126 pp.

Platunova, A.P., 1975. "Lithospheric" segments in the Kola Peninsula and their marginal suture zones. In: The geology and deep structure of the Eastern Baltic Shield. Leningrad, Nauka, pp. 84–102.

Polishchuk, V.D. and Polishchuk, V.I., 1978. Metamorphic complexes in the basement to the Kursk magnetic anomaly (KMA) basin. In: V.B. Dagelyasky and L.P. Bondarenko (Editors), Metamorphic complexes in the basement to the Russian Platform. Leningrad, Nauka, pp. 131–155.

Polkanov, A.A., 1937. Summary of the geology and petrology of the NW Kola Peninsula, pt 2. Moscow, Acad. Sci. USSR. 318 pp.

Polkanov, A.A., 1939. Pre-Quaternary geology of the Kola Peninsula and Karelia or the easternmost part of the Fennoscandinavian crystalline shield. Proc. 17th Int. Geol. Congress, Vol. 2. Leningrad and Moscow, GONTI, pp. 27–58.

Polkanov, A.A. and Gerling, E.K., 1960. The use of K–Ar and Rb–Sr methods in determining the age of Precambrian rocks in the Baltic Shield. Trudy LAGED (Proc. Lab. Precambrian Geol.), Acad. Sci. USSR, Vol. 9.

Polovinkina, Y.I., 1940. Stratigraphic scheme for the Ukrainian Precambrian. Sovetskaya Geologiya, No. 5–6.

Polovinkina, Y.I., 1953. Stratigraphy, magmatism and tectonics of the Precambrian in the Ukrainian SSR. Trudy LAGED (Proc. Lab. Precambrian Geol.), Acad. Sci. USSR, Vol. 2, pp. 127–130.

Polovinkina, Y.I., 1954. Volcanosedimentary and magmatic complexes in the Ukrainian crystalline massif. Trudy VSEGEI (Proc. All-Union Geol. Inst.), Vol. 1. Moscow, Gosgeoltekhizdat, 94 pp.

Polunovsky, R.M., Panov, B.S. and Lavrinenko, L.F., 1983. Some problematical questions in the Precambrian stratigraphy of the Azov massif. In: Stratigraphy of Precambrian formations in the Ukrainian Shield. Abstracts, 1st Regional Strat. Conf., Dnepropetrovsk, 1983. Kiev, pp. 73–77.

Polyak, E.A., 1968. Geological structure of the Pechenga belt. In: Geology and deep structure of the Eastern Baltic Shield. Leningrad, Nauka.

Postnikova, I.Y., 1977. The Upper Precambrian of the Russian Platform and its oil content. Moscow, Nedra, 222 pp.

Predovsky, A.A., 1980. Reconstruction of conditions of sediment genesis and volcanism in the Early Precambrian. Leningrad, Nauka, 152 pp.

Predovsky, A.A., Fedotov, Z.A. and Akhmedov, A.M., 1974. Geochemistry of the Pechenga complex (metamorphosed sediments and volcanics). Leningrad, Nauka, 139 pp.

Predovsky, A.A., Petrov, V.P. and Belyayeva, O.A., 1967. Rare element geochemistry of Precambrian metamorphic series. Leningrad, 180 pp.

Priyatkina, L.A. and Sharkov, E.V., 1979. Geology of the Lapland deep fault. Leningrad, Nauka, 128 pp.

Pushkarev, Y.D., Kravchenko, E.V. and Shestakov, G.I., 1978. Geochronological reference points in the Precambrian of the Kola Peninsula. Leningrad, Nauka, 135 pp.

Pushkarev, Y.D., Ryungenen, G.N., Shestakov and Shurkina, L.K., 1979. Granitoids older than 2800 Ma in the Kola Peninsula. In: The oldest granitoids of the Baltic Shield. Apatity, Kola Branch, Acad. Sci. USSR, pp. 18–43.

Puura, V.A. and Sudov, Y.A., 1975. On the question of zones of platform-type tectonic activity in the southern part of the Baltic anticline and associated metallogeny. In: Precambrian metallogeny. Leningrad, pp. 125–127.

Puura, V.A., Vaker, R.M. and Klein, V.M., 1983. The crystalline basement of Estonia. Moscow, Nauka, 208 pp.

Radchenko, M.K., 1984. Geological structure and setting of gabbro-anorthosite intrusions in the structure of the Verkhneponoy block. In: Geology and history of formation of Precambrian structures in the Kola Peninsula. Apatity, Kola Branch, Acad. Sci. USSR, pp. 33–38.

Robonen, V.I., Rybakov, S.I. and Svetova, A.I., 1975. Palaeovolcanic reconstructions of Lower Proterozoic volcanic structures in Karelia. Sovetskaya Geologiya, No. 8, pp. 135–140.

Rundqvist, D.V. (Editor), 1981. Ore content and geological formations in the structure of the Earth's crust. Leningrad, Nedra, pp. 18–37.

Rundqvist, D.V., 1982. The use of evolutionary patterns in time of mineral formations in metallogenic forecasting investigations. Paper presented at 6th meeting of the All-Union Mineralogical Society, 26 Jan. 1982. ZVMO (Proc. All-Union Min. Soc.) No. 4, Pt 3, pp. 407–421, Leningrad, Nauka.

Rundqvist, D.V., 1984. Time factor in ore genesis. (Evolution and distribution patterns of mineral deposits). Global tectonics and metallogeny, Vol. 2, No. 3/4, pp. 169–257 [in English].

Rundqvist, D.V. and Popov, V.E., 1980. Patterns of ore formation and emplacement of useful minerals. In: Metallogeny of the Eastern Baltic Shield. Leningrad, Nauka, pp. 213–228.

Rundqvist, D.V. and Popov, V.E., 1981. Some general features of the geological structure and metallogeny of nickeliferous volcanic belts. In: Problems in petrology related to nickel-sulphide formations. Moscow, pp. 204–212.

Rundqvist, D.V. and Sokolov, Y.M., 1986. Major metallogenic epochs in the Precambrian of the European part of the USSR. Int. Conf. Metallogeny of Precambrian, IGCP Project 91, Geol. Survey Czechoslovakia. Prague, pp. 93–109 [in English].

Rusakov, N.F., Rusakova, A.F. and Litovsky, D.I., 1979. Sapphirine-bearing schists in a gneiss-schist sequence in the Saltychan anticlinorium. In: Petrology and correlation of crystalline complexes in the East European Platform. Kiev, pp. 132–154.

Ryabenko, V.A., 1970. Main features of the tectonic structure of the Ukrainian Shield. Kiev, Naukova Dumka, 124 pp.

Ryabenko, V.A., 1983. Lithostratigraphic complexes and geological structures of the Ukrainian Shield. Geol. Zhurnal, Vol. 43, No. 2, pp. 59–70.

Rybakov, S.I., 1982. Volcanosedimentary and metamorphic factors in the formation of pyrite deposits in the Baltic Shield. In: Metallogeny of Karelia. Petrozavodsk, Karelian Branch, Acad. Sci. USSR, pp. 40–56.

Rybakov, S.I., 1980. Metamorphism of Early Precambrian volcano-sedimentary formations in Karelia. Petrozavodsk, 134 pp.

Rybakov, S.I. and Svetova, A.I., 1981. Volcanism of Archaean greenstone belts in Karelia. Leningrad, 152 pp.

Safronova, G.P., 1982. Metal ore prospects in the Tikshozero intrusion with reference to new types of alkaline and essentially carbonate rocks. In: Metallogeny of Karelia, Petrozavodsk, pp. 143–161.

Sakko, M., 1960. Varhais-Karjalisten metabaasien radiometrisia zirconilkia (Zircon radiometric dates on East Karelian metabasics). Geologi, Vol. 23, No. 9, pp. 117–118 [in Finnish].

Salop, L.I., 1973. General Precambrian stratigraphic scale. Leningrad, Nedra, 309 pp.

Salop, L.I., 1983. Geological evolution of the Earth during the Precambrian. Berlin, Springer Verlag, 459 pp. [in English].

Salye, M.Y., 1983. Retrograde metamorphism — the main phase of ore genesis during pegmatite formation. In: The geology and genesis of pegmatites. Leningrad, Nauka, pp. 30–59.

Salye, M.Y., Batuzov, S.S. and Dusheyko, S.I., 1985. Geology and pegmatites of the Belomorides. Leningrad, Nauka, 251 pp.

Salye, M.Y. and Glebovitsky, V.A., 1976. Metallogenic signatures of pegmatites in the eastern Baltic Shield. Leningrad, Nauka, 188 pp.

Saranchina, G.M., 1972. Precambrian granitoid magmatism, metamorphism and metasomatism — a case study from Lake Ladoga and other regions. Leningrad, 128 pp.

Semenenko, N.P., Bordunov, I.N. and Polovko, N.I., 1978a. Iron-chert formations in the Ukrainian Shield, Vol. 2. Kiev, Naukova Dumka, 367 pp.

Semenenko, N.P., Ladiyeva, V.D. and Bordunov, I.N., 1978b. Iron-chert formations in the Ukrainian Shield, Vol. 1. Kiev, Naukova Dumka, 327 pp.

Semenenko, N.P., Ladiyeva, V.D. and Boyko, V.L., 1982. Metabasic and keratophyric aluminosilicate formations in the central Ukrainian Shield. Kiev, Naukova Dumka, 376 pp.

Semeneko, N.P., Ryabokon, S.M. and Bordunov, I.N., 1979. Ultrabasic associations in the central Ukrainian Shield. Kiev, Naukova Dumka, 426 pp.

Semenenko, N.P., Tkachuk, L.G. and Shcherbak, N.P., 1964. Regional geochronological scale for the Ukrainian Shield and its folded margins. In: Absolute age of geological formations. 22nd session Int. Geol. Congress, Soviet Contributions, Problem No. 3, Moscow, Nauka, pp. 325–328.

Semenenko, N.P., Vinogradov, A.P., Komlev, L.V. and Tugarinov, A.I., 1965. Geochronological map of the Ukrainian Precambrian. In: Precambrian geochronology of the Ukraine. Kiev, Naukova Dumka, pp. 5–15.

Semenenko, N.P., 1958. Precambrian stratigraphy of the Ukrainian crystalline massif. In: Geology of the USSR, Vol. 5, Ukraine and Moldavia. Moscow, Gosgeoltekhizdat.

Semikhatov, M.A., Aksenov, Ye.M. and Bekker, Yu.R., 1979. Subdivisions and correlation of the Riphean in the USSR. In: Upper Proterozoic stratigraphy of the USSR (Riphean and Vendian). Leningrad, Nauka, pp. 6–42.

Sergeyev, S.A., Arestova, N.A., Levchenkov, O.A. and Yakovleva, S.Z., 1983. U–Pb isotopic age for the Semch gabbro-diorite intrusion, Karelia. Izvestiya Akad. Nauk SSSR, Ser. Geol., No. 11, pp. 15–20.

Sergeyev, S.A., Lobach-Zhuchenko, S.B., Levchenkov, O.A. and Yakovleva, S.Z., 1985. U–Pb zircon dating of a grey gneiss complex. In: Problems in the isotopic dating of metamorphic and metasomatic processes. Alma-Ata.

Sergeyeva, E.I., Timofeyev, B.V., Sergeyev, A.S. and Kukharenko, A.A., 1971. On the age and relationships of the Tersk and Turyin Formations, southern shore of the Kola Peninsula. Doklady Akad. Nauk SSSR, Vol. 200, No. 4, pp. 941–943.

Sharkov, Y.V., 1984a. Anorthositic associations in the Kola Peninsula. In: Terrestrial and lunar anorthosites. Moscow, Nauka, pp. 5–61.

Sharkov, Y.V., 1984b. Continental rift-type magmatism in the Lower Proterozoic of the Kola-Karelia region. Geotectonics, No. 2, pp. 37–50.

Shatsky, N.S., 1946. Main structural features and evolution of the East European Platform. Izvestiya Akad. Nauk SSSR, Ser. Geol., No. 1.

Shcherbak, N.P., 1975. Petrology and geochronology of the Precambrian in the western Ukrainian Shield. Kiev, Naukova Dumka, 270 pp.

Shcherbak, N.P. (Editor), 1978. Catalogue of isotopic dates for rocks in the Ukrainian Shield. Kiev, Naukova Dumka, 222 pp.

Shcherbak, N.P., 1980. Precambrian stratigraphy and geochemistry of the Ukrainian Shield. In: Precambrian. 26th session, Int. Geol. Congress, Soviet contributions. Moscow, Nauka, pp. 126–131.

Shcherbak, N.P. (Editor), 1983. Geological map of the crystalline basement to the Ukrainian Shield, 1 : 500,000 scale. Explanatory note TsTZ Mingeo UkrSSR, 102 pp.

Shcherbak, N.P., 1984. Problems in stratigraphic and isotope-geochronological correlations of chert-banded ironstone formations in the Early Precambrian of the Ukrainian Shield. In: Problematical questions in the stratigraphy of chert-banded ironstones of the Ukrainian Shield. Preprint, IGFM Acad. Sci. UkrSSR, Kiev, pp. 4–9.

Shcherbak, N.P., Bartnitsky, Y.N. and Lugovaya, I.P., 1981a. Isotope geology of the Ukraine. Kiev, Naukova Dumka, 245 pp.

Shcherbak, N.P. and Bibikova, E.V., 1984a. Stratigraphy and isotope geochronology of the early Precambrian in the Ukrainian, Baltic and Aldan Shields. Geol. zhurnal, Vol. 4, No. 3, pp. 81–99 [in Russian and English].

Shcherbak, N.P. and Bibikova, E.V., 1984b. Early Precambrian stratigraphy and geochronology of the USSR. Proc. 27th Int. Geol. Congress. Section C.05, Precambrian Geology, abstracts, Vol. 5, pp. 3–14. Moscow.

Shcherbak, N.P., Bibikova, E.V., Zhukov, G.V. and Makarov, V.A., 1982. Isotopic dating of palaeovolcanics in the Konko-Verkhovtsev series in the Central Dneiper region. Doklady Akad. Nauk UkrSSR, Ser. B, No. 11, pp. 29–33.

Shcherbak, N.P., Sollogub, V.B. and Usenko, I.S., 1981b. Evolution of continental lithospheric composition in the early stages — the Precambrian of the Ukrainian Shield as an example. In: Petrology and ore content of the lithosphere. Leningrad, Nedra, pp. 12–14.

Shcherbak, N.P., Yeliseyeva, G.D. and Levkovskaya, N.Y., 1978. Geological and radiogenic age of the Klesov Group and the Osnitsky complex. Geol. Zhurnal, Vol. 38, No. 4, pp. 28–43.

Shcherbakov, I.B., 1975. Petrography of Precambrian rocks in the central Ukrainian Shield. Kiev, Naukova Dumka, 279 pp.

Shcherbakov, I.B., Siroshtan, R.I. and Shcherbakova, T.G., 1982. Main metamorphic patterns in the Ukrainian Shield. In: Abstracts of 4th Interdepart. meeting on problems of metamorphic ore genesis and the 5th All-Union symp. on metamorphism. Vinnitsa, pp. 216.

Shcherbakov, I.B., Yesipchuk, K.E. and Orsa, V.I., 1984. Granitoid associations in the Ukrainian Shield. Kiev, Naukova Dumka, 192 pp.

Shukolyukov, Y.A. and Bibikova, Y.V. (Editors), 1986. Methods in isotope geology and the geochronological scale. Moscow, Nauka, 226 pp.

Shulga, P.L., Furtes, V.V. and Lapchik, F.E., 1982. Palaeozoic sediments in the Belokorovich graben-syncline. Geol. Zhurnal, Vol. 42, No. 2, pp. 120–132.

Shurkin, K.A., Gorlov, N.V., Salye, M.Y., Dook, V.L. and Nikitin, Y.V., 1962. The Belomoride complex of Northern Karelia and the south-west Kola Peninsula. Geology and pegmatite content. Moscow and Leningrad, Acad. Sci. USSR, 306 pp.

Shurkin, K.A., Mitrofanov, F.P. and Shemyakin, V.M., 1980. Early Precambrian igneous associations in the USSR, Vols 1–3. Moscow, Nedra, 285, 283, 266 pp.

Shurkin, K.A., Shemyakin, V.M. and Pushkarev, Y.D., 1974. Geology and magmatism in the junction zone between the Belomorides and the Karelides — the White Sea–Karelia deep fault. Leningrad, Nauka, 183 pp.

Shvarts, G.A. and Pitade, A.A., 1980. Geological structure and composition of gneisses in the Bratsk synclinorium. Geol. Zhurnal, Vol. 40, No. 5, pp. 20–28.

Simonen, A., 1960. Plutonic rocks of the Fennoscandian Svecofennides in Finland. Bull. Comm. Géol. Finlande, No. 189, 101 pp. [in English].

Simonenko, T.N. and Tolstikhina, M.M., 1968. Block structure of the folded basement of the European USSR. Geotectonics, No. 4, pp. 37–53.

Siroshtan, R.I., Shcherbakova, T.G. and Kravchenko, 1982. Metamorphism of sapphirine-bearing rocks in the Ukrainian Shield. Vinnitsa, pp. 218–219.

Sitdikov, B.S., 1982. Early Precambrian metamorphic complexes in the crystalline basement of the eastern Russian Platform within the Volga-Kama anticline. In: Geology, petrology and correlation of crystalline complexes in the European USSR. Leningrad, Nedra, pp. 147–158.

Sitdikov, B.S. and Polyanin, V.A., 1976. Petrography, geological structure and conditions of formation of the crystalline basement to the Tatar arch. In: Geology, petrology and metallogeny of crystalline formations in the East European Platform, Vol. 1. Moscow, Nedra, pp. 115–119.

Sivoronov, A.A., Berzenin, B.Z. and Malyuk, V.I., 1981. Early Precambrian metamorphosed volcanogenic formations in greenstone belts in the Ukrainian Shield — structure and composition. Geol. Zhurnal, Vol. 41, No. 5, pp. 20–28.

Slyusarev, V.D., Kravchenko, A.N. and Kozlova, N.Y., 1975. A new type of alkaline magmatism in Northern Karelia. In: Information summary for 1974. Petrozavodsk, pp. 19–25.

Slyusarev, V.D. and Kulikov, V.S., 1973. Geochemistry of Proterozoic basic-ultrabasic magmatism. Leningrad, 104 pp.

Slyusarev, V.D., Pekurov, A.V. and Bogachev, A.I., 1982. Metallogeny of Archaean greenstone belts in the Urosozero-Vygozero region. In: Metallogeny of Karelia. Petrozavodsk, Karelian Branch, Acad. Sci. USSR, pp. 92–125.

Sobolev, D.N., 1936. Precambrian stratigraphy and tectonics of the Ukrainian crystalline plate. Problems of Soviet geology, Vol. 6, No. 9, pp. 786–807.

Sobotovich, E.V., Grashenko, S.N., Aleksandruk, V.M. and Shats, M.N., 1963. Determining the age of ancient rocks using the lead-isochron and isotopic-spectral methods. Izvestiya Akad. Nauk SSSR, Ser. Geol., No. 10, pp. 3–15.

Sokolov, B.S., 1979. The Vendian. Principles of identification, boundaries and status in the geological time scale. In: Upper Proterozoic stratigraphy of the USSR (Riphean and Vendian). Leningrad, Nauka, pp. 43–61.

Sokolov, V.A., 1972. Geological evolution of the Middle Proterozoic in Karelia. Geotectonics, No. 5, pp. 61–75.

Sokolov, V.A., 1980. Jatulian formation of the Karelian ASSR. In: A. Silvenoinen (Editor), Jatulian geology in the eastern part of the Baltic Shield. Proc. Finnish-Soviet symp., 21–26 Aug. 1979, Rovaniemi, pp. 163–194 [in English].

Sokolov, V.A. (Editor), 1984. Precambrian stratigraphy of Karelia. Petrozavodsk, 88 pp.

Sokolov, V.A., Goldobina, L.P., Ryleyev, A.V., Satsuk, Y.I. and Heiskanen, K.I., 1970. Geology, lithology and palaeogeography of the Jatulian in Central Karelia. Petrozavodsk, 363 pp.

Sokolov, V.A., Robonen, V.I., Rybakov, S.I., Svetova, A.P., Kulikov, V.S., Golubev, A.I., Svetova, A.I. and Goncharova, L.I., 1978. Volcanic edifices in the Proterozoic of Karelia. Leningrad, 168 pp. Sokolov, Y.M. and Kratz, K.O., 1984. Metallogenic impulses of endogenic activity in the Earth's crust during the Precambrian. In: Early Precambrian metallogeny of the USSR. Leningrad, Nauka, pp. 4–14.

Sollogub, V.B. and Chekunov, A.A., 1975. Deep structure and evolution of the Earth's crust. In: Problems of Earth physics in the Ukraine. Kiev, Naukova Dumka, pp. 118–141.

Sollogub, V.B., Kharitonov, O.M. and Chekunov, A.A., 1980. Deep structure of the East European Platform from geophysical data. Geofiz. Zhurnal, Vol. 2, No. 6, pp. 26–35.

Staritsky, Y.T. (Editor), 1981. History of development and mineral genesis of the Russian Platform cover. Trudy VSEGEI, new series, Vol. 308, Leningrad, Nedra, 224 pp.

Stenar, M.M., 1973. Deformation periods in the Belomoride complex. In: Stages in the Precambrian tectonic evolution of Karelia. Leningrad, Nauka, pp. 9–18.

Stenar, M.M. and Volodichev, O.I., 1970. Relict granulite facies regional metamorphism in the western White Sea region. In: Regional metamorphism and metamorphic ore formation. Leningrad, Nauka, pp. 137–142.

Stepanov, V.S., 1981. Basic magmatism in the Precambrian of the western White Sea region. Leningrad, Nauka, 216 pp.

Storchak, P.N. (Editor), 1984. Geology and metallogeny of the Precambrian of the Ukrainian Shield. Set of maps, 1 : 1,000,000 scale, Explanatory Note, Vol. II, Kiev, 150 pp.

Strygin, A.I., 1978. Petrology and ore formations of the Precambrian in the Ukrainian Shield. Kiev, Naukova Dumka, 258 pp.

Stupka, O.S., 1980. Main stages in the formation of the Earth's crust and structures in the southern margin of the East European Platform during pre-Riphean time. Geotectonics, No. 4, pp. 3–17.

Sudovikov, N.G., 1939a. Petrology of the western White Sea region (Granitization of the Belomorides). Trudy Len. Geol. Upr. (Mem. Leningrad Geol. Surv.)

Sudovikov, N.G., 1939b. Review of Precambrian stratigraphy, tectonics and igneous activity in Karelia. In: Stratigraphy of the USSR, Vol. 1, Precambrian. Moscow and Leningrad.

Sudovikov, N.G., Glebovitsky, V.A., Sergeyev, A.S., Petrov, V.P., Kharitonov, A.L., 1970. Geological evolution of deep zones in mobile belts. The north Ladoga region. Leningrad, 228 pp.

Suslova, S.N., 1971. Lithological composition of Talya tundra gneisses and schists, Kola Peninsula. In: Problems in Precambrian lithology. Leningrad, Nauka, pp. 84–108.

Suslova, S.N., 1984. Structure and composition of the granulite complex in the Kola Peninsula. Sovetskaya Geologiya, No. 2, pp. 77–89.

Svetov, A.P., 1979. Platform-type basaltic volcanism in the Karelides. Leningrad, Nauka, 208 pp.

Svetova, A.I., 1979. Lithological composition and palaeogeographical conditions of accumulation of volcanosedimentary formations in the Hautavaara zone of Southern Karelia. In: Early Precambrian geology of Karelia. Petrozavodsk, pp. 49–70.

Sviridenko, L.P., 1974. Metamorphism and granite formation in the Early Precambrian of Western Karelia. Proc. Inst. Geol. Karelian Branch, Acad. Sci. USSR, Vol. 21. Leningrad, 156 pp.

Sviridenko, L.P., Semenov, A.S. and Nikolskaya, L.D., 1976. The Kaalam gabbro and tonalite intrusion. In: Precambrian basic-ultrabasic intrusive complexes in Karelia. Leningrad, pp. 127–140.

Systra, Y.I., 1978. Structural evolution of the Belomorides in the western White Sea region. Leningrad, Nauka, 167 pp.

Timofeyev, V.V., 1969. Proterozoic spheromorphids. Leningrad, Nauka, 145 pp.

Tugarinov, A.I. (Editor), 1978. Geochronology of the East European Platform and the junction between the Caucasus-Carpathian systems. In: Proc. 19th session of the Commission for determining the absolute age of geological formations. Moscow, Nauka, 306 pp.

Tugarinov, A.I. and Bibikova, E.V., 1980. Geochronology of the Baltic Shield using zircon data. Moscow, Nauka, 129 pp.

Tugarinov, A.I., Zykov, S.I. and Bibikova, E.V., 1963. Age determination of sedimentary rocks by the lead–uranium method. Geokhimiya, No. 3

Turchenko, S.I., 1978. Metallogeny of metamorphic sulphide ore deposits in the Baltic Shield. Leningrad, Nauka, 120 pp.

Usenko, I.S., 1960. Basic and ultrabasic rocks in the west Azov region. Kiev, Acad. Sci. UkrSSR, 220 pp.

Usenko, I.S., Belevtsev, R.Y. and Shcherbakova, T.G., 1980. Rock-forming garnets in the Ukrainian Shield. Kiev, Naukova Dumka, 176 pp.

Usenko, I.S., Kalyayev, G.I. and Belevtsev, R.Y., 1976. Metallogenic signature of geological formations in the Ukrainian Shield. In: Geology, petrology and metallogeny of crystalline rocks in the East European Platform, Vol. 2. Moscow, Nedra, pp. 246–248.

Usenko, I.S., Shcherbakov, I.B. and Siroshtan, R.I., 1982. Metamorphism of the Ukrainian Shield. Kiev, Naukova Dumka, 308 pp.

Valeyev, R.N., 1978. Aulacogens in the East European Platform. Moscow, Nedra, 153 pp.

Vardanyants, L.A., 1964. Geological map of the crystalline basement to the Russian Platform, scale 1 : 2,500,000. Moscow, Nedra.

Velikoslavinsky, D.A., Birkis, A.P. and Bogatikov, S.A., 1978. The anorthosite-rapakivi granite association of the East European Platform. Leningrad, Nauka, 293 pp.

Venediktov, V.M, Glevassky, Y.B. and Golub, E.N., 1979. Rock-forming pyroxenes in the Ukrainian Shield. Kiev, Naukova Dumka, 228 pp.

Vetrin, V.R., 1984. Geological and geochemical features of the oldest granitoids in the Kola Peninsula. In: Natural associations of Archaean grey gneisses: geology and petrology. Leningrad, Nauka, pp. 113–123.

Vetrin, V.R., 1983. Primary composition and petrochemistry of the oldest granitoids in the Kola Peninsula. In: Precambrian igneous complexes in the north-east Baltic Shield. Apatity, Kola Branch, Acad. Sci. USSR, pp. 17–32.

Vetrin, V.R., Belkov, Y.I. and Pushkarev, Y.D., 1977. Age of granitoids in the margin of the Ust-Ponoy structure, Kola Peninsula. Doklady Acad. Sci. USSR, Vol. 237, No. 6, pp. 1434–1437.

Vinogradov, A.N. and Vinogradova, G.V., 1981. The geochemical type of primary crustal granites. In: The oldest granitoids of the USSR. Grey gneiss complex. Leningrad, Nauka, pp. 49–59.

Vinogradov, A.P., Tarasov, L.S. and Zykov, S.N., 1959. Isotopic composition of ore leads in the Baltic Shield. Geokhimiya, No. 7.

Vinogradov, A.P. and Tugarinov, A.I., 1964. Precambrian geochronology of the eastern Baltic Shield using data from the Pb–U–Th absolute age dating method. In: Precambrian geology and geochronology. Proc. LAGED (Laboratory for Precambrian Geology), Vol. 19. Moscow and Leningrad, Nauka, pp. 185–204.

Vinogradov, L.A., Bogdanova, M.N. and Yefimov, M.M., 1980. The granulite belt in the Kola Peninsula. Leningrad, Nauka, 208 pp.

Volodichev, O.I., 1975. Facies of metamorphism in kyanite gneisses — the Belomoride complex as an example. Proc. Inst. Geol., Karelian Branch Acad. Sci. USSR, Petrozavodsk, Vol. 25, 170 pp.

Voronovsky, S.N., Ovchinnikov, L.V. and Ovchinnikov, L.N., 1981. Geochronology of granitic pegmatites in the Karelia-Kola region. In: Problems in geochronology and isotope geology. Moscow, Nauka, pp. 64–80.

Vrevsky, A.B., 1980. Komatiites from the Early Precambrian Polmos-Poros belt, Kola Peninsula. Doklady Akad. Nauk SSSR, Vol. 252, No. 5, pp. 1216–1219.

Vrevsky, A.B., 1985. Petrology and geodynamic conditions of formation of a komatiite-tholeiite series in the Polmos-Poros greenstone belt. In: Petrology and criteria for ore content in Precambrian mafic-ultramafic associations in the Kola-Karelia region. Apatity, Kola Branch, Acad. Sci. USSR, pp. 64–73.

Vrevsky, A.B. and Kolychev, Y.A., 1986. Tectonic evolution and metallogeny of Archaean greenstone belts in the Kola Peninsula. In: Geology and ore prospects in the basement to ancient platforms. Leningrad, Nauka.

Yaroshchuk, M.A., 1983. Iron ore formations in the Belotserkov-Odessa metallogenic zone. Kiev, Naukova Dumka, 169 pp.

Yaroshchuk, M.A., Fomin, A.B., Kogut, K.V. and Kucherok, Y.Y., 1982. Relationship between chert-banded ironstones in the SW Ukrainian Shield and komatiitic and tholeiitic basalts. Geol. Zhurnal, Vol. 42, No. 1.

Yeliseyev, N.A., Kushev, V.G. and Vinogradov, D.P., 1965. A Proterozoic intrusive complex in the east Azov region. Moscow, Nauka, 204 pp.

Yevdokimov, B.N., Goryainov, P.M. and Schleifstein, B.A., 1978. Regional metamorphism of banded ironstone-chert formations in the Imandra iron ore region. In: Geology and mineral resources of the Kola Peninsula. Apatity, Kola Branch, Acad. Sci. USSR, pp. 84–93.

Yezhov, S.V., 1973. Stages in the formation of a migmatite-gneiss complex in the Allarechka ore field, NW Kola Peninsula. Izvestiya VUZ, Geol. i Razvedka, No. 9, pp. 48–52.

Yudin, B.A., 1980. A gabbro-labradorite association in the Kola Peninsula and its metallogeny. Leningrad, Nauka, 169 pp.

Zagorodny, V.G., Mirskaya, D.D. and Suslova, S.N., 1964. Geological structure of the Pechenga volcano-sedimentary group. Leningrad, Nauka, 218 pp.

Zagorodny, V.G., Predovsky, A.A. and Basalayev, A.A., 1982. The Imandra-Varzuga zone of the Karelides: geology, geochemistry and evolution. Leningrad, Nauka, 280 pp.

Zagorodny, V.G. and Radchenko, L.T., 1983. Early Precambrian tectonics of the Kola Peninsula. Leningrad, Nauka, 97 pp.

Zak, S.I., 1980. Ultramafic associations in the Kola Peninsula. Leningrad, Nauka, 160 pp.

Zak, S.I., Slyusarev, V.D. and Bogacheva, A.I., 1975. Ultramafic belts in the Karelia-Kola region. Doklady Akad. Nauk SSSR, Vol. 221, No. 6, pp. 1395–1398.

Zander, V.N., Tomashunas, Y.I. and Berkovsky, A.N., 1967. Geological structure of the basement to the Russian Platform. Leningrad, Nedra, 124 pp.

Zapolnov, A.K., Neyelov, A.N., Lobach-Zhuchenko, S.B. et al., 1980. Comparative analysis of basement tectonics in ancient cratons of the USSR. In: K.O. Kratz and F.P. Mitrofanov (Editors), Problems of early Precambrian tectonics. Leningrad, Nauka, pp. 21–37.

Zapolnov, A.K. and Neyelov, A.N., 1984. Precambrian of the USSR. In: The oldest rocks of the USSR. Leningrad, Nauka, pp. 8–14.

Zaytsev, Y.S., Ivanov, A.B. and Lebedev, I.P., 1978. Early Precambrian metamorphic complexes in the SW part of the Voronezh crystalline massif. In: Metamorphic complexes in the basement to the Russian Platform. Leningrad, Nauka, pp. 115–130.

Zhdanov, V.V. and Malkova, T.P., 1974. Iron ore deposits in zones of regional basification. Leningrad, Nedra, 198 pp.

Zoubek, V., Krylova, M.D. and Losert, J., 1979. Leptinites in the Karelia-Kola part of the Baltic Shield and the Bohemian Massif. In: An attempt at correlating igneous and metamorphic rocks. Moscow, Nauka, pp. 5–72.

Part II. Precambrian of the Siberian craton

Introduction

HISTORY OF INVESTIGATION AND TECTONIC STRUCTURE

The Siberian craton occupies a vast area in Central Siberia: between the Yenisey valley in the west, the Verkhoyan and Sette-Daban ranges in the east, the North Siberian Lowlands in the north and the East Sayan and Baikal Highlands in the south. The Precambrian basement to the craton crops out at the surface in the south-east: the Aldan Shield and the Dzhugdhzur-Stanovoy region, in the north: the Anabar Shield, and in the south-west: the Sayan and other marginal elevations (the Kan–Pre-Sayan region and the Baikal-Patom belt). In the rest of the territory the basement is hidden beneath a thick sedimentary cover and is exposed in only a few rare boreholes in the south-western and central parts of the craton. The craton is surrounded on all sides by Phanerozoic fold belts of various ages. Along most of the western and south-western borders are the fold structures of the Baikalides, while there are Hercynian and Mesozoic fold belts along the south-eastern and eastern edges. The position of the craton boundaries has so far not been unequivocally determined in a few regions. For example, the south-western boundary has recently been the subject of heated debate. Only a few authors now include the Baikal mountain region in the basement (Fedorovsky, 1985) and consequently the Baikal-Patom belt, which was previously regarded as a Phanerozoic belt (Salop, 1982). The Dzhugdzhur-Stanovoy region is also frequently considered to be a Phanerozoic fold belt. The basis for treating it as an integral part of the basement is that we have now demonstrated that it has a common history with the Aldan Shield throughout geological time.

The earliest work on the geology and tectonics of the Siberian craton was carried out at the end of the 19th century by Kropotkin (1875) and Chersky (1886). In the first quarter of the 20th century, Suess (1909), De Loué (1913) and Obruchev (1923) devoted their research to this region. The term "Siberian craton" was first introduced into the literature by Borisyak (1923). These early publications laid the foundations for modern ideas on the structure of the craton. Shatsky's 1932 tectonic structural scheme is still important today. This showed the outlines of the Siberian craton for the first time, and these boundaries are accepted almost without change today.

The structure, composition and age of Precambrian rocks in the Aldan and Anabar Shields and the Sayan Highlands have been actively investigated since the 1930s. The results of this work have been published by Obruchev (1949), Pavlovsky (1964), Frolova (1951), Dzevansky and Sudovikov (1958), Salop (1978, 1982, 1983), Neyelov (1968) among many others.

The overall geological structure of the Siberian craton is reflected in the tectonic maps compiled under the leadership of Shatsky (1956), the Precambrian tectonic map of Siberia (ed. Kosygin), the tectonic map of Siberia and the Far East (ed. Fotiadi) and the tectonic map of Eurasia (ed. Yanshin).

A new stage in the study of the Siberian craton began with the carrying out of regional geophysical investigations, in the first place aeromagnetic surveys. The accumulated geophysical data were used widely in constructing geological and tectonic maps and schemes by many workers, including Fotiadi (1967), Bulina (1970), Kropotkin et al. (1971), Grishin et al. (in Kontorovich et al., 1981), Mironyuk (1961) and a team working on the tectonic map of the USSR basement (Nalivkin, 1974).

The tectonic schemes for the Siberian craton basement constructed by various authors differ appreciably from one another. One reason is the multiplicity of interpretations of both the geophysical data and the structural and age relationships between the principal Precambrian terrains in the exposed territories. Another reason is the use of different tectonic concepts in analysing the materials — either geosynclinal or plate tectonics (Markov, 1978).

Earlier attempts at constructing tectonic schemes relied mainly on aeromagnetic survey data, since other geophysical methods such as deep seismic sounding, electrical exploration and detailed gravimetric surveys have been employed in Siberia only recently. From these new data bases it emerges that the basic results agree mostly with the regional tectonic scheme advanced by Grishin et al. (in Kontorovich et al., 1981).

Structure of the Precambrian basement

Due to the incomplete geophysical investigation and a total lack of geological data from the central part of the craton, we can present only the broadest outlines of the basement structure at this point in time. The block pattern for the basement structure of the Siberian craton (Fig. II-1) was drawn up from an interpretation of the regional gravity and magnetic fields. As a preliminary, the physical fields were transformed in order to identify those components which reflected basement influences. With this in mind, the gravity field was simplified by removing the components due to the sedimentary cover and mantle inhomogeneities. The averaged field of the remaining gravity anomalies, reflecting basement effects, is shown in

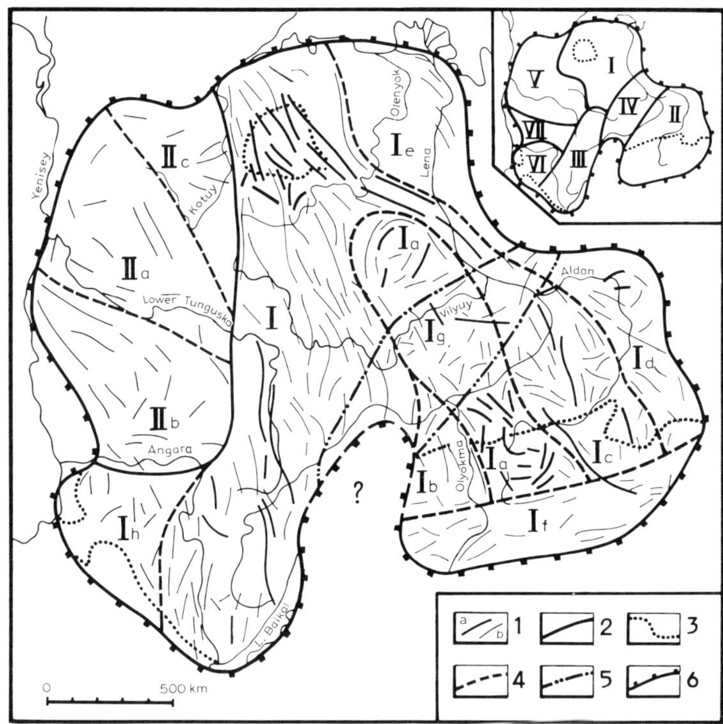

Fig. II-1. Block structure of the Precambrian basement to the Siberian craton. Major Precambrian crustal blocks: *I*-Aldan-Anabar-Angara, *II* = Tungusska. Younger blocks: *Ia* = Olyokma, *Ib* = Iyengr (W Aldan), *Ic* = Timpton-Dzheltulin (E Aldan), *Id* = Batomga, *Ie* = Olenek, *If* = Stanovoy, *Ig* = Taseyev; *IIa* = N Tungusska, *IIb* = S Tungusska, *IIc* = Kotuy. Phanerozoic blocks: *Ih* = Vilyuy; exposed Precambrian crystalline rocks: *An* = Anabar Shield, *Ald* = Aldan Shield. *D–St* = Dzhugdzhur-Stanovoy terrain; marginal outcrops of the basement: *K1* = S Yenisey, *K* = Pre-Sayan (Biryusa and Sharyzhalgay). *1* = magnetic anomaly axes (*a* = high intensity, *b* = low intensity); *2* = boundary between major Precambrian crustal blocks; *3* = boundary of exposed Precambrian; *4* = boundary between younger smaller Precambrian crustal blocks; *5* = boundary of Phanerozoic blocks; *6* = edge of Siberian craton; *7* = area of crustal block II; *8* = area of crustal block *I* (compiled by L.Ye. Shustova). *Inset*: Sketch map of surface relief of the basement to the Siberian craton. *I* = Anabar anticline, *II* = Aldan anticline, *III* = Angara-Lena anticline, *IV* = Vilyuy syncline, *V* = Tungusska syncline, *VI* = Sayan-Yenisey syncline, *VII* = Markin saddle. (Compiled by V.K. Pyatnitsky.)

Fig. II-2. The magnetic field in turn was simplified by removing numerous extremes caused by individual bodies, sedimentary cover sequences and plateau basalts belonging to the middle Palaeozoic to early Mesozoic Tunguska complex. Figure II-1 shows magnetic anomaly axes which reflect the structure and composition of the craton basement.

The Siberian craton basement is a complex assemblage of blocks varying in dimensions and configuration. Two of these are major crustal blocks, I — the Aldan-Anabar-Angara, embracing the central, south-western and

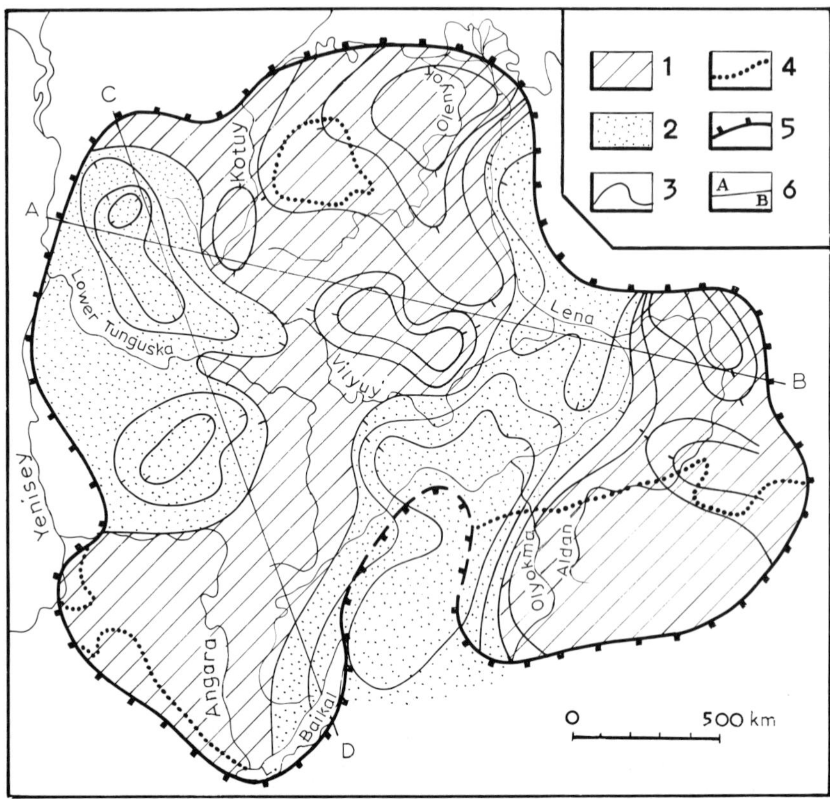

Fig. II-2. Sketch map of averaged remanent gravity anomalies, reflecting the influence of the consolidated crust of the Siberian craton (compiled by L.Ye. Shustova); effects of sedimentary cover and deep mantle structures removed. *1* = regions of higher field; *2* = regions of lower field; *3* = gravity isolines; *4* = surface exposure of crystalline rocks; *5* = edge of Siberian craton; *6* = DSS profile lines

eastern parts of the craton, and II — the Tunguska, which occupies its north-western part.

These blocks have different physical field characteristics. Block I typically has high gravity and magnetic fields, while block II is lower in both respects. In addition, they differ in the thickness of the sedimentary cover. In block I, the basement is buried to a depth of 1 km, and in block II to 4 km.

Different field intensities show that the rocks in the Precambrian basement to block I must have high magnetization and density, while in block II the values are lower. The results of laboratory studies on the physical properties of these rocks have shown that the average density of rocks from the Aldan and Anabar Shields equals 2.75 g/cm^3 (Lotyshev, 1979; Malyshev, 1977). In block II, numerous boreholes in the area between the Stony Tunguska and Lower Tunguska rivers have exposed granite-gneisses with a low density equal to 2.62–2.65 g/cm^3 (Zamarayev, 1967).

Over most of the Aldan-Anabar-Angara block, the magnetic field consists of a system of strip anomalies, striking nearly N–S and NNW for many hundreds of kilometres. In the Anabar Shield in the north, the intense strip anomalies correspond to Lower Archaean high-grade metamorphic rocks, the Daldyn, Upper Anabar and Khapchan Groups. The anomalies radiate out from the Anabar Shield to the south, south-east (with a break in the river Vilyuy basin) and south-west. The south-eastern branch extends as far as the eastern part of the Aldan Shield, where it corresponds to Lower Archaean complexes, the Timpton and Dzheltulin Groups. The south-western branch enters the territory of the Irkutsk amphitheatre (the Taseyev massif) and ends in the Sayan basement high. Such well-expressed gravity and magnetic field characteristics in the Aldan-Anabar-Angara block and the correspondence between these fields and the physical properties of lower Archaean rocks in the Aldan and Anabar Shields would seem to infer that block I consists entirely of formations similar to the lower Archaean complexes in the Aldan and Anabar Shields.

The vast Aldan-Anabar-Angara block is subdivided into smaller units. Block Ia — the Iyengr or West Aldan — has a very high and intensely variable magnetic field. The mosaic pattern and high anomaly intensity reflects the complex structural form of Early Archaean high grade metamorphic rocks (the Iyengr Group) and numerous younger granites. It is worth repeating that the elongate strip anomalies are a property of the block I magnetic field. Two branches of these anomalies engirdle the central part of the Iyengr block with its mosaic of intense anomalies. A certain difference in the configuration of anomalies in the western and eastern branches was the reason for identifying the Timpton-Dzheltulin or East Aldan block (Ic) as a separate entity. To the west of the Iyengr block (Ia) lies the Olyokma block (Ib), with its characteristic quiet magnetic field, the north being lower and the south higher. Against this relatively quiet magnetic background, there are individual north–south striking intense magnetic maxima, especially in the east. The extreme south-eastern Batomga block (Id) and north-eastern Olenek block (Ie) show certain similarities in geophysical characteristics. In the south of the Aldan-Anabar-Angara block is the Stanovoy block (If), which displays a characteristically low magnetic field with east–west magnetic maxima standing out sharply against this background.

In the central part of the region between the rivers Lena and Vilyuy, N–S and NW magnetic anomaly strips are cut almost at right angles by anomalies belonging to the Vilyuy block (Ig), which is characterized by a low magnetic field with E–W and NE magnetic maxima. This block extends from the Patom Highlands in the north-east, and its subdued field merges with the low Verkhoyan field. If the block I basement on the whole consists of Archaean rock complexes, and if the break-up into smaller blocks also occurred in the Precambrian, which is indicated by a Precambrian age for

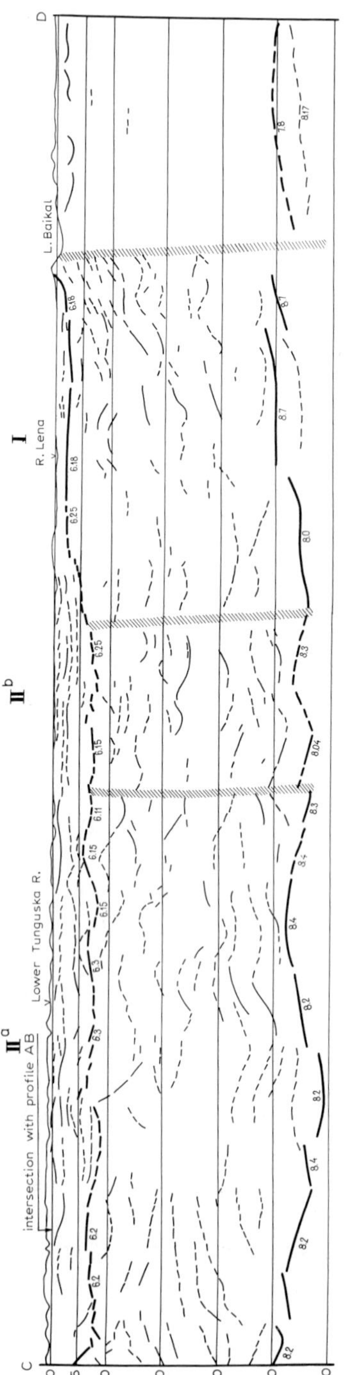

Fig. II-3. Seismic sections along lines *A–B* and *C–D* (see Fig. II-2). *1* = seismic boundaries; *2* = basement surface, M boundary and boundary velocities from seismic data; *3* = boundaries between crustal blocks from gravity data (see Fig. II-1)

marginal fault zones around the Aldan Shield, then the formation of the Vilyuy block structure must therefore belong to a much younger period. The almost north–south strike of the Aldanides-Anabarides is expressed very weakly here, even in places where the basement is not buried very deeply. The observed intense E–W and NE-striking anomalies coincide with the strike of supposed Palaeozoic rift structures (Smelov, 1986). Geophysical data also indicate that block Ig is rift-like in nature. The gravity field is low, indicating low density rocks in the block. Seismic sections along profiles AB and CD (Fig. II-3) clearly show that the crustal thickness in the Vilyuy block changes sharply and in general is attenuated. The M-boundary rises to 32 km, while over the entire Siberian craton its average value is around 40 km. The low gravity field and thinned crust are geophysical evidence for a rift structure.

The south-western part of this enormous block I is treated as an independent block (Ih), the Taseyev massif. The crystalline basement differs here from all other parts of the Aldan-Anabar-Angara block in that it is buried to a great depth, in some places to 7 km. In this block we appear to be dealing with the least eroded section of the Earth's crust, which is also the reason for the less clearly expressed magnetic field characteristics. There are no magnetic anomalies here with elongate axes, and the field intensity is very low.

The magnetic field resulting from the composition and structure of the crystalline basement in block II, the Tunguska block, differs markedly from that of block I, the Aldan-Anabar-Angara block. Here, extensive linear anomalies are absent and the magnetic field intensity is also much lower. The major Tunguska block is subdivided into a number of smaller blocks on the basis of gravity and magnetic field characteristics. Block IIa has a low-intensity magnetic field; anomaly axes are mainly E–W, occasionally NW-striking. Block IIb has a more intense magnetic field with anomaly axes characteristically striking E–W and NE. Both blocks IIa and IIb have subdued gravity fields, which is in accordance with low-density granite-gneisses recovered from deep boreholes. Block IIc has a high gravity field, probably indicating fairly dense rocks in the basement. Unfortunately, there are no boreholes anywhere in this block.

The inhomogeneous nature of the Tunguska block is also very clearly defined by seismic characteristics. Subdivisions of the block are quite evident in seismic sections along profiles AB and CD (Fig. II-3).

The crustal blocks in the Siberian craton enumerated above are separated by inter-block boundaries which are distinguished mainly on the basis of zones of maximum gravity field gradients and zones of rapid strike swings and sudden changes in the nature of magnetic anomalies. From the seismic data, these boundaries are identified from changes in wave properties and from sharp changes in the relief of seismic boundaries, but sometimes also from seismic areas with steep angles of dip. These geophysical features as a rule are taken to be evidence of deep faults. This is confirmed in the Aldan Shield by geological field work. The boundaries between blocks Ia, Ib, Ic, Id and If in this shield are irrefutably deep faults.

Extrapolating the results of interpreting geophysical data over the entire unexposed territory of the craton, we may surmise that the identified crustal blocks are also separated by deep fault zones beneath the sedimentary cover. We note in this regard that judging from the available seismic sections the fault zones which separate crustal rock assemblages do not always appear clearly at the surface.

The structures in the sedimentary cover of the Siberian craton do not display any clear link with the basement block structure which originated in the Precambrian. The boundaries of even major structures in the cover, such as the Tunguska syncline and the Anabar anticline, do not coincide with block boundaries (Fig. II-1 and inset).

* * *

The results of analysing geological and geophysical data obtained from exposed and unexposed territories in the basement to the Siberian craton are shown on the regional tectonic sketch map (Fig. II-4). On this map the tectonic provinces are identified using the same criteria as in the East European craton. We have found that Archaean mobile belts predominate in the Siberian craton and we assume that Late Archaean internal processes were responsible for completing the formation of crustal structure in these belts. Subsequent processes, including those operating in the Early Proterozoic, had local effects and utilized previously formed structural elements. The sketch map shows Early Proterozoic structures as narrow belts.

The interpretation proposed on this sketch map for the tectonic evolution is a break with tradition and is not always accepted by researchers working in specific regions. This is particularly the case with the Batomga block, which is defined in the regional summary of the Aldan Shield as a granite-greenstone terrain.

As far as particular structural features are concerned, the Siberian craton basement differs significantly from that of the East European craton. Even allowing for the fact that several differences arise from different levels of investigation and a variety of interpretations, there are nevertheless a number of divergent facts which have been proved for each time segment:

– in the Phanerozoic, the thickness of the sedimentary cover is greater on the Siberian craton than the East European craton;

– in the Late Precambrian, Early Riphean aulacogens are far fewer in Siberia;

– in the Early Precambrian in particular crust-forming processes were completed earlier in Siberia (Rudnik et al., 1969).

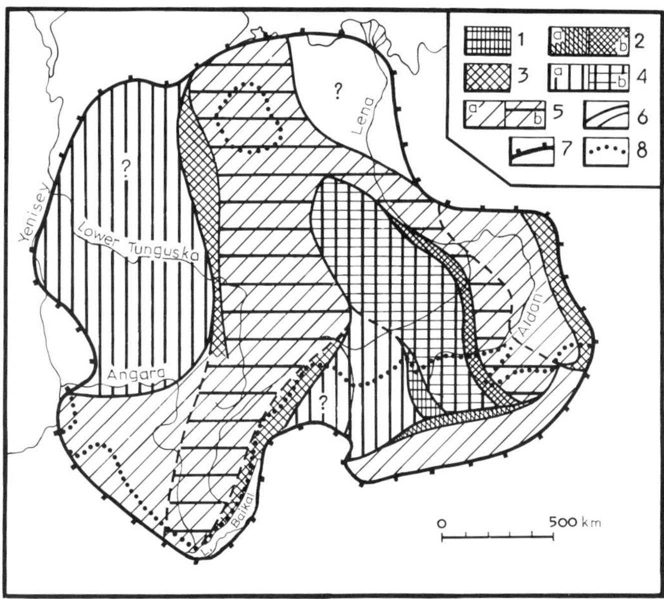

Fig. II-4. Sketch map of tectonic regions of the Precambrian basement of the Siberian craton (after V.Ya. Khiltova). *1–3*: Early Proterozoic mobile belts: *1* = with geosynclinal precursors, *2* = without geosynclinal precursors (a: shallower, b: deeper sections), *3* = belts with completed orogenic complexes; *4–5*: Archaean mobile belts (epi-Archaean cratons): *4* = granite-greenstone belts (*a*: in upper sections, *b*: in lower sections), *5* = granulite-greenstone belts (*a*: in upper sections, *b*: in lower sections), *6* = boundaries between geostructures; *7* = craton margin; *8* = shield boundaries.

Among Archaean mobile belts in the Siberian craton basement, granulite-greenstone terrains occupy a greater area that granite-greenstone belts. The largest and best studied region in East Siberia — the Aldan Shield and the adjoining Dzhugdzhur-Stanovoy province — may be referred to different types of Archaean mobile regions, separated by relatively narrow Early Proterozoic mobile belts. Accordingly, the different blocks that make up the shield at the present erosion level may be regarded as different sections through Archaean crustal provinces (Fig. II-5). For example, the Olyokma block (Ib), the western part of the shield and the Batomga block (Id), its eastern part, combine upper and mid-crustal sections. A characteristic of these sections is the preservation of Late Archaean mobile belts, greenstones and, or paragneisses with fairly thick supracrustal sequences associated with basement complexes. These two extreme blocks possess a number of distinguishing features:

(1) The blocks differ in tectonic pattern — linear structures are a feature of the Batomga block, while in the western part of the Olyokma

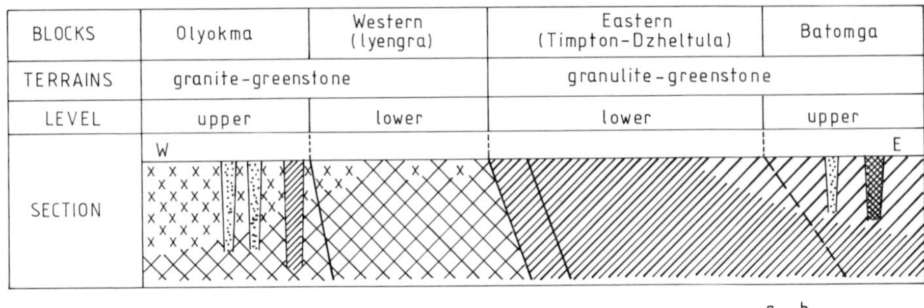

Fig. II-5. Section showing major structure of the Aldan Shield (after V.Ya. Khiltova). *1* = Early Proterozoic and late Archaean mainly terrigenous troughs; *2* = Late Archaean volcanic troughs; *3* = granite-gneiss terrains; *4* = granulite-gneiss terrians, low-*P* granulites; *5* = granulite-gneiss terrains, high-*P* granulites (*a*: in lower section, *b*: in upper section).

block, unaffected by Early Proterozoic processes, the pattern is more whorl-shaped.

(2) They differ in the nature of Archaean metamorphism, which is moderate gradient in the Batomga and high gradient in the western part of the Olyokma block, as shown by recent work.

(3) They also differ in geophysical field characteristics.

Those blocks which constitute the Aldan granulite-gneiss terrain, the Iyengr in the west and the Timpton-Dzheltulin in the east, are deep sections through different lower Archaean mobile regions. These sections consist of lower Archaean (Katarchaean?) supracrustal assemblages and their basement complexes, metamorphosed at granulite facies. We have documented a large number of Archaean and Late Archaean intrusive rocks; in the Iyengr block these are mostly acid, but basic in the Timpton-Dzheltulin block. The two blocks differ in a number of other respects. In the Timpton-Dzheltulin block the tectonic structures have a marked linear nature, granulites are high-pressure types, basic and ultrabasic rocks are numerous and widespread and thrust faulting is common. In the Iyengr block, dome structures predominate, granulites are low-pressure types, and most rocks are of acid composition.

While there are essential differences in the pairs of blocks examined, there is nevertheless a similarity in the most important features of internal evolution during the Late Archaean: between the Batomga and Timpton-Dzheltulin on the one hand, and the Iyengr and Olyokma on the other. Typical features characterising the Batomga and Timpton-Dzheltulin blocks allow them to be considered as different sections of one granulite-greenstone type terrain, whereas the Iyengr and Olyokma blocks belong together in a granite-greenstone terrain.

A linear fold zone, the Nuyam or Idzhek-Sutam zone, may be taken as a boundary suture between these types of Archaean mobile regions — the Olyokma-Iyengr granite-greenstone terrain and the Batomga-Timpton granulite-greenstone terrain.

The Dzhugdzhur-Stanovoy province (the Stanovoy block) is situated to the south of the zone called in the literature the Stanovoy suture (Fedorovsky, 1985) or the South Aldan deep fault (Neyelov, 1980), and is also an Archaean granulite-greenstone terrain. Its tectonic structure was formed by Stanovoy or more likely late Stanovoy processes, with an age of around 2400 Ma. Early Proterozoic processes, localized in narrow zones, did not alter the structural provinces, since they did not create a new structural pattern in the region. They developed locally and utilized structural directions that originated in the Late Archaean.

For the Dzhugdzhur-Stanovoy province we may note all the characteristics belonging to granulite-greenstone terrains in upper crustal sections.

(1) Preservation of supracrustal assemblages, which had completed their development in the Late Archaean, by 2500 ± 100 Ma.

(2) Linear structural plan.

(3) Moderate to low gradient type of metamorphism during Late Archaean events. In mid-crustal sections, amphibolite facies rocks belonging to the kyanite-sillimanite facies series were formed, and in lower crustal sections, high-pressure granulites. These granulites are presently exposed mostly in zones where, at present erosion levels, deep sections of the Archaean crust are elevated. These zones are Early Proterozoic mobile belts, formed at the margins of different types of Archaean provinces, for example the Pre-Stanovoy belt (the Stanovoy suture).

Thus, despite the debate about the interpretation of geological and geochronological data from the Aldan Shield, which was affected by processes from Early Archaean (Katarchaean) to Proterozoic, crust-forming structures are thought to have ended over most of the shield by the Late Archaean, i.e. by 2500 ± 100 Ma ago. The western part resembles the granite-greenstone terrains of the East European craton, although it is not identical in all respects; the eastern and southern parts are like granulite-greenstone terrains. Early Proterozoic effects were restricted locally to suture zones and did not alter the overall structure of the region, although they had a substantial effect on the closure time of isotopic systems, producing close radiometric age values, obtained by various methods, for geological formations of different ages. These age determinations probably coincide with the time of elevation of this section to higher crustal levels (higher than amphibolite facies level).

In spite of its inhomogeneous structure, the Anabar Shield can also be regarded in total as a Late Archaean mobile region. It consists mainly of Early Archaean supracrustal and igneous rocks, and Late Archaean (3000–2600 Ma) intrusions, metamorphosed in part at granulite facies. The well-expressed linear nature of the shield structure, the mainly low-gradient type of granulite facies metamorphism over most of the shield, the widespread development of Late Archaean basic and ultrabasic intrusive rocks and the common thrusts are typical features that are associated with lower sections of granulite-greenstone terrains. The occurrence of moderate-pressure granulite facies metamorphic rocks in one of the blocks in the Anabar Shield may have something to do with our present imperfect understanding of various geological processes. In particular, this could be due to the greater preservation in this block of a much older medium-pressure granulite facies metamorphism in many lower Archaean complexes, which gave way to high pressure metamorphism in the Late Archaean.

The marginal elevations in the south-west of the Siberian craton are also comparable with Archaean granulite-greenstone terrains in terms of their structure and internal evolution during the early Precambrian, the western ridges being upper- to mid-crustal sections of such terrains. They contain Late Archaean supracrustal complexes — trough-shaped greenstone belts (grabens). The eastern part of the Sayan ridge, the Sharyzhalgay block, may be a lower crustal section. As in other regions in the Siberian craton, Early Proterozoic processes in the marginal elevations are manifest in suture zones and they were particularly intense in southern and south-western ridges, where they coincide with the boundaries of the Siberian craton.

The basement highs at the margins can be traced in terms of structure and evolution beneath the platform cover (Zamarayev, 1967; Pavlov and Parfenov, 1973). Individual highs, separated at present by Late Precambrian to Phanerozoic platform sediments, form parts of the single Taseyev massif (Neyelov, 1980), or the Angara platform (Salop, 1982).

Early Proterozoic tectonic provinces fall into two categories of belt, as in the East European craton. Those without geosynclinal precursors, such as the Pre-Stanovoy belt, separate mobile regions of the Aldan Shield itself from the Dzhugdzhur-Stanovoy granulite-greenstone terrain. Another example of this category of belt may be the Idzhek-Sutam linear fold belt, which separates various types of Archaean terrains across the whole Aldan Shield.

Mobile belts — those with geosynclinal precursors — include the eastern part of the Olyokma province, which is a junction zone between various sections through Archaean granite-greenstone terrains. During the Early Proterozoic, this was the site where graben structures were initiated, the growth of which concluded with folding, metamorphism and igneous activity. The region which lies to the west of Lake Baikal belongs to this category of tectonic structure (see Fig. II-4). This is the Baikal-Patom belt. Most workers accept that its constituent metamorphic complexes are Early Proterozoic in age, while the final crust-forming event in the belt is a volcanic-plutonic association (the Akitkan suite and associated granites), which yields age dates of 1700 Ma and younger (Manuylova, 1968).

Chapter 1

THE ANABAR SHIELD

Introduction

The Anabar Shield is the exposed basement to the Siberian Craton and geographically coincides with the Anabar Plateau of the Central Siberian Plateau. The crystalline rocks of the shield are overlain at the margins by gently-dipping Riphean sediments. The eastern and western edges of the shield are determined by flexural warps in the sedimentary cover. These flexures are caused by the sudden downthrow of the basement along faults. Within 30–40 km of the edge of the sedimentary cover, the basement is downfaulted by 2–3 km; southwards, the basement surface is gradually buried and 500–700 km from the edge of the cover, it is around 3 km deep.

The Anabar Shield was discovered during the Khatanga expedition, led by Tolmachev and Backlund in 1905.

Investigations of the Anabar Shield during the years 1946–1950 showed that its constituent rocks form a complex Archaean volcano-sedimentary assemblage, intruded by granites, anorthosites and ultrabasic rocks, then thought to be Early Proterozoic (Rabkin, 1959). The sequence of metamorphic rocks was divided into three groups (from below upwards): the Daldyn, Upper Anabar and Khapchan, consisting of pyroxene-plagioclase gneiss and schist together with marble, calciphyre, hypersthene-garnet and other paragneisses. The Upper Lomuy Group was also described by Rabkin (1959); this consists of amphibolite facies paragneiss and amphibolite. It was also established that the metamorphic rock complexes form broad belts striking north–west and reflecting the fold structure of the shield. Granulite facies rocks predominate: ultrametamorphic enderbites and charnockites, pyroxene-plagioclase schists and meta-ultramafics. Also present are quartzite, banded ironstone, marble and calciphyre. Amphibolite facies rocks are less important and where present they crop out in NW-striking shear zones as biotite and hornblende-biotite gneisses, amphibolites, migmatites and granitized rocks containing relict granulite facies mineral assemblages. The anorthosite, gabbro and granitoid intrusions are restricted to these particular zones.

Research carried out in the period 1956–1961 shed light on the question of the geology and petrology of the granulite complex of the shield, which on the basis of lithological criteria (Lutz, 1964) is subdivided into a sequence of formations: the Daldyn spilite-keratophyre volcanic Group gives way to the

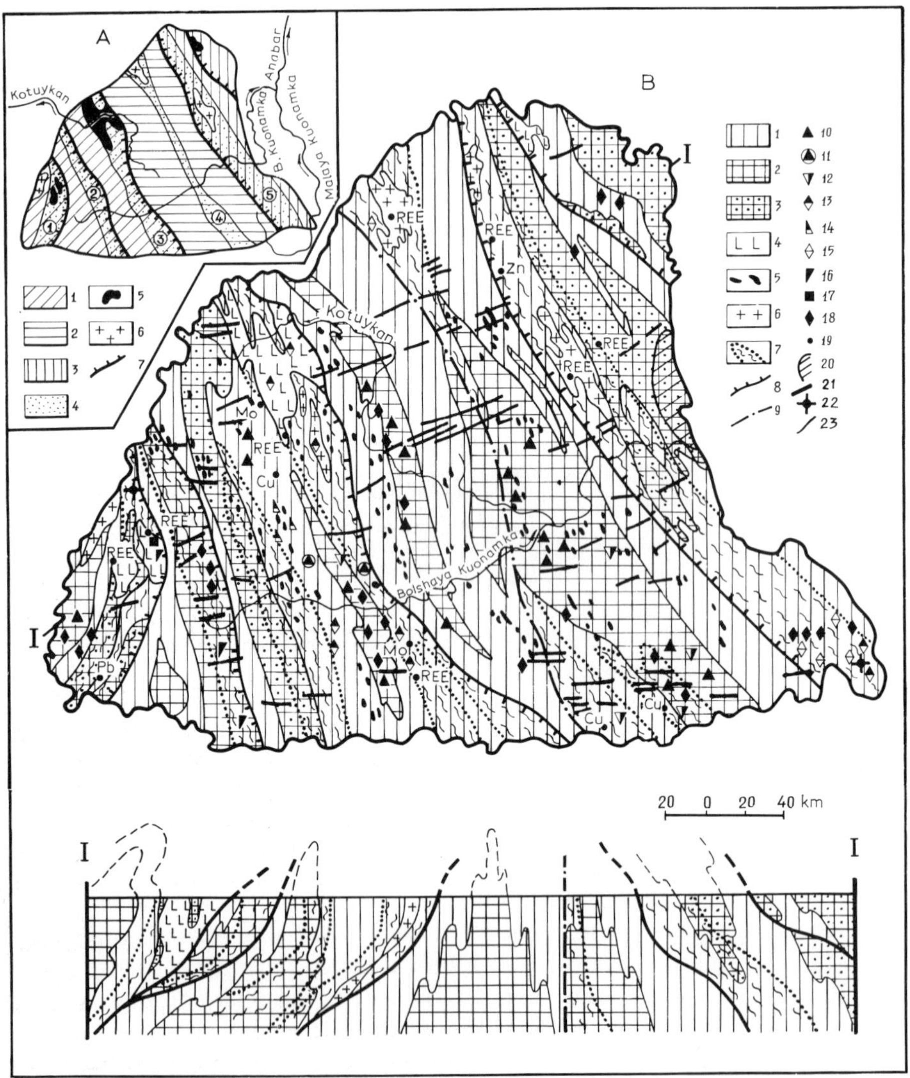

Fig. II-6. A. Sketch map showing tectonic regions of the Anabar Shield (after Vishnevsky, 1978). Early Archaean metamorphic complexes: *1* = Magan, with medium *P–T* granulites, *2* = Central Anabar with high *P–T* granulites, *3* = Khapchan, with relatively low *P–T* granulites. *4* = zones of superimposed early Proterozoic amphibolite facies metamorphism and ultrametamorphism (numbers on sketch: *1* = Magan, *2* = Lamuy, *3* = Kotuykan-Monkhoolin, *4* = Kharap, *5* = Billyakh); *5* = anorthosites, *6* = granitoids, *7* = deep-level thrusts, *8* = block boundaries. B. Geological structure and mineral deposits of the Anabar Shield. Archaean provinces (supracrustal assemblages): *1* = sialic (enderbite-gneiss) (Upper Anabar Group), *2* = sialic-femic (Daldyn Group and Ambardakh assemblage of the Upper Anabar Group), *3* = sialic with large volume of terrigenous-carbonate rocks (Khapchan Group and Vyurbyur assemblage of Upper Anabar Group), *4* = anorthosites and gabbro-anorthosites, *5* = meta-ultramafics. Early Proterozoic province: *6* = granitoids, *7* = migmatites, granites and amphibolite facies retrogressed rocks. *8* = deep-level thrusts,

Upper Anabar sandstone-shale terrigenous Group and finally the Khapchan carbonate-flysch Group. It has also been suggested that the Upper Lomuy Group is a granitized and amphibolite facies polymetamorphosed part of the Upper Anabar Group. The block structure of the shield was recognised later by Vishnevsky (1978), individual blocks being at different metamorphic grades and separated by shear zones (Fig. II-6). These shear zones display amphibolite facies metamorphism, accompanied by partial melting and granitoid magmatism. At the same time, the stratigraphic scheme being employed then was shown to be entirely provisional, but was confirmed in later publications (Rozen, 1981).

Tectonic structure

Seismic results from the Vorkuta-Tiksi deep seismic profile (Sokolov and Bilibina, 1981) which crosses the Anabar Shield in the north indicate that basement rocks are buried stepwise to a depth of some 4 km. Intracrustal seismic boundaries within the Anabar Shield form discontinuous surfaces dipping outwards to the west and east from the centre of the shield to the periphery.

A major characteristic of the tectonic structure of the shield is the clearly expressed block pattern owing to the presence of NW-striking shear zones and superimposed metamorphism. Five such major zones have been mapped in the shield (Fig. II-6A) which possess a number of characteristic features: (1) a length of hundreds of kilometres and a width of a few tens of km; (2) overlapping pattern of crustal slices and a complex thrust-nappe mechanism of formation; (3) widespread granitization and amphibolite facies retrograde metamorphism; (4) the presence of deep-level igneous rocks: anorthosite, norite, gabbro, also granitoids and pegmatites; (5) a protracted period of formation for the majority of the zones, from the Late Archaean to the end of the Early Proterozoic. These shear zones divide the Anabar Shield into a number of relatively small blocks which are grouped together into three major blocks, the Central Anabar, the Magan and the Khapchan.

The Central Anabar block consists of the Daldyn and part of the Upper Anabar Groups of mafic schists. Charnockites are locally present as well as granitoids of igneous and partial melt origin.

The Magan block is separated from the Central Anabar by the largest tectonic zone of superimposed metamorphism — the Kotuykan-Monkhoolin

---

9 = Anabar deep fault. Mineral deposits: *10* = iron (magnetite-pyroxene quartzites), *11* = chrome in meta-ultramafics, *12* = nickel in meta-ultramafics, *13* = muscovite in pegmatites, *14* = garnet, *15* = phlogopite, *16* = apatite, *17* = titanium, *18* = graphite, *19* = non-ferrous and rare metals, *20* = part of Kuonam kimberlite field, *21* = diabase dykes, *22* = alkali-ultramafic intrusion and carbonatites, *23* = geological boundaries.

— and is in turn divided into smaller blocks by similar zones. The dominant rocks in the Magan block belong to the Upper Anabar Group, with some Daldyn Group rocks also present. Intrusive granitoids and migmatites are restricted to superimposed tectonic zones and in this block they have gabbro-anorthosite bodies associated with them.

Most of the Khapchan block is occupied by Khapchan Group rocks. Pyroxene-plagioclase gneisses, referred to the Upper Anabar Group, are present.

A characteristic of the Anabar Shield is its intense magnetic field and marked elongate magnetic anomalies, defined by alternating narrow and wide strips, respectively highly and weakly magnetised. A wide zone of intense magnetic anomalies occurs in the Central Anabar block, approximately coinciding with the Daldyn Group outcrop area. Here the gravity field is also high, corresponding to the high density of these rocks (Grishin and Lotyshev, 1974). In the north-east (the Khapchan block) and within the Magan block, relatively low magnetic and gravity fields are typical. In some instances, positive magnetic anomalies are restricted to NW-trending tectonic zones, with abundant granitization and granitoid magmatism. It is also noteworthy that where Upper Anabar Group rocks are present, we observe less intense isometric magnetic anomalies with narrow bands of intense anomalies, doubtless due to the magnetic properties of two-pyroxene schists, which occur as lenses amidst the plagioclase gneiss and enderbite unit.

A clear picture of the Anabar Shield emerges, with a broad domal structure in which the oldest rocks of the Daldyn and Upper Anabar Groups crop out in the centre of the shield. They give way to Upper Anabar Group and Khapchan Group carbonate-terrigenous rocks on the north-eastern and south-western margins of the shield. This overall structure is complicated by minor synclines and anticlines which form a series of elongate structures striking NW for 200–300 km and only a few tens of kilometres wide. The major mapped structural forms — folds a few kilometres or tens of metres across — sometimes also show up clearly on aerial photographs. These are isoclinal folds, accompanied by the squeezing out of plastic material in fold closures, which shows that the folds formed during a period of regional metamorphism. On top of this, minor structural forms are highly variable: boudinage and minor shear folds, lineations and shear zones accompanied by migmatization and late blastomylonitization, especially well shown in zones of superimposed metamorphism.

The structure of the Anabar Shield is complicated by various generations of tectonic breaks, including fault zones, thrusts, crush belts, shear zones and tectonic displacement zones. Some of these are old structures which originated during the Archaean and became reactivated during the Early Proterozoic period of evolution in the shield. The youngest structures are E–W faults which can be traced by Late Proterozoic and Mesozoic dolerite dyke trends (Maschak, 1973). Broadly speaking, the structure of the shield can be represented as a series of major blocks consisting of stratified Archaean sedimentary, volcanic and intrusive rocks, metamorphosed at granulite facies. The blocks are separated by younger (Early Proterozoic) zones of tectonothermal reworking. These formed at the sites of deep-seated thrust-type lineaments with amphibolite facies metamorphism, granitoid magmatism and partial melting. The deepest erosion level is seen in the northern parts of these zones, which is precisely where we find anorthosite, gabbro and charnockite bodies. There is a sharp rise in the amount of granitization in the southern parts of these zones. At the southern margin of the Khapchan zone (*4* on Fig. II-6) we have an assemblage of amphibolites, hornblende-biotite-plagioclase gneisses, quartzites and quartzose sandstones without relict granulite facies minerals (Vishnevsky, 1978; Lutz, 1964). The age of the quartzose sandstones is 2300–2400 Ma, from a K–Ar determination carried out by Krylov (1962). On this basis the deep fault zones could be Early Proterozoic or Late Archaean basins, filled with sedimentary sequences, progressively metamorphosed at amphibolite facies and subjected to granitization.

From what has been presented above, it follows that the structure of the Anabar Shield formed over a long time interval as the result of repeated deformation of various kinds and that it consists of three structural domains: (1) lower Archaean-granulite facies rocks, the Daldyn, Upper Anabar and Khapchan Groups metamorphosed under a variety of PT conditions; (2) upper Archaean, consisting of plutonic igneous associations — gabbro-anorthosite intrusions and bodies of charnockite and enderbite; (3) lower Proterozoic, consisting of reworked Archaean rocks showing retrograde metamorphic effects, granitoids and relict supracrustal rocks showing prograde amphibolite facies metamorphism.

Alkaline ultramafic intrusions and carbonatites, referred to the Riphean (Shpunt et al., 1982), are found locally in Late Archaean–Early Proterozoic NW-striking tectonic zones, e.g. the south-western part of the Billyakh zone. East–west regional fracture zones control the emplacement of diabase and dolerite dykes in the shield, ranging in age from Early Riphean to Late Palaeozoic and Mesozoic (Maschak, 1973). Kimberlite bodies occur in the east, forming part of the Kuonam field of early Mesozoic age.

Geochronology

Most of the radiometric age determinations on Anabar Shield rocks were carried out in the 1960s using the K–Ar method (Table II-1). U–Th–Pb dating has been done on whole-rock samples as well as on monazite and zircon from pegmatites and migmatites (Stepanov, 1974). Most of the K–Ar determinations on micas from granitoids and pegmatites gave an age of 2120–1750 Ma, which dates an epoch of granitization

TABLE II-1

Isotopic age of the Anabar Shield

No.	Target	Material	Method	Age (Ma)	Refs
1.	Upper Anabar Group	Amphibolites, amphibole	K-Ar	2570-2300	1, 2
2.	Granite pegmatites and granitoids	Micas	K-Ar	2120-1750	1, 2, 3
3.	Two-pyroxene schists gneisses, pegmatites	Uraninite thorite, monazite	U-Th-Pb $^{207}Pb/^{206}Pb$	3150-160 3550±500	1, 2, 4
4.	Upper Anabar Group enderbitoids	Zircon	U-Pb isochron	2890±20	5
5.	Anorthosites	Zircon	Pb-Pb thermochron	2770-270	6
6.	Diabase dykes	Whole rock	K-Ar	1570-1506 1412-1400 1318-1250 242±5	7
7.	Kimberlites, NE part of Shield	Zircon	Fission track	250-170	8

References: 1 = Krylov et al. (1963); 2 = Rabkin and Vishnevsky (1968); 3 = Tarasov et al. (1963); 4 = Stepanov (1974); 5 = Vishnevsky (1978); 6 = Sukhanov et al. (1984); 7 = Mashchak (1973); 8 = Komarov and Ilupin (1978).

and retrograde metamorphism. Older K–Ar dates (2570–2300 Ma) were obtained on amphiboles from Upper Anabar schists.

$^{207}Pb/^{206}Pb$ ratios in uraninites, thorites and monazites from two-pyroxene schists, gneisses and migmatites show wide variations in age from 3150 to 1600 Ma (Stepanov, 1974), although most fall into the 1800–2000 Ma interval, which is in good agreement with K–Ar determinations. Some workers (Musatov et al., 1981) consider that the upper age boundary for supracrustal rocks in the shield is 3550 ± 500 Ma (on monazite from a pegmatite cutting Khapchan amphibole gneisses), using the U–Th–Pb method (Krylov et al., 1963). According to them, this figure dates the earliest episode of partial melting, suggesting that the primary rocks of the metamorphic series in the shield are older than this date. Radiometric dating by the U–Pb isochron method on zircons from Upper Anabar Group hypersthene-plagioclase gneisses (enderbites, Bibikoba et al., 1985) yields a value of 2890 ± 20 Ma. A thermochron determination on zircon from a gabbro-anorthosite intrusion in the Kotuykan-Monkhoolin zone (Fig. II-6) showed that this suite belongs to the Late Archaean, the date obtained being 2770–2700 Ma (Sukhanov et al., 1984).

Younger igneous rocks of the shield, represented by dolerite and diabase dykes, have been dated by the K–Ar method (Maschak, 1973) and show a broad age spectrum (Table II-1). We can identify an Early Riphean dyke swarm with an age of 1570–1400 Ma, a Late Riphean swarm at 1318–1250 Ma, as well as early Mesozoic dykes at 242 ± 5 Ma. Kimberlites are 250–170 Ma old, from fission track determinations on zircons (Komarov and Ilupin, 1978).

Tectonic domains

The *Early Archaean domain* consists of supracrustal assemblages in the form of stratigraphic sequences of varied volcanic and terrigenous rocks, metamorphosed at granulite facies. On a lithological basis they can be subdivided into a number of assemblages (Table II-2), which are described in detail elsewhere (Vishnevsky, 1978; Babkin, 1959; Anon., 1986). We simply point our here that the Daldyn Group is presumed to lie at the base of the succession; it consists of two-pyroxene schist (c. 60%) and gneiss, alternating with hypersthene-plagioclase gneiss-enderbitoids (c. 30%) containing meta-ultramafics (peridotite and pyroxenite) and magnetite quartzite in the form of individual bands and lenses up to hundreds of metres long. There are also bands of leucocratic granulite (garnet-plagioclase gneiss) and pyroxene-magnetite schist (c. 10%). The Upper Anabar Group consists of two-pyroxene and hypersthene plagiogneiss (c. 40%) — enderbitoids, amphibole–two-pyroxene gneiss (c. 30%) with bands of garnet and biotite-garnet plagiogneiss (c. 15%) and two-pyroxene schist (c. 15%). The stratigraphic relations of these two groups are unknown.

Rozen (1981) considers that the Daldyn metabasite-plagiogneiss (enderbite) association is volcanic in origin, while the Upper Anabar Group is an association of plagiogneisses or enderbites proper. Since charnockite also features as a typical member of the latter association, then it may be defined as enderbitic-charnockitic, so that in the proposed scheme (Fig. II-7) we have two separate groups: (1) a sialic Upper Anabar enderbite-charnockite group; and (2) a sialic-femic group, comprising the Daldyn Group and the Ambardakh Formation of the Upper Anabar Group.

The stratigraphically higher Khapchan Group consists in the main of garnet gneiss and leucocratic granulite plus thin marble bands alternating with hypersthene-plagioclase gneiss (enderbitoids) and calciphyre (diopside-plagioclase and scapolitic types). This group is easily identified due to the first appearance of marble bands up to 200 m thick. According to Rozen, it is a carbonate-gneiss association which in the scheme of Fig. II-7 is included as part of the Vyurbyur carbonate-gneiss Formation in the Upper Anabar Group as a group of sialic formations containing substantial volumes of carbonate terrigenous rocks (Table II-2). The total thickness of the section is estimated to be about 10 km.

Reconstructing the primary composition of the high grade metamorphic rocks in the Anabar Shield is exceedingly difficult, but attempts have been

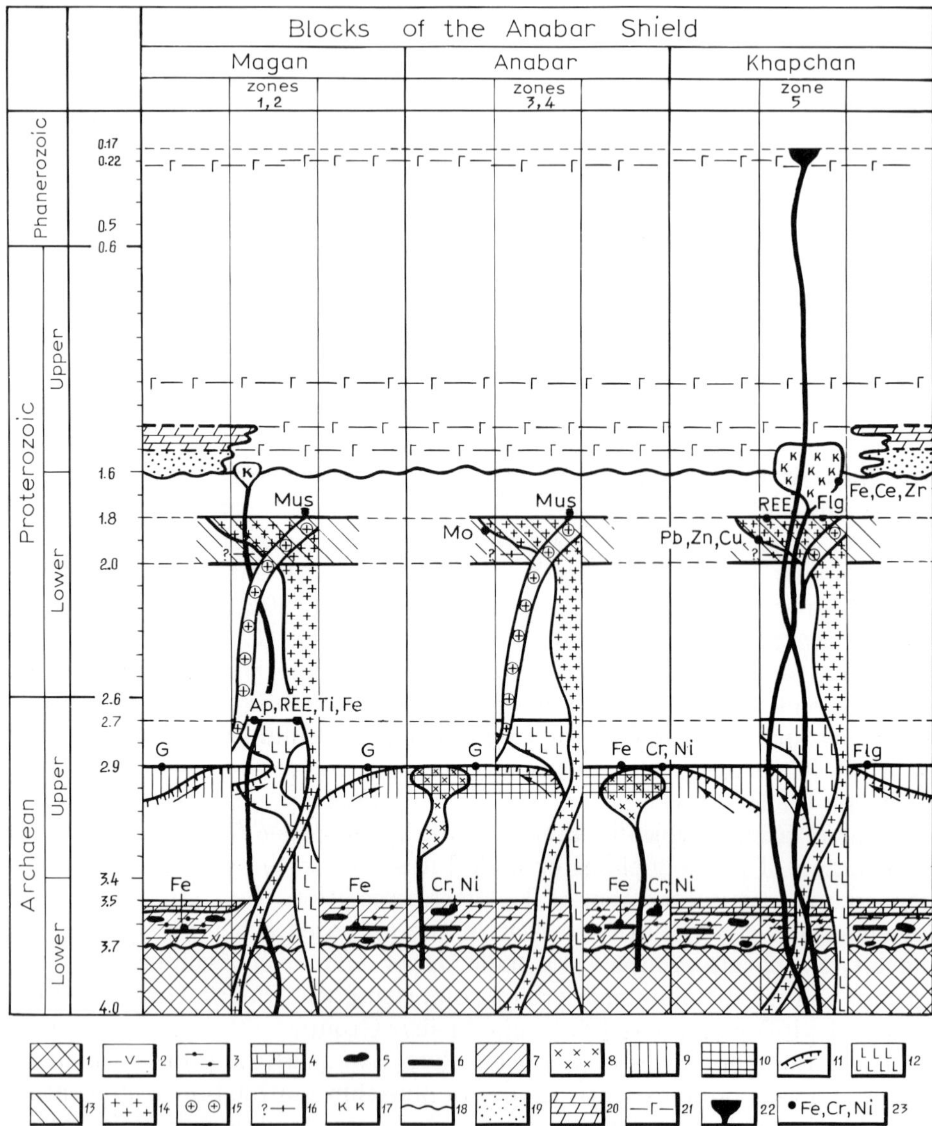

Fig. II-7. Diagrammatic evolution of the crust in the Anabar Shield. *1* = Undifferentiated basement, early Archaean supracrustal assemblages; *2* = Daldyn Group (basic granulites, enderbites), *3* = Upper Anabar Group (basic and acid granulites, enderbitoids), *4* = Khapchan Group (marble, paragneiss, acid granulites), *5* = meta-ultramafics (komatiites), *6* = ferruginous quartzites. *7* = medium-*P*, high-*T* granulites, *8* = enderbites, charnockites, *9* = medium *P–T* granulites, *10* = high *P–T* granulites, *11* = deep-level thrusts, *12* = gabbro-anorthosites, *13* = superimposed amphibolite facies metamorphism and ultrameta-morphism, *14* = granitoids, *15* = pegmatites, *16* = possible early Proterozoic supracrustal assemblage, *17* = alkali-ultrabasic rocks; *18* = unconformity; *19* = Mukun Formation (mostly sandstones), *20* = Billykah Formation (mostly limestones), *21* = gabbro and diabase dykes; *22* = kimberlites, *23* = mineral deposits: Fe, Cr, Ni, Pb, Zn, Cu, Mo, Ce, Zr, Musc (muscovite), Ap (apatite), REE, Phl (phlogopite), Grp (graphite).

TABLE II-2

Composition and subdivision of Archaean supracrustal rocks of the Anabar Shield

Group	Formation, after Vishnevsky (1978), Bilibina & Sokolov (1984)	Rock composition, after Vishnevsky (1978), Bilibina & Sokolov (1984)	Thickness (m), from Vishnevsky (1978)	Rock association from Rozen (1981) Lutz (1984)	Group of rocks in this work
Khapchan	Billekh-Tamakh	Leucocratic and mesocratic garnet, biotite-garnet-graphite gneiss Sillimanite-garnet schist and gneiss, sometimes with graphite and cordierite	2500-3000	Carbonate-gneiss	Sialic with large volumes of terrigenous and carbonate rocks
	Khaptasynnakh	Hypersthene, two-pyroxene and garnet-hypersthene plagiogneiss. Biotite, biotite-garnet schist and gneiss, calciphyre, marble	2000-2500	Carbonate-flyschoidal	
	Vyurbyur	Hypersthene and two-pyroxene plagiogneiss, biotite-garnet and garnet-sillimanite-cordierite schist, calciphyre, marble	2000-2500	Plagiogneiss (enderbite proper)	
Upper Anabar	Eymi	Hypersthene and two-pyroxene plagiogneiss, biotite-garnet-hypersthene, sillimanite-biotite-garnet, garnet-sillimanite-cordierite schist, quartzite with sillimanite and garnet	3000	Volcanogenic-terrigenous	Sialic
	Ambardakh	Hypersthene and two-pyroxene plagiogneiss	2000-3000		
Daldyn	Kilegir	Meso- and melanocratic two-pyroxene plagiogneiss, quartzite, ferruginous quartzite, pyroxene schist, garnet-pyroxene-magnetite schist, cordierite-sapphirine-sillimanite schist, hypersthene-garnet plagiogneiss	3000-5000	Metabasite-plagiogneiss (enderbitic)	Sialic-femic
	Bekelekh	Meso- and melanocratic two-pyroxene plagiogneiss, pyroxene schist, hypersthene plagiogneiss. Biotite-garnet gneiss, garnet-pyroxene-magnetite schist, pyroxenite, peridotite	1500-2000	Volcanogenic	

made, based on the petrochemistry of the major rock types. The Daldyn and Upper Anabar mafic schists are analogues of tholeiitic basalts, while the pyroxene-plagioclase gneisses (enderbitoids) are dacite and andesite analogues (Lutz, 1984; Rozen, 1981). The petrochemical data shown on variation diagrams (Fig. II-8) demonstrate that these rocks form an unbroken volcanic series of basalt-andesite-dacite composition. On the basis of the Peacock calc-alkali index of 64 and 60, this series corresponds to two of its branches: calc and calc-alkali (Lutz, 1984). Meta-ultramafics, varying in composition from peridotite to pyroxenite (previously regarded as intrusives by Lutz, 1964 and Rabkin, 1959) are now included in the volcanogenic series of stratiform formations, while chemically they correspond to komatiites (Lutz, 1984). The majority of the Daldyn Group rocks together with ferruginous quartzites may be treated as a volcanogenic komatiite-

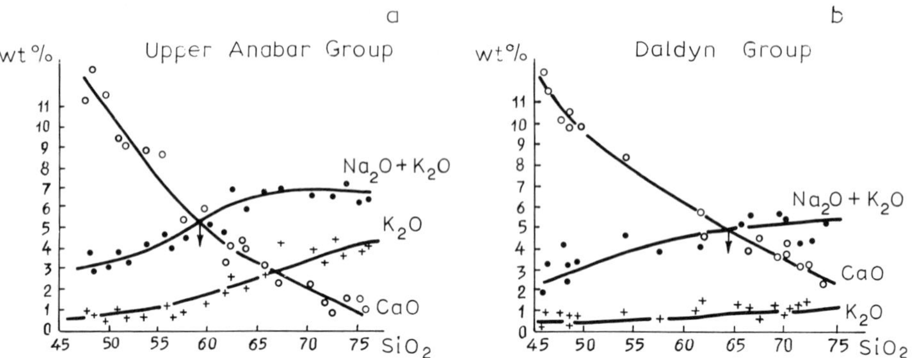

Fig. II-8. Variation diagrams for rocks of the Upper Anabar (a) and Daldyn (b) Groups (after Lutz, 1974).

tholeiite-basalt association. Upper Anabar Group rocks plus garnet-biotite gneisses, which are treated as continental clastic rocks (Rozen, 1981), form a volcanic-terrigenous (basalt-dacite with continental clastics) association.

In the upper part of the series we see the pattern changing, with these formations being replaced in a regular manner by the Khapchan carbonate-flyschoid Group. The association of marble with garnet and biotite gneiss forms a characteristic petrochemical group, corresponding to terrigenous rocks of the greywacke family (Rozen and Dimroth, 1982). This rock group is exposed in the Magan and Khapchan blocks (Fig. II-6). The Central Anabar block comprises sialic and sialic-femic supracrustal rock groups.

As far as particular granulite facies metamorphic features are concerned, these blocks also differ from one another in terms of their deep crustal level metamorphic mineral assemblages. Typomorphic parageneses in the Central Anabar block are hypersthene with sillimanite and quartz (figures show iron content): Hyp_{28-36} + Sill + Q, Gr_{41-43} + Hyp_{22-32} + Saph + $Cord_{9-15}$, as well as Hyp_{15-41} + $Saph_{7-25}$ + $Cord_{10-20}$ + Sill + Bt_{20-27} and Gr_{44} + Hyp_{20} + $Saph_{25}$ + $Cord_{14}$ + Bt_{22} in alumina-rich rocks.

The following parageneses are also typical: Gr_{49-58} + Hyp_{25-29} + $Cord_{18-20}$ + Bt_{24-30}; Gr_{55-65} + $Cord_{14-22}$ + Sill; Gr_{58-60} + Sill + Bt; Gr_{53-64} + Cord + Bt_{27-50} which from the constituent mineral compositions indicate high metamorphic pressures and temperatures. The complex contains abundant occurrences of deepest level garnetiferous mafic schists, eclogite-like garnet-clinopyroxene, garnet–two-pyroxene and garnet-hypersthene rocks with the following mineral associations: Gr_{57-58} + MP_{20-30} + Hyp_{40-45} + Pl + Q; Gr_{70-77} + MPy_{33-65} + Pl_{40-80}; Gr_{50-60} + Hyp_{30-42} + Pl_{36-60}; Gr_{68-84} + Hyp_{47-61} + MPy_{33-65} + Pl_{40-56} and Gr_{53-65} + $Amph_{40-48}$ + Pl_{30-45}.

The metamorphic conditions prevailing in this complex, as far as we can deduce from the mineral assemblages in CaO-rich and CaO-poor rocks (Vishnevsky, 1978; Lutz, 1974) have fairly wide limits and correspond to the garnet-hypersthene-sillimanite and the sillimanite-biotite-garnet-orthoclase subfacies of the granulite facies. Metamorphism in the Central Anabar block was therefore at high or medium pressure granulite facies conditions. Parameters for the thermodynamic regime of metamorphism in this block obtained from mineral pair phase equilibria have widely varying limits: T = 820–950°C and P = 8–11 kbar (Vishnevsky, 1978). These wide limits for the parameters obviously reflect metamorphic heterogeneity, related to two granulite facies metamorphic episodes. Evidence for this is the discovery of the association spinel-garnet-quartz in high-alumina rocks which allows us to infer an earlier granulite facies metamorphism in the spinel-quartz subfacies at relatively low pressure but higher temperature.

One important feature of rocks in the Magan block is the complete set of formations which are fundamentally volcanics of tholeiite-andesite and andesite-rhyolite composition, giving way to continental clastic-volcanic, clastic, carbonate and calc-silicate rocks.

We have found examples of practically all the typical metamorphic rock types for the Anabar Shield within the confines of the Magan block, although we have noticed significant differences in the mineral parageneses of the rocks. In general, the aluminous gneisses do not have assemblages found in the granulite facies pressure and temperature extremes, such as hypersthene + sillimanite + quartz or hypersthene + sapphirine + cordierite. The usual mineral assemblages are Hyp_{40} + $Cord_{25}$ + Gr_{64} + Bt_{36}; Hyp_{30-45} + $Cord_{20-30}$ + Bt_{30-40}; Gr_{60-65} + Cord + Sill + Bt; Gr_{68} + Sp + Bt + Q; Gr_{60-65} + Cord + Bt_{27-35}, showing that in the Magan block the rocks belong to the garnet-hypersthene-cordierite-orthoclase and sillimanite-biotite-garnet-orthoclase subfacies. Mafic schists also contain two-pyroxene-amphibole-plagioclase assemblages, and garnetiferous varieties have also been found among these, e.g.: Gr_{71-85} + CPx_{44-55} + Hyp_{50-65} + Pl_{30-46}; Gr_{70-80} + Hyp_{50-60} + CPx_{43-50} + $Amph_{42-60}$ + Pl.

For calc-silicate rocks, the typical paragenesis is Di + Serp + Pl + Mus + Q, while marbles and calciphyres have Di + Fo + Ca.

Mineral assemblages of CaO-rich and CaO-poor rocks in this complex are stable at medium pressures and in a fairly restricted temperature interval. They are comparable with the higher pressure sillimanite-garnet-cordierite-orthoclase field and the lower temperature field of the garnet-hypersthene-cordierite-orthoclase subfacies. Metamorphic conditions for this complex have been estimated at T = 780–850°C and P = 7–8.5 kbar. Alumina-rich rocks contain relict spinel-garnet-sillimanite-quartz and spinel-quartz assemblages. Obviously, these also indicate an earlier high-temperature shallow metamorphism which proceeded under spinel-quartz subfacies conditions.

Most of the rocks cropping out in the Khapchan block belong to the sialic group with significant amounts of carbonate-clastic rocks (the Khapchan Group) and part of the Upper Anabar Group belonging to the sialic group

(Table II-2). This block differs from the others in having a preponderance of high-alumina, calc-silicate and carbonate rocks.

Mineral assemblages in the high-alumina rocks are mainly in the sillimanite-garnet-cordierite-orthoclase and, in part only, the sillimanite-biotite-garnet-orthoclase subfacies of the granulite facies. The most typical parageneses are: Gr_{61} + Cord; Gr_{65} + $Cord_{25}$ + Sill + Bt_{40}; Gr_{65-70} + Sill + Bt_{50-55}, and also Gr_{60-85} + Bt_{36-61} and Gr_{65-72} + Hyp_{45-65}. Additional phases in each association are plagioclase, potash feldspar and quartz.

Calc-silicate and carbonate rocks include the following parageneses: Ca + Di + Sc + Q + Ort; Ca + Di + Sc + Plg + Ort; Dol + Ca + Fo + Sp; Ca + Di + Sc + Gross + Woll.

The mineral associations in CaO-poor and CaO-rich rocks from the complex belong in the main to the middle part of the sillimanite-garnet-cordierite-orthoclase subfacies and only partly to the sillimanite-garnet-biotite subfacies. The metamorphic pressure was 5.5–7.5 kbar and the temperature 750–820°C. Taking the P–T conditions together with the mineral assemblages, we conclude that the granulite facies metamorphism of the Khapchan Group could be polycyclic. The rock complex in this block underwent extensive granitization and superimposed amphibolite facies metamorphism which is particularly clear in the Billyakh zone (*5* on Fig. II-6).

The lower Archaean supracrustal rocks, represented by the Daldyn and Upper Anabar Groups, were metamorphosed twice. The earlier phase, accompanied by partial melting, and dated by the U–Th–Pb method at 3550 Ma, proceeded at uniformly high temperatures of approximately 1000–1100°C and a pressure of 7–8 kbar. The second granulite phase metamorphism took place at much higher pressures (9–11 kbar), while high temperatures were maintained (800–950°). In the Magan block, repeated granulite facies metamorphism occurred at moderate pressures and temperatures, 750–850° and 7–9 kbar, affecting the earlier spinel-quartz subfacies assemblages. In the Khapchan block, granulite facies metamorphic conditions were moderate temperature and pressure (T = 750–850°C and P = 5.5–7.5 kbar), possibly for both episodes.

The *upper Archaean domain* comprises igneous rocks which crop out mainly in thrust-type linear zones. It was during this period of evolution in the shield that the second granulite metamorphic episode referred to above occurred in the Magan and Central Anabar blocks and a monofacies granulite metamorphism in the Khapchan block. Closely related to the second metamorphic episode are early partial-melt phenomena, expressed as injection migmatites with a granite-charnockite and enderbite composition. There are also a number of closely related charnockite and alaskite intrusions, often several kilometres long and up to 200 m wide. In the Magan zone the maximum dimensions attained by one of these bodies is 22 km long by 2.5 km wide. Bodies such as this crop out close to the Billyakh zone. Most rocks of this type contain monazite, zircon, chevkinite and euxenite which are responsible for the rare-metal–rare-earth metallogenic signature.

The largest linear thrust belts, to which are restricted superimposed amphibolite facies metamorphism and granitization, include gabbro-anorthosite intrusions. These bodies have been mapped in the Magan block and the Kotuykan-Monkhoolin zone and in the northern part of the Khapchan block. Although our understanding of these massifs is fairly limited, an investigation of the largest ones in the Kotuykan-Monkhoolin zone (Sukhanov et al., 1983) suggest that the rock association is mangerite-anorthosite, similar to rocks in typical mangerite-anorthosite intrusions such as the Kalara and Adirondacks types. A special feature of the composition of massifs in the Anabar Shield is that in addition to gabbro-anorthosite, gabbro-norite-anorthosite and pyroxenite, individual areas also contain jotunite and mangerite (subalkaline-diorite and diorite containing hypersthene). In spite of the anorthosites and associated rocks having been partly or completely retrogressed at amphibolite facies, their petrographic features provide evidence for a link with granulite terrains and consequently a Late Archaean age (Sukhanov et al., 1984). Gabbro-anorthosite massifs are sheet-like bodies deformed into asymmetric folds in which it is possible to establish the elements of layering from gabbro to gabbro-norite.

The *lower Proterozoic domain* consists mainly of retrograde metamorphic rocks formed after granulite associations in thrust-type shear zones at amphibolite and epidote-amphibolite facies. Within individual zones, such as Kotuykan-Monkhoolin and Khapchan, retrogression preceded intense blastocataclasis and blastomylonitization and in some outcrop areas the newly formed rocks underwent metamorphic differentiation with the formation of banding and a penetrative schistosity.

In other zones, such as Lamuyka and Billyakh, granulite facies rocks were less intensely reworked and retrograde metamorphism in them shows up mainly as the replacement of granulite assemblages by correspondingly shallower-level and lower temperature amphibolite facies assemblages. Metamorphic differentiation of retrogressed rocks is not seen in these zones, although there is always a second schistosity expressed as mineral growth. These rocks are not the product of prograde metamorphism but instead resulted from polyphase Early Proterozoic reworking of granulite facies terrains within the zones mentioned above. Relict granulite facies minerals are commonly found in these rocks. Retrograde rocks include biotite, amphibole, biotite-garnet and garnet-amphibole gneiss and plagiogneiss, also amphibole, garnet-amphibole and biotite-amphibole schists (sometimes with equilibrium pyroxene), quartzite, marble, calciphyre and calc-silicate rocks. Aluminous and high-alumina rocks contain the following characteristic assemblages: Gr_{74} + Sill + And + Sp + Pl_{47} + Ort + $Cord_{34}$

TABLE II-3

Ore shows in the Precmabrian of the Anabar Shield

Useful mineral; deposit, show	Tectonic setting; blocks & zones after Kitsul et al. (1973)	Ore-forming and genetic types of mineralization	Geological and mineralogical features: Host mineralized rocks	Shape of deposit, morphology of ore bodies	Type of ore	Ore minerals (minor minerals)	Age of host rock and mineralization
Iron ore shows	Central Anabar block	Metamorphic, magnetite quartzite	Daldyn Group gneiss and granulite	Conformable lenticular-layer deposits	Densely disseminated	Magnetite (ilmenite, haematite, pyrite, chalcopyrite)	Archaean
Shows of apatite, titanomagnetite; phosphorus, titanium, iron	Magan block	Apatite-titanium-iron ore (apatite-titano-magnetite)	Anorthosite, gabbro, gabbro-norite	Lenticular, vein-type bodies	Vein-type dissemina-tions, massive	Titanomagnetite, apatite, ilmenite, (pyrrhotite, chalcopyrite)	Archaean
Copper-nickel ore shows; copper, cobalt	Central Anabar block	Copper-nickel-sulphide; magmatic, associated with ultrabasic bodies	Ultrabasic bodies	Schlieren	Disseminated	Pyrrhotite, pentland-ite, chalcopyrite (pyrite, magnetite)	Archaean
Molybdenite ore shows; molyb-denum	Magan block	Molybdenite (pegmatitic, hydrothermal)	Pegmatites	Veins, pockets	Disseminated	Molybdenite, muscovite	Proterozoic
Phlogopite shows	Khapchan block, Kotuykan-Monkho-olin zone	Phlogopite; metasomatic	Calciphyres, diopsidic metasomatites	Schlieren, pockets	Pockets	Phlogopite	Archaean
Muscovite shows	Khapchan block, Kotuykan-Monkho-olin zone	Muscovite; pegmatitic	Pegmatites	Veins	Pockets	Muscovite	Proterozoic
Graphite shows	Magan and Khapchan blocks	Metamorphic	Daldyn & Upper Anabar gneisses and schists	Conformable and lenticular deposits	Disseminated, pockets	Graphite	Archaean

(only in the Kotuykan-Monkhoolin zone); Gr_{80} + $Cord_{35}$ + Bt_{56} + Sill + Q (in the Lamuyka zone) and Gr_{73-100} + Bt_{50-86} for garnet-biotite and garnet gneisses and plagiogneisses, and the assemblage Gr_{87-97} + $Amph_{75-88}$ + Bt_{75-85} for garnet-hornblende and garnet-hornblende-biotite gneisses.

Mafic schists display two types of paragenesis: MPy + Amph + Pl + Q and Amph + Pl + Q, and for garnetiferous varieties Gr_{79} + MPy_{45} + $Amph_{63}$ + Pl_{50} and Gr_{75-90} + $Amph_{50-60}$ + Pl_{45-65} + Q.

Retrograde mineral assemblages in marbles and calciphyres are Tr + Ca + Di ± Q and Act + Ca + Sc + Pl + Q.

From their mineral assemblages, all these rock types belong to the amphibolite facies and only partly to the epidote-amphibolite facies (only within the Kotuykan-Monkhoolin zone), corresponding to temperature steps of 650–800° and 500–600°C.

The magnitude of pressure during the formation of retrograde rocks has been determined from phase diagrams for the coexisting mineral pairs garnet-biotite, garnet-amphibole and garnet-cordierite-biotite; for the Lamuyka zone it is 4.0–5.0 kbar, for the Billyakh zone 6.7–7.0 kbar, but the garnet-biotite and garnet-amphibole pairs give 6.0–7.0 kbar for these same assemblages in all the zones.

Useful minerals

The tectonic domains of the Anabar Shield have only a limited number of useful mineral shows (Table II-3).

Mineral deposits associated with lower Archaean tectonic domains are metamorphic in origin, the principal group being metamorphic iron and graphite deposits (Fig. II-6). Iron ore shows are represented by magnetite quartzites and magnetite–two-pyroxene-garnet schists, forming lenticular bodies up to 50 m thick and from 2.5 to 4 km long and containing from 28–33% to 60–80% of iron oxides in Daldyn and Upper Anabar Group basic granulites. Graphite shows are restricted to the same metasedimentary units and form graphitic members with up to 20% graphite. In addition, meta-igneous chromite and copper-nickel ore mineralization is associated with meta-ultramafics. Ore mineralization of the metamorphic class is also represented by metamorphic-metasomatic garnet occurrences in the form of lenticular bodies containing up to 60–70% garnet in aluminous rocks from the Upper Anabar Group in the Magan block.

The magmatic formations that constitute the upper Archaean domain are characterised by a rare-metal–rare-earth signature, a feature found in charnockites and alaskites, and there is also apatite-ilmenite-titanomagnetite ore mineralization in gabbro-norite-anorthosite intrusions (Sukhanov and Ruchkov, 1984).

Zones of superimposed metamorphism, blastomylonitization and granitoid magmatism, constituting the Early Proterozoic domain, are characterised by metamorphic-metasomatic and hydrothermal ore mineralization. Ore shows of metamorphic genesis are represented by phlogopite, strontium- and boron-rich apocarbonate metasomatites which crop out in the south of the Billyakh zone, also muscovite and rare-metal pegmatites. Mineral deposits of hydrothermal genesis in superimposed zones are localized in mylonite and shear zones. They include metasomatites associated with granitoid intrusions, accompanied by sulphide mineralization with copper, molybdenum, lead and zinc.

Crustal evolution in the Anabar Shield

The upper part of the Earth's crust in the Anabar Shield to a depth of 12–14 km consists of granulite complexes, represented by Early Archaean volcanogenic and terrigenous rocks, affected by amphibolite facies polyphase metamorphic processes, accompanied by intense granitization and granitoid magmatism during the Early Proterozoic. These processes are manifest in zones which divide the shield into blocks.

The Early Archaean crust of the Anabar Shield was in all probability a combination of tonalite-gneiss terrains and greenstone belts, metamorphosed at granulite facies in the Early Archaean. This association at the present time is represented by Upper Anabar enderbite-gneiss formations and highly metamorphosed komatiite-basalt and basalt-andesite-dacite assemblages of the Daldyn and part of the Upper Anabar series. These assemblages form two belts close together in the central part of the shield, 450–500 km long by 120–150 km wide (Fig. II-6). A fragment of a third belt may be seen in the south-west of the shield (the Magan block), where we have mapped zones of dome-shaped structures consisting of enderbite-gneiss associations, 50–250 km across.

Both components of this association — greenstone belts and tonalites — underwent several episodes of metamorphism, igneous activity and deformation during the Archaean and Early Proterozoic. Resulting from these processes, all the rocks were compressed into tight isoclinal folds and were affected by blastomylonitization, imparting to the shield a predominantly north–west oriented structure and in outward appearance an "interleaving" of meta-igneous and metasedimentary rocks.

The overall impression is that the history of formation of the Anabar Shield progressed in a number of distinct phases (Fig. II-7).

(1) The lower Archaean domain formed on an "undetermined basement", a pre-existing primary cover at the end of the pre-geological or beginning of the geological stage of Earth evolution (4.0–3.7 Ga), which has been hypothetically reconstructed by modelling the crustal composition of the Anabar Shield below the 12–15 km thick "stratified layer" (Musatov et al., 1981) from experimental data and theoretical considerations (Christensen and Fountain, 1975; Goodwin, 1981). Early Archaean terrains were

formed during the period 3.7–3.5 Ga ago and underwent two granulite facies metamorphic events at relatively high pressures and temperatures. Metamorphism was accompanied by the earliest occurrence of partial melting, with which the formation of ancient pegmatites is associated. These have an age of 3550 Ma, on monazite using the U–Th–Pb method. It was these processes which were responsible for forming the ancient crust of the Anabar Shield, part of the extensive Anabar-Aldan granulite-gneiss terrain which in turn is a constituent part of the basement to the Siberian craton.

(2) The Late Archaean (3.0–2.7 Ga) episode saw the formation of deep-level thrusts, probably reflecting early horizontal crustal movements at granulite and partly amphibolite facies level. Simultaneously, in tectonically stressed zones, there was the intrusion of plutonic magmas, forming gabbro-anorthosite, charnockite and enderbite bodies.

(3) During the Early Proterozoic (2.0–1.8 Ga) episode, complexities were produced in the continental crustal structure of the region. Initially, the Archaean structure was fragmented into a number of major blocks along deep faults which had as their precursors Late Archaean tectonic zones. These movements were accompanied by cataclasis of the Archaean terrains and blastomylonitization in thrust-nappe type tectonic zones with associated superimposed amphibolite facies metamorphism at medium to low pressure. During a later stage, intense granitization of rock assemblages and intrusion of granitoids and pegmatites took place in tectonic zones. These particular zones were the likely sites for the formation of sedimentary basins, filled with Early Proterozoic supracrustal assemblages, the relicts of which are now preserved only in the southern part of the Khapchan belt.

(4) The Riphean period signifies the beginning of a fundamentally new episode in the evolution of the Earth's crust in the Anabar Shield. Rifting occurred initially as older inherited sites became tectonically reactivated. This is particularly clearly seen in the south of the Billyakh zone (Fig. II-6), where they are accompanied by the intrusion of Early Riphean bodies of alkaline-ultrabasic composition and carbonatites with titanomagnetite–rare-earth mineralization. Subsequently, at the end of the Early Riphean and in the Middle Riphean, continental platform sediments were deposited, represented by clastics and carbonates of the Mukun and Billyakh Formations with rare volcanic layers of basaltic composition and gabbro and diabase dykes, dated at 1570–1400 Ma and 1318–1250 Ma (Mashchak, 1973).

(5) The Mesozoic stage in the evolution of the shield was characterized by a new phase of magmatic activity, seen in the shape of early Mesozoic dolerite and kimberlite dykes.

Chapter 2

THE ALDAN SHIELD

The Aldan Shield is the largest basement elevation in the Siberian Craton. It extends for a distance of some 1200 km and is 270–350 km across. The shield is bounded to the south by the Stanovoy suture, which marks the junction between Early Precambrian crystalline rocks of the Shield and the Dzhugdzhur-Stanovoy province, an epicratonic orogen. To the north and east, basement rocks are overlain by flat-lying continental platform sediments of Vendian to Palaeozoic and/or Riphean age. The exceptionally wide distribution of granulite facies rocks and fold phases over an area of 600 × 300 km in the centre of the shield which took place under these conditions (lower temperature effects are seen only in electron probe studies of mineral grain boundaries) sets the Aldan Shield apart from other shield areas in the Soviet Union.

Our present understanding of the lower Precambrian geology and petrology of the Aldan Shield stems from work carried out by Korzhinsky (1936, 1939), who first defined the Iyengr, Timpton (charnockites) and Dzheltula granulite facies metamorphic rock assemblages and confirmed a younger age for the amphibolite facies metamorphism in the Stanovoy Ridge. This initial scheme was enlarged upon and developed by the researches of Glukhovsky, Dzevanovsky, Dook, Katz, Kudryavtsev, Lazko, Minkin, Mironyuk, Mokrousov, Neyelov, Petrov, Sudovikov, Frolova, Fedorovsky, Frumkin, Cherkasov, Shpak and others. Important contributions to metamorphic petrology were made by Drugova, Kitsul, Marakushev, Smelov and Shkodzinsky. Detailed structural work on the metamorphic assemblages was undertaken by Balagansky, Bogomolova, Dook and Kharitonov. By the 1970s, tectonic regional mapping schemes had identified the following: the Aldan crystalline massif — a core domain that was stabilized in the Early Archaean and younger fold systems, separated from it by deep faults, the Olyokma, Batomga and Stanovoy, in which lower Archaean granulite facies terrains underwent structural reworking at amphibolite facies conditions (Drugova and Neyelov, 1960; Sudovikov et al., 1965; Dodin et al., 1968; Kosygin et al., 1964). In more recent schemes the Olyokma and Batomga systems are referred to granite-greenstone terrains which give way to the west and east to the Aldan granulite-gneiss terrain (Fig. II-9A).

The Olyokma-Aldan-Chara-Aldan lithospheric plate (Markov, 1978), the

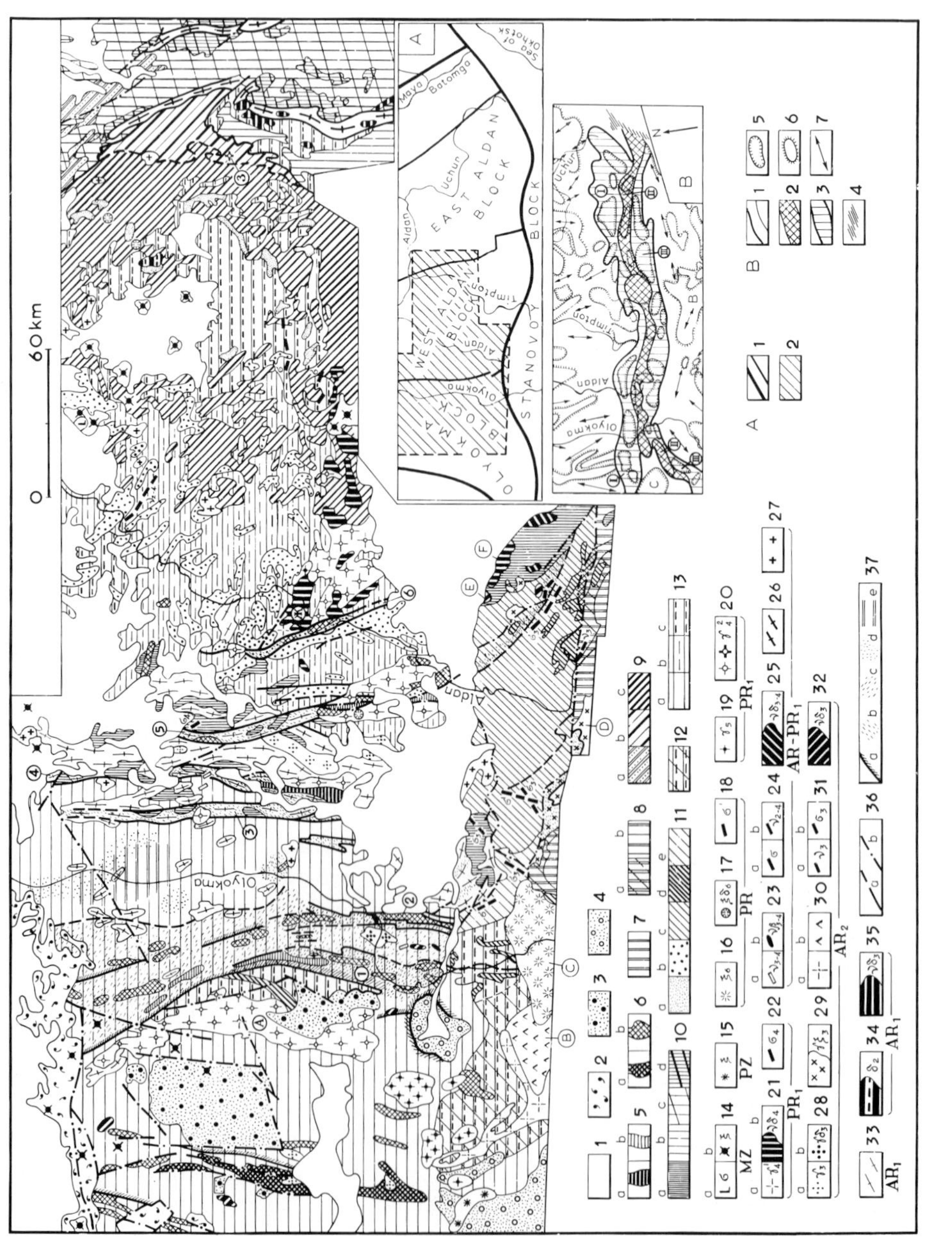

0 60 km
Olyokma
Aldan
WEST ALDAN BLOCK
EAST ALDAN BLOCK
OLYOKMA BLOCK
STANOVOY BLOCK
Uchur
Timpton
Maya
Batomga
Sea of Okhotsk
MZ
PZ
PR
PR₁
AR-PR₁
AR₁
AR₂

Fig. II-9. Geological map of the central Aldan Shield (simplified from Dook et al., 1986; Mitrofanov and Moskovchenko, 1985). *1* = V, Pz, Mz sediments; *2* = platform sediments, R; *3* = Uguy Group RR_1; *4* = Udokan Group RR_1; *5* = volcanic belts (*a* = Subgan Group) and orogenic complexes (*b* = Tasmiyeli Group) in Lower Proterozoic belts; *6* = upper Archaean greenstone belts, rift phase (*a* = Olondin Group) and cratonic phase (*b*-Tungurcha Group and analogues); *7* = Stanovoy Group; *8* = Stanovoy Group with inliers of Kurultin Group (*a*) and Chuga Formation (*b*); *9–12*: lower Archaean Aldan Supergroup: *9* = late-stage assemblages in late provinces (*a* = Kyurikan, *b* = Idzhek, *c* = Fyodorov); *10* = early-stage assemblages in late provinces [*a* = Chuga, *b* = Seym, *c–d*: Kholbolokh (*c* = Sutam Formation, *d* = Ampardakh Formation)]; *11* = assemblages in early provinces [*a, b* = Kurumkan (*a*: in SE, *b*: in Central and SW subprovinces), *c* = Zverev, *d* = Kurultin, *e* = Zverev and Kurultin Groups (undiff.)]; *12* = Kurultin Group and enderbite-gneisses of the infracrustal complex; *13* = infracrustal complex (*a* = grey gneisses of Olyokma granite-greenstone terrain with small inliers of Chuga Formation and Olyokma Group supracrustal rocks, *b* = granite-gneiss with enderbite relicts, *c* = enderbite-granite-gneiss and enderbite-gneiss complexes); *14–35*: intrusive complexes: *14* = Mesozoic (*a* = dunites and peridotites, incl. kimberlitic, *b* = syenites, alkaline syenites, incl. porphyritic); *15* = Palaeozoic alkaline syenites and granosyenites (synnyrites); *16* = aegirine-diopside syenites and granosyenites, Tass massif; *17* = diorites-monzonites; *18* = serpentinite belts; *19–22*: Lower Proterozoic complexes: *19* = porphyritic granites and subalkaline granites (biotite and two-mica with tourmaline and fluorite), cutting rocks of Tasmiyeli, Udokan and Yarogu Groups; *20* = charnockites of the W Aldan block (*a*) cutting biotite granites and subalkaline granites of the Olyokma and W Aldan blocks (*b*); *21* = gneissose tonalites (*a*) and metagabbros-diorites (*b*) of the Ungra complex; *22* = metapyroxenite and metaperidotite bodies and belts; *23–27*: undifferentiated $AR–PR_1$ complexes: *23* = metagabbros-norites (drusites) of Borsala-Nelyuka zone (*a*), metadiabases, metagabbro-diabases of Temulyakit-Tungurcha zone (*b*); *24* = meta-ultrabasic (*a*) and metagabbro (*b*) belts; *25* = metagabbro-diorites; *26* = gneissose subalkaline granites and granosyenites of the Idzhek-Nuyam and Tyrkanda fault zones; *27* = undifferentiated alaskite granites of the Aldan Supergroup; *28–32*: Upper Archaean complexes: *28* = gneissose subalkaline diopside-hornblende granites (*a*) and Ust-Oldongso gneissose quartz diorites and tonalites (*b*); *29* = Kabakta gneissose monzonite-syenites, granosyenites and alkaline granites; *30* = Kalara mangerites, granosyenites (*a*), anorthosites, gabbro-anorthosites (*b*); *31* = metagabbro (*a*) and meta-ultrabasic (*b*) belts; *32* = metagabbro and subalkaline metagabbro; *33* = gneissose granites and subalkaline granites; *34–35*: Lower Archaean complexes: *34* = metadiorites, *35* = subalkaline metagabbros and metagabbro-diorites of the Idzhek-Nuyam fault zone; *36* = fault zones with recrystallized rocks at granulite and amphibolite facies (*a*) and greenschist facies (*b*); *37* = shear zones and blastomylonite zones (dense ornament) and zones of retrogression (light ornament) of V (*a*), IV (*b*), III–IV (*c*) cycles of activity; zones of intense migmatization (*d*) and Olyokma grey gneisses (*e*) containing inliers of granulite facies mafic schists. Rock index subscripts (e.g. σ_3, δ_2) correspond to cycle of crust-forming internal processes. "Soft" fault zones (numbers): *1* = Temulyakit, *2* = Tungurcha, *3* = Borsala-Nelyuka, *4* = Amga, *5* = Reutov zone, *6* = Aldan-Kiliyera. Intrusions (letters): *A* = Charodokan, *B* = Kalara, *C* = Tass, *D* = Kabatka, *E* = Tenevoy, *F* = Khonchegrin, *G* = Ungra, *H* = Ust-Idzhek. A: Sketch map showing tectonic regions. *1* = province and subprovince boundaries; *2* = location of Fig. II-9 on sketch. B: Sketch map of regions in gravity field of Stanovoy suture zone (after A.M. Alakshin and L.P. Karsakov, Tikhookeanskaya Geologiya, No. 3, 1985; pp. 76–95). *1* = contours of linear anomaly zone; *2, 3* = linear anomalies relative to high and low fields, respectively; *4* = boundary gravity step; *5, 6* = local anomalies, minima and maxima, respectively; *7* = dominant strike of anomalies. Anomaly provinces: *A* = Aldan, *B* = Stanovoy, *C* = Vitim-Olyokma. Linear anomalies: *I* = Chara-Tyrkan, *II* = Kalara-Maya, *III* = Kolakan-Maya.

Stanovoy province and the Stanovoy suture separating them are clearly distinguishable in the physical fields (Fig. II-9B). The boundary between the Aldan granulite-gneiss terrain and the Olyokma granite-greenstone terrain is a strip up to 70 km wide containing five nearly north–south zones with higher gravity gradients and chains of magnetic anomalies. We know far less about the boundary between the East Aldan and Batomga blocks and the entire Batomga block, which is regarded as either a granite-greenstone terrain (Petrov et al., 1978) or a granulite-greenstone terrain.

Much discussion in the literature has centred on the age of rocks in the Aldan Shield. They were first studied by the K–Ar method and the closure time for the K–Ar system was found to be 1900–1700 Ma. Rb–Sr isotopic systems were also closed at around 2000 Ma ago (Gorokhov et al., 1981). Most stratigraphic schemes quote older age dates of around 4 Ga, obtained by K–Ar on diopsides (Manuylova, 1968) and by the whole-rock Pb–Pb method (Rudnik et al., 1969), although in the specialist literature on radiometric dating it is pointed out that initial isotopic ratios for elements were disrupted in the intervals 2800–2500 Ma and 2100–1200 Ma ago (Gerling et al., 1979; Iskanderova et al., 1980). Work done in regions where the Aldan massif is in contact with the Stanovoy fold system has led to the conclusion that deformation and metamorphism affected the territory of the Aldan Shield more than once (Dook, 1977; Dook and Kitsul, 1975; Dook et al., 1975; Balagansky, 1979; GSA, 1983), with superimposed metamorphism reaching granulite facies conditions.

The polyphase nature of deformation and metamorphism in tectonic cycles was also demonstrated in the central part of the Aldan Shield (Dook et al., 1975).

The lack of unconformities in the great majority of the separate rock groups identified in the Aldan Shield demanded the use of criteria from field and laboratory studies on folding and metamorphism and their interrelations in order to analyse the sequence of geological events. The age status of rock complexes (Fig. II-9) was established from their relationships with crust-forming events. Structural age scales established for areas where detailed structural mapping had been completed were correlated by following key surfaces (folded faults and axial surfaces of linear folds). Identical structural-metamorphic parageneses were followed and this turned out to be a basis for identifying zones and domains of different evolutionary stages.

The main geological boundary was taken as the time of formation of the Stanovoy fold belt which cuts provinces and metamorphic belts in the Aldan granulite-gneiss terrain. From the data available, the following stages in structural evolution were identified: Aldanian–Early Archaean (cycles I^1, I^2, I and II); Olyokma-Stanovoy or early Stanovoy–Late Archaean (III); late Stanovoy or Stanovoy proper–Early Proterozoic (IV^1); Ungrian (IV^2); and Udokanian (V)–Early Proterozoic.

On the basis of available geological and geophysical information, the Aldan Shield is divided into the Aldan granulite-gneiss terrain, which includes the West Aldan and East Aldan block, and to the west and east of this the Olyokma and Batomga granite-greenstone terrains — the Olyokma and Batomga blocks, respectively. The enormous variation in the amount known about each block has determined the way in which we have presented our description of the Aldan Shield. We start with the Aldan granulite-gneiss terrain since we are able to follow the Early Archaean stage of development best of all there. We then describe the Olyokma and Batomga terrains, which contain Late Archaean and Early Proterozoic supracrustal rocks.

In the Aldan Shield, the Early Proterozoic was a period of transition to a sub-platform regime, continental clastic sediments of this age being found only at the margins of the shield, so therefore we have presented the particular features of this phase using the Udokan and Tasmiyeli Groups in the Olyokma terrain and tectonically active zones in the West Aldan block as a case study.

1. THE ALDAN GRANULITE-GNEISS TERRAIN

Early Archaean evolution

During the Early Archaean stage, all the supracrustal assemblages of the Aldan Supergroup (megacomplex) were formed; these are known as the Iyengr, Timpton and Dzheltula Groups (Korzhinsky, 1939; Mironyuk et al., 1971), as well as the Kurultin and Zverev Groups on the flanks of the Aldan Shield (Katz, 1962; Drugova and Neyelov, 1960). The differences in rock compositions and associations in the Aldan Supergroup both laterally and in section, reflecting the primary structural and petrographic inhomogeneity in the region of formation (Fig. II-10) have been described by many workers (Anon., 1981; Dook et al., 1986). Tectonic domains have been identified at two time levels in the supergroup (Dook et al., 1986). An unstratified complex of tonalitic and granitic composition — enderbite- and charnockite-gneiss, plagiogneiss and granitic gneiss, occupying around 50% of the area (Fig. II-9), which had previously been included as part of various groups and formations in the Aldan Shield — is defined as the infrastructure complex. Its boundaries are known precisely and new tectonic domains have been identified (Fig. II-10).

The composition of lower Archaean lithological groups was examined in sections, on separate macrorhythms and data from ore sampling in an area. The interpretation of petrochemical data was done using classification diagrams for sedimentary and volcano-sedimentary rocks (Neyelov, 1980; Predovsky, 1970; Golovenok, 1977) and statistical methods (factor analysis and cluster analysis), allowing us to reconstruct the primary nature of metamorphic rocks on the basis of compositional differentiation trends, specific for sedimentary and igneous rocks and for rocks altered during metamorphism.

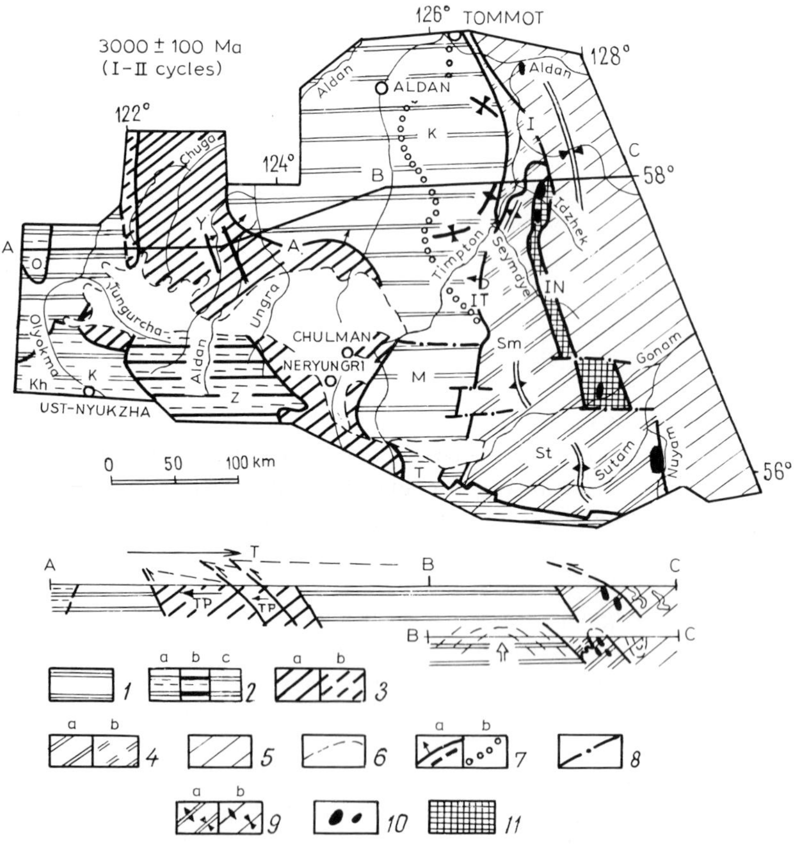

Fig. II-10. Sketch map of tectonic divisions in the Olyokma-Aldan region during evolutionary cycles I–II (after Dook et al., 1986). *1–2*: early Provinces: *1* = Central Aldan, *2* = Olyokma-Kurultin-Zverev (*a*) and Zverev subprovince (*b*), *c* = Olyokma infracrustal complex; *3–5*: late provinces: *3* = West Aldan (*a*) and its allochthon (*b*), *4a* = Idzhek-Sutam, *5* = East Aldan zone and its allochthon (*4b*); *6* = boundary of Mesozoic sediments; *7* = province boundaries (*a*), *b* = eastern margin of Kurumkan quartzite outcrop; *8* = late fault zone; *9* = axial surfaces of cycle II folds: *a* = anticlines and synclines, *b* = folds surrounding Lower Timpton dome; *10* = metagabbros and metadiorites; *11* = granitoids in Idzhek-Sutam fault zone. Letters: *Ya* = Yarogu, *A* = Aldan, *Ti* = Timpton thrust, *IN* = Idzhek-Nuyam, *I* = Idzhek fault zone; blocks: *O* = Olomokit, *Kh* = Khanin, *K* = Kurulta, *Z* = Zverev, *T* = Tangrak, *N* = Nimnyr, *M* = Melemken, *Sm* = Seym, *St* = Sutam.

Early Archaean tectonic domains (AR_1^1)

Representative structures of this time are the Olyokma-Kurultin-Zverev (South Aldan) and Central Aldan domains which differ widely in the environments of formation of their rock assemblages.

The *Olyokma-Kurultin-Zverev (South Aldan) domain* is subdivided into Western (Kurultin), Central (Zverev) and Eastern subdomains on the basis of rock associations and composition. Subdomain boundaries inherited Early Archaean structural trends and are oriented oblique to the strike of major Late Archaean fold structures (Fig. II-10).

The Kurultin subdomain (basin of the river Olyokma) consists of rocks of the unstratified infrastructure complex (enderbite-gneiss, charnockite-gneiss and granite-gneiss), with intermittent outcrops, in the form of complex folds, of Kurultin Group rocks, mainly garnet-biotite-plagioclase gneiss with lenses and bodies of two-pyroxene-plagioclase schist. Mafic schists are also found among rocks of the infrastructure complex. The structural pattern of the Kurultin Group is defined by the presence of three isoclinal fold systems accompanied by at least two episodes of shearing. This has resulted in repeated extremely fine-scale interleaving of all the rock types mentioned.

Typical features in the infracomplex are a uniform schistosity or occasionally a finely banded texture due to differentiation into melanocratic and leucocratic bands. These are interpreted as the result of anatectic processes and metamorphic differentiation. In terms of their chemistry and REE distribution (Fig. II-11), the enderbite-gneisses are analogous to grey gneisses

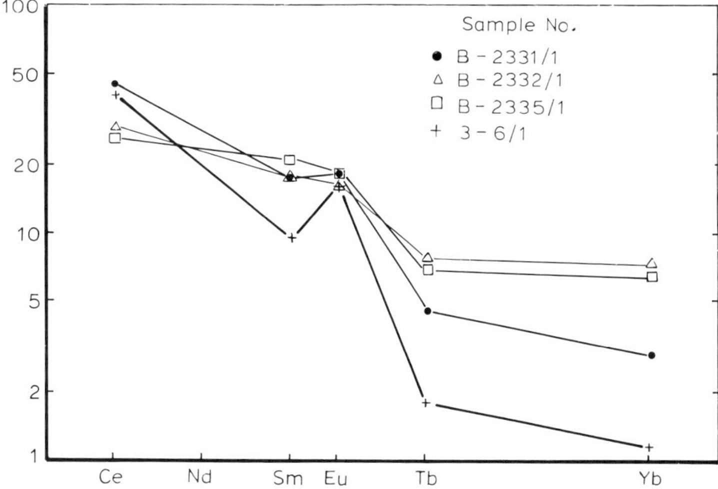

Fig. II-11. Chondrite-normalized REE distribution in enderbites from the Kurulta complex.

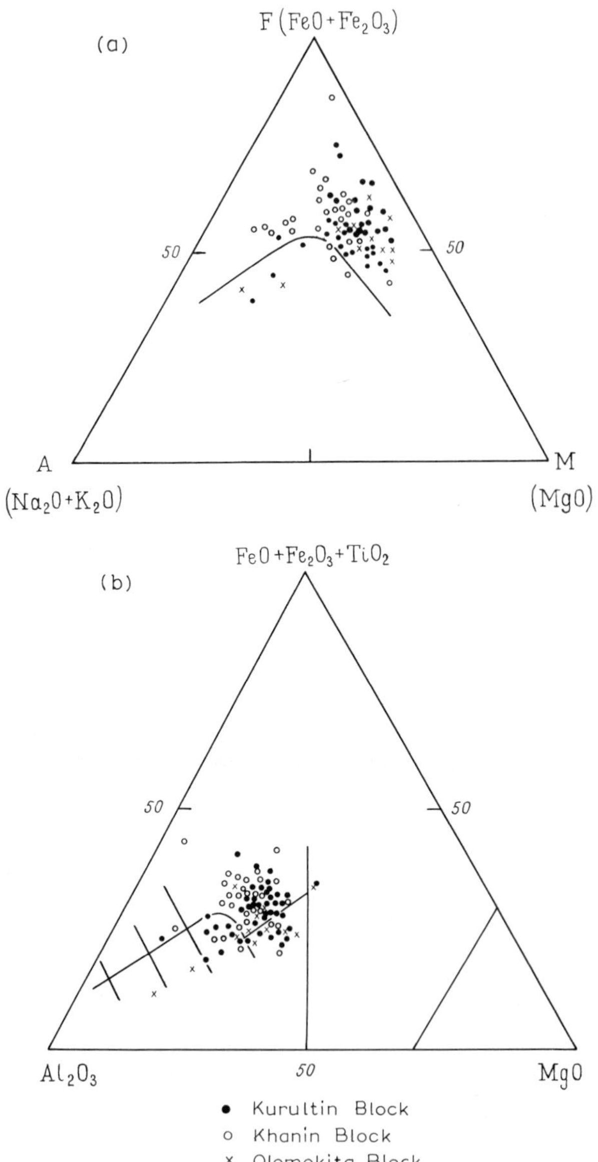

Fig. II-12. *AFM* diagram and Al_2O_3–(FeO + Fe_2O_3 + TiO_2)–MgO for schists in the infracrustal complex and the Kurulta Group.

from around the world (Condie, 1983; Barker, 1983) and distinguished from them by lower Si, Rb, Th and U contents.

The mafic schists which appear to be inclusions among rocks of the infrastructure complex, occupy around 15% of the area. Compositionally

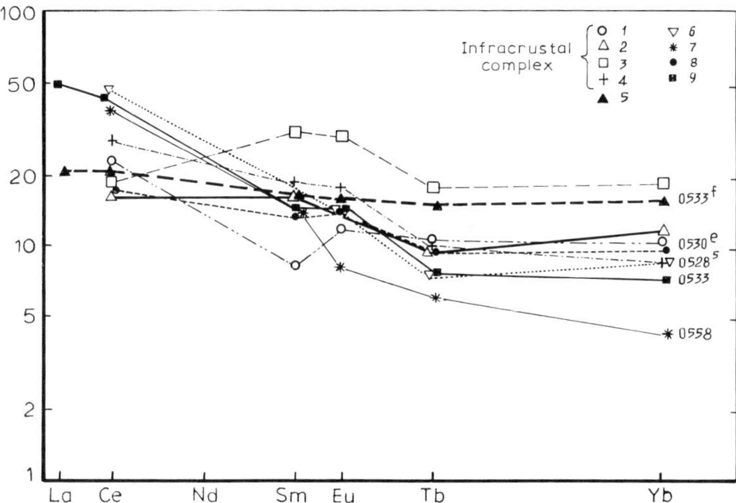

Fig. II-13. Chondrite-normalized REE distribution in mafic schists from inliers in the infracrustal complex (*1–4*) and from the Kurulta Group (*5–9*) (after Dook et al., 1986).

they correspond to tholeiitic basalts with low to medium alumina and high Ca contents. A few rare varieties are low-Ti, subalkaline and high-alumina basalts (Figs. II-12, II-13).

The Kurultin Group consists of the following rock association: garnet-biotite-plagioclase gneiss 60–70%, sometimes with sillimanite and cordierite, less usually garnet-hypersthene-biotite-plagioclase gneiss; two-pyroxene (± amphibole ± garnet) plagioclase schist 15%, with thin lenses of chert containing garnet or magnetite, calc-silicate rocks and very rare eulysite-type rocks. In the eastern part of the Kurultin subdomain, at its boundary with the Zverev subdomain, occurrences of the association two-pyroxene (± amphibole) plagiogneiss and plagioschist (sometimes with garnet) with subordinate lenses of garnet-biotite-plagioclase gneiss have been found. But here we may be dealing with tectonic interleaving of Kurultin and Zverev Group rocks.

Fine migmatitic banding (0.5–3 cm) is a common feature in the garnet-biotite plagiogneisses, oriented parallel to the foliation. Thicker leucocratic bands (garnetiferous granulites) sometimes cross-cut the banding in garnet-biotite-plagioclase gneisses. On very rare occasions it is possible to observe in these same gneisses graded-bedding type asymmetric banding. Mafic schists occur in both uniformly schistose and banded varieties, sometimes with exceptionally fine-scale banding (1–5 cm). This is either alternation of relatively leucocratic and melanocratic bands or bands rich in hypersthene, diopside or garnet.

The petrochemistry of 67 rock varieties in the Kurultin subdomain has been studied. On the component diagram (Fig. II-14/1) we have isolated rock groups whose average composition and inferred primary nature are shown in Table II-4/1. Here and subsequently, rock group numbers coincide with field numbers on the component diagrams. Mafic schists (1) are compositionally equivalent to tholeiites. The more magnesian of these (1a) most often occur amongst rocks of the infracomplex (1b), usually alternating with garnet-biotite gneiss. Compositions with higher Al and Na contents (1c) represent the most leucocratic bands in banded crystalline schists.

Group 2 combines various types of low-Al melanocratic schists, rich in hypersthene (2a), diopside (2b) or garnet (2c). They appear to be products of metamorphic differentiation. Garnet-biotite (± hypersthene) gneisses are compositionally little differentiated. From the relict graded bedding we may infer that these rocks were originally sediments — greywacke-siltstones (3) and greywacke-sandstones (5) of the K–Na series. The latter have been

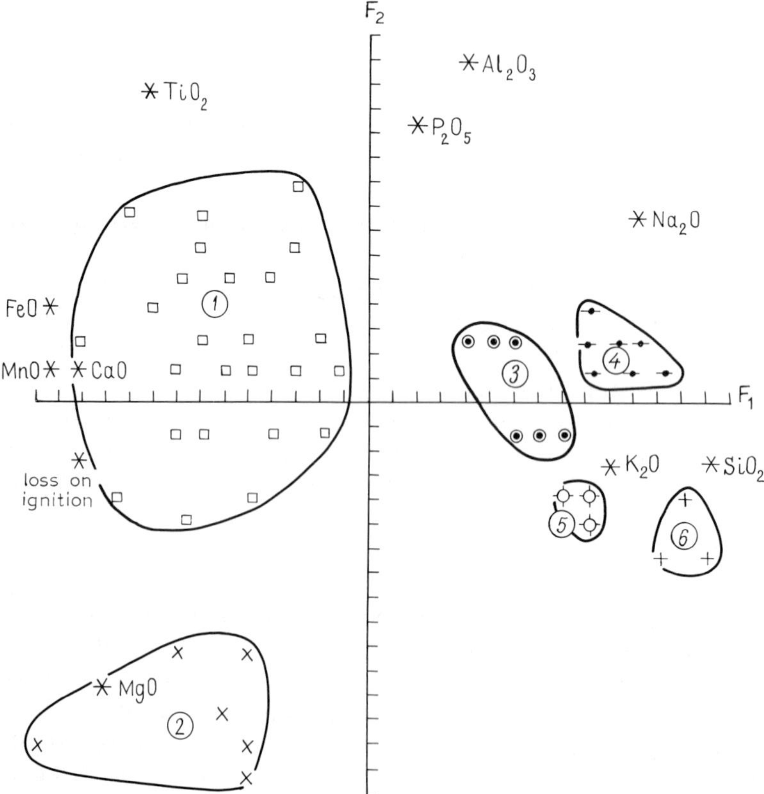

Fig. II-14/1. F_1–F_2 component diagram for rocks of the Kurulta subprovince. Oxide symbols with an asterisk (*) are factor loadings, other symbols are factor values. The ratio of factor loading to factor value scales is 6.0.

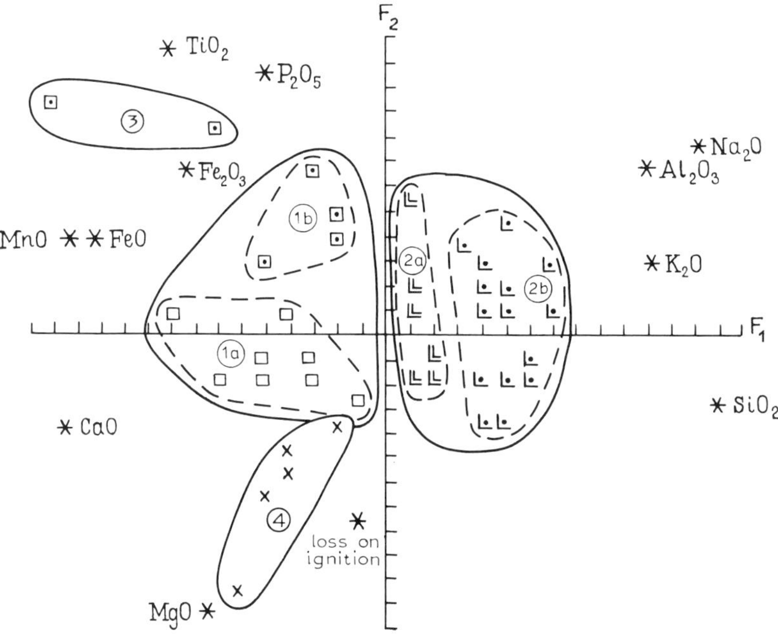

Fig. II-14/2. F_1–F_2 component diagram for mafic schists of the Zverev sub-province.

affected by anatexis. Rocks from the infracomplex (4) are distinguished by their high Na content and resemble the enderbites which occur in wide areas of the infrastructure. Garnetiferous granulites (6) stand out on account of their high K content and are probably autochthonous and subautochthonous granites.

Excluding the enderbites from the Kurultin assemblage and referring them to the infracomplex together with those rocks that are the result of anatexis and metasomatism, we reach the conclusion that the rock association of the Kurultin Group includes garnet-biotite (± hypersthene) gneisses-siltstone greywackes and mafic schists of tholeiitic composition.

The Zverev subdomain is made of the Zverev Group, which alternates with Kurultin Group rocks on the south-eastern flank. The Zverev Group consists of two-pyroxene and two-pyroxene-hornblende (± garnet) plagiogneiss and schist, amongst which we find individual lenses of calc-silicate rocks, calciphyre and chert with diopside and plagioclase. Common amongst these are diorite-gneisses with 10–20% quartz, for which both banded, uniform and massive textures are typical. Mapping in the diorite-gneiss field has shown up lenticular bodies of hypersthene plagiogneiss, usually homogeneous but sometimes indistinctly banded, the bands being garnet-hypersthene, two-pyroxene and hypersthene-hornblende gneiss and schist.

Kurultin Group rocks in the form of tectonic lenses of various shapes

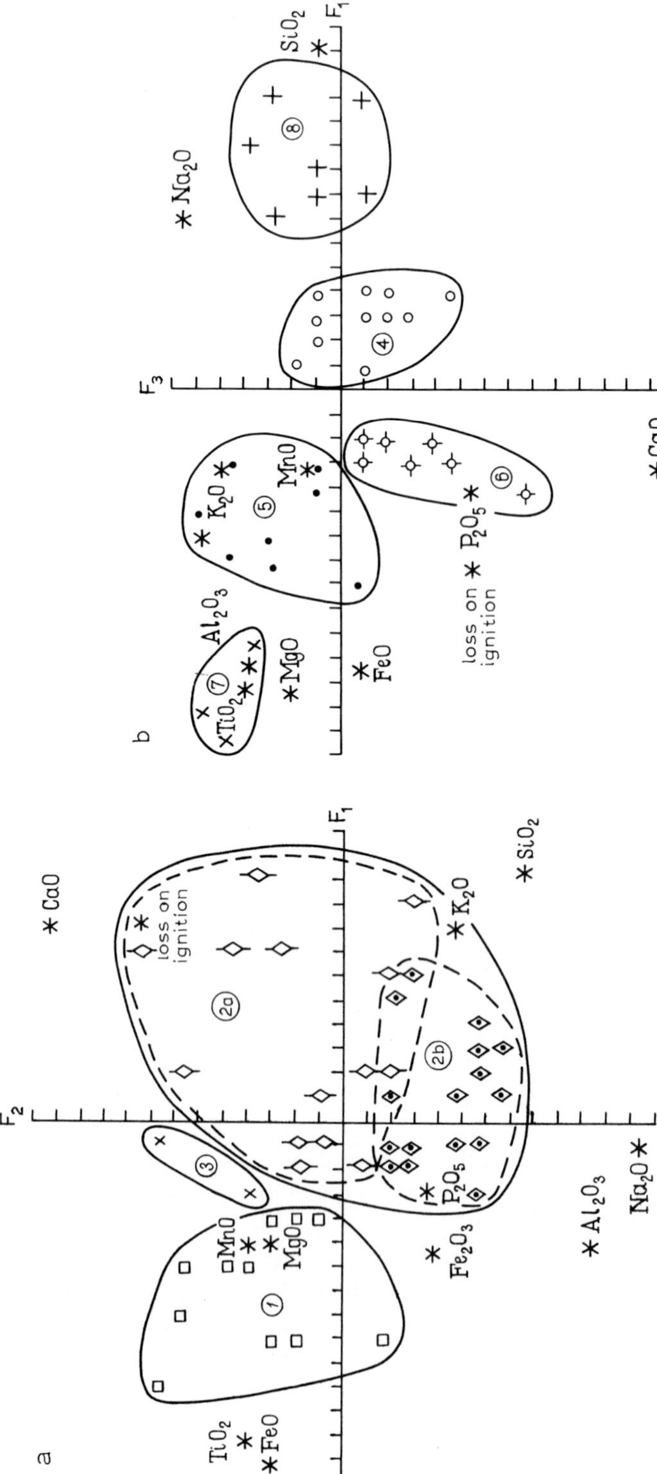

Fig. II-14/3. F_1–F_2 component diagram for rocks of the Kholbolokh assemblage, river Kholbolokh.

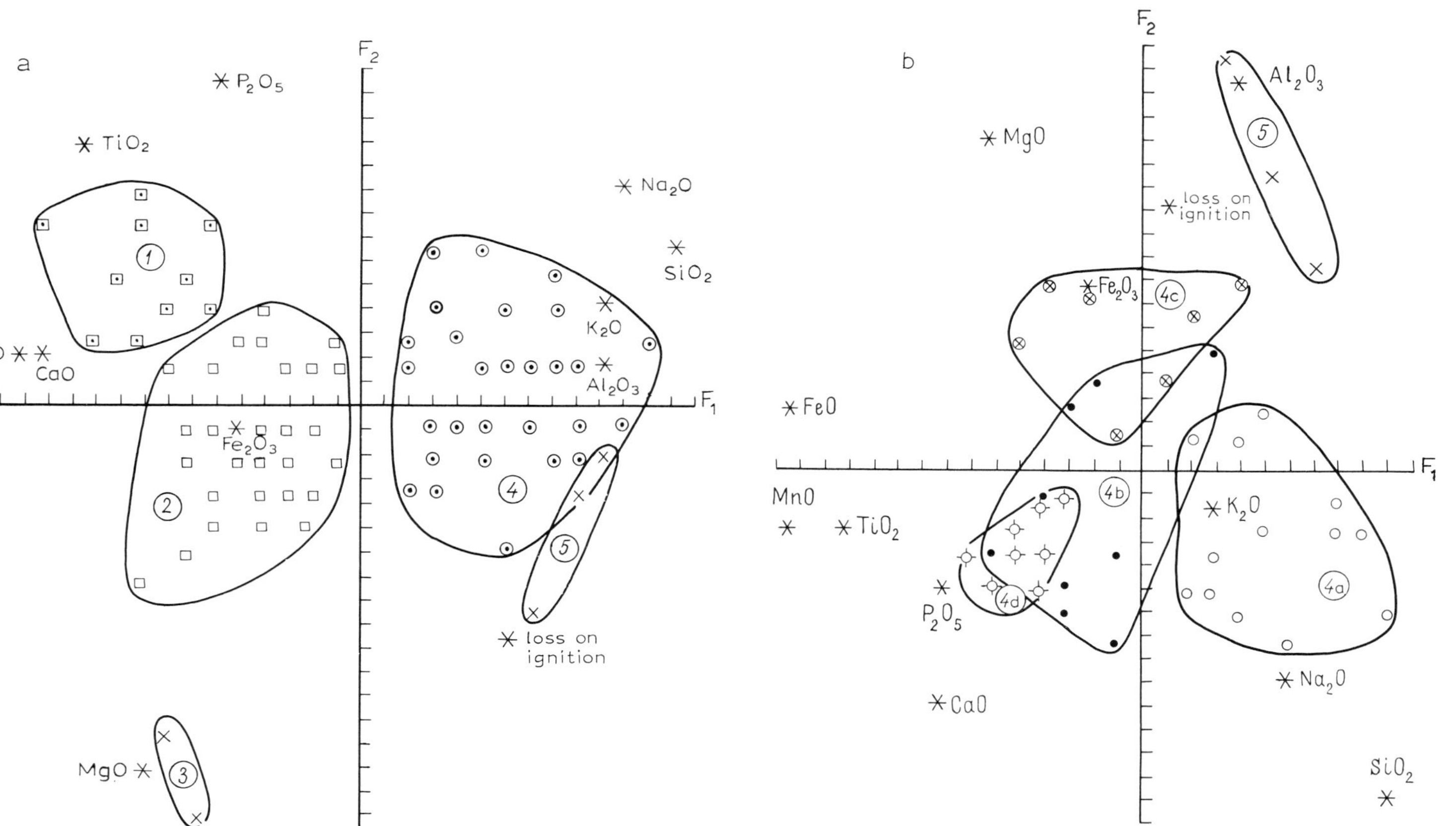

Fig. II-14/4. F_1–F_2 component diagram for rocks of the Seym assemblage.

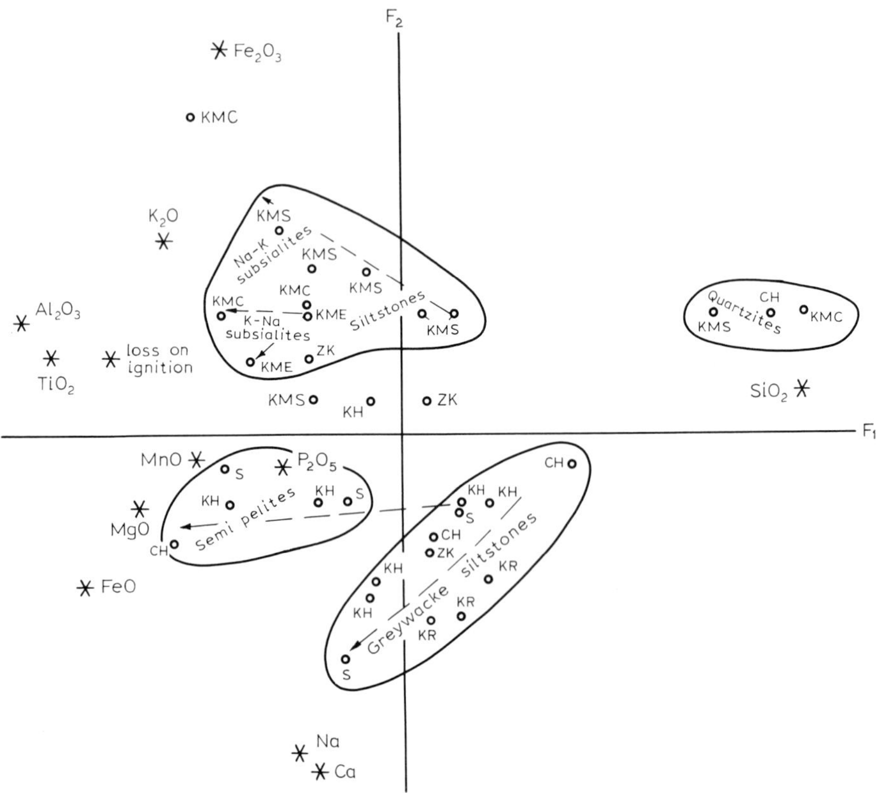

Fig. II-14/5. F_1–F_2 component diagram for combined samples of average compositions of gneisses and quartzites in the Aldan Supergroup.

and dimensions (0.5 × 5 km; 10 × 10 km; 20 × 1 km) have been mapped on the south-eastern flank of the subprovince (Fig. II-9). The morphology of the bodies and the complexly folded nature of their boundaries result from polyphase isoclinal folding. The compositions of rocks in the Zverev subdomain are shown in Table II-4/2 and mafic schist compositions are also shown on diagrams (Figs. II-14/2 and II-15). The mafic schists clearly divide into two types: (1) melanocratic low-alkali (tholeiites, the Kurultin Group) and (2) two-pyroxene schist and hypersthene plagiogneiss with high Si, Al and alkalis and low Ti [subalkaline basalts (2a) and andesites (2b), the Zverev Group]. Tholeiites are differentiated from more magnesian compositions (1a) to those rich in Fe^{3+}, Ti and P (1b). Old dykes (3) belong to the extreme differentiates of this series and were probably intruded at the time when the later West Aldan province was initiated (AR_1^2; see below). Ultrabasic rocks are located in the high-Mg compositional field (4). Garnet-biotite plagiogneisses are associated with tholeiite-type mafic schists (the Kurultin Group). We consider that the gneisses represent

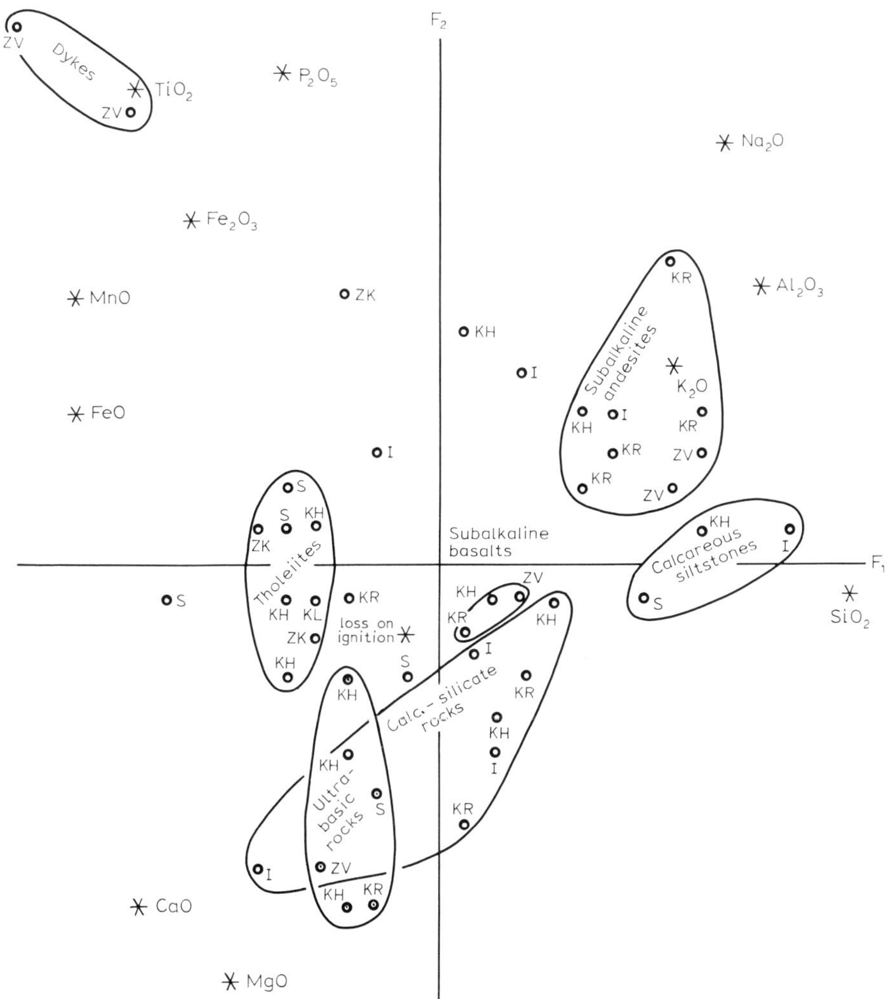

Fig. II-14/6. F_1–F_2 component diagram for combined samples of average compositions of mafic schists and calc-silicate rocks.

greywacke-siltstone sedimentary rocks, due to the presence of relict bedding and interleaving with calc-silicate rocks in particular cases (Neyelov, 1980). Variations in the composition of these rocks, from Na–Ca (5b) to Si–K (5a) probably reflect anatectic differentiation. The higher Al content in the gneisses (5c) is most often associated with secondary processes producing sillimanite enrichment.

The *Central Aldan domain* consists of the Kurumkan supracrustal assemblage (Upper Aldan and Nimnyr Formations of the Iyengr Group), which has been mapped among enderbite-, charnockite- and granite-

TABLE II-4/1–3

1. Average compositions of rocks (wt%) in the Kurultin subdomain

Gp	Composition	No. of samples	SiO_2	TiO_2	Al_2O_3	Fe_2O_3	FeO	MnO	MgO	CaO	Na_2O	K_2O	P_2O_5	l.o.i.	Total
1a	Two-pyroxene amphibole schists (tholeiites)	11	46.48 2.79	1.12 0.46	14.46 1.43	3.79 1.70	9.84 1.50	0.22 0.06	8.09 0.96	10.96 1.25	2.14 0.30	0.64 0.45	0.09 0.07	2.39 0.72	100.24
1b	Two-pyroxene amphibole schists (tholeiites)	9	48.30 2.8	1.03 0.44	14.55 1.03	2.9 0.72	10.33 1.81	0.3 0.07	6.37 0.93	11.1 2.7	2.13 0.57	0.21 0.33	0.07 0.05	1.30 0.51	98.60
1c	Two-pyroxene amphibole schists (high-Al types)	3	47.79 20.37	0.63 0.16	17.66 1.43	2.16 0.36	7.86 0.44	0.11 0.05	8.13 0.39	10.77 0.05	2.89 0.33	0.80 0.18	0.02 0.04	1.67 0.15	100.50
2a	Diopside-hypersthene schists	4	50.90 4.00	0.37 0.21	5.97 1.84	2.15 1.93	9.06 1.05	0.23 0.04	20.88 3.34	7.15 3.49	1.10 0.20	0.68 0.73	0.02 0.01	1.43 0.26	99.93
2b	Amphibole-diopside schists	1	51.36	0.12	4.78	1.58	4.45	0.12	13.91	20.81	0.93	0.07	0.001	2.4	100.12
2c	Garnet-diopside-amphibole schists	1	46.50	1.38	6.04	2.97	15.46	0.30	10.63	13.72	0.52	0.17	0.13	0.61	99.43
3	Biotite-garnet plagiogneisses (greywacke-siltstones)	11	63.02 2.24	0.67 0.25	16.3 0.83	1.07 1.11	6.16 2.03	0.10 0.06	2.58 0.55	3.91 1.36	3.22 0.14	1.68 0.84	0.13 0.12	0.92 0.40	99.80
4	Biotite-hypersthene plagiogneisses (infracomplex)	7	68.02 2.60	0.39 0.12	16.08 0.68	0.74 0.51	2.96 0.75	0.02 0.01	1.38 0.72	4.37 0.67	4.31 0.28	1.05 0.53	0.12 0.05	0.39 0.11	99.83
5	Biotite-garnet and biotite-hypersthene gneisses, affected by anatectic processes	3	73.39 2.31	0.48 0.12	12.77 1.29	0.30 0.13	3.53 0.56	0.02 0.01	1.98 0.49	2.71 0.49	2.52 0.42	1.49 0.32	0.02 0.02	0.54 0.28	99.75
6	Garnetiferous granulites	3	74.76 1.02	0.07 0.07	13.21 0.27	0.13 0.08	2.08 0.80	0.07 0.05	0.27 0.06	1.24 0.15	2.47 0.28	5.47 1.30	0.03 0.03	0.12 0.16	99.91

2. Average compositions of rocks (wt%) in the Zverev subdomain

1a	Two-pyroxene schists (±ga + amph) (tholeiitic basalts, low alkalis, Na)	8	46.79 4.47	1.58 0.75	14.70 0.91	3.91 2.2	11.08 3.04	0.30 0.06	6.57 2.02	10.56 2.17	1.84 0.67	0.39 0.23	0.12 0.06	2.20 0.41	100.06
1b	Two-pyroxene schists (±ga ±amph)(moderately alkaline basalts with higher Fe^{+3}, Ti, P)	4	46.65 1.71	2.55 0.93	15.69 2.14	6.24 3.76	9.20 2.44	0.17 0.02	4.47 2.22	9.45 1.02	3.19 0.46	0.72 0.48	0.48 0.36	0.76 0.40	99.84
2a	Two-pyroxene and two-pyroxene amphibole schists (subalkaline basalts, K-Na)	6	50.84 2.56	0.85 0.29	15.84 1.58	3.77 2.47	6.98 2.34	0.14 0.04	6.65 0.49	9.14 0.57	3.08 0.70	1.14 0.29	0.08 0.05	1.67 0.65	100.18
2b	Hypersthene, hypersthene-amph. plagiogneisses (subalkaline andesites, K-Na)	14	57.08 2.53	0.85 0.34	16.70 1.81	2.94 1.45	5.38 1.63	0.11 0.07	4.12 0.81	6.15 1.38	3.75 0.52	1.13 0.33	0.21 0.18	1.48 0.97	99.90
3	Two-pyroxene (±ga±amph) schists (basalts with high Ti, Fe^{+3}, P; dykes)	2	41.65	3.53	9.62	11.18	12.69	0.39	5.84	10.53	2.04	0.57	0.82	1.13	100.12
4	Two-pyroxene schists (bodies of ultrabasic rock)	5	48.82 2.10	0.52 0.23	9.20 2.83	3.24 1.33	7.54 1.39	0.16 0.02	15.01 2.22	10.85 2.17	1.29 0.71	0.60 0.18	0.06 0.08	2.04 1.11	99.35
5a	Leucocratic biotite-garnet gneisses, moderately alkaline	16	67.28 3.96	0.50 0.19	15.24 1.53	1.04 0.56	4.35 2.00	0.07 0.05	2.24 0.93	1.77 0.69	2.69 0.63	3.41 1.21	0.03 0.02	1.05 0.48	99.69
5b	Biotite-garnet gneisses (greywacke siltstones, tuffites)	10	64.29 3.54	0.63 0.12	15.45 0.80	2.79 1.04	3.21 1.74	0.06 0.04	2.33 0.69	4.11 1.44	3.56 0.53	1.85 0.76	0.12 0.09	1.18 0.55	99.58
5c	Sillimanite-garnet-biotite gneisses	3	61.48 2.60	0.65 0.15	21.30 1.11	0.77 0.20	4.50 1.62	0.03 0.02	1.98 0.58	1.39 0.42	2.33 0.77	3.53 1.55	0.03 0.02	1.77 0.37	99.93

3. Average compositions (in wt%) of rocks from the Kurumkan Group, Central subdomain (stratotype)

I	Quartzites (ultra-silicites)	5	96.01 1.55	0.03 0.05	1.68 1.27	0.39 0.46	0.67 0.58	—	0.29 0.08	0.32 0.21	0.11 0.08	0.28 0.21	0.01	0.01	98.80
II	Bi-ga quartzitic gneisses (ferruginous quartzites)	1	77.26	0.66	9.55	0.48	3.48	0.03	2.42	1.33	1.90	1.84	—	1.00	99.95
III	Bi-cord-sill quartzitic gneiss	1	70.12	1.17	11.86	3.03	3.20	0.06	4.19	tr	0.01	4.06	tr	2.62	100.41
IV	Ga-cord-sill-bi gneiss (ferruginous siltstone)	1	66.23	1.22	15.75	1.85	4.76	0.09	1.76	0.65	1.22	3.30	—	2.46	99.29
Va	Bi-sill±cord±ga gneisses (semipelites)	9	61.38 0.62	0.93 0.16	19.11 1.14	2.91 1.23	5.96 0.60	0.08 0.02	2.23 0.35	1.32 0.36	1.70 0.68	3.00 1.00	0.05 0.06	0.63 0.90	99.30
VIa	Ga-bi-cord-sill gneiss (pelites, Na-K)	4	56.76 1.33	1.01 0.16	22.35 1.51	3.84 1.05	5.87 0.93	0.10 0.03	2.33 0.58	1.52 0.80	1.77 0.79	2.38 1.46	0.06 0.04	1.04 1.45	99.03
VIb	Bi-cord-sill±ga gneisss (high-K pelites)	4	58.49 1.49	1.00 0.16	19.65 1.77	5.21 1.71	3.86 0.53	0.06 0.02	3.01 0.86	0.35 0.27	0.74 0.39	5.81 0.39	0.05 0.03	2.33 0.94	100.56

TABLE II-4/4–6

4. Average compositions (in wt%) of rocks from the Kurumkan Group, North-Eastern subdomain

Gp	Composition	No. of samples	SiO_2	TiO_2	Al_2O_3	Fe_2O_3	FeO	MnO	MgO	CaO	Na_2O	K_2O	P_2O_5	l.o.i.	Total
I	Quartzites (ultrasilicites)	4	88.57/4.30	0.24/0.24	2.94/2.15	1.55/1.98	4.07/2.44	0.05/0.03	0.71/0.69	0.10/0.08	0.24/0.18	0.89/0.65	—	0.48	99.84
II	Quartzitic gneisses (ferruginous quartzites)	4	73.13/2.29	0.72/0.19	6.66/1.13	4.37/2.89	9.24/1.50	0.23/0.18	3.62/0.89	0.38/0.16	0.19/0.11	0.74/0.63	—	0.03	99.31
IV	Ga-bi-cord±sill gneisses	7	69.19/1.57	0.59/0.14	14.41/1.61	2.42/2.23	3.89/1.56	0.33/0.45	3.50/1.87	0.65/0.62	0.44/0.42	2.71/1.16	0.02/0.01	0.67/0.06	98.82
Va	Bi-ga-cord-sill gneisses (semipelites)	2	64.27	0.82	18.16	3.04	3.16	0.03	2.28	0.90	2.25	4.33	—	—	99.24
Vb	Ga-bi-cord gneisses (ferruginous semipelite)	1	60.84	0.54	17.05	—	9.93	1.97	4.91	0.89	0.29	2.70	0.02	0.44	99.58
Vc	Garnet-biotite gneiss (ferruginous semipelite)	1	57.61	0.56	14.52	4.28	13.85	0.53	4.78	1.35	0.52	1.18	0.05	0.85	100.08
VIa	Ga-bi-cord-sill gneisses (low-Fe pelites, Na-K)	8	59.00/1.98	1.04/0.27	21.00/1.87	4.21/2.63	5.71/1.54	0.10/0.13	3.07/0.60	1.11/0.75	0.78/0.42	2.61/1.03	0.03/0.04	1.03/0.29	99.69
VIb	Ga-bi-cord-sill gneisses (low-Fe pelites, K-Na)	8	58.54/3.01	1.03/0.20	19.97/1.56	4.20/1.17	6.20/2.48	0.10/0.03	2.48/0.67	2.35/1.25	2.74/1.15	1.38/0.47	—	—	98.99
VIIa	Bi-cord-sill±ga gneisses (illite-hydromica subsialites, Na-K)	6	54.00/2.44	1.21/0.13	27.24/4.01	5.27/2.30	6.10/1.08	0.09/0.03	2.31/0.75	0.91/0.44	0.59/0.35	1.43/0.65	—	0.86	100.01
VIIb	Bi-cord-sill±ga gneiss (illite-hydromica subsialites, K-Na)	2	53.51	0.98	24.52	9.84	2.86	0.07	2.52	1.00	2.65	1.43	—	0.30	99.68

5. Infracrustal complex

Gp	Composition	No. of samples	SiO_2	TiO_2	Al_2O_3	Fe_2O_3	FeO	MnO	MgO	CaO	Na_2O	K_2O	P_2O_5	l.o.i.	Total
1	Ga+bi±cord±sill gneisses (pelites, Na-K)	16	58.66/1.95	0.83/0.19	21.80/1.96	4.36/2.25	5.48/1.37	0.09/0.05	2.27/0.53	1.14/0.27	1.78/0.98	2.80/0.93	0.10/0.03	1.35/0.50	100.66
2	Ga+bi±cord±sill gneisses (pelites, Na-K)	7	58.66/1.95	0.83/0.19	21.80/1.96	4.36/2.25	5.48/1.37	0.09/0.05	2.27/0.53	1.14/0.27	1.78/0.98	2.80/0.93	0.10/0.03	1.35/0.50	100.66
3	Ga+ga+sill+cord gneiss (pelites, K-Na)	4	55.44/1.33	0.96/0.16	23.26/1.51	4.53/1.05	8.01/0.93	0.10/0.03	2.01/0.58	1.03/0.80	3.37/0.79	1.19/1.46	0.07/0.04	0.80/1.45	100.77
4	Ga+bi+cord-sill gneisss (hydromica subsialites, K-Na)	5	50.63/2.75	0.97/0.21	24.26/1.42	4.98/1.15	7.88/1.85	0.21/0.06	3.18/0.79	1.86/0.51	2.43/0.67	3.27/0.85	0.08/0.03	0.66/0.47	100.41
5	Ga-cord-sill-bi gneiss	2	45.85	1.67	25.80	4.38	11.17	0.21	3.84	2.64	2.33	0.74		0.40	99.03

6. Average compositions (in wt%) of Kurumkan Group rocks, South-Western subdomain, Melemken block

Gp	Composition	No. of samples	SiO_2	TiO_2	Al_2O_3	Fe_2O_3	FeO	MnO	MgO	CaO	Na_2O	K_2O	P_2O_5	l.o.i.	Total
VIa	Ga+bi±cord±sill gneisses (pelites, Na-K)	11	58.66/1.95	0.83/0.19	21.80/1.96	4.36/2.25	5.48/1.37	0.09/0.05	2.27/0.53	1.14/0.27	1.78/0.98	2.80/0.93	0.10/0.03	1.35/0.50	100.66
VIb	Ga+ga+sill+cord gneiss (pelites, K-Na)	3	55.44/1.61	0.96/0.29	23.26/1.76	4.53/3.76	8.01/4.28	0.10/0.01	2.01/0.42	1.03/0.34	3.37/1.07	1.19/0.78	0.07/0.01	0.80/0.07	100.77
VIIa	Ga+bi+cord-sill gneisss (hydromica subsialites, K-Na)	6	50.63/2.75	0.97/0.21	24.26/1.42	4.98/1.15	7.88/1.85	0.21/0.06	3.18/0.79	1.86/0.51	2.43/0.67	3.27/0.85	0.08/0.03	0.66/0.47	100.41
VIIb	Ga-cord-sill-bi gneiss (ferruginous subsialites)	1	45.85	1.67	25.80	4.38	11.17	0.21	3.84	2.64	2.33	0.74		0.40	99.03

gneisses (the infrastructural complex) as fragments of complexly polyphase folded structures, from a few hundreds of metres to 15 × 30 km (Figs. II-9, II-10). We have identified three types of succession and three corresponding subdomains: Central, North-Eastern and South-Western.

The section in the Central subdomain is taken as a stratotype. It

TABLE II-4/7–9

(7) Average compositions (in wt%) of Kholbolokh assemblage rocks, river Kholbolokh

Gp	Composition	No. of samples	SiO_2	TiO_2	Al_2O_3	Fe_2O_3	FeO	MnO	MgO	CaO	Na_2O	K_2O	P_2O_5	l.o.i.	Total
1	Hyp-amph, two-pyroxene-amph schists (tholeiitic basalts)	11	46.23 3.10	1.70 0.64	14.83 1.41	2.1 1.89	12.46 1.98	0.21 0.06	7.34 1.90	10.83 1.41	1.88 0.52	0.56 2.42	0.15 0.14	1.49 0.95	99.73
2a	Calc-silicate rocks	14	51.40 3.61	0.67 0.24	14.32 1.79	2.00 1.52	6.03 1.98	0.16 0.08	6.17 2.41	13.57 2.73	1.80 0.84	1.38 1.30	0.12 0.13	1.99 1.15	99.60
2b	Two-pyroxene-amph±ga schists (calcareous siltstones)	18	52.03 2.02	0.701 0.24	16.10 1.67	1.82 0.89	7.31 1.62	0.16 0.03	6.59 1.63	9.17 1.18	2.98 0.70	1.25 0.81	0.17 0.14	1.12 0.73	99.39
3	Two-pyroxene-amph schists	2	49.46	0.9	10.25	2.44	8.08	0.08	14.76	11.34	1.04	0.40	0.11	1.35	100.27
4	Ga-bi plagiogneisses (greywacke-siltstones)	13	67.79 2.63	0.56 0.19	14.73 1.92	1.02 0.76	4.82 1.19	0.08 0.03	1.94 0.46	3.32 1.32	2.73 0.73	1.99 0.96	0.13 0.18	1.09 0.48	100.17
5	Ga-bi±cord±sill gneisses (semipelites)	8	59.80 3.47	0.76 0.17	17.97 1.89	1.31 0.67	6.75 1.35	0.09 0.04	2.89 0.85	3.05 1.27	2.71 0.59	3.23 0.99	0.11 0.07	1.19 0.57	99.86
6	Ga-hyp-bi gneisses (calcareous siltstones)	6	63.82 1.67	0.63 0.05	15.09 1.92	0.83 0.67	6.01 0.75	0.17 0.15	2.58 0.32	4.92 1.22	2.42 0.82	1.52 0.84	0.14 0.06	0.72 0.53	98.82
7	Sill-ga-bi gneisses with secondary sillimanitization	3	52.35 1.41	0.95 0.18	23.86 4.00	2.78 2.17	6.78 1.65	0.1 0.03	3.54 0.96	2.39 0.79	2.40 1.49	3.27 2.02	0.08 0.02	1.62 0.34	100.11
8	Cord-ga-bi gneisses, altered by granitization	8	72.39 1.68	0.27 0.10	13.48 0.94	0.99 0.64	2.09 0.73	0.03 0.01	0.96 0.52	2.67 1.43	3.75 0.96	2.59 1.70	0.04 0.06	0.68 0.4	99.91

8. Average compositions (in wt%) of Kyurikan Group rocks (Western sector)

Gp	Composition	No. of samples	SiO_2	TiO_2	Al_2O_3	Fe_2O_3	FeO	MnO	MgO	CaO	Na_2O	K_2O	P_2O_5	l.o.i.	Total
1	Biotite gneisses (±ga±hyp) (greywacke siltstones, tuffites)	9	64.58 5.13	0.75 0.32	15.32 2.06	0.96 0.64	4.67 0.99	0.09 0.03	1.92 0.60	4.70 2.05	3.59 0.60	2.02 0.83	0.22 0.09	0.81 0.55	99.63
2	Ga-bi, amph-bi gneisses (ferruginous siltstones)	4	61.78 2.27	0.69 0.21	16.50 1.05	3.44 2.58	5.08 1.01	0.18 0.12	2.08 0.74	4.47 1.81	3.60 0.65	1.39 0.58	0.22 0.11	0.44 0.37	100.23
3	Diopside (±scapolite) calc-silicate rocks	16	54.49 4.8	0.76 0.24	12.23 2.68	0.76 0.29	4.81 1.38	0.11 0.06	4.69 3.74	15.81 3.16	1.25 0.78	0.92 0.60	0.28 0.15	3.63 3.29	99.73
4	Calciphyres	4	30.52 4.31	0.50 0.30	7.38 2.34	0.97 0.68	2.35 0.51	0.08 0.01	2.23 0.71	31.63 3.60	1.01 0.63	1.11 0.66	0.19 0.06	21.53 3.79	99.60
5	Two-pyroxene, di-amph schists (calcareous, ferruginous pelites)	2	48.6	1.31	17.92	3.49	8.82	0.21	4.34	12.64	1.25	0.37	0.14	1.15	100.17
6	Two-pyroxene-amph (±biotite) plagiogneisses (andesites)	4	57.03 1.71	0.91 0.37	15.89 0.26	2.03 0.81	7.23 2.21	0.16 0.05	4.18 1.17	7.37 1.63	3.12 0.48	0.65 0.25	0.12 0.09	0.97 0.48	99.66
7	Two-px-amph, di-amph, hyp-amph schists (subalkaline basalts)	10	48.52 1.60	0.81 0.18	14.33 1.01	2.35 0.61	8.61 0.91	0.16 0.02	8.69 1.11	10.52 1.25	3.25 0.94	0.80 0.42	0.17 0.29	1.09 0.36	99.30
8	Hypersthene-plagioclase schists	2	53.77	2.18	15.82	3.62	8.62	0.19	2.32	7.40	2.90	0.73	0.70	1.87	100.35

9. Average compositions (in wt%) of Seym assemblage rocks

Gp	Composition	No. of samples	SiO_2	TiO_2	Al_2O_3	Fe_2O_3	FeO	MnO	MgO	CaO	Na_2O	K_2O	P_2O_5	l.o.i.	Total
1	Two-px schists with high Fe, Ti, P (low-alkali basalts, sodic)	8	49.76 2.14	2.65 0.47	13.18 0.54	1.48 1.27	14.04 1.67	0.02 0.03	5.78 0.69	8.62 1.69	1.44 0.79	0.43 0.37	0.33 0.13	0.08 0.14	98.04
2a	Two-px (+amph) schists (low-alkali tholeiitic basalts, sodic)	8	48.76 2.58	1.86 0.15	13.18 1.335	2.56 1.75	12.25 2.01	0.24 0.03	7.52 2.84	9.69 2.72	1.81 0.71	0.75 0.84	0.08 0.09	0.49 0.64	99.21
2b	Two-px schists with biotite (picrito-basalts, associated with anatectites)	5	47.79 2.03	0.72 0.31	12.54 3.57	1.04 0.38	10.38 1.32	0.16 0.02	13.70 2.81	9.14 1.01	0.77 0.45	1.56 0.61	0.10 0.07	0.46 0.17	98.36
3	Ol-hyp-amphibole schists (ultrabasic rocks)	3	45.13 2.23	0.62 0.82	5.21 1.46	2.74 1.45	10.82 0.67	0.21 0.03	27.98 5.12	4.22 3.11	0.35 0.19	0.15 0.09		1.23 1.52	98.66
4a	Bi-ga (±hyp±cord) gneisses (greywacke siltstones, moderately alkaline, K-Na)	14	66.51 3.91	0.39 0.21	15.88 1.55	0.95 0.51	4.52 1.53	0.05 0.02	2.48 1.08	2.65 1.10	3.18 0.64	2.31 1.02	0.05 0.08	0.79 0.30	99.74
4b	Hyp-ga-bi gneisses (greywacke siltstones, tuffites, moderately alkaline, K-Na)	10	61.64 2.78	0.842 0.29	14.65 1.72	0.93 0.61	7.76 1.78	0.08 0.02	4.99 1.91	2.76 1.15	2.24 0.95	2.30 1.09	0.21 0.13	0.91 0.68	99.35
4c	Cord-ga-bi (±hyp±sill) gneisses (semipelites, mod. alk., K-Na)	7	55.18 2.60	0.82 0.36	19.36 1.40	1.14 1.17	8.27 1.31	0.09 0.05	5.88 2.77	1.85 0.60	2.14 1.04	3.11 1.40	0.06 0.08	1.54 1.15	99.48
4d	Bi-hyp (±di) gneisses (calc-ferruginous siltstones)	8	58.11 2.67	0.81 0.16	15.88 1.24	0.74 0.61	7.91 0.93	0.11 0.02	5.06 1.17	5.97 1.23	2.70 0.79	0.87 0.38	0.14 0.07	0.45 0.21	98.80
5	Ga-bi-cord-hyp gneisses (subsialites)	3	52.38 2.23	0.12 0.19	29.10 1.27	1.90 0.75	4.15 1.18	0.03 0.02	7.01 2.46	1.11 0.71	2.2 1.44	0.84 0.47	0.67 0.02	1.27 0.56	100.15

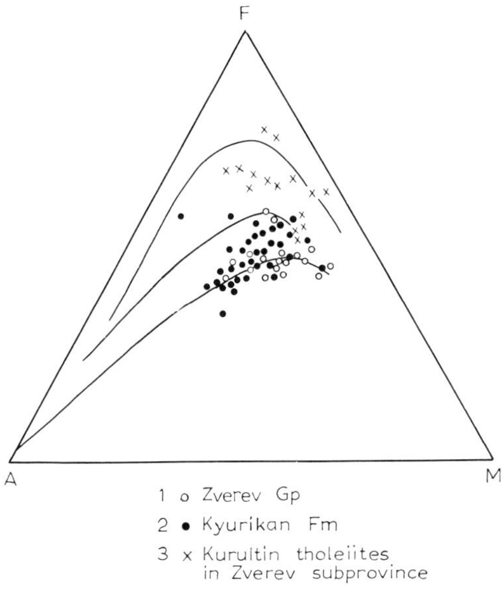

Fig. II-15. *AFM* diagram for volcanics in the sub-alkaline andesite-basalt series.

comprises an association of quartzites and high-alumina gneisses. The high-alumina gneiss members are often restricted to synformal fold cores and appear to be higher that the quartzites in the section. In the North-Eastern subdomain, the Kurumkan Formation is made of high-alumina gneiss only, and immediately overlies rocks of the infracrustal complex (known as the Nimnyr Formation). In the South-Western subdomain, the assemblage consists of quartzite (around 50% of the section), high-alumina gneiss, lenses of calc-silicate rocks, calciphyre, banded ironstone and diopside quartzite. In the north of the subdomain, rock outcrops are bounded by the Aldan and Yarogu thrusts and occur in allochthonous contact with the younger AR_1^2 province (the Amedichi Formation; Dook et al., 1986). Fine banding is characteristic in the quartzites, which in part at least reflects primary sedimentary bedding.

In all the subdomains, the high-alumina gneisses have varieties with garnet, cordierite, biotite and sillimanite and a fine migmatitic banding is common (migmatites resulting from differential anatexis).

Within the supracrustal and infracrustal rocks in the Central Aldan domain, basic plagioclase schists occupy about 5–10% of the area. These are two-pyroxene, sometimes with biotite and amphibole; diopside-amphibole, and amphibolites. Compositionally they are quite uniform in the Central and North-Eastern subdomains and are regarded as volcanics in the tholeiite series (Velikoslavinsky, 1976; Velikoslavinsky and Rudnik, 1983; Berezkin and Kitsul, 1979; Bulina and Spizharksy, 1970). From their U and

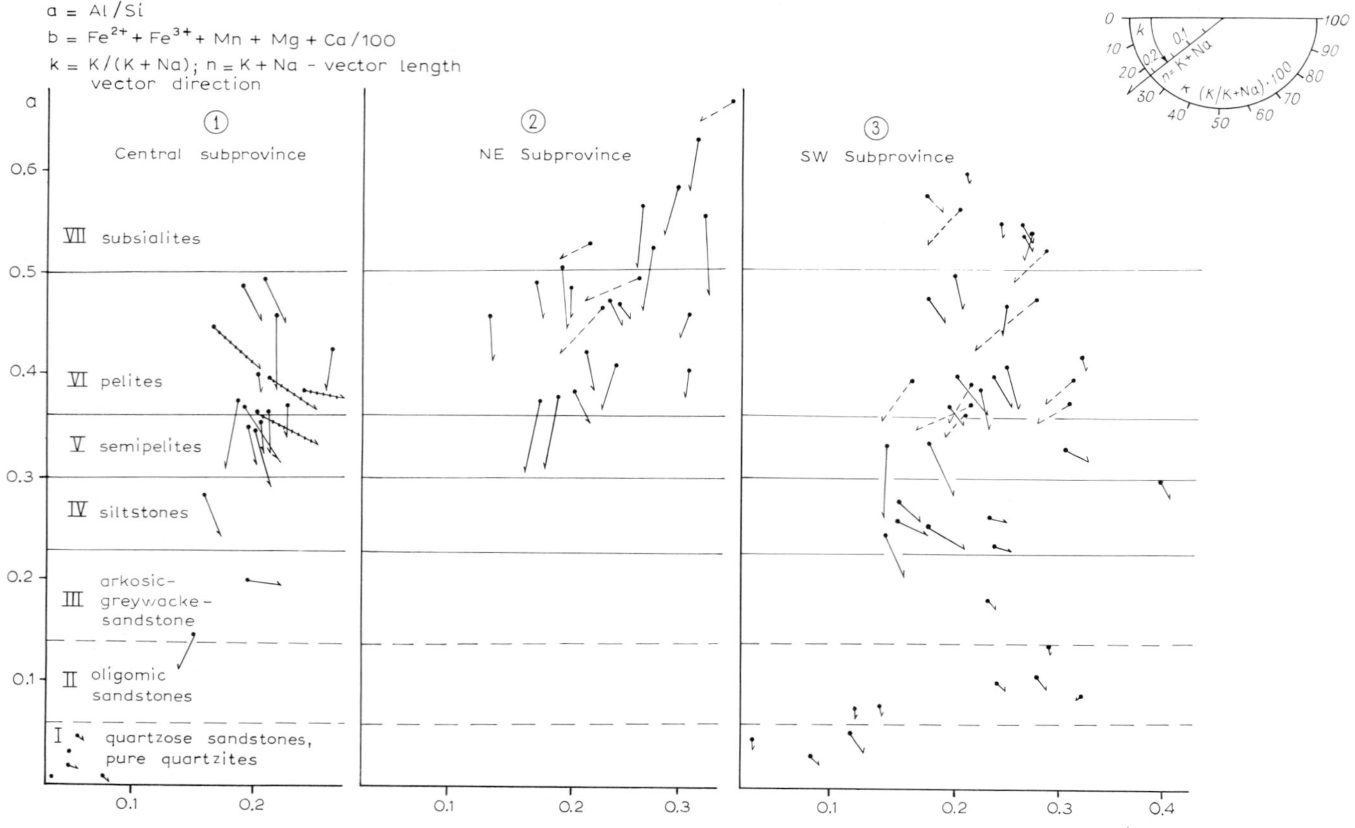

Fig. II-16. "*ab*" diagram for rocks of the Kurumkan assemblage (after Neyelov, 1980). Solid lines on arrows show vectors for rocks in Na–K series, broken line: K–Na series, dotted line: essentially K rocks.

Th ratios, they are divided into continental and oceanic tholeiites (Vaganov, 1981). Velikoslavinsky and Rudnik (1983) have compared them with oceanic basalts. According to data obtained by Travin (1977), the mafic schists are lithologically and geochemically unrelated to the host gneisses (traps?). Rare earth studies suggest that they belong to tholeiites from plutonic magmas (Lobach-Zuchenko et al., 1984). For the South-Western subdomain (the Melemken block) we have observed that the schist compositions are inhomogeneous. As well as tholeiites, we also have varieties with high Na and Al contents, similar to subalkaline basalts in the Zverev subdomain. We have mapped areas containing higher concentrations of mafic schists in the NW of the Central Aldan domain, within which we have also found ultrabasic rocks. They are of different ages and are compositionally heterogeneous. Bodies of basic and ultrabasic rocks cut across lithological boundaries discordantly as well as the contacts between supracrustal rocks and rocks in the infra-complex. Field observations (Glebovitsky and Sedova, 1986) show that some bodies cut the migmatitic banding and are themselves metamorphosed at granulite facies conditions.

The chemical composition of Kurumkan Formation supracrustal rocks is shown diagrammatically on Fig. II-16 and in Tables II-4/3, II-4/4 and II-4/6. Rock groups are separated on their Al/Si ratios in accordance with Neyelov's (1977, 1980) classification.

In the Central subdomain (the stratotype region), the Kurumkan Formation consists of rocks with extreme compositions: monomineralic quartzite (pure silica rocks, field I) and high-alumina gneiss (siltstone and shale, fields V–VI). The Kurumkan assemblage in the North-Eastern subdomain comprises only high-alumina rocks — pelites and hydromica subsiallites (fields VI–VII). Rocks in the South-western subdomain (the Melemken block) form a continuous series on the "*ab*" diagram from pure silica rocks (I) to hydromica subsiallites (VII). However, the quartzites (chert?) in this case have characteristically high Al and Fe contents, as distinct from the stratotype region. Here too we observe that in the transition from silica to high-alumina rocks (clays), the magnitude of the alkalinity vector scarcely changes. In practical terms, there is a decrease in the Si content at the expense of an increase in Al, while all the other characteristic properties of clays are maintained (low Na and Ca and high K). The impression gained is that there were two depositional trends (silica and clay), giving mixed rock compositions in fields III and IV.

A characteristic feature of the high-alumina rocks in all the subdomains is the Na–K profile of sediments, a very low Ca content and high Fe with a high degree of oxidation. However, K–Na compositions, also enriched in Ca, do occur among the high-alumina rocks in the North-Eastern and especially the South-Western subdomains. The existence of such rocks may be evidence that rocks with a montmorillonite composition are present among hydromica clays. High alumina in clays is usually attended by an increase in Fe and Ti. High-alumina rocks in the South-Western subdomain (the Melemken block) stand out on account of having the lowest alkalis and highest melanocratic colour index.

There are currently two points of view concerning the primary nature of the silica rocks, although neither is as yet sufficiently well-reasoned: quartzose sands derived from weathering crusts (Dzevanovsky, 1958; Kulish, 1973) and chemogenic sediments (Frolova, 1951; Salop, 1982). The latter view is supported by our investigations of trends in the petrochemical differentiation of Kurumkan Formation compositions in the Melemken block which indicate the existence of two discrete rock groupings: silica and clay.

Differences in composition between rocks of the Kurumkan Formation and the infracomplex are strikingly obvious from the example in the North-Eastern subdomain where the rock association consists of high-alumina gneisses only, while rocks in the infracomplex are hypersthene-biotite (± garnet) enderbite- and charnockite-gneisses. Using factor analysis, we have established that they are not related to a single differentiation series. Similarly, high Al and Fe contents are typical of the high-alumina gneisses, the Fe being in a high oxidation state (Table II-4/4), while for rocks in the infracomplex we have higher Ca, Na and Mg (Table II-4/5).

In the Central subdomain, the volume of charnockite-gneiss and gneissose granite (granite-gneiss) increases, while gneissose granite dominates in the South-Western subdomain. These rocks were previously considered to be metasomatic nebulitic migmatites and granites, derived from mafic schists. From the REE distribution pattern (Figs. II-17, II-18) we can identify gneissose granites (granite-gneisses) with a marked negative europium anomaly (Ce_N/Yb_N = 15.5–30, Tb_N/Yb_N = 1.2–4.9 — group 1; and Ce_N/Y_N = 2.6–5.7, Tb/Yb = 0.69–1.3 — group 2) and a compositionally close rock (sp.

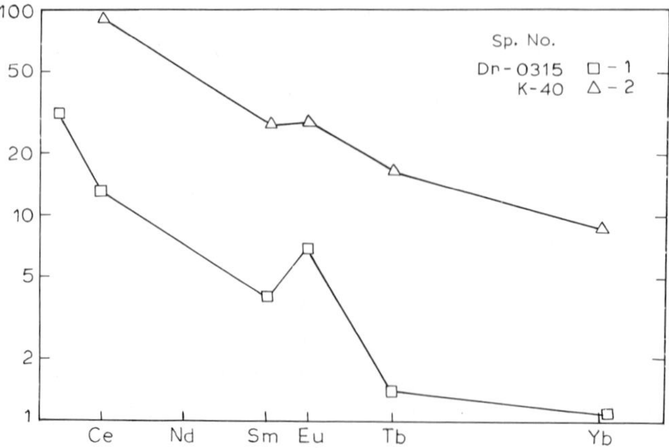

Fig. II-17. Chondrite-normalized REE distribution in enderbite-gneisses of the infracrustal complex (after Dook et al., 1986).

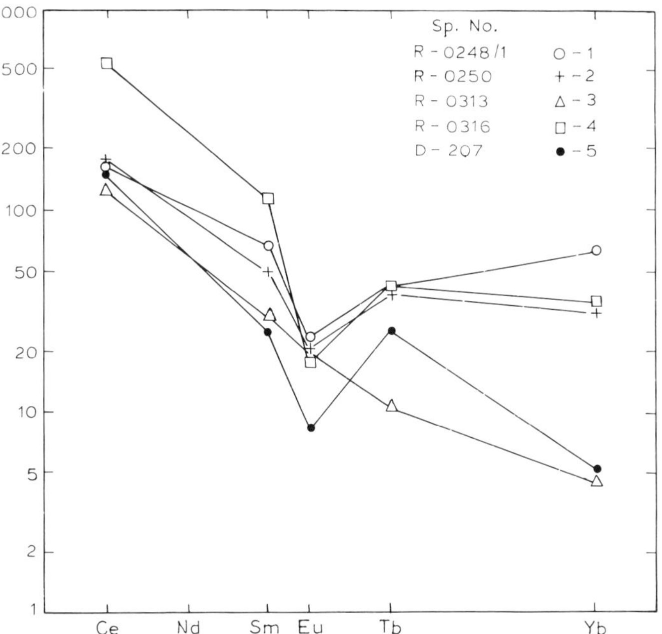

Fig. II-18. Chondrite-normalized REE distribution in granite-gneisses of the infracrustal complex of the Central Aldan province (after Dook et al., 1986).

R-0313), with a REE distribution pattern similar to Archaean grey gneisses in composition, i.e. tonalite-trondhjemites. Since enderbite-gneisses and group 1 granite-gneisses lie in the same position on partial melt trends and since the granite-gneisses have a high initial $^{87}Sr/^{86}Sr$ ratio equal to 0.723, we conclude that the granite gneisses formed from the partial melting of rocks with a tonalite composition, followed by fractional crystallization (Dook et al., 1986). In this case, enderbite-gneisses (sp. Dr-0315d, Fig. II-17) with anomalously low REE content and positive europium anomaly may be considered as a restite.

Late Archaean tectonic domains (AR_1^2)

AR_1^2 domains form a system of structures with a north–west strike: the Eastern Aldan, Idzhek-Sutam and Western Aldan domains (Fig. II-10). Each domain is characterised by an active tectonic environment. Tholeiites, originating from plutonic magmas, were intruded throughout this period in the Central Aldan domain during periods of crust-forming activity.

The Eastern Aldan domain is bounded to the west by the Idzhek-Sutam deep fault. Its eastern boundary with the Batomga block follows the Ulkan fault zone. Supracrustal assemblages are represented by the Kholbolokh assemblage and the Kyurikan Formation (Dook et al., 1986). The episode of folding which separates them has been taken to be the boundary between cycles I and II (Dook and Kitsul, 1975; Dook et al., 1979; AGU, 1983).

The Kholbolokh assemblage (Dzheltulin Group) consists of two spatially isolated rock associations (Fig. II-9). In association 1 (the Sutam Formation), which accounts for around 80% of the assemblage's volume, garnet-biotite-plagioclase gneisses are the dominant rocks, containing thin bands of garnet-hypersthene, hypersthene and diopside-hypersthene plagiogneisses and there are a few places where we see members consisting of rhythmic alternations of these rocks. The most typical rhythms are garnet-biotite plagiogneiss → hypersthene (± garnet)-biotite plagiogneiss → diopside-hypersthene plagiogneiss. Calc-silicate rocks appear in the upper part of the rhythm but are very rare. Occasionally in the garnet-biotite plagiogneisses we observe fine asymmetric lamination, expressed as a gradual increase in garnet content from the base to the top of a rhythm. The rhythms are up to a few tens of centimetres thick, and we have also mapped macrorhythms 100–150 m thick. Around 7–10% of association 1 consists of garnet-biotite gneiss with sillimanite and/or cordierite. They occur as thin bands amongst garnet-biotite plagiogneisses and are often associated with silica rocks, sometimes with garnet or garnet + plagioclase. Mafic schists make up 5–7% of the volume of this association.

Rocks in association 2, mapped as outcrops measuring from 10 × 100 m to 10 × 30 km, form an independent assemblage for which we have proposed the term Ampardakh Formation. They account for 20% of the volume of the Kholbolokh assemblage and are characterised by a somewhat different, more calcic composition in the layers and rhythms. The following rhythm is typical for association 2: hypersthene-biotite (rarely garnet-biotite) plagiogneiss → hypersthene (± diopside)-biotite plagiogneiss → two-pyroxene plagioclase schist → calc-silicate rock (sometimes calciphyre). The last two occasionally alternate with each other in thin bands. On the eastern limb of the synformal structure known as the Sutam or Gonam-Dzheltulin synclinorium, it has been possible to establish that association 2 occupies a lower stratigraphic position, based on evidence from the macrorhythms.

Migmatitic banding and accompanying metamorphic differentiation are found in both rock associations 1 and 2. The banding occurs only to a minor extent in rocks transitional to plagioclase schists (with 10–20% quartz), where we usually see injection-type migmatites instead. Zones of feldspathization and abundant migmatite injection (strips from 3–10 m to 20–30 m wide) are found in the vicinity of "soft" faults. The formation of migmatites in garnet-biotite-plagioclase gneisses was accompanied by the obliteration of the asymmetric bedding structures. Sillimanite-rich zones have been mapped in the gneisses, cutting obliquely across the migmatitic banding and the foliation which is parallel to it. These sillimanite zones contain around 50% quartz and lenses of secondary quartzites are also present.

In the stratotype region — the upper reaches of the river Kholbolokh — we analysed separately two series of rock samples with very different SiO_2 contents: mafic schists, together with calc-silicate rocks and gneisses. Using cluster analysis on factor values, rocks from the first series were divided into two major groups. Group 1 — hypersthene (± diopside ± garnet)-hornblende schists with high Fe and Ti contents, corresponding to alkali-poor tholeiites in composition (Fig. II-14/3, Table II-4/7). These rocks form bodies and boudins in garnet-biotite-plagioclase gneisses. They do not maintain a regular position in the section and are probably orthoschists of volcanic and possibly intrusive origin. Group 2 has both two-pyroxene-hornblende (± garnet) schists and calc-silicate, diopside-carbonate and diopside-scapolite rocks and differs from group 1 in having only half the amount of Fe and Ti and higher SiO_2. Within the group, the calc-silicate rocks (2a) are differentiated from the schists (2b) on the basis of lower Na and Al and higher Ca. Both types are members of rhythms, suggesting that they might be para-rocks. A third, small group (3) of two-pyroxene-hornblende schists is distinguished by its low Al and high Mg content. These schists form thin hypersthene-rich melanocratic bands among group 1 rocks and may be the products of metamorphic differentiation.

Judging by their field appearance, there is no doubting the primary sedimentary nature of the gneisses. The type of differentiation of these rocks is defined by two factors (Fig. II-14/3):

$$F_1 = \frac{\mathrm{Si}_{91}(\mathrm{Na}_{44})}{\mathrm{Ti}_{77}\mathrm{Mg}_{74}\mathrm{Fe}^{2+}_{70}\mathrm{Al}_{69}} 34\% \quad \text{and} \quad F_2 = \frac{(\mathrm{Na}_{40})}{\mathrm{Ca}_{86}}\ 14.7\%$$

Factor F_1 may be interpreted as the differentiation of rock compositions from sandy-silty (Si, Na) to clayey (Ti, Mg, Fe, Al). In this series there are two separate major rock groups (4 and 5), the composition of which does not essentially change during metamorphism. The tendency to increased alumina in group 7 (sillimanite-garnet-biotite gneisses) increases with the process of secondary sillimanitization. The primary composition of group 8 rocks (garnet-biotite-two-feldspar gneisses) is altered by anatexis.

Using factor F_2 we can identify a Ca-branch for sediments — group 6, whose end member is calciphyre.

The composition of rocks in the Kholbolokh assemblage may be summarised as follows. Clastic terrigenous rocks, represented by moderately alkali K–Na varieties, are scarcely differentiated on alumina content from greywacke-siltstones (garnet-biotite-plagioclase gneisses) to semipelites (sillimanite ± cordierite-garnet-biotite gneisses). There is a simultaneous increase in K, which is a regular feature for sedimentary differentiation. The second branch of differentiation is manifest mainly in an increase of Ca in the rocks, due to the appearance of the following rock series: biotite-hypersthene, garnet-hypersthene-biotite, diopside plagiogneisses, two-pyroxene-amphibole (± garnet) paraschists and calc-silicate rocks. The calc-silicate rocks are characterised by variable Mg and Ca contents, with Ca sharply dominant over Mg and they are gradually transitional to carbonate rocks which in this case were probably background sediments. All these rock varieties form an integral part of sedimentary rhythms.

The concluding period in the evolution of the Eastern Aldan province is represented by rocks of the Kyurikan Formation. It differs from the Sutam Formation in containing rocks of the andesite-basalt series, a more varied rhythmically bedded structure, more frequent alternations of layers which, as we shall show later, are slightly different in composition. In the region around the mouth of the river Timpton, the formation is characterised in Western and Eastern (stratotype) portions, separated by outcrops of Sutam Formation rocks. By examining the rhythmic layering in detail, we have found that there is a gradual thickening of the succession towards the inner portion of Kyurikan Formation outcrops (Dook et al., 1986).

The most complete rhythms are represented by the following rock succession in which there is a regular increase from base to top in Ca and other melanocratic components (in brackets we show the percentage of that rock variety in the Western area succession): biotite plagiogneiss (8%) and garnet-biotite plagiogneiss (29%) → hypersthene-biotite plagiogneiss (4%) → diopside (± hypersthene ± amphibole) plagiogneiss (1%) and hypersthene (± diopside ± amphibole)-plagioclase schist (16%) → diopside (± amphibole)-plagioclase schist (4%) → calc-silicate rocks, calciphyre and marble (14%). Individual layers vary in thickness from a few centimetres to tens of centimetres to a few metres. Garnet gradually increases towards the top in the gneisses, then as the rocks become more melanocratic and the Ca-content increases, the garnet content gradually drops off. Judging by the nature of the rhythms, we may infer a primary sedimentary origin for these rocks. There are sometimes thin reaction rims of diopside-amphibole-feldspar rock at the contact with calciphyres. Intermittently banded hypersthene (± amphibole) or melanocratic garnet-biotite plagiogneiss, low in quartz, account for 19% of the volume of the formation and 5–7% schists, with neither of these types occupying a strictly defined position in the rhythmic sequence. These are probably ortho-rocks, corresponding in composition to andesites and basalts.

The chemical composition of rocks in the Kyurikan Formation is considered using the Western area as an example (Fig. II-19, Table II-4/8).

Gneisses fall into field IV, greywacke-siltstones (or tuffites) and several analyses which differ in having higher Al and Fe are in field V — semipelites. Both rock groups have characteristically high alkalis (moderately-alkaline in Neyelov's (1980) classification). Typically, they all lie exclusively within the igneous rock trend, suggesting that they are tuffites, close in composition to igneous rocks.

Calc-silicate rocks vary in the amount of CaO they contain, hence there

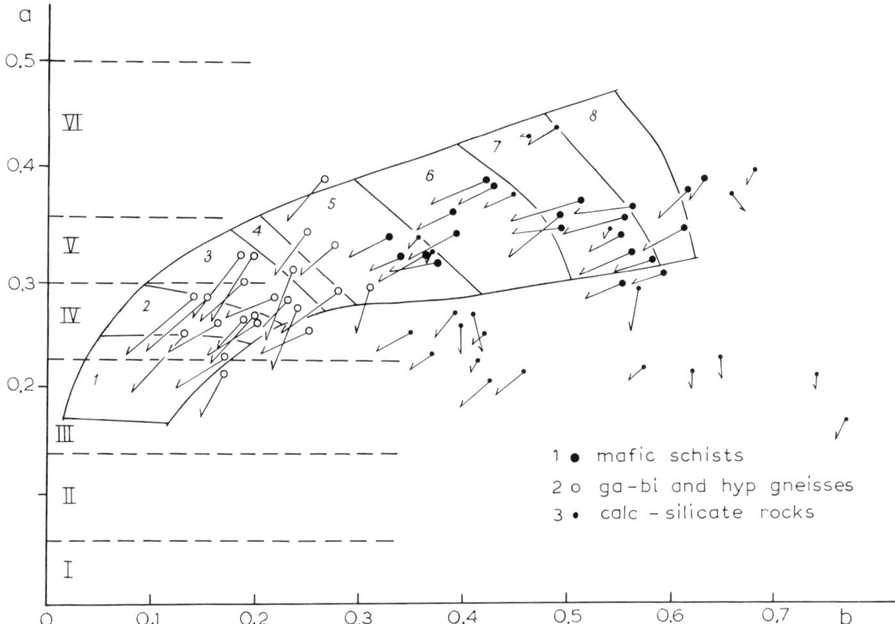

Fig. II-19. "*ab*" diagram for rocks of the Kyurikan assemblage, western sector. Italicized numbers show fields of volcanic rocks: *1* = liparites, *2* = liparitic dacites, *3* = dacites, *4* = andesitic dacites, *5* = andesites, *6* = anesitic basalts, *7* = basalts. Other symbols as on Fig. II-15.

is a scatter of points on the "*ab*" diagram along the *b*-axis. Some of those with higher Al content ($\alpha > 0.4$) are probably lime-muds, usually restricted to the upper parts of rhythms. These rocks all have low alkali contents.

The remaining rocks — two-pyroxene-amphibole-plagioclase gneisses (andesites) and two-pyroxene-amphibole schists (basalts) — form a single series within the igneous rock trend (fields 5–8). They do not occur in any particular or regular order in the rhythmic banding sequence.

In presenting the broad features of the Kyurikan Formation as a whole, we should point out that its distinguishing feature is the high alkali content of all the clastic terrigenous and volcanogenic rocks due to high Na, which is approximately 1% higher than in the Kholbolokh assemblage rocks. Volcanic rocks typically have low Ti. We have encountered individual chemical analyses of basic schists with anomalously high (for the formation) Ti, Fe and P and low Mg (Table II-4/8).

Tholeiites, common in other assemblages, are not seen in the Kyurikan Formation. The fine interbanding between gneisses and calc-silicate rocks points to two sedimentary trends: a formation of high-Na clastic terrigenous rocks (tuffites) at the expense of weakly differentiated material, alternating with typical volcanic basalts and andesites and a formation of carbonate rocks which appear to have been background deposits in the sedimentary basin.

Formational analogues of part of the rocks in the Eastern Aldan domain are rocks of the Idzhek Formation in the Timpton Group. The formation crops out as a tectonic sheet (Figs. II-9, II-10) covering the northern part of the Central Aldan and Idzhek-Sutam domains (the latter is described below). To the east the Idzhek fault (which is traced by following a line of granitoids and high temperature blastomylonites) separates Idzhek Formation rocks from the Eastern Aldan domain.

In the stratotype region, the river Idzhek estuary, 60–70% of the formation volume is made up of association 1, hypersthene (± diopside-amphibole) plagiogneiss alternating with thin bands and lenses of diopside-plagioclase (± scapolite), diopside and calc-silicate rocks, garnet-biotite and garnet-hypersthene-biotite gneiss and plagiogneiss. These rocks form thin rhythmically banded members with increasing Ca-content from base to top. As the carbonate-rich portion of the section increases in thickness, it is possible to map out bodies — complex fold fragments — in which diopside-amphibole-plagioclase schists predominate (association 2, 20% of the formation volume); they alternate with diopside (± scapolite)-plagioclase schists and calc-silicate rocks and contain individual bands of garnet-biotite and hypersthene-garnet-biotite-plagioclase gneiss. Association 3 forms major mappable bodies of garnet-biotite (± cordierite) gneiss alternating with garnet-hypersthene-plagioclase gneiss with subordinate mafic schists and calc-silicate rocks. This association is often referred to as the Uluncha Formation (Frumkin, 1961).

Diopside-bearing rocks are found amongst the calc-silicate rocks, whilst garnetiferous silica rocks occur amongst the garnet-biotite-plagioclase gneisses. In the south-eastern part of the Idzhek Formation outcrop area, magnetite chert and associated calc-silicate rocks are present. Two-pyroxene-amphibole, hypersthene-amphibole and two-pyroxene schists occur in all associations of this formation, up to 10–20% of its volume. They sometimes have a uniform texture and are interpreted as igneous rocks. Charnockite-gneisses have been mapped throughout the outcrop area, forming bodies up to 1 × 7 km. In a number of cases, we have established a link between them and cross-cutting zones oriented sub-parallel to the Idzhek fault. Migmatitic and metamorphic banding are characteristic for rocks of the Idzhek Formation. The mafic schists correspond in composition to subalkaline basalts (and partly subalkaline andesites).

From its terrigenous-carbonate rock association and subalkaline basic magmatism, the Idzhek Formation is analogous to formations in the late stage of evolution of the Eastern Aldan province (Dook et al., 1986).

The Idzhek-Sutam domain has well-expressed tectonic boundaries, traceable as intensely sheared granulite facies “pencil gneisses”. The zone can be traced southwards as far as the Stanovoy suture. Metamorphic equivalents of the rocks that form this zone are known in the Chogar and Dzhugdzhur blocks in the Dzhugdzhur-Stanovoy domain (Karsakov, 1978; 1980).

The Idzhek-Sutam domain consists of supracrustal rocks belonging to the Seym assemblage and some 40–50% of enderbite-gneisses belonging to an unstratified infracrustal complex. The northern part of the zone is covered by allochthonous rocks of the Eastern Aldan domain — the Idzhek allochthon (Dook et al., 1986).

Eighty percent of Seym assemblage outcrops consist of association 1, represented by alternating garnet-biotite gneiss and plagiogneiss with high-alumina sillimanite- and/or cordierite-bearing garnetiferous gneiss (with bands of hypersthene ± garnet-cordierite gneiss), and also hypersthene-biotite, two-pyroxene and diopside-hornblende meso- and melanocratic plagiogneiss. Melanocratic plagioclase gneiss accounts for 20% of the association 1 volume. Additionally, this association contains lenses of silica rocks (with garnet, plagioclase and magnetite), calc-silicate rocks and diopside-forsterite calciphyre. Everywhere we find thin bands and lenses up to a few tens of cm thick of basic and ultrabasic two-pyroxene and two-pyroxene-hornblende schists with occasional hypersthene and olivine-two-pyroxene schists also. These make up 5–10% of the volume of this assemblage.

Fitfeen to twenty percent of the Seym assemblage outcrops consist of association 2 diopside, hypersthene, diopside-hypersthene and garnet-hypersthene plagiogneisses alternating with mafic schists. The schists themselves make up around 30% of the association volume, some of them having high Al_2O_3 contents (up to 16.8%). Bands or thin beds of garnet-cordierite-biotite gneiss and calc-silicate rocks are also present in association 2.

On the F_1–F_2 component diagram (Fig. II-14/4), mafic schists (1, 2, 3) are distinguished from gneisses (4, 5) on the basis of factor F_1 (melanocratic index) for all the rocks in the assemblage. Using factor F_2, we can separate out ultrabasic rocks (3) from the mafic schists and they themselves further subdivide into two groups: group 2, in which the rock composition varies from tholeiite (2a; Table II-4/9) to picritic basalt (2b) and group 1, which differs from group 2 in having lower Mg and higher Fe, Ti and P. The opposition of these elements may be interpreted as the Fennerov differentiation trend, which is confirmed by the *AFM* diagram (Fig. II-20).

The REE distribution in mafic schists in the Idzhek-Sutam province (Fig. II-21) indicates heterogeneity in light REE ratios, close to the chondritic distribution of heavy REE, and the presence of rocks with an unfractionated REE distribution.

F_1, the major factor in gneisses (Fig. II-14/4b), characterises the differentiation of compositions on melanocratic index (Si, Na against Fe^{2+}, Mn, Ti) which probably reflects the amount of endogenic impurities and may be evidence for these rocks being tuffs. Factor F_2 reflects sediment

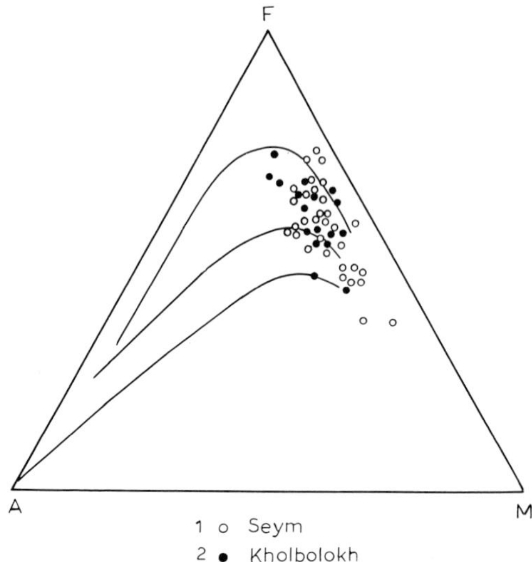

Fig. II-20. *AFM* diagram for volcanics in the Seym and Kholbolokh assemblages.

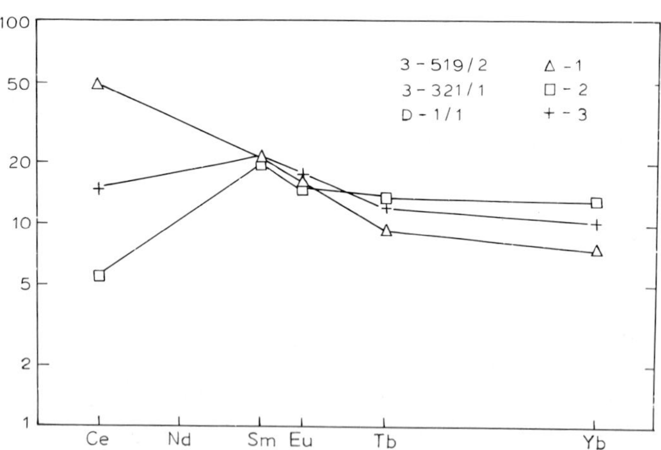

Fig. II-21. Chondrite-normalized REE distribution in mafic schists in the Idzhek-Sutam zone (after Dook et al., 1986).

differentiation according to maturity (Al, Mg — the clay component; and Si, Na, Ca — the sand-silt component).

Rocks in field (4d) differ from all the gneisses by having twice the amount of Ca. These are probably sediments with a carbonate admixture. They usually occur amongst garnet-biotite gneisses, often in contact with calc-silicate rocks. But we have to bear in mind the fact that the composition of these rocks is also close to andesites.

One characteristic feature of all the rocks in the Seym assemblage is the high Fe and Mg. Tholeiites in this zone typically have the highest colour index, the lowest alkali content, the shallowest conditions for the formation of magmas in the Aldan Supergroup and contain varieties transitional to ultrabasic rocks (picrito-basalts).

The Fyodorov Formation occurs in the Central Aldan domain (AR_1). Previously, it was considered to be the uppermost formation in the Iyengr Group (Dzevanovsky, 1958). Compared with other assemblages, it shows fewer deformation episodes and on this basis it has been referred to the upper domain (Dook and Kitsul, 1975; Dook et al., 1975), corresponding to the concluding stage in the evolution of late (AR_1^2) provinces (Dook et al., 1986). Cherkasov (1978) and Frumkin (1981) exclude it from the Iyengr Group.

Association 1 which predominates, accounting for 70% of the outcrop area of this assemblage, comprises amphibole (biotite, diopside ± hypersthene) plagiogneisses and occasional schists, plus thin bands and lenses of diopside and phlogopite-diopside rocks, calciphyres and hypersthene plagiogneisses. Diopside-amphibole schist, calc-silicate rocks and calciphyres forming association 2 contain productive phlogopite deposits in the shape of complex fold fragments (up to 0.5 × 15 km) within association 1 rocks. Basic schists and plagioclase gneisses correspond to basalts, andesite-basalts and subalkaline to calc-alkaline andesites, as can be seen from petrochemical and geochemical data. Fyodorov Formation volcanics form a single uninterrupted series from basalts to dacites inclusive, with a Bowen differentiation trend (Berezkin and Kitsul, 1975). Diopside plagiogneisses and schists correspond to terrigenous and chemogenic sedimentary rocks with tuffaceous impurities. The structure and composition of the Fyodorov Formation are discussed further in Velikoslavinsky (1976), Velikoslavinsky and Rudnik (1983), Petrova and Smirnova (1982), Reutov (1981) and Cherkasov (1978).

The contact between the Fyodorov and Kurumkan Formations often shows signs of being a tectonic boundary — such as the juxtaposition of different rock associations belonging to the Kurumkan and Fyodorov Formations at the contact surface. Observations such as this, together with the fact that there are no intrusive analogues of Fyodorov volcanics within Kurumkan rocks, the conformable nature of the boundary between the Fyodorov Formation and Idzhek allochthon suggest that the Fyodorov Formation is in an allochthonous position. The Idzhek and Fyodorov Formations could possibly belong to a terrain which formerly lay between the Idzhek-Sutam and Eastern Aldan zones and was obliterated during the formation of deep-level thrusts in the Idzhek-Nuyam fault zone.

The Western Aldan domain is bounded to the east by the Aldan thrust surface and to the west its constituent rocks are partially found in allochthonous juxtaposition with the Olyokma granite-greenstone terrain (Dook et al., 1986). The Western Aldan domain boundaries typically show a hiatus in the way metamorphic regimes change, and igneous rocks from deeper lithospheric levels make an appearance. Localities where tectonic boundaries were rejuvenated (the Amga and Aldan-Kiliyer zones) are marked by Lower Proterozoic blastomylonites at amphibolite and epidote-amphibolite facies. The domain contains rocks of the Chugin Formation, which was first recognised by Mironyuk (1961) as an independent subdivision of the Upper Aldan Formation (Iyengr Group). In addition to this, it includes outcrops of various stratigraphic subdivisions of the western part of the Aldan Shield (Dook et al., 1986). Chugin supracrustal rocks occur as isolated outcrops among gneissose granites. The following paragenesis is typical for the formation: garnet-biotite (± sillimanite) and amphibole (± diopside) plagioclase gneiss and schist, calc-silicate rocks, calciphyre and silica rocks (often with diopside, amphibole and plagioclase, sometimes magnetite). From the few chemical analyses available, the garnet-biotite gneisses belong to weakly differentiated greywacke-siltstones with a K–Na profile. Amphibole-diopside gneisses are close in composition to greywacke-siltstones but have a higher Ca content. It can be seen from Fig. II-14/5 that the average compositions in the Chugin assemblage are described by the greywacke-type rock differentiation trends that are typical for the majority of assemblages in the Aldan Supergroup. High-alumina rocks with sillimanite and sometimes corundum are uncommon; they have varied alkali contents and in most cases appear to be products of acid leaching. Silica rocks correspond in composition to pure silica rocks and cherts, with characteristically high Ca and sometimes Fe, as is the case with the gneisses. Migmatized garnet-biotite-plagioclase gneisses are sometimes altered during shearing to lenticular-banded or uniform rocks containing two feldspars. Mafic schists make an important contribution (from 10 to 30%) to all the rock associations. They include tholeiites, but the majority of chemical analyses correspond to K–Na subalkaline basalts, with andesites to a lesser degree. We have encountered rocks with anomalously high TiO_2 (3.29%) and P_2O_5 (1.4%) contents, similar to early dykes in the field where the Zverev Group is developed. Rocks belonging to the subalkali series are mainly localised along the western and eastern tectonic boundaries of the province under discussion. From the occurrence of tholeiites and subalkaline basic rocks we may infer that early and late evolutionary stage rock associations are present in the Western Aldan zone.

Discussion of results

The average compositions of the commonest rocks in the Aldan Supergroup have been studied using factor analysis (Fig. II-14/5, 6). A quartzite group was identified on the basis of factor:

$$F_1 = \frac{\mathrm{Si}}{\mathrm{Al}_{83}\mathrm{Ti}_{82}\mathrm{Fe}^{2+}_{80}\mathrm{Mg}_{58}\mathrm{Mn}_{41}\mathrm{Fe}_{40}}\ 42\%$$

Factor:

$$F_2 = \frac{Fe^{3+}_{67}(K_{30})}{Ca_{92}Na_{78}} \ 21\%$$

divides the gneisses into two genetic groups: highly-differentiated sediments-siltstones, pelites and semipelites with a Na–K profile and high Fe^{3+} (the Kurumkan Formation) and rocks in the K–Na series — weakly differentiated greywackes (all other assemblages).

The greywacke group has two differentiation branches: one based on the degree of maturity in the direction of an increasing clay component, enriched in femic elements (reaching only siltones), the other branch being calcic (increasing amount of carbonate admixture in the sediments). While possessing general tendencies, the greywackes in each group have their own specifics, related to features of igneous rocks associated with them and depositional conditions. For example, sediments in the Seym assemblage in the Idzhek-Sutam domain are the most melanocratic, the Kholbolokh assemblage (Sutam and Ampardakh Formations) the most calcic, and the Kyurikan the most sodic.

High-alumina gneisses and quartzites in the Kurumkan Formation have a K profile for the sediments and high degree of Fe oxidation, with the Fe content being quite high. The high oxidation state of Fe precisely here is not peculiar to the gneisses in the infracomplex and the mafic schists which underwent the same metamorphism as the Kurumkan Formation. The compositional features of the Kurumkan Formation are thus more likely to be related to depositional conditions.

The average composition of mafic schists, including in this sample several groups of hypersthene gneisses and calc-silicate rocks, using factor:

$$F_1 = \frac{Si_{88}Al_{63}Na_{60}K_{49}}{Fe^{2+}_{75}Mn_{73}Ca_{59}Fe^{3+}_{49}Mg_{40}} \ 42\%$$

represents an acid–basic differentiation trend, divided into rocks of tholeiitic and subalkaline series. Tholeiites are typical for the Kurultin Group, the Kurumkan Formation and the Kholbolokh and Seym assemblages. The subalkaline andesite-basalt series typifies the Zverev Group and the Kyurikan and Idzhek Formations (Fig. II-14/6). Tholeiites in the Aldan Supergroup are low in alkalis and high in Fe and Ti compared to rocks in the subalkaline series. In rocks belonging to the subalkaline series Na, K and Al are high. From factor F_2 tholeiitic basalts are differentiated from their Fe, Ti and P contents. Rocks with the highest amounts of these elements are present in the Kurultin Formation and the Seym assemblage. Extreme Ti-differentiation is seen in the composition of AR_1 dykes found in the Zverev subzone. Factor F_2 also allows us to divide the subalkaline andesite-basalt series into andesites and basalts as well as indicating the differentiation of calc-silicate rocks and terrigenous-carbonate rocks. The calc-silicate rocks vary from Mg–Ca compositions to alkali-alumina due to increased terrigenous or more likely endogenic impurities.

An analysis of sedimentary rock compositions shows that for the Early Archaean of the Aldan Shield, the commonest rocks are weakly differentiated essentially greywacke-type sodic sediments, their composition being closely dependent on that of volcanic rocks associated with them. This is especially marked in the case of the Seym and Kyurikan assemblages. Besides greywackes, high-alumina sediments with a K profile also existed, with a high degree of Fe oxidation (the Kurumkan Formation), which may be evidence for significant amounts of free oxygen in the Earth's atmosphere from the very earliest stages in geological history.

During AR_1^2 times the evolution of volcanism was clearly developed, as expressed by a change from tholeiites to a subalkaline sodic andesite-basalt series.

We therefore see that, beginning in the Archaean, differentiation of tectonic regimes and sedimentary environments existed in the Aldan Shield.

During the early stage (AR_1^1) of the formation of the Aldan Supergroup in the province, there were highly distinctive tectonic regimes: cratonic for the Central Aldan tectonic domain and active for the Olyokma-Kurultin-Zverev tectonic domain. Where they are most intimately in contact (the Melemken block in the Central Aldan domain), we can detect in rocks belonging to the Kurumkan Formation a transition to an active tectonic environment. In addition to high-alumina gneiss and quartzite with a Na–K profile (semipelite, pelite and psammite), the Kurumkan Formation also has ferruginous and Mn–Fe chert, calc-silicate rocks, pelites with a K–Na profile and basic volcanics. This all provides indirect evidence that the data from the province are the same age.

All the domains belonging to the later stage (AR_1^2) are characterised by an active tectonic environment with the formation of tholeiitic series initially, followed by the andesite-basalt series plus large amounts of redeposited volcanic material. AR_1^2 domains usually have undifferentiated K–Na profile terrigenous rocks and minor amounts of carbonates. Two differentiation types have been established for the sedimentary rocks, corresponding to two depositional trends: from greywacke-siltstones to melanocratic silty-clay sediments enriched in Al, Fe and K (most clearly seen in the Idzhek-Sutam province) and a differentiation branch from greywackes to calc-silicate and carbonate rocks (especially characteristic for the Eastern Aldan domain). The higher Fe in volcanics and sediments, the shallow level of tholeiitic magma generation and the presence of picrito-basalts and picrites are characteristic features of the Idzhek-Sutam domain.

Clastic terrigenous rock associations (greywacke-siltstones, close in composition to andesites), carbonate and calc-silicate rocks, subalkaline basalts, large volumes of subalkaline series andesites and the absence of tholeiitic basic rocks are typical for assemblages in the final evolutionary stage of late tectonic domains.

Metamorphism

Peak metamorphic conditions were attained after the accumulation of sedimentary assemblages in late domains of AR_1^2 age and characterise that stage in the evolution of the region, which is expressed structurally in the formation of isoclinal fold systems IIF_{1-3} (Dook et al., 1975, 1979; Balagansky, 1979; GSA, 1983). The boundaries of domains and metamorphic belts often coincide, indicating that the major structures in the shield developed along directions inherited from earlier periods. This section on metamorphism relies heavily on the extensive work done over many years by Kitsul (Kitsul and Shkodzinsky, 1972; Dook et al., 1986).

In the Southern Aldan (Olyokma-Kurultin-Zverev) belt, metamorphism occurred in the biotite-sillimanite-garnet-orthoclase subfacies of the low-gradient granulite facies (Table II-5). The Western subzone of the Southern Aldan domain corresponds to the Kurultin, Olomokita and Khanin blocks, the Central to the Zverev block and the Eastern to the Tangraka block (Fig. II-10). In those blocks in the Western subzone, $N_{Mg^{ga}}$ reaches 37.8–38.4 in parageneses with Sill + Cord in the Kurultin block and 36.7 in the Olomokita block. Cordierite is rare in the Kurultin block and then only as a mineral forming reaction rims around garnet grains in sillimanite gneisses. Two generations of cordierite are present in the Khanin and Olomokita blocks, an earlier one in the matrix and a later one as reaction rims around garnet. From Ga–Cord equilibrium, we obtain a pressure of 7.2 kbar and temperature of 730°C in the Olomokita block (Dook et al., 1986) which in fact characterises a retrogressive metamorphic event ($N_{Mg^{ga}}$ in the middle portion of a grain is 33.0, which is less than the maximum value in these particular assemblages in this block).

TiO_2 in biotite is up to 5.11–6.30 wt% in the Kurultin block, 5.45–6.0 in the Khanin block and up to 4.54–5.31 in the Olomokita block.

The amphibole in schists (Table II-5) is hornblende with up to 2.79 wt% TiO_2 in the Khanin and 2.32 in the Olomokita block. The Na_2O content in hornblendes is 2.11 and 1.41 wt% in the same blocks, respectively.

The paragenesis Ga + Cpx is found in the Kurultin block, with two-pyroxene equilibrium temperatures of 880–890°C (Dook et al., 1986). Garnet is rare in mafic schists from the Khanin block, while in the Olomokita block the paragenesis Ga + Cpx +/or Amph is not found. $N_{Mg^{ga}}$ in mafic schists from the Kurultin block reaches 27.8–28.8 with $N_{Mg^{cpx}}$ 58.6–63.9 and $N_{Mg^{opx}}$ 46.8–56.2; in the Khanin block $N_{Mg^{ga}}$ is 17.0 with $N_{Mg^{opx}}$ 43.9 and $N_{Mg^{amph}}$ 47.4. From Ga–Bi equilibrium the temperature in the Kurultin block was 800–660°C, in the Olomokita block 790–760°C and

TABLE II-5

Mineral associations in assemblages belonging to the Aldan Supergroup (Early Archaean, cycle II) and estimated metamorphic conditions

Rock group	Southern Aldan Province	Central Aldan Province	Western Aldan Province	Eastern Aldan Province	Idzhek-Sutam Province Northern part	 Sutam block
Gneisses, quartzites, enderbites	Sill+ga$_{40}$+bi+Q+Ksp Hyp$_{62}$+bi$_{64}$+pl+Q±Ksp ±ga$_{36}$ Ga$_{41}$+bi$_{67}$+pl+Q±Ksp Sill+ga$_{41}$+bi$_{57}$+Ksp+cor	Hyp+cor+ga$_{37}$+pl+Ksp+Q Sill+ga$_{35}$+cor+pl+Ksp+Q +bi Cor+ga$_{33}$+bi$_{54}$+pl+Ksp+Q Ga+bi+pl±Ksp+Q Hyp+ga+bi+pl+Ksp+Q ***In SW subprovince:*** Sill+bi+ga$_{29}$+pl+Ksp+Q	Sil+bi+pl+Q±Ksp±ga$_{23}$ Sill+cor+bi+pl+Ksp+Q +ga$_{24}$ Cor+bi+pl+Ksp+Q±ga Hyp+bi+pl+Q±ga Ga+bi+pl+Ksp+Q Sil+cor$_{59}$+ga$_{21}$+bi+pl+Q	Sill+ga$_{40}$+pl+Ksp+Q±bi Sill+cor$_{79}$+ga$_{47}$+pl+Ksp+Q±bi Hyp+ga+pl+Ksp+Q±bi Hyp+pl+Q±Ksp±bi Ga+bi+pl+Q±Ksp	Hyp+cor+pl+Q±ga$_{43}$±bi±Ksp Cor+ga$_{39}$+pl+Q±bi±Ksp Sill+cor+ga$_{41}$+bi+pl+Q±Ksp Ga+pl+Q±bi±Ksp Ga+sill+pl+Ksp+Q±bi Hyp+bi+pl+Q±di±Ksp Ga$_{45}$+cor$_{80}$+sill+pl+Q±Ksp Ga+Q+hyp±di±mt Hyp+di+Q+mt	Hyp$_{79}$+ga$_{51}$+pl±Ksp+Q Ga$_{48}$+bi$_{85}$+pl+Q±Ksp Ga$_{48}$+Sill+pl+Ksp+Q±bi$_{75}$ Hyp$_{78}$+sill+ga$_{60}$+pl+Q+Ksp±bi$_{78}$ Hyp$_{64}$+cor$_{81}$+ga$_{49}$+pl+Q+Ksp±bi$_{78}$ Ga+hyp+di+pl+Ksp+Q±bi Cor+sap+pl+hyp
Mafic and ultramafic schists	Opx$_{65}$+cpx$_{71}$+am$_{64}$+pl±Q Opx+am+bi+pl±Q±ga Cpx+am+pl+ga Opx+cpx+pl+bi±Q Ga$_{45}$+opx$_{69}$+cpx$_{72}$+am$_{66}$ +pl+mt±sp Ol+opx+cpx+sp±am	Opx+cpx+pl+bi±Q Opx$_{75}$+cpx$_{82}$+am+bi+pl±Q Cpx+pl±am±Q Am+pl	Opx+cpx$_{67}$+am$_{58}$+pl Opx+cpx+bi+pl+Q Opx+am+pl+Q±bi Cpx+am+pl±Q Am+pl±bi±Q±Ksp	Ga$_{42}$+opx$_{63}$+cpx$_{69}$+pl+am$_{55}$ Opx$_{42}$+cpx$_{55}$+pl±am Opx$_{52}$+cpx$_{69}$+bi+pl±ca±am Opx+am+pl±bi Cpx+am+pl Opx+cpx±am+sp Ol+opx+cpx±am+sp	Opx+cpx+pl±am Opx+cpx+pl±Q±bi Ol+opx+cpx+am+sp	Ga$_{44}$+opx$_{69}$+cpx$_{77}$+am$_{98}$+pl Opx+cpx+pl±Q
Calc-silicate & carbonate rocks	Di+pl+Q+sph+ga	Di+pl+scap+Q+ca+sph Di+pl+Q+scap+sph±Ksp Ca+dol+for+sp±di Ca+for+sph±di Gross+cpx+pl$_{61}$+scap$_{74}$ +Q+ca+sph±Ksp ***In Fyodorov Fm:*** Woll+gross+cpx+pl$_{39}$ +scap$_{59}$+Q+ca+sph Woll+pl$_{69}$+scap$_{88}$+Q+ca sph±cpx Andr+cpx+pl+Q+mt	Di+pl+Q+sph Gross+di+pl+sph Ca+for+di+phl Ca+di+phl	Ga+cpx+pl+Q+ca±scap±sph Cpx+pl+sph±scap±Q Ca+for+di±phl±sp	Ga+di+pl+ca+sph±scap Di+pl+sph±scap	Ca+for+di Ca+for+di+ga+sp
Estimate of peak metamorphic P-T conds. Two-px equilib. T in brackets	800°C (900°C); >7.5 kbar	800°C (900°C); 7.5 kbar ***in SW subprovince:*** 800°C (900°C); 8.2 kbar	800°C (850°C); 6.8 kbar	830°C (930°C); 8.7 kbar	***North:*** 880°C (930°C); 8.7 kbar. ***South:*** 880°C; 9.0 kbar	≥900°C; ≥9.0 kbar

in the Khanin block 750–740°C (Dook et al., 1986) — these figures represent retrograde stage mineral growth and superimposed effects. Opx–Cpx and Ga–Bi equilibria yield similar temperatures of 870 and 860°C, respectively, for the Olomokita block.

The rocks in the Zverev subzone form a block of the same name, and also form a number of outcrops to the south in adjacent structures belonging to the Stanovoy domain. In the Zverev and Kurultin Formations, the amphibole is hornblende with up to 3.06 TiO_2 and Na_2O up to 3.17 wt%. $N_{Mg^{ga}}$ is 39.7–44.9 and $N_{Mg^{cpx}}$ is 68.0–72.5; $N_{Mg^{opx}}$ 63.4–68.8; $N_{Mg^{amph}}$ 60.0–65.6. Cor + Q reaction rims around garnet in sillimanite-bearing gneiss are associated with a retrograde metamorphic event. In the Sill + Cord (reaction) paragenesis, $N_{Mg^{ga}}$ reaches 39.6–41.0; the TiO_2 content in Bi is up to 5.02–5.54. From two-pyroxene equilibrium data (Dook et al., 1986), T = 880–920°C. Garnet-biotite equilibria, with rare exceptions (860°C), reflect conditions of retrogressive and superimposed episodes of mineral growth (780–680°C).

In the Eastern subzone of the Southern Aldan domain (the Tangrak block), $N_{Mg^{ga}}$ is up to 34.4 in parageneses with Sill; TiO_2 in Bi is up to 4.92–5.36. Hornblende has up to 2.60 TiO_2 and up to 2.5 wt% Na_2O. $N_{Mg^{ga}}$ reaches 30.4–37.8 for $N_{Mg^{cpx}}$ = 61.7–70.2 and $N_{Mg^{opx}}$ = 53.0. Temperatures of mineral equilibria are 820–930°C from Opx–Cpx and 750–800°C from Ga–Bi. In the southern part of the block, superimposed kyanite-sillimanite facies series amphibolite facies metamorphism is most intensely developed.

The lowest gradient and/or deepest metamorphic conditions were attained in the Zverev (Central) subdomain. Moving away from it to blocks in the Kurultin subdomain, we see a gradual transition to shallower and/or higher gradient granulite facies conditions (Cor + Q, Hyp + Pl rims around garnet), while the later superimposed metamorphism took place under variable conditions, depending on the different blocks and different parts of these blocks.

In the Central Aldan domain (Fig. II-10), metamorphism was at high-gradient granulite facies conditions (in the Nimnyr block) with a transition to lower gradient regimes in its South-Western subdomain, the Melemken block (Table II-5). In Hyp + Cord, Sill + Cord and cordierite-bearing assemblages in the North-Eastern subdomain, the $N_{Mg^{ga}}$ maximum is 37.3–34.8. In the Central subdomain, due to the widespread late amphibolite facies metamorphism, high-Mg garnets in these particular assemblages are rare, and $N_{Mg^{ga}}$ = 32.7–31.7. In both subdomains, regressive zoning in garnets and cordierite-quartz rims have been found. The maximum parameters using Ga–Cord equilibria are 800–820°C, 7.6–7.3 kbar in the North-Eastern subdomain and 800–830°C, 7.6–7.2 kbar in the Central subdomain, determined on garnets with Mg-contents of 25.2–24.8 and 23.3–22.9, respectively, and are unlikely to reflect peak metamorphic conditions since the Mg-content at the centre of grains in the same specimen is 28.9

(Dook et al., 1986). TiO_2 in biotite is up to 4.63–5.70 in the North-Eastern and 4.09–5.20 wt% in the Central subdomain.

Only garnet-free assemblages are found in the mafic schists belonging to the Kurumkan and Fyodorov Formations, which are stable in subfacies of shallow and intermediate depth. Two-pyroxene equilibrium temperatures for the Kurumkan Formation are 870–900°C, and 820–890°C for the Fyodorov Formation. In all parts of the Nimnyr block, parageneses in calc-silicate and carbonate rocks from the Fyodorov Formation are similar: Woll + $Plag_{69}$ + $Scap_{88}$ + Q + Cal + Sph ± Cpx, Gross + Cpx + $Plag_{61}$ + $Scap_{74}$ + Q + Cal + Sph, Andr + Cpx + Pl + Q + Mt, Cal + For + Sp ± Di, Cal + Dol + For + Sp. Wollastonite and grossular show no signs of having formed later. In the south (the Melemken block) and west (the Aldan thrust) of the South-Western subdomain, $N_{Mg^{ga}}$ maxima are close and in parageneses with Cord and Hyp respectively reach 39.9 and 42.7; with Cord + Sill 35.8 and 38.1; with Cord 33.3 and 32.7; with Sill 28.9 and 28.3. Al_2O_3 in Hyp in assemblages with Cord is 7.41–5.34 wt%. TiO_2 in biotite is up to 5 wt%. Ga–Cord equilibrium parameters in sillimanite-bearing gneiss in the Melemken block were determined for $N_{Mg^{ga}}$ = 33.3; 34.5 and gave 800°C, 8.2 kbar, while in the Western part of the subdomain in an assemblage with Hyp, the values were 790°C and 8.6 kbar for $N_{Mg^{ga}}$ = 42.7. In both parts of the subdomain we find the widespread development of parageneses and mineral compositions characteristic of lower Ga–Cord equilibrium parameters in sillimanite gneiss: 780–690°C and 7.5–6.1 kbar in the Melemken block and 650°C, 5.9 kbar in the west of the subdomain. The two-pyroxene equilibrium temperature is 880–930°C. In the suture zone separating the Western Aldan and Central Aldan domains (the Aldan thrust), we find the assemblage Ga + Cpx + Plag. The following assemblages occur in the frontal part of the Aldan thrust: Opx + Cpx + Ga + Amph + Plag ± Q; Amph ± Bi + Ga + Plag ± Q; Ga + Opx + Amph + Plag; Ga + Amph + Bi + Cumm + Plag + Q. In associations with Opx and Amph, $N_{Mg^{ga}}$ = 16.0 for $N_{Mg^{opx}}$ 40.9 and $N_{Mg^{amph}}$ = 51.5. In associations with Amph, $N_{Mg^{ga}}$ = 19.2 for $N_{Mg^{amph}}$ = 39.2. The two-pyroxene equilibria temperature is 930°C.[1]

The following associations are found in calc-silicate and carbonate rocks: Di + Plag ± Scap + Q ± Cal + Sph; Di + Plag ± Kfs + Q + Scap + Sph; Gross + Di + Plag ± Scap + Q ± Cal + Sph; Cal + For + Di + Sph; Cal + For + Di + Sph; Cal + Di + Phlog. Relict amphibole (reddish-brown to brown hornblende) has up to 2.15 wt% TiO_2 and up to 2.12 wt% Na_2O. There are examples of parageneses which contain in addition to relict high-temperature minerals, common hornblende, cummingtonite or gedrite in K_2O-undersaturated quartz-bearing gneisses, highly ferruginous

---

[1] In the Kurumkan Formation, situated in the Aldan thrust zone, Lavrenko (1957) noted the presence of armoured sillimanite in orthoclase in hypersthene-cordierite-biotite and hyoersthene-biotite gneisses.

garnet ($N_{Mg^{ga}}$ = 12.3–12.1), biotite with average titanium, newly formed muscovite, epidote, fibrolite and sometimes andalusite. Cor + Q and Hyp + Plag symplectites and reaction rims are common around garnet grains, also green hornblende replacing brown, gedrite and biotite replacing hypersthene and cross-hatched microcline replacing orthoclase-perthite. Peak metamorphic conditions were at granulite facies, in a transitional regime from high to low gradient with the formation of parageneses typical of increased crustal depths, found in the most deeply eroded part of the allochthon. Subsequently, there was widespread retrogression and superimposed amphibolite facies metamorphism.

The Western Aldan metamorphic belt (Table II-5) forms a west- and south-facing arc some 50 km across and corresponds to the Western Aldan domain (Fig. II-10).

The maximum $N_{Mg^{ga}}$ in parageneses with Sill and/or Cord is 23.2–23.8, Ga–Cord equilibrium parameters in a paragenesis with Sill are 780–800°C and 6.8 kbar, obtained for $N_{Mg^{ga}}$ = 21.1; 20.5 and the TiO_2 content in biotite of up to 4.34 wt% is the lowest maximum for this value in metasediments belonging to the Aldan Supergroup. Hypersthene + oligoclase assemblages are absent. Metamorphism is non-uniform. To the west of the Yarogu thrust (Fig. II-10), two-pyroxene varieties are missing from the mafic schists. In the Yarogu thrust zone, in the western (Pritoplenny) sheet (relative to the autochthon), gedrite is seen to be replaced by hypersthene-cordierite symplectite with expulsion of an ore phase. Two-pyroxene assemblages in this sheet are found only in the western part, close to the boundary with the Olyokma block. A similar picture emerges for the Chuga Formation which occurs in an allochthonous position in the Olyokma granite-greenstone terrain. Here, two-pyroxene mafic schists crop out only in the western, most deeply eroded part of the sheet (in the river Tas–Khoyko basin). In this region the mafic schists and Ca-rich plagiogneisses have these assemblages: Ga + Opx + Amph + Plag + Q; Ga + Cpx + Amph + Plag; Ga + Amph ± Bi + Plag ± Q. $N_{Mg^{ga}}$ in mafic schists is 21.3 for $N_{Mg^{opx}}$ = 44.9, $N_{Mg^{cpx}}$ = 60.3 and $N_{Mg^{amph}}$ = 53.1.

The stability of the biotite + sillimanite paragenesis and the appearance of garnet in association with clinopyroxene and amphibole in mafic schists in the fronal part of the allochthons demonstrate the transitional nature of metamorphic conditions from low- to high-gradient. In the less-deeply eroded eastern parts of tectonic sheets, metamorphic conditions were transitional between granulite and amphibolite facies, while in more deeply eroded western (frontal) parts of allochthons in garnetiferous and non-garnetiferous parageneses in mafic schists and plagiogneisses, mineral compositions are typomorphic for the granulite facies (i.e. $N_{Mg^{amph}}$ = 61.9–61.8; TiO_2 = 2.46; N_2O = 2.19 wt%; $N_{Mg^{opx}}$ = 67.4–67.5; $N_{Mg^{cpx}}$ = 74.6–67.1). The two-pyroxene equilibrium temperature is 840–870°C to 900°C. Superimposed retrogressive amphibolite facies metamorphism is widespread. New mineral growth comprises: green and blue-green hornblende, cummingtonite, gedrite, low-Ti biotite, fibrolite, muscovite and ferruginous garnet ($N_{Mg^{ga}}$ = 15.8, 16.0 to 5.7 and 4.1). In the Eastern Aldan metamorphic belt (the Eastern Aldan domain, Fig. II-10), metamorphism was at medium- to low-gradient granulite facies conditions (Table II-5).

The following features are noteworthy: the absence of Hyp + Cor and Bi + Cor + Ga parageneses; wide distribution of Bi + Sill ± Ga, Ga + Opx + Cpx and Ga + Cpx assemblages; in the assemblage Sill + Cor + Ga, cordierite often shows signs of late-stage growth. In the northern part of the belt: TiO_2 in biotite is up to 5.0–5.43 wt%; the maximum $N_{Mg^{ga}}$ in parageneses with Cor and/or Sill is 39.4–36.9; Ga–Cor equilibrium parameters are 830°C and 8.7 kbar, obtained for $N_{Mg^{ga}}$ = 36.9; the temperature from two-pyroxene equilibria is 850–930°C, including 910–930°C in garnetiferous schists. Ga, Cor and Bi compositions often reflect retrograde metamorphic conditions (Cor–Q reaction textures), the parameters of which are 750°C and 7.3 kbar; 660°C and 6.1 kbar; 650°C and 5.7 kbar from Ga–Cor equilibria in sillimanite gneisses (Dook et al., 1986). Ga–Bi equilibrium temperatures, including in those samples with highest maxima for Ga–Cor equilibria values, reflect retrograde conditions in the metamorphic evolution. In mafic schists, $N_{Mg^{ga}}$ = 18.7–19.4; for $N_{Mg^{cpx}}$ = 46.5; $N_{Mg^{opx}}$ = 35.4–40.6; $N_{Mg^{amph}}$ = 40.7. Hornblende is found with a TiO_2 content of up to 2.34 and Na_2O up to 2.50 wt%.

In rocks from the Idzhek allochthon, mineral assemblages are similar to those above. In assemblages containing Cord and/or Sill, $N_{Mg^{ga}}$ reaches 34.5–35.1; the maximum parameters for Ga–Cord equilibria are 830°C and 8.7 kbar. Two-pyroxene equilibrium temperatures are 880°C, obtained in garnet-bearing assemblages. The TiO_2 content in hornblende is up to 2.2 wt%, $N_{Mg^{ga}}$ is up to 21.7 for $N_{Mg^{cpx}}$ = 61.0, $N_{Mg^{opx}}$ = 56.6 and $N_{Mg^{amph}}$ = 51.4. The assemblage Bi–Cord–Ga is found in the allochthon, and is probably a reflection of the metamorphic evolution of rocks which occurred subsequent to their emplacement in the Idzhek-Sutam province and metamorphic zone.

The low-gradient character of this metamorphism, as outlined above, is preserved over the entire outcrop area of the Eastern Aldan belt (the Eastern metamorphic zone; Kitsul and Shkodzinsky, 1972). However, in the centre of the belt, $N_{Mg^{ga}}$ increases to 42.6–39.2 in assemblages with Cord and Sill. The TiO_2 content in biotite is up to 6.0 wt%, but 3.38 wt% in amphibole. In the southernmost part of the belt, close to the Stanovoy suture, $N_{Mg^{ga}}$ = 47.0 for $N_{Mg^{cpx}}$ 79.7 in a paragenesis with Cord and Sill. In mafic schists, $N_{Mg^{ga}}$ 41.9–53.1 for $N_{Mg^{cpx}}$ = 69.4–73.5; $N_{Mg^{opx}}$ = 62.7–68.4; $N_{Mg^{amph}}$ = 55.2, but here we also encounter the paragenesis Ga + Cord + Hyp + Bi + Plag + Kfs + Q.

At the end of the metamorphic event in the Eastern metamorphic zone, there was a transition to shallower and/or higher-gradient granulite facies conditions.

In the Idzhek-Sutam province, peak metamorphic conditions (Fig. II-10) in the northern and central portions (the Seym block) corresponded to boundary conditions between the biotite-cordierite-garnet-orthoclase and hypersthene-cordierite-garnet-orthoclase subfacies of the granulite facies. Compared with other rocks in analogous metamorphic regimes in the North-Eastern subdomain of the Central Aldan domain, metamorphism occurred at deeper-level and higher-temperature conditions (see Table II-5 for parageneses). The maximum $N_{Mg^{ga}}$ value in assemblages containing Hyp + Cord, Sill + Cord and Cord = 43.2, 41.0 and 39.2 respectively. The TiO_2 content in biotite is 5.6–6.93 wt% and *f* is up to 1.95 wt%. In hypersthenes found in assemblages with cordierite, the Al_2O_3 content is 7–8.5 wt%.

Characteristic features are the absence of garnet in mafic schists, instability of the Bi + Sill paragenesis (sillimanite is usually armoured in cordierite), and wide distribution of a stable Hyp + Cord paragenesis. Maximum *P–T* values for Ga–Cord equilibria in sillimanite-bearing gneisses from the north of the block are 840–880°C and 8.7–9 kbar, obtained for $N_{Mg^{ga}}$ = 36.5 and 33.6. The close agreement in Mg contents for Ga and Cord obtained by chemical analyses and by microprobe analyses (Anon., 1983) and the low variation in Mg content in one and the same specimen testify to their mainly homogeneous composition.

The rocks along the western and eastern boundaries and in the north of the Seym block experienced intense shearing, with the formation of pencil schistosity or rodding (Dook et al., 1979; GSA, 1983). The same mineral assemblages are seen in both schistose rocks and unaffected varieties. Sillimanite gneisses contain poorly expressed Cord + Q symplectites, while Hyp–Ca–Cord gneisses have Hyp + Cord symplectites. In gneisses with Sill and Cord, $N_{Mg^{ga}}$ drops to 32.0 and 26.5. PT parameters for Ga–Cord equilibria are 760–800°C and 8.1–7.1 kbar, obtained for $N_{Mg^{ga}}$ = 32.7; 26.5. In the Timpton thrust zone, rodded gneisses show amphibolite facies retrogression (Dook et al., 1975, 1979, Mitrofanov and Glebovitsky, 1985). Well-developed Cord + Q symplectites are formed around garnet and spinel, cross-hatched Kfs appears, and there is secondary andalusite. Andalusite is also found in rocks belonging to the Idzhek allochthon which has been emplaced here. $N_{Mg^{ga}}$ decreases to 21.5; 18.2. Bi and Amph compositions correspond to amphibolite facies (Dook et al., 1975). Ga–Cord equilibrium PT parameters are 700°C and 5.7 kbar (Dook et al., 1986).

The Sutam block is situated in the southern part of the Idzhek-Sutam domain. It is separated from the Seym block by the Atugey-Nuyam graben, filled with upper Precambrian and Mesozoic non-metamorphic sediments. Some of the highest temperature and deepest crustal rocks are exposed in the Sutam block — hypersthene-sillimanite facies (the Sutam depth facies, Marakushev, 1965), with the Hyp + Sill assemblage first discovered and described by Kadensky (1960). The block is bounded to the west by the Davangra-Khugda graben, consisting of non-metamorphic late Precambrian and Mesozoic rocks, to the east by the Idzhek-Nuyam fault and to the south by the Stanovoy suture.

Mineral assemblages belonging to three subfacies of the granulite facies have been identified in the Sutam block (Fig. II-10): hypersthene-sillimanite, hypersthene-cordierite-garnet and biotite-sillimanite-garnet-orthoclase, which is characteristic for PT metamorphic conditions close to the Hyp + Sill + Bi + Ga + Plag + Kfs + Q invariant assemblage point (Marakushev and Kudryavtsev, 1965). There is a voluminous literature on the metamorphism of the Sutam block: Kadensky (1960), Marakushev (1965), Kastrykina (1976), Korikovsky and Kislyakova (1975), Kitsul and Shkodzinsky (1972), Dook et al. (1986), Anon. (1983). A noteworthy feature is the absence of the assemblage Bi + Cord + Ga + Plag + Kfs + Q among prograde assemblages, related to the transition between this assemblage and a Hyp + Sill assemblage for a garnet Mg-content of 50, the maximum attained for quartz-bearing rocks. In Hyp + Sill parageneses, relict $N_{Mg^{ga}} > 50$ (limiting values = 58–60). These garnets are found in equilibrium with hypersthenes containing 7.1–10.4 wt% Al_2O_3, $N_{Mg^{hyp}}$ 73–80 and high-Mg, high-f biotites ($N_{Mg^{bi}}$ = 80–84, $f \geq 2.4$ wt%).

Cordierite-free Hyp + Sill + Ga gneisses are found only as relicts. Between Hyp and Sill in cordierite-bearing rocks we find cordierite rims in quartz-bearing gneisses and cordierite-sapphirine, sometimes cordierite-spinel rims in quartz-free assemblages. Cord + Hyp symplectitic borders are present around garnet, as well as rare hypersthene rims around biotite. In such cases, $N_{Mg^{ga}}$ decreases from the centre to the edge of a grain and becomes comparable with $N_{Mg^{ga}}$ in Hyp + Cord + Ga gneisses. Hypersthene in Cord + Hyp rims is more ferruginous than that in the matrix (Korikovsky and Kislyakova, 1975). The association Hyp + Sill + Ga transforms into the association Hyp + Cord + more ferruginous Ga, typomorphic for the Hyp–Cord–Ga–Orth subfacies.

Hyp–Cord–Ga–Orth subfacies assemblages are found in the same outcrops as Hyp–Sill subfacies assemblages. The limiting $N_{Mg^{ga}}$ value is 49; $N_{Mg^{hyp}}$ = 62–68, $N_{Mg^{bi}}$ = 66–70 and $N_{Mg^{cord}}$ = 80–81. Cord + Q symplectites occur around garnet grains in gneisses and there are plagioclase reaction rims between Ga and Bi and textures indicating biotite replaced by hypersthene. In grain centres $N_{Mg^{ga}}$ = 49, and 42–36 at the edges. High $N_{Mg^{ga}}$ in reaction Hyp, Bi and Cord and low $N_{Mg^{ga}}$ compared to $N_{Mg^{ga}}$ of these same minerals in gneisses with mineral equilibrium ratios, a lower alumina content in newly-formed hypersthenes and lower titanium content in biotites prove that these mineral transformations occurred during a retrograde episode in the metamorphic evolution, with a decrease in temperature (redistribution of Mg from garnet and Fe from cordierite to the contact mineral). Sill + Cord + Ga ± Bi gneisses are spatially associated with the first two rock groups. In varieties without reaction textures, $N_{Mg^{ga}}$ = 46–43; $N_{Mg^{cod}}$ = 79–80; $N_{Mg^{bi}}$ = 67–69. In varieties with reaction textures (cordierite-quartz symplectites, cordierite rims around sillimanite), $N_{Mg^{ga}}$ is 46 at grain centres and 39–20 at the edges. Cordierites are also chemically inhomogeneous. The occurrence of early and late mineral assemblages in the rock groups examined results in a wide scatter of PT values for Ga–Cord equilibria. The highest values for each group, in the order in which they were described, are: 920°C, 10.8 kbar; 947°C, 10.6 kbar; 850°C, 9.6 kbar. Lower estimates have been found for all groups, down to 590°C and 6.4 kbar. In Ga + Sill ± Bi + Plag + Kfs + Q parageneses, $N_{Mg^{ga}}$ is up to 47.6 and $N_{Mg^{bi}}$ is up to 75.6. The assemblage Cord + Ga ± Bi + Plag + Kfs + Q formed only during the retrograde stage of metamorphic evolution.

Two-pyroxene equilibrium temperatures in plagioclase-bearing schists are 730–970°C (Wood and Banno, 1973) and 780–940°C (Perchuk, 1977). The wide scatter of temperature estimates reflects the metamorphic evolution of rocks in the Sutam block, expressed in mafic schists in the replacement of garnets by Hyp + Plag symplectites, Cpx by Hyp or Amph, in the emergence of reverse zoning in plagioclase, in the formation of Ga rims at the contact between Plag and Opx and Cpx (Kitsul et al., 1973). This is the only place in the Aldan granulite-gneiss terrain where we encounter the assemblage Cal + For + Di + Ga + Sp (Kitsul and Shkodzinsky, 1976). In mafic schists the maximum $N_{Mg^{ga}} = 44.1$; $N_{Mg^{cpx}} = 77.4$; $N_{Mg^{opx}} = 69.0$; $N_{Mg^{amph}} = 58.01$. Hornblende contains up to 3.25 wt% TiO_2 and up to 2.50 wt% Na_2O.

The presence of highly-magnesian Ga, Hyp, Cord and Bi, with Mg at its upper limits, and highly-aluminous Hyp, high-Ti and high-F Bi, assemblages containing Hyp with Sill and Hyp with Cord in quartz-bearing rocks, the wide distribution of sapphirine-bearing varieties of these in quartz-free rocks, together with the assemblage Ga with Cpx and Amph in mafic schists all indicate that high temperature–high pressure granulite facies conditions were attained during a prograde metamorphic episode. Considering that the subsequent metamorphic evolution lead to a change in phase relationships and mineral compositions, we must take the parameters for the peak metamorphic conditions as being not less than $T \geq 900$°C and $P \geq 9$ kbar.

Two trends emerge in the metamorphic evolution of the Sutam block: low-gradient for the evolution of Hyp + Sill gneisses in the south, characterised by embryonic reaction and symplectite textures, and high-gradient for the evolution of Ga + Cord + Sill gneisses in the central and northern parts of the block, characterised by the intense development of reaction and symplectite textures in Hyp + Sill gneisses (Dook et al., 1986). Kyanite is a possible mineral in assemblages belonging to the low-temperature part of the low-gradient trend (Perchuk, 1983). In the Seym block, P–T parameters for Ga–Cord equilibria which arose during the peak metamorphic episode and during the episode when rodded gneisses formed, and during the amphibolite facies retrogressive metamorphic event, form a single evolutionary trend, which falls within a higher-temperature range relative to the trend in the northern part of the Sutam block (Dook et al., 1986).

Typically, individual metamorphic regimes correspond to late tectonic domains (AR_1^2). In the Central Aldan cratonic zone (AR_1^1) which was active during this period, rocks were metamorphosed in regimes of AR_1^2 structures which bound it in the NE in the Idzhek-Sutam domain, but to the SSW in the regime of the Western Aldan domain. During AR_1^2 times, the AR_1^1 mobile zone continued to be active, evidence for this being the autonomous low-gradient metamorphic regime in the Southern Aldan domain. It is worth noting that in the Idzhek-Sutam and Eastern Aldan domains which differ in metamorphic regime, lower-gradient and deeper-level metamorphic conditions are found in their southern portions, but each zone retains its own individual features. This fact points to the presence during AR_1^2 or AR_2 time of a zone of maximum downward movement at the edge of the Chara-Aldan lithospheric plate (in the region of the Stanovoy suture) in which AR_1^2 provinces were involved, with their own distinctive metamorphic conditions.

Deformation at the end of the Early Archaean

In the concluding stage of Early Archaean evolution, after the attainment of peak pressure conditions of prograde metamorphism and the formation of migmatites, deep-level thrusts formed which are clearly marked on textural and structural grounds and particular metamorphic features. The major thrusts have been mapped at the boundaries of tectonic domains showing different metamorphic regimes; these are the Timpton and Aldan thrusts (Fig. II-10). Within the thrust zones, where nappe sheets have been translated from east to west, the rocks are intensely sheared under retrograde granulite facies conditions (cf. schistose rocks in the Timpton thrust zone). The imposition of schistosity led to the obliteration of earlier folds (IIF_{1-3} folds, Dook et al., 1979; AGU, 1983) and in the Aldan thrust zone this was accompanied by the accumulation of conglomeratic rocks, known as the Upper Aldan conglomerate formation (Kulish, 1973). At the boundaries between the Eastern, Central and Western Aldan tectonic domains and within the Western Aldan tectonic domain (the Yarogu thrust), metamorphic surfaces were distorted (Fig. II-10), increasingly in the western part of the domain during the formation of Early Proterozoic thrusts with a similar displacement vector. Rocks in the Western Aldan domain were thrust over the Olyokma block (the infracomplex of the Olyokma granite-greenstone terrain). Superimposed metamorphism in rocks of the allochthon according to Smelov correspond to the peak metamorphic conditions in the Olyokma infracomplex (Smelov, 1986); while cycle III granites (AR_2) in the western part of the granulite-gneiss terrain which occur amongst granulite facies rocks, carry intrusive tonalite xenoliths metamorphosed at amphibolite facies and typical of the Olyokma granite-greenstone terrain which lies 20 km to the west.

The major fold structures in the Aldan Supergroup, preserved in a least altered state only in the centre of the Aldan granulite-gneiss terrain, formed at the end of the episode. At the boundary between the Eastern Aldan and Western Aldan blocks, a system of linear folds developed, encompassing the entire Idzhek-Sutam domain — the Gonam-Sutam anticlinorium (Dook et al., 1975, 1979; AGU, 1983) and the adjacent part of the Eastern Aldan province — the Sutam synclinorium. Major oval structures (100–150 km in diameter) are characteristic features in the interior regions of the Eastern Aldan and Central Aldan domains. These structures in the centre of the Aldan Shield have been known since the 1950s, after the work of Minkin, who referred to them as the Lower Timpton domes. The cupolas developed separately and independently as local gravity uprises, prior to the formation of several isoclinal fold systems and after the inception of the Timpton thrust, more or less simultaneous with but slightly later than the formation of the Gonam-Sutam anticlinorium (Dook et al., 1979; AGU, 1983). The subsequent intervening period when the Aldanides were consolidated took place during the final cycle II folding — vertical flexures and accompanying shear zones at retrograde granulite facies conditions.

Recently, metabasic dykes have been found in the north of the Idzhek-Sutam province, cutting $\text{II}F_{1-3}$ folds and the leucosome of $\text{II}M_{1-3}$ migmatites (A.B. Kotov, pers. commun.). The dykes were sheared and recrystallized during the time of formation of the Timpton thrust and major $\text{II}F_4$ folding (retrograde granulite facies pencil-rodded gneisses). During this period, mobilization and/or crystallization of granitoid melts are atypical. Melts had probably already originated by the time the thrusts formed. We may conclude from this that the thrusts, rodded gneisses, major Aldanide folds ($\text{II}F_4$) and late cycle II flexures all represent the deep crustal expression of an orogenic stage of evolution.

Isotopic dating of the Early Archaean complexes in the Aldan granulite-gneiss terrain has been carried out using zircons from hypersthene tonalite-trondhjemite gneisses in the infracrustal complex of the Central Aldan province and from mafic schists in the Olyokma-Kurultin-Zverev (Southern Aldan) domain.

In the Central Aldan domain on the river Aldan, hypersthene tonalite-trondhjemite gneisses of the supposed basement to the Kurumkan Formation have been found (sp. Dr-0315, Fig. II-17), for which the U–Pb age of oldest generation of zircons falls in the interval 3.4–3.7 Ga (Mitrofanov and Moskovchenko, 1985). The dated zircons are milky white, prismatic, of hyacinth habit with finely zoned cores. It is uncertain whether the age date corresponds to the time of metamorphism or to the age of the initial tonalites. The Pb–Pb age (by the evaporation technique) of second generation zircons extracted from the same samples (yellow, sub-idiomorphic and

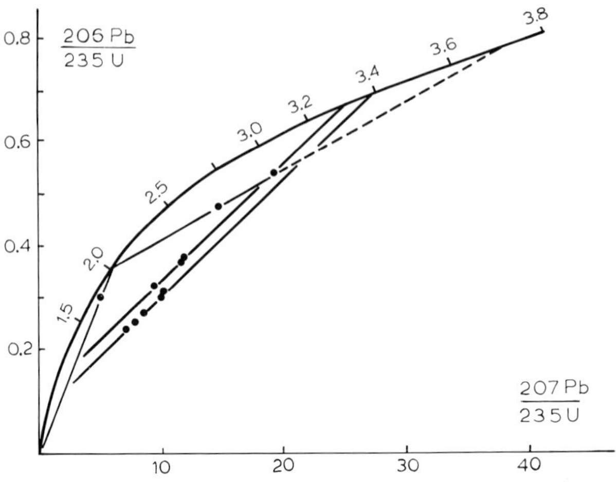

Fig. II-22. $^{206}Pb/^{235}U$–$^{207}Pb/^{235}U$ concordia diagram for plagiogneisses from rocks of the infracrustal complex of the supposed basement to Archaean supracrustal rocks.

rounded crystals and aggregates forming overgrowths on first generation zircons) is 3.0 Ga. During the lower Proterozoic, the outcrop area of the supposed basement was subjected to intense fold movements at granulite facies. The U–Pb age of third generation zircons (pinkish-lilac, transparent, rounded, with high RI) from the same specimens is 2.0 Ga, which corresponds to the age of the lower intersection of the U–Pb dyscoria with the concordia plot (Fig. II-22). Zircon ages of up to 3.0 Ga have been obtained here also using the Pb–Pb method (the evaporation technique) on zircons from mafic schists (inclusions in plagiogneisses). Zircons from quartzites (the Kurmukan Group) have an age of 2.53.0 Ga (Pb–Pb method, evaporation technique).

The minimum age for zircons (pink, prismatic, hyacinth habit with zoned prismatic cores) from granulite-facies metamorphosed tholeiites in the Olomokita and Kurultin blocks and the Olyokma-Kurultin-Zverev province is 3.3 Ga (Anon., 1986). This is either the lower age limit for the crystallization of the tholeiites or the time of granulite facies metamorphism.

Late Archaean evolution

The rocks in the Southern Aldan (Olyokma-Kurultin-Zverev) and Western Aldan domains and the southern areas of the Central Aldan, Idzhek-Sutam and Eastern Aldan domains were involved in an episode of active reworking at the boundary between the Early and Late Archaean. The northern parts of these zones formed the relatively stable Aldan massif which during this period appeared to undergo predominantly vertical uplift movements of up to 10 km in amplitude, from geobarometry studies (Glebovitsky and Sedova, 1986). The Olyokma-Stanovoy active transition zone was initiated at the boundary between the Aldan massif and the Dzhugdzhur-Stanovoy domain (Fig. II-23) and was separated from the Dzhugdzhur-Stanovoy domain by the Stanovoy suture. As a result of repeated movements on the Stanovoy suture, Late Archaean faults are preserved fragmentarily (the upper reaches of the river Aldan) in places where Late Archaean tectonic boundaries, deformed by NW folds, were deflected from later boundaries by the almost exactly east–west-trending structures of the suture (Figs. II-9, II-23). The suture separates the Stanovoy geoblock, which appeared during the Late Archaean stage of evolution of the region during low-gradient, high-pressure metamorphism, from the Olyokma-Aldan domain which at this time was characterised by higher-gradient metamorphism.

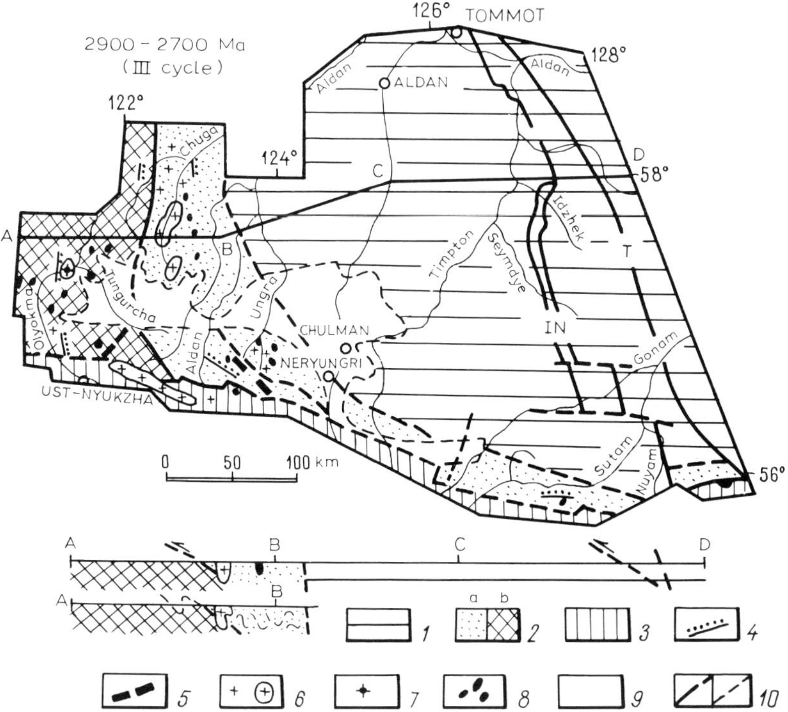

Fig. II-23. Sketch map of tectonic divisions in the Olyokma-Aldan region during evolutionary cycle III. *1* = stable provinces (Aldan massif); *2* = active provinces: *a* = Olyokma-Stanovoy, *b* = Olyokma subprovinces; *3* = Dzhugdzhur-Stanovoy mobile belt; *4* = axial surfaces of cycle III folds; *5* = zones of ultrabasic bodies; *6* = granites and granosyenites; *7* = Ust-Oldonga granodiorite and tonalite intrusion; *8* = basic and ultrabasic bodies; *9* = Mesozoic sediments; *10a* = tectonic boundaries, *10b* = boundary of Mesozoic sediments. Letters: *IN* = Idzhek-Nuyam, *T* = Tyrkanda fault zones.

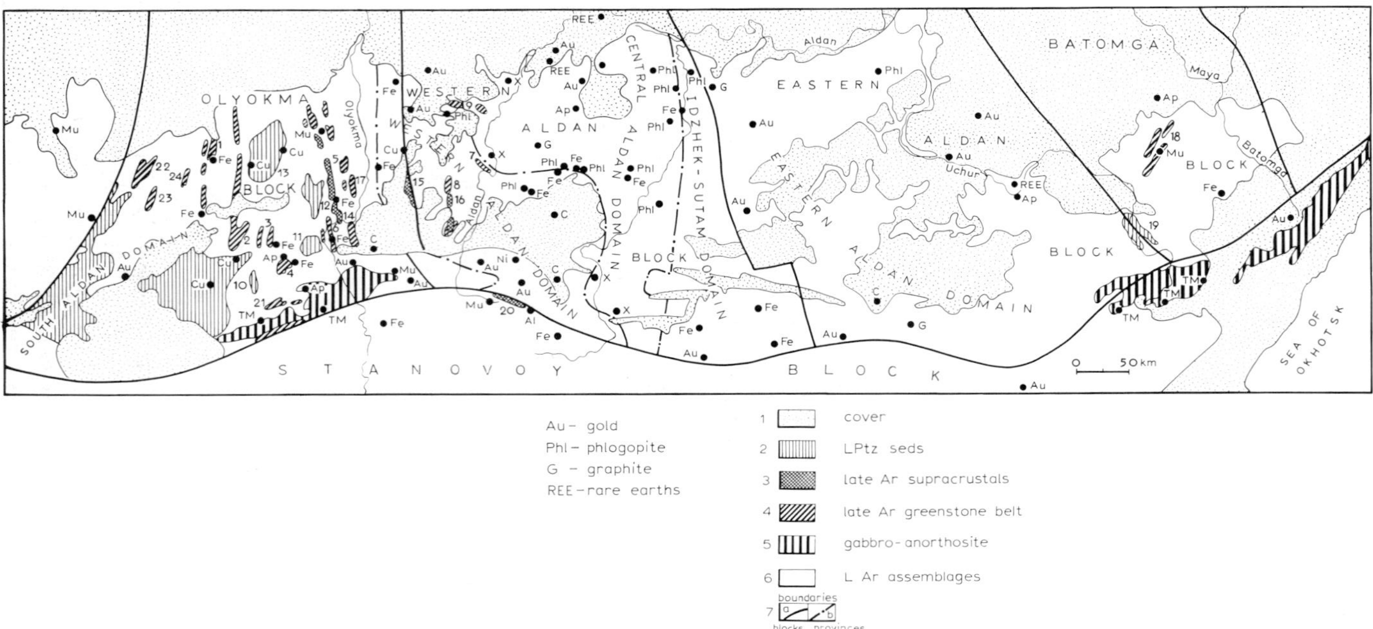

Fig. II-24. Sketch map of the Olyokma-Aldan region showing position of useful minerals, ore shows and supracrustal complexes of Late Archaean–Early Proterozoic age. *1* = platform cover and Quaternary sediments; *2* = Lower Proterozoic terrigenous subcratonic sediments; *3* = Late Archaean (Early Proterozoic) supracrustal assemblages, Tasmiyeli level; *4* = Late Archaean greenstone belts; *5* = gabbro-anorthosite intrusions; *6* = Early Archaean assemblages on Aldan Shield; *7* = terrain (*a*) and province (*b*) boundaries; *8* = mineral deposits and ore shows: *Au* = gold, *Phl* = phlogopite, *G* = graphite, *REE* = rare earth elements, *Fe* = iron, *TM* = titanomagnetite, *C* = coal, *Mu* = muscovite, *Ap* = apatite, *X* = rock crystal, *Cu* = copper, *Ni* = nickel, *A* = corundum, sillimanite. Supracrustal belts (numbered): *1* = Tarynakh, *2* = Olondo, *3* = Itchilyak, *4* = Byrylakh, *5* = Tasmiyeli, *6* = Syrylyr, *7* = Sagara, *8* = Bulgunyakhtakh, *9* = Balaganakh and Sogolokh, *10* = Udokan, *11* = Lower Khanin, *12* = Oldongso, *13* = Uguy, *14* = Temulyakit, *15* = Subgan, *16* = Yarogu, *17* = Tungurcha, *18* = Chumikan, *19* = Ulkan, *20* = Chulman, *21* = Kalar, *22* = Saymagan, *23* = Torochan, *24* = Chara.

The Olyokma-Aldan domain may be subdivided into two subdomains which were tectonically active: the Olyokma subdomain (Olyokma block) and the Fore-Olyokma subdomain (the western part of the Western Aldan block) (Fig. II-23). To the north of the Chulman Mesozoic basin, the boundary between the subdomains is represented by a system of thrusts of Archaean and Early Proterozoic age, by which the Aldan granulite-gneiss terrain was translated over the Olyokma granite-greenstone terrain. To the south of the Mesozoic basin, the boundary can be provisionally traced by means of a north–west-striking upper Archaean thrust which splays off from the Stanovoy suture, separating the Kurultin and Zverev blocks of the Southern Aldan domain. Here, as in the north, the eastern (Zverev) block is thrust over the western (Kurultin).

In the allochthonous nappe front, deep-level upper Archaean metamorphic rocks are obliterated; they are represented in Zverev Group mafic schists by reactions of repeated eclogitization.

The further evolution of active tectonic processes in the Dzhugdzhur-Stanovoy mobile region subsequently led to the northwards overthrusting of the Kurultin and Zverev blocks (Dook et al., 1986). In the Zverev and Sutam blocks, Late Archaean metamorphism was accomplished at granulite facies conditions, in the Kurultin block at granulite-amphibolite facies boundary conditions and in the remainder of the Fore-Olyokma subdomain at high-temperature, high-gradient amphibolite facies conditions.

A single arcuate linear fold system in the Olyokma-Fore-Stanovoy active zone is oriented subparallel to the boundary of the Aldan massif, with a gradual swing in the strike of fold structures from east–west to near north–south (Figs. II-9, II-23). Tight folds give way to more open forms going towards the direction of the Aldan massif. This is especially clearly established in the east–west part of the structure (Dook, 1977; Dook et al., 1979; AGU, 1983; Balagansky, 1979), oriented parallel to the Dzhugdzhur-Stanovoy mobile region. Rocks at various metamorphic grades and upper Archaean deep-level nappes are deformed into a single linear fold system.

A characteristic feature seen in the western part of the Western Aldan block is the presence of upper Archaean greenstone belts, analogous to belts in the eastern part of the Olyokma block. The Bulgunyakhtakh and the Balaganakh greenstone belts have been mapped out (these are the Balaganakh, Sagar and Sogolokh fragments, *7*, *8*, *9* on Fig. II-24).

In the best-preserved Bulgunyakhtakh belt, the succession consists of the following members: (1) garnet- and sillimanite-bearing schist alternating with biotite schist; in some cases sillimanite-bearing varieties predominate, in others, biotite; (2) pure silica rocks with isolated thin bands of sillimanite schist; (3) ore-free and ferruginous silica rocks alternating with hornblende schist. Thin calc-silicate and diopsidic calciphyre horizons are present. The hornblende schists are compositionally similar to tholeiites, slightly more alkaline than the tholeiites in the greenstone belts belonging to the Olyokma granite-greenstone terrain. Garnet-biotite (± sillimanite) and microcline gneisses correspond to a K–Na group of similar rocks in the greenstone belt in the east of the Olyokma block and typically have high MgO, CaO and FeO contents. The belts are located in "soft" fault zones, >130 km long by 5–10 km wide (*5*, *6* on Fig. II-9). The juxtaposition of greenstone belt metasedimentary assemblages with Aldan Supergroup granulites took place during the formation of Early Proterozoic deep-level thrusts in the Ungra crust-forming cycle (IV^2). It was during this particular episode that greenstone belt assemblages in the Western Aldan block reached maximum *P–T* metamorphic conditions and the major observable fold groups in them developed (Dook et al., 1986).

Asymmetric metamorphic zoning is observed in the major fragments of the Bulgunyakhtakh belt (the Reutov zone). Metapelites in the western parts of the fragments contain armoured relict staurolite in plagioclase, muscovite and cordierite. The amount of sillimanite-bearing associations increases in the eastern part of exposures, and γ_4^2 injection migmatites make an appearance. The commonest mineral assemblage is Ga + Bi ± Musc + Plag + Q; other less common ones are: Ga + Bi + Musc + Sill + Plag + Q [Staur], Ga + Bi + Cord + Plag + Q [Staur], Ga + Sill + Cord + Bi + Q, Bi + Musc + Sill + Micr + Plag + Q. $N_{Mg^{ga}}$ = 9–12, $N_{Mg^{bi}}$ = 30–39. In hornblende schists: Hb + Bi + Plag + Q, rarely Ga + Hb + Cumm + Plag + Q. The parageneses indicate metamorphism under biotite-muscovite gneiss subfacies conditions. *P–T* parameters for Ga–Bi equilibria in the presence of muscovite or sillimanite are 640–670°C and 4–6 kbar; using Ga–Cord equilibrium, 670°C and 4.1 kbar.

In a zone lying to the east of the Aldan-Kiliyera zone (*6* on Fig. II-9) — the Sagar, Balaganakh and Sogolokh fragments — metamorphism occurred under conditions transitional between biotite-sillimanite-orthoclase and garnet-cordierite-orthoclase subfacies of the amphibolite facies. Mineral assemblages in metapelites are: Ga + Bi + Sill + Plag + Kfs + Q, Bi + Cord + Plag ± Kfs + Q, Ga + Bi + Cord + Plag + Q, Ga + Bi + Cord + Sill + Plag + Kfs + Q. $N_{Mg^{ga}}$ = 12–15, $N_{Mg^{bi}}$ = 37–42, $N_{Mg^{cord}}$ = 49. In hornblende schists: Hb + Plag + Q, rarely Ga + Hb + Bi + Plag + Q. In the easternmost part of the Sogolokh fragment, *P–T* parameters reached 650°C and 4.6 kbar from Ga–Bi equilibria data, and 680–700°C, 4.6–5.6 kbar from Ga–Cord equilibria.

The Late Archaean stage of evolution was characterised by intrusive magmatic activity. In the Stanovoy suture zone, extensive swarms of basic and ultrabasic dykes were emplaced *en échelon* (ν_3, σ_3, Fig. II-9). In the west of the suture (the Zverev block) and along a north–west-striking splay fault separating the Olyokma subzone, they form dextral *en échelon* swarms, oriented oblique to the strike direction of tectonic boundaries. In the Sutam block they form a dextral *en échelon* swarm, striking from near-E–W to NE. The dykes cut second-cycle folds and migmatitic banding and are themselves deformed by third-cycle folds; they are altered to amphibolite (in the upper reaches of the river Tungurcha) and granulite facies (Zverev and Sutam blocks) ultramafic and mafic schists. The Stanovoy suture can be traced from anorthosite and mangerite bodies. One of the largest bodies is the Kalara intrusion (*A* on Fig. II-9).

Intrusive igneous complexes formed in the tectonically active part of the Western Aldan block. These are differentiated intrusions of subalkaline gabbro and gabbro-diorite, sometimes including bodies of ultrabasic rock. In the northern part of the Zverev block, they were metamorphosed under conditions transitional between amphibolite and granulite facies and on this basis they are distinguishable from surrounding high-temperature granulites. Associated with these bodies are subalkaline gneissose granite, syenogranite and aegirine-bearing alkaline syenite. Gabbro and gabbro-diorite bodies are known in the Bulgunyakhtakh greenstone belt. There are numerous bodies of gneissose Bi + Hb + Cpx normal and subalkaline series granite (γ_3) in the west of the Aldan granulite-gneiss terrain (in the river Chuga basin; Fig. II-9). All the intrusive rock suites referred to here were involved in the Late Archaean deformation cycle (III). The Kabaktan massif formed in the Stanovoy suture zone (*C* on Fig. II-9) — gneissose syenogranite and subalkaline granite, 70 × 7 km in size. The massif contains gneissose diorite and riebeckite- and aegirine-bearing gneissose rocks.

In the north of the Zverev block the best known massif is the Tenevoy — gabbro-amphibolite containing bodies of ultrabasic rock, calciphyre and calc-silicate rock (*D* on Fig. II-9). Compositionally, the rocks of this massif are pyroxenite, peridotite, subalkaline gabbro, gabbro-diorite, diorite and rare plagioclasite. These rocks were previously included in the Burpali massif and were considered by Gabyshev and Gabyshev to be its lower basic portion. Compositions of upper Archaean intrusive rocks are given in Dook et al. (1986).

Early Proterozoic evolution (cycles IV and V)

During the Early Proterozoic, a zone of intense tectonomagmatic reworking formed in the west of the Aldan granulite-gneiss terrain (Fig. II-25), referred to as the Reutov, Aldan-Kiliyera and Ungra-Dyos-Melemken zones. The initiation of this activity was marked by the intrusion of differentiated gabbro-diorite-tonalite ($\gamma\delta_4$) and minor ultrabasic bodies (σ_4).

The most intense magmatism took place in the Ungra-Dyos-Melemken zone — the Ungra complex (*F* on Fig. II-9). The zone is over 200 km long and up to 40 km wide in its middle portion. The largest bodies, which measure up to 30 × 20 km, are concentrated in an area where the zone takes a knee-shaped flexure. The opening of cavities is in agreement with dextral thrust displacements along a suture zone. Intrusive bodies are found within the tonalite gneiss and granite-gneiss infracrustal complex

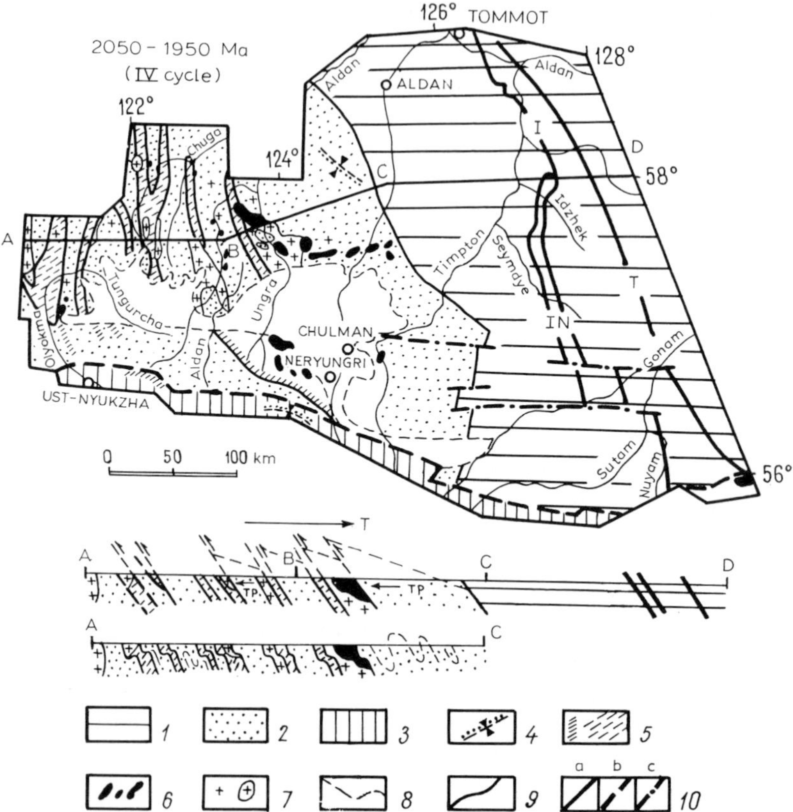

Fig. II-25. Sketch map of tectonic divisions in the Olyokma-Aldan region during evolutionary cycle IV (after Dook et al., 1986). *1* = stable region; *2* = active region; *3* = Stanovoy zone of Dhugdzhur-Stanovoy mobile belt; *4* = axial surfaces of cycle IV folds; *5* = zones of intense shearing and blastomylonitization; *6* = gabbro-diorites and diorites; *7* = granites; *8* = boundary of Mesozoic sediments; *9* = boundaries of shear zones and active zones; *10a* = fault zones, *10b* = edge of Dzhugdzhur-Stanovoy mobile belt, *10c* = late faults.

and in rocks belonging to the Kurumkan and Fyodorov Formations. In spite of deformation and high-temperature, high-gradient amphibolite facies metamorphism, which altered the intrusive rocks to hornblende and biotite-hornblende (± diopside) schist, dioritic gneiss and tonalite gneiss, it is not uncommon to find relict igneous textures — gabbroic and blasto-ophitic and minerals — lath-like plagioclase and very rare rhombic and monoclinic pyroxene. The temperature of metamorphism increases to the east (brown high-titanium hornblende appears in the metamorphic parageneses), while the degree of schistosity decreases. Tonalite makes up a significant part of the complex, and plagioclasite is also found. Rock compositions are presented in Dook et al. (1985).

In rocks belonging to the Ungra complex, the foliation is deformed by relatively tight folds with subvertical axial surfaces, which here are the major IV-cycle mapped folds and the first folds which deform the schistosity in intrusive rocks. Folds belonging to cycles II and III are absent from Ungra complex rocks but are seen in the country rocks; also missing are the early folds which are found in assemblages belonging to the Bulgunyakhtakh and Balaganakh greenstone belts (Dook et al., 195). Cycle IV^2 folds are superimposed on the western and north-western parts of the Lower Timpton Archaean dome structure. In the north-west of the dome, IV^2 folds were not accompanied by any clear expression of retrogression in Aldan Supergroup rocks. On the contrary, there is in fact evidence of polyphase granulite facies metamorphism (Glebovitsky and Sedova, 1986). Between the metamorphic peaks of these two events, there was a decompression of around 3 kbar due to uplift of the block. The new metamorphic event was accompanied by burial of the area by at least 10 km.

The time of emplacement of the Ungra complex is taken as a reference event in the Early Proterozoic epoch of tectonomagmatic activity. The U–Pb age of zircons from a gabbro with well-preserved igneous plagioclase is 2040 $\pm$ 20 Ma. Metamorphic changes are close in time (20 Ma) to the origin of the rocks in the complex and occur within the limits of analytical accuracy.

The Ungra complex is similar to the Khonchengra massif in the north of the Zverev block (*E* on Fig. II-9) in terms of chemistry, metamorphic features and relationship to deformation. It stretches out in a near east–west direction, concordant with the strike of cycle III and IV structures. The massif consists of gabbro-diorite, diorite and quartz diorite.

Minor ultramafic bodies, metamorphosed at amphibolite facies conditions, are found in the north of the Reutov zone and belong to this time interval.

The concluding stage of Early Proterozoic activity (cycle IV^2) is marked by the formation of numerous deep-level thrusts, with east to west nappe transport (Fig. II-25). Rocks of the Aldan Supergroup and cycle IV^2 intrusive rocks were affected by blastomylonitization and shearing in fault zones under the same facies conditions as the prograde metamorphism of assemblages in the Bulgunyakhtakh and Balaganakh greenstone belts: T = 650–700°C and P = 4.5 kbar (Smelov, 1986). Granitoid magmatism appeared simultaneously with the metamorphic peak, mainly in suture zones (γ_4, Fig. II-9). Medium- to finely-porphyritic granites γ_4 cut Ungra complex rocks. Foliation is only weakly expressed in the granites and is oriented parallel to the axial surfaces of tight folds which deform their contacts with the host rocks. The granites contain relict primary igneous granitic textures. Compositionally they are normal and subalkaline granites, with K_2O predominating over Na_2O. Intense IV^2 cycle folding, producing tight folds, occurs sporadically in the Olyokma-Aldan region and is mainly restricted to suture zones which developed as shear zones after the formation of thrusts. The Ungra complex differentiated intrusions suffered shearing in these zones. Tectonic contacts and planar fabrics in rocks of the same age are deformed by north–south folds oriented parallel to the strike of the suture zones.

Structures in the Western Aldan block formed later than those in the Dzhugdzhur-Stanovoy province. In the north of the Stanovoy block, near east–west folds (IV^1) — the major mappable folds in this region — are cut by a belt of roughly north–south-striking planar structures, which may be a possible continuation of the Reutov zone and/or the Ungra-Dyos-Melemken zone (IV^2). Here, only the schistosity in greenschist blastomylonites is parallel to the suture.

In the Aldan granulite-gneiss terrain, cycle V structures have been found in the Western Aldan block, where they are represented by the Yarogu Group (zone *5*, Fig. II-9; *16* on Fig. II-24), metadiabase, metagabbro-diabase and granite (β_5, $\nu\beta_5$, γ_5). The Yarogu Group consists of micaceous quartzite, metapsammite and mica schist, sometimes with andalusite and skeletal cordierite. Carbonate horizons, metagrits and conglomerates containing quartzite and granite pebbles are also present. For the quartzite, metapsammite and mica schist, K_2O typically predominates over Na_2O. Hornblende schist is found (showing relict porphyritic texture with intersertal texture in the groundmass), corresponding in composition with alkali basalt and andesite-basalt, acid meta-effusive rocks and tuffaceous sandstone. The Yarogu Group is in tectonic contact with the surrounding γ_4 granites, the Chuga Formation and Bulgunyakhtakh greenstone belt assemblages and is situated in a narrow graben separating them. The Yarogu Group is cut by γ_5 granites which contain xenoliths of blastomylonitized γ_4 granites. Rocks of this group were metamorphosed at dolomite-quartz and tremolite grades of the andalusite-sillimanite facies series (Dook et al., 1986).

Compositionally, the metagabbros and metadiabases are olivine dolerites, dolerites and occasional ore-rich subalkaline olivine dolerites (with up to 10% titanomagnetite). Granites belonging to γ_5 generation are finely-porphyritic biotite and two-mica types with fluorite and tourmaline; they form elongate bodies up to 10×3 km. These granites belong to the subalkaline series. Primary gneissose facies are present at the margins.

In the Western Aldan block the Proterozoic evolutionary stage concludes with the intrusion of a diorite-monzonite suite (stocks and dykes) and extensive congo-diabase dyke swarms.

Structural evolution of the Aldan granulite-gneiss terrain

The granulite complexes in the terrain have undergone extensive and complex tectonic evolution. Identification of the first tectonic cycle is based on structural and stratigraphic discordances at the base of the Kyurikan and Fyodorov Formations. The fold phase which separates them was taken as the boundary between cycles I and II (Dook and Kitsul, 1975; Dook et al., 1979).

The second tectonic cycle was accompanied by the main prograde metamorphism, represented by a series of isoclinal folds, alternating with open folds, flexures and block movements. Cycle II peak metamorphic conditions corresponded with the formation of folds belonging to the first and second deformational phases, which preceded the formation of thrusts and pencil-rodded gneisses. IIF_1 and IIF_2 folds deform migmatitic banding and the junctions between petrographic rock varieties; the axial surfaces of these folds had a subhorizontal attitude by the beginning of phase IID_4. One group of early folds had a north–west axial strike, a second group north–east.

A number of zones and regions emerge, based on particular structural features and time of formation of the major mappable folds.

In the central part of the Western Aldan block, the main structural elements are major isometric IIF_4 structures, isometric or oval in plan (the Lower Timpton dome, for example). In the east of this block and in the west of the Eastern Aldan block, linear IIF_4 folds are developed.

The IIF_4' fold system developed at the boundary between the Western and Eastern Aldan blocks initially had subvertical axial surfaces and subhorizontal NNW-striking hinges. IIF_4 folds are tight, with interlimb angles of less than 40°. The IID_4 episode saw the completion of the Gonam-Sutam anticlinorium and the Sutam synclinorium, which define the main structural features of the Timpton and Gonam river basins. The Seym assemblage and the infracrustal complex occur in the core of the Gonam-Sutam anticlinorium, while the Fyodorov, Idzhek and Kyurikan Formations are found on the limbs. In the central and northern parts of this structure, pencil-rodded gneisses, which formed at the IID_3–IID_4 boundary (shearing related to thrusting) and the formation of IID_4 phase folds are found throughout the lower complex and in the upper complex at its contact with the lower. The lineation in the rodded gneisses trends north–west. Upright IIF_4' folds (with gently NW-plunging hinges) are the first folds in the pencil-rodded gneisses. IIF_4'' folds — close recumbent folds with NW-striking hinges (IIF_4' and IIF_4'' hinges are coaxial) — are developed in a 15–20 km wide zone on the eastern flank of the Lower Timpton dome [a region where there is mutual interference between the Lower Timpton dome and the Gonam-Sutam anticlinorium (AGU, 1983; Dook et al., 1979)].

Detailed mapping within the Lower Timpton dome revealed associated folds (a few km across) of the same generation as IIF_4 folds in the Gonam-Sutam anticlinorium.

IIF_4 folds, which complicate the domal structure, surround the core of the dome. The second cycle concludes with open folds with subhorizontal axial surfaces (IIF_5) and flexures which gradually give way to associated steeply-dipping shear zones. Since the folds with horizontal axial surfaces originate during vertical compression of rock layers, and flexural folds and steep shear zones reflect vertical displacement of blocks under conditions of horizontal extension, all these structures are referred to a single deformational episode. IIF_5 shear zones consist of granulite facies minerals. In the north of the Gonam-Sutam anticlinorium they have a subvertical attitude; the majority dip to the north-west, others to the south-east. Near north–south flexures and shear zones in the SW of the Lower Timpton dome occupy the same position in the deformation sequence. Blocks close to the core of this positive structure are uplifted, indicating that block movements are related to the final stages in the formation of the structure. Shear zones cut granites belonging to the second cycle.

Tectonic movements at the beginning of the third cycle were accompanied by the intrusion of alkali gabbro, mafic and ultramafic rocks in the Stanovoy suture zone and in the south of the Western and Eastern Aldan blocks. These intrusive bodies took part in third cycle deformation events, which were intensely expressed only in the southern marginal part of the blocks, close to their boundary with the Dzhugdzhur-Stanovoy fold belt. Tight folds formed in the Zverev block during deformational stage IIID_1 (Dook, 1977), with initially subhorizontal axial surfaces. It is clearly evident that a penetrative mineral schistosity developed in basic and ultrabasic dykes during this episode. The IIIF_1 fold interlimb angle is 35–50° for minor and major folds.

IIIF_2 folds are minor, with dipping axial surfaces. IIIF_3 folds are the main third cycle mappable folds in the Zverev and Sutam blocks. Their axial surfaces are subvertical, with a NW strike. IIIF_3 hinges show a wide scatter in original orientation due to the inhomogeneous superposition of deformed planar structures. IIIF_3 folds become less tight as we move northwards away from the southern margin of the blocks. Minor fold forms maintain the same degree of tightness, with interlimb angles of 60 to 90°. In the south of the Zverev block, close to the Stanovoy suture, early cycle IV structures are non-linear upwarps which deform IIIF_3 fold axial surfaces. Simultaneously, steeply-dipping planar fabrics were folded into open folds with subhorizontal axial surfaces and interlimb angles of 120–150°. IVF_2 folds have vertical axial surfaces all striking east–west and with interlimb angles of 100 to 150°. IVF_3, IVF_4 and IVF_5 open folds are found everywhere, with hinges striking N–S, NE and NW respectively and with vertical axial surfaces. The cycle ends with open folds with subhorizontal axial surfaces. In the west of the Western Aldan block, numerous examples of tight NW-striking folds are to be found in zones of tectonomagmatic reworking.

2. THE OLYOKMA GRANITE-GREENSTONE TERRAIN

Early Archaean evolution

The Kurultin Group formed during the Early Archaean stage of development of the Olyokma terrain; it occurs in the form of a series of tectonic blocks around the periphery of the Olyokma grey gneiss terrain, and consists of biotite and hornblende-biotite plagiogneiss of tonalite-trondhjemite composition and intrusive granitoids, among which are found rare garnet-biotite (± sillimanite) gneiss and quartz gneiss, alternating with garnet-biotite plagiogneiss and high-Ca schist (the Olyokma Group) as bands, lenses and fragments of complex folds, which are the probable age equivalents of the Chuga Formation. Migmatization is common in gneisses of the infracomplex, which has vein material with the composition of plagioclase granite to granodiorite, less commonly syenogranite.

The plagioclase gneiss in the infracomplex is compositionally similar to "grey gneisses" from other parts of the world. Typically, these have a tonalite-trondhjemite trend and calc-alkali differentiation, with 2.5–5 times as much Na_2O as K_2O and around 15–17% Al_2O_3 (Dook et al., 1986). The gneisses are characterized by a strongly fractionated REE distribution with light REE enrichment and heavy REE depletion (Fig. II-26), high Sr contents — $380–530 \times 10^{-4}\%$, and low U concentrations — $0.44–1.67 \times 10^{-4}$ (Kovach et al., 1984). Particular compositional and textural features suggest that the plagiogneisses of the infracomplex are high-alumina type tonalite-trondhjemite series rocks. They display obvious geochemical heterogeneity (e.g. sp. B-2372 and B-2373, Fig. II-26).

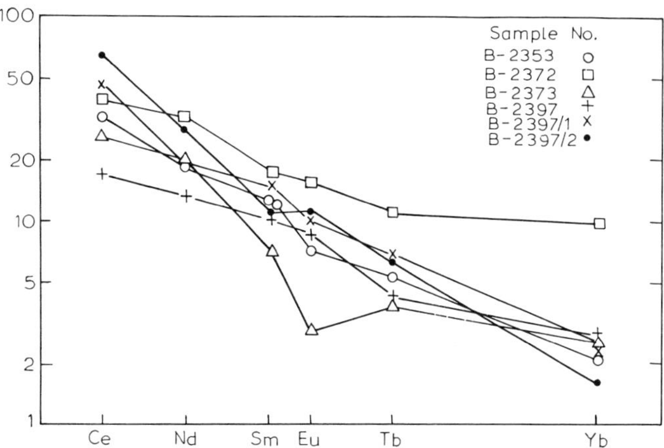

Fig. II-26. Chondrite-normalized REE distribution in plagiogneisses of the Olyokma infracrustal complex (after Dook et al., 1986).

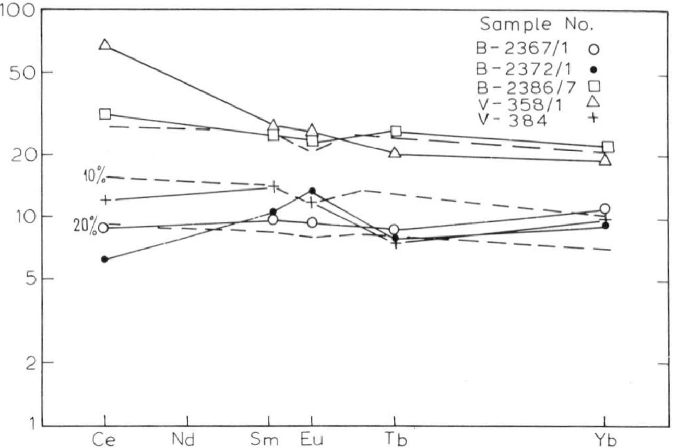

Fig. II-27. Chondrite-normalized REE distribution in mafic schists from inliers in the Olyokma infracrustal complex (after Dook et al., 1986). Lower broken lines: model REE distribution for 10% and 20% melting of upper mantle plagioclase peridotite. Upper broken line: model REE distribution for 20% melting of plagioclase peridotite with subsequent fractional crystallization.

Mafic schist inclusions (Amph + Bi + Plag + Q, Cpx + Amph + Plag ± Q) correspond in composition to tholeiites (Fig. II-27). They contain metatholeiites with a flat REE distribution curve, slightly depleted in LREE, and tholeiites with a fractionated REE distribution with LREE enrichment and a flat distribution curve for HREE, close to "tholeiites I" and "tholeiites II", respectively (Condie, 1981). The first group of metatholeiites could have formed by direct melting of upper mantle plagioclase lherzolite, already depleted in LREE, while the tholeiites of the second group would appear to have undergone fractional crystallization (Kovach et al., 1984). The petrochemical and geochemical heterogeneity of the Olyokma infracomplex rocks would therefore appear to be due to the presence among the plagiogneisses of both relicts of older unaltered material and tonalite remobilizates, restite-like rocks.

Peak metamorphic conditions (high-temperature amphibolite facies, from the amphibole composition and evidence of widespread plagioclase migmatization) in the Olyokma complex were attained during the period of cycle II–III folding. The commonest assemblages are Amph + Pl ± Q, Amph + Bi + Pl + Q ± Kfs, Di + Amph + Plag, and Bi + Pl + Q + Kfs, but they yield less useful information concerning metamorphic conditions and *P–T* parameters. The amphibole is green hornblende with low Al_{VI} 0.329/0.114 (average of 16) in gneisses and 0.399/0.158 (average of 8) in schists (formula units); the Ti content in f.u. is 0.115/0.023 and 0.109/0.041, respectively (Dook et al., 1986). $N_{Mg^{amph}}$ = 47–70, $N_{Mg^{bi}}$ = 50–69. From the Mg ratio in coexisting biotites and amphiboles, the pressure has been estimated at

5–6 kbar (Anon., 1986). The low Al_{VI} content in Amph and the complete absence of assemblages with Ga and Amph indicate high-gradient type metamorphism. Post-migmatitic alkali metasomatism in the gneisses of the complex was expressed in the formation of biotite- and hornblende-bearing quartz-plagioclase porphyroblastic rocks of granitic composition. K–Na metasomatism occurred during the very last stages of ultrametamorphism. The porphyroblastic rocks of granitic composition are most strongly developed in "soft" faults which bound terrains to the north-east and south-west in which we find schist lenses with the assemblages Ga $\pm$ Hyp + Cpx + Amph + Pl amongst biotite- and biotite-hornblende-plagioclase gneisses. The concordant U–Pb age date of 2.8 Ga obtained from a zircon crystalline phase from plagiogneisses in the Olyokma infracomplex corresponds to the time of Late Archaean reworking of the complex and post-crystallization resetting of the zircon U–Pb system. An older age for the complex is obtained on the basis of discordant values for a minimum Pb–Pb age of 2.92 Ga.

Late Archaean greenstone belts

Crustal extension and movement of crustal blocks led to the initiation of Late Archaean greenstone belts.

The Olyokma granite-greenstone terrain contains the following belts (from west to east): Tarynakh (1), Olondo (2), Byrylakh and Itchilyak (3) and Tungurcha (fragments 6, 12, 14, 17), included in rocks of the Olyokma complex (Fig. II-24).

The Tungurcha belt consists of tectonically disrupted and disconnected fragments of a single sedimentary basin, at least 70 km wide, containing the Tungurcha Formation (Fig. II-28). The fragments are concentrated in zones striking approximately north–south and dipping gently eastwards. The main mappable folds are also north–south-trending, usually overturned to the west. These tight folds appear to be the first structures to deform both the tectonic contacts between the Tungurcha Formation and rocks belonging to the Olyokma infracrustal complex, and the foliation of blastomylonites in "soft" fault zones. They are preceded by early F_{1-2} folds in the Tungurcha Formation.

The most complete section of the Tungurcha Formation is in the Syrylyr fragment on the right bank of the Olyokma river opposite the mouth of the river Khani (Mironyuk et al., 1971; Dook et al., 1986). According to Timofeyev and Bogomolova (1986), the formation consists of the following members, from bottom to top: (1) carbonate and calc-silicate rocks (>500 m); (2) interlaminated calc-silicate rocks, quartzite, chert and mica schist (350 m); (3) black graphitic two-mica-quartz schist, often with garnet, staurolite and kyanite, more rarely with andalusite and sillimanite. Northwards along strike, mafic schist, ore-free quartzite,

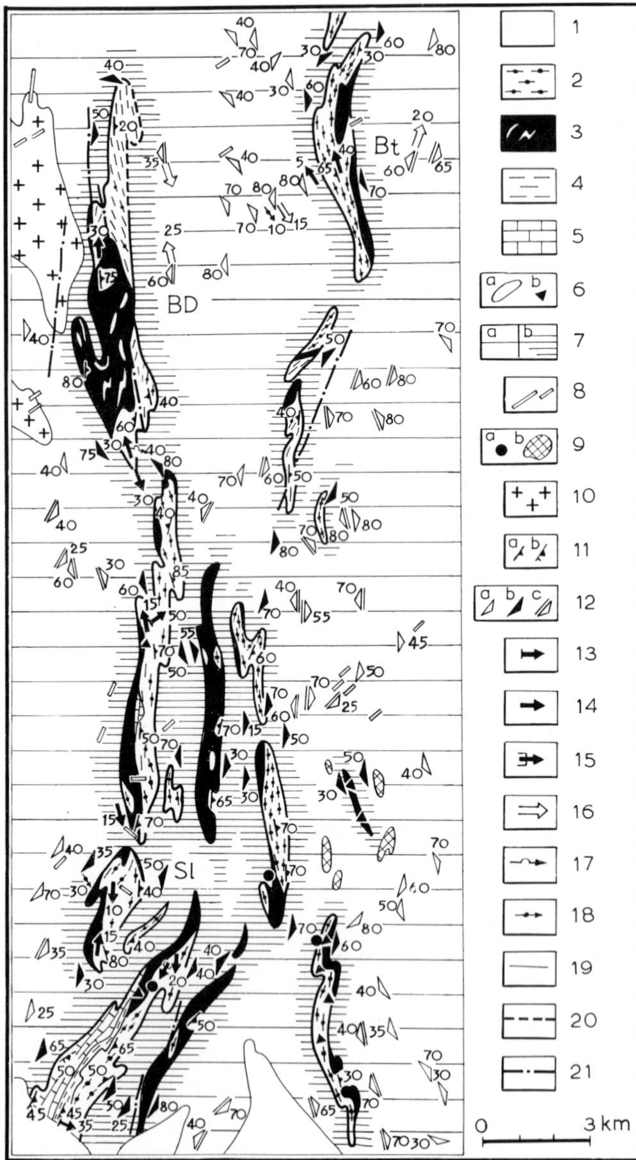

Fig. II-28. Geological map of the region between the rivers Syrylyr and Bolshaya Dagda (compiled by L.M. Bogomolova and V.F. Timofeyev using material from V.I. Berezkin, A.Ye. Breydo, V.L. Dook, L.N. Malkov, S.S. Rozhin and A.P. Smelov). *1* = Quaternary and Mesozoic; *2–6*: Tungurcha greenstone belt (*2* = aluminous schists (with ga, st, cord, sill, ky), *3* = interbedded qz-mica schist, mafic schist, quartzite and Fe quartzite, *4* = qz-mica schist and microgneiss, *5* = marble and calciphyre, *6* = ferruginous quartzites: *a* = lenses, *b* = occurrence); *7* = bi, hb-bi plagiogneisses and migmatites of infracomplex (*a*), and in epidote-amphibolite facies shear zones (*b*); *8* = metaconga-diabase, greenschist facies; *9* = meta-ultrabasics (*a*), metagabbros, metagabbro-diorites (*b*) of Amnunaktin complex, at

ferruginous chert and garnet ($\pm$ cummingtonite) schist appear in the middle member. Ferruginous chert, garnet and cummingtonite-garnet schist form bands up to 30 m thick at the boundary between mafic and aluminous schists and appear to be banded ironstones.

The mafic schists are compositionally basalts. From their $SiO_2/(Na_2O + K_2O)$ ratio, they are normal or rarely subalkaline basalts. Present among the volcanics are tuffites with a very slight excess of SiO_2, K_2O and Na_2O. Rocks with a high alkali content are restricted to marginal parts of bodies. Mafic schists in the Tungurcha Formation belong to the tholeiite series, the main components being tholeiite and olivine tholeiite. The tholeiitic differentiation type is clearly established: MgO decreases as $FeO + Fe_2O_3$, TiO_2, MnO and P_2O_5 decrease, while SiO_2, Al_2O_3, CaO, Na_2O and K_2O remain almost constant. Solitary lenticular and sheet-like bodies of monomineralic amphibolite up to 1 m thick are present among the metatholeiites and correspond compositionally to picrite and picrito-basalt or their intrusive equivalents. The metatholeiites possess an unfractionated REE distribution.

The quartzites are cherts and oligomict psammites belonging to Na–K–high-K families. Banded ironstones and cummingtonite-garnet schists are substantially differentiated with respect to alumina, they are high in Fe and Mg and to a lesser extent Ca and Mg, they are low in alkalis, with the alkali ratio varying from $Na_2O > K_2O$ to $K_2O > Na_2O$ which, together with their alumina differentiation probably indicates the composition of terrigenous impurities. Microcline gneiss, mica-plagioclase-quartz schist and psammite form a family of moderately alkaline rocks of the K–Na and Na–K series, with $FeO > Fe_2O_3$. Schists with high-alumina minerals are compositionally differentiated, from siltstones to subsiallites. Characteristics are moderate alkali values and sharp fluctuations in the alkali ratios, which have a predominantly Na–K profile. The appearance of K–Na varieties is obviously due to the presence of impurities and (or) the erosion of sodic volcanics. Compositional points for metapelites on the Al_2O_3–K_2O–MgO diagram lie in the compositional field of hydromica clays and clays with a hydromica-montmorillonite composition. Subsiallite compositions are shifted in the

---

epidote-amphibolite facies, ^{207}Pb–^{206}Pb zircon age $\geq$2.9 Ga; *10* = porphyritic biotite and two-mica granites; *11* = S_{1-3} schistosity (*a*), coincident with S_0 (*b*) in Tungurcha Group; *12* = fabrics in infracomplex plagiogneiss (*a*), in epidote-amphibolite facies shear zones (*b*), banding and coincident foliation in migmatites (*c*); *13-16*: minor folds (*13* = with axes along dip of axial surface, *14* = NNW, NW with gently plunging axes, *15* = with subhorizontal axial surface, *16* = NNW, NW in infracomplex plagiogneisses and migmatites); *17* = crenulation; *18* = quartz lineations in schistose gneisses of infracomplex; *19* = geological boundaries; *20* = faults accompanying schistosity at epidote amphibolite facies conditions; *21* = zones of greenschist retrogression. *BD* = Bolshaya Dagda sector; *Bt* = Botolkoy sector; *Sl* = Syrylyr sector.

direction of the field of kaolinitic clays. Calc-silicate rocks correspond in composition to calcareous sandstones and/or silica rocks (Timofeyev and Bogomolova, 1986).

The moderate thickness, formational composition, overall nature of the succession and the petrochemistry of these rocks allow us to refer the Tungurcha belt to platform-phase greenstone belts, according to the classification of Groves and Batt (1984).

The Olondo belt has a quite different formational composition and succession type. The belt is situated in the central part of the Olyokma granite-greenstone terrain (Fig. II-24). Its rocks are localised in a narrow synform striking approximately north–south, with steeply-dipping limbs and steeply-plunging hinges (these are the major mappable folds). The synform folds the foliation, metamorphic banding and axial surfaces of early folds with flat-lying hinges (Fig. II-29).

Gabbro and ultramafic intrusions were emplaced simultaneously with and immediately preceding the main mappable folds. Surrounding the Olondo belt are rocks of the Olyokma infracrustal complex-biotite and hornblende-biotite gneisses with thin amphibolite streaks and lenses. Among these rocks there is a significant volume of tonalite-granodiorite bodies, analogous to the granitoids which cut the Olondo Group, and younger than a group of tonalites which were emplaced into fine-grained tonalitic gneisses. Contacts between the Olondo belt and the surrounding rocks are tectonized in most cases. At the eastern tectonic contact, there is a small outcrop (a few hundreds of metres across) of rhythmically layered metasandstones and metasiltstones containing thin bodies of intermediate composition metamorphosed effusive rocks (an analogue of the PR_1 Udokan Group), overthrust onto rocks of the Olondo Group. The western contact is less disturbed.

At least 90% of the Olondo Group consists of metavolcanics. On the western limb of the synform, close to the contact, the supposed lower part of the succession consists of basic and ultrabasic metavolcanics, which are gradually replaced by intermediate and acid volcanics eastwards (upwards in the succession?). Petrochemical data (Fig. II-30) allow three series of metavolcanics to be identified, totalling some 3 km in thickness: komatiitic (ultrabasic), tholeiitic (basic — Condie's group I tholeiites) and calc-alkaline (intermediate to acid composition). This last series contains thin metasedimentary horizons, compositionally equivalent to siltstones and shales and polymict and greywacke-sandstones — weakly differentiated immature sediments (Sochava, 1986). The combination of volcanic and sedimentary rocks, the great abundance of volcanics in the succession and the common occurrence of komatiites suggest that the Olondo belt has similarities with rift-phase greenstone belts (Groves and Batt, 1984).

Gabbroid and ultramafic intrusions form subconcordant bodies and dykes. Gabbroids which are concordant with the host rocks (foliation and

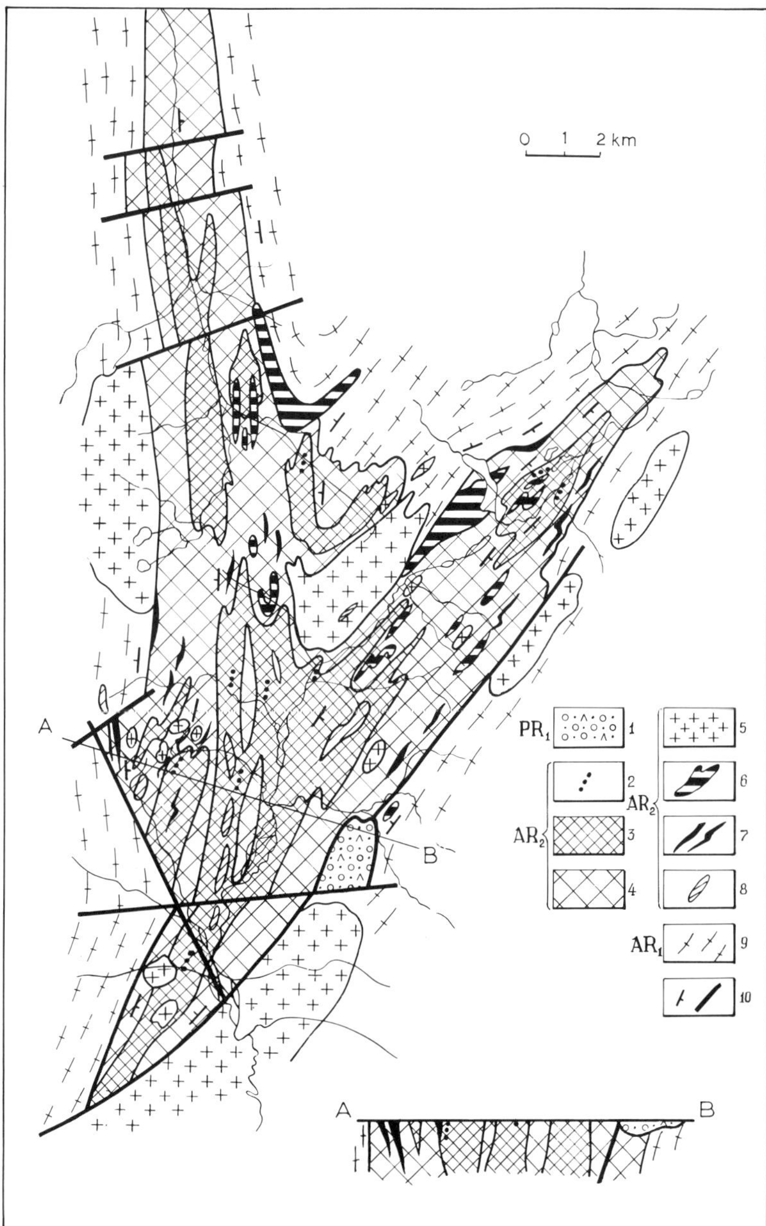

Fig. II-29. Geological map of the Olondo greenstone belt (compiled by G.M. Drugova, using material from A.L. Kharitonov, S.A. Bushmin and Ye.B. Lvova). *1* = Udokan metasandstones; *2* = thin metasediments; *3* = intermediate and acid (with minor basic) metavolcanics; *4* = basic metavolcanics; *5* = granites and granodiorites; *6* = ultrabasic intrusions; *7* = komatiite lavas; *8* = metagabbroids; *9* = granite gneisses with thin amphibolites, Olyokma complex, Early Archaean; *10* = dip and strike, tectonic breaks.

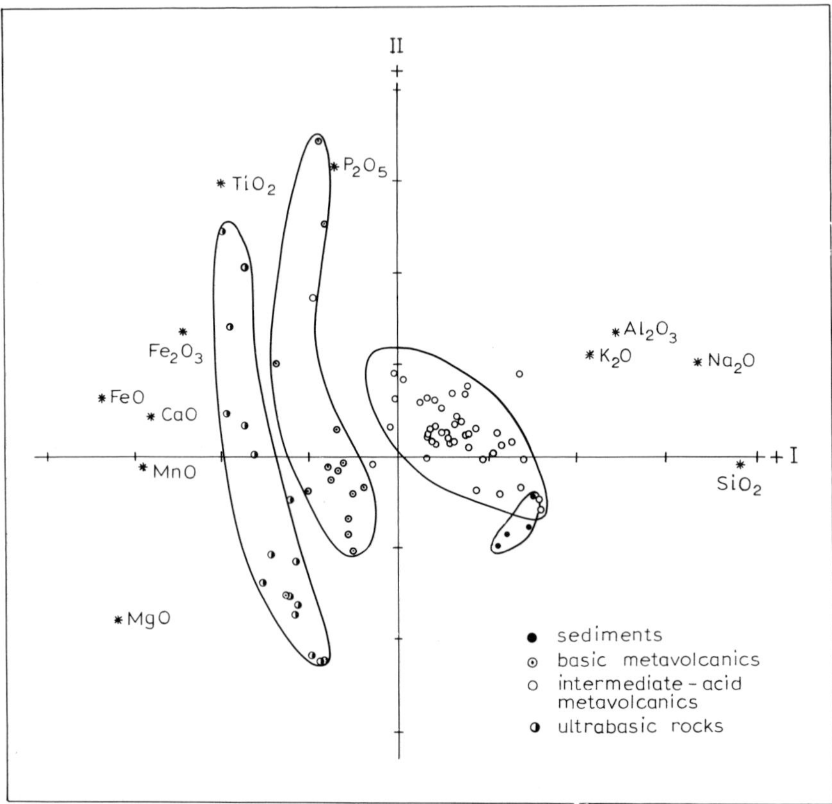

Fig. II-30. F_1–F_2 component diagram for rocks of the Olondo Group.

banding in them is usually concordant) are recognisable from relict gabbroic textures and a more massive and more homogeneous appearance than the host metavolcanics. The gabbroids are visually indistinguishable from the metavolcanics in marginal parts of the Olondo belt, in zones subjected to intense metamorphic and structural reworking. Ultrabasic intrusions are mostly situated in the north of the belt (Fig. II-29). They consist of subconcordant bodies, occasionally obliquely cross-cutting, up to 500–700 m thick. Dobretsov has found multiple asymmetric textures (repeated four times) in the largest body: medium- to coarse-grained cumulate textures → fine-grained spinifex-like textures (harrisitic) → coarse-grained serpentinized harrisitic textures. Asymmetric rhythms also repeat four times in the section: dunite (Ol + chrome-Sp, intercumulate Cpx + Opx) → peridotite and lherzolite → olivine pyroxenite. According to Pukhtel and Drugova, pyroxenites are developed in the marginal zone of the massif. The asymmetric structure would appear to have been caused by repeated injection of pulses of melts (slight openings of the magma chamber under a continued extensional regime). Dunite predominates among the intrusive bodies, with harzburgite and lherzolite present to a much lesser extent. Minor bodies and the margins of major bodies are altered to talc-serpentine and talc-serpentine-carbonate rocks. Ultrabasic intrusions and gabbroids of this group are found close to the Olondo belt amongst rocks of the Olyokma infracrustal complex. Intrusions of ultrabasic rocks are exposed in the Itchilyak belt, where they also occur amongst rocks of the Olyokma infracomplex.

The type of section in the greenstone belts found in the Olyokma granite-greenstone terrain changes both to the east and the west. The Tarynakh belt which is situated to the west of the Olondo synform closely resembles the Tungurcha belt in terms of its formational composition. Compared to it, the Tarynakh belt has much greater volumes of a volcanogenic-ironstone-chert formation (the Tarynakh ironstone deposit). The highest-gradient metamorphism — T = 550–570°C, P = 3.5–4.5 kbar — is characteristic for the Olondo belt and for the Byrylakh belt some 50 km to the south-east [675°C, 4.5 kbar, from garnet-biotite equilibria (Neyelov, 1983)]. Metamorphism in the Olondo belt is non-uniform, varying from epidote-amphibolite to amphibolite facies. This inhomogeneity is complicated by changes that arose after the complex had attained peak metamorphic conditions. Mafic metavolcanics have the following parageneses in the epidote-amphibolite facies zone: Hbl + Cumm + Pl + Q, Act + Bi + Pl + Q, Hbl + Pl + Epi + Q; and in acid to intermediate metavolcanics — Musc + Bi + Pl + Q. In plagiogneisses (Ga + Bi + Pl + Q), $N_{Mg^{ga}}$ = 8.3–12.5, $N_{Mg^{bi}}$ = 42.1–55.4; in the assemblage Ga + Bi + St + And + Pl + Q, $N_{Mg^{ga}}$ = 12.3, $N_{Mg^{bi}}$ = 49.2. In the amphibolite facies zone, mafic metavolcanics have the following mineral assemblages: Cpx + Hbl + Pl + Q, Hbl + Bi + Pl + Q, rarely Ga + Hb + Cumm + Pl + Q. In this last paragenesis, $N_{Mg^{ga}}$ = 9.2, $N_{Mg^{amph}}$ = 24.5, for F_{tot} = 78.5; in amphibole, Al_{VI} = 0.92, Al_{IV} = 1.83 f.u. The assemblage Bi + Pl + Mic + Q is found in acid metavolcanics. The assemblage Trem + Act + Pl + Q is present in metagabbros in the epidote-amphibolite facies zone, while Hbl + Pl is found in the amphibolite facies zone.

Non-uniform kyanite-sillimanite facies series metamorphism is characteristic for the Temulyakit-Tungurcha zone (the Tungurcha greenstone belt). Fragments of different metamorphic facies are exposed in the Temulyakit-Tungurcha zone, due to translations along faults and subsequent differences in the depth of erosion (Smelov, 1986). In the Syrylyr and Tungurcha belts (Fig. II-28), metamorphism corresponds to the kyanite-biotite-staurolite subfacies of the epidote-amphibolite facies. In the extreme eastern part of the Syrylyr window and in the Yayelakh fragment, the rocks belong to the biotite-muscovite gneiss subfacies (staurolite-garnet-cordierite-biotite subfacies in Glebovitsky's facies scheme). The associations Ga–St–Bi, St–Bi and Ga–Bi schists (with muscovite as a rule) are very common in metapelites belonging to the staurolite subfacies. The commonest paragenesis in quartzites and quartz schists is Bi + Musc + Pl + Q; and in quartz schists and two-mica schists — Musc + Bi + Chl + Pl + Q. In mafic schists which alternate with metapelites, we commonly find the assemblage Hbl + Pl ± Bi ± Epi + Q + Ore, less commonly Ga + Hbl ± Epi + Pl + Bi + Q and Ga + Hbl + Cumm + Pl + Q. $N_{Mg^{ga}}$ varies from 11 to 13 in assemblages containing an Al_2SiO_5 polymorph. The TiO_2 content in biotite reaches 1.43 wt%.

The most characteristic compositional feature of hornblendes, which allows us to distinguish them from hornblendes in the Olyokma infracomplex, is their high Al_{VI} content — 0.766–1.11 f.u., a lower Ti content and a lower total Ca + Na + K. From their TiO_2 and F_{tot} ratio, they fall in the epidote-amphibolite field. Similar features are found in hornblendes in the Tungurcha complex metagabbroids which occur amongst grey gneisses. Trails of garnet grains appear at the boundary between hornblende and plagioclase in the metagabbroids. $N_{Mg^{ga}}$ in Tungurcha schists is 7.4, with $N_{Mg^{amph}}$ = 34 and TiO_2 = 0.24–0.27 wt%.

The biotite-muscovite gneiss subfacies is represented in the extreme eastern exposure of the Syrylyr fragment. In metapelites the mineral assemblage is Ga + Bi + Sill + Musc + Pl + Q [St]; in K_2O undersaturated rocks: Ga + Bi + Cord + Sill + Pl + Ky + Q [St]. Cordierite is found as armoured relicts in Ga, Pl and Cord and St is not contiguous with quartz. $N_{Mg^{st}}$ is 15 in muscovite-bearing assemblages, and 21 in those with garnet-cordierite. $N_{Mg^{ga}}$ = 12, Mg and the TiO_2 content in biotite is 1.93–2.18 wt%. Stable parageneses in K_2O-undersaturated rocks in the Yayelakh fragment are: Ga + Ged + Bi_{41} + Pl + Q; Ga_{92} + Bi + Qtz + Ore; Ga_{88} + Bi_{57} + Sill + Q + Ore. In metabasics belonging to the biotite-muscovite gneiss subfacies, the assemblages are: Hbl + Bi ± Cumm + Pl + Q; Hbl + Cpx + Pl + Q; and rarely Ga + Hbl + Bi + Pl + Q.

P–T parameters in the staurolite subfacies, in assemblages with Ky, are 500–600°C and 4.5–6 kbar from Ga–Bi and Ga–St equilibria. The scatter of calculated *T* and *P* values in the rocks investigated is due to compositional inhomogeneities in minerals from grain to grain and inhomogeneous grain compositions, especially for garnet. Zoning in garnets is the progressive type-CaO and MnO decrease while MgO increases towards grain edges. $N_{Mg^{bi}}$ varies from 43.4 in the groundmass to 44.6 at the contact with garnet. In the assemblage Ga + Bi + St + Sill(?br) + Musc + Pl + Q, the lowest PT parameters, 525°C and 4.7 kbar, are obtained from Ga_{core}–St_{core}. Ga–Bi equilibria yield from 550°C and 5.0 kbar, Ga_{core}–Bi_{matrix}, to 600°C and 6.0 kbar, Ga_{rim}–Bi_{rim}, showing that metamorphism was prograde.

For the biotite-muscovite gneiss subfacies, *P–T* parameters in assemblages with Sill are: T = 600°C, P = 5.5–6.0 kbar, based on Ga–Bi equilibrium. The relationships between *P* and *T* in both subfacies are evidence of a constant metamorphic regime, belonging to the kyanite-sillimanite facies series. *P–T* metamorphic parameters in the Syrylyr fragment increase from west to east. High-T conditions also characterise the easternmost Yayelakh fragment.

In order to elucidate the metamorphic evolution and to correlate mineral parageneses with deformational events, the orientations of St, Cord, And, Sill, Ky, Musc, Bi and Q were studied in the assemblages Ga + St + Bi + Musc + Pl + Q, Ga + St + Bi + Musc + Pl + Q + Ky, Ga + St + Ky + Sill + Cord + Bi + Pl + Q, Ga + St + Bi + Musc + ?br + Pl + Q and And + Musc + Bi + Pl + Q.

Structurally equilibrated mineral assemblages, i.e. those in the same stress field, were identified according to the conditions set out in Kozhevnikov (1982), Lazarev and Kozhevnikov (1973).

Using mineral interrelations and microtextural studies, we have established that the structural paragenesis with F_{1-2} folds is represented by Musc + Q and Musc + Ky + Q. In the biotite-muscovite gneiss subfacies, the eastern part of the Syrylyr fragment and the Tungurcha window with F_{1-2} folds corresponded to the Ky–Bi–St subfacies. In the western and central parts of the Syrylyr fragment this was attained only during the period of formation of the major F_3 folds, which deform the tectonic contact surface of the Syrylyr fragment. The orientation of St, Bi, Cord, Ky, Sill and Musc are here correlated with F_3 folds. In the Botolkoy fragment, the Sill–Bi–St subfacies of the epidote-amphibolite facies corresponds to F_3 fold formation, while in the Yayelakh and the east of the Syrylyr fragments the Si–Bi–Musc subfacies of biotite-muscovite gneisses corresponds to this event. Peak metamorphic conditions were attained in the Tungurcha Group after it was emplaced as a wedge into the migmatitic gneisses of the Olyokma infracomplex.

The orientation of Bi, Musc and And was investigated in the south of the Tungurcha fragment. Schistosity, expressed as oriented Bi and Musc, is folded by VF_1 folds. Andalusite crystals form a maximum of [001] axes, coinciding with VF_1 fold hinges. Here, andalusite-sillimanite facies series metamorphism is manifest locally and from its relationship to deformation it corresponds to events in the Tasmiyeli-Udokan stage which were accompanied by prograde metamorphism of comparable regime in the Tasmiyeli and Udokan Groups.

A transition to higher-gradient conditions in the Tasmiyeli-Udokan stage is also observed in the Olondo Group. The relatively higher-gradient conditions of the later metamorphism are here indicated by the relatively high Al content in hornblendes — 0.92 f.u. (in the assemblage Ga + Hbl + Cumm + Pl + Q) and by the P–T parameters calculated on the basis of Ga–Bi equilibria, i.e. T = 500°C, P = 3.5 kbar; T = 570°C, P = 3.5 kbar, for early and late parageneses, respectively.

The Tyrynakh belt which is situated in the west (Fig. II-24), is similar to the Tungurcha belt in terms of metamorphic regime and evolution of regimes. These facts all suggest that the Olondo belt is a fragment of the central rift zone of the Olyokma granite-greenstone terrain, characterised by the maximum development of basic and ultrabasic magmatism and the highest heat flow.

Transitional peak metamorphic conditions were attained in intervening belts, such as the Itchilyak fragment: $T = 550°C$, $P = 4.5$ kbar (garnet-biotite equilibria).

Evidence for Late Archaean internal processes is widespread in the Olyokma infracrustal complex. Tonalites were remobilized during this period. The largest massif of quartz diorite and granodiorite-tonalite, 15 × 7 km, is located in the region around the mouth of the river Oldongso (Fig. II-9). It consists of clinopyroxene-hornblende-, hornblende- and biotite-bearing gneissose rocks with relict igneous plagioclase and patches of granitic texture.

Petrochemical and geochemical features of the rocks in this massif suggest that it may possibly have formed by the remobilization of plagiogneiss in the infracomplex. Relative to older tonalites, these are enriched in LREE, Th, Zr, Hf, Ba, Cs and U (Kovach et al., 1984). Compared with the country rock gneisses, the rocks of the massif show a much more indistinct foliation, which was involved in fewer deformational episodes. Xenoliths of the host grey gneiss are present. In the fault zone that forms a margin to outcrops of the Tungurcha Group, the rocks of the massif were sheared during a period corresponding to when the Tungurcha Group reached peak metamorphic conditions. It follows that this massif may be considered as having formed during the concluding phase of the Late Archaean crust-forming cycle (γ–$\gamma\delta_3$).

Simultaneously with the attainment by the Tungurcha Group and its age equivalents of peak metamorphic conditions, two types of retrograde rocks formed in the Olyokma grey gneiss complex; these have been best studied in the Temulyakit-Tungurcha zone (Smelov, 1986). The first consists of intensely sheared rocks, localized in "soft" fault zones which bound the Tungurcha Group (Figs. II-9, II-28). The second type of retrogressed rocks show no substantial tectonic reworking and occupy extensive areas close to shear zones. Gneisses and schists in shear zones are altered to medium-grained finely-foliated rocks with a schistosity that coincides with that in the prograde metamorphic rocks of the Tungurcha Group. Sheared rocks (blastomylonites) with augen texture are taken to be metamorphosed acid effusives (metaporphyrites and leptites).

In the western part of the Aldan granulite-gneiss terrain, greenstone belt assemblages were metamorphosed in a rather higher-gradient regime and at higher temperatures compared to the Temulyakit-Tungurcha zone. Peak metamorphic conditions in greenstone belt assemblages were attained here during the time of formation of lower Proterozoic deep-level thrusts in the Ungra (IV^2) cycle. The later elevation of the Western Aldan block in comparison to the Olyokma block within the zone of high temperature metamorphism may have resulted in the asynchronous formation of morphologically and metamorphically similar fault zones. They could have been earlier (cycle III) in the Olyokma granite-greenstone terrain.

Only the greenstone belts in the Olyokma granite-greenstone terrain have been dated. Zircon ages from calc-alkali series acid to intermediate volcanics in the Olondo belt have been determined by the U–Pb isochron method at 2960 ± 70 Ma (Neyelov, 1983). The following age dates have been obtained using the Pb–Pb method (evaporation technique): 2900 ± 50 Ma on zircons from aluminous schists in the Olondo Group, and 2800 ± 50 Ma from Tungurcha Group schists (the Syrylyr fragment). Zircons showing signs of transport gave an age of 3020 ± 100 Ma (aluminous schists from the Tungurcha Group). Bibikova obtained a date of 2700 Ma for zircons from apophyses of the Ust-Oldongso intrusion (γ–$\gamma\delta_3$) by the U–Pb method on a rock with well-preserved igneous textures. A Pb–Pb (evaporation technique) age for zircons from rocks of the intrusion itself is also 2700 Ma. The age of ferruginous chert in the Tarynakh belt (the southern group of deposits), obtained by Neymark using the lead-isochron method, is 2700 ± 50 Ma. Figurative points for amphibolites and magnetite chert "lie" on an isochron with a model age of 2650 Ma. Results obtained from the isochron agree with data from Wetherill's model of 2720 ± 80 Ma.

The nature of folding is characteristically non-uniform throughout the area. In the north and east, there are linear north–south folds with steep axial surfaces overturned to the west and flat hinges. In the south the major folds formed under conditions of local compression. In the cores of dome-shaped structures of this generation, in rocks of the activated basement, there are folds developed with subhorizontal axial surfaces. On the limbs of the domes they give way to tight isoclinal folds with vertical axial surfaces (Anon., 1986). These deformation types are best observed in the region of the Byrylakh syncline (Fig. II-24). The formation of these domes concluded the Archaean stage of evolution of this terrain, with the appearance of a structural paragenesis typical of granite-greenstone terrains: linear structures of supracrustal belts and dome-shaped elevations in remobilized assemblages of the granitized basement.

Early Proterozoic evolution

In the Olyokma granite-greenstone terrain, a suite of minor gabbro-amphibolite bodies in the Temulyakit-Tungurcha zone, cutting the schistosity in the Late Archaean Ust-Oldongso intrusion, is of Early Proterozoic age. Relict ophitic texture is preserved in the gabbro-amphibolites. From their composition and differentiation type, they are close to Tungurcha Group tholeiites, differing slightly in higher Na_2O and especially K_2O contents. Metagabbro-dolerite dykes with relict subophitic texture, elongate prismatic plagioclase, orthorhombic and monoclinic pyroxene are encountered in the Borsalin-Nelyuka zone. Highly xenomorphic quartz is found very occasionally. Similar rocks, but with tiny amount of olivine, are sometimes described by the term "drusites". Chemically, these rocks correspond to normal tholeiitic gabbro and leucogabbro.

The disconnected nature and type variety of lower Proterozoic structures in the Olyokma granite-greenstone terrain is illustrated in the compositional and structural features of assemblages belonging to the Tasmiyeli-Udokan level.

Lower Proterozoic metasedimentary assemblages in the south of the Olyokma block have been described as the Udokan Group (Mironyuk et al., 1971; Fedorovsky, 1985; Sochava, 1986), in the Koda-Udokan basin and the Lower Khanin graben-syncline; in the north is the Uguy Group in the Oldongso and Uguy graben-synclines.

Udokan Group clastic terrigenous sediments in the Kodar-Udokan basin were metamorphosed twice (Korikovsky and Kislyakova, 1975; Kudryavtsev and Kharitonov, 1976). The earlier metamorphic episode was low-temperature, reaching only the chlorite-sericite subfacies of the greenschist facies, and is related to a burial episode. Such metamorphism is characteristic for the majority of rocks in the central part of the structure. The later metamorphic episode belongs to the concluding Early Proterozoic cycle and is related to the growth of dome structures and high heat flow at the contact between Udokan rocks with those of the reactivated basement. As a result, high-gradient metamorphic zoning formed, with the appearance of andalusite-staurolite-biotite-muscovite, andalusite-biotite-muscovite, sillimanite-biotite-muscovite and sillimanite-biotite-microcline zones. Temperatures increased rapidly to 400–600°C at a pressure of 2–3 kbar in the epidote-amphibolite facies zone.

Metamorphism in the Lower Khanin graben-syncline was of a different type, being lower-gradient that in the Kodar-Udokan basin, with the temperature not exceeding 500°C at a pressure of 4 kbar.

The metamorphic grade in Uguy Group rocks in the Oldongso and Uguy graben-synclines did not exceed low-temperature greenschist facies.

On the basis of the different level of structural and metamorphic effects seen in rocks belonging to the Udokan and Uguy Groups and differences in their chemical compositions, sediment thicknesses and nature of the successions, we conclude that these assemblages are different in age, with the Udokan Group occupying a lower stratigraphic position.

Both the Udokan and Uguy Groups consist in the main of terrigenous rocks or their metamorphic products — interbedded sandstone, siltstone and shale. Carbonate rocks play a significant role only in the Butun and Atbastakh Formations of the Udokan group and in the Namsalin Formation of the Uguy Group.

A general trend in the composition of terrigenous rock associations during the Early Proterozoic consists in a gradual change from weakly differentiated associations with higher Na_2O, MgO, CaO and FeO to K-type differentiated associations with $Fe^{3+} > Fe^{2+}$. This trend reflects a change in the composition of the source area, the type of vulcanism and the intensity of weathering processes during the formation of mature continental crust in this region (Sochava, 1986).

The Tasmiyeli basin is an orogenic type. Tasmiyeli Group molasse sediments (Petrov, 1974, 1976) are localised in a north–south graben-type superimposed structure; they are usually regarded as an assemblage that accumulated during the final stages of upper Archaean greenstone belt formation. The graben is bounded to the west and east by blastomylonite zones (Fig. II-31). Blastomylonitization affected not only the Tasmiyeli Group and the surrounding rocks, but also granites (γ_5). The succession thickens from east to west and from west to east, which is interpreted as a major tight F_1 anticline with a northward-plunging hinge (Timofeyev and Bogomolova, 1986). F_1 folds deform bedding S_0 and a bedding-parallel schistosity S_0'. An S_1 schistosity is developed parallel to F_1 axial surfaces. During D_2, rare open minor folds with subhorizontal axial surfaces formed. F_3 folds are accompanied by an axial planar subvertical schistosity. Granites occupy a cross-cutting position relative to F_1 folds and in the granites a steep schistosity is developed, parallel to S_3. The origin of tectonic boundaries of the Tasmiyeli Group is a D_3 or later event.

The group, which is over 900 m thick, has six rhythmically arranged rock members — sandstones, metasiltstones and metapelites.

Metasandstones are characterized by low and moderate alkalis. Metapelites have moderate alkalis and in composition diagrams they fall in the field of hydromica and montmorillonite clays. The supposed volcanics of the Tasmiyeli Group are, according to Petrov (1974, 1976) and Berezkin (in Dook et al., 1986), represented by hornblende schists which are similar to metabasics that cut the Tasmiyeli Group; these correspond to andesites, andesite-basalts and basalts.

Zonal metamorphism was accomplished in the andalusite-sillimanite facies series regime (Smelov, 1986). The metamorphic grade increases from west to east from the biotite (biotite zone) and almandine-chlorite-chloritoid (garnet zone) subfacies of the greenschist facies to the staurolite-chlorite (staurolite zone) subfacies of the epidote-amphibolite facies (Fig. II-32). Using Ga–Bi and Ga–St equilibria, T = 480–550°C and P = 2.2–2.6 kbar.

The Subgan terrigenous-volcanogenic assemblage was first recognised as an independent member of the Aldan Supergroup by Mitich in 1945, who referred it to the Proterozoic. It is usually regarded as belonging to AR_2 greenstone belts. Outcrops of Subgan Group rocks are restricted to the Amga fault zone which separate the Aldan granulite-gneiss and Olyokma granite-greenstone terrains and can be traced as a major zone of blastomylonites. The group consists predominantly of basic metavolcanics-amphibole ortho-schist and two-mica schist with andalusite. Petrochemically, the alu-

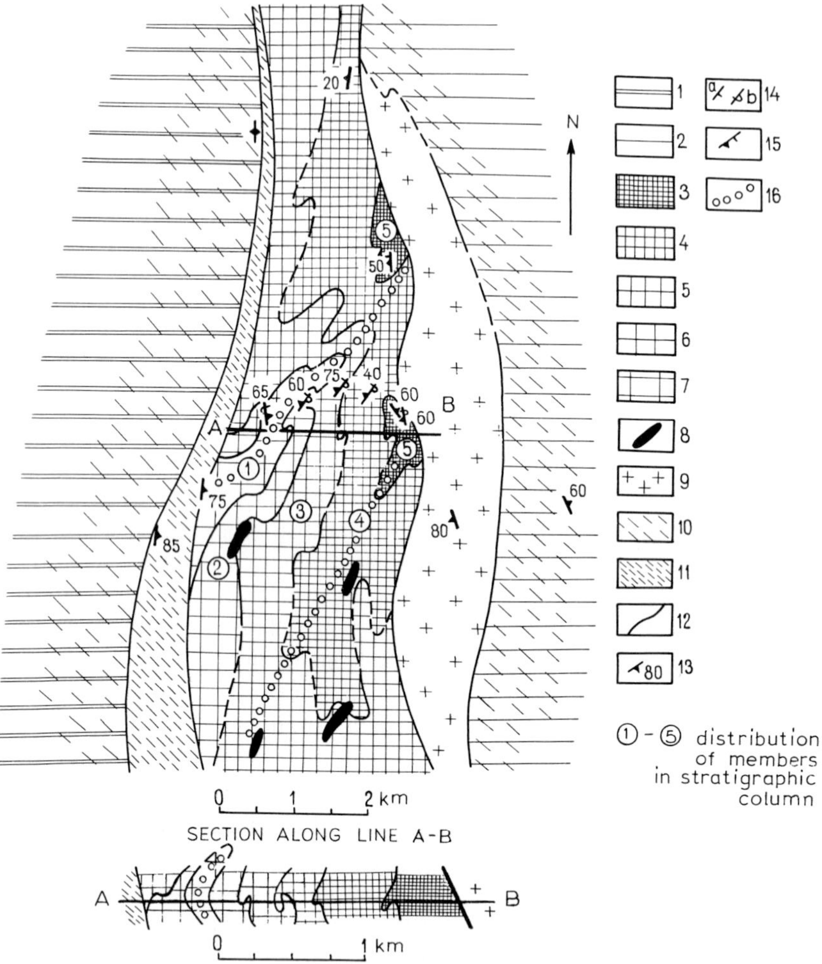

Fig. II-31. Structural sketch map of the Tasmiyeli segment [after L.M. Bogomolova and V.F. Timofeyev (Dook et al., 1986)]. *1* = enderbites, gneisses; *2* = tonalites; Tasmiyeli Group members: *3* = first, *4* = second, *5* = third, *6* = fourth, *7* = fifth and sixth; *8* = metadiabases; *9* = two-mica granites; *10* = greenschist and epidote amphibolite facies shear zones; *11–12* = greenschist facies mylonites; *13* = schistosity; *14* = bedding; *15* = bedding//schistosity; *16* = fold axial surfaces. Numbers in circles: order of members in stratigraphic column.

minous schists are characterised by a predominance of K_2O over Na_2O and they correspond to hydromica clays, while the hornblende schists represent high-temperature tholeiitic basalts.

In the Yarogu structure, outcrops of supracrustal rocks belonging to the Yarogu Group are restricted to a narrow graben separating outcrops of Bulgunyakhtakh Group rocks, and consist of micaceous quartzite, metasandstone, metaconglomerate, high-alumina shales and carbonate rocks.

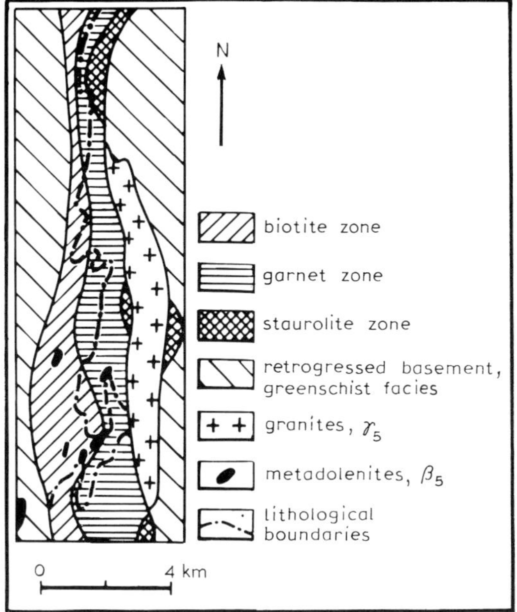

Fig. II-32. Sketch map of metamorphism in the Tasmiyeli graben [after A.P. Smelov (Dook et al., 1986)].

Metamorphism in the Subgan Group is zonal, having progressed in a high-gradient regime of the andalusite-sillimanite series. The metamorphic grade increases from west to east, from the andalusite-biotite-staurolite subfacies to biotite-muscovite gneiss facies. Within the confines of the Yarogu graben, metamorphism took place under high-gradient conditions at dolomite-quartz and tremolite grades with an increase in metamorphic grade from east to west (Dook et al., 1986).

Metamorphic evolution in the Aldan-Olyokma region is expressed as a general decrease in metamorphic grade, related to lesser internal crustal activity, and a progressive increase from early to later stages of evolution in the contrast between metamorphic regimes (Fig. II-33). Differentiation in metamorphic regimes can be seen even as far back as the Early Archaean and consists of a decrease in the granulite facies metamorphic gradient from the centre of the region towards the south-west, south and east. This is probably related to intense magmatic basification of the crust around the periphery of the Early Archaean craton, which also caused to varying degrees a marked tendency towards downward movements of a denser crust and a decrease in metamorphic gradient. A minimum gradient has been estimated for the Kurultin and Zverev complexes; for the Kurultin it is 26–28°/km, which approximates to the geothermal gradient in the oldest granulite terrains in West Greenland. During the Late Archaean,

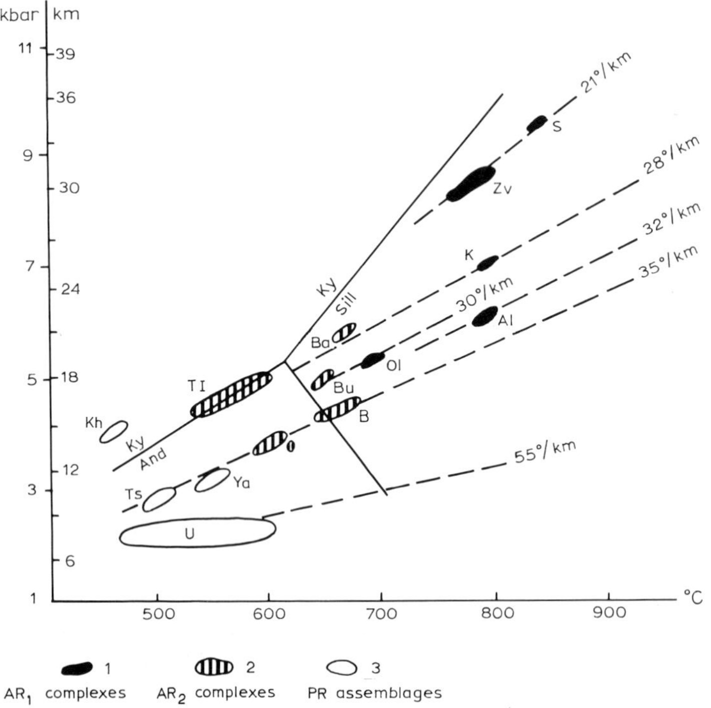

Fig. II-33. Conditions of metamorphism in terrains of different age in the Olyokma-Aldan region (after G.M. Drugova). *1* = Lower Archaean terrains [*Al* = granulites of Central Aldan province, *K* = of Kurulta and Zverev terrains in South Aldan province, *Ol* = of Olyokma terrain, *S* = granulites of Sutam subprovince (southern part of Idzhek-Sutam province)]; *2* = Upper Archaean terrains (*O* = Olondin terrain, *B* = Bulgunyakhtakh Group, *TI* = Tungurcha Group and Itchilyak assemblage); *3* = Proterozoic assemblages (*U* = Udokan Group, *Kh* = assemblages of Lower Khanin ridge, *Ts* = Tasmiyeli Group, *Ya* = Yarogu Group).

the metamorphic regime in supracrustal assemblages in the Olyokma block varies in the disjointed structures as a result of laterally changing tectonic environments during the formation period of granite-greenstone terrains and finally in the lower Proterozoic, sub-platform conditions that characterise the Olyokma block there is a sharp drop in metamorphic grade and a significant increase in the contrast between the metamorphic regimes: in the Udokan Group, the geothermal gradient was c. 56°/km, but in the grabens restricted to fault zones, it dropped to 28–30°/km. In the Aldan granulite-gneiss terrain during the Proterozoic, there was a final high-temperature metamorphic event in the central, southern and eastern parts of the terrain, under granulite and retrograde amphibolite facies conditions.

3. THE BATOMGA GRANITE-GREENSTONE TERRAIN

The Batomga granite-greenstone terrain (Batomga block) occupies the eastern part of the Aldan Shield and is in contact with the Eastern Aldan block along the Ulkan fault zone. Three complexes of different ages have been delineated in the Batomga zone: Omna, Batomga and Chumikan (Glebovitsky, 1975; Petrov et al., 1978; Anon., 1977). The oldest complex, the Omna, consists of schist and gneiss belonging to the Omna supracrustal group, pre-metamorphic basic and ultrabasic intrusions, charnockite, enderbite and tonalite. Rocks of the Omna complex were initially metamorphosed at granulite facies and they form areas within outcrops of the younger Batomga Group. The composition of the Omna complex is very similar to that of the Kurultin complex. Mineral assemblages in the schists of the Omna complex contain two pyroxenes in association with amphibole and garnet, while calciphyres contain forsterite and diopside.

The Batomga complex includes the Batomga Group, early-orogenic basic intrusions, granitic gneisses and granitoids. The Batomga Group consists of diopside-bearing hornblende-plagioclase schist, amphibolite, garnet amphibolite, garnet-sillimanite gneiss, quartzite, diopsidic calciphyre and marble. In the west of the zone, in the divide between the Udokan and Maymakan rivers, banded paragneisses predominate, together with quartzites and calc-silicate rocks; the main rocks cropping out in the east are mafic orthoschist with metadiorite and metagabbro. The rocks were metamorphosed under high-temperature amphibolite facies conditions (almandine-amphibolite facies). Partial melting was more intense in the west, where all the rocks are migmatized, with outcrop areas of hornblende-biotite granitic gneiss, similar in character to the infracrustal complex in the Olyokma block. In the eastern part of the Batomga zone, partial melt processes are less well seen due to the predominantly basic composition of the rock units.

The Chumikan complex consists of the supracrustal Chumikan Group, plus metagabbroids and a suite of biotite granites that cuts the group. This complex forms a 60 km NE-striking belt some 10–12 km wide (*18* on Fig. II-24) in the lower reaches of the river Chumikan. The supracrustal group has a composition typical for belts in the Olyokma terrain: porphyroid and porphyrite metavolcanics, mica schist, quartzite, quartz schist and tremolite marble. Metamorphism is zonal, increasing from greenschist facies in the centre of the belt to low-temperature, high-gradient amphibolite facies at the margins.

The structural similarity with the Olyokma block increases with polyphase deformation and metamorphism of blocks of older rocks in the Omna complex (Kurultin complex in the Olyokma region) at amphibolite facies conditions or epidote-amphibolite facies conditions in individual zones, related to folding and metamorphism of the Chumikan greenstone terrain. The Batomga complex occupies the same structural position as the

Olyokma; however, although the Olyokma Group resembles the Batomga in composition, it is present in the Olyokma terrain in a reduced form. We assume that the outcrop areas of granite-gneiss in the Batomga complex and in the Olyokma terrain contain portions of reworked tonalitic basement.

Geological history

An analysis of the material presented above shows that even at the very earliest stages of evolution, the lithosphere was differentiated laterally into separate terrains which had developed in quite different ways. This differentiation is expressed as differences in sedimentation conditions in the various separate provinces, in particular features concerning the manifestation and composition of early magmatism, in the thermodynamic regimes of metamorphism, and in the volume and composition of granitoids. Three major periods of crustal growth in the basement to the Siberian Platform have been identified in the Early Precambrian history of the Aldan Shield: Early and Late Archaean and Early Proterozoic, subdivided into a number of sequential evolutionary cycles.

The Early Archaean epoch resulted in the formation of all the major structural domains. This epoch embraces two growth periods. The earlier one is related to the accumulation in early provinces (AR_1^1) of the Kurumkan (Iyengr) assemblage in the middle of the shield and the Kurultin and Zverev assemblages around the margins. Deposition of these assemblages, at least in the central and western parts of the shield, took place on an older sialic basement, consisting of enderbitic and tonalitic gneiss, whose source material age is 3.7 Ga. Sedimentation and magmatism in the earlier period took place in conditions of lateral tectonic inhomogeneity — more stable at the centre of the shield and less stable at the margins, resulting in associated lithological differences in the composition of Early Archaean assemblages in the central and marginal parts of the shield. This is particularly noticeable in the Kurumkan Formation (Iyengr Group) which has a quartzitic gneiss composition, compared to the Kurultin and Zverev complexes in which mafic schists and basic intrusive rocks play a much more pronounced role. Accumulation of the Kurumkan assemblage characterised cratonic conditions in the centre of the region. The relatively mobile regions at the shield margins possess a low degree of differentiation in the sedimentary assemblages which accumulated there, and they contain saturated igneous rocks.

A lower geothermal gradient for the succeeding regional metamorphism at the edges of the shield appears to be connected with crustal basification around the margins of the early craton as a consequence of the more rapid burial of denser crust. High-gradient regimes of granulite and amphibolite facies regional metamorphism which took place during the later period of the Early Archaean epoch (AR_1^2) in the central part of the Western

Aldan block are replaced in the Southern and Eastern Aldan provinces by lower-gradient events, which reach a minimum in the Zverev complex. The geothermal gradient in these terrains approximates to the minimum Archaean geotherm (Fig. II-33). Regional granulite facies metamorphism at the centre of the Aldan Shield was accompanied by partial melting and intense granitization, where as a result of a density inversion in the granitized systems, gigantic dome-shaped structures formed, of the Lower Timpton cupola type.

Nappe-thrust structures and the Central Aldan linear fold belt formed at the boundary between the Western and Eastern Aldan blocks. The age of the supracrustal assemblages which accumulated during the Early Archaean period has not been firmly established. On the basis of radiometric data obtained from gneisses and schists in the Olyokma and Kurultin terrains, this age is $\geq$3.0 Ga or $\geq$3.3 Ga, respectively.

The Late Archaean epoch (cycle III) was a major stage in the evolution of the Aldan Shield. Relatively stable conditions prevailed in the central part of the shield. Crust-forming internal processes were displaced to marginal zones, where granite-greenstone terrains originated: the Olyokma in the west and the Batomga in the east. This cycle was marked by the subsequent formation of supracrustal assemblages in individual spatially isolated structures of slightly different age, distinguished from one another in terms of lithology and types of internal processes. In the Olyokma granite-greenstone terrain, supracrustal rocks constitute the Olondo, Borsala and Tungurcha Groups; the Bulgunyakhtakh and Balaganakh Groups in the pre-Olyokma active zone; and the Chumikan Group in the Batomga block. There are lateral changes in the quantity and composition of volcanic rocks and in their relationship with sedimentary rocks in various linear belts within the granite-greenstone terrain. The earliest event in this cycle was the initiation of the Olondin structure in the centre of the Olyokma block. The basement to this belt was the Olyokma complex with segments of reworked older basement. The Olondin structure can be considered a typical greenstone belt on account of its steep synclinal structure and the composition of its volcanic and intrusive rocks. East and west of the Olondo belt are the Borsala and Tungurcha Groups, which occur in isolated linear structures, forming a series of belts concentrated along linear tectonic zones — the Chara-Tokk and Temulyakit. The volume of basic metavolcanics in these belts decreases westwards and eastwards away from the Olondo structure. The regional metamorphic gradient also changes laterally, decreasing westwards and eastwards from the Olondo belt. Remobilization and granitization took place in the granite-gneiss and gneiss of the greenstone belt basement, with the formation of dome structures.

The South Aldan deep fault and the development of the Stanovoy fold belt were initiated at the start of the Late Archaean. Large numbers of basic and ultrabasic intrusions were emplaced along the fault zone, including the layered gabbro-anorthosite intrusions of the Kalar and Dzhugdzhur massifs. As far as cycle III is concerned, basic dykes are seen only in the southern part of the Aldan massif, where they were involved in the third deformational cycle. Folds appear only in the immediate vicinity of the Stanovoy zone. The degree of tightness in the folds decreases rapidly northwards.

The lower Proterozoic epoch encompasses two cycles, IV and V. Cratonization begins to be seen in the Olyokma and Batomga granite-greenstone terrains. The foundations of the broad Kodar-Udokan basin were laid down on an Archaean basement in the Olyokma block and the basin was filled by the Udokan Group and a number of more minor graben-synclines, for example the Lower Khanin. Most of the sediments are terrigenous clastics and their regular succession indicates a gradually stabilizing tectonic environment and a transition to a sub-platform regime. The lower part of the Udokan Group succession contains a significant proportion of greywackes, suggesting that they formed from the weathering of volcanics. From geophysical data, the Udokan Group is underlain by "light" rocks, the physical properties of which correspond with those of the granite-gneisses in the Olyokma complex. Metamorphism of the Udokan rocks proceeded in two stages, the earlier zonal one of which was related to thermal uprise in the activated Archaean basement during the formation of dome structures under an exceptionally high geothermal gradient. Porphyritic granite intrusions are a characteristic feature and they are often restricted to dome cores. The Uguy and Oldongso Groups are other assemblages which formed during the lower Proterozoic epoch and the constituent rocks in the Uguy and Oldongso belts can be distinguished on the basis of their low metamorphic grade.

A whole series of zones of tectonic activity, recrystallization and remobilization formed in the west of the Aldan massif during the lower Proterozoic epoch, together with Early Proterozoic plutonic belts, such as the Ungra gabbro-tonalite belt, metamorphosed at high-temperature amphibolite facies in a high gradient regime. Thin assemblages of weakly metamorphosed Yarogu Group supracrustal rocks are present in tectonic wedges. Intense granite-forming processes in tectonically active zones (Reutov, Aldan-Killiyera, Ungra-Dyos-Melemken) belong to this epoch. The main deformation type in these active zones was the formation of nappes and shear folds (horizontal compression). We may assume the existence of similar zones at the eastern contact of the Aldan massif as well.

The Proterozoic epoch saw the final cratonization of the Aldan Shield and the formation of near-horizontal platform-type Riphean sedimentary assemblages, as well as variegated Vendian deposits. Internal activity was expressed in the intrusion of dykes and gabbro-diabase, diabase porphyrite and monzodiorite massifs and the formation of the Saligdar apatite deposit.

TABLE II-6

Typical ore deposits and shows in the Aldan Shield

Mineral deposit or ore show, useful mineral	Tectonic setting	Ore-formational and genetic type of mineralization	Geological and mineralogical features: Host rocks or mineralized rocks	Shape of deposit, ore body morphology	Type of ore	Mineral composition of ores (minor minerals) and hydrothermally altered rocks	Age of host rock	Age of mineralization
Tessegey (Temulyakit-Tasmiyeli zone); iron	Olyokma block; greenstone belt	Magnetite quartzites; metamorphic	Bi-amph and amph gneisses, amphibolites	Conformable layered and lenticular deposits	Densely disseminated to massive	Magnetite, quartz (haematite, pyrrhotite)	Late Archaean	
Yagindya (Sutam group of deposits); iron	Western Aldan block; Idzhek-Sutam province	Hyp-qz-magnetite; metamorphic	Hyp and hyp-ga gneisses	Lenticular and vein-type bodies	Ditto	Magnetite, quartz, hypersthene (haematite, pyrrhotite, pyrite)	Early AR	Archaean to Early PR (?)
Tayezhnoye (Tayezhnoye group of deposits); iron	Western Aldan block	Magnetite; deep skarns	Px, amph, bi-amph gneisses and schists	Lenses and layer-type deposits	Massive, pocket disseminations	Magnetite (pyrite, pyrrhotite, ludwigite)	AR-Early PR	Early PR
Imalyk, Tarynakh; iron	Olyokma block; greenstone belt	Magnetite quartzites; metamorphic	Bi and amph gneisses, amphibolites	Conformable layer deposits	Disseminated to massive	Magnetite, martite, haematite (pyrrhotite)	Late Archean	
Nelyukin; iron	Ditto	Magnetite quartzites; metamorphic	Bi-amph gneisses and schists	Ditto	Ditto	Magnetite (haematite, pyrrhotite)	Ditto	
Davangrin; iron	Western Aldan block	Haematite; sedimentary	Sandstones, siltstones, shales	Ditto	Disseminated (oolitic); nodular oolitic	Haematite (pyrite, gold)	Late Proterozoic	
Chiney; iron, titanium, vanadium, copper	Olyokma block	Titanomagnetite; magmatic copper sulphide ore related to gabbro-norites	Gabbro-norite, gabbro	Lenses, pockets, vein-like bodies	Disseminated massive	Titanomagnetite, chalcopyrite (magnetite, ilmenite, pyrrhotite, etc.)	Early Proterozoic	
Udokan; copper (Naminga-Chitkandin group of deposits)	Ditto	Cupriferous sandstones; sedimentary–hydrothermal, metamorphosed	Sandstones, siltstones	Conformable layered deposits	Disseminated, rarely vein-like disseminations	Chalcosite, bornite, chalcopyrite, pyrite, pyrrhotite (magnetite, titanomagnetite, sphalerite, etc.)	Early Proterozoic	
Katugin zone; rare metals, rare earths	Ditto	Rare-metal; metasomatic (albitite, hydrothermal) related to granites of Kalar massif	Granite contact with schists	Irregular-shaped ore deposits, stockworks	Disseminated, vein-type disseminations	Pyrochlore, zircon (fluorite, scheelite, pyrite, molybdenite, sphalerite, etc.)	Early Proterozoic	
Ore shows in Ulkan zone; rare metals	Batomga block	Rare-metal; metasomatic-greisen, albititic, related to alkali granite intrusions	Endocontact zone of granites	Stockworks predominate	Ditto	Wolframite, cassiterite, molybdenite (sphalerite, chalcopyrite, columbite, etc.)	Proterozoic	
Gold ore shows, Kabaktan-Tungurcha zone	Olyokma and Western Aldan blocks	Gold-sulphide-quartz; hydrothermal	Greenschist facies retrogressed gneisses & schists	Cross-cutting veins	Disseminated, clusters, pockets	Pyrite, galena, chalcopyrite, sphalerite, gold	Early AR	PR-MZ
Ore shows in Atugey-Nuyam zone; gold	Western Aldan block, late Proterozoic basin	Auriferous conglomerates and gravels; sedimentary	Conglomerates, gravels and sandstones	Bedded deposits (conglomerate cement)	Disseminated, vein-type disseminations	Magnetite, ilmenite, rutile, gold	Late Proterozoic	
Central Aldan group of deposits; gold	Western Aldan block	Gold-sulphide-quartz; hydrothermal-metasomatic	Limestones and dolomites of the Yudomian	Layered and lenticular deposits, veins, mineralized crush zones	Disseminated vein-type disseminations	Pyrite, chalcopyrite, sphalerite, galena, haematite, gold	Vendian	Mezozoic
Emeldzhak; (deposits in Emeldzhak-Tayezhnoye zone); phlogopite	—Ditto—	Phlogopite; deep skarns	Carbonate and calc-silicate rocks in Fyodorov Fm	Pockets, cross-cutting and bedding-parallel veins	Cavities	Phlogopite (magnetite, apatite, pyrrhotite, chalcopyrite, etc.)	Early AR	Early PR
Tene-Olyokma muscovite, rare-metal zone	Olyokma block	Rare-metal, muscovite; pegmatitic	Gneisses and schists	Pegmatite veins	Cavities	Muscovite (pyrite, chalcopyrite, galena, molybdenite, columbite)	Archean, Early Proterozoic	
Rock crystal, Central Aldan	Western Aldan block	Quartz-crystal; hydrothermal	Quartzites	Veins, stockworks, pipe-type deposits	Cavities	Rock crystal (haematite, fluorite, calcite, barite, zeolites, etc.)	Early AR	Early PR
Seligdar deposit; apatite	—Ditto—	Apatite-dolomite; hydrothermal–metasomatic	Schists, gneisses, granite-gneisses	Stock-like deposits	Disseminated, vein-type disseminations	Apatite, dolomite (quartz, martite, haematite, calcite)	Early AR	Proterozoic

The Palaeozoic epoch was marked by the deposition of Lower Cambrian horizontal carbonate successions and the intrusion of various granitoids, the most important of which are the alkaline granitoids in the Kodar-Udokan foredeep.

During the Mesozoic, Jurassic sediments were deposited in broad basins on the Aldan Shield; these include coal measures, accompanied by hypabyssal intrusions of alkaline rocks.

Particular groups of useful mineral deposits correspond to different stages in the evolution of the region. Their distribution in the Olyokma-Aldan region is shown in Figure II-24 and brief characteristics are outlined in Table II-6.

Chapter 3

DZHUGDZHUR-STANOVOY PROVINCE

The Dzhugdzhur-Stanovoy province is situated to the south of the Aldan Shield proper. Its southern boundary is the Phanerozoic Mongolia-Okhotsk belt, while to the west it is bounded by the Baikal and Patom belts (Fig. II-34). The tectonic development of the terrain is represented by several cycles of internal activity which began in the Early Archaean, reached a culmination in the Late Archaean and were completed by the start of the Early Proterozoic when the terrain was cratonized. However, even in later periods the quasi-platform tectonic regime was interrupted by plutonic and tectonic activity, particularly in the south where the most intense processes coincided with the Cimmerian orogeny.

A particular feature of the province is its high degree of mobility, which is expressed in the frequent alternation of cycles of activity and the

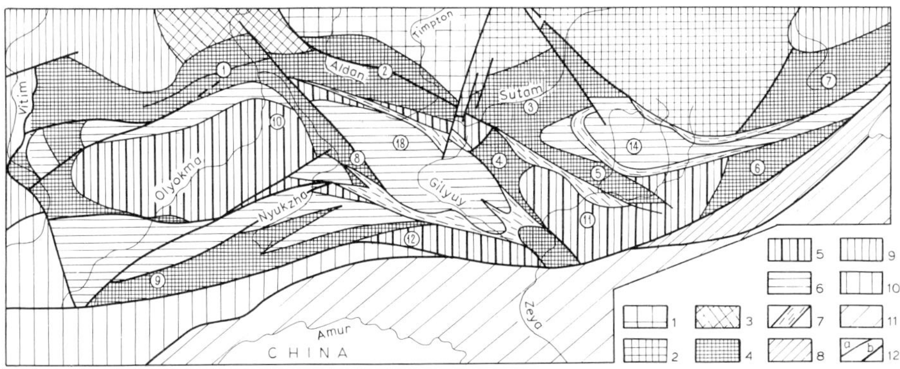

Fig. II-34. Sketch map showing structure of the southern margin of the Siberian craton (compiled by N.I. Moskovchenko). Aldan Shield: *1* = W Aldan granuilte-gneiss terrain, *2* = E Aldan granulite-gneiss terrain, *3* = Olyokma granite-greenstone terrain. Dzhugdzhur-Stanovoy fold belt: *4* = polychronous granulite blocks (numbers in circles: *1* = Kurultin, *2* = Zverev, *3* = Sutam, *4* = Bryansk, *5* = Tokk-Sivakan, *6* = Chogar, *7* = Dzhugdzhur, *8* = Larbin, *9* = Mogochin); *5* = amphibolite facies Stanovide blocks (numbers in circles: *10* = Nyukzha, *11* = Zeya, *12* = Urikan); *6* = sialic type (numbers in circles: *13* = Ilikan, *14* = Kupuriy); *7* = Lower Proterozoic suture troughs; *8* = Stanovides buried by younger sedimentary cover. Baikal fold belt: *9* = Baikal-Patom and Barguzi systems and regions covered with Middle and Upper Proterozoic sediments; *10* = Upper Vitim system; *11* = Mongolia-Okhotsk fold belt; *12* = boundaries of structural zones: *a* = block boundaries, expressed as syntectonic structural zones, *b* = edge of Dzhugdzhur-Stanovoy province.

spatial inhomogeneity of processes. This produced a variety of structural-lithological complexes or terrains at all time levels, and a heterogeneous structure both vertically and laterally.

Structural features and tectonic subdivisions

The folded-block structure of the Dzhugdzhur-Stanovoy province formed during the Archaean and Early Proterozoic. Geological cycles were accompanied by a southward shift of active tectonic zones from an older stablized core, the Aldan granulite-gneiss terrain. These processes occurred at different times and under different conditions both laterally and vertically, thus producing a complex geological and tectonic structure, which is extremely varied at the present erosion level.

For many years, the main principles for subdividing and correlating the Precambrian complexes of the Dzhugdzhur-Stanovoy province were based on differences in the metamorphism of the rocks. There were two points of view then concerning its geological structure and history. According to one of these, the province was made up of Early Archaean granulites which were correlated with the Aldan granulites but differed from them in being reworked several times in lower-temperature regimes, beginning from the Late Archaean (Korzhinsky, 1939; Korikovsky, 1967). According to another, amphibolite-facies metamorphosed supracrustal rocks represent an independent Late Archaean Stanovoy Group (Dzevanovsky, 1958; Sudovikov et al., 1965; Moshkin, 1958), with the granulites which formed in the Early Archaean being the basement to this group. Lower Proterozoic rocks, metamorphosed at epidote-amphibolite and greenschist facies were considered to be assemblages that formed in later trough-type structures.

At the present time, tectonic schemes for the subdivision of the Dzhugdzhur-Stanovoy province always reflect three major early Precambrian evolutionary periods, independent of any concepts concerning the geological structure, extent and age of terrain boundaries. Granulite blocks have been identified, traditionally referred to as Early Archaean basement inliers. Most of the region consists of rocks of the second structural stage, represented by upper Archaean formations of the Stanovoy complex. The third stage comprises lower Proterozoic basin infill assemblages.

In compiling the structural scheme for the Dzhugdzhur-Stanovoy province (Fig. II-34), account was taken of differences in rock metamorphism. Here, however, we have used results of the interrelations between structure and metamorphism, isotope geochemistry and lithoformational analysis which have shown that differences also occur within each structural stage. The tectonic subdivision of the Dzhugdzhur-Stanovoy province was therefore made on the basis of the sum total of all the factors peculiar to terrains of different ages in each structural belt.

In recent years, a marked inhomogeneity in the composition and structure of blocks consisting of granulites has been demonstrated. The fact that the granulites in the south of the Aldan granulite-gneiss terrain and the Dzhugdzhur-Stanovoy province are highly mafic has been known for a long time, and that they formed at high *P* (Dook, 1977; Karsakov, 1978; Kastrykina, 1983; Marakushev and Kudryavtsev, 1965; Moskovchenko et al., 1985; Panchenko, 1985), to which have been added recent data proving these features to be due in large measure to internal processes being active in the Late Archaean. The highest-pressure granulites formed at precisely this time (Dook, 1977; Shukolyukov and Bibikova, 1986; Moskovchenko et al., 1985; Shuldiner et al., 1983). The changing nature of the petrogenetic conditions of the granulites and their spatial distribution deserve special attention. They prove that the Dzhugdzhur-Stanovoy province was tectonically differentiated even during the Early Archaean. For example, the deepest-level Early Archaean granulites in the Aldan evolutionary cycle are localized along the "Stanovoy suture" and comprise the Sutam, Tokk-Sivakan and Chogara blocks (Fig. II-34). They can be traced as a continuation into the Dzhugdzhur-Stanovoy province of the linear structure of the Aldan granulite-gneiss terrain — the Idzhek-Sutam zone (Part II, Chapter 2). This zone divides the Dzhugdzhur-Stanovoy province into two parts — western and eastern — in which the terrains at each of the three structural stages are distinct. In particular, the oldest granulites in the western part differ in being more mafic and having formed at lower pressure.

The highest pressure granulites are restricted to the southern part of the Zverev and Sutam blocks, the Larba block, that is to internal belts within the Dzhugdzhur-Stanovoy province.

Amphibolite facies rocks also display heterogeneity and various times of formation. Some of the amphibolite facies rocks were reworked in the Late Archaean Stanovoy cycle, which may correlate with the early granulites. Femic rocks seen within the Nyukzha and Zeya blocks (lying to the west of the Idzhek-Sutam lineament) belong here. The femic evolutionary trend of these blocks is also preserved during the formation stage of the rocks of the second structural stage, which consists of the Stanovoy Group, corresponding to the subsequent Late Archaean amphibolite facies processes. The femic blocks are separated by ensialic belts whose basement is represented by primary crustal tonalitic gneisses (Kratz et al., 1984; Shuldiner et al., 1983). Examples are the Ilikan block and, in the east of the Dzhugdzhur-Stanovoy province, the Kupurin block. Differences in composition and internal evolution at the early stages in these blocks determined the various trends in how they subsequently developed.

Some of the rocks at amphibolite facies metamorphic grade are currently described as lower Proterozoic trough complexes. Their constituent volcanogenic and volcano-sedimentary formations occur in a number of sutures, restricted to block boundaries or other older inhomogeneities, gravitating towards the nearly north–south central dividing zone of the Dzhugdzhur-Stanovoy province (Fig. II-34). These complexes shifted southwards with time, accompanied by a change in formational characteristics and a reduction in metamorphic grade. The sediments in these particular troughs were previously referred to as the Tukuringra complex (Sudovikov et al., 1965) and more recently the Gilyuy, Odolgo and other complexes.

Starting in the Late Archaean, there was a reworking of the older structural plan of the Dzhugdzhur-Stanovoy province and the establishment of northwest to approximately east–west strike directions. By the end of the Early Proterozoic these structures had obscured the trends of the boundaries of older crustal inhomogeneities. Differences in the tectonic regime of the various episodes, together with the displacement of zones of activity and alterations in the structural plan during progressive development, led to an exceptionally heterogeneous structure, which was further complicated by post-folding fault movements and plutonic activity of different ages during the long orogenic episode. This implies that the identification of blocks and their boundaries is somewhat provisional.

The best known region is the Olyokma-Gilyuy interfluve in the west of the Dzhugdzhur-Stanovoy province — the Stanovoy fold belt (Fig. II-35). Its northern boundary passes south of the Chulman depression with Jurassic coal measures. As a result of the uneven development of the territory, it was subdivided many years ago into a number of structural-formational zones (Sudovikov et al., 1965; Kastrykina, 1983; Shuldiner et al., 1983). The northern zone is transitional to Aldan structures and includes the Zverev and Kurultin granulite blocks in the Sutam metamorphic belt (Glebovitsky, 1975). Its evolution began in the Early Archaean and by the Late Archaean it was a relatively stable crustal segment, having undergone intense reworking by internal processes, connected with the development along the southern boundary of the zone of a system of volcanogenic trenches during Stanovoy time. The central Stanovoy zone, despite the presence of an older basement, owes its structure and composition to several impulses of crust-forming processes which occurred at the Archaean-Proterozoic boundary. It is divided by the Larba granulite block into the femic Nyukzha block and the sialic Ilikan block. Due to the more intense action of internal processes during the Early Proterozoic, the southern margin of the Ilikan block is sometimes delineated as a separate subzone, included as a constituent of the southern Tukuringra zone (Sudovikov et al., 1965). The Urkan block of metabasics belongs here too, together with the granulites and amphibolites of the Mogocha block. These blocks are distinguished not only on the basis of repeated plutonic and tectonic alteration during the Late Precambrian and Phanerozoic stages, but also by having older terrains. Therefore it is more correct to consider the Dzheltulak suture trench to be the boundary

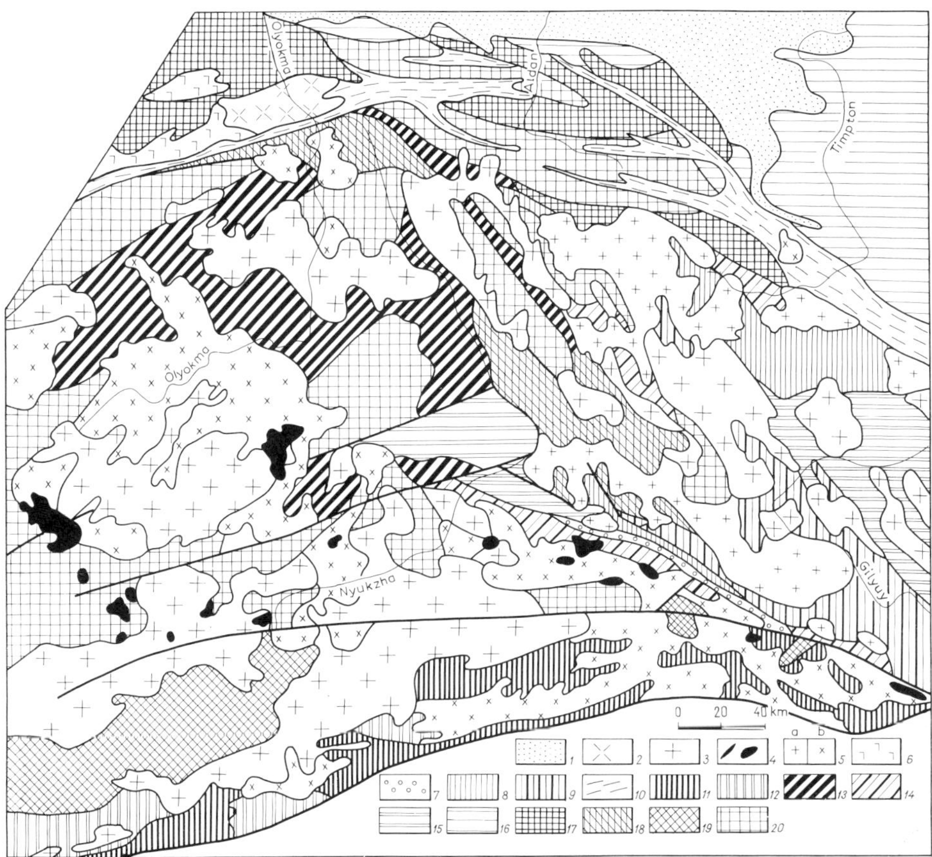

Fig. II-35. Geology and structure of the western part of the Dzhugdzhur-Stanovoy fold belt (compiled by N.I. Moskovchenko). *1* = Mesozoic sediments. Early Proterozoic intrusions: *2* = syenite and granosyenite, *3* = granitoids, *4* = basic and UB. Late Archaean intrusions: *5* = late- and post-orogenic granitoids: *a* = K–Na and K series, Ilikan block, *b* = Na and K–Na series, Nyukzha, Zverev and Urkan blocks; *6* = gabbro-anorthosite-enderbite series. Lower Proterozoic supracrustal assemblages: *7* = carbonate-terrigenous schists, conglomerates, marbles, *8* = volcanogenic-terrigenous gneisses, amphibolites. Upper Archaean assemblages: *9* = bimodal volcanogenic amphibole and biotite schists, gneisses, amphibolites, *10* = mafic-ultramafic-quartzite, *11* = amphibolites and hb-bi schists, *12* = as *11*, strongly granitized. Archaean assemblages: *13* = mafic-ultramafic amphibolites, ga-hb and px schists, *14* = as *13*, in late Archaean zones of intense reworking, *15* = qz schists, gneisses, quartzites, *16* = quartzites, amphibolites, hb-px schists. Polyphase Archaean orthocomplexes: *17* = px–hb ± ga mafic granulites (schists) in Zverev and Kurultin blocks, *18* = as *17*, in Larbin block. *19* = Stanovoy granulites in zones of subsequent reworking; *20* = kinzigite-gneiss and granite-gneiss complexes with relicts of orthogneisses from older sialic crust.

of the central and southern zones; the suture formed throughout the Late Archaean–Early Proterozoic and completed its structural and metamorphic history at the end of the Early Proterozoic. Each of these tectonic zones and blocks is distinguished primarily on the basis of the composition, structure and evolution of its constituent terrains.

Structural-lithological terrains

The best studied terrains are those in three blocks, the Nyukzha, Zverev and Ilikan (Table II-7). Lower Archaean terrains, which form the lower structural stage in each block, include granulites and amphibolite facies rocks. Essential differences between the granulites allow the granulite terrains to be subdivided into the *Zverev*, *Kurultin* and *Larba*. There are two high-temperature amphibolite facies terrains, the *Elgakan* mafic-ultramafic complex, and the *Chilchin* gneiss complex.

The granulite terrains can be traced uninterruptedly in the Sutam belt and along the boundary between the Nyukzha and Ilikan blocks. The Sutam belt, in which the oldest formations are the Zverev and Kurultin complexes, forms an independent suture zone.

Three formations can be distinguished in the granulite terrains. They are usually regarded as part of the succession of a single supracrustal group, the Zverev Group (Katz, 1962; Neyelov and Sedova, 1963). On the whole, they remain constant within all the granulite outcrop areas. The oldest are mafic and ultramafic schists, the Imangrakan Formation. The second formation, the Zverev, includes gneisses, both aluminous and acid, as well as granulite, kinzigite and rarer quartzite. At the top of the section comes the Kurbalikita Formation of varied gneisses with pyroxene, amphibole and biotite, together with subordinate amphibolites and mafic rocks.

The three-fold structure of the terrains is explained in another way by adherents of the plutono-metamorphic nature of granulites (Moskovchenko et al., 1985). Granulite source materials are tonalitic orthogneisses and mafics, the ratio between the two being variable in this ancient bimodal association (Kratz et al., 1984). The basic rocks in the Zverev complex belong to the andesite-basalt series, in which the basic component predominates and which is characterized by a high alumina content. Sometimes it is inclined towards tholeiites (Fig. II-36). Tholeiites are characteristic of segments containing the most widespread plagiogneisses, for example the Kurultin block, while the andesite-basalt series is typical of basification zones. As the basic rocks evolve, the plagioclase gneisses lose their homogeneous texture, and the first banding appears, folded by early F_1^{zv} folds at the same time as boudinage of basic bodies. A second generation of basic rocks is represented by intrusions which cut the earlier structure; they have a layered structure and a wide range of compositions, corresponding to the tholeiite series. They differ markedly from older tholeiites in having higher iron content and lower alumina and alkalis. The oldest basic rocks in the Larba complex belong to the tholeiite-komatiite series.

TABLE II-7

Structural-lithological provinces

Northern Province (Sutam belt)	Central Province (Stanovoy)		Southern Province (Tukuringri)
Zverev & Kurultin blocks	Nyukzha block	Ilikan block	Urkan block
Vulcanism: basalt-andesite-liparite formation	Dykes of trachybasalt-trachyliparite formation Intrusions of diorite-granodiorite series		Vulcanism: tranchyandesite-trachyliparite formation Terrigenous assemblages
Local greenschist facies retrogression and blastomylonitization			Local deformation Low-temperature metamorphism Basic vulcanism during formation of eugeosynclinal assemblages Yankan complex
Magmatism and metasomatism during episode of stablization Granite-granosyenite series; low-T alkaline and acid metasomatism Late Stanovoy granites Local metamorphism, deformation Chulman complex Metabasic dykes Beginning of stabilization of block Horizontal schistosity and blastomylonitization	Granodiorite-granite series Layered basic and ultrabasic intrusions	Alkaline and acid metasomatism High-gradient zonal metamorphism Volcano-sedimentary Dzheltulak complex Horizontal movements Synorogenic granitoids Folding, low-gradient zonal metamorphism up to amphibolite facies Volcanogenic and terrigenous assemblages Late Stanovoy, Gilyuy and other complexes Horizontal schistosity, basic dykes	Multiple plutono-tectonic processes, predominantly granitic magmatism
Early Stanovoy granites Horizontal tectonic movements Folding, metamorphism at almandine-amphibolite facies, formation of autochthonous granites Vulcanism, tholeiite series Early Stanovoy complex		Granite emplacement (analogues of Early Stanovoy granites) Folding, medium-gradient metamorphism at amphibolite facies, granite formation Commencement of stabilization of block	Deformation and metamorphism up to granulite facies Mogocha complex
Magmatism: layered intrusions of gabbro-anorthosite-enderbite-mangerite series Metamorphism: zonal, medium- and low-gradient, up to granulite facies Deformation with predominantly horizontal tectonic movements Vulcanism: tholeiites, komatiites, andesite-basalts Kholodnikan complex	Enderbites, charnockitic gneisses Folding and low-gradient metamorphism Multiple basic volcanic activity, andesite-basalt series Deformation, high-gradient metamorphism, granulite and amphibolite facies Volcanism (?) komatiite-tholeiite series	Temporary stablization of block	
Dykes of basaltic and peridotitic komatiites Folding, medium- and low-gradient metamorphism, granulite facies Basic intrusions (gabbro-leucogabbro) Folding, metamorphism (?) Andesite-basaltic and tholeiitic magmatism Zverev and Kurulta complexes	Larba and Elgakan complexes	Folding, metamorphism, granite formation Trachyto-basalt dykes Gneissic and quartzitic-gneissic assemblages Chilchin complex	Folding, high-gradient amphibolite facies metamorphism Vulcanism of andesite-basalt series Urkan complex

Early Archaean, 3.3 Ga		Late Archaean, 3.3-2.7 Ga	
Aldan Shield		Early Stanovoy cycle	
1st stage	2nd stage	1st stage	2nd stage

Early Proterozoic, 2.7-2.0 Ga		Late Proterozoic	Geological epoch
Late Stanovoy cycle			Cycle and stage
1st stage	2nd stage		of development

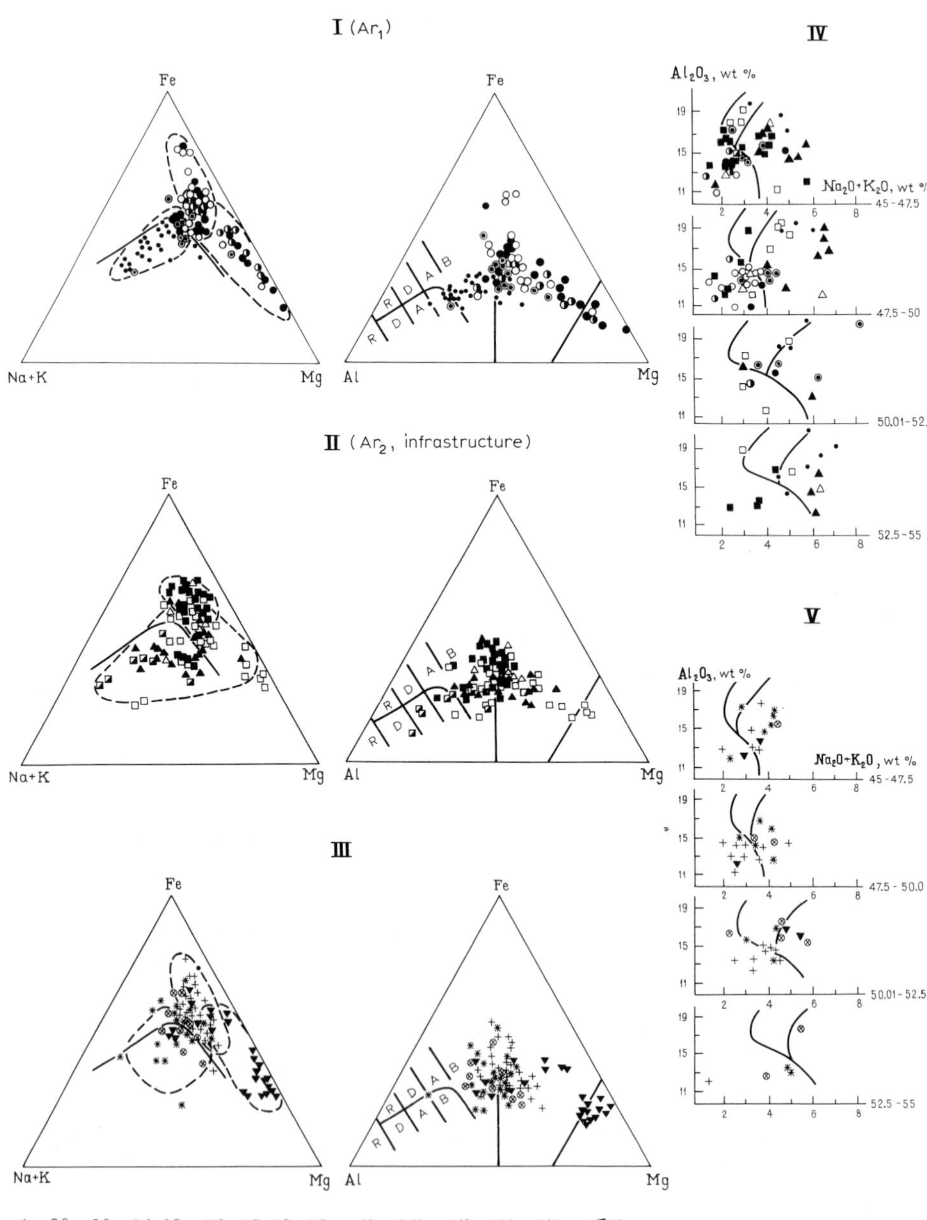

Fig. II-36. Petrochemical classification diagrams for basic rocks (after Kuno, 1968). I. (Na + K)–Fe–Mg and Al–Fe–Mg diagrams for Early Archaean period. Oldest mafic rocks: *1* = Zverev and Urikan blocks, *2* = bodies dispersed among grey gneisses. Mafic rocks of Aldan cycle: *3* = Zverev, *4* = Larbin, *5* = Nyukzha blocks. II. Diagrams for complexes in infrastructure during Late Archaean peridod in: *6* = Zverev, *7* = Larbin, *8* = Nyukzha blocks. Proterozoic basic dykes in: *9* = Zverev, *10* = Larbin blocks. III. Diagrams for mafic rocks in complexes of second and third structural stages: *11* = Early Stanovoy, *12* = Kholodnikan, *13* = Ilikan, *14* = Late Stanovoy; *15* = compositional fields for groups of mafic rocks of various ages. IV, V: Stepwise diagrams for groups of mafic rocks of various ages in infrastructure and second–third structural stages, respectively.

Zones of basic magmatism, referred to as the Imangrakan "formation", are distinctive on account of being highly mafic. Remobilization of plagiogneisses occurred at zone boundaries, culminating in the appearance of displaced anatectic masses. They mark the beginning of acid granulites with the simultaneous formation of kinzigites and aluminous gneisses. This association is usually referred to as the Zverev "formation". It separates zones of basification (the Imangrakan "formation") from areas of plagiogneiss with minor amounts of basic magmatism (the Kurbalikita "formation").

The granulite terrains were initiated by crust-forming processes active during the Early Archaean period. Their radiometric age is not younger than 3.4 Ga, according to isochron determinations on zircons from mafic schists in the Kurultin complex (Shuldiner et al., 1983). During the Late Archaean, the composition and structure of the granulites became complicated owing to rock-forming processes during the development of rocks of the second structural stage, which were completed by 2.7 Ga ago.

The start of the second period is characterized by the intrusion of basic and ultrabasic dykes, separated from the earlier layered gabbroids by a second fold episode — F_2^{zv}. The dykes correspond in composition to basaltic and peridotitic komatiites (Fig. II-36). Subsequent plutonic activity took place under syntectonic conditions and evolved against a background of repeated folding and metamorphism of the plutonic rocks. Basic magmatism was renewed several times during this period. In particular, it was accompanied by the formation of east–west vertical structures, F_4^{zv}, marking the initial imposition of early Stanovoy trenches along the southern boundary of the Zverev block (Fig. II-35). A particularly strong impulse of igneous activity corresponds to the episode of formation of F_3^{zv} horizontal structures, which accompanied the appearance of layered gabbro-anorthosite intrusions, the largest being the Kalara massif in the west of the terrain. Fractionates of these magmas belong to the calc-alkali series, producing minor intrusions of enderbite and charnockite, the latest of these being 2.7 Ga old. Their petrochemical signature indicates retrogression of the granulites, which experienced granitization under the action of highly-alkaline fluids in the conditions existing in the infrastructure. Retrogression was completed by the emplacement of tonalites and Na-alaskites, the intrusion of which was controlled by steep north–south trends of F_5^{zv} deformational events. Granite formation was terminated by F_6^{zv} folding, accompanied by intense horizontal movements, regional schistosity development and blastomylonitization, flattening of all contacts and planar orientation and tectonic interleaving of rocks formed at different crustal levels.

The major mappable folds formed in the period from the end of the Late Archaean to the start of the Early Proterozoic, their trends being nearly north–south (F_7^{zv}), nearly east–west (F_8^{zv}) and north-east (F_9^{zv}). Simultaneously, low-temperature recrystallization of the rocks took place along or overprinted older planar orientations. These then controlled late tholeiitic and andesitic dykes. The history is completed by the intrusion of normal-alkali (late Stanovoy) granites and high-alkali granites.

Granulite petrogenesis is related to a sequence of plutono-metamorphic processes, synchronous with the first three episodes of folding in the Aldan and early Stanovoy cycles, Early and Late Archaean respectively (Dook and Kitsul, 1975; Dook, 1977; Shuldiner et al., 1983). Early granulites are mostly represented by two-pyroxene assemblages without garnet, which formed at temperatures of 900–1100°C (Karsakov, 1978; Kastrykina, 1983). Such high-temperature regimes are confirmed by sapphirine + quartz parageneses in the presence of hypersthene and sillimanite, also garnet with hypersthene and sillimanite in the Sutam and Chogar blocks. These define the lower pressure limit, which even at the very start of mineral formation did not drop below 9 kbar. The formation of garnet in two-pyroxene schists, typical for these pressures, occurs under lower-temperature conditions, simultaneously with amphibolization in a wide P–T interval: T = 970–780°C and P = 8–12 kbar, determined from garnet equilibria (X_{Mg} = 0.25–0.50) with clinopyroxene and amphibole (Moskovchenko et al., 1985). The change in P–T values corresponds to two lines of the geothermal gradient (Fig. II-37). The lower temperature gradient characterises structures caught up

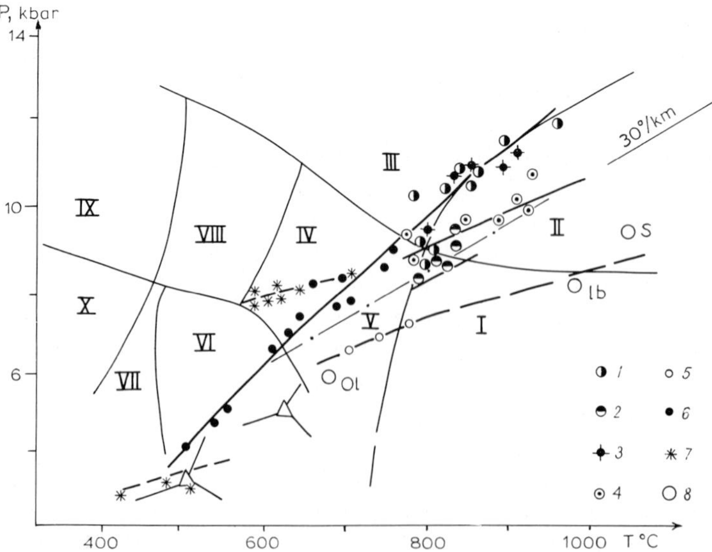

Fig. II-37. P–T parameters and geothermal gradients during the formation of terrains of various ages. Granulites: *1* = Zverev, *2* = Kurultin, *3* = Larba, *4* = Sutam and Chogar; periods of medium-grade metamorphic conditions: *5* = Early Archaean, *6* = Late Archaean, *7* = Early Proterozoic. Metamorphic facies: *I* = ga-cord, *II* = ga-hyp-sill (ga-two-px schists), *III* = eclogite-pyrgarnite, *IV* = ga-ky-ged, *V* = ga-cord-ged, *VI* = qz-st schists, *VII* = blueschists (glaucophane-greenschist), *VIII* = zoisite-hb apo-eclogites, *IX* = zoisite-glauc apo-eclogites, *X* = ga-law-glauc schists.

in internal processes in the early Stanovoy cycle — the southern margin of the Zverev and Sutam blocks and the Larba block. The low gradient line is related to an immediate transition to points of early parageneses of the Stanovoy complex, reflecting a single zonation during Late Archaean progressive metamorphism (Shuldiner et al., 1983; Moskovchenko et al., 1985; Panchenko, 1985) with an age of 2.7 Ga (Shukolyukov and Bibikova, 1986; Shuldiner et al., 1983). Basic garnetiferous granulites of other structures correspond to the higher gradient line, with its flat incline towards the temperature axis, at the end of which are situated high-temperature parageneses of the Chogar and Sutam blocks. Medium-pressure granulite assemblages in Aldan structures also correspond to this same line. Gradient lines meet near the values T = 750–775°C, P = 8–9 kbar. Assemblages which originated during retrogression of granulites as well as those which enter a single zoning pattern with the Late Archaean prograde amphibolite facies correspond to these conditions.

The amphibolite rocks also contain recognisable polychronic complexes, comparable in all evolutionary stages with the granulites and which have undergone Late Archaean high-temperature reworking. One of these is the femic Elgakan complex in the Nyukzha block (Kastrykina, 1983). From its composition and sequence of events, it correlates with the granulites in the Larbin complex (Moskovchenko et al., 1985). The original pre-metamorphic rocks of the complex were metabasics with a high degree of oceanicity of tholeiitic and komatiitic composition (Fig. II-36). After several deformational episodes ($F^{el}_{1,2,3}$), accompanying early high-gradient metamorphism and granitization, a second generation of metabasics appeared. They broadly correspond to tholeiites, but as their silica content decreases, their compositions shift into the high-alumina and alkali basalt field. Metamorphism at T = 710–750°C and P = 8–9 kbar led to the formation of garnetiferous amphibolites and eclogitic rocks which crop out on the flanks of Late Archaean granulite complexes (Panchenko, 1985; Bilibina and Shaposhnikov, 1976). Synmetamorphic folding and granitization ended with intense horizontal movements and the formation of recumbent folds (F_4^{el}), cleavage and blastomylonitization, which concluded the early Stanovoy cycle. Subsequent internal processes include the intrusion of several different age swarms of tholeiite and alkaline basalt dykes, and polyphase granitoid magmatism — mostly Early Proterozoic granitoids.

Sialic terrains show a different pattern in their evolution. These are the Chilchin assemblage and in the east the Zeya assemblage and the Ud-May Group. The composition of the oldest sialic rocks and the degree of their metamorphism change across the strike of the folded terrain. In the west, the Chilchin complex has a basement of homogeneous, massive plagiogneiss around the periphery of the Nyukzha block (Fig. II-35). Their monotonous structure is broken by the first deformational episode, accompanied by early generation of basic rocks and the beginnings of anatexis in gneisses adjacent to the basic rocks. As migmatization progresses in the substrate, garnet, sillimanite, cordierite and other aluminous minerals appear. As in the Elgakan complex, zonal metamorphism occurs during the second episode, synchronous with basic magmatism, in the almandine amphibolite facies temperature interval, reaching the granulite facies boundary with T = 700–730°C and P = 7–8.5 kbar (Kastrykina, 1983; Moskovchenko et al., 1985).

In the Ilikan block, plagioclase gneisses are associated with thick quartzite horizons and sheet-like bodies of alkaline metabasics. Metamorphism is lower grade here, as a result of which quartzite-gneiss associations often identify with the upper parts of the Stanovoy Group succession. However, the more complex nature of deformation and granite formation show that at least some of the gneisses belong to older formations in the block and have an age in excess of 3.0 Ga (Anon., 1986). In the Kupuri block, the carbonate-gneiss Zeya assemblage is the age equivalent of the Chilchin gneisses. Its non-uniform metamorphism reached medium-pressure granulite facies (Karsakov, 1980). Calcareous gneisses and schists in the Ud-May Group on the other hand differ in their high-pressure metamorphism. Judging from the development of zoisite-kyanite and garnet-clinopyroxene assemblages (Kozyreva et al., 1985), they formed in a medium temperature interval at a pressure in excess of 8 kbar.

The upper Archaean complex (second structural stage) comprises supracrustal assemblages and related plutono-metamorphic rock associations. The complex is subdivided into a number of groups and formations. The upper Archaean stratotype is the Nagorny Formation of the Stanovoy Group in the upper reaches of the river Timpton. The succession consists of uniform meta-tholeiites with an initial age of 3.3 ± 0.15 Ga (Neymark et al., 1981). Their structural and metamorphic evolution commenced during the time of retrogression of the Zverev complex, synchronous with F_4^{zv}–F_5^{zv} deformation. Pre-granite metamorphism of the complex took place under almandine-amphibolite facies conditions, with T = 650–670°C and P = 7–8 kbar, that is with the same geothermal gradient as the Late Archaean zonation in the infrastructure. A succession of processes begins with this event, including the formation of nebulous and *lit-par-lit* migmatites, and concluding with the emplacement of Early Stanovoy tonalites, the intrusion of which was controlled by horizontal structures, correlated with F_6^{zv}. These processes have not been established in other parts of the Stanovoy Group succession, creating the major structural features of that portion of rocks, which is identified as the *Early Stanovoy complex*. According to U–Pb isochron determinations on zircon from granitized amphibolites, the Early Stanovoy cycle ended no later than 2.7 Ga ago (Neymark et al., 1981). Within the Ilikan block, upper Archaean supracrustal rocks and related plutonometamorphic associations are grouped together as the *Ilikan complex*. Its basement consists of metavolcanics, compositionally equivalent to picrite, basalt, andesite and dacite. Basic and intermediate compositions are inclined towards higher alumina types. Sodic-type alkali basalts are also present. In addition, there are small outcrops of metapelite and quartzite, including ferruginous types. Metamorphism took place within the amphibolite facies temperature interval. It has been established that it belongs to the kyanite-sillimanite type only for low-temperature fields. Particular features of granitization are low CO_2 and highly alkaline fluids with high f_{O_2}, which limits the formation of garnetiferous assemblages and promotes the growth of magnetite.

Within the internal zones in the Zverev block, north of outcrops of the Early Stanovoy formation stratotype, upper Archaean rocks are represented by metavolcanics of the *Kholodnikan complex*. The basement to this succession includes tholeiitic-komatiitic series volcanics, which are overlain by andesite-basalt series metavolcanics (Moskovchenko, 1983; Krasnikov, 1984). Compositionally they differ markedly from all the other volcanogenic assemblages and are comparable only with dykes (second generation in the infrastructure of the block) and early orthogranulites and metabasics of the Larba and Elgakan complexes (Fig. II-36). Metamorphism proceeded under medium- to high-temperature gradient conditions, with T = 500–600°C and P = 4–5 kbar. The sequence of structural forms and mineral forming processes turns out to be more complex than in the Early Stanovoy complex. Deformation and metamorphism of the Early Stanovoy cycle in the north of the Dzhugdzhur-Stanovoy province probably commenced slightly earlier than in southern regions. During the culmination of Early Stanovoy processes, second generation troughs had already been initiated, the rock successions of which are conventionally correlated with third cycle lower Proterozoic formations.

The lower Proterozoic complex constitutes the third structural stage. It consists of volcanogenic-terrigenous formations which differ in formational characteristics and in the grade and conditions of metamorphism. Supracrustal rocks formed in a series of troughs, one replacing the other in time as they migrated southwards. Belonging to the earliest of these are structures restricted to the northern and southern boundaries of the Zverev block. Along the northern boundary they are filled with high-alumina basalt and andesite-basalt associated with quartzite and metamorphosed in a low-gradient regime, with T = 600°C and P = 7 kbar. These rocks are referred to as the *Chulman complex*. On the southern boundary, their equivalents are gneisses and minor amphibolites which had previously been included in the Stanovoy Group.

From metamorphic conditions they are indeed close to Early Stanovoy formations. However, being in an autochthonous position relative to the former, these rocks are distinct in possessing a simpler history of deformation and granite formation, processes superimposed on the Early Stanovoy complex (Moskovchenko et al., 1983). On this basis, they are recognised as an independent *Late Stanovoy complex*, characterised by andesite-basalt volcanics and gneisses, including aluminous varieties. Metamorphism at T = 630–650°C and P = 8 kbar was accompanied by the development of granites of normal and later high K-alkalinity, forming a late Stanovoy group (Sudovikov et al., 1965).

Within the internal zones of the Ilikan block, equivalents of the late Stanovoy formations are units of gneiss, quartzite and amphibolite of medium metamorphic grade, identified separately as the Odolgo, Agikan and other formations of the Burpala complex. Geologically they occupy an intermediate position between the Ilikan complex and the later Dzheltulak complex. These assemblages, together with part of the lower units in the Dzheltulak complex, are sometimes grouped together as the *Gilyuy complex* (Karsakov, 1980). Supracrustal rocks in the complex are represented by metamorphosed clastic sediments. Andesite-basalt series metavolcanics predominate in other structural zones. Kyanite-sillimanite type zonal metamorphism corresponds to the temperature interval T = 500–600°C, with P = 4–5 kbar (Kastrykina, 1983). Some workers consider that this metamorphism is the only one to have affected the *Dzheltulak complex* and that supracrustal sediments cap the section of the upper structural stage. The volcanogenic-terrigenous section of the complex is represented by two assemblages of zonally metamorphosed rocks, including aluminous schists, quartzites and basic, intermediate and acid metavolcanics. Clastic sediments, including intraformational meta-conglomerate, occupy a prominent position. Volcanics and quartzites decrease from west to east, while carbonates increase (Kastrykina, 1976; Yelyanova et al., 1985). Metamorphic zoning within the Dzheltulak complex approaches isobaric (Fig. II-37) and with increasing temperature it shifts into the andalusite stability field. Rocks in all the metamorphic zones are cut by granites belonging to the Tyndin, Igam and other complexes, distinguished by high alkalinity. The Igam complex embraces a wide spectrum of magmatic and metasomatic rocks. Metasomatites, restricted to tectonic zones, include rocks of sodic and potassic evolutionary series, as well as concluding acid leaching processes (Yelyanova et al., 1985). In addition to the layered basic and ultrabasic rocks of Mount Lukinda, they comprise the distinguishing features of end-Stanovoy magmatism in the time interval 2.2–2.1 Ga ago. These processes mark the extensive Dzheltulak suture, which separates the Ilikan block from the Urkan block. They also characterize the Gilyuy-Armazar lineament, in which a series of plutonic intrusions appears, continuing right up into the Mesozoic.

The latest supracrustal assemblages are rocks of the Yankan complex — weakly metamorphosed diabases and schists. They are considered to be either late Precambrian or end-Proterozoic formations, or to mark the start of the Phanerozoic history of the Mongolia-Okhotsk fold belt. This period of development was accompanied by polyphase tectonic movements, forming a complex small-scale block-mosaic structure on the southern margin of the Siberian Platform.

Tectonic evolution of the Dzhugdzhur-Stanovoy province

The first inhomogeneities in the Dzhugdzhur-Stanovoy province were initiated in the period prior to 3.5 Ga ago (Fig. II-38). They are expressed in the varied composition of primary crustal tonalite gneisses. In structural zones subsequently inherited by mafic granulites, the orthogneisses are seen to contain lower silica and alkalis, but higher basics, alumina and total iron (Kratz et al., 1984). They are also characterized by the appearance of andesite and aluminous tonalite. Early events include the metamorphism of plagiogneisses, related to their first magmatic basification. The early basic rocks of the Zverev and Kurltin complexes also belong to this stage, and are represented by tholeiite and high-alumina basalt with high Na-alkalinity. The same magmatic type is found in the Urkan block to the south.

These processes are separated from the succeeding stage of igneous activity, which has a tholeiitic trend, by an episode of folding. Basic igneous activity developed on a regional scale, characterised by komatiitic melts, associated with metatholeiites with maximum level of oceanicity, developed in the substrate of the Larba and Elgakan complexes. Their outcrop area in the Nyukzha block forms an oval, separating symmetrically arranged zones of andesite-basalt magmatism. The Zeya block may also represent a similar structure. The margins of the oval consist of remobilized masses of tonalitic orthogneiss. In zones where they are best preserved, initial magmatism is highly variable but has a tholeiitic trend; it is accompanied by an intense display of rheomorphism of older sialic masses. The Olyokma is the largest of the blocks of slightly broken sialic crust, which later evolved according to the type of granite-greenstone terrains.

There is no unique interpretation of the interrelations between Early Archaean terrains. Most data indicate that andesitic-basaltic magmatism represents an initial destructive phase of ancient crust, while the komatiitic-tholeiitic type represents a mature stage in the formation of the terrain. Another interpretation holds that an initial event could be the formation of ocean-type magmas, filling a number of broad ovals. In this case, the zones of andesitic to basaltic magmatism are secondary and developed as a result of the destruction of isostatic equilibrium at the boundaries between sialic and simatic crustal segments. Oval structures are characterised by a high geothermal gradient and the absence of any signs of folding. In linear zones, on the other hand, low-gradient metamorphism appeared in a shear stress environment, with the formation of the first granulites with Sutam facies parageneses, corresponding to the lowest temperature gradient for the Aldan cycle.

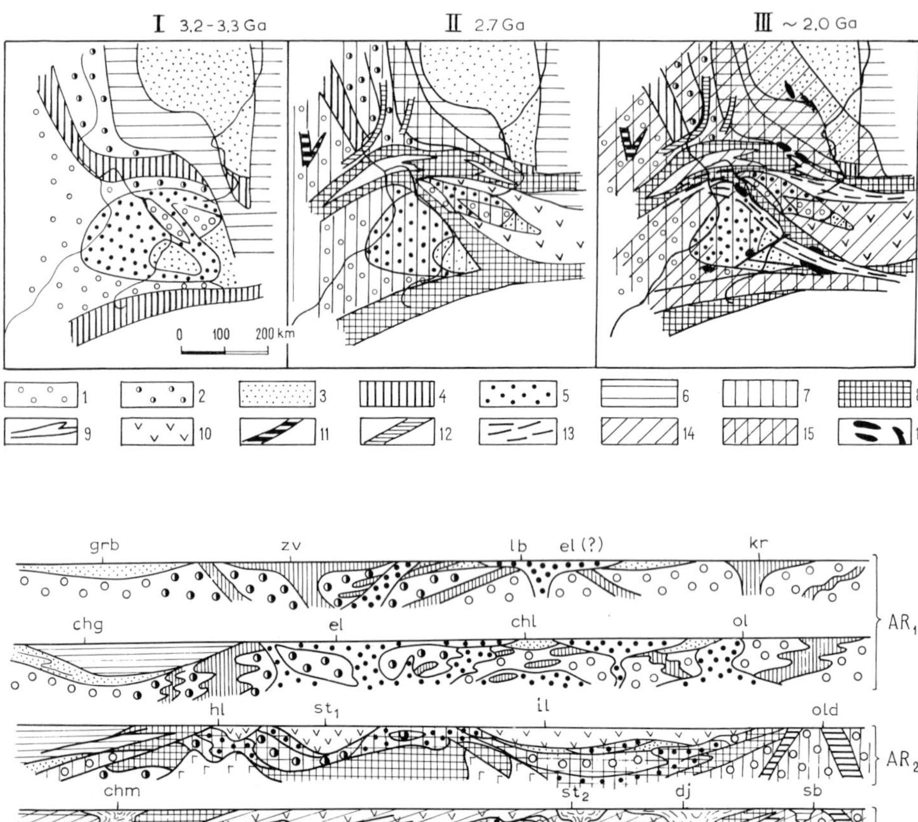

Fig. II-38. Sequential evolution of early Precambrian structural belts and complexes in the Aldan-Olyokma-Stanovoy region. Relicts of old sialic crust: *1* = tonalitic, *2* = andesitic; oldest provinces: *3* = stable, with quartzitic-gneiss-schist protoplatform cover, *4* = belts of early andesite-basalt magmatism, *5* = regional komatiite-tholeiite magmatism, *6* = first mobile belts of Aldan Shield; Early Stanovoy (Late Archaean) provinces: *7* = amphibolite facies belts, *8* = granulite facies belts, *9* = first generation volcanogenic basins with Kholodnikan greenstone assemblages, *10* = second generation volcanogenic basins with tholeiitic and bimodal complexes, *11* = Olondo greenstone belt, *12* = paragneiss belts of Aldan Shield; *13* = Early Proterozoic basins and zones of intense Late Stanovoy reworking; *14* = greatest intensity of igneous activity and plutono-tectonic alteration of concluding phase and subsequent orogenic processes; *15* = as *14*, in zones of continuous distribution of plutonic rocks of various ages; *16* = zones of maximum distribution of basic magmatism. Provinces: AR_1 (*grb* = Gorbylakh (Iyengr), *zv* = Zverev, *kr* = Kurulta, *lb* = Larba, *el* = Elgakan, *chl* = Chilchin, *chg* = Chuga, *ol* = Olyokma); AR_2 (*hl* = Kholodnikan, st_1 = Early Stanovoy, *il* = Ilikan, *old* = Olondo); PR_1 (st_2 = Late Stanovoy (Gilyuy, Odolgo), *chm* = Chulman, *dj* = Dzheltulak, *sb* = Subgan).

Lateral inhomogeneities which originated in the Early Archaean ended at around 3.4–3.3 Ga ago with the separation of three tectonic provinces — the Aldan, Olyokma and Stanovoy — the evolutionary paths of which diverged during their subsequent evolution. Naturally, with the appearance of these provinces, their junction zones began to develop as independent structures. The rocks of the Sutam and Nyukzha belts owe much of their compositional features to Early Archaean processes. However, the major compositional, petrogenetic and structural features arose here during the second Late Archaean period. It encompasses the age interval of 3.2–2.7 Ga, when there was a change in tectonic regime in all three provinces (Fig. II-38). During this period, the Aldan and Olyokma provinces were characterised by more stable conditions, experiencing brittle faulting, along which in particular the Olondo and other greenstone belts formed, and their attendant paragneiss structures. In the Dzhugdzhur-Stanovoy province, intense crustal activity is related to the Early Stanovoy cycle. In the lower structural stage the cycle commences with faulting and the intrusion of parallel swarms of tholeiite and basaltic komatiite dykes. Slightly later, gabbro-anorthosite and calc-alkali series plutons were emplaced. They are restricted to the central and southern parts of the Sutam belt, which in time was transformed into the South Aldan deep fault. High-temperature metamorphism is associated with the intrusions, and granulite facies conditions were once more attained, this time at higher pressures. In the Nyukzha belt during this period there was uplift of deep-level asthenoliths of andesitic-basaltic composition, with the subsequent intrusion of enderbite, andesite and charnockite. The associated zonal metamorphism and granulite facies mineral transformation ended at 2.7 Ga.

The second structural stage, represented by Late Archaean metavolcanics, also reflects tectonic mobility and differentiation of the region during this period, as deduced from the composition and evolution of various associations. The earliest manifestations of volcanism are seen in the north, where the Kholodnikan ultramafic-mafic-siliceous complex and meta-tholeiites of the Early Stanovoy complex crop out, which on the basis of structural complexity and degree of involvement in internal processes are equivalent to the bimodal leptinite amphibolite association of the Ilikan complex. This forms ensialic basins, covering older gneisses and quartzites of the Ilikan block.

All three types of volcanogenic association belonging to the Early Stanovoy period differ in metamorphic conditions. The bimodal Ilikan complex was metamorphosed under relatively high-gradient conditions. Northwards, in the direction of the South Aldan fault, the temperature gradient decreases, but metamorphism of Early Stanovoy volcanics proceeded at the same pressures as granulite mineral growth in the infrastructure. On the northern flank of the Sutam belt, the gradient increases again going farther into the stable Aldan province, and the assemblages in the Kholodnikan complex were altered at relatively low pressure, while high-pressure regimes appear in the subsequent generation of basins.

During progressive evolution, tectonic processes became more and more localized in the junction zones between heterogeneous blocks (Fig. II-38). Low-gradient transformations which affected the infrastructure and suprastructure of the Stanovoy system and ended 2.7 Ga ago with horizontal tectonic movements, predetermined the initiation of a new generation of basins with andesitic to basaltic volcanism and clastic terrigenous sedimentation. They mark the next cycle, which after a brief rest period emerged as a strong impulse of internal activity, peaking at 2.5–2.6 Ga ago. This is the time of metamorphism and folding of the Chulman and Late Stanovoy complexes, restricted to narrow zones, inherited from the strike direction of Late Archaean granulite-mafic belts. Along the boundary of the Sutam belt there is evidence for the lowest-gradient metamorphism, which reached high-temperature amphibolite facies. In the southern zone, parallel to the Nyukzha belt, a reduced vertical temperature gradient also operated, although the metamorphic grade in the rocks is not high. Igneous activity in the form of basic dyke intrusion was widespread outside the basins.

The Early Proterozoic cycle ended with the intrusion of huge volumes of Late Stanovoy granites, a change in the sense of tectonic movements and overthrusting of older deeper-level complexes onto the rocks of the Early Proterozoic basins. Associated with this episode of compression and thrust stacking are complementary extension zones, filled with terrigenous clastic formations of the Dzheltulak complex. The overall tendency towards regional uplift is maintained during this episode and is accompanied by the final intrusion of high-alkali granites at 2.0–2.2 Ga, and by high-gradient zonal metamorphism. These processes affected the Aldan and Olyokma provinces which prior to this had relatively stable regimes. Plutonic regimes became established at 2.0 Ga; in the Aldan province these were accompanied by low-pressure polyphase granulite facies metamorphism.

Metallogenic features

The polyphase history of development of the complexes of various ages is reflected in the distribution of useful minerals. The metallogenic signature has a mixed nature. During the early stages of each cycle, especially the Aldan and Early Stanovoy, it had a mantle origin. During the concluding cycles, crustal sources played a progressively more important role. Late Stanovoy superimposed plutono-tectonic processes had the greatest significance for ore generation.

The lower structural stage, in which the major compositional and structural features were shaped by Early Archaean and Late Archaean processes, controls the emplacement of ore belonging to an apatite-titanomagnetite

TABLE II-8

Typical deposits and ore shows

Deposit, ore show; useful mineral	Tectonic setting	Type of mineralization: ore-formational and genetic	Geological and mineralogical features				Geological age of	
			Host or mineralized rocks	Shape of deposit,	Ore type	Ore mineral composition (minor minerals)	Host rock	Mineralization
Sivakan (Bomnak iron-ore zone); iron	Tokk-Sivakan & Zeysky blocks	Ferruginous quartzites; metamorphic	Gabbro amphibolites	Lenticular deposits	Densely disseminated	Magnetite (haematite, pyrrhotite)	Archaean	
Gayum; phosphorus, titanium, iron	Dzhugdzhur block	Apatite-titano-magnetite; magmatic, related to Dzhugdzhur anorthosite intrusion	Anorthosites, gabbro-norites, norites	Bedded-lenticular deposits	Predominantly disseminated, veins, massive	Titanomagnetite, ilmenite, apatite (magnetite, pyrrhotite)	Proterozoic	
Katuga rare-metal zone; rare earths, gold	Kurulta block	Rare-metal; metasomatic (albititic, hydrothermal) related to Kalara alkali granite intrusion	Contact zone between granites and schists and gneisses	Irregular deposits, stockworks	Disseminated, vein-type disseminations	Pyrochlore, zircon, orthite (fluorite, scheelite, molybdenite, sphalerite, etc.)	Proterozoic	
South Stanovoy auriferous zone; copper, gold	Ilikan and Zeysky blocks	Gold-quartz; hydrothermal	Greenschist facies retrogression of metamorphosed Archaean rocks	Veins	Disseminated	Gold (pyrite, pyrrhotite, chalcopyrite)	Proterozoic	
Dzhultulak gold and copper bearing zone; copper, gold	Dhzeltulak suture	Gold-chalcopyrite; volcanogenic-sedimentary	Graphitic schists, blastomylonites	Sheet-like deposits	Disseminated, vein-like	Chalcopyrite (galena, sphalerite, pyrite, scheelite, gold)	Proterozoic	
North Stanovoy auriferous zone; gold	Zverev block	Gold-quartz; gold-sulphide-quartz; hydrothermal	Retrogressed mafic schists, dykes and granitoid stocks	Veins and stockworks	Disseminated	Pyrite, chalcopyrite (gold, molybdenite, etc.)	Early PZ	Early PZ, Mesozoic
East Stanovoy auriferous zone (Maymakan ore field); gold	Kupuriy block	Gold-quartz with sulphides; hydrothermal	Greenschist facies retrogressed Pre-cambrian meta-morphic rocks	Veins	Disseminated	Pyrite, pyrrhotite, chalcopyrite, gold	Early PZ	Early PZ, Mesozoic
Prishilkin gold-molybdenum zone	Mogocha block	Gold-sulphide-quartz	Greenschist facies retrogressed Pre-cambrian meta-morphic rocks and granitoids	Veins	Disseminated	Pyrite, chalcopyrite, molybdenite, gold, etc.	Archaean-Early PZ	Proterozoic, Mesozoic

ore formation. Ore shows formed during the concluding stages of the Late Archaean cycle, when layered rocks belonging to the gabbro-anorthosite-enderbite series were metamorphosed for a second time. In zones enriched in iron and titanium, the most inert components, rich magnetite and titano-magnetite ore bodies originated. In areas where calcite, the most mobile element, was redeposited, apatite and other associated mineralization became localized.

Metavolcanics belonging to the second structural stage, represented by komatiite-tholeiite and andesite-basalt series, are associated with gold-sulphide mineralization.

Frequent episodes of basic magmatism, penetrating the infrastructure of belts, controlled sulphide-nickel ore mineralization, beginning in Early Stanovoy times. Granitoids and related metasomatites, marking a prolonged orogenic regime in the Dzhugdzhur-Stanovoy province from the start of the Early Proterozoic, have a gold-rare metal signature. Table II-8 presents a brief summary of the mineral deposits and ore shows that characterise the province.

Chapter 4

BASEMENT HIGHS AROUND CRATON MARGINS

Within the Siberian Platform area we have in addition to the traditionally identified ancient basement structures of the Aldan and Anabar Shields, marginal elevations of the basement in the south-west which also belong to this category. They form the exposed southern part of the Taseyev massif (Neyelov, 1968) or the Angara platform (Salop, 1982). The marginal highs are the Pre-Sayan — to the north of the East Sayan mountain range, and the South Yenisey (the Kan microcraton) — to the south of the river Angara. These inliers, as will be demonstrated below, possess common evolutionary features and can be regarded as the early segments of a single Kan–Pre-Sayan Early Precambrian mobile region. The region west and north-west of Lake Baikal, whose Early Precambrian formations constitute the Baikal-Patom belt, is a similar structure.

1. THE KAN–PRE-SAYAN TERRAIN

This terrain extends from the SW shores of Lake Baikal westwards along the East Sayan mountain range as far as the river Yenisey. Its northern boundary is the river Angara in the north-west. The Pre-Sayan elevation has been studied in most detail both geologically and geochronologically and its major features will be described in more detail than those of the South Yenisey block.

Pre-Sayan basement inlier

The Pre-Sayan elevation has as its southern margin the Main Sayan fault, separating Pre-Sayan from the East Sayan fold belt (Fig. II-39). Towards the north-east the basement elevation is buried beneath a sedimentary platform cover consisting of Late Proterozoic Riphean sediments (the Karagas and Oselkov Formations).

The area has been studied by many research workers. One of the first was Obruchev (1949) who established the major tectonic features of the region. Korzhinsky (1939) laid the foundation for in-depth petrological studies. Further research was carried out by teams from the Academy of Sciences' Leningrad Institute of Precambrian Geology and Geochronology,

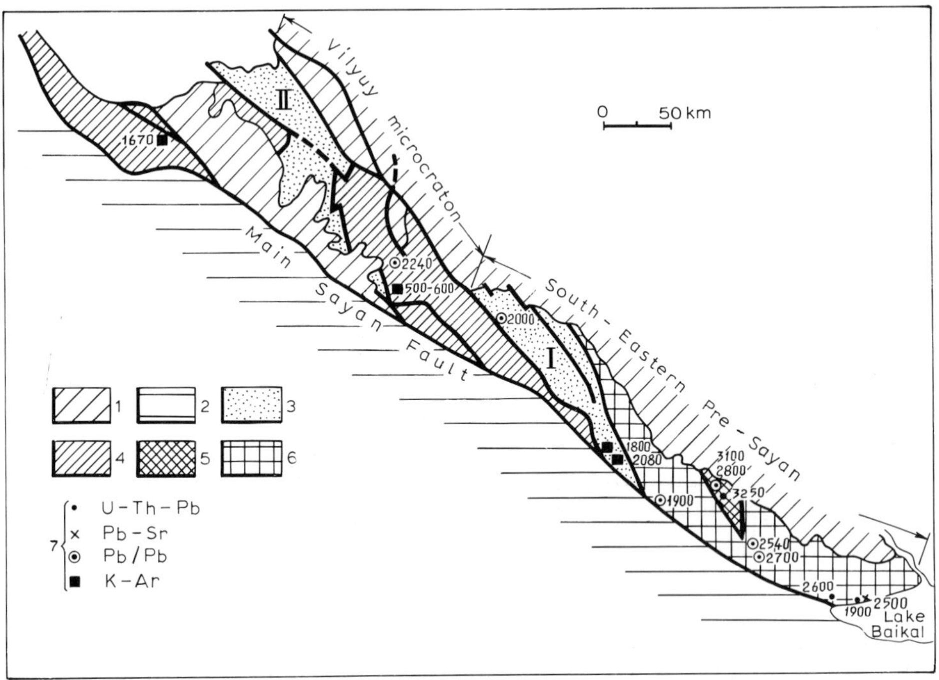

Fig. II-39. Sketch map showing tectonics and geochronology of the Pre-Sayan basement high. *1* = Phanerozoic and Riphean sediments; *2* = Precambrian of East Sayan fold belt (Baikalides); *3* = Early Proterozoic sediments of Urik-Iy (*I*) and Tumanshet (*II*) grabens; *4* = Late(?) Archaean in Biryusa microcraton; *5* = Late(?) Archaean in Onot graben; *6* = Early Archaean; *7* = radiometric age dating sample site.

the Irkutsk Institute of the Earth's Crust of the Siberian Branch of the Academy of Sciences, the All-Union Geological Institute, the Regional Geological Survey and the Novosibirsk Institute of Geology and Geophysics, also part of the Siberian branch of the USSR Academy of Sciences. Workers from the Precambrian Institute (Polkanov and Obruchev, 1964) concentrated particular attention on the petrology of Precambrian igneous and metamorphic rocks, their geochronology, stratigraphy and tectonic relationships. Geologists at the All-Union Geological Institute (VSEGEI) (Dodin et al., 1968; Konikov and Travin, 1983) made important contributions to the solution of problems in the stratigraphy and geological structure of the region. Teams of geologists from the Irkutsk Regional Geological Survey (Shirobokov and Sezko, 1979; Poletayev et al., 1979) not only carried out a systematic survey of the region, but also made detailed thematic studies of a number of aspects of the geology. More recently, detailed studies of the geology and petrology of the Pre-Sayan region have been undertaken by geologists from the Institute of the Earth's Crust (Sezko, 1972; Petrova and Levitsky, 1984; Kuznetsova, 1981; Yeskin et al., 1979), the Siberian

Research Institute for Geology, Geophysics and Mineral Resources (SNIIGIMS) (Sizykh, 1978; Sizykh and Shafeyev, 1976) and the Novosibirsk Institute of Geology and Geophysics (Pavlov and Parfenov, 1973; Nozhkin, 1985).

The Pre-Sayan marginal elevation consists of a number of blocks which differ in geological structure, composition and age. During the 1950s and '60s the status of the southern boundary of the microcraton was the subject of heated debate. A number of workers considered that only the Sharyzhalgay block was a marginal basement elevation, and referred other blocks to the East Sayan younger fold belt.

In the present work we take the Main Sayan fault to be the south-western boundary of the Pre-Sayan marginal elevation. In this case, the block is taken to include the south-eastern Pre-Sayan region and the Biryusa and Kan-Sayan microcratons.[1]

The Pre-Sayan microcraton, the major component of which is the Sharyzhalgay block, differs markedly in geophysical characteristics from other Precambrian blocks in Eastern Sayan and even from other blocks in Pre-Sayan (Pavlov and Parfenov, 1973; Savinsky, 1972). In the Sharyzhalgay block, average ratios of remanent to inductive magnetism in igneous rocks is higher than in rocks in other blocks, 3.03 as against 1.09; variational changes in density are within the range 2.6–3.22 g/cm.

Magnetic field patterns for the Sharyzhalgay block and adjacent territories prove that on the whole it is homogeneous, although the eastern part of the block (east of the river Kitoy) is characterised by a marked positive magnetic field, with an intensity of 1300–1700 nT and major north–south linear anomalies. In the south of the block, the anomalies possess a north-west strike, parallel to the marginal suture of the microcraton. The strike of geophysical anomalies often coincides with that of the crystalline rocks.

Of prime importance in the geological structure of the Pre-Sayan microcraton are the south-eastern Pre-Sayan and Biryusin areas which differ in composition and age and are therefore usually treated separately.

South-eastern Pre-Sayan area

This is one of the best studied parts of the region. It consists of three terrains of different age, forming a variety of tectonic structures. The oldest (Sharyzhalgay) terrain is exposed over the greater part of the region. It has been brought to the surface as a system of blocks, which are here grouped together under the name "Sharyzhalgay block". An upper Archaean and a lower Proterozoic complex are exposed in grabens, the Onot and Urik-Iy,

[1] In this chapter we use the same terminology for tectonic structures that has been widely employed in the regional and more general literature, although it does not always satisfactorily reflect the tectonic components.

TABLE II-9

Radiometric age of Precambrian rocks in the Pre-Sayan inlier

Rock collected (mineral)	Method	Age, Ma	Laboratory Author
Lower Archaean (Sharyzhalgay) complex			
Biotite-hypersthene gneiss, shore of Lake	Rb-Sr	2530±260	PRECAMB Baikal I.N. Krylov I.M. Gorokhov
Charnockite, Lake Baikal	Rb-Sr	2540±160	
Biotite granite, Lake Baikal	Rb-Sr	2530±35	
Pegmatitic granite, Lake Baikal	Rb-Sr	1970	
Zircon from two-pyroxene-hornblende plagiogneiss and biotite-hypersthene gneiss	U-Pb	2700-2600	GEOCHEM Ye.V. Bibikova
Zircons from charnockites and enderbites	U-Pb	2050-1950	—ditto—
Zircons from sillimanite schist in the basin of the river Kitoy (coll. by V.Ya. Khiltova)	Pb-Pb evaporation technique	2720±120	PRECAMB A.A. Nemchin
Zircon from banded calciphyre, same loc.	—do—	2430±10	—ditto—
	—do—	2530±20	—ditto—
		2460±20	—ditto—
Upper Archaean (Onot) complex			
Zircon from pegmatitic granite, left bank of the river Onot (coll. by V.Ya. Khiltova)	-do-	3024±35 2820±80 2820±80	PRECAMB I.M. Morozova
Zircon from tonalite in the Onot complex	U-Pb	3250±100	GEOCHEM Ye.V. Bibikova
Lower Proterozoic (Urik-Iy) complex			
Zircon from granite (Sayan complex), left bank of river Barbitay (collected by S.A. Reshetova)	Pb-Pb thermo-emission	1960±60	PRECAMB I.M. Morozova
Zircon from granite (Sayan complex), watershed between rivers Uda and Ognit (collected by S.A. Reshetova)	—do—	1870±30	—ditto—
Ditto, river Shablyk	—do—	2220±180	—ditto—
Biryusa microcraton			
Zircon from kyanite schist, basin of the river Biryusa (Tepsa complex)	—do— —do—	2220±20 1980±20	PRECAMB A.A. Nemchin

PRECAMB = Institute of Precambrian Geology and Geochronology, USSR Academy of Sciences, Leningrad; GEOCHEM = Institute of Geochemistry, USSR Academy of Sciences, Moscow.

respectively (Fig. II-39). Geochronological studies have been carried out in the region at different times, by a variety of methods: K–Ar, U–Pb, Rb–Sr and Pb–Pb, both whole rock and on zircons. Many age dates have already been published (Goltsman et al., 1982; Manuylova, 1968; Volobuyev et al., 1976, 1980). Table II-9 shows the results of recent research in geochronology.

The *Sharyzhalgay block* consists of a lower Archaean supracrustal complex and associated igneous rocks of the same age, as well as Late Archaean and Early Proterozoic igneous rocks (Fig. II-40).

The *lower Archaean Sharyzhalgay terrain* has for a long time been

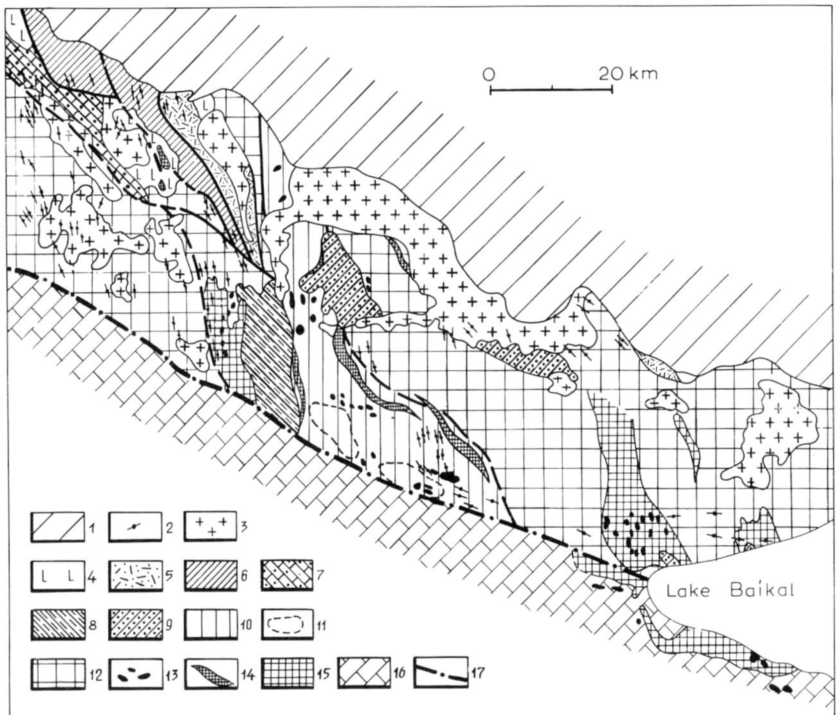

Fig. II-40. Geological sketch map of South-Eastern Pre-Sayan (modified after Kuznetsova, 1981). *1* = platform cover; *2* = 1200 Ma Nertsin dyke swarm; *3* = AR_2–PR_1 granitoids; *4* = AR_2 gabbros; *5* = AR_2–PR_1(?) volcanosedimentary assemblage; *6* = upper volcanosedimentary formation in Onot graben (AR_2^2, Sosnovy Bayts); *7* = lower volcanosedimentary formation in Onot graben (AR_2^1, Kamchadal); *8, 9* = AR_1 carbonate-terrigenous assemblages (upper part of Sharyzhalgay complex); *10* = volcanosedimentary rocks within granite-gneiss; *11* = maximum development of AR_1 supracrustals; *12* = granite-charnockites with fragments of Sharyzhalgay supracrustals; *13* = basic and UB rocks; *14* = high-*P* granulites (undiff.); *15* = high-*P* granulite (diaphthorite); *16* = Precambrian of East Sayan fold belt; *17* = Main Sayan Fault.

correlated with the oldest rocks in Eastern Siberia. Geochronological studies have determined only the complex polycyclic history of the rocks, since the available age dates do not confirm an Early Archaean age. Despite this, however, the Sharyzhalgay complex may be referred to the lower Archaean, a consequence not only of field evidence but also geochronological data from the region as a whole, as will be demonstrated below.

The complex consists of several formations with different rock compositions, the interrelations between which have not been determined unequivocally. It is divided into three formations or groups which crop out in a number of ways: (1) a group of predominantly mafic schists with thin bands of various gneisses; (2) a group of predominantly gneisses; and (3) a group consisting of schists, including high-alumina varieties, marble, gneiss and amphibolite.

The total thickness of the succession has been provisionally estimated at up to 12 km. Relationships between the first two groups have long been a matter of debate. The fact is that workers have attempted to group all the biotite and hornblende gneisses together into a single formation, although it has been noted in so doing that for certain regions the mafic schists and gneisses underlie the gneiss group (Yelizaryev, 1962), while in other regions the opposite is the case (Polkanov and Obruchev, 1964). These contradictions are probably due to the fact that the authors have observed relations between different geological formations in the different regions. In one case the gneissose rocks represent a basement complex and in another they are supracrustal gneisses on top of group 1. At the moment it is almost universally accepted that the oldest group in the supracrustal assemblage is group 1, while the other two lie progressively higher up in the succession.

The mafic schists in group 1 are usually considered to be metamorphosed basic volcanics. Compositionally they correspond to basalt or leucobasalt and rare subalkaline olivine basalt (Fig. II-41). In the (Na_2O + K_2O)–MgO–FeO triangle, mafic schist compositions fall in the field of tholeiitic basalts (Fig. II-42).

Many basalts are high-iron types with higher alumina. In this formation, eulysite and banded ironstone crop out in a few areas together with the mafic schists.

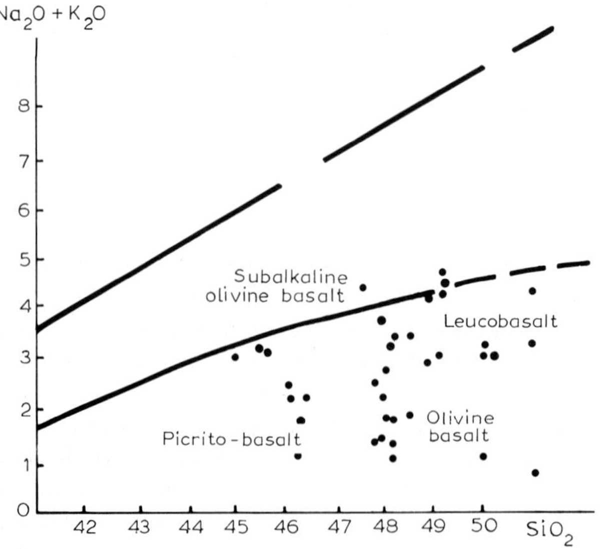

Fig. II-41. Chemical composition of mafic schists in the Sharyzhalgay complex, ($Na_2O + K_2O$)–SiO_2 coordinates.

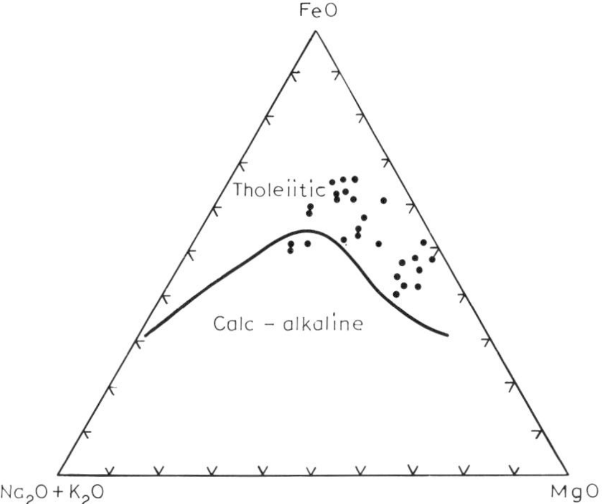

Fig. II-42. Chemical composition of mafic schists in the Sharyzhalgay complex, FeO–MgO–($Na_2O + K_2O$) coordinates.

The next group 2 contains a large number of biotite and hornblende gneisses, petrochemically similar to intermediate and acid igneous rocks, possibly volcanics.

The two lower groups are particularly widespread in the eastern part of the Sharyzhalgay block. The topmost group 3 is most fully represented in the west (in the middle reaches of the river Kitoy and the upper reaches of the river Malaya Belaya). Group 3 is particularly varied in composition — gneiss, amphibolite, marble lenses and horizons, calciphyre, ash-grey micrognеiss, sillimanite schist and quartzite with granulitic texture. Each rock type has a varied thickness over the outcrop area, particularly the marble and aluminous schist. Carbonate rocks for example vary from 700 m to 100 m or less.

Many of the rock types listed above are obviously sedimentary in origin. The amphibolites present especially in the upper part of the succession are para-rocks. The ash-grey coloured micrognеisses are probably primary acid effusive rocks. Many types of para-rocks in the group show evidence for deposition in shallow-water: cross-bedding, terrigenous material present among carbonate rocks, a high amount of detrital zircon and heavy fraction minerals, and a large quantity of graphite in aluminous rocks (Khiltova, 1971).

Areas occupied by supracrustal rocks within the Sharyzhalgay block amount to 20–30%, sometimes as much as 40–50% in certain regions of the exposed area. They are situated among strongly migmatized rocks and granite-gneiss, although they themselves are sometimes only weakly affected by migmatization, particularly the upper formations in the complex. The supracrustal rocks increase in volume to the west.

The structure of the polyphase metamorphosed and deformed lower Archaean terrain is complex and has so far not been unambiguously deciphered. The Sharyzhalgay block is broken by a near north–south fault system into a series of smaller-scale blocks (Prokofyev, 1966). In the western part (the Kitoy and Bulun blocks), major N–S-striking linear folds crop out, sometimes with axial surfaces overturned to the south-west.

In the eastern part of the Sharyzhalgay block (the Irkut block), there are major dome-and-basin shaped folds of various orders. The typical overall pattern for the area is an alternation of more or less extensive segments of the two types (Yeskin et al., 1979) with flat-lying rocks in each, separated by steep isoclinal folds. Boundaries between segments with different structure are usually sharp, and only rarely diffuse.

Three deformational and three metamorphic episodes have been identified as being responsible for the formation of the overall structure of the Sharyzhalgay block (Yeskin et al., 1979). Most of the rocks are granulites, more or less retrogressed. The upper formation in the complex is least metamorphosed, at conditions transitional between granulite and amphibolite facies. The granulites in the Sharyzhalgay complex belong to two facies series: medium- and high-pressure. There is a strong spatial link between granulites in the second facies series and basic-ultrabasic intrusions, sometimes layered, which are also frequently metamorphosed at granulite facies conditions. Instances where the second metamorphism is superimposed on the first suggest that there were two granulite facies metamorphic episodes separated by a time interval.

Typical mineral assemblages for high-pressure granulites are hypersthene with sillimanite, with garnet present in two-pyroxene schists and gneisses. Kuznetsova (1981) has shown that the pyrope content in garnet in these rocks is up to 46%. High-pressure granulites include not only basic rocks but also garnetiferous enderbite and charnockite, in which biotite has up to 5.0% TiO_2 and hypersthene up to 5.6%; in the enderbites, garnet has 52% pyrope molecule. There are occurrences among the garnet-hypersthene-sillimanite gneisses of thin bands of quartz-sapphirine-spinel associations. P–T conditions for granulite formation have been estimated to be P = 10–11 kbar and T = 750–800°C (Kuznetsova, 1981).

Upper Archaean Onot Terrain. Rocks which may be assigned to this terrain occupy tiny areas in south-eastern Pre-Sayan. Rocks of the Onot graben undoubtedly belong here and at present we also provisionally consider that a gneiss unit, the Arkhut Formation, is an integral part. The gneisses have been mapped in the south-western part of the Sharyzhalgay block. The terrain has been studied in detail both geologically and geochronologically in the Onot graben.

The graben has a NW to NNW strike and is situated within the Sharyzhalgay terrain, with which it is in tectonic contact. The age of its constituent rocks has long been a subject of debate and ideas varied widely, from Late Proterozoic to Archaean. Radiometric age dating has been carried out on these rocks only recently. The K–Ar method has determined the closure age of isotopic systems to be 1800–1700 Ma. U–Pb dates on the granite-gneiss, known in the literature as the Onot granitoid complex, unexpectedly depicted the oldest age for any of the Pre-Sayan rocks — 3250 ± 50 Ma (Bibikova et al., 1982). Later, in the Precambrian Institute laboratories, dating work was done on some granites located among relatively weakly metamorphosed rocks (on the left bank of the river Onot). The granites possess a pegmatitic texture and are frequently blastomylonitized. Three varieties of zircon were found. Dates of 3024 ± 35 Ma and 2820 ± 80 Ma were obtained from two of these, using the Pb–Pb thermo-emission method (an interpretation of the geochronological data for Pre-Sayan, including age dating results from rocks in the Onot graben, is presented below).

Most research workers divide the supracrustal rocks of the Onot graben into two formations: the (lower) Kamchadal and the Sosnovoye Baytso. The lower formation occupies the south-western part of the graben, the upper the north-eastern part (Fig. II-40).

The lower formation includes a variety of rock units (from the bottom up): amphibolite alternating with quartzite and magnetite-hornblende schist; talcose magnesite, dolomite and dolomitized limestone; the upper part has hornblende gneiss and schist.

The upper formation — Sosnovoye Baytso — consists of amphibolite with banded ironstone members which are interleaved with mica and garnet-mica schist. Schist and microgneiss predominate near the top of the formation. Most workers consider that the amphibolites in the complex are basic volcanics, among which concordant and slightly cross-cutting gabbro-diabase dykes are present. The microgneisses are primary felsite-porphyries (Shirobokov and Sezko, 1979).

Petrochemical data are scarce, but from what is available we may make the following remarks: amphibolites, especially in the upper formation, belonging to basic tholeiitic volcanics (SiO_2 around 50%), are characterized by low K_2O (<1%), total alkalis 2.5–3.5 wt%, and 1–1.6% titanium.

Amphibole-mica schists are abundant in the upper formation; their SiO_2 content is 60%, total alkalis up to 7.5% and the alkali ratio approximates to 1.

The rocks within the graben are deformed into folds trending roughly north–south. Thrusts are a common and characteristic feature of the south-western margin. Syntectonic and late tectonic basic and ultrabasic rocks are of frequent occurrence in the Onot graben (Belsk and Khoyta-oka complexes).

The basement to the Onot supracrustal rocks has not been established with any certainty. We may only surmise that it is leucocratic gneissose granite with a blastomylonitic texture. These granite-gneisses in other regions of Pre-Sayan are assigned by researchers to the second group of the Kitoy complex (Suloyev, 1950). In the Onot graben the granites were blastomylonitized at amphibolite facies conditions. Their petrochemical characteristics are: high SiO_2 content (>72%), significant alkalis (around 8%), an alkali ratio (Na_2O/K_2O) < 1, and total FeO of 2–3% (2 analyses).

Supracrustal rocks belonging to the two age groups in the Onot graben are cut by frequent granitoids, the earliest of which are referred to by all workers as the Onot complex. Compositionally they correspond to tonalite and granodiorite (SiO_2 = 68–70%), total alkalis are around 6.0% and the Na_2O/K_2O ratio approximates to 2; total FeO is around 4%. The granites always have a gneissose texture, although in fact igneous textures are often well-preserved. They are restricted to the central and marginal areas of the graben, where they cut rocks of the Kamchadal and Sosnovoye Baytso Formations. They have also been mapped in rocks of the Sharyzhalgay complex.

Onot granites are cut by microcline granites which are usually correlated with a granitoid group in the Early Proterozoic Sayan complex. Since no detailed research has been carried out on the age of the microcline granites in the Onot graben, this correlation is purely an analogy at present.

The rocks of the terrain are zonally metamorphosed from epidote-amphibolite to high-temperature amphibolite facies in the kyanite-sillimanite facies series (moderate-gradient type of metamorphism).

From the widespread development of dolomite and magnesite in the lower formation, it has been proposed that sedimentation occurred in shallow-water conditions in a relatively quiet environment, as distinct from the upper formation which contains amphibolite and ferruginous quartzite, assigned to a jaspilite formation, and significant volumes of fragmental rocks. The sedimentary basin had larger dimensions than the present outcrop of the Onot complex, although it is scarcely likely to have encompassed the whole of the Pre-Sayan region, since the Arkhut Formation, with which the rocks of the Onot graben are correlated on an age basis, was deposited under different conditions.

The presence of large volumes of amphibolite and basic intrusive rocks among the supracrustal assemblages, the association of amphibolite (basic volcanics) with microgneiss (felsite-porphyry), and an Archaean age for the terrain, mean that there are close parallels between the rocks of the Onot graben and Archaean greenstone belts. However, widespread carbonate development and smaller volumes of basic volcanics compared to such belts as the Barberton greenstone belt in South Africa, reflect the individuality of the Late Archaean Pre-Sayan terrain, which probably formed in an environment similar to that of the Isua belt.

In terms of its age situation, the second group of upper Archaean rocks (the Arkhut Formation) is analogous to the rocks of the Onot graben. It overlies the Sharyzhalgay complex and in several regions there is a probable erosive contact. At the base of the formation and its analogues, rocks occur which certain workers believe to be basal conglomerates, while others claim they are tectonites. Although the nature of these rocks is still indeterminate in most regions, there are some cases where conglomerates are present at the base of the formation, characterized by poorly sorted fragmental material. There are frequent occurrences of flattened pebbles of pyroxene and amphibole gneiss, marble, mica-feldspar gneiss and granite-gneiss.

Arkhut Formation rocks are fairly uniformly metamorphosed at amphibolite facies, in the kyanite-sillimanite facies series.

Lower Proterozoic Urik-Iy Terrain. In south-eastern Pre-Sayan, this forms a single relatively major structure, the Urik-Iy graben (Fig. II-43). It is situated between the Sharyzhalgay block and the Biryusa microcraton, forming a north–west-striking belt over 40 km wide. The graben is in tectonic contact with older complexes. Younger upper Proterozoic sediments, which in fact belong to the platform cover, overlie the complex in the north of the belt.

An Early Proterozoic age for the Urik-Iy complex has been reliably ascertained from radiometric data on cross-cutting granites. These granitoids form part of the Sayan complex, which intrude the lower part of the succession in the Urik-Iy graben and have a K–Ar date of 1890 Ma (Manuylova, 1968). A Pb–Pb thermo-emission (evaporation technique) date on zircons is 1900–2200 Ma (see Table II-9).

The granites, as mentioned above, cut only the lower part of the complex. Folding and metamorphism of these rocks are related to the time of origin of the granitoids. In accordance with the basis for determining the boundaries of structural-lithological terrains adopted in this book, only the lower part of the rocks in the Urik-Iy graben should be assigned to the lower Proterozoic terrain. However, since workers in this region have traditionally treated different parts of the stratigraphic section as a single unit (despite the fact that they are separated by folding and granite formation), we have also referred them to a single terrain, especially since the age of the upper part of the succession is greater than 1800 Ma, i.e. the entire supracrustal assemblage in the Urik-Iy graben has an Early Proterozoic age. The lower part of the succession, which would be defined as the lower subcomplex according to most workers, reflects the mobile development of the belt; the upper subcomplex, which combines marine and continental molasse formations, is the orogenic part. The basement to the supracrustal assemblage is not exposed and we can only surmise that the assemblage rests with erosive unconformity on Archaean rocks. There is at present no complete agreement on either the stratigraphic sequence or the internal structure of the graben. In this work, we have followed the interpretation proposed by Konikov and Travin (1983).

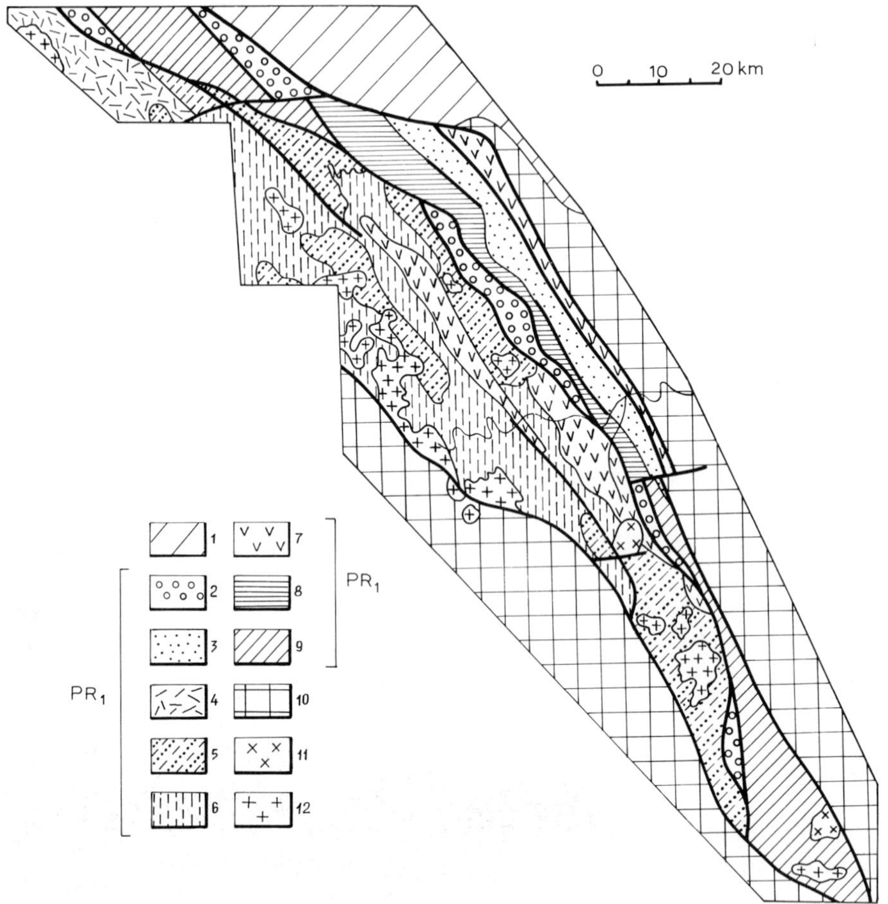

Fig. II-43. Geological sketch map of the Urik-Iy graben (simplified from Konikov and Travin, 1983). *1* = Phanerozoic sediments; *2–9*: Early Proterozoic rocks (*2* = sandstone, conglomerate, tuff, *3* = Ingashi Formation (metasandstone, metaconglomerate, phyllite), *4* = Shablyk Formation (dacitic and liparitic porphyry and tuff, andesitic and basaltic tuff), *5* = Kirey Formation (greywacke metasandstone, schist, metaconglomerate), *6* = oligomict metasandstone, phyllite (greenschist–low-*T* amphibolite facies), *7* = Daldarmi Formation (metadiabase, pillow lava, lava breccia), *8* = Ognoy Formation (quartzose metasandstone, graphitic schist), *9* = Arshan Formation (phyllitic dolomitic schist, metasandstone)); *10* = Archaean rocks; *11* = Gunik syenite; *12* = Sayan granite.

The lower subcomplex is subdivided into a series of lithologically distinct formations. The oldest of these crops out in the north-eastern part of the graben and consists of finely laminated argillaceous and carbonate-clay shale, dolomite and argillaceous dolomite, 2600 m thick. In the south, phyllite, chlorite, hornblende-biotite, two-mica and graphite-mica schist belong to this stratigraphic level.

Above lie argillaceous-carbonate sediments of the Ognoy Formation:

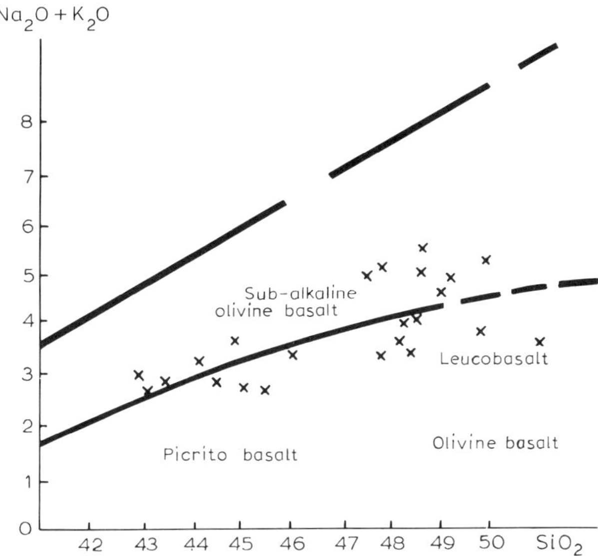

Fig. II-44. Chemical composition of basic volcanics in the Shablyk Group (PR_1), ($Na_2O + K_2O$)–SiO_2 coordinates.

creamy quartzite, graphitic-silty shale, various sandstones and volcanic sheets. At the base of the formation are lenticular intercalations of quartzite, coarse arkose and conglomerate, up to 15 m thick. The overlying Shablyk Formation contains massive pillow diabase, basic lava breccia with magnetite quartzite members in the lower parts of the section, and magnetite-quartz-chlorite schist. The succession of the lower subcomplex concludes with phyllitic schist, sandstone and slate. The total thickness is up to 4000 m.

The Shablyk Formation contains the highest volume of volcanics, commencing with acid varieties and on the whole the volcanics belong to a highly differentiated series with a trachytic slope (Fig. II-44). Compared to volcanics in older terrains, we see an increase in the average value of total alkalis and K_2O. Sedimentary rocks in this formation are sometimes characterized by high sediment maturity; sedimentary breaks and unconformities are seen between individual formations.

Fragments of underlying rocks have been found in the sediments belonging to the lower Proterozoic upper subcomplex, including Sayan granites and their basement rocks (Semeykin, 1972). Overall, the upper subcomplex (the Ingashi and Yermosokh Formations) contains coarsely fragmental rocks, with redbeds and grey-coloured assemblages near the top.

Rocks belonging to the complex were zonally metamorphosed in the andalusite-sillimanite facies series. The highest grade metamorphic rocks (amphibolite facies) crop out in the south of the graben, coinciding with the

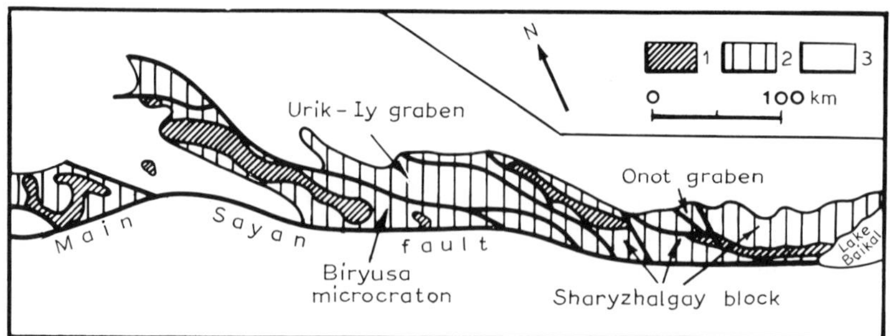

Fig. II-45. Distribution of basic and ultrabasic intrusions in the Pre-Sayan inlier (after Abramovich et al., 1979).

outcrop area of granitoids. However, there is no direct relationship between an increase in metamorphism and individual granite intrusions. The Sayan granitoids, as mentioned previously, cut only rocks in the lower subcomplex and had already begun to be eroded by the start of the formation of the upper subcomplex. Particular compositional features of the supracrustal rocks in the Urik-Iy graben and the association of granitoids with the Main Sayan fault zone set this terrain apart from other older terrains and make it more like a Precambrian structure belonging to the sub-platform regime.

Structures in the Pre-Sayan region are characterized by widespread basic and ultrabasic intrusions (Fig. II-45). Major massifs are restricted to the Onot graben, suggesting that many are related to the Late Archaean growth of the region.

Different varieties of basic volcanism affected the Pre-Sayan terrains of various ages. In the Sharyzhalgay lower Archaean terrain, basic volcanics belong mostly to high-Fe basalts with increased Al_2O_3 contents. If the assumption is correct that the gneisses interleaved with them also formed from volcanics, then we can accept that the lower Archaean volcanic series was andesitic-basaltic. The upper Archaean volcanic series of the Onot graben is bimodal rhyolite-basalt, while the volcanics in the lower Proterozoic terrain as a whole form a highly differentiated series with a trachytic slope.

Each of the above terrains in the region has its own particular associated suite of useful minerals. Associated with the Sharyzhalgay terrain are groups of ferruginous quartzite which occur in an assemblage containing two-pyroxene schist, and known in the literature as the Kitoy Group (Shirobokov and Sezko, 1979). In the same complex there is also a sillimanite deposit (Dabady) and phlogopite shows.

The following useful minerals are associated with the upper Archaean terrain: banded ironstone deposits (the Onot group), restricted to the upper formation (Sosnovy Bayts), magnesite deposits, occurring among the middle and upper portions of the lower formation (Kamchadal) and its carbonate formations. Gabbro-diabase dykes cutting rocks of the Kamchadal Formation have associated cobalt mineralization in pyrite-rich seams (Poletayev et al., 1979).

Ferruginous quartzites belong to the lower Proterozoic terrain, associated with its sedimentary rocks. Rare-metal pegmatites are found in the outcrop area of the Sayan granites.

In concluding this extremely brief description of the southwestern Pre-Sayan region, we shall consider radiometric age dates since it is obviously the case that, despite the overwhelming U–Pb and Pb–Pb dates falling within the range 2.6–2.7 Ga, the Sharyzhalgay complex is assigned to the lower Archaean and the Onot to the upper, although it is precisely within the confines of upper Archaean belts that dates of 3250 ± 100 Ma were obtained by the U–Pb method and 3100 ± 20 Ma on zircons by the Pb–Pb thermo-emission method (evaporation technique). The situation makes sense only if the following is taken into account: internal crust-forming processes in the two early terrains of the Pre-Sayan region (lower and upper Archaean) were ubiquitous and it was these very processes that determined the internal crustal structure of the region. Early Archaean processes were preserved in relict associations of metamorphism and folding. The results of Late Archaean history are most complete and widespread and because of this we find that most of the radiometric dates fall into the time interval 2.7–2.6 Ga. Concerning the oldest date (3250 ± 50 Ma), which was obtained from a granite cutting Onot assemblage supracrustal rocks, then the most logical way to reconcile modern geochronological data and field relations would seem to be to treat it as reflecting the age of remobilization of basement granites, which may even have been in a partially molten state, a fact which has also been responsible for the cross-cutting intrusive relations with the Onot supracrustal assemblage. In this case, the 3.25 Ga age should more properly be regarded as dating the crust-forming processes of the Sharyzhalgay and not the Onot complex.

The same conclusion can be reached regarding the 3.1 Ga date. This was obtained from a partially blastomylonitized granite, as noted above, occurring as a lenticular body among relatively weakly metamorphosed rocks in an area where thrusts are widespread.

The proposed treatment of very old dates links the commonly-obtained dates of 2.6–2.7 Ga obtained in the region with such field evidence in the Sharyzhalgay complex as: (1) superposition of NNW structures, peculiar to the time of formation of the Onot graben structure, on early flat-lying structures with a NW strike, reflecting the Early Archaean structure of the Sharyzhalgay complex; and (2) later superposition of high-pressure granulites on older medium-pressure granulites.

The preservation of the oldest age dates in the Onot graben is related to the fact that it represents the upper part of the Archaean crust, having maintained this position since the end of the Late Archaean. Early Proterozoic internal processes, seen in lower crustal sections and dated by various methods (K–Ar, Rb–Sr and U–Pb), did not alter U–Pb systems in upper crustal sections and their effect is seen only on K–Ar dates.

The Late Archaean radiometric age of charnockites and granulites from lower crustal sections, obtained by a variety of methods, suggests that firstly, the last regional metamorphism in the region was a Late Archaean event and secondly, the lower Archaean (Sharyzhalgay) terrain was a constituent of the deep crust at this time, i.e. it belonged to the infrastructure. Its component parts occupied different vertical positions in the crustal section, which can be observed thanks to the block structure of the region. Late Archaean internal processes appeared in certain parts of the Sharyzhalgay complex as superimposed high-pressure granulite assemblages and elsewhere as migmatites. The high-pressure granulites which crop out today as bands (Fig. II-40) in which migmatization effects are extremely rare, probably reflect deep sections of the Late Archaean crust, while the migmatized Early Archaean granulites are mid-crustal

TABLE II-10

Main geological events in south-eastern Pre-Sayan

Age, Ma	Event	
350	Nepheline syenites	Platform sediments
600±50	Alkali gabbroids	
	Intrusion of alpine-type ultrabasics	
1200	Nersin basic dyke complex	
1700±50	Acid volcano-plutonic association	
	(Subluk volcanics)	
	local erosion	
1800	Urik complex (upper subcomplex)	
2200	Folding, granite formation, low-P metamorphism	
	Urik-Iy complex (lower subcomplex)	
	Erosion	
2600±100	Intense folding, granite formation, high-P metamorphism	
	Sosnovoye Baytso Group	
2800±100	Folding, granite formation	
	Kamchadal Group	
	Erosion	
3200±100	Intense folding, granite formation, low-P metamorphism	
	Sharyzhalgay complex	
	(Kitoy Group and its analogues)	
3400±100	Granite formation, folding (?)	
	Sharyzhalgay complex	
	(Zhidoy and Zogin Groups and their analogues)	
	Sialic crust	

sections. Both types of secondary processes probably proceeded at the same time as the supracrustal rocks in the Onot terrain were metamorphosed.

The Early Proterozoic stage of internal processes, dated by U–Pb and K–Ar methods, is geologically more locally represented in comparison with Late Archaean events, especially in upper crustal sections, due to the preservation of Late Archaean structures and mineral associations. The major geological events in Pre-Sayan are shown in Table II-10.

The Biryusa microcraton

The term "Biryusa microcraton" has been widely used in the literature for the north-western part of the Pre-Sayan basement high.

The microcraton stretches north-west from the Urik-Iy graben for a distance of more than 400 km; it has a maximum width of 80 km. Its south-western margin is defined by the Main Sayan fault and coincides with the edge of the Siberian Platform. Its north-eastern margin is the Biryusa fault. In the west, the craton is overlain by middle Palaeozoic sediments belonging to the Agul basin. The Biryusa microcraton consists of three tectonic elements (Fig. II-46): the Biryusa block, in which the oldest rocks are exposed; the Tumanshet graben containing Lower Proterozoic rocks; and the Palaeozoic Agul basin.

The *Biryusa block*, consisting mainly of Archaean rocks including mica pegmatites, has been studied in some detail, relatively speaking, and accounts for the main area of the microcraton. The history of investigation is detailed in a number of publications (e.g. Dibrov, 1958). The primary structure of the block is defined as a major syncline, complicated by several brachyanticlines, and gently dipping monoclines on the limbs. It was later thought (Dibrov, 1958) that the metamorphic rocks formed the north-eastern limb of an anticlinorium, consisting of a system of disharmonically alternating open and close isoclinal folds.

Geologists from Institutes in Irkutsk began to make a detailed study of the geological structure and stratigraphy of the rocks in the Biryusa block as a whole from the late 1960s onwards (Sezko, 1972; Sizykh and Shafeyev, 1976). This work formed the basis for more detailed schemes of the geological structure and for refining the stratigraphy of the supracrustal assemblages.

The monoclinal nature of Precambrian rocks in this region and the absence of sharp structural breaks were responsible for the fact that for a long time, researchers treated the rocks in the Biryusa microcraton as forming a single stratigraphic complex. Nowadays, three terrains have been identified in the Biryusa microcraton: the lowermost Biryusa complex proper (or the Shelma Formation, Sizykh and Shafeyev, 1976), the Tepsa complex and the Tumanshet complex. The last two form a system of anticlines and synclines running from the south-eastern margin to its

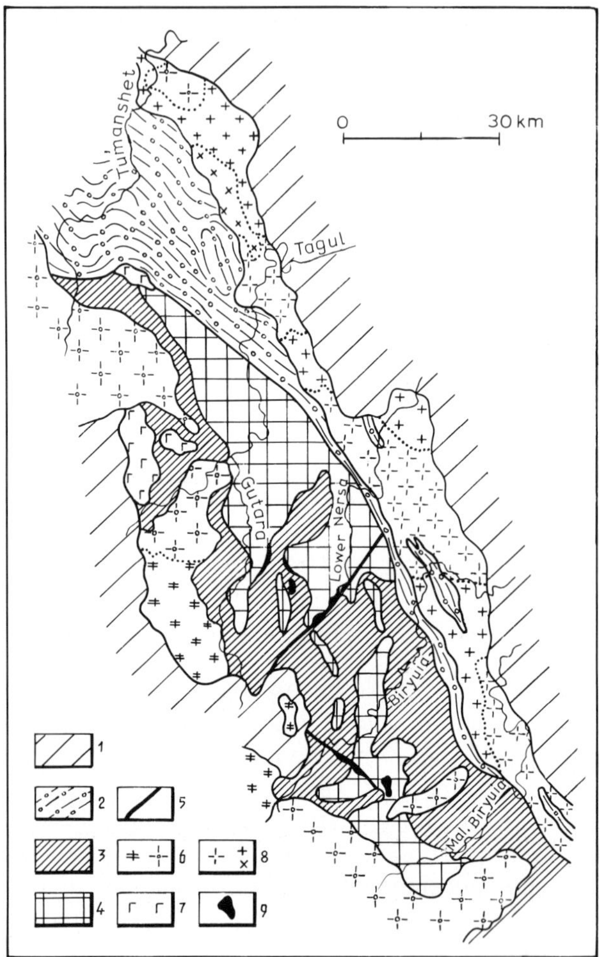

Fig. II-46. Geological sketch map of the Biryusa microcraton (simplified from Sizykh, 1978). *1* = Late Precambrian; *2* = Tumanshet complex PR_1; *3* = Tepsa complex AR_2^2; *4* = Biryusa complex AR_2^1; *5* = faults and geological boundaries; *6* = subalkaline granites, PZ_2; *7* = gabbroids, PR_2; *8* = Sayan granitoids, PR_1; *9* = ultrabasic rocks, AR_2–PR(?).

north-western termination (Fig. II-46). Synclines containing the Tepsa complex are represented by strongly flattened isoclinal folds with roughly NS-striking axial surfaces which are usually overturned to the NE. Their hinges plunge at 5–8° (in rare cases 10–12°) to the NNE.

Structural breaks or gaps of any kind between the Biryusa and Tepsa complexes are unknown. Their breakdown is based on differences in metamorphic and structural histories — these are more complex in the case of the Biryusa complex (Sizykh, 1978).

Biryusa lower Archaean(?) terrain. The biggest outcrop area is in the north-west of the Biryusa block. In the south-east, the complex crops out in anticlinal structures and is quantitatively subordinate to the upper or Tepsa complex. In the north-west the Biryusa complex is overlain transgressively by the younger lower Proterozoic terrain (Sezko, 1972). The age of this complex, as for all the others, is mainly determined from data on the overall evolution of the Pre-Sayan shield area. The rocks in the Biryusa microcraton are quite unsuitable for radiometric age dating purposes. The two major boundary faults exerted a powerful internal influence even during the Palaeozoic. This is why many K–Ar determinations on biotites and muscovites give values around 500 Ma (Polkanov and Obruchev, 1964).

Pb–Pb dating by the thermo-emission method (evaporation technique) on two zircon generations from the terrain gave values of 1960 $\pm$ 20 Ma and 2220 $\pm$ 20 Ma (Table II-9), showing the effects of the abundant intrusions of the Sayan granitoid complex on these rocks.

The Biryusa complex is rather monotonous in composition and it can be subdivided only into members; from the base upwards these are: (1) biotite and garnet-biotite gneiss with thin hornblende and hornblende-biotite gneiss; (2) hornblende-biotite gneiss with thin amphibolite sheets and hypersthene-biotite gneiss; (3) predominantly biotite gneiss with thin bands of amphibolite, quartzite and marble; and (4) monotonous gneisses.

Two metamorphic episodes have been established in the Biryusa complex. During the first episode, the rocks were metamorphosed at medium-pressure granulite facies conditions and during the second superimposed event metamorphism was at amphibolite facies or transitional to high-pressure granulite facies. During the second event, garnetiferous gneiss and amphibolite formed; the pyrope molecule in the garnet reaches 40%. Eclogitic associations have been found at a number of localities in the region (Dibrov, 1958; Kratz et al., 1984) and it is likely that they formed during the second episode. Partial melting also took place then, restricted to individual parts of the complex.

The *upper Archaean Tepsa terrain* attains its maximum outcrop area in the south-eastern part of the Biryusa microcraton, although small outcrops can be found all along its southern margin. As has already been pointed out, the terrain consists of narrow synclinal structures striking approximately north–south. The Tepsa complex has not been dated radiometrically, nor is there any direct geological evidence concerning its age. However, it has now been established that firstly the Tepsa complex is older than the Tumanshet and secondly that the Tumanshet complex is definitely correlated with the Urik-Iy complex in south-eastern Pre-Sayan, for which an Early Proterozoic age is based on radiometric data.

The Tepsa complex is assigned on this basis to the upper Archaean. The succession of supracrustal rocks in the complex is dominated by para-rocks. Representatives are high-alumina kyanite schist, kyanite-muscovite schist, calciphyre and marble containing significant numbers of amphibolites.

TABLE II-11

Typical deposits and ore shows

Deposit, ore show; useful mineral	Tectonic setting	Type of mineralization: ore-formational and genetic	Geological and mineralogical features				Geological age of host rocks and mineralization
			Host or mineralized rocks	Shape of deposit, ore body morphology	Ore type	Ore mineral composition (minor minerals)	
			South-eastern Pre-Sayan				
Baikal (Kitoy group of deposits); iron	Sharyzhalgay block	Magnetite related to ferruginous quartzites; metamorphic	Biotite-pyroxene, biotite and amphibole-pyroxene gneisses	Conformable layered and lenticular deposits	Densely dissemin-ated to massive	Magnetite (haematite, pyrrhotite)	Early Archaean
Kitoy; aluminium	Sharyzhalgay block	Sillimanite, in schists metasedimentary	Carbonate rocks, biotite and biotite-amphibole gneisses	Conformable layered deposits	Massive	Sillimanite (graphite, rutile, ilmenite)	Early Archaean
Sosnovy Bayts; iron	Onot graben	Haematite-magnetite related to ferruginous quartzites; metasedimentary	Amphibole-mica and garnet-mica schists	Conformable layered deposits	Densely dissemin-ated to massive	Magnetite, haematite (pyrrhotite)	Late Archaean
Onot; talc, magnesite	Onot graben	Talc; metasomatic, related to magnesites	Dolomites, magnesites	Steeply-dipping layered deposits	Massive	Talc, magnesite	Late Archaean
Savin; magnesite	Onot graben	Magnesite; metasomatic	Acid and intermediate effusives, talc-chlorite and biotite schists	Conformable layered deposits	Massive	Magnesite (talc, chlorite, pyrite)	Late Archaean
Urik-Iy pegmatite belt	Urik-Iy graben	Rare metal; pegmatitic	Ga-and-qz-bi schists	Pegmatitic, conformable and cross-cutting veins	Clusters, block	Columbite (fluorite)	Early Proterozoic
Bolshaya Tagna; fluorite	Urik-Iy graben	Fluorite in carbonatites related to Bolshetagnin alkaline ultrabasic intrusion	Carbonatites	Subvertical, pipe-like (?) bodies	Disseminated, vein-type, massive	Fluorite, calcite, haematite (magnetite, pyrite)	Late Precambrian
Belozimin; apatite, iron	Urik-Iy complex (graben)	Magnetite-apatite in carbonatites related to the Belozimin alkaline ultra-basic intrusion	Carbonatites	Subvertical, pipe-like (elliptical) bodies	Disseminated, vein-type disseminations	Magnetite, apatite (barite, baddleyite, fluorite)	Late Precambrian
			Biryusa microcraton				
Tepsa (Biryusa group of deposits); muscovite	Biryusa block	Muscovite in pegmatites	Two-mica schists and amphibolite gneisses	Layered, cross-cutting sheets and veins	Clusters, block	Muscovite (microcline, quartz, plagioclase)	Late Archaean (?)
Kukshero; iron	Biryusa microcraton	Iron ore, ferruginous quartzites; metasedimentary	Two-mica biotite and amphibole gneisses, schists	Conformable layered	Densely disseminated to massive	Magnetite, haematite (pyrrhotite, pyrite)	Late Archaean

The complex is provisionally subdivided into the following units, depending on which particular rocks predominate and the way in which they are interleaved (from the base upwards): (1) two-mica, garnet-two-mica schist with kyanite, >300 m thick; (2) amphibolite, amphibole and garnet-amphibole plagiogneiss; thin bands of calciphyre, marble, kyanite-staurolite and muscovite schist are common in this unit, which is up to 500 m thick; and (3) muscovite and kyanite schist and marble. Amphibolites occurring in this unit, which is up to 700 m thick, are often represented by para-rocks and show gradual transitions with carbonate rocks.

According to the latest information, the petrographic complexes in the Biryusa block can be viewed as independent stratigraphic units, the development of which reflects various crust-forming events in the region. However, it is difficult to establish the extent to which the internal processes in these complexes were separated in time. There is no doubt that the supracrustal rocks of the second complex were deposited on deeply eroded rocks of the first complex, which thus acted as a basement complex for them. The basement complex exemplifies all the internal crust-forming processes of the upper Archaean. For example, higher metamorphic pressures determined the formation of eclogitic associations. Basic and ultrabasic rocks, related in time to the second complex, were most frequently emplaced into the basement complex.

One of the characteristic features of the Biryusa block is the development of mica pegmatites. The pegmatites in the Biryusa block formed at the same time as the internal processes in the Tepsa complex were active. This conclusion is based on a structural study of the Tepsa complex, an analysis of the location of zones of high mica pegmatite concentration (Sizykh, 1978) and the mutual relationships between metamorphic processes and pegmatites. If the Tepsa complex has been correctly assigned to the Late Archaean, then the pegmatites are also of this age.

The *lower Proterozoic Tumanshet terrain* is exposed in the graben of the same name, situated in the north-east of the Biryusa microcraton, although different authors quote different outcrop areas for the complex (see Fig. II-39); it has a weathered base and is unconformable on both the older complexes. Its supracrustal rocks consist of up to 4200 m of quartzite, sandstone, schist, rare conglomerate and grit.

The rocks are zonally metamorphosed from greenschist to low-temperature amphibolite facies in the sillimanite-andalusite facies series. There is no direct evidence concerning the age of the complex. Most workers correlate the Tumanshet complex with that of the Urik-Iy graben, in which case it must be assigned to the lower Proterozoic.

Characteristic deposits of useful minerals associated with the terrains in the Pre-Sayan basement high are shown in Table II-11.

South Yenisey basement high

This elevation (the Angara-Kana microcraton) is situated in the south of the Yenisey ridge, south of the river Angara. To the north it is in fault contact with the Baikalides of the Yenisey ridge, and to the east and south it is covered by the Riphean to Palaeozoic sedimentary platform cover. To the west, its boundary is a deep fault zone along which both the basement high and the Yenisey fold belt are in contact with the West Siberian Platform.

Significant contributions to the geology of this region have been made by Yu.A. Kuznetsov, S.V. Obruchev, G.I. Kirichenko, Ye.K. Kovrigina, V.I. Volobuyeva, A.V. Vorobyov and many specialists from the Krasnoyarsk regional geological survey.

The subdivision of the Yenisey ridge into two different structural categories — the Baikalide fold belt and an ancient platform basement — is somewhat provisional. A number of workers consider the South Yenisey high to be the basement to the Baikalides of the Yenisey ridge. Reasons for assigning this region to an ancient cratonic structure are: (1) outcrops in the east of the microcraton of sub-platform Riphean sediments, and (2) an early Precambrian age for the internal processes responsible for forming the crustal structure. The South Yenisey high consists mainly of Archaean high-grade metamorphic rocks.

The *lower Archaean terrain* is known by the term Kan Group, which for many years has been correlated with the oldest rocks in Eastern Siberia. The Kan Group is subdivided into the Atamanov and Kuzeyev assemblages, totalling some 8000 m.

The lower part of the succession is the Atamanov assemblage, consisting of biotite, garnet-biotite and biotite-garnet-sillimanite gneiss with thin quartzite bands. Hypersthene diorite and garnet-hypersthene plagiogneiss occur among these rocks. The upper part of the succession is the Kuzeyev assemblage, consisting of hypersthene and two-pyroxene schists containing thin bands of ferruginous quartzite. The development of charnockite and anorthosite is associated with the evolution of this lower Archaean terrain. According to a few workers (Musatov et al., 1983), the upper unit — the Kuzeyev assemblage — forms a NNW-striking belt some 200 km long by 30–40 km wide.

Two granulite facies metamorphic events have been established in the Kan Group as a whole, the later one being high-pressure, with P = 10–9 kbar and T = 800°C (Glebovitsky, 1975).

The *upper Archaean terrain* is represented by rocks of the Pre-Yenisey belt (Nozhkin, 1985). Only the rocks in the lower part of the belt belong to the upper Archaean complex, while the upper part is lower Proterozoic. A solution to this problem requires further geochronological research and at present, on the basis of analogy with adjacent East Sayan regions, the entire

TABLE II-12

Characteristic ore deposits and ore shows in the Archaean Angara-Kan basement inlier

Mineral deposit or ore show; useful mineral	Type of mineralization: ore-forming and genetic	Host or mineralized rocks	Shape of deposit, ore body morphology	Type of ore	Mineral composition of ores and minor minerals	Age of host rocks and mineralization
Metlyakovo; molybdenum	Quartz-molybdenum; hydrothermal, related to Late Proterozoic Lower Kan granite intrusion	Granite	Cross-cutting quartz veins	Disseminated, cluster-type disseminations	Molybdenite (pyrite, chalcopyrite)	Late Proterozoic
Bogunay; gold	Gold-sulphide-quartz; hydrothermal	Pyroxene gneisses, granulites	—ditto—	Disseminated	Pyrite, chalcopyrite, sphalerite, galena, gold, sericite, chlorite	Early Archaean, Late Proterozoic
Kuzeyev; gold	Gold-sulphide-quartz; hydrothermal, related to Late Proterozoic Posolni granite intrusion	Biotite and amphibole gneisses, biotite and amphibole migmatites	Cross-cutting quartz veins, accompanied by gold placers	Disseminated	Pyrite, galena, sphalerite, chalcopyrite, pyrrhotite, gold (wolframite, scheelite)	Late Proterozoic
Bargin; muscovite, rare metals, ceramic raw materials	Muscovite-ceramic; pegmatitic	Pyroxene gneisses, granulites	Concordant vein bodies predominate	Clusters, block	Muscovite, microcline, quartz	Late Archaean
Sredneshilkin; muscovite, rare metals	Muscovite; pegmatitic related to granites	Biotite and amphibole-biotite gneisses	Cross-cutting quartz veins	—ditto—	Muscovite (microcline, quartz)	Late Archaean
Tarak; rare earths	Placers related to Archaean Tarak granite intrusion	Granites			Zircon, ilmenite, magnetite	Recent - Quaternary

rock complex in the Pre-Yenisey belt is assigned to the upper Archaean terrain.

The lower unit in the complex, the Yudin assemblage, includes amphibolite, feldspathic amphibolite, garnetiferous gneiss and schist, 1500–1700 m thick. The assemblage is intruded by stratiform metagabbroid sills, gneissose granite bodies and *lit-par-lit* veins of gneissose aplite and pegmatite. The amphibolites in this assemblage correspond to low-titanium and low-iron basalts, among which have been found thin komatiite flows. The gneisses are considered to be metavolcanics, identified petrochemical varieties being liparite, rhyodacite and dacite. The basic metavolcanics in the assemblage have low (lower than clark values) contents of lithophile elements: uranium (0.1–0.6 ppm), thorium (0.2–1.0 ppm), also gold (0.5–3.5 mg/t). Acid metavolcanics are characterized by low background values of gold (1–2 mg/t), uranium (0.5–1.5 ppm) and thorium (2.4–5.5 ppm), i.e. lower than the usual radiochemical background for andesites, which suggests that the acid volcanics are differentiates of a basic magma (Nozhkin, 1985).

The upper unit in the belt is the Predivin assemblage, with mafic and ultramafic micro-amphibolite near the base; quartzite with amphibolite members and amphibole plagiogneiss occur higher in the succession. The upper part of the assemblage consists of quartzite and schist, with a horizon of acid metavolcanics, plagioclase porphyry and amphibolite.

The upper part of the succession contains significant volumes of metagreywackes with magnetite-bearing quartz-sericite schist and hematite-magnetite quartzite horizons. Large amounts of ultramafics are a significant feature of the belts as a whole. The basic metavolcanics in the upper assemblage belong to high-titanium and high-iron varieties. Acid volcanics are sodic dacite and liparite.

An Archaean age (2.6 Ga) for the complex has been established on the basis of radiometric dating of pegmatite veins which occur among rocks of Early and Late Archaean age (Nozhkin, 1985).

Characteristic ore deposits and shows for the South Yenisey basement high are shown in Table II-12.

2. THE BAIKAL-PATOM HIGHLANDS

Introduction

The Baikal-Patom fold belt is part of a system of heterogeneous fold structures occupying a wide area in Eastern Siberia, known in Soviet geological literature as the Baikal mountains (Salop, 1964). This region was considered as the tectonotype Baikalide terrain since it has Late Proterozoic age representatives among the majority of the stratified metamorphic assemblages. However, work carried out here in the last 10 years and summarised by Fedorovsky (1985) has shown that these assemblages have a lower Proterozic age, proving that the tectonic stabilization of the Earth's crust in the NW Baikal mountain region occurred not in the Late Precambrian but at the end of the Early Proterozoic.

The Baikal-Patom belt is situated in the north-western part of the Baikal mountains, occupying an area in excess of 200,000 km^2, stretching from the central part of Lake Baikal in the south, to the river Vitim basin in the north-east. Here, the formationally varied lower Proterozoic complexes form a series of linear structural zones, each characterized by an individual evolutionary history and metallogeny. Within the confines of this belt (Fig. II-47) are inliers consisting of Late Archaean rocks (2.9–2.6 Ga), forming the reworked basement to the belt, and linear fold zones — Early Proterozoic clastic terrigenous and volcanogenic trench structures, together with an Early Proterozoic volcano-plutonic belt. To the north-west and in the east, the Baikal-Patom fold belt is surrounded by the Late Riphean sedimentary cover of the East Siberian Platform.

Ideas about the geological evolution of the belt and the entire Baikal mountain region were formulated in fundamental works by Salop (1964) and Obruchev (1949) during the period 1930–1960. Their views were further developed during survey work carried out by a large team of geologists from Irkutsk in 1960–1970. The Precambrian Institute in Leningrad has published several substantial works on the tectono-metamorphic evolution of the Baikal mountain region.

The belt has a complex structure, a factor which contributes to the controversy surrounding the age of its constituent terrains and the time of formation of the tectonic structure of the belt. Accordingly, the following outline of the tectonic structure and the description of the evolution of the region reflect elements of this controversy. In considering the age of the terrains, we have used as a data base previously published and new work on isotope geochronology that agrees with research in structural geology.

Tectonics

At the present time it has been established that the formation of the Precambrian lithosphere in the Baikal-Patom belt occurred during two major cycles which constitute its structural stages — Archaean (>2.6 Ga) and Early Proterozoic (>1.65 Ga), corresponding to which are regionally disposed structural-lithological terrains. A characteristic of the Baikal-Patom belt is the presence of elongate north–east-striking blocks of the Archaean terrain, forming inliers (Fig. II-47) — the Baikal, Chuya-Tonod, Nechera and Muya. These inliers occur within the Lower Proterozoic protogeosynclinal complex, forming trench strucutures: the volcanogenic (eugeosynclinal) Baikal-Vitim, the terrigenous (mio- and micto-geosynclinal

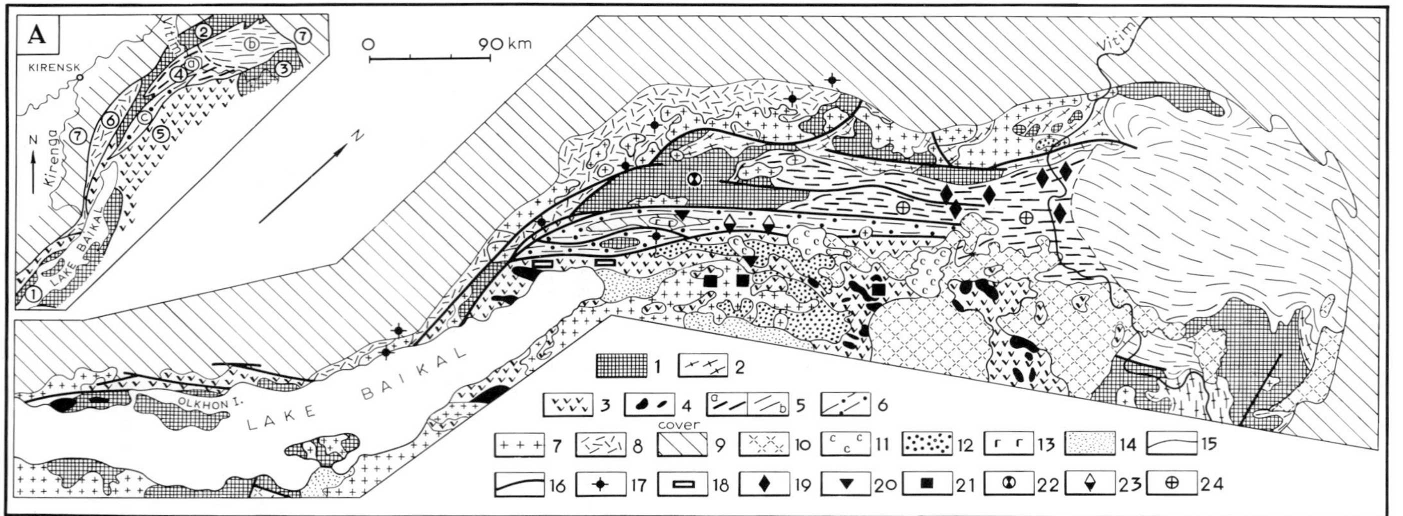

Fig. II-47. Sketch map showing geological structure of the Baikal-Patom fold belt (compiled by S.I. Turchenko and Yu.M. Sokolov). Archaean provinces: *1* = Chuya and Olkhon Groups (two-pyroxene schist, gneiss, quartzite, calciphyre, marble, amphibolite); *2* = gneissose granite and migmatite. Early Proterozoic provinces: *3* = Nyurunduka, Muya and Sarmi Groups (intermediate and acidic metalava, amphbolite, schist, quartzite, ferruginous quartzite); *4* = metagabbro, metaperidotite; *5* = carbonate-terrigenous metasediments: ga-bi, bi-st, ky-sill schists, graphitic schist, quartzite, calc-silicate rocks, amphibolite; in high-*T* zones: migmatite, gneissose granite of Mama-Oron complex, pegmatite; in greenschist facies zones: metapelite, black schist; metasandstone, grit, limestone (*a* = Mama-Teptorga Group, *b* = Tonodo-Bodaybo Group); *6* = Olokita Group (metabasalt, metarhyolite, carbonate-graphite, sulphide-graphite schists and volcanogenic-graphitic rocks, amphibolites); *7* = Primorye, Kuandi, Amondrak post-metamorphic granitoids. Early Proterozoic volcano-plutonic complex: *8* = Akitkan Group (conglomerate, sandstone, rhyolite, quartz porphyry, tuff). *9* = East Siberian platform cover: Baikal Group (Middle and Upper Riphean) — quartzite, dolomite, limestone, sandstone, siltstone, shale; *10* = Konkudero-Mamakan Palaeozoic granitiod complex; *11* = nepheline syenite; *12* = Vendian sediments: Kholodnin, Orkolikan Groups (conglomerate, sandstone, siltstone); *13* = Dovyren layered gabbro and peridotite intrusion; *14* = Quaternary sediments, Upper Angara basin; *15* = geological boundaries; *16* = major faults; major ore deposits and ore shows: *17* = pyrite-polymetallic and lead-zinc, *18* = iron, *19* = muscovite pegmatites, *20* = copper-nickel, *21* = titanomagnetite, *22* = facing stone (cordierite), *23* = rare-metal and rare-metal-muscovite pegmatites, *24* = alumina (kyanite). A: Structural sketch map of Baikal mountains. Archaean terrains: *1* = Baikal, *2* = Chuya-Tonodo, *3* = Nechera and Muya. Basins in Early Proterozoic terrain: *4* = terrigenous (*a* = Mama, *b* = Bodaybo, *c* = Olokita), *5* = Baikal-Vitim volcanogenic, *6* = Early Proterozoic Akitkan volcano-plutonic belt, *7* = Late Riphean sedimentary cover assemblage on East Siberian craton./dwars

TABLE II-13

Isotopic age of zircons from structural terrains in the Baikal-Patom belt

Geological setting	Structural-lithological terrain	Method	Age, Ga
North Baikal mountains	Konkudero-Mamakan granitoids, tectono-magmatic stage of activity	Pb-Pb evapor. techn.	0.38, 0.31
	Lower Proterozoic complex (Mama-Bodaybo, Tonod-Bodaybo, Patom Groups)	—ditto—	1.8±0.1
	Muscovite pegmatites in lower Proterozoic complex	—ditto—	1.8±0.1
North-West Baikal region	Mama-Oron granites in lower Proterozoic complex	—ditto—	1.9±0.1
	Rare-metal pegmatites in lower Proterozoic complex	—ditto—	1.9±0.1
	Iron-quartz metasomatites (Tyskoye deposit) in lower Proterozoic complex	—ditto—	1.8±0.1
West Baikal region	Granites of Primorye complex	U-Pb isochron	1.9±0.1
North Baikal mountains, North-West Baikal region	Gabbros cutting rocks of the Chuya Group in the Archaean complex	Pb-Pb evapor. techn.	2.28±0.1
	Archaean structural terrain (Chuya Group)	—ditto—	2.42, 2.29, 2.26

or basinal) Mama and Bodaybo, and the volcanogenic-terrigenous Olokita (riftogene). The Akitkan volcano-plutonic proto-orogenic structure formed during the Early Proterozoic in the evolution of the north-western margins of the belt. The Late Riphean was characterized by the formation of a sedimentary platform cover. The upper age boundary for the formation of Precambrian terrains is defined on the basis of fossiliferous Vendian to Lower Cambrian sediments (the Orkolikan and Kholodnin Groups). Intense tectono-magmatic activity occurred in the Palaeozoic along the south-eastern margin of the Baikal-Patom belt, expressed as granitoid intrusions (the Konkudera-Mamakan complex with an age of 400–300 Ma).

The latest data on zircon age determinations from rocks in this belt are presented in Table II-13 and Figure II-48.

Structural-lithological terrains

The *Archaean terrain* within the Baikal basement high (Fig. II-47) consists of the Olkhon Group, which is correlated with the Early Archaean Sharyzhalgay Group of Pre-Sayan (Yeskin et al., 1979; Petrova and Levitsky, 1984). Within the Chuya-Tonod, Nechera and Muya inliers, it accounts for

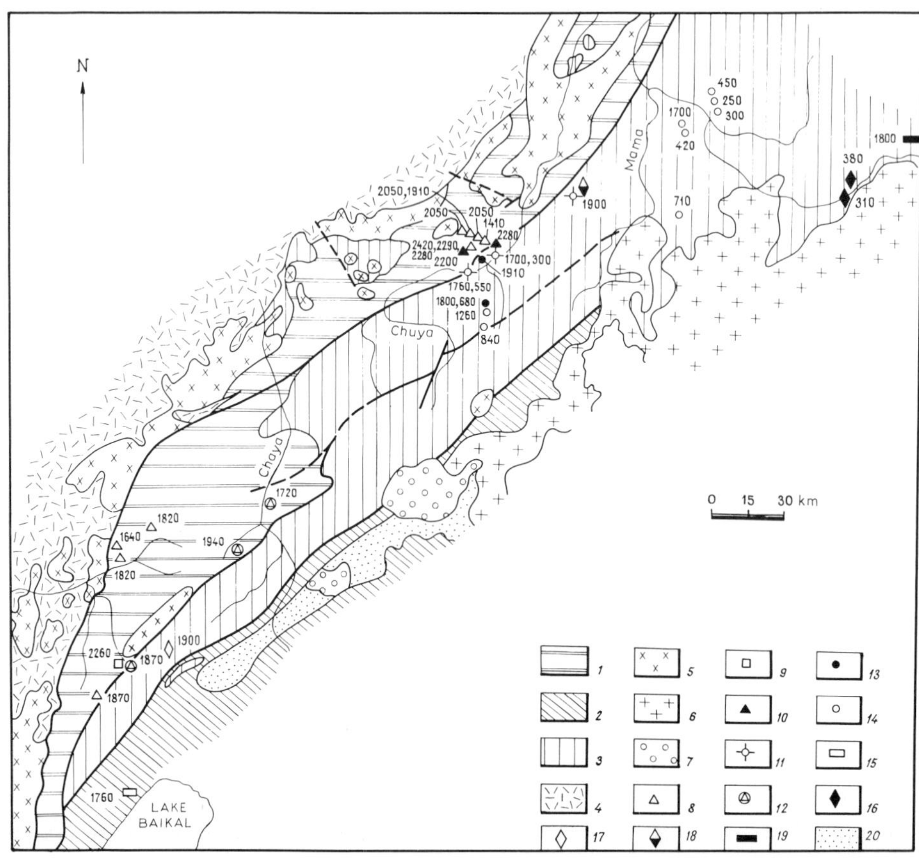

Fig. II-48. Map showing distribution of isotopic dates in the Baikal-Patom belt. *1* = Chuya Group (Archaean): polymigmatites and amphibolite facies retrogressed granultes, amphibolites and gneissose granites. Lower Proterozoic: *2* = Baikal-Vitim volcanogenic trench. Pyurunukan and Muya Groups (metadiabase, metagabbro, amphibolite facies schists); *3* = Olokita-Mama-Bodaybo terrigenous basins. Mama-Bodaybo, Olokita, Patom Groups (carbonate-terrigenous metasediments). *4* = Lower Proterozoic, Akitkan Group (acid-intermediate lava and tuff, subvolcanic granitoids, rapakivi granite); *5* = Early Proterozoic granitoids; *6* = Palaeozoic granitoids; *7* = nepheline syenites. Isotopic age dates from zircons, Pb–Pb thermochron method and composition of dated rocks: *8* = Chuya Group schists, migmatite, gneissose granite, *9* = cord-ged-phlog metasomatites, Chuya Group, *10* = metagabbro cutting Chuya Group, *11* = Teptorga and Patom Groups, schist, *12* = muscovite-rare metal pegmatites, Abchad pegmatite field, *13* = plagioclase pegmatites, North Baikal muscovite province, *14* = plagioclase-microcline pegmatites, same province, *15* = quartzites, Tya iron ore deposit, *16* = Palaeozoic granitoid dykes, *17* = gneissose granite, Mama-Oron complex, *18* = Proterozoic granites (Primorye etc.), *19* = Patom schists; *20* = Vendian.

significant areas or is seen as major boudins and relicts among metamorphic associations of the lower Proterozoic stage of rejuvenation. These rocks are grouped together within the Patom and North Baikal Highlands as the

Chuya Group, correlated with the Olkhon Group in the South Baikal region and with the Sharyzhalgay Group in Pre-Sayan, respectively.

The Archaean terrain in all mapped regions consists of two-pyroxene schist, pyroxene-magnetite quartzite, marble, calciphyre, garnet-cordierite-sillimanite gneiss and associated charnockite. The most typical assemblages are: hypersthene–diopside–plagioclase (± garnet ± spinel ± quartz); and brown hornblende–plagioclase (An_{48-54})–diopside–hedenbergite (± garnet ± spinel). For metapelites, parageneses with cordierite and sillimanite (cordierite–garnet–hypersthene–plagioclase–biotite–quartz), and for calc-silicate rocks: diopside–carbonate–scapolite–plagioclase (± quartz ± garnet ± Kfs). It is worth drawing attention to the fact that although granulites sometimes also form horizons up to a few hundreds of metres thick (West Pre-Baikal region) and of considerable extent, the characteristic outcrop pattern for granulites in all the basement elevations is still that they occur in areas of polyphase amphibolite facies metamorphism and repeated granitization. They always show clear signs of later dislocations and metamorphism of lower Proterozoic age. Additionally, the Archaean terrains contain areas of gneissose granite and polymigmatite.

A potassium-argon age for brown amphibole from Chuya Group gneissose granites is 3.46 ± 0.07 Ga, while a Pb–Pb age on zircon from a migmatite leucosome in the same group is 2.42 Ga. Zircons from gabbro cutting Chuya Group rocks is 2.28 Ga (Pb–Pb method, the evaporation technique). Furthermore, cordierite metasomatites, associated with Chuya Group rocks, also have an age close to this (2.26 Ga by the Pb–Pb method). Granite gneisses in the Chuya block give an age of 2.36 Ga using the U–Pb isochron method. Whole-rock U–Pb isotopic ages for Olkhon Group rocks are 2.36 ± 0.1 Ga according to work published by Yeskin et al. (1979).

The *Early Proterozoic terrain* contains supracrustal and igneous formations, restricted to trench structures: the Baikal-Vitim, Olokita, Mama and Bodaybo. In the first of these, the main constituents are bimodal metavolcanics, altered to schist and amphibolite, as well as thin bands and lenses of quartzite and ferruginous quartzite. These rocks include major mappable and minor bodies of metagabbro and meta-ultramafics. The supracrustal rocks in the Baikal-Vitim belt are recognised under the names Sarmi Group in the Southern Pre-Baikal region, the Nyurundukan Group in the Western Pre-Baikal region, and the Muya Group in the Northern Pre-Baikal region and the Muya highlands. As a rule, rocks of these groups are metamorphosed at epidote-amphibolite and low-temperature amphibolite facies. In terms of the petrography and the close association between tholeiitic metavolcanics and layered gabbro and ultramafic bodies, this structural zone is very similar to eugeosynclinal zones and is sometimes referred to as a greenstone belt or as an ophiolite (Fedorovsky, 1985).

However, the biggest outcrop areas in the Baikal-Patom belt are formed by terrigenous (Mama and Bodaybo belts) and volcano-terrigenous (Olokita belt) complexes (Fig. II-47). Rocks in the Mama and Bodaybo belts (particularly the former) occur as tectonic nappes on a partially rheomorphosed Archaean basement and have broadly similar types of stratigraphic section. The complex groups together members and formations of zonally metamorphosed sediments, identified as the Tonod-Bodaybo, Patom, Teptorgin and Mama-Bodaybo Groups. On the whole, these groups are characterised by the same type of succession; their total thickness is 10–11 km, consisting of bedded carbonate-clastic terrigenous rocks in which it is possible to trace over considerable distances horizons of metapelitic, metapsammitic, metacarbonate and metaconglomerate lithofacies, the initial composition and textural features of which can be recognised in all the regional metamorphic zones. Metamorphic zoning in the Mama and Bodaybo belts belongs to the kyanite-sillimanite type and is the best studied example of Early Proterozoic low-gradient regime metazonal complexes in Siberia. In the Mama structural zone, which is distinguished by a well-expressed linear fold structural pattern, metamorphic isograds cut the sequence of rock layering and can be clearly observed along strike. Furthermore, metamorphism in this structural zone reaches high-temperature amphibolite facies, and goes as far as migmatite generation, areas of intense granitization and pegmatite intrusion, plus the formation of rheomorphic granites (the Mama-Oron complex).

In summary, the zonal metamorphism of the lower Proterozoic terrain has the following characteristics: (a) local change in amphibolite facies *P–T* parameters from low-gradient to high-gradient regime, leading to a change in the regime of acid-alkali metasomatic processes during retrograde metamorphism with an associated change in metallogenic signature of pegmatites, from muscovite-bearing to rare-metal types; (b) broad greenschist facies zones and closely-spaced garnet, staurolite and kyanite isograds; (c) formation of granite-gneiss domes and migmatite zones during the final stages of granitization; and (d) retrogression with renewed acid-type post-migmatite metasomatism, which is the main reason for the formation of muscovite in the pegmatites.

Another feature of this zone is the frequent tectonic interleaving of the substrate with slices of the rheomorphosed granite-gneiss basement, most commonly encountered in zones of granitization. This feature is reflected in the specific metallogeny of the Mama zone, which is abundant muscovite. On the whole, the lithology of rocks in this zone and the metamorphic and magmatic regime are comparable with miogeosynclines or deep-water parts of the continental slope within the passive margins of an ancient protocontintent.

In the Bodaybo structural zone, despite great similarities with the Mama zone in terms of lithology and type of succession, there are differences in outcrop areas of graphitic and sulphide-graphite schist, coarse-grained sandstone and conglomerate, and in the absence of zones of granitization and high-temperature amphibolite facies metamorphism. Metamorphic zoning also belongs here to the kyanite-sillimanite type and is represented by greenschist and staurolite facies zones. Another characteristic feature for this zone is the appearance of dome-like structures, combined with linear and open fold forms, which is in sharp contrast to the Mama zone. The nature of the lithology and fold forms, together with particular features of low- and medium-T zonal metamorphism where the isograds are close to stratigraphic horizons, indicate that this zone had a different geodynamic regime — close to micto-geosynclinal environments or shelf regions.

The Olokita structural zone borders on the south-western margin of the Baikal-Vitim volcanogenic trench and in individual localities the successions are similar. Volcanics are common near the base, with the development of a basalt-rhyolite bimodal association. The remainder of the succession of the volcanogenic-terrigenous complex, which is assigned to the Olokita Group (correlated with the Mama-Bodaybo), in addition to volcanics displays a characteristically widespread development of carbonate-graphite, quartz-graphite and sulphide-graphite rocks, quartzite and lydite (black jasper). The rocks are zonally metamorphosed, up to kyanite-sillimanite facies series of the amphibolite facies. Evidence for the riftogene nature of this structure is based on particular features of the succession, the linear nature of the structure, the evolution of metabasalt composition from sub-alkaline olivine tholeiite to tholeiite and quartz tholeiite in association with rhyolite, and the presence of carbonaceous formations.

The clastic and volcano-terrigenous assemblages in this structural zone are cut by numerous stocks of syn-metamorphic granitoids belonging to the Mama-Oron complex (1.9 Ga) and later post-metamorphic granitoids (1.8 Ga). Microphytofossil evidence indicates that the period of sedimentation of the rocks in the metamorphic zonal complex was lower Proterozoic. An isotopic age of 1.9 ± 0.1 Ga has been obtained by the Pb–Pb method on zircons from rocks in the Mama-Bodaybo Group, and finally the time of formation of pegmatites and post-metamorphic granites is 1.8 Ga. These data indisputably confirm a lower Proterozoic age for the clastic terrigenous rocks in the trench structures under discussion, allowing them to be correlated confidently with lower Proterozoic terrains in other regions of East Siberia (Fedorovsky, 1985).

Early Proterozoic volcano-plutonic complex. A volcano-plutonic complex has been identified within the north-western marginal zone of the Baikal-Patom belt, consisting of a proto-orogenic zone which formed in the lower Proterozoic and completed its protogeosynclinal development. This structure is known as the Akitkan volcano-plutonic belt, comprising the Akitkan Group. The belt has an elongate linear shape and stretches for 800 km, with a width of 10–100 km. It consists of volcanogenic-terrigenous assemblages, the constituents being a lower unit of conglomerate and sandstone, a thick suite of trachybasalt-trachyandesite and trachydacite-rhyolite, and an upper unit of volcanoclastic sediments, tuffo-conglomerate and acid effusives. The volcanic rocks in the belt are accompanied by subvolcanic and hypabyssal intrusions of granodiorite and diorite, granite and syenite porphyry and rapakivi granite which formed at the same time. Formation of the volcano-plutonic complex was accompanied by tectonic breaks of various orders, associated with which are metasomatic effects (albitization, chloritization and silicification).

The lower age boundary of the Akitkan Group is defined by the overlying basal conglomerates resting with angular unconformity on lower Proterozoic phyllitic schist (the Sarmi Formation) and on a granitoid weathering crust (the Primorye complex), with an age of 1.9 ± 0.1 Ga. The upper age boundary of the group is defined by the stratigraphic superposition on its rocks of carbonate-terrigenous platform sediments (the Baikal Group). The radiometric age of the volcanic rocks in the Aktikan Group has been estimated at 1700 ± 20 Ma by the Rb–Sr isochron method (Manuylova, 1968).

The Late Riphean platform assemblage (Baikal Group) is restricted in the geotectonic plan to the Lena basin, a tectonically activated part of the cover assemblage on the East Siberian Platform. Individual portions of the assemblage form relicts within the Olokita zone, where they are represented by dolomite and limestone. Associated with these is the Dovyren layered gabbro-peridotite intrusion, which formed during the Late Riphean episode of tectono-magmatic activity.

The Baikal-Patom fold belt thus includes individual basement inliers, made of Archaean granulite and granite-gneiss, Proterozoic volcanogenic and terrigenous trenches, forming protogeosynclinal belts and a proto-orogenic volcano-plutonic structure. The south-eastern margin of the fold belt was affected by Palaeozoic activity, expressed as the intrusion of the Kondukero-Mamaka batholith complex.

Useful minerals

The heterogeneous structure of the Baikal-Patom belt has determined the variety of the metallogenic signatures of its structural zones and therefore characterizes the variety of ore deposits and useful mineral shows.

Only very small deposits are associated with the Archaean terrain. Formations in the Olkhon Group contain manganese metasedimentary deposits (Tsagan-Zaba), localized in crystalline limestones, also the Ust-Angin marble and graphite deposit. Within the Chuya Archaean inlier there are outcrops of iron-magnesium metasomatic rocks containing gem-quality cordierite. In addition, the Archaean terrain of the entire Baikal mountain region has prospects for iron-ore and phlogopite deposits, since its constituent rocks are correlated with the Sharyzhalgay Group in the Pre-Sayan region, which contains similar unique deposits.

An essentially femic geochemical spectrum of deposits is associated with the Baikal-Vitim volcanogenic trench. Here basic and ultrabasic rocks have associated deposits of asbestos (Molodyozh), nephrite (Param) and a number of copper-nickel (Chaya), titano-magnetite (Orkolikan and Gorbylyak) and chromite ore shows. Individual copper pyrite shows are restricted to metabasites in the Nyurundukan and Muya Groups. In the lower metabasalt-andesite part of the succession in this structure (the Nyurundukan Group) and in associated terrigenous rocks, there are ferruginous quartzite deposits (Tya and Yelizaveta).

The lower Proterozoic was the main metallogenic cycle in the Baikal-Patom belt, associated with the intense expression of internal processes and manifest in the terrigenous and terrigenous-volcanogenic complexes of the Mama and Olokita zones. In the first of these, muscovite and quartz-feldspar pegmatites formed during this episode, associated with kyanite-sillimanite facies series metamorphism. Rare-metal pegmatites and the iron-magnesium metasomatites with industrial stones, referred to above, formed in the Chuya basement high in zones where metamorphism was at andalusite-sillimanite facies series. The lower Proterozoic activity had a significant effect on the remobilization of ore-forming components during the formation of pyrite-polymetallic ore deposits and the ironstone-chert formations. In all probability, it was at this particular time that the deposits localized along external boundaries of Archaean microcratons originated. Ore formations associated with retrogressed rocks originated here: granulated quartz, rare-metal and rare-earth pegmatites (the Ukuchikta and Nechera zones).

The Kholodnin pyrite-polymetallic ore deposit and a number of analogous ore shows are restricted to the Olokita structural zone, which is riftogene in nature. Typical ores in the deposit are concordant sheets and impregnation-veinlet types. Economic ore bodies have a zonal structure, expressed as copper pyrite ores localized at the base and lead-zinc and pyrite ores at the top of the deposits. Banded ores were metamorphosed at amphibolite facies and altered to gneissose, banded crystalline and coarsely disseminated types. Subsequently, pockets and veinlets of galena-sphalerite ores formed in the layered ore bodies at the same time as quartz-muscovite and quartz-kyanite metasomatic rocks originated.

The emplacement of the Early Proterozoic volcano-plutonic complex was accompanied by dislocations with breaks in continuity; these had associated metasomatic effects (albitization, chloritization and silicification) on the volcano-sedimentary and intrusive rocks and on the rocks forming the basement to the belt. These hydrothermal-metasomatic changes here have associated rare-metal and copper-lead-zinc mineralization, restricted to volcano-tectonic structures (calderas and central-type volcanic systems). Ore bodies are represented by two types. The first are cross-cutting veins, pipes, veinlet-lenticular and pocket-type bodies localized in fractured portions of automagmatic breccias, lava and tuff-pyroclastic formations and in the external contact zones of individual lava flows. The second consists of sheet-like and lenticular bodies of disseminated and veinlet-disseminated ores, localized along gently-dipping dislocations.

A trachydacite-liparite rock association in the upper part of the section includes ore shows of lead, zinc, copper, fluorite and rare metals, restricted to palaeovolcanic structures. In this belt, ore-bearing metasomatic rocks also develop from blastomylonites, formed from rocks in the Chuya Group, and expressed in the formation of polyphase zones of microclinization, albitization, greisenization, silicification and fluoritization. Scheelite, tin-tungsten, arsenic-mercury-silver and lead-zinc mineralization are associated with these. In addition to this sequence, contact zones of plutons also contain copper-polymetallic and arsenic-bismuth mineralization.

Late Riphean platform-type sediments have associated copper-nickel ore mineralization in the Dovyren gabbro-peridotite intrusion (the Yoko and Rybachye ore shows).

Chapter 5

PRECAMBRIAN OF THE COVER

Late Proterozoic sediments in the Siberian craton form a number of independent structural stages in the composition of the platform cover. Reliably established Lower Riphean sediments are known in a restricted number of sections in the Uchur-May region and in the north-east of the craton. Middle and Upper Riphean sediments are widely distributed and comprise different structural units; they may fill major intracratonic basins and overlap onto the slopes of adjacent elevated blocks, and separate deep aulacogen-type troughs; or they may form thick successions in pericratonic basins and foredeeps, closely associated with neighbouring folded epigeosynclinal regions (Fig. II-49). The highest upper Proterozoic horizons (Yudomian) form part of the basal platform succession as a thick cover over practically the entire platform. Stratigraphically, the Yudomian sediments appear to correspond to the Kudash and Vendian of the East European craton.

By far the best studied Riphean and Yudomian successions are those around the craton margins where they are exposed at the surface. In central regions of the craton, upper Proterozoic sediments are buried under a thick Phanerozoic sedimentary cover. Here, Late Proterozoic sections are known exclusively from deep borehole data obtained in recent years mostly from southern regions of the craton: the Vilyuy basin, the Nep-Botuob dome, the Kamo arch and in Pre-Sayan. However, at present there are insufficient data to allow any detailed stratigraphic subdivisions or to identify structural-facies regions in the centre of the craton.

Riphean sediments in the Uchur-May region on the SE margin of the craton are among the best studied. Two types of succession are distinguished here, associated with the Yudoma-May pericratonic foredeep and the Uchur-May basin at the eastern termination of the Aldan Shield.

In the Yudoma-May foredeep, the most complete sections, commencing with the Lower Riphean, are known from its northern part, along rivers cutting through the Ulakhan-Bom and Kyllakh ranges on the right bank of the Aldan. At the base of the river Khanda section (Belaya river) is the Pioneer Formation, which has a volcanogenic-terrigenous red-bed sequence at the base, succeeded by essentially dolomitic rocks. The overlying

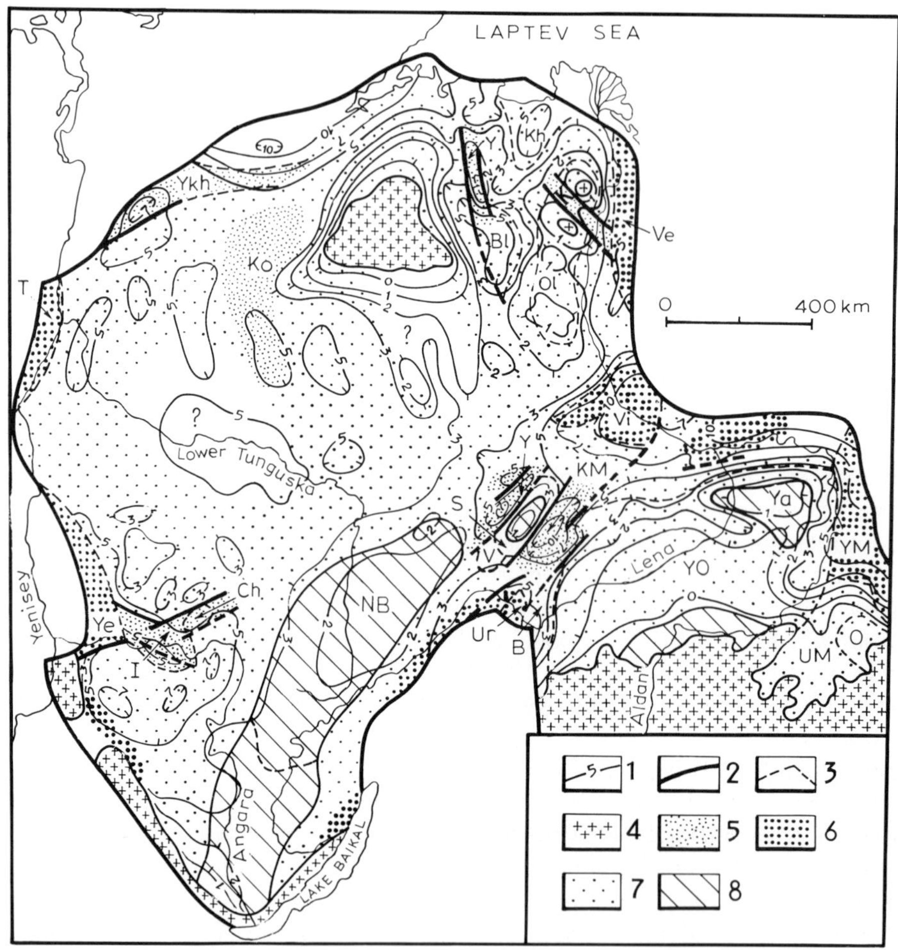

Fig. II-49. Sketch map showing distribution of Riphean sediments in the Siberian craton (compiled from Grishin et al., 1982; Malich et al., 1984; Gusev et al., 1985; Kontorovich et al., 1981). Basement relief (depths 0, 1, 2, 3, 5, 7 and 10 km below sea level) taken from basement map of Siberian craton (Kontorovich et al., 1981). *1* = depth to top of basement (km), *2* = major faults, *3* = structures in cover, *4* = basement outcrops, *5* = Riphean aulacogen sediments, *6* = Riphean pericratonic basin sediments, *7* = Riphean intracratonic basin sediments, *8* = Riphean missing. Pericratonic basins: *Ye* = Fore-Yenisey, *T* = Yenisey-Turukhan, *V* = Verkhoyan, *Vi* = Vilyuy, *YM* = Yudoma-May, *Ur* = Ura. Aulacogens: *I* = Irkineyev, *YKh* = Yenisey-Kheta, *Ko* = Kotuy, *K* = Kyutingda, *Km* = Kempendyay, *Y* = Ygyatta. Other structures: *NB* = Nep-Botuob anteclise, *Ch* = Chadobets dome, *Bl* = Bilir basin, *Kh* = Khastakh basin, *Ol* = Olenyok arch, *S* = Suntar arch, *V* = Upper Vilychan dome, *B* = Beryozova basin, *YO* = Yakutsk-Olyokma scarp, *Ya* = Yakutsk arch, *UM* = Uchur-May basin, *O* = Omnya arch.

terrigenous-carbonate assemblages belong to the Khanda Group. These rocks contain an increased amount of finely-disseminated carbonaceous material, imparting a dark grey colour to the rocks (Gusev et al., 1985). Lower Riphean sediments have been deformed and altered under conditions close to greenschist facies, so that the overlying Middle and Upper Riphean assemblages are in contact with Khanda Group sediments above an angular unconformity.

Middle to Upper Riphean sediments occur throughout the basin. Similar but much thinner sequences are also described from the adjacent north-eastern part of the Uchur-May basin along the river May. Four groups are usually distinguished among these sediments, corresponding to major sedimentary cycles. The lowest, Aimchan Group, consists of sandstone (at the base) and carbonate shale. The carbonate portion of the succession is made up of grey stromatolitic dolomite. The group (or its analogues) reaches 500–700 m in thickness. Kerpyl Group sediments are usually unconformable on earlier sediments. The lower part of the group consists of sand and shale, the upper carbonate. The carbonate part of the sequence constitutes a marker unit, the Malgina or Duga Formation of mottled limestone and the Tsipanda or Fira Formation of pale-coloured massive stromatolitic dolomite (Table II-14). Two stromatolite associations are described from the dolomites in the Aimchan and Kerpyl Groups (the Svetlyy and Tsipanda respectively), proving a Middle Riphean age for the host sediments (Semikhatov et al., 1979).

Above come the Lakhanda Group carbonate sediments (or its equivalents). As is the case with the Kerpyl Group, a conformable relationship is observed only in the axial part of the trough, while to the west the sediments were preceded by erosion, in places accompanied by a bauxite-bearing weathering crust. The lower formation, the Neryuyen, consists of stromatolitic limestone and dolomite with thin mottled shale and siltstone. It is overlain by limestone and dolomite of the Ignikan Formation. The age of the upper (Nelkan) member of the Neryuyen Formation is well-constrained by numerous age determinations on glauconite which fall in the 920–970 Ma range (Semikhatov and Serebryakov, 1983). The Lakhanda Group is 500–600 m thick, while its analogues in the axial region of the palaeotrough reach 1350 m in thickness.

There are sudden changes in the stromatolitic assemblages in the Lakhanda Group succession. The Lakhanda stromatolite association from the Neryuyen Formation with massive baicalia, conophytons and yakutophytons, belongs to a Middle Riphean assemblage. The question of whether or not there is also a change in the microphytolith assemblages at this same horizon has not been finally settled (Semikhatov et al., 1979).

This creates particular difficulties in comparing the Western Urals and Uchur-May Riphean reference sections. On the one hand, K–Ar dates for the lower boundary of the Karatau and Lakhanda Groups are clearly similar.

TABLE II-14

Riphean assemblages of the Siberian craton

Age	Uchur-Maya basin	Yudoma-Maya depression	Ura aulacogen	Fore-Yenisey basin	Pre-Yenisey (Turukhansk) basin	Marginal slopes of the Anabar massif	Udzha aulacogen	Kyutingda aulacogen	Fore-Verkhoyan basin (Tuora-Sis, Kharaulakh)
U. Riphean	Lakhanda Group: Ignikan Fm *c* Neryuyen Fm *c*	Uy Group *t* Gren Fm *c* Setmuyal Fm *c* Vil Fm *c*	Zhuya Gp *t*	Tungusik Group: Oslyan Gp *tc* Kirgitey Fm *tc* Shuntar Fm *t* Potoskuy Fm *ct*	Dolomites, limestones Shorikhin Fm *c* Burovoy Fm *c* Derevnin Fm *ct*	Yusmastakh Fm: Upper member *c*	Udzha Fm *tvr* Khapchanyr Fm *tc*	Khaypakh Fm: Upper member *tc* Lower member *ct*	Sietachan Fm *tc* Neleger Fm *c*
M. Riphean	Kerpyl Group: Tsipanda Fm *c* Malgina Fm *c* Totta Fm *t* Aimchan Group *ct*	Fir Fm *c* Duga Fm *c* Muskel Fm *t* Bik Fm *t* Svetly Fm *ct* Taly Fm *t*	Valyukhta Group *t* Barakun Group *t* Bolshepatom Group *t* Balagannakh Formation *t*	Sukhopit Group: Aladyin Fm *c* Kartochka Fm *c* Pogoryuy Fm *t* Uderey Fm *t* Gorbilok Fm *t* Kordin Fm *t*	Sukhotungusik Fm *c* Linok Fm *c* Bezymyany Fm *t*	Yusmastakh Fm: Lower member *c*	Unguokhtakh Fm *tv*	Debengda Fm *tc* Arymas Fm *ct*	Eselekh Fm *c* Ukta Fm *tr*
L. Riphean	Uchuri Group: Ennin Fm *tc* Omakhta Fm *tc* Gonam Fm *tc*	Khanda Group: Belorechka Fm *tc* Dima Fm *tc* Tryokhgornaya Formation *tc* Pioneer Fm *tc*				Kotuykan Fm *c* Mukun Fm *tr*	Ulakhan-Kurunga Fm *tc*	Kyutingda Fm *c* Sygynakhtakh Fm *tr*	

Note: sediment types — *c* – carbonate, *tc* – terrigenous-carbonate, *ct* – carbonate-terrigenous, *t* – terrigenous, *tv* – terrigenous-volcanic, *r* – redbeds

This is also strengthened by particular features in stratigraphic similarities between Upper Riphean Kipchak and Tangaur sediments from the Western Urals and the Lakhanda and overlying Uy Groups in the Uchur-May region. On the other hand, the change in Middle to Upper Riphean stromatolite assemblages in Siberia occurs higher in the succession, at approximately the level corresponding to the base of the Katav Group in the Urals succession. This state of affairs led Semikhatov to propose that the boundary between the Middle and Upper Riphean should be moved up and fixed according to the Siberian palaeontological model (and not the Western Urals model).

The Riphean succession in the Yudoma-May basin is capped by sediments belonging to the Uy Group, a 3–4 km thick suite of grey and greenish sandstone, siltstone and shale, transgressive over the Lakhanda Group. Coarse polymict red sandstone appears towards the top in south-eastern sections. The rocks are noticeably dislocated. Using a variety of techniques, age dates of 640–740 Ma were obtained for cross-cutting central-type ultrabasic and alkaline intrusions (Semikhatov and Serebryakov, 1983).

To the west of the Yudoma-May basin, the Riphean assemblages decrease in thickness stepwise along faults and west of the river Aldan the section takes on a platform character. South-west of the foredeep, Riphean sediments crop out in the Uchur-May basin on the east margin of the Aldan Shield.

Lower Riphean sediments in the western part of the Uchur-May basin are represented by red and grey sandstone and grit, the Gonam Formation (80–140 m thick) and grey chemogenic and biogenic dolomite, sandstone and siltstone, the Omakhta (220–220 m) and Enni (140–150 m) Formations. Numerous age determinations on glauconite from the lower part of the Gonam Formation fall in the interval 1450–1520 Ma (Semikhatov and Serebryakov, 1983). The above-named formations are usually grouped together as the Uchur Group. Yudomian sediments follow, above an unconformity.

The greater part of Lower Riphean sediments wedge out on the Omnya elevation in the centre of the Uchur-May basin. Here at the base of the section are 250 m of almost exclusively Ennin Formation terrigenous clastic sediments (with rare thin dolomite horizons) and Middle Riphean sediments, missing from western regions of the Uchur-May basin: siltstone and shale belonging to the Omnya Formation (300 m) and limestone and dolomite of the Malgina and Tsipanda Formations (approximately 300 m). The Malgina limestones are often bituminous (Bogolepov and Votakh, 1977).

Riphean sediments on the northern slopes of the Aldan Shield in the Lena and Amga river basins form the lower platform cover almost the entire distance from Yakutsk to Olyokminsk, gradually giving way on approaching the Patom Highlands to sediments in the Beryozova basin and the Ura palaeotrough (the Ura anticline in the modern structural plan).

Grey-coloured Upper Riphean sediments (Dikimda Formation) are found over most of the Yakutsk-Olyokminsk step. The rocks are dolomite with thin marls, gypsum beds and argillite and a horizon of clastic rocks at the base, with bitumen shows. The sediments become gradually thicker, from a few tens of metres in the central regions of the platform to 200 metres and over in the lower reaches of the rivers Chara and Olyokma.

Farther west, within the confines of the Beryozova basin, the Upper Riphean succession thickens rapidly, increasing downwards by the stepwise addition of progressively older assemblages. The Upper Riphean is represented by the Dikimda Formation in deep boreholes in the Beryozova basin, and the Middle Riphean by the Imalyk Formation (Kontorovich et al., 1981). The Dikimda Formation attains 400 m in thickness, while the Imalyk (red conglomerate and sandstone with thin marls, shale and siltstone) is 20–70 m.

Riphean successions around the margins of the Patom Highlands and in the Ura anticline which juts far into the main body of the platform are distinguished by having greatly increased thicknesses. For example, in the Ura anticline and the west flank of the Beryozova basin in the Zhuya fault zone, the Riphean contains a large number of formations, making four or five groups, corresponding to four major depositional cycles, with a gradual transition in each cycle from coarse-grained terrigenous clastic sediments to carbonates, often with a marked sedimentary break at the base.

The lowest sedimentary cycle comprises the Balaganakh Group (>1400 m), which consists of terrigenous clastics at the base, gradually giving way upwards to a unit of sandstone, shale and limestone, the Mariinsk Formation. The next cycle is a combination of the Bolshepatom (Dzhemkukan and Moldoun Formation) and Barakun carbonate-terrigenous Groups, totalling 1700–2700 m in thickness. The top cycle of the Middle Riphean consists of the Valyukhta Group — around 1000 m of green siltstone and limestone. Overlying sediments form the Upper Riphean Zhuya Group. These rocks are up to 2000 m thick and also form a transgressive cycle, comprising at the base mottled siltstone with dolomite members and mostly limestone at the top.

In all the groups listed above the fragmental material everywhere contains large volumes of redeposited volcanoclastics. The Valyukhta Group is seen to contain higher amounts of altered alkaline-basic volcanic material (Gusev et al., 1985). Coarse clastics are present in the Bolshepatom Group, which contains blocks and debris of pre-Riphean rocks.

Post-sedimentary changes as a rule did not exceed deep epigenesis. More pronounced changes, approaching greenschist facies conditions, are seen in pre-Valyukhta rocks at certain localities in the Ura trough (aulacogen).

The position of these groups in the stratigraphic column rests to a large extent on microphytolith determinations. The Mariinsk and Dzhemkukan Groups contain assemblage II microphytoliths. Assemblage III fossils have been extracted from limestones in the Chencha Formation, upper Zhuya Group (Prokofyev, 1966). Stromatolitic assemblages are mainly represented by endemic Riphean sediments in western regions of the craton.

Geophysical data indicate a continuation of the Ura aulacogen into central regions of the craton, where the Kempendyay aulacogen (Grishin et al., 1982; Malich et al., 1984) is supposed to occur, and continues farther until it joins with the Vilyuy aulacogen of the eastern margin of the Siberian craton (Fig. II-49).

Riphean sediments in western regions of the craton. The Siberian craton is fringed to the west by the Late Proterozoic Yenisey fold belt with complete and in many regions well-studied Riphean miogeosynclinal successions. In the areas around the lower reaches of the rivers Stoney Tunguska and Angara, Riphean sediments overstep onto the platform, forming the Pre-Yenisey (Turukhan) and Fore-Yenisey pericratonic basins (Grishin et al., 1982).

In the lower Angara region (the Fore-Yenisey pericratonic basin), the Riphean succession is represented by several kilometres of Middle and Upper Riphean sediments in clearly-expressed cycles. Middle Riphean Sukhopit Group sediments form two major sedimentary cycles. The lower cycle, around 4 km thick, begins with basal conglomerate members, while higher up there are siltstones alternating with phyllites. The overlying 2.5 km thick cycle commences with alternating siltstone and quartzite units and concludes with the Kartochka Formation — mottled limestone (a 300–500 m marker horizon) and the Aladyin Formation — dolomite, 500–800 m thick. Fossils are found only in the Kartochka limestone (endemic stromatolites belonging to the genus *Malginella*).

Upper Riphean sediments form the next two major sedimentary cycles. An erosive surface is seen in some places at the base of the Tungusik Group, while in more southerly outcrops the group immediately overlies pre-Riphean formations. The lower part of the succession of this group contains shale, quartzose sandstone and dolomite — the Potoskuy Formation (up to 1500 m). Above come the Shuntar Formation shales (up to 2000 m), with an erosive base in places, and the Kirgitey Formation — up to 1100 m of shale and carbonate. Numerous conophytons, yakutophytons, baicalia and other Lakhanda assemblage stromatolites have been determined in the upper part of the Potoskuy Formation. Stromatolite genera in the Kirgitey Formation successions (the Sery Klyuch and Dadykta members) show a sharp change (Semikhatov, 1974).

Above lie sediments of a further major (1.6–4 km) transgressive cycle, consisting of shale, sandstone and limestone below and muddy limestone and dolomite above. They are taken together as the Oslyan Group. The stromatolites present most likely belong to the Upper Riphean Minyar assemblage.

Riphean sediments are exposed at the surface in the Chadobets elevation, east of the Fore-Yenisey basin. From geophysical data and deep drilling, this zone belongs to the major Irkineyeva-Chadobets aulacogen.

Here, the Riphean is divided into a number of formations which broadly correlate with sediments belonging to the Sukhopit and Tungusik Groups, but are thinner. An exception is the lowermost unit of limey-argillaceous shale (the Chadobets Formation, over 50 m thick), which can probably be assigned to the Lower Riphean. The overlying Semyonov Formation is transgressive, with a conglomeratic member at the base. The formation consists of fine-grained sandstone, siltstone and argillaceous-siliceous shale with glauconite, totalling over 1000 m in thickness. The Semyonov Formation, together with the Dolchikov Formation (350 m of banded limestones and subordinate thin siltstones) are usually correlated with the Sukhopit Group. Glauconite from the Semyonov Group has been dated at 1290 Ma (Bogolepov and Votakh, 1977)

The overlying formations, which total some 1400 m in thickness, are usually correlated with the Tungusik Group succession. We point simply to the presence of a number of unconformities in the upper part of the succession, also red and grey coarse- to medium-grained sandstones, the Togona Formation (up to 300 m), which concludes the Riphean succession.

Analogues of various parts of the Tungusik Group have been found in boreholes NE of the Chadobets elevation within the Katanga saddle. For example, the Vanavara borehole has uncovered grey dolomite with thin mottled siltstone dipping at 15–30°. Equivalent units of dolomite, limestone and shale were found in boreholes in the Kamo arch, NW of the Chadobets elevation. In the lower reaches of the Stoney Tunguska river, in the Lebyazhi field, there are dark grey and black muddy dolomite, limestone, marl and argillite, comparable in a number of features with Shuntar Formation rocks of the Tungusik Group (Kontorovich et al., 1981).

In more easterly regions, on the NW and E limbs of the Nep-Botuob anticline, the thickest and most complete Upper Riphean succession is found in the Verkhnevilyuchan field. Here, the lower parts of the succession consist mainly of oligomict and quartzose sandstones of the Verkhnevilyuchan Formation (210 m). Above are grey argillaceous-carbonate (with darker thin siltstone beds) and carbonate rocks of the Bochugunor Formation (210 m). Overall, the Riphean in the NE part of the Nep-Botuob anticline increases in thickness from 0 to 420 m (Kontorovich et al., 1981).

In the north-west of the platform, in the Turukhan region, the broad features of the section in the Fore-Yenisey basin are preserved. In the sections studied here, analogues of the upper part of the Sukhopit, Tungusik and Oslyan Groups are known. The correlation of the sections is confirmed by equivalence of marker units — mottled carbonates belonging to the Linok and Kartochka Formations. The Linok, Sukhotungusik, Derevnin and Burovaya Formations, corresponding to the carbonate portion of the

Sukhopit Group and the lower (pre-Kirgitey) part of the Tungusik Group, contain stromatolites close in composition to those of the Lakhanda assemblage. Age determinations on glauconite from the upper part of the Burovaya Formation give values of 890–920 Ma. Immediately above the Burovaya Formation, as well as immediately above the Shuntar Formation in the Fore-Yenisey succession, massive Upper Riphean stromatolites appear, replacing the Lakhanda assemblage stromatolites (Semikhatov, 1974).

Riphean sediments in northern regions of the Siberian craton. In the north of the platform, the Riphean succession commences with the Lower Riphean on the Anabar massif, the Udzha elevation (aulacogen) and on the eastern slope of the Olenek arch — in the Kyutingda aulacogen, linked to the Pe-Verkhoyan pericratonic basin.

On the western and eastern slopes of the Anabar massif, Mukun Group sediments overlie weathered Archaean gneisses. The rocks (up to 690 m on the western slope) are mostly red quartzose and quartz-feldspar sandstones with shallow-water sedimentary structures. Basalt lava flows occur at the base on the western slopes of the Anabar massif. Shpunt et al. (1982) have found thin beds of silicified alkaline pyroclastics in the sandstones; pyroclastics with trachy-rhyolite compositions have also been found. Above come stromatolitic dolomites with thin shales and siltstones of the Kotuykan Formation, up to 220 m in the east and 520 m in the west.

In the Kyutingda aulacogen, the Lower Riphean is made up of the Sygynakhtakh Formation (grey sandstone and conglomerate) and the Kyutingda Formation — up to 380 m of dolomite. In the Udzha aulacogen, the Ulakhan-Kurunga Formation corresponds to the upper part of the Lower Riphean; the well-known Riphean succession in this zone begins with this formation. The formation, which attains a visible thickness of 500–550 m, is distinguished by having a significant volume of volcanogenic-sedimentary rocks, interbedded with algal dolomites. The Kotuykan, Kyutingda and Ulakhan-Kurunga Formations are reliably correlated with one other from their stromatolite assemblages (Semikhatov, 1974; Semikhatov and Serebryakov, 1983). Although endemic forms are abundant, the stromatolite assemblages on the whole can be correlated with Lower Riphean assemblages in the Urals and Uchur-May sections and are accompanied by assemblage I microphytoliths (Lower Riphean). These data are confirmed by numerous K–Ar age determinations on glauconites from the upper horizons of the Mukun Group (1400–1500 Ma), the Sygynakhtakh Formation (1430–1470 Ma) and from various horizons in the Kotuykan and Kyutingda Formations (1300–1390 Ma).

Middle and Upper Riphean sediments in North Siberia on the whole represent a single assemblage and yield to strict stratigraphic subdivision with great difficulty. On the slopes of the Anabar massif we find terrigenous-carbonate sediments belonging to the Yusmastakh Formation. On the western slope, stromatolites have been identified from brown dolomites in the lower member (up to 210 m thick) which closely resemble the Midde Riphean Svetla assemblage in the Uchur-May region. Grey organic dolomites in the upper member (up to 400 m) have yielded assemblage III microphytoliths. The stromatolites are mostly represented by endemic forms. Throughout the succession there are sub-volcanic sheet-like bodies of trachydolerite and trachybasalt lava flows. Three transgressive cycles are provisionally recognised in the formation, from thin beds and lenses of fragmental rocks. K–Ar determinations on glauconite from the base of the lower cycle give values of around 1300 Ma, and 1120–1140 Ma from the upper (Semikhatov and Serebryakov, 1983). The boundary between the Middle and Upper Riphean would seem to fall in the middle of the upper member. No more than 120 m of the formation has been preserved on the eastern slope of the Anabar massif, due to pre-Yudomian erosion. A volcanic structure with explosive trachyte breccias has been mapped here among the dolomites.

The Udzha aulacogen contains an unbroken Middle and Upper Riphean section, the Unguokhtakh (Middle Riphean), Khapchanyr and Udzha Formations. The volcano-sedimentary Unguokhtakh Formation, which is up to 600 m thick, contains basic volcanoclastic material. The Khapchanyr Formation, 300 m thick, consists of interbedded siltstone, limestone and dolomite. The siltstones contain large volumes of redeposited alkaline-basic volcanic material. Numerous stromatolites have been identified in the Khapchanyr Formation, including jacutophyton, conophyton, tungussia and baicalia. A quite different stromatolite assemblage, with inzeria and jurusania, is present in the lower part of the Udzha Formation. According to Semikhatov et al. (1979), the host sediments of the Khapchanyr and lower part of the Udzha Formations can be broadly correlated with the Lakhanda Group. Geologists from Yakutsk (Gusev et al., 1985) treat the Khapchanyr Formation as belonging to the Middle Riphean. The Upper Riphean Udzha Formation is mainly represented by volcanogenic-terrigenous redbeds. Stromatolitic dolomite horizons, up to 50 m thick, are seen at the base and top of the formation.

Middle to Upper Riphean sedimentary successions are exposed by rivers cutting the Kystyk plateau on the eastern slope of the Olenyok arch. The most complete sections are found in the Kyutingda aulacogen. Here we have the Arymas, Debengda and Khaypakh Formations, each of which represents an independent transgressive cycle. Khaypakh Formation sediments are missing to the south of the Kyutingda aulacogen, possibly as the result of erosion in pre-Yudomian time. Middle Riphean sediments overlie Kyutingda Formation dolomites without apparent unconformity, directly on top of an eroded crystalline basement surface.

The Arymas Formation (270–380 m thick) consists of alternating siltsone and glauconite-quartz sandstone with thin dolomites at the base and top and alkaline-basic, alkaline and basic lavas. The Debengda Formation (460 m) consists of interbedded dolomite, sandstone and siltstone; dolomite dominates the succession on the NW slope of the Olenyok arch. Endemic stromatolite forms are widespread in both formations, although from the predominance of conophytons, jacutophytons, colonnella and the presence of baicalia, the stromatolite assemblage has definite similarities with the Lakhanda and older Uchur basin assemblages (Semikhatov et al., 1979).

Sediments belonging to the overlying Khaypakh Formation were deposited above an unconformity. The lower member — coarse-grained clastic sediments at the base and dolomite with siliceous inclusions above — is up to 90 m thick. The upper member consists of rhythmically alternating mottled sandstone, siltstone and dolomite, up to 110 m. Substantial amounts of volcanoclastic material is seen in both members.

K–Ar age determinations on glauconite vary with monotonous regularity from 1220 Ma in the lower part of the Arymas Formation to 1040 Ma at the top of the Debengda Formation, and from 1010 Ma at the base of the lower member of the Khaypakh Formation to 900 Ma in the upper member (Semikhatov and Serebryakov, 1983). The stromatolite assemblage from the upper member, although also represented in the main by endemic forms, is nevertheless quite distinct from the Arymas–lower Khaypakh assemblage. This suggests that here, as in the Uchur-May region, as well as in the Fore-Yenisey basin, the change in stromatolite assemblages occurs above the accepted boundary between the Middle and Upper Riphean at 1050 $\pm$ 50 Ma.

Riphean sections increase east of the Olenyok arch, with a sharp decrease in the percentage of clastic terrigenous formations. For example, in the Tuora–Sis Range successions (right bank of the lower Lena), within the Pre-Verkhoyan pericratonic basin, terrigenous formations predominate only in the lower volcanogenic-terrigenous Ukta Formation red beds (up to 200 m), immediately overlying the crystalline basement. Above come limestones and dolomites — the Eselekh Formation (500 m), correlated with the Debengda Formation, Neleger Formation dolomites (250 m) and the Sietachan mottled terrigenous-carbonate Formation (400–500 m). These last two formations correlate with the lower and upper members of the Khaypakh Formation. A similar succession is also preserved in the northern regions of the Kharaulakh Range.

Yudomian sediments

The topmost upper Precambrian stratigraphic subdivision in East Siberia is the Yudomian. In the East European craton, the Kudash(?) and Vendian correspond to the Yudomian. In the Siberian craton, Yudomian sediments together with the Lower Palaeozoic form an independent structural stage, separated from the underlying stage in most cases by a marked break in sedimentation, and in many cases by a pronounced angular unconformity. The formation of the thick sedimentary platform cover over the entire cratonic region commenced in Yudomian time, except for the highest regions of the Aldan and Anabar domes and a few major arches.

The Yudomian assemblage consists in the main of carbonate and terrigenous-carbonate sequences. Two subdivisions are often identifiable, corresponding to separate sedimentary cycles.

In the Yudoma-May basin and in the east Aldan Shield, the Yudoma Group has two formations. The lower formation, the Aim, consists of sandstone (mostly quartzose), argillite and black limestone and dolomite, bituminous in places. The sandstones are replaced by quartz grits in a number of cases. Overall the colour changes from lighter at the base to darker at the top. The thickness varies from a few metres to 400 m. As a rule, deposition of the sediments was preceded by deep erosion and only in a few sections in the Yudoma-May basin is it possible to observe a gradual transition to Upper Riphean Uy Group sediments. The upper formation, known as the Ustyudoma, is 20–460 m thick and comprises mainly pale saccharoidal dolomite with subordinate sandstone at the base and limestone at the top. Yudomian dolomites contain assemblage IV microphytoliths and a specific stromatolite assemblage (Semikhatov et al., 1979). An Rb–Sr age determination on argillites from the lower part of the Aim Formation gave a value of 632 ± 20 Ma (Semikhatov and Serebryakov, 1983).

Yudomian sections in northern regions of the Siberian craton have much in common with those of the south-east. On the western slopes of the Anabar massif, sandstones and siltstones belonging to the Starorechensk Formation and the lower member of the Ustkotuykan Formation, totalling some 300 m, are assigned to the Yudomian. In the Udzha aulacogen, there are siltstone, sandstone and grit in the Tomptor Formation and dolomite in the Lower Turkut Formation. On the slopes of the Olenyok basement high, essentially carbonate sediments belonging to the Khorbusuonk Group and the lower part of the Kesyusa Formation (c. 500 m) are referred to the Yudomian. Finally, the Kharayutekh dolomite Formation and the basal sandstone beds of the Tyusser Formation (400 m) in the Pre-Verkhoyan basin belong to the Yudomian.

All these sections contain assemblage IV microphytoliths and numerous stromatolites. Upper parts of the succession, containing fossil chyolites *Anabarites trisulcatus* Miss., are known as the Nemakit-Daldyn horizon (lowest Kesyusa Formation, lower member of the Ustkotuykan Formation, etc.). The possibility that this horizon is wholly or partly Cambrian in age is currently being widely discussed in the literature (Shishkin, 1979; Kontorovich et al., 1981).

In southern regions of the platform, west of the Uchur basin, the Yudomian gradually decreases, mainly through the lower horizons thinning out. On the NW slope of the Aldan Shield and on the Yakutsk-Olyokma step, rocks belonging to the Porokhtakh Formation are assigned to the Yudomian; they are unconformable on Upper Riphean carbonate rocks. At the base of the formation is a basal mottled terrigenous member with thin marls and dolomite, while higher up come dolomite and sandy dolomite with thin gypsum horizons in the west and limestone in central regions of the Yakutsk-Olyokma step (the Sinsk zone). The formation increases in thickness westwards, reaching 530 m at the edge of the Beryozova basin.

The Yudomian in the Beryozova basin is considered to be part of the Macha Group: the Dzerba quartz sandstone Formation at the base, around 300 m thick, and the Tinnov Formation — up to 500 m of dolomite and limestone, usually bituminous, containing assemblage IV microphytoliths. This assemblage, together with a distinctive endemic stromatolite assemblage, is traceable up the sequence into the lower member of the Nokhtuy Formation. This has led a number of workers in the region to increase the age of the upper boundary of the Yudomian also (Bogolepov and Votakh, 1977; Semikhatov et al., 1979). Dzherba and Tinnov Formation sediments can be followed over the entire northern near-platform margin of the Patom Highlands.

All over the south-west of the platform, within the Nep-Botuob dome, the Angara-Lena step, in the Pre-Sayan region and the Kamo arch (lower reaches of the Stoney Tunguska river), are outcrops of the Mota Formation or its analogues. It has three members (Kontorovich et al., 1981) and is equivalent to the Dzherba and Tinnov sediments. The lowest member, which is up to 360 m thick in the Bratsk region, consists of grey or reddish-brown quartz-feldspar or quartz sandstone, finely laminated siltstone and shale, often bituminous. At the top of the member is the Parfyonov marker horizon — poorly cemented grey sandstone with thin shale and siltstone towards the top, with which are associated numerous oil and gas shows. The middle member (70–130 m) consists of compact grey dolomite and dark grey quartz sandstone, siltstone and shale. The top member consists of dolomite and limestone, in places with thin saline dolomite, anhydrite and rock salt. This part of the formation is actually often assigned to the Cambrian.

Sediments belonging to the Mota Formation overstep various Riphean horizons or the crystalline basement. In various regions around the edge of the craton, the formation lies on top of the eroded surface of the Ushakov or Olkhina Formations, which are also usually assigned to the Yudomian assemblage. The Mota Formation has been shown to lie conformably above Ushakov Formation rocks (up to 120 m) only by drilling in the Bratsk region. The Ushakov Formation consists basically of greywacke, grit and conglomerate. Carbonate-terrigenous sediments of the Olkhina Formation (up to 400 m) crop out in the extreme south of the region and are found in boreholes in the Irkutsk region and other areas.

The lower and middle members of the Mota Formation have been shown to belong to the Yudomian on the basis of the nature of the

TABLE II-15

Typical deposits and ore shows

Deposit or ore show; useful mineral	Tectonic setting	Type of mineralization	Geological and mineralogical features				Geological age of	
			Host rocks or mineralized rocks	Shape of deposit, ore body morphology	Ore type	Ore mineral composition (minor minerals)	Host rock	Mineralization
Uchuri ore zone; iron	Uchuri basin	Sedimentary iron ore	Red quartzose sandstone	Conformable patches	Oolitic	Haematite	Early Riphean (Gonam Fm)	
Omni ore zone; iron	Eastern slopes of Omni elevation	Sedimentary iron ore	Quartzose sandstone	Conformable horizon	Massive	Haematite	Middle Riphean (Totti Fm)	
May ore zone; iron	May basin	Sedimentary iron ore	Quartzose sandstone	Conformable lenticular deposits; sandstone cement	Massive, disseminated	Limonite, haematite	Middle Riphean (Aimchan Group)	
May ore zone; iron	May basin	Sedimentary iron ore	Mottled sandstone and clay	Conformable lenses and interbeds	Oolitic and massive	Haematite, chamosite, siderite (sphalerite)	Late Riphean (Lakhanda Group)	
Sardana ore region (Sardana, Uruy and Perevalnoye deposits); lead, zinc	Yudoma-May pericratonic basin	Lead-zinc sulphide, stratiform and epigenetic hydrothermal	Dolomite	Flattened pipes, tabular-shaped, ribbon-like deposits	Disseminated	Sphalerite, galena (pyrite, marcasite, arsenopyrite)	Late Vendian (Ust-Yudoma Formation)	Late Vendian to early Palaeozoic (?)
Tabornoye & others; lead, zinc, fluorite	Cis-Baikal basin	Galena-sphalerite-fluorite stratiform sedimentary & hydrothermal metasomatic	Dolomitic limestone, talc-carbonate rocks	Multi-level conformable bedded and lenticular deposits	Vein-like disseminated, massive	Sphalerite, galena, fluorite (pyrite, arsenopyrite)	Middle Riphean (Goloustensk, Uluntay Formations)	
Sukahrikhina copper-bearing zone; copper (zinc, lead)	Pre-Yenisey (Turukhansk) basin	Stratiform; metasedimentary	Mottled sandstone, siltstone, limestone	Conformable horizons	Disseminated	Chalcopyrite, bornite, chalcocite, sphalerite, galena	Vendian (Sukharikhina Formation)	
—Ditto—	—ditto—	Cupraceous sandstone	Sandstone, siltstone	Conformable horizons	Disseminated	Chalcopyite, pyrite, bornite, chalcocite	Vendian (Izluchina Formation)	
Teya copper-bearing zone; copper	Fore-Yenisey basin	Cupraceous sandstone	Red sandstone	Conformable horizons	Disseminated	Chalcocite, bornite, chalcopyrite	Vendian (Taseyeva Formation)	
Ushakov copper-bearing zone; copper	Cis-Baikal basin	Cupraceous sandstone	Red sandstone, siltstone	Conformable horizons	Disseminated	Bornite, chalcopyrite (malachite)	Vendian (Ushakov Formation)	
Gornostil ore show; bauxites	May basin	Bauxitic, weathering crust on dolomites	Ferruginous-kaolinitic rocks, shales	Layer-like deposits	Massive	Diaspor (haematite, limonite, goethite, galloisite)	Middle Riphean (Tsipanda Fm)	M.–Late Riphean
Pre-Sayan ore region; phosphorites	Pre-Sayan basin	Phosphoritic	Sandstone, dolomite, breccia	Conformable layers	Disseminated	Francolite, apatite	Late Riphean (Karagas Group, upper fms)	
Cis-Baikal ore region; phosphorites	Cis-Baikal basin	Phosphoritic; sedimentary	Limestone, dolomite	Conformable horizons	Disseminated	Apatite	Middle Riphean (Uluntay Formation)	
Nikolayev and others (Nizhneudinsk group of deposits); manganese	Pre-Sayan basin	Remnant, remnant-infiltration	Arkosic sandstone, clastic dolomite	Layered bodies; weathering crust on sandstone & dolomite	Compact (in ore nests), earthy, sooty	Psilomelane, pyrolusite, vernandite	M. Riphean (Karagas Gp)	Cenozoic ?
Arshan ore show	—ditto—	Sedimentary	Sandstone	Lenses	Disseminated, vein-like layers	Hausmannite (braunnite, rhodochrosite), haematite	Middle Riphean (Tagul Fm, Karagas Gp)	
Peshcherny sector, etc. (Cis-Baikal talc province)	—ditto—	Talc; hydrothermal-metamorphic	Graphitic-clay shale, limestone, dolomite	Layered bodies	Massive	Talc	Middle Riphean	
Sarmi; phosphorus	—ditto—	Phosphoritic; metasedimentary	Graphitic-clay shale, limestone	Layered bodies	Massive, brecciated	Apatite (collophane)	Late Riphean	

stromatolite assemblage and the IV microphytolith assemblage. Pebbles in Ushakov conglomerates contain identifiable assemblage III microphytoliths (Semikhatov et al., 1979).

The Yudomian around the western margin of the platform is mainly clastic redbeds containing certain features of an orogenic assemblage. In southern regions of the Yenisey ridge (lower reaches of the river Biryusa), most of the Yudomian is represented by the Taseyeva Group, which lies on various Riphean horizons with angular unconformity. The group has variable thickness, which can reach 2.5–3.0 km in the most complete sections. It is divided into three formations, the lower and upper being red sandstones and grits and the middle (Chistyakov) being a marker unit 80–320 m thick of grey sandstone, argillite and rare dolomite. Stromatolites are practically unknown in these sediments, and microphytolith assemblages also have distinctive features. A Yudomian age for the Taseyeva Group has been determined from broad field relations and inter-regional correlations (mainly with the Nemchan Group in the NE Yenisey ridge) and is strengthened by K–Ar dates on glauconites (635 Ma) from the Chistyakov Formation.

The Ostrov Formation (150–550 m) follows, which in the extreme eastern regions oversteps progressively older sediments. The section commences with a conglomerate horizon which grades upwards into quartz or oligomict sandstone with thin dolomite at the top. Impressions of the soft-bodied fossil *Cyclomedusa* sp. have been found in the Ostrov Formation.

Farther north, analogues of the Taseyeva Group and Ostrov Formation are identified by the term Nemchan Group, the lower part of which has been dated at 690–635 Ma by glauconites (Semikhatov et al., 1979).

Data concerning the most important mineral resources associated with Riphean sediments in the Siberian craton are shown in Table II-15.

The pattern of Late Proterozoic structural stages in the Siberian craton and its margins is in many respects quite different from that of the East European craton.

In the East European craton, pericratonic basins developed not only around the margin, but also as broad embayments which overstepped the platform limits, where they are subsequently transformed into aulacogens. In the Middle(?) to Late Riphean, the aulacogens rapidly evolved and formed a continuous network in the central and eastern regions of the platform; Riphean sediments are practically absent from the cells between the branches. Aulacogen growth ceased at the beginning of the Vendian and only Upper Vendian (Valday) sediments form the platform cover.

In contrast, the Middle to Late Riphean aulacogens in the Siberian craton grew sluggishly, mainly at craton margins, and did not form a connected network. Additionally, many of the syneclises and major basins at this time were filled with basin-type sediments, resembling platform cover sediments in many characteristics. The true cover began to form at the start of the Yudomian, which seems to correspond to the start of Kudash (still pre-Vendian *s.s.*) time. Associated with this period was the onset of a major marine transgression which embraced practically the entire territory to the north and west of the Aldan Shield. This would appear to provide evidence for a much greater degree of consolidation of the Siberian craton at the beginning of the Riphean, compared to the East European craton. It is probably also related to the fact that tectonic activity was much less intense in the Siberian craton during the preceding Early Proterozoic growth stage.

Chapter 6

MAJOR FEATURES OF PRECAMBRIAN METALLOGENY

The metallogeny of the Siberian Platform has been studied by Yu.G. Staritsky, N.S. Malich, Ye.V. Tuganova, V.M. Terentyev, T.V. Bilibina, Yu.V. Bogdanova, P.M. Khrenov and V.G. Kushev, who have published extensively. This chapter will briefly deal with the following aspects: (1) general features of Precambrian metallogeny in the Siberian craton; (2) ore-bearing structures; (3) metallogenic zoning; and (4) major metallogenic epochs. These questions are all mutually dependent and intimately associated with one another and taken together they give an impression of the most important time–space patterns controlling the emplacement of ore deposits.

General features of Precambrian metallogeny of the Siberian craton

As major "typomorphic" useful minerals in the Precambrian of East Siberia, mention must be made of copper, polymetallic and rare-metal deposits, noble element mineralization, antimony, and to a lesser degree iron and titanium, and the exceptionally varied spectrum of non-metallic minerals: muscovite, phlogopite, apatite, asbestos, rock crystal and granular quartz, marble, asbestos and aluminous raw material. Ore mineralization types found here include the Udokan cupriferous sandstones; pyrite-polymetallic deposits in blastomylonites occurring in Early Proterozoic trough complexes (the Kholodna deposit); metasomatic phlogopite deposits among magnesian skarns (the Sharyzhalgay basement high and the Aldan granulite-gneiss terrain); and typical iron-silica metasomatites (the Tayezhnoye deposit). Magnetite-phlogopite deposits are widely developed, also apatite (the West Aldan province and the Seligdar deposit) in the metamorphic units of the Fyodorov Formation (AR_1^2) — a result of primary carbonates being much commoner in supracrustal assemblages in Siberia, compared to the European part where they are less widespread. Gold deposits associated with black shale formations are also more typical of the Siberian Precambrian.

One feature common to the metallogeny of both the East European and Siberian cratons is the presence of ferruginous quartzite, titanomagnetite ore mineralization associated with anorthosite and gabbroids, ceramic and muscovite pegmatites, rock crystal and rare metal mineralization.

Characteristic differences can also be observed in the rare-earth mineralization. In the European part, we have mainly rare-metal pegmatites and "apogranites" (the Perzhan zone in the Ukraine) and skarn-greisen mineralization related to rapakivi granites. In Siberia, on the other hand, we have principally alkaline metasomatites in deep fault zones, "apogranites" (the Katuga zone in the west Baikal region) and mineralization associated with orogenic volcano-plutonic complexes that formed at the end of the Early Proterozoic cycle (the Akitkan and Ulkan belts).

Many of the particular metallogenic features seen in Precambrian belts in Siberia are determined by processes that were active in the Proterozoic and again in the Mesozoic, forming gold and polymetallic deposits as well as apatite-iron-titanium-rare-earth mineralization (Jurassic ultrabasic-alkali intrusions and carbonatites in the Western and Central Aldan blocks).

Major ore-bearing structures

The most significant ore-bearing tectonic zones in the Precambrian of the Siberian craton are linear sutures of various types and ages: (1) Late Archaean greenstone belts with banded ironstones (the Chara-Tok group of deposits in the Olyokma block with iron reserves exceeding 8 billion tonnes); (2) deep fault zones which evolved over an extensive period (AR_2–PR_1) with associated metamorphic, igneous and metasomatic rocks along them; (3) Proterozoic trenches, grabens and aulacogens; and (4) Riphean pericratonic and aulacogen-type structures around the craton margin.

The metallogenic signature of greenstone belts is defined in the first place by banded ironstones; in places there are signs of gold mineralization and ore occurrences of non-metallic raw materials — talc, asbestos and apatite. Late Archaean to Early Proterozoic deep fault zones have an important metallogenic significance; blastomylonite, gabbro-anorthosite and granite complexes and metasomatic rocks have developed along them. Major gabbro-anorthosite bodies have associated apatite-titanomagnetite mineralization, seen in the Kalara, Dzhugdzhur and other minor massifs in the junction zones of the Aldan Shield, the Dzhugdzhur-Stanovoy belt and in Anabar linear fault zones.

A fundamentally different type of deep fault zone, AR_2–PR_2 age, has Na and K alkali metasomatites developed along them (Rudnik et al., 1970). In the first case, these are zones of albitites, quartz-oligoclase metasomatites with accompanying rare-metal and tungsten (scheelite) mineralization. In the second case the mineralization is associated with K-metasomatites. Narrow linear zones of granitization and alkali metasomatism are cogeneric with other active structures: regional-scale zonal granulite and amphibolite metamorphism and granitization of AR_2 and PR_1 age, which will be discussed below.

The most important Early Proterozoic ore-bearing structures are the Olokita and NW Baikal-type trenches. The polygenetic pyrite-polymetallic deposits among volcanogenic-terrigenous-carbonate formations and barite-polymetallic deposits developed here have no analogues in the Precambrian of European USSR. The Early Proterozoic basins in Siberia are also ore-bearing structures. The most interesting is the Udokan basin with its cupriferous sandstones; its structural position is defined by the margins of the Baikal-Patom and Olondo belts and the Stanovoy block.

The volcano-plutonic belts which were emplaced at the end of the Proterozoic cycle and occur along the rear margins of mobile belts are also typical suture-type Precambrian ore-bearing structures in the Siberian craton. The best known of these is the Akitkan in the W and NW Baikal region with fluorite, polymetallic and rare-earth mineralization. Close to the eastern margin of the Aldan Shield is the Ulkan belt, a similar type, with complex rare-metal mineralization. The formation of these structures at the boundary between the Early and Late Proterozoic fixes the most important boundary in geological history. Noting the wide distribution of orogenic volcano-plutonic associations in the Siberian Platform, it is impossible not to compare this fact with the sharply reduced occurrence of typical rapakivi granites which are close in age and which are abundant in the Baltic and Ukrainian Shields.

While emphasising the important significance of linear suture zones in controlling ore emplacement — deep fault zones, trenches, aulacogens, volcano-plutonic belts etc., it is important at the same time also to draw attention to the common occurrence of "regional" metallogeny in the Aldan Shield, related to Early Archaean supracrustal assemblages.

The most typical is widely distributed phlogopite, magnetite and apatite mineralization in the Fyodorov Formation (AR_1^2) in belts surrounding granite-gneiss domes. Among these are iron ore deposits: Tayezhnoye and Dyos-Legriyer; phlogopite: Emeldzhak; and apatite: Seligdar. In addition to the important role of lithologic control — the development of mineralization among calc-silicate rocks and calciphyres in the Fyodorov Formation — retrograde metamorphism and metasomatism were significant processes which manifest themselves in both the Late Archaean and Early Proterozoic.

A further type of regional mineralization in the Aldan Shield — rock crystal deposits in the Melemken and Nimnyr blocks (see Fig. II-10) — also demonstrates the important ore-controlling significance of the lithological factor and late-stage metasomatic effects. Deposits of this type are controlled by quartzite horizons in the Kurumkan Formation (AR_1^1). Both groups of deposit are also controlled by zones of metamorphism and metasomatism which accompanied the later metamorphism of granite-gneiss domes and leucogranite intrusions of Early Proterozoic age. There are regional shows of graphitic mineralization and high-alumina raw material — sillimanite schists — within the Aldan Shield and the Stanovoy belt.

Of the lower Proterozoic regional ore-bearing tectonic structures, mention must be made of the zone controlling the distribution of the North Baikal muscovite pegmatite province. Here, the ore controlling zone is a geanticlinal elevation of Early Proterozoic age which formed on Archaean sediments of the Chuya microcraton. In such an examination we can see the fundamental similarity in the evolution of the Mama mica province in the Baikal-Patom belt in Siberia and the White Sea province in the Baltic Shield. The common feature in this case is the long period over which the structures developed on the same site (AR–PR_1) with repeated kyanite-sillimanite facies series metamorphic episodes.

The most productive ore-bearing structures of Riphean age are synclines and pericratonic basins which grew around the periphery of the Siberian craton and penetrated farther in along deep faults of Archaean to Proterozoic age. An example is the Yudoma-May pericraton, which bounds the platform to the south-east. Here, carbonate units (R_3–V) contain stratiform polymetallic deposits (Sardana, Uruy and Perevalnoye); cupriferous sandstones are also present, and there are indications of bauxites and phosphorites.

Oolitic iron ores are found as ore shows in the Early Riphean on the northern limb of the Aldan syncline.

Metallogenic zoning

The most general scheme of metallogenic zonation for the Precambrian of the Siberian craton is defined by particular features of its tectonic structure. In a sense the platform is surrounded on all sides by Ripheid zones which are characterized in the first place by stratiform Pb and Zn deposits, fluorite in carbonate rocks, which are atypical for the pre-Riphean folded basement of the craton. At first sight this is general concentric zoning, but under more detailed scrutiny it turns out to be more complex — the western (Yenisey ridge) and eastern (Verkhoyan and Sette-Daban) regions are essentially different in tectonic style and have their own metallogenic features.

The major (first order) zoning in the Siberian Platform is therefore defined by the distribution of belts around a pre-Riphean basement of systems of marginal structures, including pericratonic depressions and transition zones bordering on territories consolidated in the Early Proterozoic. Lead and zinc (with fluorite and barite) deposits, ore shows and geochemical anomalies, localized along the Yenisey ridge, the NW Baikal region (Tabor and Novo-Anay deposits), SW Yakutia (Kylakh ore zone, Malich et al., 1984), and the Sette-Daban and Taymyr mountain ranges constitute the Siberian polymetallic belt, which is several thousand kilometres long (Bilibina and Sokolov, 1980). In most cases, ore genesis in this belt is defined by sedimentogenic-katagenic and hydrothermal-metasomatic ore types (Sidorenko, 1982) — a feature which is not characteristic for pre-Riphean structures in the East European platform.

The internal structure of the pre-Riphean basement to the Siberian craton is much less well studied than the East European. Few boreholes have reached the basement beneath the sedimentary cover, implying that it is valid to speak about metallogenic zoning only with reference to the exposed segments of the basement — the Anabar and Aldan Shields and adjacent Baikal and Stanovoy regions (including Dzhugdzhur), as well as the south-eastern basement highs. There are a number of conflicting views regarding the structural status of these territories within the Siberian craton. As well as the tectonic regional scheme referred to above, another scheme is significant in metallogenic analysis, compiled by Glukhovsky et al. (1983) from satellite imagery and incorporating geophysical data from magnetic and gravity surveys. They have noticed several first-order concentric zonal structures or "nuclei" in Siberia: Kheta-Olenyok, Angara, Aldan-Stanovoy, Vilyuy, and others, which they consider to be relicts of ancient structures which arose during a nucleation phase of development.

Taking into account these authors' publications, the Anabar Shield is the NE fragment of the Kheta-Olenyok nucleus, and the SW basement highs are the marginal part of the Angara nucleus, which is essential for understanding the overall metallogenic zonation of the craton.

We shall now deal in more detail with the metallogenic zonation of the Anabar and Aldan Shields and the SW basement highs (second order metallogenic zonation).

The basic features of metallogenic zonation in the Anabar Shield are defined by the combination of two major elements: (1) metallogeny of blocks made of lower Archaean supracrustal rocks, metamorphosed at granulite facies; and (2) metallogeny of NW-striking parallel fault systems (suture zones) of AR_2–PR_1 age which segment the Anabar Shield into a series of blocks. These suture zones control the emplacement of ore-bearing intrusive, metasomatic and metamorphic formations.

The metallogeny of the blocks is defined by ore-bearing primary volcanogenic and sedimentary-volcanogenic assemblages (>3.4 Ga), metamorphosed to two-pyroxene schist, hypersthene gneiss and enderbite. Typically they have banded ironstone, ferruginous two-pyroxene-garnet schist, graphite deposits, rare-metal pegmatites and indications of copper-nickel mineralization in minor meta-ultramafic bodies associated with the supracrustal rocks. Shield blocks did not experience the effects of subsequent Late Archaean and Early Proterozoic tectono-magmatic activity and as a result the granite-gneiss dome structures, so typical of most Precambrian regions, do not appear in them. By virtue of this, the metallogeny of suture zones (inter-block junction zones) is more interesting and varied. Here the main ore-bearing formations were Late Archaean to Early Proterozoic. They are represented by ore-bearing intrusive, metasomatic and metamorphic rocks, which taken together form a single evolutionary series from gabbro-anorthosite to tonalite and leucogranite. This also defines their ore mineral content — the association apatite–REE–Ti–Fe in anorthosites; Pb, Zn, Cu, Au, Mo in moderately acid granitoids and alkali metasomatites. For the gabbro-anorthosites we have an association that at first sight is an unusual collection of elements — Au, Cu, Zn and Pb.

A crustal type of metallogeny is observed in the narrow marginal parts of suture zones among activated rocks of the Early Archaean blocks — micaceous and rare-metal pegmatites. Having noted these two major elements in the metallogeny of Anabar (Archaean blocks and AR–PR_1 suture zones), it is essential to draw attention to a further structural-metallogenic feature. On analysing formations, we see a certain younging of terrains and mineralization from west to east. The youngest Khapchan Formation for example crops out in the extreme east of the region; it consists of marble and graphitic rocks, AR_1 charnockite and enderbite are characteristically absent and there is only one granulite facies metamorphic episode at relatively low pressure. The composition of the rocks in the formation in the eastern (Khapchan) block determines the important role played by graphite and phlogopite mineralization and mica-bearing pegmatites close to the suture zone on its southern flank. Secondly, it is in precisely this eastern Khapchan block that alkali-ultrabasic rock complexes were emplaced at the end of the early and start of the Late Proterozoic with accompanying Fe, Ce, Zr mineralization.

If in future this pattern of younging seen in the structural-lithological terrains and the metallogeny for the east of the Anabar Shield also emerges for the western Magan block, then it will be possible to refer to zoning of the Anabar Shield, with high-pressure granulite cores at the centre and younger lower-grade metamorphic zones around the periphery, complicated by a series of sutures in the form of linear fault zones.

Metallogenic zoning in the exposed part of the Aldan Shield, including the Dzhugdzhur-Stanovoy and Olyokma terrains and the Baikal-Patom belt reflects most completely the general features of the distribution of mineralization in Early Precambrian structures (Fig. II-50). Here the metallogenic zoning is determined by the combination of three main elements: (1) metallogeny of ancient Archaean structures, rejuvenated in subsequent epochs (Central Aldan, the Chara "microcraton", the Olyokma terrain, and others); (2) suture zones with a long history of development (AR_2–PR), forming an exceptionally complicated network within the Dzhugdzhur-Stanovoy and Olyokma blocks; and (3) linear fold zones — the Baikal-Patom mobile belt.

The overall metallogenic zoning seen in the SE exposed part of the Siberian craton can be compared with that seen in the Karelia-Kola region, the outcrop areas being exactly the same size. If we ignore the easternmost Batomga block, which is least well known but is thought to belong to a granulite-greenstone terrain, then for the remaining basement outcrops

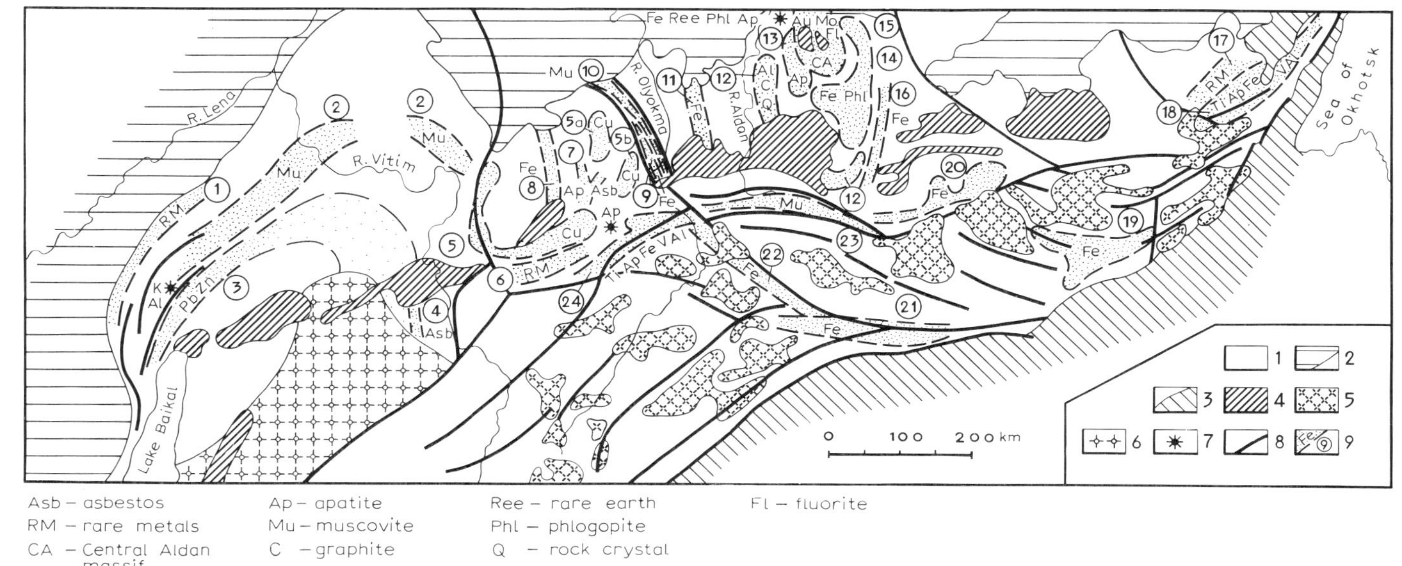

Fig. II-50. Metallogeny of the Vitim-Aldan region (compiled from material by Yu.V. Bogdanov, L.I. Krasny and N.S. Malich). *1* = Early Precambrian; *2* = edge of platform cover; *3* = edge of Mongolia-Okhotsk belt; *4* = Precambrian metamorphic complex and granites of various ages; *5* = Mesozoic-Cenozoic basins; *6* = Mesozoic granitoids; *7* = central-type alkaline complexes; *8* = major faults; *9* = major metallogenic zones (numbers in circles): *1* = Akitkan, RM, *2* = Mama-Chuya, Musc, *3* = Olokita, Pb and Sm, *4* = Molodyezhnoye, Asb, *5* = Kodar-Udokan, Sm, *6* = Katugin, RM, *7* = Olondo, Ap, Asb, *8* = Chara-Toka, Fe, *9* = Middle Olyokma, Fe, *10* = Temulyakit, Musc, *11* = Nelyuka, Fe, *12* = Upper Aldan-Timpton, Q, Al, C, *13* = Seligdar, Ap, *14* = Aldan-Timpton, Fe, Phlog, *15* = Central Aldan, Au, Mo, Phlog, *16* = Seym, Fe, *17* = Ulkan, RM, *18* = Dzhugdzhur, T, Ap, Fe, V, Al, *19* = Upper Zeya, Fe, *20* = Daurkachan-Usman, Fe, *21* = Getkan, Fe, *22* = Nyukzha, Fe, *23* = Chulman, Musc, *24* = Kalara, T, Ap, Fe, Al, Zr. *Ch* = Chara microcraton. *CA* = Central Aldan massif. *RM* = rare metals, *Asb* = asbestos, *Ap* = apatite, *Musc* = muscovite, *Phlog* = phlogopite, *Q* = rock crystal, *C* = graphite, *Fl* = fluorite.

we see an east to west progressive younging of structures and age of mineralization, analogous to that seen in the Baltic Shield (from north to south).

The Eastern Aldan block has AR_1 supracrustal assemblages, high-pressure AR_1–AR_2 granulite facies metamorphism, Na-series granites with K-granites localized along narrow zones, weakly expressed Proterozoic tectono-magmatic activity, weak retrograde metamorphism, and low ore content.

The Western Aldan block has AR_1^1–AR_1^2 supracrustal assemblages showing low-pressure granulite facies metamorphism, intensely altered in subsequent metamorphic tectono-magmatic episodes of activity, abundant granite-gneiss dome formation, K-series granites, and intense endogenic mineralization — phlogopite, iron, apatite, rock crystal and high-alumina raw material.

The Olyokma block has granite-greenstone terrains, AR_1^1 and AR_1^2 volcano-sedimentary supracrustal assemblages, AR_2 amphibolite and epidote-amphibolite facies metamorphic rocks, intensely developed PR_1 internal processes (in the east) with abundant granitoids, and several mineralization epochs — ferruginous quartzites, gold shows, plus talc and asbestos.

The Baikal-Patom belt is a Proterozoic mobile belt with a regular pattern of alternating zones, reminiscent of Phanerozoic fold belts.

Elements of concentric zoning within Archaean structures have previously been emphasised in publications by Moralev and Glukhovsky, using the Aldan crystalline massif and the Chara microcraton as a case study. In the Aldan massif for example the configuration of zones of distribution of metamorphic-metasomatic magnetite and phlogopite mineralization forms an arc surrounding to the south and east the core of the massif — the Lower Timpton dome. Distribution zones of apatite metamorphic-metasomatic mineralization enhance this arc, surrounding the core to the west. The central core itself has younger Jurassic granitoids with gold mineralization.

The Early Archaean Chara craton is folded on its eastern side by the Archaean Chara-Tok banded ironstone zone and to the SW by the Early Proterozoic Kodar-Udokan copper-bearing zone, associated with an arcuate basin.

Ring structure fragments and corresponding concentric metallogenic zonation are also observed in the Dzhugdzhur-Stanovoy terrain. The cores of such ring structures often expose leucocratic gneiss and granulite, and there are frequent occurrences of highly aluminous schist and granitoids, including those of Phanerozoic age, associated in the main with Jurassic activity (Fig. II-50). These sectors contain Au, Zn, Pb, Mo and rare-metal mineralization. Melanocratic rocks increase in importance around the periphery of ring structures; these are mafic schist, amphibolite, marble, etc., with Fe, Ti, apatite, phlogopite and other mineralization.

The second characteristic element of metallogenic zonation are long-lived (AR_2–PR_1) linear suture zones of deep crustal origin, with a typical complex metallogeny — Fe, Ti, apatite, REE, Mo, Au, Cu and rare metals.

The suture zones in the region under consideration, as distinct from the Anabar Shield, are more varied with not only intrusive and metasomatic "cross-cutting" formations but also "concordant" stratified sedimentary-volcanic assemblages localized in them. This type of metallogenic zoning is best expressed within the Dzhugdzhur-Stanovoy terrain. Its broad features are somewhat reminiscent of a structure referred to in ore geology as a "horsetail" — two subparallel undulose curved east–west regional thrust-nappe systems (the Stanovoy and Tukuringra faults) and numerous NW-striking splay faults feathering out from them, forming a branching pattern not unlike a horsetail. These systems have associated AR_2 and PR_1 intrusive, metasomatic and volcano-sedimentary complexes. Situated in lenticular blocks of the Archaean basement between them, fragments of ring structures are preserved with Jurassic granitoids in the cores. The metallogeny of the Stanovoy block has yet to be studied in detail. One of its most interesting ore-bearing structures is the Kholodnikan greenstone belt (Krasnikov, 1984) which evolved over a long interval (AR_1^2–AR_2), with signs of gold mineralization, directly related to the Stanovoy suture zone. Also associated with this zone is Siberia's biggest gabbro-anorthosite complex, the Kalar-Dzhugdzhur, with its Fe–Ti–V, apatite and rare-earth mineralization. Alkali metasomatites and apogranites with rare-metal mineralization (the Katuga deposit) are found in the western continuation of this structure and on its flanks. North–south fracture systems contain upper Archaean ferruginous quartzites (the Sutam, Upper Zeya and other iron ore regions).

Small fields of muscovite pegmatite occur in the marginal parts of suture zones and in remnant undeformed blocks among the intensely altered rocks of such zones in addition to rare-metal pegmatites.

The third element in metallogenic zoning is defined by the growth of a system of parallel metallogenic linear zones, complicated by thrust and nappe-type oblique and transcurrent faults. This element of metallogenic zoning is expressed most fully in the North-West Baikal region. From an axial suture, represented by a Proterozoic ophiolite belt with asbestos and talc deposits and copper-nickel ore shows towards the periphery, in the NW there is a regular pattern, typical of mobile belts, in the zones on the side of the block which had been consolidated earlier: the Olokita eugeosynclinal schist type with pyrite-polymetallic and barite-polymetallic deposits; the Chuya geanticline and a miogeosynclinal type of cover, in which are localized economically important mineral deposits — muscovite pegmatites and gold-bearing formations in the Bodaybo region; and an outer arc, the Akitkan marginal volcano-plutonic belt with tin, fluorine, gold and polymetallic mineralization.

We can present only an outline of the metallogenic zoning for the South Yenisey and Pre-Sayan basement highs in the SW of the Siberian craton.

If we consider the Kan–Pre-Sayan terrain as an example, we see a certain similarity in its metallogenic zoning with that of the Dzhugdzhur-Stanovoy terrain. Here, the basic elements of zoning are defined by a major NW-striking subparallel fault system and associated minor oblique faults splaying off from major structural sutures. These regional splay systems control the sites of AR_2 and PR_1 linear trench zones and their metallogeny. Primary crustal relicts containing lower Archaean supracrustal assemblages are preserved in the blocks between these zones.

The metallogeny of the blocks typically takes the form of Archaean banded ironstones, such as the Kitoy group of deposits, high-Al sillimanite schists (e.g. the Dabod deposit), and others. The main factors determining the sites of this mineralization are firstly the primary lithological control exercised by the assemblages and secondly subsequent metamorphism and metasomatism. The Kitoy group of banded ironstones for example is restricted to the gneisses and schists in the Sharyzhalgay complex; the Biryusa muscovite pegmatites belong to the AR_2 Tepsa Formation — high-Al rocks from kyanite-sillimanite metamorphic facies series.

The metallogeny of late-stage trench complexes can be subdivided into a number of subtypes, depending on the composition of the sedimentary and volcanic rocks of the trenches. The most important are the first type — lower Archaean greenstone belts in the South Yenisey high with gold and sulphide mineralization. The second type — lower Archaean trenches — volcanogenic-terrigenous-carbonate (Onot graben) with banded ironstones, magnesite (Sava deposit) and talc (Onot deposit). The third type — trenches with marine and continental molasse sediments represented by the Early Proterozoic Urik-Iy graben, where the mineralization consists of rare-metal pegmatites related to end-Early Proterozoic granitoids and Palaeozoic (PZ_1–PR_2) rare-earth-apatite-fluorite carbonatites of the Ziminsky complex, intruded along a roughly N–S belt cutting the Urik-Iy graben.

It is obvious from the above that in broad terms the metallogenic zoning in the SW Precambrian basement highs of the Siberian craton is principally defined by major lower Archaean and in part Early Proterozoic tectonic zones. These zones broke the hypothetical Angara craton, consolidated in the Early Archaean, into separate blocks around its margins, each with its own individual metallogenic features. This type of metallogenic zoning could thus be called grid or lattice type.

Metallogenic epochs

Major metallogenic epochs have been described in outline by Bilibina, Terentyev, Kazansky, Mironyuk and Salop.

Judging from the results of isotopic studies on ores, metasomatic rocks and ore-bearing intrusions, the major metallogenic epoch of the Siberian craton is Early Proterozoic, 2.2–1.8 Ga. This is the only epoch widely documented in both Archaean granite-greenstone and granite-gneiss terrains, and in the Baikal-Patom mobile belt. A more detailed investigation has established that this age in the Anabar and Aldan Shields and in the SW marginal elevations corresponds in most cases to the termination of the processes which were responsible for forming the mineral deposits. They began to form much earlier than this, during preceding geological epochs.

For example, the Central Aldan phlogopite, magnetite and apatite deposits are dated isotopically as Early Proterozoic but field relations, in particular their connection with late Subgan granitoids, indicate that they are Late Archaean to Early Proterozoic.

Data obtained from isotopic dating of apatite deposits in the Aldan Shield are particularly indicative. For example, it has been established that the ores in all three different genetic types of apatite deposit have the same age: 1850 ± 20 Ma (Pb–U–Th and K–Ar methods). The following analyses were made in this study: metamorphic apatite shows from the Early Archaean Fyodorov Formation (AR_1); hydrothermal-metasomatic deposits from the same formation, but associated with younger (PR_1) granitoids surround granite-gneiss domes (the Seligdar deposit); and also hydrothermal-metasomatic types in fault zones associated with gabbro-syenite intrusions (the Khanin deposit).

These and other data suggest that the interval 2.2–1.8 Ga was a period mainly of repeated metamorphism, granitoid protoactivation, remobilization of ore material and regeneration of many previously-formed deposits.

Field relations and geochronological data from the Baikal-Patom belt are fundamentally different. Here we have a scatter in the age of ore emplacement, from PR_1 to R_3. For example, two ages emerge for the active formation of pyrite-polymetallic deposits in the Olokita zone — Early Proterozoic and Early Riphean; but the mica deposits in the Mama pegmatite fields are PR_1. The mineral deposits of the Early and Late Archaean metallogenic epoch in the Baikal region are unique — they are high-Al schists and Fe–Ti–Cu–Ni shows related to gabbro-pyroxenites in remnant central massifs.

Research workers find different numbers of metallogenic epochs in the Precambrian evolutionary history of the Siberian craton. The most detailed scheme (Anon., 1983) has identified four epochs in the Archaean alone: Iyengr (AR_1^1 > 3.4 Ga), Fyodorov or Timpton-Dzheltulin (AR_1^2 = 3.4–3.0 Ga), early and late Subgan (in the interval 3.0–2.6 Ga).

We have adopted the simplest scheme of periodicity in the Early Precambrian history, which identifies three Archaean (AR_1^1, AR_1^2, AR_2) and two Proterozoic (PR_1^1, PR_1^2) epochs. However, it should be emphasised that in establishing the boundaries between epochs and assigning specific deposits to one or other epoch, numerous difficulties arise, as is the case for the

East European craton. On the one hand, they are related to the spread of age boundaries in different blocks and the lack of agreement between metallogenic boundaries and accepted stratigraphic boundaries; on the other hand, they are related to the long growth history of the deposits. We now present a brief review of the characteristics displayed by the major Early Precambrian metallogenic epochs.

AR_1^1 (>3.4 Ga)

The metallogeny of this epoch is principally defined by metasedimentary assemblages in the Fe–Al–Mn triad, during the metamorphism of which major quartz-pyroxene-magnetite deposits formed (the Baikal deposit in the Pre-Sayan region, the Kholodnikan deposit in the Dzhugdzhur-Stanovoy terrain, the Sutam deposit in the south Aldan Shield, and a number of minor deposits in the Anabar Shield), economic accumulations of kyanite-sillimanite ores (the Anabar Shield, Yenisey ridge and Pre-Sayan region) and deposits of abrasive and industrial garnets, also sapphirine (the Anabar Shield). In the Western Aldan block, deposits of dinas-rock (refractory quartzose claystone) [the Melemken and Nimnyr structural-metallogenic zones (Bilibina and Shaposhnikov, 1966)] formed during granulite facies metamorphism, while graphite, copper-nickel and pyrite ore shows are seen practically everywhere.

For this metallogenic epoch, as is the case for the East European craton, typomorphic deposits include minor shows of ilmenite ores belonging to the titanium-iron formation, associated with gabbro-norite and copper-nickel and nickel-chromium formations in ultramafics. Among deposits that are rare for the Precambrian is the Chaynyt corundum deposit in the Zverev tectonic domain.

AR_1^2 (3.4–3.0 Ga)

One characteristic feature of this epoch is the almost ubiquitous development of stratified volcano-sedimentary formations, including the essential role played by carbonate, siliceous and highly carbonaceous rocks. They served as the source of and favourable environment for the formation of the ore material for many of the deposits which originated in subsequent metamorphic and igneous cycles. In particular, assemblages of this age have associated deposits of the skarn-magnetite formation, diopside-phlogopite rocks, apatite and graphite mineralization, and high-Al gneisses. The only syngenetic mineralization appears to be ferruginous quartzites (the Magan region in Aldan).

AR_2 (3.0–2.6 Ga)

The most typical ore formations for this epoch are those associated with greenstone terrains. Belonging to this period are the ironstone deposits of the Chara-Tok, Tungurcha, Nelyuka, Khanin, Sivaka, Olyokma and

Stanovoy provinces, as well as chromium copper and nickel ore shows, asbestos and talc associated with ultramafics in greenstone belts. The gabbro-anorthosite complexes of the Kalara, Dzhugdzhur and other massifs evidently formed in this epoch (from 3.0 to 2.6 Ga), with their associated titanium, iron, vanadium, apatite and rare-metal ore mineralization. These massifs have a complex metamorphic-metasomatic origin and are fundamentally different from the gabbro-anorthosite complexes of layered intrusions. According to Bibikova, the main granulite facies metamorphic episode in the Aldan region occurred in the interval 3.1–2.7 Ga. Granite formation, directly associated with this metamorphic episode and its retrograde stages, is linked by several workers to the formation of skarn bodies with iron ore, phlogopite, apatite and rock-crystal mineralization. In addition, isotope data for metasomatic rocks and granites provide more grounds for assigning these to Early Proterozoic formations. Rare-metal pegmatites are atypical for this epoch.

PR_1; PR_1^1 (2.6–1.8; 1.65 Ga)

The epoch at the start of the Early Proterozoic (PR_1), analogous to the early Karelian in Europe, is expressed very indistinctly in Siberia. Economically valuable mineralizations include cupriferous sandstones in Udokan and possibly also the copper-nickel and iron-titanium ore shows in the Chiney and other basic intrusions. The main mineralization formed in the concluding stages of the Early Proterozoic epoch in the interval 2.2–1.8 Ga, and in a number of structures, in particular the Akitkan orogenic-volcanic belt in the Baikal region, also continued to develop actively, judging from the isotopic age determinations, up to 1.6 Ga or possibly 1.4 Ga.

Varied types of deposits originated during the 2.2–1.8 Ga time interval, the most important of which are as follows: (1) pegmatites belonging to this epoch are practically ubiquitous quartz-feldspar, muscovite, rare-metal and rare-earth types; (2) iron-ore and phlogopite deposits in magnesian skarns; (3) varied genetic types of apatite deposit, as mentioned previously; (4) rock crystal deposits in quartz veins; (5) pyrite-polymetallic deposits and noble metal mineralization related to black schist formations in the Baikal region; and (6) rare-metal, fluorite and polymetallic mineralization in orogenic volcanic belts.

Recent findings allow us to note the following broad patterns in the evolution of mineralization in the Siberian Precambrian:

(1) The evolution of Precambrian mineral deposits in the Siberian craton took place over a prolonged time interval and often stretched over two or three epochs due to the frequent remobilization of ore elements, changes in ore mineral composition during metamorphism and later overprinting by intrusive-metasomatic processes.

(2) Periodicity in the evolution of mineralization of several scales is a characteristic feature, expressed in the appearance of particular repetition

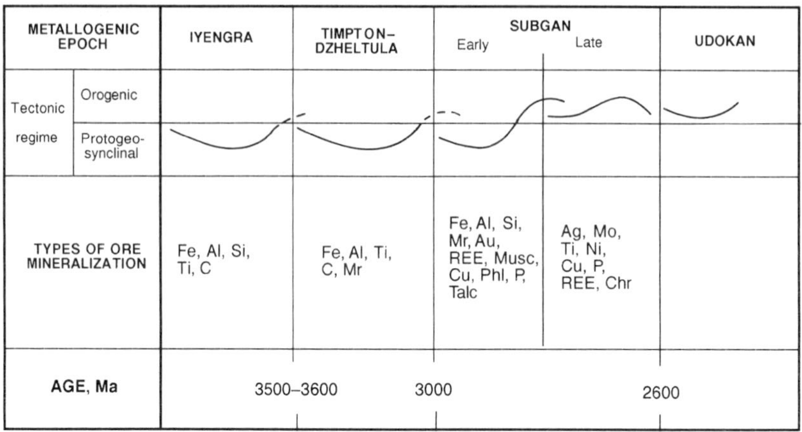

Fig. II-51. Relationships between various associations of useful minerals in the geological evolution of the Aldan Shield (from Sokolov and Bilibina, 1984).

(in general features) of the evolutionary sequence of ore types — initially associated mainly with supracrustal sedimentary and volcanic assemblages, further by metamorphic processes and then orogenic and activation magmatism. Relationships between the different natures of ore formations and ore deposits change in a regular fashion from earlier to later periods. This position can be illustrated using the Olyokma and Stanovoy blocks of the Aldan Shield as an example (Fig. II-51).

(3) The characteristic alternation between relatively poorly productive periods with periods of maximum distribution of mineralization. Many workers have previously noted this inhomogeneous intensity in the appearance of mineralization. Marakushev and Kudryavtsev (1965) have emphasised that the main age dates, 2600, 1800 and 1000 Ma, are of global significance.

(4) Of the specific features characterising the evolution of mineralization in the Precambrian history of the shields in the Siberian craton, we may draw attention to the greater expression (compared to European shield areas) of the Early Archaean metallogenic epoch. This allows the Early Archaean epoch to be divided into two: Katarchaean (>3.4–4.0 Ga) and Early Archaean (3.4–3.0 Ga).

(5) Attention should also be drawn to the important significance for the Siberian craton of Riphean metallogeny. In the general case, the Siberian craton is surrounded by Riphean to Vendian actively evolving tectonic belts. As a result of this, Riphean tectono-magmatic activity manifest itself within the shields and basement highs of the Siberian craton much more completely than in the East European craton. This feature is seen most clearly when we compare the Svecofennides of Karelia and their age equivalents in the Baikal region. The widespread development of Early

Cambrian formations alongside Riphean endogenic processes served, as we know, as the basis for assigning this province to Ripheids ("Baikalides").

In conclusion, we would emphasise that, unfortunately, lack of sufficient accurate age dates from structural-lithological terrains and ores makes it extremely difficult to analyse the evolution of mineralization with time, and at present prevents us from accurately correlating internal and external processes in the different blocks of the Siberian craton.

REFERENCES

(All references are in Russian unless otherwise stated)

Abramovich, G.Ya., Svirina, I.F. and Volynets, Yu.N., 1979. Igneous complexes of Eastern Sayan. In: Major features of the geology of Eastern Sayan. Irkutsk, East Siberian Publishers, pp. 37–49.

Aksyonov, Ye.M., Vafin, R.F. and Khaydarov, R.A., 1985. Comparative formational analysis of trench structures in the western part of the Patom-Aldan-Stanovoy shield. In: Precambrian trench structures in the Baikal-Amur region and their metallogeny. Novosibirsk, Nauka, pp. 111–116.

Alakshin, A.M. and Korsakov, L.P., 1985. Deep structure of the Stanovoy suture zone. Tikhookean. Geologiya, No. 3, pp. 76–86.

Andreyeva, Ye.D., Bogatikov, O.A. and Borodayevskaya, M.B., 1981. Classification and nomenclature of igneous rocks. Moscow, Nedra, 160 pp.

Anon., 1968. Geological structure of the USSR. Vol. 4, Moscow, Nedra, 504 pp.

Anon., 1977. Structure of the basement beneath the sedimentary cover of the Siberian platform. Trudy Inst. Geol. and Geophys., SO AN SSSR, No. 378, pp. 207–214.

Anon., 1981. Geology of the Yakutsk ASSR. Moscow, Nedra, 300 pp.

Anon., 1983a. Correlation of endogenous processes and their role in the Early Precambrian crustal development of SE Siberia. In: Profiles of orogenic belts. Geodynamics series. American Geophysical Union, Washington D.C., and Geol. Soc. America, Boulder, Colorado, Vol. 10, pp. 145–204 [in English].

Anon., 1983b. Tables of chemical compositions and crystallochemical formulae of minerals from metamorphic rocks and granitoids of the Aldan Shield. Yakutsk, Yakutsk Branch, SO AN SSSR, 358 pp.

Anon., 1986a. Evolution of the early Precambrian lithosphere in the Aldan-Olyokma-Stanovoy region. Leningrad, Nauka.

Anon., 1986b. Structure of the Earth's crust in the Anabar Shield. Moscow, Nauka.

Aranovich, L.Ya., 1983. Biotite-garnet equilibrium in metapelites. Reviews of physico-chemical petrology. Moscow, Nauka, pp. 121–136.

Aranovich, L.Ya. and Podlessky, K.K., 1981. An experimental study of garnet + sillimanite + quartz = cordierite equilibrium. Doklady Akad. Nauk SSSR, Vol. 259, No. 6, pp. 1440–1443.

Balagansky, V.V., 1979. Early Precambrian internal processes in the junction zone between the Aldan Shield and the Stanovoy belt. In: Geodynamic researches, Soviet Radio, No. 5, pp. 30–45.

Barker, F. (Editor), 1983. Trondhjemites, dacites and associated rocks. 490 pp. [in English].

Berezkin, V.I. and Kitsul, V.I., 1979. Two differentiation trends in metavolcanics of granulite and amphibolite complexes in the central Aldan Shield. In: Problems in petrogenesis and ore mineralization. Correlation of internal processes. Abstracts of 3rd East Siberian Regional Petrography Meeting, Irkutsk, 40 pp.

Bibikova, Ye.V., Belov, A.N., Gracheva, G.V. and Rozen, O.M., 1985. Upper age boundary of granulites from the Anabar Shield. Izvestiya Akad. Nauk SSSR, Ser. Geol., No. 8, pp. 19–24.

Bibikova, Ye.V., Khiltova, V.Ya., Gracheva, T.V. and Makarov, V.A., 1982. Age of greenstone belts in the Pre-Sayan region. Doklady Akad. Nauk SSSR, Vol. 267, No. 5, pp. 1171–1174.

Bilibina, T.V. (Editor), 1976. Geology and metallogeny of shield areas in ancient cratons. Leningrad, Nedra, 338 pp.

Bilibina, T.V. and Shaposhnikov, G.N. (Editors), 1976. Geological formations and metallogeny of the Aldan Shield. Leningrad, Nedra, 337 pp. (Trudy VSEGEI, New Series, Vol. 276).

Bilibina, T.V. and Sokolov, Yu.M. (Editors), 1975. Metallogeny of the Precambrian. Leningrad, Nedra, 197 pp.

Bilibina, T.V. and Sokolov, Yu.M. (Editors), 1979. Features of Precambrian metallogeny of the USSR. Leningrad, Nauka, 123 pp.

Bilibina, T.V. and Sokolov, Yu.M. (Editors), 1980. Regional metallogeny of the Precambrian of the USSR. Leningrad, Nedra, 115 pp.

Bilibina, T.V. and Sokolov, Yu.M. (Editors), 1984. Precambrian stratiform mineral deposits. Geology, genesis and metallogeny. Leningrad, Nedra, 125 pp.

Bogdanov, Yu.V. (Editor), 1981. Metallogenic map of the Baikal-Amur Mainline railway region. Leningrad, VSEGEI, 138 pp.

Bogolepov, K.V. and Votakh, O.A. (Editors), 1977. The ancient cratons of Eurasia. Novosibirsk, Nauka, 312 pp.

Bulina, L.V. and Spizharsky, T.N., 1970. Basement heterogeneity of the Siberian craton. In: Siberian Tectonics, Vol. 3, Tectonics of the Siberian craton. Moscow, Nauka, pp. 54–60.

Cherkasov, R.F., 1978. The Archaean of the Aldan Shield in the river Timpton valley stratotype region. In: Precambrian stratigraphy and sedimentary geology of the Far East. Vladivostok, DVNTs AN SSSR, pp. 19–49.

Cherkasov, R.F., 1979. Archaean of the Aldan Shield. Moscow, Nauka, 159 pp.

Christensen, N.Y. and Fountain, D.N., 1975. Constitution of the lower continental crust based on experimental studies of seismic velocities in granulites. Geol. Soc. Am. Bull., Vol. 86, pp. 227–236 [in English].

Condie, K., 1983. Archaean greenstone belts. Amsterdam, Elsevier, 390 pp. [in English].

Davydov, Yu.V., 1975. Riphean carbonate sediments of the SE Siberian craton and its margins. Novosibirsk, Nauka, 127 pp.

Dedeyev, V.A. (Editor), 1974. Structure of the basement to cratonic regions of the USSR. Leningrad, Nauka, 400 pp.

Demina, L.I. and Yeskin, A.S., 1974. Eclogites in the western Baikal region. In: Problems in geology, geochemistry and geophysics in exploring for deposits of useful minerals and in engineering geological surveys. Moscow, Moscow Univ. Publ., pp. 59–69.

Dibrov, V.Ye., 1958. Geological structure of the Gutar-Biryusi mica-bearing region. Voronezh, Voronezh Univ. Publ., 128 pp.

Dodin, A.L., Konikov, A.Z., Mankovsky, V.K. and Taschchilov, A.F., 1968. Stratigraphy of Precambrian rocks of Eastern Sayan. Moscow, Nedra, 280 pp.

Dook, V.L., 1977. The Aldan crystalline massif and the Stanovoy fold belt. In: Structural evolution of metamorphic complexes. Moscow, pp. 27–47.

Dook, V.L., 1984. Diapiric megastructures (ovals) in ancient shields. The Lower Timpton structure of the Aldan Shield. In: Geological survey of regions showing Precambrian dome structures. Leningrad, Nedra, pp. 44–53.

Dook, V.L., Balagansky, V.V. and Zergenezov, A.N., 1975. Deformation sequence in Archaean rocks in the Sutam block. In: Structural and metamorphic petrology of the Precambrian of the Aldan Shield. Yakutsk, pp. 19–41.

Dook, V.L. and Kitsul, V.I., 1975. An investigation of the Precambrian of the Aldan Shield by structural and metamorphic petrology. In: Structural and metamorphic petrology of the Precambrian of the Aldan Shield. Yakutsk, pp. 5–19.

Dook, V.L., Kitsul, V.I. and Berezkin, V.I., 1979. Lower Precambrian structures and metamorphism of the Aldan Shield in the Timpton and Sutam river basins. In: Geodynamic research. Results of Int. Geophys. Projects, No. 5, pp. 7–29.

Dook, V.L., Kitskul, V.I. and Petrov, A.F., 1986. The early Precambrian of Southern Yakutia. Ed. Dobretsov, N.L. Moscow, Nauka, 280 pp.

Dook, V.L., Pavlov, S.N., Verkhalo-Uzkiy, V.N. and Gusakova, I.N., 1985. Geology and petrology of the Ungra metagabbro-tonalite complex. In: Early Precambrian of the Aldan Shield and its margins. Leningrad, Nauka, pp. 20–34.

Dook, V.L., Salye, M.Ye., Baykova, V.S., 1975. Structural and metamorphic evolution and phlogopite content of granulites in the Aldan region. Leningrad, Nauka, 266 pp.

Drugova, G.M. and Neyelov, A.N., 1960. Polymetamorphism of Precambrian rocks in the southern part of the Aldan Shield and the Stanovoy range. Trudy Lab. Precamb. Geol., Vol. 11, pp. 142–216.

Dzevanovsky, Yu.K., 1958. The Archaean metamorphic complex of the Aldan Shield. In: Proc. Interdepartmental meeting on unified stratigraphic schemes for Siberia. Moscow and Leningrad, pp. 37–42.

Fedorovsky, V.S., 1985. Lower Proterozoic of the Baikal mountains, geology and conditions of formation of the continental crust in the early Precambrian. Moscow, Nauka, 200 pp. (Trudy GIN AN SSSR, Vol. 400).

Fotiadi, E.E., 1967. Major tectonic features of the structure of Siberia in the light of regional geological and geophysical information. Trudy SNIIGGIMS, Regional Geology Series, Vol. 57, pp. 30–57.

Fotiadi, E.E., Grishin, M.P. and Neyelov, A.N., 1974. Early Precambrian fold systems in the basement to the Siberian craton. In: Structure of the basement to cratonic regions of the USSR. Leningrad, Nauka, pp. 99–113.

Frolova, N.V., 1951. Conditions of sedimentation during the Archaean. Trudy Irkutsk Univ., Ser. Geol., Vol. 2.

Frumkin, I.M., 1961. New data on the Archaean stratigraphy of the Aldan Shield. In: Abstracts of an Interdepartmental meeting on developing stratigraphic schemes for the Yakutsk ASSR. Leningrad, VSEGEI.

Frumkin, I.M., 1984. Metallogenic epochs and provinces in the Archaean of the Aldan Shield. In: Metallogeny of the early Precambrian of the USSR. Leningrad, Nauka, pp. 198–202.

Gerling, E.K., Iskanderova, A.D., Levchenkov, O.A. and Mikhaylov, D.A., 1970. On the age of marbles in the Dzheltulin and Iyengr Groups of the Aldan Shield from the U–Pb isochron method. Doklady Akad. Nauk SSSR, Vol. 194, No. 6, pp. 1397–1400.

Glebovitsky, V.A. (Editor), 1975. Metamorphic belts of the USSR. Leningrad, 55 pp.

Glebovitsky, V.A. and Sedova, I.S., 1986. The Aldan megablock. In: Evolution of the early Precambrian lithosphere of the Aldan-Olyokma-Stanovoy region. Leningrad, Nauka.

Glukhovsky, M.Z., Katz, Ya.G. and Moralev, V.M., 1983. The nuclei of the world's continents. Izvestiya Vuzov, Geologiya i Razvedka, No. 8, pp. 14–19.

Golovenok, V.K., 1977. High-alumina assemblages in the Precambrian. Leningrad, Nedra, 268 pp.

Goltsman, Yu.V., Bibikova, Ye.V. and Bairova, E.D. et al., 1982. Geochronology of granites in the Primorsk complex of the SW Baikal region from U–Pb, Rb–Sr and K–Ar data. Izvestiya Akad. Nauk SSSR, Ser. Geol., No. 1, pp. 5–15.

Goodwin, A.M., 1981. Precambrian perspectives. Science, Vol. 213, No. 4503, pp. 55–61 [in English].

Gorokhov, I.M., Dook, V.L. and Kitsul, V.I., 1981. Rb–Sr systems of Precambrian polymetamorphic complexes. Izvestiya Akad. Nauk SSSR, Ser. Geol., No. 8, pp. 5–17.

Grishin, M.P. and Lotyshev, V.I., 1974. Density characteristics and gravity effects of the sedimentary cover of the Siberian platform. Nauch. Trudy SNIGGIMS, Vol. 196, pp. 22–29.

Grishin, M.P., Lotyshev, V.I. and Surkov, V.S., 1982. Structure of the basement to the Siberian craton and its effect on the formation of sedimentary basins of Riphean age. In: Geophysical methods in regional geology. Novosibirsk, Nauka, pp. 102–110. (Trudy IGIG SO AN SSSR, Vol. 543).

Groves, D.I. and Batt, W.D., 1984. Factors controlling the heterogenous distribution of metallogenic associations in Archaean greenstone belts: the West Australian shield as a case study. In: 27th Int. Geol. Congress, Precambrian Geology. Moscow, Nauka, pp. 133–143 [in English].

Gusev, G.S., Petrov, A.F. and Gradkin, G.S. et al., 1985. Structure and evolution of the Earth's crust of Yakutia. Moscow, Nauka, 247 pp.

Iskanderova, A.D., Neymark, L.A. and Rudnik, V.A., 1980. Geological formations in the region of the Baikal-Amur Mainline railway. Trudy VSEGEI, New Series, Vol. 307, pp. 123–138.

Kadensky, A.A., 1960. Magnetite ore mineralization in the Sutam region. In: Iron ores of southern Yakutia. Moscow, Akad. Nauk SSSR.

Karasik, V.M., 1984. A study of seismic velocities in the western part of the Siberian craton. Obzor VIEMS, Moscow, 52 pp.

Karsakov, L.P., 1978. Deep-level granulites: the Chogar complex in the Stanovoy fold belt of Eastern Siberia. Moscow, Nauka, 149 pp.

Karsakov, L.P., 1980. Tectonic setting of deep-level granulites and the structure of the lower crust. In: Tectonics of Siberia. Vol. 8, Methodological problems in tectonics and problems in tectonic regional mapping. Novosibirk, Nauka, Siberian Branch, pp. 69–74.

Kastrykina, V.M., 1976. Mineralogy of metamorphic rocks in the Archaean Sutam complex of the Aldan Shield. In: Modern methods of petrological research. Moscow, Nauka, 1976, pp. 56–76.

Kastrykina, V.M., 1983. Metamorphism of the central part of the Dzhugdzhur-Stanovoy fold belt. In: Early Precambrian metamorphism in the region of the Baikal-Amur Mainline railway. Leningrad, pp. 140–163.

Katz, A.G., 1962. Archaean stratigraphy of the SW Aldan Shield. Trudy VAGT, Vol. 8, pp. 90–92.

Kazansky, V.I., 1983. Metallogeny of the early Precambrian. VINITI. Progress in science and technology. Series on Ore Deposits, Vol. 13, 93 pp.

Khiltova, V.Ya., 1971. Highly aluminous rocks of the Kitoy Group, Eastern Sayan — lithology and conditions of formation. In: Problems of Precambrian sedimentology. Leningrad, 1971, pp. 96–108.

Khrenov, pp.M. (Editor), 1983. Geological map of the Baikal area, scale 1 : 1,000,000. SNIIGGIMS.

Khrenov, pp.M. and Shafeyev, A.A. (Editors), 1981. Metallogeny of the Precambrian. Irkutsk, 380 pp.

Kitsul, V.I. and Shkodzinsky, V.S., 1972. Inertness and mobility of components in granulite facies metamorphic and migmatite-forming conditions and criteria for their subdivision: an example from the Aldan Shield. In: Summary maps and general problems in metamorphism. Novosibirsk, Inst. Geol. Geophys., Vol. 2, pp. 81–91.

Kitsul, V.I., Petrov, A.F. and Zedgenizov, A.N. Structural-lithological provinces of the Aldan Shield. In: Major tectonic complexes of Siberia. Novosibirsk, IGG SO AN SSSR, pp. 16–31.

Kitsul, V.I. and Shkodzinsky, V.S., 1976. Granulite facies of the Aldan Shield. In: 25th Int. Geol. Congress. Precambrian Geology. Soviet contributions. Stratigraphy and Sedimentology. Moscow, Nauka, pp. 275–286.

Kitsul, V.I., Shkodzinsky, V.S. and Zedgenizov, A.N., 1973. Physico-chemical analysis of garnet formation in granulite facies mafic schists. In: Petrology of the granulite facies in the Aldan Shield. Moscow, Nauka, pp. 4–28.

Klimov, L.V., 1978. High-pressure granulite metamorphism in deep-level Sutam-type early Precambrian complexes. In: Cyclicity and directional trends in regional metamorphic processes. Leningrad, pp. 33–56.

Komarov, A.N. and Ilupin, I.P., 1978. New fission-track data on the age of kimberlites in Yakutia. Geokhimiya, No. 7, pp. 1004–1014.

Konikov, A.Z. and Travin, L.V., 1983. Lower Proterozoic stratigraphy of the Urik-Iy graben, Presayan region. In: F.P. Mitrofanov and A.A. Shafeyev (Editors), Precambrian stratigraphy of Central Siberia. Collection of papers. Akad. Nauk SSSR, Geology, Geophysics and Geochemistry. Scientific Council for Precambrian Geology. Leningrad, Nauka, 168 pp.

Kontorovich, A.E., Surkov, V.S. and Trofimuk, A.A. (Editors), 1981. Geology of oil and gas in the Siberian Platform. Moscow, Nedra, 552 pp.

Korikovsky, S.P., 1967. Metamorphism, granitization and post-magmatic processes in the Precambrian of the Udokan-Stanovoy zone. Moscow, 298 pp.

Korikovsky, S.P., 1979. Facies of metamorphism of metapelites. Moscow, Nauka, 262 pp.

Korikovsky, S.P. and Fedorovsky, V.S., 1971. Geological relations of rocks in the Udokan Group and the trench complex, SW Aldan Shield. Sov. Geologiya, No. 10, pp. 120–124.

Korikovsky, S.P. and Kislyakova, N.G., 1975. Reaction textures and phase equilibria in hypersthene-sillimanite schists of the Sutam complex, Aldan Shield. In: Metasomatism and ore mineralization. Moscow, Nauka, pp. 314–342.

Korzhinsky, D.S., 1936. Petrology of the Archaean complex of the Aldan Shield. Trudy TsNIGRI, Vol. 86, 76 pp.

Korzhinsky, D.S., 1937. Depth dependence of mineral formation. ZVMO (Proc. All-Union Min. Soc.), Pt 66, No. 2.

Korzhinsky, D.S., 1939. Precambrian of the Aldan Shield and the Stanovoy Ridge. In: Stratigraphy of the USSR, Vol. 1. Moscow and Leningrad, pp. 349–366.

Kosygin, Yu.A. (Editor), 1984. Tectonic nature of geophysical fields of the Far East. Moscow, Nauka, 200 pp.

Kosygin, Yu.A., Basharin, A.K. and Volontey, G.M., 1964. Precambrian tectonics of Siberia. Novosibirsk, 128 pp.

Kovach, V.P., Kitsul, V.I. and Vaganov, pp.A., 1984. Rare elements in plagiogneisses and metabasites of the Olyokma Group, Aldan Shield. In: Natural associations of Archaean grey gneisses. Leningrad, Nauka, pp. 84–94.

Kovrigina, Ye.K., 1977. Tectonics of the Angara-Kana part of the Yenisey ridge. Trudy VNIGI, pp. 24–40.

Kozhevnikov, V.N., 1982. Conditions of formation of structural-metamorphic parageneses in Precambrian complexes. Leningrad, Nauka, 184 pp.

Kozyreva, I.V., Avchenko, O.V. and Mishkin, M.A., 1985. Deep-level metamorphism of late Archaean volcanic belts. Moscow, 164 pp.

Krasnikov, N.N., 1984. The Kholodnikan greenstone belt in the structure of the Aldan massif. Doklady Akad. Nauk Sssr, Vol. 275, No. 4, pp. 938–941.

Krasny, L.I. (Editor), 1979. Structural elements in the region of the Baikal-Amur Mainline railway and their mineralogical features. Leningrad, Nedra, 96 pp. (Trudy VSEGEI, New Series, Vol. 303).

Kratz, K.O. (Editor), 1984. Fundamentals of the metallogeny of Precambrian metamorphic belts. Leningrad, Nauka, 340 pp.

Kratz, K.O., Moskovchenko, N.I. and Shemyakin, V.M., 1984. Grey gneisses in the Aldan massif and its margins: geological setting, associations, petrogenesis. In: Natural associations of Archaean grey gneisses. Leningrad, pp. 63–72.

Kropotkin, pp.N., Valyayev, B.M., Gafarov, R.A., Solovyov, N.A. and Trapeznikov, Yu.A., 1971. Lower crustal tectonics of ancient cratons in the northern hemisphere. Trudy Geol. Inst. Akad. Nauk SSSR, Vol. 209, 390 pp.

Krylov, A.Ya., Vishnevsky, A.N. and Silin, Yu.I., 1963. Absolute age of rocks in the Anabar Shield. Geokhimiya, No. 12, pp. 1140–1144.

Krylov, I.N., Gorokhov, I.M., Kutyavin, E.P., Melnikov, N.N, Varshavskaya, E.S., 1977. On the problem of dating polymetamorphic assemblages in the Shayzhalgay Group, SW Baikal region. In: Geological interpretation of data from geochronology. Abstracts. Irkutsk, pp. 42–43.

Kudryavtsev, V.A., 1974. Upper Archaean and Lower Proterozoic stratigraphy of the Aldan Shield. In: Precambrian and Palaeozoic of North-Eastern USSR. Magadan, pp. 29–31.

Kudryavtsev, V.A., Akhmetov, R.N. and Biryulkin, G.V., 1975. Precambrian structural-lithological terrains in the Temulyakit-Tungurchi fold belt, Aldan Shield. In: Precambrian geology and tectonics of the Far East. Vladivostok, DVND AN SSSR, pp. 49–60.

Kudryavtsev, V.A. and Kharitonov, A.L., 1976. Structural relations of middle Precambrian complexes in the east of the Udokan ridge. Doklady Akad. Nauk SSSR, Vol. 227, No. 5, pp. 1195–1198.

Kudryavtsev, V.A. and Nuzhnov, S.V., 1981. Upper Archaean structures of the Aldan Shield. Geologiya i Geofizika, No. 6, pp. 28–37.

Kulish, Ye.A., 1973. Highly aluminous lower Archaean metamorphic rocks of the Aldan Shield and their lithology. Khabarovsk, 370 pp.

Kushev, V.G., 1986. Greenstone basins (trench complexes) of Eastern Siberia in a system of Archaean cratons and Proterozoic mobile belts. In: Precambrian trench structures of the Baikal-Amur region. Novosibirsk, Nauka, pp. 28–34.

Kuznetsov, V.A. (Editor), 1985. Precambrian trench structures in the Baikal-Amur region and their metallogeny. Novosibirsk, 199 pp.

Kuznetsova, F.V., 1981. The granulite complex of the SW Baikal region. Novosibirsk, Nauka, Siberian Branch, 182 pp.

Lavrenko, Ye.I., 1957. Paragenetic relations in alumina-rich schists and gneisses of the Aldan complex. ZVMO (Proc. All-Union Min. Soc.), Pt. 86, No. 1.

Lazarev, Yu.I. and Kozhevnikov, V.N., 1973. A structural-petrological study of granitization. Leningrad, 124 pp.

Letnikov, F.A. (Editor), 1981. Geology of granulites: Field guide to the August, 1981 Baikal Excursion, Int. Symp. for Int. Geol. Congr. "Archaean Geochemistry" and "Precambrian metallogeny" Projects. Irkutsk, Akad. Nauk SSSR, Siberian Branch, East Siberian Filial, 98 pp.

Lobach-Zhuchenko, S.B., Dook, V.L. and Krylov, I.N., 1984. Geological and geochemical types of associations of tonalite-trondhjemite series in the Archaean. In: Natural associations of Archaean grey gneisses. Leningrad, Nauka, pp. 17–51.

Lotyshev, V.I., 1979. Density inhomogeneities of the upper mantle of the Siberian craton. In: Advances in petrophysics and geology in Siberia. Novosibirsk, pp. 20–28.

Lutz, B.G., 1964. Petrology of the granulite facies of the Anabar massif. Moscow, 124 pp.

Lutz, B.G., 1974. Petrology of deep zones in the continental crust and upper mantle. Moscow, Nauka, 304 pp.

Lutz, B.G., 1984. Mantled grey gneiss domes of the Anabar massif and the Aldan Shield. Byull. Mosk. Ob-va Ispytateley Prirody. Otd. Geolgoii. Vol. 58, No. 2, pp. 27–35.

Malich, N.S., 1975. Tectonic evolution of the Siberian platform sedimentary cover. Moscow, Nedra, 215 pp.

Malich, N.S. et al. (Editors), 1974. Geological formations of the pre-Cenozoic cover of the Siberian craton and their ore content. Moscow, Nedra, 279 pp. (Trudy VSEGEI, New Series, Vol. 194).

Malich, N.S. Mironyuk, Ye. pp. and Turchanova, Ye.V., 1984. Geological structure of the Siberian craton. 27th Int. Geol. Congress. Geology of the Soviet Union. Colloq. K.01, Abstracts Vol. 1, Moscow, Nauka, pp. 35–43.

Malyshev, Yu.F., 1977. Geophysical studies of the Precambrian of the Aldan Shield. Moscow, Nauka, 127 pp.

Manuylova, M.M. (Editor), 1968. Precambrian geochronology of the Siberian craton and its folded margins. Leningrad, Nauka, 331 pp.

Marakushev, A.A. and Kudryavtsev, V.A., 1965. The hypersthene-sillimanite paragenesis and its petrological significance. Doklady Akad. Nauk SSSR, Vol. 164, No. 1, pp. 179–183.

Markov, M.S. (Editor), 1978. Tectonics of the basement to the East European and Siberian cratons. Moscow, Nauka.

Mashchak, M.S., 1973. Petrochemistry of diabase and dolerite dykes of various ages in the south of the Anabar Shield. In: Geology and geochemistry of basic rocks in the eastern part of the Siberian craton. Moscow, pp. 78–86.

Mironyuk, Ye.P., 1961. Some new data on the Precambrian stratigraphy of the junction zone between the Olyokma and Iyengr Groups in the Aldan Shield. In: Abstracts. Meeting to develop stratigraphic schemes of the Yakutsk ASSR. Leningrad, Nauka, pp. 13–15.

Mironyuk, Ye.P., 1967. New data on the stratigraphy of the Precambrian of the NW margins of the Stanovoy ridge. In: Precambrian of eastern regions of the USSR. Leningrad, Nedra, pp. 165–172.

Mironyuk, Ye. pp., Lyubimov, B.K. and Magnushevsky, E.L., 1971. Geology of the western part of the Aldan Shield. Moscow, Nedra, 238 pp.

Mirnoyuk, Ye.P. and Zagruzina, I.A., 1983. Tectonic domains ("provinces" or "geoblocks") of Siberia and stages in their formation. In: Tectonics of Siberia. Vol. 2. Crustal structure in the east of the USSR in the light of modern tectonics concepts. Novosibirsk, Nauka, 206 pp.

Mitrofanov, F.P. (Editor), 1979. General problems concerning the subdivision of the Precambrian of the USSR. Leningrad, Nauka, 163 pp.

Mitrofanov, F.P. and Glebovitsky, V.A. (Editors), 1985. Internal PT regimes in the formation of the crust and ore formation in the early Precambrian. Leningrad, Nauka, 287 pp.

Mitrofanov, F.P. and Moskovchenko, N.I. (Editors), 1985. The early Precambrian of the Aldan massif and its margins. Leningrad, Nauka, 182 pp.

Mitrofanov, F.P. and Shafeyev, A.A. (Editors), 1983. Precambrian stratigraphy of Central Siberia. Leningrad, Nauka, 167 pp.

Miyashiro, A., 1961. Evolution of metamorphic belts. Journal of Petrology, Vol. 2, No. 3, pp. 277–311 [in English].

Moshkin, V.N., 1958. Precambrian stratigraphy: the Stanovoy and Dzhugdzhur mountain ranges. In: Geological structure of the USSR. Vol. 1. Moscow, pp. 128–130.

Moskovchenko, N.I., Krasnikov, N.N. and Semenov, A.P., 1983. Internal evolution of terrains in the junction zone between Aldanides and Stanovides. In: Metamorphism of the Precambrian in the region of the Baikal-Amur Mainline railway. Leningrad, pp. 97–127.

Moskovchenko, N.I., Semenov, A.P. and Verkhalo-Uzkiy, V.N., 1985. Granulite complexes in the Stanovoy fold belt. In: Early Precambrian of the Aldan Shield and its margins. Leningrad, pp. 121–144.

Musatov, D.I., Fedorovsky, V.S. and Mezhelovsky, N.V., 1983. Archaean tectonic regimes and geodynamics. Obzor VIEMS. General and Regional Geology and Geological Mapping. 42 pp.

Musatov, D.I., Levitova, F.N. and Chernyshov, N.M., 1981. Deep structure of the Anabar Shield. Obzor VIEMS. General and Regional Geology and Geological Mapping. Vol. 6, pp. 1–17.

Neyelov, A.N., 1968. Precambrian palaeotectonics of the Siberian craton and some patterns in the evolution of Precambrian mobile belts. In: Precambrian geology. 23rd Session of the Int. Geol. Congress. Leningrad, pp. 41–51.

Neyelov, A.N., 1977. Chemical classification of sedimentary rocks in the study of Precambrian metamorphic complexes. In: Lithology and geochemistry of the early Precambrian. Apatity, pp. 96–105.

Neyelov, A.N., 1980. Petrochemical classification of metamorphosed sedimentary and volcanic rocks. Leningrad, Nauka, 100 pp.

Neyelov, A.N. (Editor), 1983. Metamorphism of the Precambrian in the region of the Baikal-Amur Mainline railway. Leningrad, 231 pp.

Neyelov, A.N. and Sedova, I.S., 1963. The western part of the Stanovoy range. In: Stratigraphy of the USSR. Lower Precambrian. Asiatic part of the USSR. Moscow, pp. 264–285.

Neymark, L.A., Iskanderova, A.D., Chukhonin, A.P., 1981. A U–Pb Archaean age for metamorphic rocks in the Stanovoy ridge. Geokhimiya, No. 9, pp. 1386–1397.

Nikolayevsky, A.A., 1968. Deep structure of the eastern part of the Siberian craton and its margins. Moscow, Nauka, 181 pp.

Nozhkin, A.D., 1985. Early Precambrian trench complexes and their metallogeny in the SW of the Siberian craton. In: Precambrian trench structures in the Baikal-Amur region and their metallogeny. Novosibirsk, Nauka, Siberian Branch, pp. 34–46.

Obruchev, S.V., 1949. Tectonics of the western part of the Sayan-Baikal Caledonian fold belt. Doklady Akad. Nauk SSSR, Vol. 68, No. 5.

Obruchev, S.V., Neyelov, A.N., Nikitina, L.P. and Manuylova, M.M., 1967. Precambrian and Cambrian stratigraphy and correlation of Central Siberia. Novosibirsk, pp. 238–246.

Panchenko, I.V., 1985. Geology and evolution of metamorphism of lower Precambrian complexes of the Stanovoy range. Bladivostok, 150 pp.

Pavlov, Yu.A. and Parfenov, L.M., 1973. Deep structure of the East Sayan and South Aldan terminations of the Siberian craton. Novosibirsk, Nauka, Siberian Branch, 110 pp.

Pavlovsky, Ye.V. and Yeskin, A.S., 1964. Composition and structure of the Archaean of the Baikal region. Moscow, Nauka, 128 pp. (Trudy Geol. Inst. AN SSSR, Vol. 110).

Perchuk, L.L., 1977. An improved two-pyroxene geothermometer for deep-level peridotites. Doklady Akad. Nauk SSSR, Vol. 233, No. 3, pp. 456–459.

Perchuk, L.L., 1983. Thermodynamic aspects of polymetamorphism. In: Metamorphic zoning and metamorphic complexes. Moscow, Nauka, pp. 21–37.

Perchuk, L.L., Lavrentyeva, I.V., Aranovich, L.Ya. and Podlessky, K.K., 1983. Biotite-garnet-cordierite equilibrium and the evolution of metamorphism. Moscow, Nauka, 196 pp.

Petrov, A.F., 1974. Upper Archaean and lower Proterozoic stratigraphy of sediments in the west of the Aldan Shield. Sov. Geologiya, No. 2, pp. 135–142.

Petrov, A.F., 1976. Precambrian orogenic complexes in the west of the Aldan Shield. Novosibirsk, Nauka, 119 pp.

Petrov, A.F., Timofeyev, V.F., Lubenovsky, V.M. and Rozhin, S.S., 1978. Precambrian stratigraphy of the Batomga block of the Aldan Shield. Bulletin N.T.I. Yakutsk, pp. 3–6.

Petrova, Z.I. and Levitsky, V.I., 1984. Petrology and geochemistry of granulite complexes in the Baikal region. Ed. Letnikov, F.A. Novosibirsk, Nauka, Siberian Branch, 201 pp.

Petrova, Z.I. and Smirnova, Ye.V., 1982. Rare earth elements in ultrametamorphic and phlogopite-forming processes — an example from the Aldan phlogopite province. In: Geochemistry of rare earth elements in internal processes. Novosibirsk, Nauka, pp. 111–128.

Poletayev, I.A., Shames, pp.I. and Shcherbakov, A.F., 1979. Geological structure of the Savin magnesite deposit. In: Major features of the geology of Eastern Sayan. Irkutsk, East Siberian Book Publishers, pp. 114–126.

Polkanov, A.A. and Obruchev, S.V. (Editors), 1964. The Precambrian of eastern Sayan. Moscow and Leningrad. Trudy Lab. Precambrian Geol., Akad. Nauk SSSR, Vol. 18.

Predovsky, A.A., 1970. Geochemical reconstruction of the primary composition of Precambrian metamorphosed volcano-sedimentary formations. Apatity, 116 pp.

Prokofyev, A.A., 1966. Evolution of the south of the Siberian craton in the Precambrian. Geology and Geophysics. Novosibirsk, Nauka, Siberian Branch, pp. 3–14.

Pyatnitsky, V.K., 1974. Basement relief and structure of the sedimentary cover of the Siberian platform. Geologiya i Geofizika, No. 9, pp. 89–98.

Rabkin, M.I., 1959. Geology and petrology of the Anabar crystalline shield. Trudy NIIGA, Vol. 87, 164 pp.

Rabkin, M.I. and Vishnevsky, A.N., 1968. The Anabar Shield. In: Precambrian geochronology of the Siberian craton and its folded margins. Leningrad, Nauka.

Reutov, L.M., 1981. Precambrian of Central Aldan. Novosibirsk, 184 pp.

Rozen, O.M. and Dimroth, E., 1982. Ancient metamorphosed greywackes at the base of the continental crust; studies of primary mineral composition — examples from Canada and the USSR. In: Sedimentary geology of Precambrian high-grade metamorphic complexes. Moscow, Nauka, pp. 155–179.

Rozen, O.M., 1981. Enderbitoids of the Anabar Shield and the grey gneiss problem. In: Ancient granitoids of the USSR: grey gneiss complexes. Leningrad, pp. 125–135.

Rozhkov, B.N., Moor, G.G. and Tkachenko, B.V., 1936. Materials on the geology and petrography of the Anabar massif. Trudy Ark. In-ta, Vol. 16. Leningrad.

Ruchnik, G.V., Bogovin, V.D. and Donets, A.I., 1977. Lead-zinc mineralization in Vendian carbonate assemblages of SE Yakutia (the Sardan ore region). Geologiya Rudnykh Mestorozhdeniy, No. 4, pp. 3–20.

Rudnik, V.A., Belyayev, G.M. and Terentyev, V.M., 1970. Patterns in the formation of quartz-feldspar metasomatic zones in regional faults. In: Problems of metasomatism. Moscow, Nedra, pp. 261–274.

Rudnik, V.A., Sobotovich, E.V. and Terentyev, V.M., 1969. On an Archaean age for rocks of the Aldan complex. Doklady Akad. Nauk SSSR, Vol. 188, No. 4, pp. 897–900.

Rundqvist, D.V. (Editor), 1981. Ore content and geological formations in crustal structures. Leningrad, Nedra, 420 pp.

Rundqvist, D.V., Kushev, V.G., Popov, V.Ye. and Sinitsyn, A.V., 1989. Types of structural-metallogenic zones of Archaean to Early Proterozoic age. In: Tectonics of the basement to ancient cratons.

Salop, L.I., 1964. Geology of the Baikal mountains. Vol. 1, 515 pp.

Salop, L.I., 1967. Geology of the Baikal mountains. Vol. 2, 700 pp.

Salop, L.I., 1973. General stratigraphic scale of the Precambrian. Leningrad, Nedra, 309 pp.

Salop, L.I., 1982. Geological evolution of the Earth during the Precambrian. Leningrad, Nedra, 343 pp. [English translation 1983, Berlin and New York, Springer Verlag, 459 pp.]

Salop, L.I., 1983. Tectonic cycles in the Precambrian. Sov. Geologiya, No. 3, pp. 37–46.

Savinsky, K.A., 1972. Deep structure of the Siberian craton from geophysical data. Moscow, Nedra, 167 pp.

Semeykin, I.N., 1972. Lithofacies characteristics of Proterozoic terrigenous sediments in the south of the Urik-Iy graben. In: Geology and gold content of Riphean and Vendian conglomerates in the southern margins of the Irkutsk amphitheatre. Irkutsk, pp. 26–45.

Semikhatov, M.A., 1974. Proterozoic stratigraphy and geochronology. Moscow, Nauka, 302 pp.

Semikhatov, M.A., Aksyonov, Ye.M. and Bekker, Yu.R., 1979. Subdivision and correlation of the Riphean of the USSR. In: Stratigraphy of the Upper Proterozoic of the USSR (Riphean and Vendian). Leningrad, Nauka, pp. 6–42.

Semikhatov, M.A. and Serebryakov, S.N., 1983. The Siberian hypostratotype for the Riphean. Moscow, Nauka, 223 pp. (Trudy GIN AN SSSR, Vol. 367).

Sezko, A.I., 1972. New data on the relationships between Precambrian metamorphic complexes of the Biryusa microcraton, Eastern Sayan. Scientific information "Geology of Eastern Siberia". Inst. of the Earth's crust. Irkutsk, pp. 11–13.

Shatsky, N.S. (Editor), 1956. Tectonic map of the USSR and adjacent countries, 1:5,000,000 scale. Gosgeoltekhizdat.

Shcheglov, A.D., 1980. Fundamentals of metallogenic analysis. Moscow, Nedra, 426 pp.

Shirobokov, I.M. and Sezko, A.I., 1979. Major features of the Precambrian stratigraphy of Eastern Sayan. In: Major features of the geology of Eastern Sayan. Irkutsk, East Siberian Book Publishers, pp. 8–36.

Shishkin, B.B., 1979. Upper boundary of the Precambrian in the north of the Siberian craton. In: Upper Proterozoic stratigraphy of the USSR (Riphean and Vendian). Leningrad, Nauka, pp. 152–155.

Shpunt, B.R., Shapovalova, I.G. and Shamshina, E.A., 1982. Late Precambrian of the north of the Siberian craton. Novosibirsk, Nauka, 226 pp.

Shpunt, B.R., Sochneva, E.G., 1981. Early Riphean carbonatites of the Anabar massif. Doklady Akad. Nauk SSSR, Vol. 289, No. 4, pp. 946–951.

Shukolyukov, Yu.A. and Bibikova, Ye.V. (Editors), 1986. Methods of isotope geology and the geochronological scale. Moscow, Nauka, 226 pp.

Shuldiner, V.I., Panchenko, I.V. and Shuldiner, I.S., 1983. Petrology of metamorphic complexes in the basin of the river Nyukzha. In: Precambrian metamorphism in the region of the Baikal-Amur Mainline railway. Leningrad, pp. 127–139.

Sidorenko, A.V. (Editor), 1982. Precambrian ore-bearing structures. Moscow, Nauka, 202 pp.

Sizykh, A.I., 1978. Structure of the Biryusa metamorphic belt and stages in its formation. In: Dynamics of the Earth's crust in Eastern Siberia. Novosibirsk, Nauka, Siberian Branch, pp. 102–112.

Sizykh, A.I. and Shafeyev, A.A., 1976. On the relationship between Precambrian complexes of the Biryusa microcraton and the Tumanshet graben. Geology i Geofizika, No. 6, pp. 16–24.

Smelov, A.P., 1986. Metamorphism of rocks of the Tungurcha and Tasmiyeli Groups and their age equivalents. In: Early Precambrian of Southern Yakutia. Moscow, Nauka, pp. 219–235, 250–255.

Smirnov, Yu.M., Glebovitsky, V.A. and Turchenko, S.I., 1975. Genetic classfication of metamorphic mineral deposits. Sov. Geologiya, No. 2, pp. 52–66.

Sokolov, Yu.M. and Bilibina, T.V. (Editors), 1981. Geology of Precambrian mineral deposits. Leningrad, Nauka, 329 pp.

Sokolov, Yu.M. and Bilibina, T.V. (Editors), 1984. Metallogeny of the early Precambrian of the USSR. Leningrad, Nauka, 247 pp.

Stepanov, L.L., 1974. Radiogenic age of polymetamorphic rocks in the Anabar Shield. In: Early Precambrian formations of the central part of the Arctic and associated mineral deposits. Leningrad, pp. 76–83.

Sudovikov, N.G., Glebovitsky, V.A., Drugova, G.M., Krylova, M.D., Neyelov, A.N. and Sedova, I.S., 1965. Geology and petrology of the southern margin of the Aldan Shield. Leningrad, Nauka, 288 pp.

Sukhanov, M.K., Bogdanova, N.G., Sumin, L.V. and Rachkov, V.S., 1984. First results of thermochron radiometric dating of ancient anorthosites. Doklady Akad. Nauk SSSR, Vol. 277, No. 3, pp. 684–688.

Sukhanov, M.K. and Rachkov, V.S., 1984. Apatite-bearing gabbro-norites of the Anabar Shield. Izvestiya Akad. Nauk SSSR, Ser. Geol., No. 12, pp. 115–118.

Sukhanov, M.K., Rachkov, V.S. and Sonyushkin, V.Ye., 1983. Anorthosites of the Anabar Shield. Izvestiya Akad. Nauk SSSR, Ser. Geol., No. 6, pp. 29–42.

Suloyev, A.I., 1950. Magmatism of Eastern Sayan. Sov. Geologiya, No. 6, pp. 16–24.

Tarasov, L.S., Gavrilov, Ye.Ya. and Lebedev, V.I., 1963. On the absolute age of the Precambrian of the Anabar Shield. Geokhimiya, No. 12, pp. 1145–1151.

Timofeyev, V.F. and Bogomolova, L.M., 1986a. The Tungurcha greenstone belt. In: Early Precambrian of southern Yakutia. Moscow, Nauka, pp. 195–202.

Timofeyev, V.F. and Bogomolova, L.M., 1986b. The Tungurcha greenstone belt and the Tasmiyeli orogenic basin. In: Early Precambrian of Southern Yakutia. Moscow, Nauka, pp. 236–244.

Tomson, I.N. (Editor), 1985. Internal structure of ore-bearing Precambrian faults. Moscow, Nauka, 167 pp.

Travin, L.V., 1977. Petrochemical and formational features of Archaean metasedimentary formations in the central part of the Aldan Shield. Litologiya i Poleznyye Iskopayemyye, No. 3, pp. 115–126.

Vaganov, pp.A., 1981. Neutron-activation studies of geochemical associations of rare elements. Moscow, Energoizdat Publishers, 112 pp.

Velikoslavinsky, S.D., 1976. Early Archaean patterns of basic volcanism in the central Aldan Shield. ZVMO (Proc. All-Union Min. Soc.), Pt. 105, pp. 48–58.

Velikoslavinsky, S.D. and Rudnik, V.A., 1983. Geochemistry of Precambrian volcanism of the Aldan Shield. Trudy VSEGEI, New Series, Vol. 323, pp. 62–83.

Vishnevsky, A.N., 1978. Metamorphic complexes of the Anabar crystalline shield. Trudy NIIGA, Vol. 184, 213 pp.

Volobuyev, M.I., Zykov, S.I. and Stupnikova, N.I., 1976. Geochronology of Precambrian formations of the Sayan-Yenisey region of Siberia. Moscow, Nauka, 1976, pp. 96–123.

Volobuyev, M.I., Zykov, S.I., Stupnikova, pp.I. and Vorobyov, I.V., 1980. Lead isotopic geochronology of Precambrian metamoprhic complexes in the SW margins of the Siberian

craton. In: Geochronology of Eastern Siberia and the Far East. Moscow, Nauka, pp. 14–30.
Wood, B.Y. and Banno, S., 1973. Garnet-orthopyroxene and orthopyroxene-clinopyroxene relationships in simple and complex systems. Contrib. Mineral. Petrol., Vol. 42, No. 2, pp. 109–124 [in English].
Yanshin, A.L. (Editor), 1964. Tectonic map of Eurasia, 1 : 5,000,000 scale. Moscow, GUGK Publ.
Yelizaryev, Yu.Z., 1962. Archaean stratigraphy of SW Baikal region. Trudy East Sib. Geol. Inst. SO AN SSSR, Vol. 5, pp. 147–151.
Yelyanov, A.A., Kastrykina, V.M. and Kastrykin, Yu.P., 1985. New data on the geological structure and metallogeny of the Dzheltulin suture. In: Precambrian trench-type structures in the Baikal-Amur region and their metallogeny. Novosibirsk, pp. 151–156.
Yeskin, A.S., Morozov, Yu.A. and Ez, V.V., 1979. Correlation of Precambrian internal processes in the Pre-Olkhon region (West Baikal region). In: Results of Int. Geophys. Projects. Geodynamic research. Correlation of internal processes in Precambrian metamorphic complexes. Moscow, Soviet Radio, pp. 62–80.
Zamarayev, S.M., 1967. Marginal structures of the southern part of the Siberian craton. Moscow, Nauka, 247 pp.
Zverev, S.M. and Kosminskaya, I.I., 1980. Seismic models of the lithosphere in the major geological structures of the USSR. Moscow, Nauka.

Part III. Precambrian in younger fold belts

Younger fold belts of various ages are found over a significant part of the USSR. The Riphean and Palaeozoic belts which separate the East European and Siberian cratons and deflect the Siberian craton from the south, are considered to be part of the Urals-Mongolia belt (Fig. III-1) (Milanovsky, 1983; Muratov, 1965). Structures in the East European craton are cut in the south by the Mediterranean belt, which stretches in an approximately east–west direction through the East Carpathians, the Crimean peninsula, the Caucasus and the margins of the Pamirs. East of the Siberian craton lie the folded terrains of the Pacific belt, which occupy the north-east USSR, the Maritime province (Primorye) and the Lower Amur province.

The Precambrian rocks in these belts are less well studied in comparison to the Precambrian of ancient cratons. This is due to their unknown geological status, since they have been affected by several episodes of tectono-magmatic activity during the late Precambrian and Phanerozoic. A whole host of problems relates to dating high-grade metamorphic assemblages lacking diagnostic fossils. Using isotopic age determination methods is complicated by the fact that radiogenic rejuvenation has taken place. Nevertheless, at present, further data, including isotopic, are being accumulated on Precambrian formations in younger mobile belts, thus allowing us to attempt an identification of the major features of their evolution compared to terrains in the basement of ancient cratons.

The main directions in which research should be carried out on problems of the Precambrian in younger fold belts were defined by Kratz et al. (1982) as a series of questions:

(1) Does the Precambrian actually exist in these younger fold belts?

(2) What is the geological status of the Precambrian in younger mobile belts?

(3) What was the Precambrian crust like in Phanerozoic fold belts?

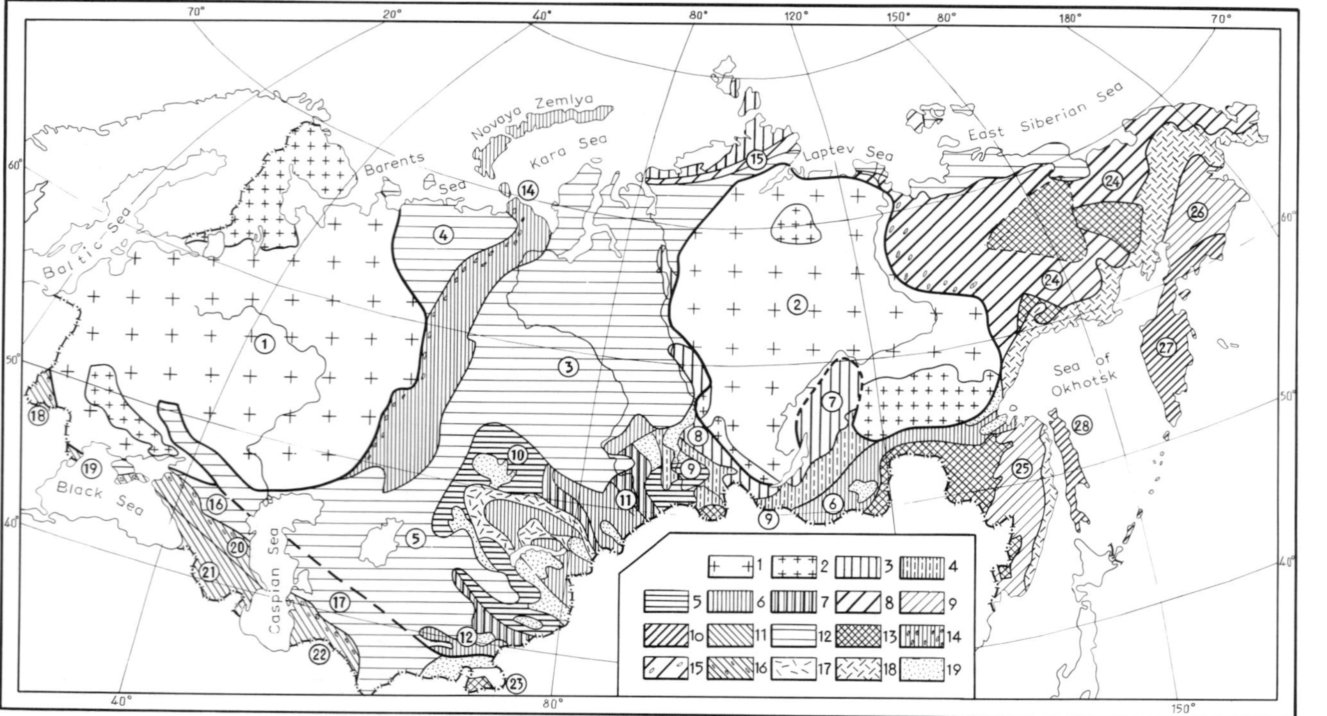

Fig. III-1. Sketch map showing younger fold belts in the USSR (Tseisler et al., 1984). Cratons: *1* = platforms, *2* = shields. Fold belts: *3* = Riphean, *4* = Salair, *5* = Caledonian, *6* = Hercynian (Variscan), *7* = Caledonian-Hercynian, *8* = Mesozoic (Cimmerian), *9* = Late Mesozoic (Laramide), *10* = Pacific (Cenozoic), *11* = Alpine; *12* = young platforms; *13* = major pre-Riphean massifs. Marginal basins: *14* = Late Palaeozoic, *15* = Jurassic-Cretaceous, *16* = Neogene-Quaternary. Marginal volcanic belts: *17* = mid-late Palaeozoic, *18* = Cretaceous-Palaeogene; *19* = Major intermontane basins. Major structures (numbers in circles): *1* = E European craton; *2* = Siberian craton; *3–15*: Urals-Mongolia belt (young platform plates: *3* = W Siberian, *4* = Timan-Pechora, *5* = N Turan; fold belts and systems: *6* = Mongolia-Okhotsk, *7* = Baikal, *8* = Sayan-Yenisey, *9* = Kuznetsk–Upper Vitim, *10* = Kazakhstan–N Tien Shan, *11* = Kazakhstan-Altay, *12* = Tien Shan, *13* = Urals, *14* = Paykhoy–Novaya Zemlya, *15* = Taymry–Severnaya Zemlya). *16–23*: Mediterranean belt (young platform plates: *16* = Scythian, *17* = South Turan; fold belts and systems: *18* = E Carpathians, *19* = Crimean, *20* = Greater Caucasus, *21* = Lesser Caucasus, *22* = Kopetdag and Greater Balkhan, *23* = Pamirs); *24–28*: Pacific belt (fold belts and systems: *24* = Verkhoyan-Chukchi, *25* = Sikhote-Alin, *26* = Taygonos-Koryak, *27* = Olyutora-Kamchatka, *28* = Hokkaido-Sakhalin).

(4) Is there any dependence between the age and nature of Phanerozoic fold belts and the age and structure of Precambrian belts?

(5) Are the ways in which Precambrian and Phanerozoic mobile belts grew comparable?

(6) Do any global patterns exist in the spatial distribution of younger fold belts?

Research along the lines suggested by these questions will permit us to understand the reasons for separating the major structures in the Earth's continental crust — ancient cratons and younger fold belts.

1. URALS-MONGOLIA BELT

In Soviet territory, the Urals-Mongolia belt is situated between the ancient platforms of the East European and Siberian cratons (Fig. III-2). To the SW, it meets the Mediterranean belt along the Gissara-Mangyshlak fault. The south-eastern part of the belt embraces South Siberia, the Baikal region, China and Mongolia, separating the Siberian craton from the Tarim and Sinian cratons. To the south-east, the Palaeozoides of the Urals-Mongolia belt are cut by Mesozoid structures of the Pacific belt. From the beginning of the Mesozoic, the area of the Urals-Mongolia belt was converted into a young epi-Palaeozoic platform.

The Timan-Pechora fold belt[1]

The term Timan-Pechora belt is understood to mean the territory on the extreme north-eastern European USSR — the modern Timan ridge and the Pechora lowlands.

There are two divergent points of view concerning the tectonic structure of this region. According to one of these, which developed from Shatsky's ideas, first put forward in the 1930s, the entire region is occupied by an independent fold system which originated from a Late Proterozoic (Riphean) geosyncline.

The second point of view goes back to ideas put forward as long ago as the end of the last century, when it was thought that in the NE part of the Pechora lowlands there existed a rigid core, surrounded by the Urals-Paykhoy fold system (Fig. III-3). This point of view had its final expression in Stille's notion of two ancient cratons, Fennosarmatia and Barentsia, separated by the narrow orthogeosynclinal belt of the Timanides and sutured together as a result of Assyntian folding. Shatsky subsequently also came round to this idea, and considered the Timan region to be an

[1] This section was written by A.K.Zapolnov.

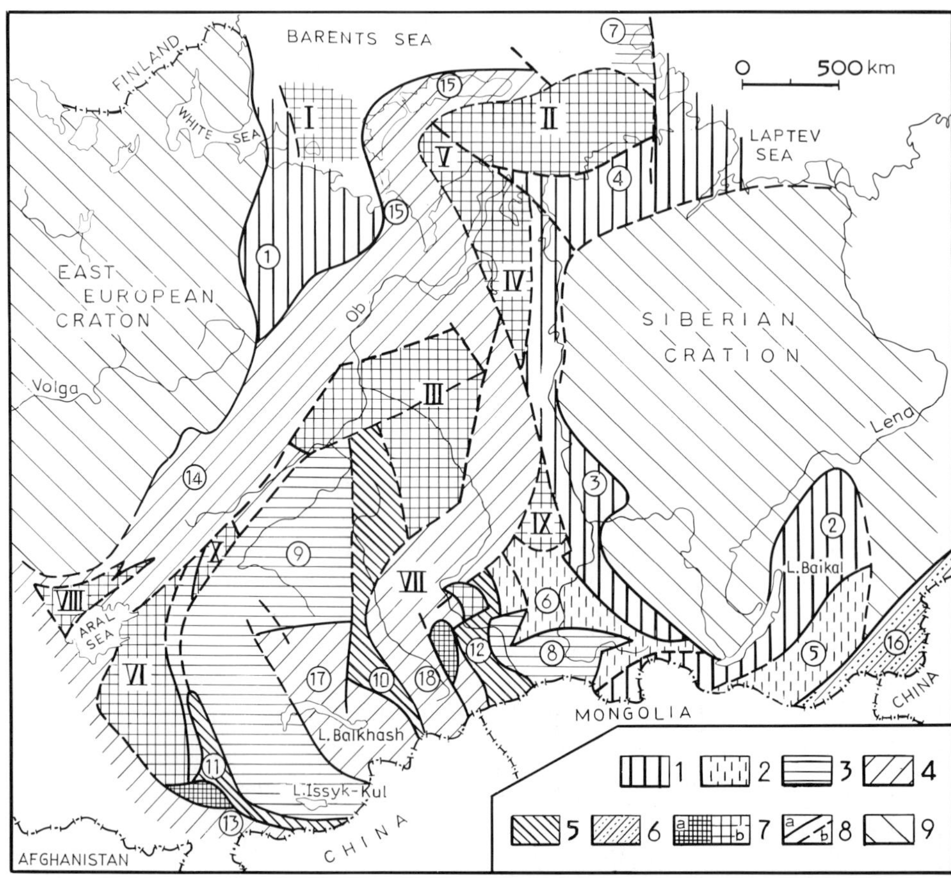

Fig. III-2. Sketch map showing major structures in the Urals-Mongolia belt (Tseisler et al., 1984; tectonic map from Anon., 1974). Fold belts: *1* = Riphean, *2* = Salair, *3* = Caledonian, *4* = Hercynian, *5* = Caledonian-Hercynian, *6* = Hercynian reworked in Mesozoic; *7* = major massifs with pre-Riphean basement (*a*: at surface, *b*: beneath Mesozoic-Cenozoic cover); *8* = boundary faults (*a*: proven, *b*: conjectured); *9* = cratons. Fold belts and systems (numbers in circles): *1–4*: Ripheides (*1* = Timan-Pechora province, *2* = Baikal, *3* = Sayan-Yenisey, *4* = Taymyr); *5, 6*: Salairides (*5* = Dzhidin–Upper Vitim, *6* = Kuznetsk-Tuva); *7–9*: Caledonides (*7* = Severnaya Zemlya, *8* = Altay–West Sayan, *9* = Kazakhstan–North Tien Shan); *10–12*: Caledonian-Hercynian (*10* = Chingiz-Tarbagatay and Salym, *11* = Central Tien Shan, *12* = Altay-Salair); *13–18*: Hercynian (*13* = S Tien Shan, *14* = Urals, *15* = Paykhoy–Novaya Zemlya, *16* = Mongolia-Okhotsk, *17* = Dzhungara-Balkhash, *18* = Ob-Zaysan). Massifs with pre-Riphean basement: *I* = Barents, *II* = Kara, *III* = Monsy, *IV* = Nadoyakh, *V* = Yamal, *VI* = Syr Darya, *VII* = Barnaul, *VIII* = Ustyurt, *IX* = Upper Kas, *X* = Turgay.

aulacogen, along which Barentsia split away from the rest of the East European craton (Shatsky and Bogdanov, 1961).

At the present time it is difficult to state a preference for one or other view with justification, since present-day exposures of Precambrian rocks

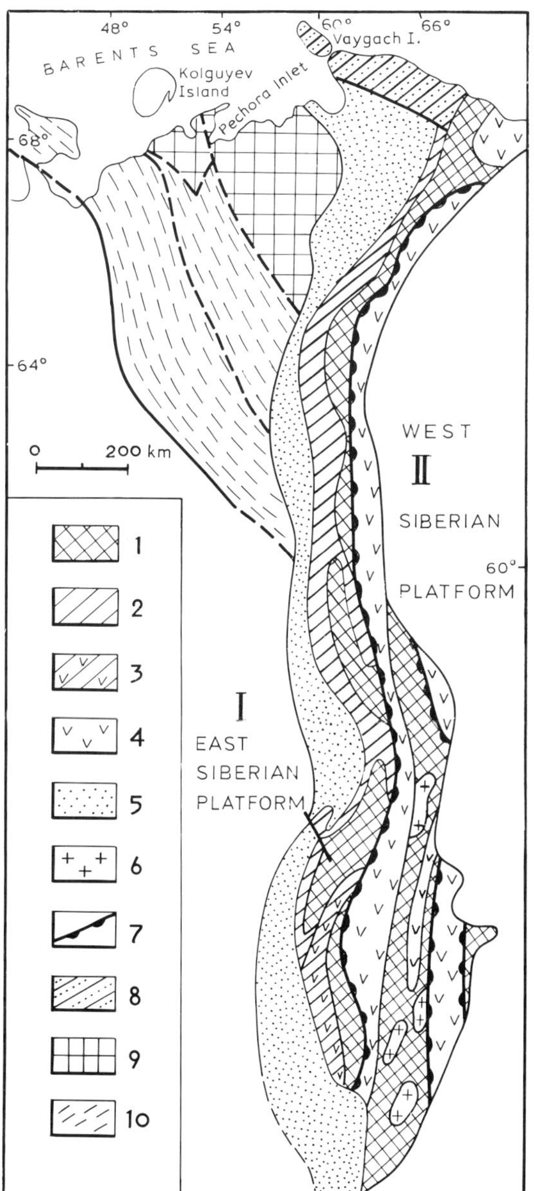

Fig. III-3. Tectonic sketch map of the Urals and the Timan-Pechora platform (Tseisler et al., 1984; tectonic map from Anon., 1974). Urals system: *1* = basal Uralide assemblage, *2* = miogeosynclinal formations of external zone, *3* = schist formation of external zone, *4* = eugeosynclinal formations of internal zone, *5* = orogenic complex, *6* = Hercynian granitoids, *7* = ultramafic belts; *8* = Palaeozoides of Paykhoy–Novaya Zemlya system; *9* = hypothetical pre-Riphean massifs; *10* = Riphean fold complex. *I* = East European craton, *II* = West Siberian craton.

are found at the surface only in the west of the province, in two disparate regions: Central and Southern Timan and the north end of Timan, and on Kanin Peninsula.

The Central and Southern Timan Region. Isolated exposures and insufficient biostratigraphic and radiometric studies on Riphean sediments of the region currently allow various interpretations to be made concerning the status of the groups and formations identified in the reference section.

In the most complete Riphean succession, at Chetlassky Kamen in Central Timan, the Middle Riphean Chetlass Group has been defined at the base of the section. Rocks of this group consist of alternating grey and greenish-grey quartzose sandstone and shale. In the upper part of the section it is possible to observe a 0.8–1.0 km rhythmic alternation of silty-muddy shale (occasionally slate), mica-clay shale, siltstone and quarzitic sandstone with grit lenses. The Chetlass Group has been estimated to be 2.5 to 3.0 km thick.

This group correlates to the east with rocks of the Kisloruchey and Vym Groups, which have an overall flyschoidal character. Individual rhythms consist of dark coloured arkosic sandstone, siltstone and muddy shale. Psammitic material is practically absent from rhythms at the base of the section, but gradually increases in progressively higher horizons. Carbonaceous material is typically absent from the upper part of the section. Numerous sills and cross-cutting dolerite bodies are present among the rocks of the formation. The total thickness of both groups exceeds 4 km.

Clastic assemblages of the Anyuga Formation overlie various horizons of the Chetlass Group with angular unconformity. These are grey, yellowish or greenish arkosic and quartzose sandstone and grit, interbedded with siltstone and muddy shale. Analogous sediments have also been identified in other regions in Central and Southern Timan (the Dzhezhim and other formations). These formations are a few hundreds of metres thick.

Carbonate-shale and carbonate units occupy a still higher stratigraphic position; they are referred to by different names, according to the stratigraphic scheme employed. The carbonate-shale unit consists of finely laminated mottled dolomite with thin marl, shale and calcareous sandstone horizons, up to 400–500 m in total thickness. A few horizons in this unit contain stromatolitic structures, mostly endemic forms, but they are nevertheless comparable with the Upper Riphean of the Bashkirian Urals (Chernaya et al., 1979). The carbonate unit (the Pavyuga Formation), which fills narrow trench-like basins, is composed predominantly of massive stromatolitic dolomite with a visible thickness of up to 700 m. The rich stromatolite and oncolite assemblage permits a definite correlation between the Pavyuga Formation and the Minyara Formation in the Upper Riphean of the Bashkirian Urals (Raaben, 1975).

Post-sedimentary changes in all the above-named Riphean assemblages usually do not exceed the stage of katagenesis and only in a few places in eastern Timan do they reach the lower greenschist facies.

Intrusive igneous activity in this region is of very limited extent. In addition to the dolerite dykes referred to above, there are also picrite, lamprophyre and carbonatite dykes. Granitoid bodies are represented by minor granite and aplite stocks, also syenites and diorites. These bodies are all post-folding. Polymetallic ore mineralization is genetically related to granitoid magmatism.

North Timan and the Kanin Peninsula. Precambrian sediments are exposed in two belts: the high ground around the Ludovaty promontories in the south-west, and the Kanin Kamen ridge in the north-east. A direct structural continuation of the Kanin Kamen ridge is the coastal region of North Timan.

Pre-Riphean rocks crop out in the Kanin Kamen ridge. They include the metamorphic Mikulkin and Tarkhanov Groups.

The Mikulkin Group consists of almandine-amphibole-biotite schist, amphibolite and paragneiss. Metamorphism reaches mid almandine-amphibolite facies (higher than the staurolite zone). The group has been estimated to be 1.5 km thick (Getsen, 1970).

The Tarkhanov Group consists of black schist with fine-grained quartzite members at the base of the section. The schists are predominantly almandine-quartz-biotite and quartz-plagioclase-amphibole in composition. The metamorphic grade does not usually exceed epidote-amphibolite facies in the kyanite-sillimanite facies series. The group has been estimated to be 3 km thick.

In North Timan, the Tarkhanov Group is correlated with the Rumyanichny Formation quartz-mica and quartz-feldspar-biotite schists.

Dark grey and greenish quartz-biotite and quartz-biotite-muscovite schists with thin quartzose sandstone beds are exposed in the watershed area of the Kanin Kamen ridge. This 1.5–2 km thick rock unit, metamorphosed under high-temperature greenschist facies conditions is considered by some workers (Dedeyev and Keller, 1986; Getsen, 1975) to be the upper unit of the Tarkhanov Group to emphasize the zonal nature of metamorphism in the outcrop area of the Mikulka and Tarkhanov Groups.

Sedimentary rocks of the Tabuyev Group and the Cape Ludovaty Formation on the Kanin Peninsula are referred to the Riphean. The succession of the Tabuyev Group is a major 3.5–3.8 km thick transgressive cycle, beginning with a quartzose sandstone unit and concluding with a carbonate-shale unit. The Cape Ludovaty Formation consists of massive dolomite, 0.7 km thick. The dolomite contains numerous stromatolites and oncolites of the Upper Riphean Minyar assemblage (determined by M.Ye. Raaben). Drilling records show that the dolomites in this formation are covered by a thick unit of red and green argillites, the Shoyna Formation, the age of which is probably Vendian (Getsen, 1970). K–Ar age determinations on sericite from analogous sediments, found in boreholes in North Timan, give a value of 565 Ma.

Igneous rocks in Riphean sediments of this region take the form of mafic and ultramafic dykes and sills, with associated sulphide-nickel ore shows. Sulphide-polymetallic shows are related to suites of granites and occur along highly fractured zones around the edges of granite and syenite intrusions.

Vendian sediments in the Timan-Pechora province have been identified only provisionally. Thin clastic assemblages of the Seduyakha Formation and volcanogenic-fragmentary rocks of the Sandivey Formation are usually taken to be Vendian. These rocks have been drilled in the last 15 years by deep boreholes in the central and southern region of the Pechora lowlands. The Sandivey Formation contains polymict sandstone and tuffo-sandstone, shale, albitophyre and trachyte. Sandivey sediments cover most of the province in the form of a blanket. From geophysical data, we estimate that the formation is 1.5–2.0 km thick. The base has nowhere been drilled.

The Taymyr fold belt

The Taymyr fold belt includes Riphean structures with an approximately east–west strike, which gradually swings round to north–east due to the deflection of the Kara massif (Fig. III-4). A horst block in the southern part of Cape Chelyuskin, composed of Archaean rocks, forms the Chelyuskin massif or microcraton.

Early Precambrian crystalline rocks are described as belonging to the Kara Group, consisting of three formations: Lower Kara, Upper Kara and Faddeyev (Pogrebitsky, 1972). The Lower Kara Formation consists of almandine-biotite gneiss, interleaved with amphibole plagiogneiss with garnet and biotite. The lower part has amphibolite and hornblende-pyroxene gneiss, while the upper part has staurolite, cordierite and graphitic gneiss. The formation crops out mainly around the shores of the Kara Sea.

The Upper Kara Formation consists of almandine-biotite plagiogneiss with subordinate hornblende gneiss and garnet schist. This formation also crops out around the Kara Sea and has an estimated thickness of 2500 m.

The Faddeyev Formation crops out around the periphery of the Kara massif and on elevated basement blocks around the shores of Taymyr Bay and the Gulf of Faddey. The lower boundary of the formation is provisionally drawn on the basis of the first appearance of a garnet-biotite-quartz schist member among garnet-biotite plagiogneiss; the schists gradually predominate upwards in the gneiss sequence. The upper part of the sequence has conglomerate with pebbles consisting of microcline gneiss with almandine, and thin marble skarn. The formation attains a thickness of 1500 m. The

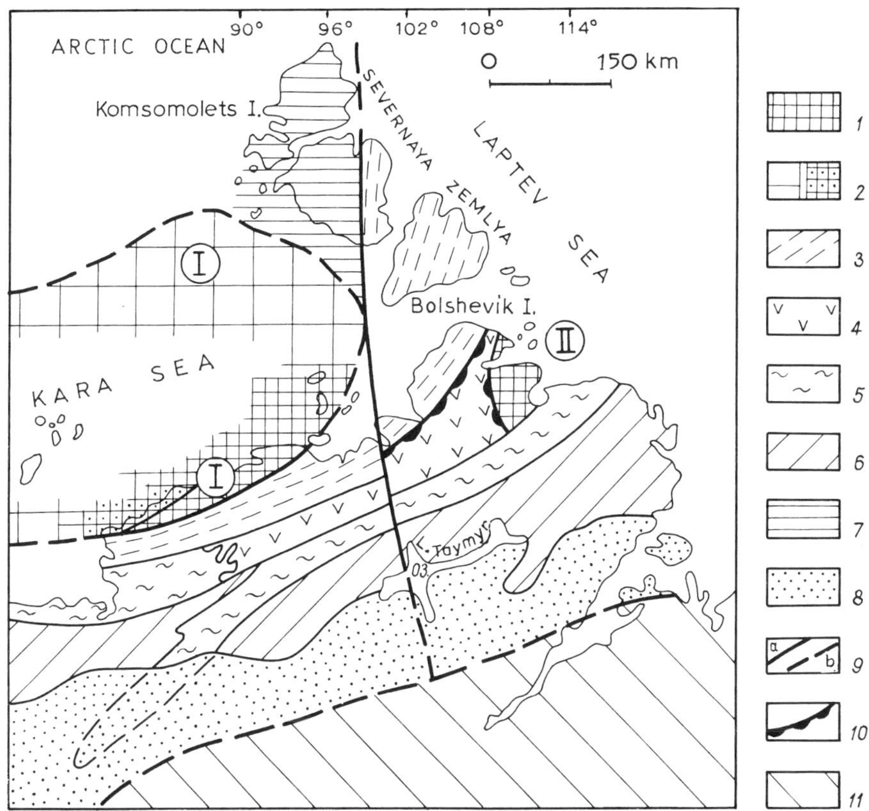

Fig. III-4. Tectonic sketch map of Taymyr and Severnaya Zemlya (Tseisler et al., 1984; tectonic map from Anon., 1974). Pre-Riphean massifs (*I* = Kara, *II* = Chelyuskin): *1* = Archaean, *2* = sedimentary cover (*a*: at surface, *b*: beneath Kara Sea); Baikalide fold system of N Taymyr: *3* = miogeosynclinal subzone, *4* = eugeosynclinal subzone, *5* = cover (top Riphean, Vendian, Lower-Middle Palaeozoic); *6* = S Taymyr zone (Upper Palaeozoic and Triassic, folded by early Mesozoic structures); *7* = Severnaya Zemlya Caledonides; *8* = Mesozoic and Cenozoic platform cover, Yenisey Estuary and Khatanga basins; *9* = faults (*a*: proven, *b*: hypothetical); *10* = ophiolite belts; *11* = Siberian craton.

Kara Group is assigned to the Archaean (Pogrebitsky, 1972), although previously these rocks were thought to be lower Proterozoic. Furthermore, there is the proposition that an independent early Precambrian terrain is missing and that the crystalline rocks of Taymyr formed during Late Proterozoic and later metamorphic episodes (Glukhovsky and Moralev, 1978). Nevertheless, it is more valid to refer these rocks to the pre-Riphean basement of the Kara massif, correlatable with assemblages in the basement of the Siberian craton.

Riphean sedimentary formations unconformably overlie the pre-Riphean structures of the craton. In this context, the major geosynclinal assemblage consists of Middle and basal Upper Riphean sediments; the molasse formations at the top of the Riphean form an orogenic assemblage; finally, the lower and middle Palaeozoic Vendian are represented by sub-platform sediments. Based on formational types, the major geosynclinal complex is divided into a northern miogeosynclinal zone and a southern eugeosynclinal zone. The miogeosynclinal zone borders on the Kara massif, and consists of Riphean terrigenous-carbonate assemblages up to 10 km thick. They are unconformably overlain by the molasse sequence in the upper part of the Riphean. Higher in the succession lies a less deformed Vendian to Cambrian flyschoidal sequence.

The eugeosynclinal zone consists of intensely deformed sediments belonging to the Middle and lower part of the Upper Riphean, comprising rocks of spilite-keratophyre and quartz-dolerite associations. Ophiolite belts are restricted to this zone. The rocks in the main geosynclinal complex are cut by granite intrusions dated at 840–820 Ma. In the south and south-west of this zone, Riphean sediments of the main geosynclinal complex are covered by molasse sequences belonging to the uppermost Riphean. The upper complex, comprising Vendian to Carboniferous subplatform formations, has a total thickness of 4–6 km.

The Precambrian of Southern Siberia

Southern Siberia contains the Sayan-Yenisey branch of the Ripheides, the Baikal Highlands and the Kuznetsk–Upper Vitim branch of the Salairides and Caledonides (Fig. III-2). Here, the boundary between the Siberian craton and the epicratonic Ripheides is somewhat provisional and in many cases is disputed. This is particularly the case for structures in the Baikal Highlands; recent work which assigns the constituent metamorphic assemblages to the Riphean or lower Proterozoic (Fedorovsky, 1985) has provoked a very lively debate. Figures III-1 and III-2 illustrate the traditional approach, in which this region is assigned to the Ripheides; Fig. III-5 on the other hand sets out the position of Fedorovsky (1985), whereby these formations are assigned to the pre-Riphean. The problem is so confused at the moment that we prefer not to examine it in this book.

The Sayan-Yenisey Ripheides

The Sayan-Yenisey province embraces Eastern Sayan and the Yenisey Ridge. In the north-east, the province borders on the Siberian craton, and in the south-west it meets the Salairides of the Kuznetsk-Tuva system (Fig. III-2). The most important assemblages in the province belong to the Lower-Middle and base of the Upper Riphean; among them are blocks and massifs consisting of Archaean and lower Proterozoic rocks (Fig. III-5).

The major structures in Eastern Sayan are the Protero-Sayan anticlinorium, with pre-Riphean rocks cropping out in the core, and the Oka and Ilchir synclinoria, which consist mainly of Lower to Middle Riphean

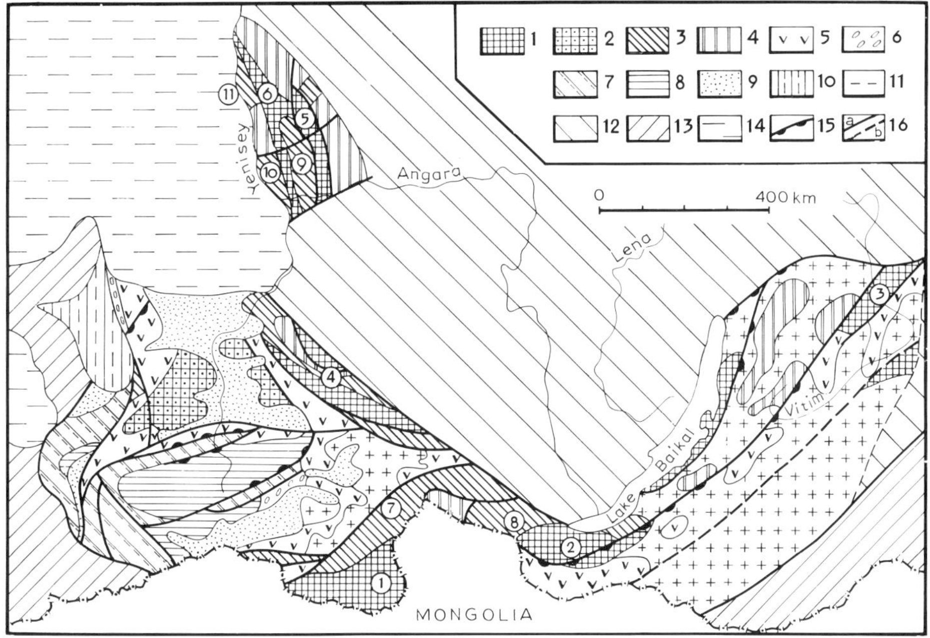

Fig. III-5. Position of Precambrian crystalline rocks in structural belts of Southern Siberia. *1* = Pre-Riphean crystalline rocks in central massifs and anticlinoria; *2* = as *1*, but beneath cover; *3* = Riphean structural storey; *4* = Upper Riphean-Vendian–Lower Cambrian structural storey. Salairides: *5* = Upper Riphean–Middle Cambrian, *6* = Upper Cambrian–Lower Ordovician. Altay–W Sayan Caledonides: *7* = Vendian–Middle Cambrian, *8* = Upper Cambrian–Silesian; *9* = Devian-Permian; *10* = Upper Palaeozoic orogenic complex; *11* = Palaeozoic granitoids; *12* = Siberian craton; *13* = Hercynides; *14* = cover of W Siberian craton; *15* = ultramafic belts; *16* = major faults (*a*: proven, *b*: hypothetical). Names of structures (numbers in circles): *1* = Sangilen massif; *2* = Khamar-Daba massif; *3* = S Muya block; *4* = Protero-Sayan anticlinorium; *5* = Central anticlinorium of Yenisey Ridge; *6* = Pre-Yenisey anticlinorium of Yenisey Ridge; *7* = Oka synclinorium; *8* = Ilchir synclinorium; *9* = Angara-Tey synclinorium; *10* = Yenisey synclinorium; *11* = Vorgov synclinorium.

rocks. There are massifs with a pre-Riphean basement: the Sangilen and Khamar-Daban (Fig. III-5), which are parts of a much larger massif, the Tuva-Mongolian (Ilyin, 1971).

The most complete Precambrian section is found in the Sangilen massif, which contains four terrains (Mitrofanov et al., 1981): the Erzin, Moren, Balyktygkhem and Naryn (Fig. III-6). The Naryn terrain is dated as Lower-Middle Riphean on fossil evidence. The other three terrains are pre-Riphean, Early Archaean, Late Archaean and Early Proterozoic, respectively.

The oldest, Erzin, terrain contains migmatite, gneiss, amphibolite, marble and quartzite, the distinguishing feature of which is the presence of relict medium-pressure granulite facies associations. A reconstruction of

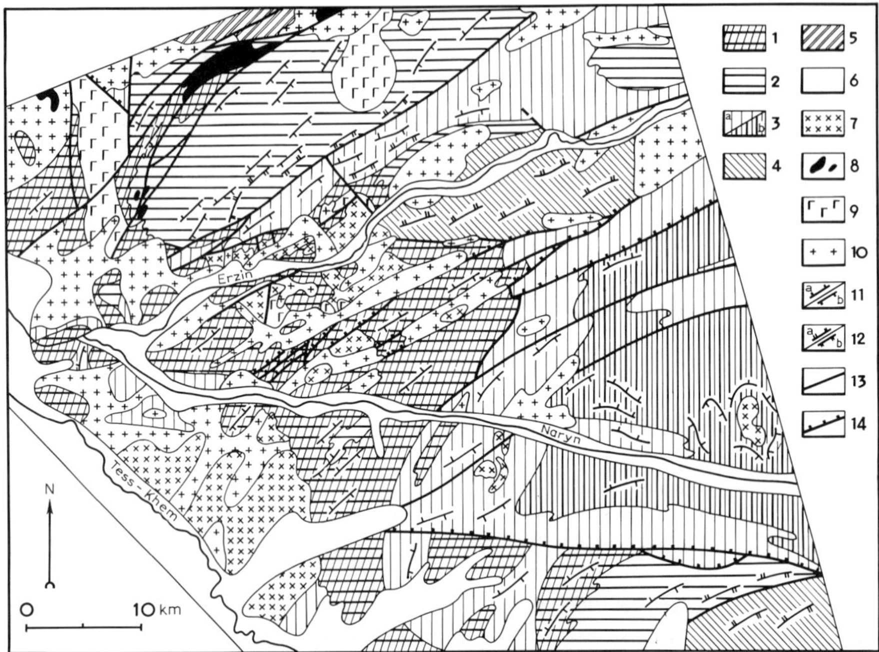

Fig. III-6. Structural map of the western part of the Sangilen massif (Kozakov, 1986). *1* = Erzin complex, Lower Archaean; *2* = Moren complex, Upper Archaean; *3* = Balyktygkhem complex, Lower Proterozoic (a = Balyktygkhem Formation, marbles, b = Chartiss Formation, marble, quartzite, gneiss); *4* = Naryn complex, schists, Lower-Middle Riphean; *5* = Lower Cambrian mafic volcanics; *6* = Cenozoic sediments; *7* = Proterozoic granitoids; *8* = Aktrovrak ultramafics (M. Camb.); *9* = Tannuol complex, diorites and gabbros (M. Camb.); *10* = Palaeozoic granitoids; *11* = orientation of pre-Riphean planar structures (*a* = gently dipping, *b* = vertical); *12* = orientation of Riphean planar structures (*a* = gently dipping, *b* = vertical); *13* = faults; *14* = thrusts.

the compositions of some of the rocks in the terrain shows the presence of analogues of volcanic and sedimentary rocks (Kozakov, 1986). Metavolcanics are divided into two groups — low-alkali tholeiite and a differentiated series from low-Ti medium-alkali tholeiite to andesite and dacite of moderate alkalinity on the K–Na slope. The sedimentary rocks are aluminous pelite (subarkose), calcareous siltstone and pelite with low alkalinity of the K–Na slope, also tholeiitic andesite-basalt tuff.

Supracrustal rocks in the Late Archaean Moren terrain, metamorphosed at high-pressure amphibolite facies, consist mainly of metavolcanics in association with ferruginous quartzite. The metavolcanics form a contrasting basalt-rhyolite series. The metasediments contain analogues of chert, oligomict quartzose sandstone and kaolinite-quartz sandstone. Moren rock associations correspond broadly to upper Archaean riftogene assemblages in the south of the Siberian craton (Olondo type of structure) which formed on a "buried" proto-metamorphic layer, reflecting the destruction of ancient sialic crust at the boundary between the Early and Late Archaean.

The composition of dykes which preceded the Early Proterozoic low-pressure amphibolite facies metamorphism corresponds to alkali K–Na andesite-basalts, which are characteristic of "mature" continental rifts. The basic supracrustal rock association of the lower Proterozoic Balyktygkhem complex is quartzite and marble. The quartzite corresponds to pure quartz sandstone, while the marbles are limestone, dolomitic limestone (0.31–5.84% MgO) and dolomite with kaolinite-hydromica impurities. Individual gneiss bands in the marbles correspond to Na–K arkose or oligomict sandstone.

A similar association of limestone and dolomite with thin quartzite and metasandstone is characteristic of continental basins, the allochthonous material of which consists of redeposition products of a deep weathering crust.

The greenschist-facies metamorphosed Naryn complex contains clastic terrigenous-carbonate rocks (limestone with thin quartz-mica schist) and terrigenous-volcanogenic rocks (sericite, sericite-chlorite and biotite schist, acid metavolcanics, micaceous quartzite, metasandstone and metaconglomerate). A stratigraphic, structural and metamorphic break with rocks of pre-Riphean complexes is seen everywhere at the base of the complex. The terrigenous-carbonate unit in the Naryn complex contains the fossil forms *Newlandia major* Walk., *N. concentrica* Walk., *Clatristroma cf tarnovskii* Posp., *Osagia undosa* Reitl., allowing it to be correlated with Middle Riphean units in Siberia (Gintsinger et al., 1979).

The Precambrian of the Tuva-Mongolian terrain ends with Upper Riphean to Vendian assemblages, which together with Lower Cambrian rocks constitute a particular complex of sub-platform (shelf) formations.

A scheme of geological processes in pre-Upper Riphean complexes in the western part of the Tuva-Mongolian massif is presented in Table III-1.

The identification of early Precambrian rocks in the geological evolution of the region is also confirmed by the first isotopic age determinations. The radiogenic age of zircons from granulite boulders in conglomerate at the base of the Moren complex in Sangilen (Mitrofanov et al., 1977), using the U–Th–Pb method, is 3100 ± 150–280 Ma (Volobuyev et al., 1976). The ages of zircon and monazite from migmatite, granite and pegmatite in Sangilen are 2100, 1970, 1960, 1880 and 1710 Ma by the same method, reflecting Early Proterozoic internal processes, while dates of 1010 ± 10, 830 ± 20, 740 ± 35 and 630 ± 20 Ma reflect Late Proterozoic activity (Volobuyev et al. 1976).

Structures in the Yenisey ridge, striking approximately north–south along the western boundary of the Siberian craton (Fig. III-5) are grouped into external and internal zones. The external zone contains the Central anticlinorium with early Precambrian rocks cropping out in the core

TABLE III-1

Scheme of geological processes in the Precambrian of the Tuva-Mongolia massif

Tectonometa-morphic cycles	Metamorphic facies, facies series	Deformation cycle: in upper structural stage (suprastructure)	Deformation cycle: in lower structural stage (infrastructure)	Tectonometamorphic reworking of the infrastructure	Igneous activity, migmatization
IV (Early-Middle Riphean)	Greenschist facies dominant, andalusite-sillimanite facies series	Complete	Reduced	Limited effects	Granites, granosyenites Early fold granodiorites
Erosion of rocks of early Precambrian structural stage, formation of Lower-Middle Riphean volcanogenic-terrigenous-carbonate assemblages (Naryn complex), rare mafic dykes					
III (Early Proterozoic)	Amphibolite facies, andalusite-sillimanite facies series	Complete	Complete	Regional reworking	Granites, pegmatites, granosyenites Migmatization in discrete zones Early fold microcline-plagioclase granites and tonalites
Formation of Lower Proterozoic quartz-carbonate and terrigenous-carbonate assemblages (Balyktygkhem complex), mafic dykes					
II (Late Archaean)	Amphibolite and granulite facies, kyanite-sillimanite facies series	Complete	Complete	Regional reworking	Granites and pegmatites Regional migmatization Early fold tonalites
Formation of Upper Archaean terrigenous-volcanogenic assemblages (Moren complex), mafic dykes					
I (Early Archaean)	Low-medium pressure granulite facies	Complete	?	?	Alaskites, charnockites Enderbite-charnockite migmatization Early fold diorites and gabbros
Formation of lower Archaean carbonate-terrigenous-volcanogenic assemblages (Erzin complex)					

and two synclinoria, composed of Riphean miogeosynclinal formations. Overlying these is a basin filled with Vendian and Cambrian rocks. In the internal zone, separated from the external zone by the Tatar fault, there are the Angara-Tey, Vorgov and Yenisey synclinoria, composed of Riphean eugeosynclinal facies, and the Pre-Yenisey anticlinorium, which has early Precambrian rocks cropping out at the surface.

The early Precambrian Tey Group (Votakh et al., 1978) consists of the Karpinsky Ridge and Penchenga Formations, which appear in anticlinoria in the northern and central parts of the Yenisey ridge (Fig. III-7).

The Karpinsky Ridge Formation (800–2000 m) consists of gneiss, migmatite, mica schist with kyanite, staurolite, garnet and sillimanite, amphibolite and quartzite. A characteristic feature is the presence of high-Al rocks with thin bands of quartzite and quartz-mica schist at the base of the section, indicating that the formation is derived from highly mature clay products with a kaolinite composition (Votakh et al., 1978). The Penchenga Formation (1700 m) overlies Karpinsky Ridge sediments with gradual transition and consists of banded graphitic marble with thin bands of schist, quartzite and amphibolite. The schists include varieties with kyanite, sillimanite and andalusite.

Some workers consider that the early Precambrian rocks of the Yenisey ridge can be divided into three stages: lower Archaean, upper Archaean and lower Proterozoic (Tseisler et al., 1984). The lower Archaean has granulite facies metamorphic rocks, while the upper Archaean has amphibolite facies rocks, cropping out in the Angara-Kan anticlinorium. The

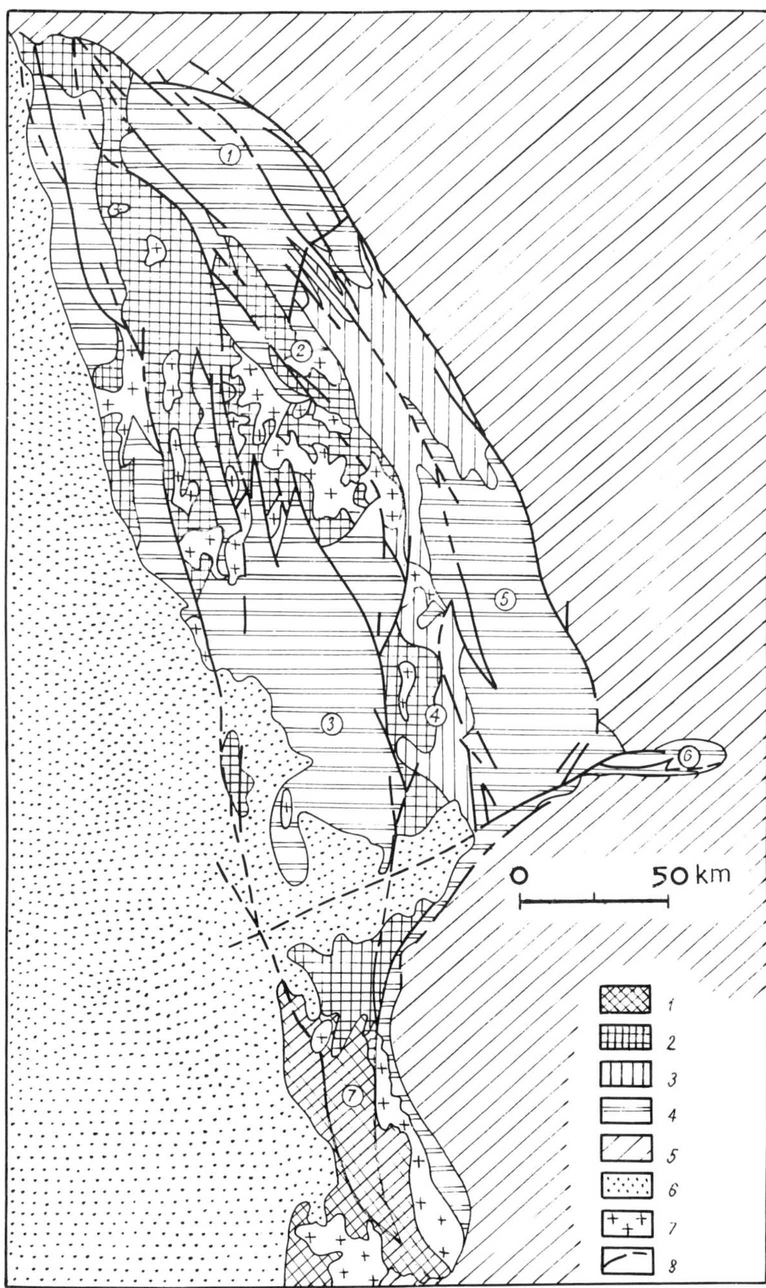

Fig. III-7. Tectonic sketch map of the Yenisey Ridge (Votakh et al., 1978). *1* = Kana Group (Archaean) in anticlinoria; *2* = Vesnin and Tey Groups (Archaean–Lover Proterozoic) in anticlinoria; *3*= Sukhopita Group (Lower-Middle Riphean) in highs; *4* = Upper Proterozoic in synclinoria; *5* = Palaeozoic; *6* = Mesozoic; *7* = granitoids; *8* = faults. Structural belts (numbers in circles): *1* = Korda-Lebyazhy synclinorium, *2* = Panimba anticlinorium, *3* = Bolshepit synclinorium, *4* = Tatar anticlinorium, *5* = Angara-Pit synclinorium, *6* = Irkineyev high, 7 = Angara-Kan anticlinorium.

Teya Group, particularly the Penchenga Formation, is considered to be lower Proterozoic, since we have an isotopic date for it in the interval 1850–1650 Ma.

The Riphean complex in the Yenisey ridge commences with the Sukhopit and Tungusik Groups. In the external zone the Sukhopit Group has slate, terrigenous flysch and carbonate formations. The Tungusik Group typically shows abrupt facies changes: in foredeeps there is a clay-carbonate unit, on internal elevations this is replaced by carbonates, and on the limbs of the Central anticlinorium it is a coarsely fragmental unit. The total thickness of the two groups in the synclinoria of the external zone is 9–10 km. In the internal zone of the Yenisey ridge, the Sukhopit and Tungusik Groups occur in synclinoria and contain spilite, dolerite, basic and intermediate amygdaloidal lava and tuff as members amongst carbonate rocks. The total thickness of the two groups increases to 12–15 km. All the units in this complex are deformed into upright isoclinal folds, which are most intense in the internal zone; the folds are more open in the external zone. Glauconite from rocks in the Sukhopit Group gives an age of 1140 Ma (Votakh et al., 1978).

The succession of the late Precambrian (orogenic) complex of the Yenisey ridge is divided into two units. The lower (the Oslyan Group in the south and the Chingasan Group in the north) comprises grey and red clastic terrigenous rocks. Hematite ores are found at the base of the Oslyan Group, which are considered to be a redeposited weathering crust. The upper unit (the Taseyev Group in the south and the Chapa Group in the north) contains more coarse-grained clastic material, cross-laminated redbeds and coarse flysch. These rocks are replaced eastwards (towards the platform) by more finely laminated sediments containing thin shallow-water carbonates with stromatolites. The succession has a maximum thickness of 6 km. Rocks in the lower unit have been dated at 850–700 Ma, and in the upper unit at 650–560 Ma.

Kuznetsk–Upper Vitim Salairides and Caledonides

This terrain occupies the central and south-eastern part of the South Siberian region. It is bounded on the west by the Kazakhstan-Altay Hercynides, on the NE by the Sayan-Yenisey Ripheides and in the east by the Mongolia-Okhotsk Hercynides. The terrain includes the Kuznetsk-Tuva and Dzhida–Upper Vitim systems (Fig. III-2), containing Middle and Upper Riphean, Vendian and Cambrian (Lower and Middle) sediments. The main Caledonian complex in the SW has Upper Cambrian, Ordovician and Lower Silurian units. The late Precambrian formations of the Caledonides in this system occur in anticlinoria (Biy-Katun, Chulyshman, Dzhebash and Kurtushuba), while the early Precambrian is known only in basement massifs.

The major massif in the Kuznetsk-Tuva system — the Shora-Batenev —

is fragmented into separate microcratons: Mrassk, Tomsk and Batenev. The lowest Precambrian is represented in the Tomsk microcraton as the Konzha and Tersa Formations (Gintsinger and Vinkman, 1978). The Konzha Formation (>4000 m) consists of granite-gneiss, amphibolite, garnet-biotite, garnet-biotite-hornblende and garnet-biotite-sillimanite gneiss with thin bands of marble, graphite-mica schist and quartzite. A K–Ar age on amphiboles from this formation gives a value in the interval 1470–1880 Ma. The Tersa Formation (>2000 m) consists mainly of marble with thin bands, sheets and lenses of quartz-graphite, hornblende, garnet-biotite-muscovite and garnet-biotite-hornblende gneiss. The age of amphiboles from this formation, based on K–Ar dating techniques, is 1740–1860 Ma (Gintsinger and Vinkman, 1978).

Higher levels in the Precambrian succession in this region contain Lower Riphean (Koltass and Tashelga Formations), Middle Riphean (Kabyrzin Formation), Upper Riphean (Kanym and West Siberian Formations) and Vendian (Belkin Formation) assemblages. Sediments provisionally referred to the Lower Riphean comprise amphibolized mafic volcanics, chloritic and siliceous schist, acid lava and pyroclastics and marble. A carbonate-type succession is more typical for the Middle Riphean, with subordinate phyllite, chert and siliceous schist members; volcanics crop out only to a limited extent and then mostly towards the base of the succession. The limestones in this formation contain numerous fossil phytoliths, including newlandiids. Upper Riphean sediments rest unconformably on Middle Riphean units. They are divided into two parts: the lower consists mainly of quartz porphyrites and volcanoclastic rocks with minor sandstone, siltstone and shale; the upper part consists basically of dolomite with limestone, chert and clay-silica shale. Sandstone, grit and conglomerate occur in places at the base of the formation. The upper formation may belong to the Vendian. The overlying Vendian Belkin Formation is conformable on the dolomite unit and has far more abundant limestone than dolomite, with various shales and carbonate-silica phosphorites also present.

The oldest rocks in the Dzhida–Upper Vitim system are most fully represented in the South-Muya microcraton (block) (Fig. III-5). Archaean formations are divided into three units, which are, from the oldest up: Kindin, Ileir and Lyunkuta. The Kindin assemblage has at the base amphibolite, two-pyroxene gneiss and biotite gneiss with thin calciphyre. The upper part has calciphyre with thin amphibolite, gneiss and plagiogneiss. Prograde metamorphism of the Kindin assemblage occurred at granulite facies conditions, with a later amphibolite facies event. The Ileir assemblage consists of two-mica and biotite gneiss, plagiogneiss and rare thin amphibolites. The Lyunkuta assemblage consists of finely alternating calciphyre, garnet gneiss and amphibolite. Regional metamorphism of the Ileir and Lyunkuta assemblages took place at amphibolite facies conditions. Lower Proterozoic assemblages in the South Muya block belong to the

Kedrov and Samodurov Formations which are mainly metasandstone, conglomerate, quartz-biotite schist and quartzose sandstone.

Kazakhstan–Tien Shan region

Structures represented in this region are the Kazakhstan-Altay and South Tien Shan Hercynide terrain and the Kazakhstan–North Tien Shan Caledonide terrain.

Kazakhstan-Altay Hercynide terrain

Structural belts in this terrain strike NW–SE from beneath the cover of the West Siberian platform into China and Mongolia. From east to west we have the separate Altay-Salair, Ob-Zaysan, Chingiz-Tarbagatay and Dzhungara-Balkhash systems (Fig. III-2).

Precambrian rocks in the Altay-Salair system crop out on the Terekta horst (Fig. III-8). The age of the assemblages making up the Terekta horst is a matter of some debate at present, although they most probably belong to the Riphean, as confirmed by a K–Ar isotopic age of 1140–750 Ma on amphiboles and phengites (Dook, 1982).

The oldest rocks in the horst, which make up the Uymona and Terekta Formations, are grouped together as the single Terekta glaucophane-greenschist complex, comprising mafic volcanics and tuff, sandstone, siltstone, grit, greywacke, argillaceous and limey shales, metamorphosed at glaucophane schist facies conditions. There is a metamorphic and structural break between Terekta rocks and the overlying assemblages of the Upper Riphean to Vendian Baratil Formation (1000 m), which consists of limestone, dolomite, mafic volcanics and chert. Higher in the section are Vendian to Lower Cambrian rocks, the Suchash Formation (3000 m), consisting of grit, sandstone, siltstone, greywacke, mafic volcanics and tuff. Rocks of both formations were metamorphosed at low-pressure greenschist facies conditions.

From an analysis of the petrochemical features of the rocks in the volcanogenic-terrigenous Terekta complex, we conclude that they are eugeosynclinal assemblages. The assemblages making up the Terekta complex are thought to lie on a pre-Riphean tonalite basement, to which are assigned the tonalitic rocks of the Turgunda massif and its surrounding gneisses and schists (Dook, 1982).

Precambrian crystalline complexes are also known in the axial zone of the Chingiz-Tarbagatay system in the cores of anclinoria. Here we have identified the Osakarov, Oshachanda and Shincharev Formations, consisting of amphibole, biotite-albite and biotite-muscovite-quartz schist, plagiogneiss, amphibolite and marble. These assemblages show evidence of polyphase metamorphism, the latest event being dated by K–Ar techniques on actinolite as 764–710 $\pm$ 70 Ma (Kisilev and Koralev, 1978).

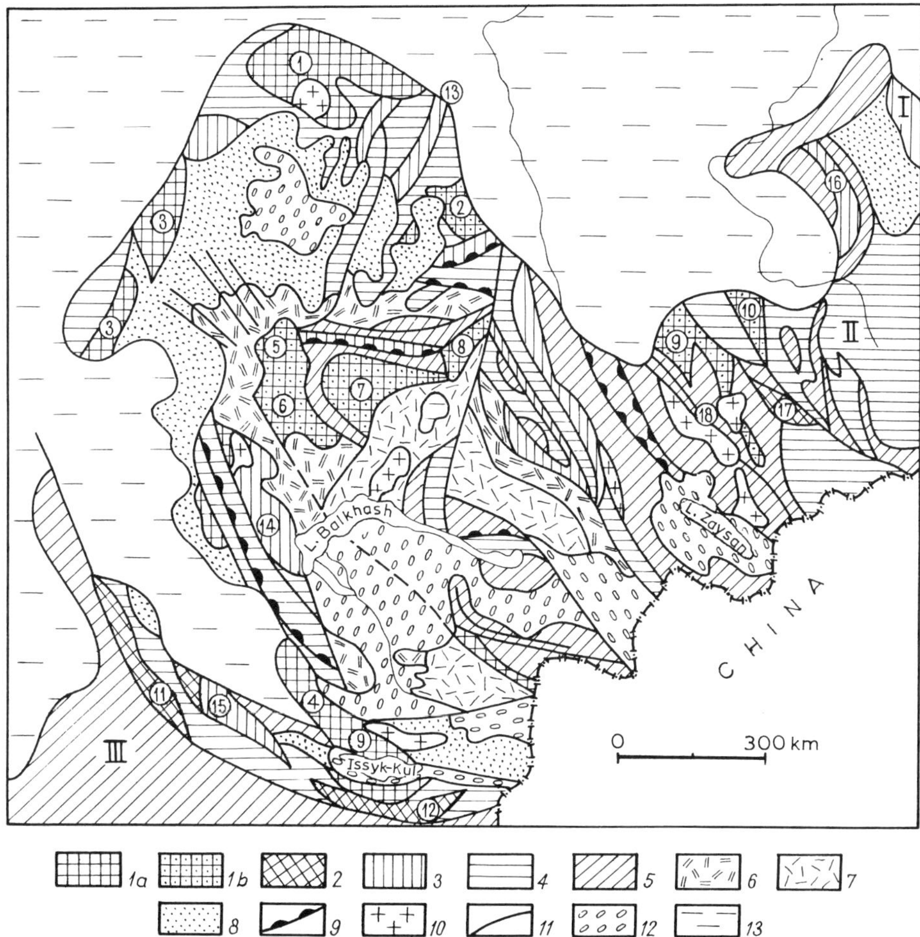

Fig. III-8. Tectonic sketch map of Kazakhstan, North Tien Shan and Altay (Tseisler et al., 1984). *1* = Central massifs (*a*: with pre-Riphean basement, *b*: with Palaeozoic cover); *2* = Riphean structural stage; *3* = Vendian-Cambrian (Salair) structural stage; *4* = Cambrian-Silesian (Caledonian) structural storey; *5* = Late Silesian–Early Carboniferous (Hercynian) structural stage; *6* = Devonian volcanic belt; *7* = Upper Palaeozoic volcanic belt; *8* = "transition complex" (Devonian-Permian); *9* = ophiolite belts; *10* = granitoids; *11* = major faults; Mesozoic-Cenozoic: *12* = internal basins, *13* = cover of epi-Palaeozoic platform. *I* = Kuznetsk-Tuva Salairides, *II* = Altay–W Sayan Caledonides, *III* = Tien Shan Hercynides. Numbers in circles: *1* = Kokchetau massif, *2* = Boshchekul massif, *3* = Ulutau massif, *4* = Muynkum massif, *5* = Atasuy microcraton, *6* = Aktau-Monti microcraton, *7* = Zhaman-Sarysuy microcraton, *8* = Karasor microcraton, *9* = Aley anticlinorium, *10* = Talitsky anticlinorium, *11* = Talass anticlinorium, *12* = Terskey anticlinorium, *13* = Yerementau anticlinorium, *14* = Buruntau anticlinorium, *15* = Makbal-Kirghiz anticlinorium, *16* = Central Salair anticlinorium, *17* = Terektin horst, *18* = Irtysh shear belt.

The Dzhungara-Balkhash system occupies south-eastern Kazakhstan (Fig. III-2). It is surrounded to the north, west and south by the Kazakhstan–North Tien Shan Caledonide terrain, while in the east it borders on the Chingiz-Tarbagatay system. The Atasu-Dzhungara (Balkhash) massif is situated in its western part, and is divided into a number of blocks (or microcratons): Atasu, Aktau-Mointa, Zhaman-Sarysuy and Karasorssa (Fig. III-8). The oldest formations, which are pre-Riphean as a rule, are overlain by a Palaeozoic sedimentary cover. In the Aktau-Mointa microcraton, a unit of porphyroids is unconformably overlain by Riphean quartzite and sericite-chlorite schist. In the SE part of the Chu-Iliy mountains, a pre-Riphean complex has been identified as part of the Zhingilda Formation, consisting of garnet-biotite-hornblende gneiss, amphibolite and garnet-biotite schist with thin bands of forsterite marble. These formations correlate with the Zerendin complex in the Kokchetau Caledonide terrain.

Kazakhstan–North Tien Shan Caledonide terrain

This terrain skirts the Dzhungara-Balkhash system on the NW, W and S (Fig. III-2), and is separated from it by a Devonian marginal volcanic belt (Fig. III-8). Its structural belts continue south-eastwards into China and in the north they are overlain by the sedimentary cover of the North Turan platform; in the south, they are separated by the Nikolayev faults from the Central and Southern Tien Shan Hercynides.

A particular feature of the tectonic structure of this terrain is the way in which the outlines of Caledonide structures are subordinate to the numerous Precambrian massifs, the best known of which are the Kokchetau, Boshchekul, Ulutau and Muyunkum (Figs. III-8, III-9). Elevated basement blocks have Archaean and lower Proterozoic rocks exposed at the surface, while down-faulted blocks have Riphean and lower Palaeozoic rocks. The terrain contains two zones, according to the assemblages in the main geosynclinal complex: an external miogeosynclinal zone in the west and an internal eugeosynclinal zone in the east, in which there are anticlinoria and synclinoria. Precambrian assemblages are also exposed in the cores of the anticlinoria.

A characteristic feature of the Precambrian in this province is the development of eclogite-gneiss terrains (Kushev and Vonogradov, 1978; Moskovchenko, 1982). These crop out within the Kokchetau and Ulutau massifs and the Makbal-Kirghizia and Aley anticlinoria (Fig. III-8). The Zerenda Group within the Kokchetau massif is identified as such a terrain, analogues of which have been found in other massifs. An extremely important fact is the discovery at the base of the Zerenda Group of relict granulite facies rocks, which were subsequently metamorphosed under lower-temperature conditions (amphibolite facies), but at higher pressures (kyanite-sillimanite series), with which is associated the formation of the

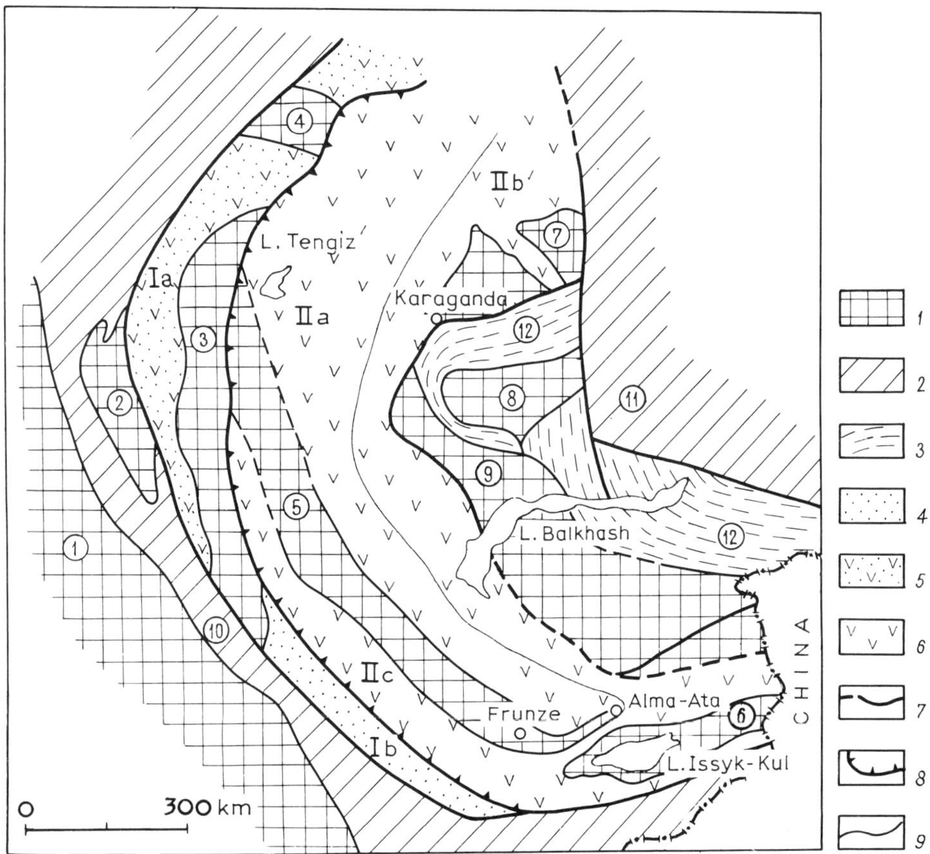

Fig. III-9. Structural domains in the Caledonides of Central Kazakhstan–North Tien Shan in the early Palaeozoic (Tseisler et al., 1984). *1* = Central massifs; *2* = Caledonian-Hercynian fold systems; *3* = Hercynian fold systems. Structural-formational zones: *4* = miogeosynclinal, *5* = mesogeosynclinal, *6* = eugeosynclinal. Boundaries: *7* = of fold systems, *8* = of external and internal zones of the Caledonides, *9* = of Caledonide subzones. Structural-formational zones: external: *Ia* = Baykonur-Ishim subzone, *Ib* = Karatau-Talass subzone; internal: *IIa* = Stepnyak-Betpakdal subzone, *IIb* = Yerementau-Chuliy subzone, *IIc* = Kirghiz-Terskey subzone. Structures (numbers in circles): central massifs: *1* = Syr-Darya, *2* = Turgay, *3* = Ulutay, *4* = Kokchetau, *5* = Muyun-Kuma, *6* = Issyk-Kul, *7* = Boshchekul, *8* = Zhaman-Sarysuy, *9* = Atasu-Dzhungara; fold systems: *10* = Central Tien Shan, *11* = Chingiz-Tarbagay, *12* = Dzhungara-Balkhash.

high-pressure eclogite-gneiss terrains proper (Abdulin et al., 1982). This suggests that the lower part of the Precambrian section in these structures contains the oldest granulite terrain, represented by a gneissose unit with thin quartzite and marble bands among which relict hypersthene gneiss has been found. This gneissose unit, the Berlyk Formation, can be assigned provisionally to the lower Archaean (Abdulin et al., 1982). Regional metamorphism of the overlying upper Archaean assemblage, represented in the main by mafic metavolcanics (the Zholdybay Formation) proceeded at high pressures. This is the first metamorphic episode to have affected rocks in the Zholdybay Formation, but for the Berlyk gneisses it is a later event. The succeeding Daulet Formation overlies the Berlyk and Zholdybay Formations and is represented by acid metavolcanics with marble horizons, which underwent prograde amphibolite facies metamorphism in the andalusite-sillimanite facies series. This metamorphism is superimposed on the underlying metamorphic rocks. Isotopic age determinations on these assemblages fall in the interval 1840–1700 Ma, i.e. they belong to the early Precambrian (Tseisler et al., 1984).

The basement to the central massifs in the Kazakhstan–North Tien Shan terrain had formed by the end of the Early Proterozoic, since Riphean sediments form the lower parts of the sedimentary cover succession as well as of mio- and eugeosynclinal foredeeps.

The most complete Riphean sections are found in Northern Tien Shan. Miogeosynclinal formations are developed in the Talass anticlinorium (Fig. III-8): limestone with stromatolites, terrigenous flysch, and carbonate-terrigenous (5–6 km thickness for the Lower to Middle Riphean, 3.5 km for the Upper). The Riphean succession in the Terskey anticlinorium is characterised by a eugeosynclinal assemblage type, and includes spilite, basalt and tuff, jasper, siliceous schist and lenses of organic limestone; the thickness is 3.5 km (Tseisler et al., 1984). The largest unconformity in Riphean sediments is seen at the boundary between the Middle and Upper Riphean. The Late Proterozoic in the Talass anticlinorium is capped by Vendian molasse redbeds.

The Middle Riphean in the central massifs is represented by dolomite and marble with rare andesite and dacite lenses; the thickness is less than 1000 m. Shallow-water marine and continental assemblages are characteristic of the Middle-Late Riphean boundary in the Central Kazakhstan central massifs (the Kokchetau Group and its analogues). The most typical is a continental quartzose sandstone formation, no more than 200–300 m thick. Upper Riphean successions are developed only to a limited extent and are represented by sandstone and acid volcanics (the Koktsiy Group).

The Southern Tien Shan Hercynide terrain

This terrain occupies the southern part of the Kazakhstan–Tien Shan region. It borders with the Mediterranean fold belt to the south along the Mangyshlak-Gissara fault and with the Central Tien Shan Caledonides to the north along the Karatau-Talas-Terskey fault. To the east it reaches into China and to the west it is hidden by the Turan platform sedimentary cover (Fig. III-1). The terrain is divided into two systems, Central and Southern Tien Shan (Fig. III-10).

The oldest rocks in Central Tien Shan are exposed in the Sarydzhas

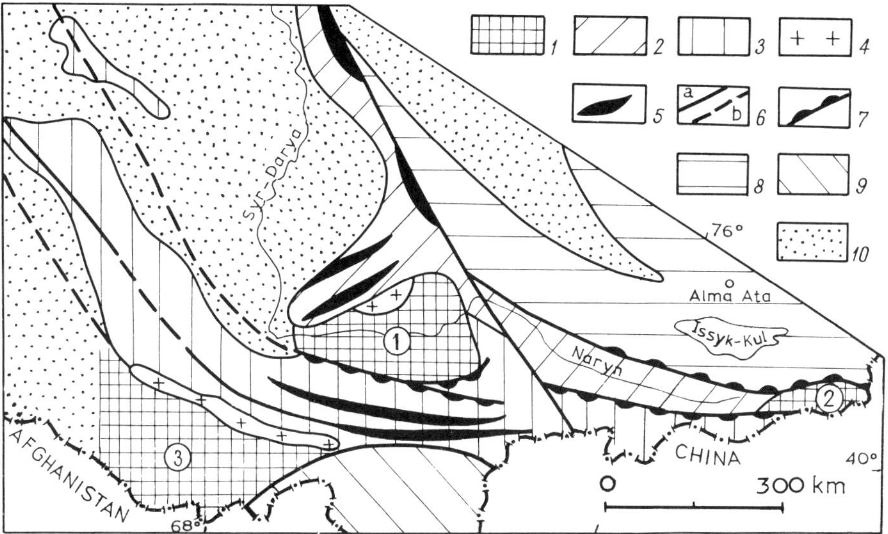

Fig. III-10. Tectonic sketch map of Hercynides in the Tien Shan province (Tseisler et al., 1984, tectonic map from Anon., 1974). *1* = Central massifs. Geosynclinal complex: *2* = Central Tien Shan system, *3* = Southern Tien Shan system; *4* = granitoids; *5* = anticlinorium axis; *6* = faults (*a*: proven, *b*: hypothetical); *7* = ultramafic belts; *8* = Caledonides of N Tien Shan; *9* = Pamir fold belt; *10* = N Turan platform cover. Massifs (numbers in circles): *1* = Fergana, *2* = Sarydzhas, *3* = Afghan-Tadzhik.

massif, where there are outcrops of biotite and garnet-biotite gneiss, amphibolite and marble, gneissose granite and migmatite which form the Kuylyu complex (Kisilev and Koralev, 1978). This can be compared with Archaean or lower Proterozoic complexes on the basis of lithology and metamorphic and partial melt phenomena (amphibolite facies, kyanite-sillimanite series) (Anon., 1977). The Kuylyu gneisses are overlain by the Greater Naryn Group above a sharp unconformity, the lower members of which are older than Middle Riphean, but overall the entire group is referred to Middle-Upper Riphean (Kisilev and Koralev, 1978; Tseisler et al., 1984). It comprises porphyritoids, siliceous-terrigenous rocks and marble and attains 2500 m in total thickness. Above comes a Vendian sand and shale unit, 3000 m thick, which has a coarse clastic base. Beyond the margins of the massif, Vendian units form part of the major geosynclinal complex.

The Southern Tien Shan fold system is situated between the Fergana and Afghan-Tadzhik massifs and is separated from them by deep faults (Fig. III-10). Structural belts in this system are mainly overlain by the Turan platform sedimentary cover. Archaean and Proterozoic formations appear at the surface along the anticlinal axis of the Zeravshan-Turkestan zone in the Karatega and Atbashi horsts.

The oldest rocks referred to the Archaean in the Karatega horst are polymetamorphosed gneiss, marble, calciphyre and pyroxene-amphibole schist, containing relict granulite facies parageneses amongst amphibolite facies rocks. An Rb–Sr isochron age for amphibolite facies gneiss gives a value of 2400–2900 Ma. The presence of kyanite-sillimanite facies series metamorphic rocks, including eclogite, is characteristic.

Atbasha Group rocks in the Atbasha horst consist of gneiss, schist, garnetiferous amphibolite, marble and eclogite (Dobretsov, 1974). These rocks form relicts among greenschist facies retrograde rocks. They typically show evidence of Late Proterozoic and Palaeozoic glaucophane schist metamorphic effects (Moskovchenko, 1982). The Southern Tien Shan glaucophane schist belts form a single system with the high-pressure Hercynide terrains of the Urals (the Maksyutov complex) and are characteristic for eugeosynclinal formations (Dobretsov, 1974).

Riphean supracrustal formations within the massifs are mainly represented by terrigenous-carbonate formations: mica schist, quartzite, dolomite and limestone with stromatolites and oncolites (Tseisler et al., 1984).

The Urals belt

The Urals belt borders on the East European craton and the Timan-Pechora plate on the west, and the West Siberian and Turan plates on the east (Fig. III-1). The belt is subdivided into the Southern, Central, Northern and Polar Urals; the south-eastern extension of the Southern Urals is Mugodzhary. The structure of the Urals is comprised of Precambrian and Palaeozoic terrains. The Variscan terrain proper — the Uralides — contains Ordovician to Carboniferous rocks, while the pre-Uralides have pre-Ordovician rocks which are exposed at the surface in anticlinoria (Fig. III-3). In this scheme, early and late Precambrian formations form independent structural stages (Fig. III-11).

Early Precambrian. The oldest pre-Riphean formations belong to the Taratash complex, which bears similarities to the oldest formations of the East European craton, especially with granulite terrains in the Sea of Azov region (Bekker, 1978). The Taratash complex crops out over an area of some 400 km^2 in the north-eastern periclinal closure of the Bashkiria anticlinorium. It was deeply eroded then unconformably overlain by Lower Riphean sediments belonging to the Aya Formation. A marked metamorphic break has also been discerned between these two — Lower Riphean rocks are practically non-metamorphic, while the rocks in the Taratash complex show evidence of granulite, amphibolite and greenschist facies processes.

The rocks in the Taratash complex include schists (hypersthene-plagioclase, hornblende, salite-plagioclase, two-pyroxene-plagioclase), gneiss (garnet-cordierite-sillimanite, hypersthene, garnet-biotite), banded iron-

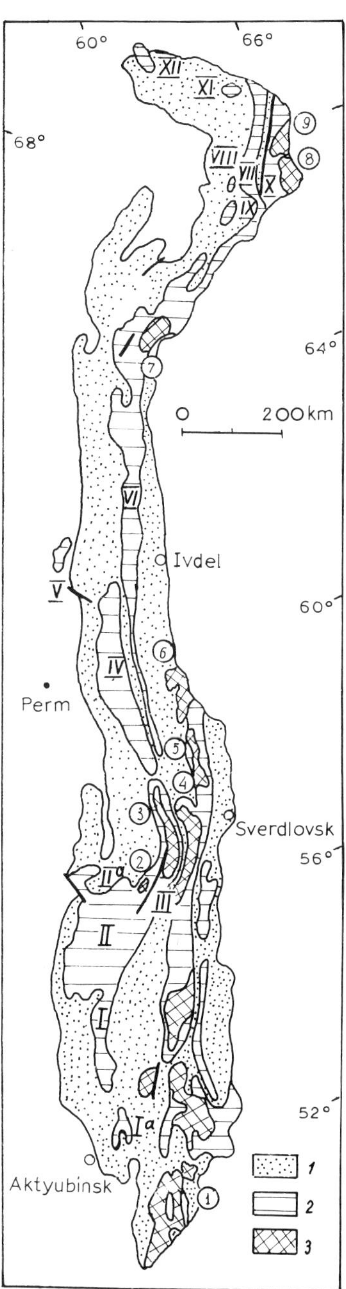

Fig. III-11. Sketch map showing Precambrian complexes in the Urals (simplified from Bekker, 1978). *1* = Uralides; *2* = Ripheides; *3* = pre-Ripheides. Riphean structures: *I* = Uraltau anticlinorium (*Ia* = Oryilek high), *II* = Bashkir anticlinorium (*IIa* = Karatau high), *III* = Zlatoust high, *IV* = Chusov anticlinorium, *V* = Polyudovo high, *VI* = Verkhnepechera anticlinorium, *VII* = Molokarsky anticlinorium, *VIII* = Yenganepey high, *IX* = Manutonyrd high, *X* = Harbey anticlinorium, *XI* = Yeduney high, *XII* = Amderma high. Major pre-Riphean complexes (numbers in circles): *1* = Mugodzhary, *2* = Taratsh, *3* = Ufaley, *4* = Sysert-Ilmenogorsk, *5* = Murza, *6* = Saldin, *7* = Lyapin, *8* = Khanmenkhoy, *9* = Murunkeus.

stone, and igneous and partial melt rocks — gneissose granite, plagiogneiss, charnockite and enderbite. The complex is cut by numerous diabase and gabbro-diabase dykes.

Dating of the internal processes manifest in the Taratash complex rocks is based on the following data. The maximum age dates are obtained using the U–Pb method on zircons (3300–2700 Ma) and K–Ar on minerals (3100–2700). These values still require confirmation by other methods, and at present we can reasonably accept the age of the granulite facies metamorphism (possibly a second event) as being around 2700 Ma, obtained using U–Pb techniques on zircons (Krasnobayev, 1980). The oldest intrusive rocks have an isotopic age of 2400 Ma, using data from the U–Th–Pb method. A value of 2000–2200 Ma has been obtained using the same method, as well as from Rb–Sr determinations on microcline. This value is related to processes of superimposed high-temperature amphibolite facies metamorphism. The final stage of Early Proterozoic crust-forming processes is marked by the intrusion of allochthonous granites dated at 1700–1800 Ma. A subsequent peak of activity is seen at 1100–1200 Ma, apparently in response to diastrophism of Grenville age within the Urals-Mongolian belt. Isotopic age values in the interval 600–300 Ma reflect the total breakdown of isotopic systems during the progress of the Caledonian and Hercynian cycles.

Pre-Riphean formations immediately to the east of the Taratash complex occur in a narrow north–south band in the Zyuratkul fault zone. They include outcrops of plagioclase and garnetiferous amphibolite and biotite gneiss grouped together as the Alexandrov amphibolite-gneiss complex.

The Ufaley granite-gneiss complex (Fig. III-11) includes units with amphibolite and hornblende-garnet gneiss showing widespread effects of migmatization. Towards the top of the section are the units of the Ukazar Formation — sericite-albite-quartz, sericite-epidote-quartz and sericite-quartz schist, conglomerate, hornblende and biotite-hornblende schist and muscovite gneiss. These assemblages are deeply eroded and unconformably overlain by Ordovician conglomerates (Bekker, 1978). Different workers have determined the age of the Ukazar Formation to be from pre-Riphean to Vendian.

Pre-Riphean formations in the Polar Urals are included in the Lyapin complex (Fig. III-11), which has two-mica and biotite gneiss with minor amphibolite and hornblende schist and muscovite gneiss with thin quartzite bands. The upper part of the section contains assemblages belonging to the Shchekuryi Formation: micaceous marble, quartzite, quartz-mica schist and leucocratic gneiss. Algae found in the marbles (*Murandavia magna* Vol. and *Nelcarella* sp.) are known from Jatulian sediments in the Baltic Shield. All the rocks were initially metamorphosed at amphibolite facies then subsequently at greenschist facies; they are overlain by Puyvi Formation sediments and cut by granites dated at 1370 Ma (Bekker, 1978).

The oldest formations in the Sysert-Ilmenogorsk complex belong to the Selyanka assemblage, consisting of graphite-biotite, garnet-biotite and sillimanite-garnet-biotite gneiss with subordinate garnetiferous amphibolite members. Age dates of 2225, 2035, 1160 and 220 Ma were obtained using the Rb–Sr method on biotite gneiss; and 1900 and 1392 Ma for migmatite (Dunayev et al., 1969). The Firsov assemblage occurs higher in the section (from 260 to 800 m), and comprises biotite and garnet-biotite gneiss, amphibolite and sillimanite-bearing quartzite with marble and calciphyre lenses. In the overlying Ilmenogorsk Formation (1000 m), amphibolites predominate with minor volumes of gneiss and quartzite. The effects of migmatization at amphibolite facies level are seen in all these assemblages. The Igish graphitic quartzite Formation overlies these above a structural hiatus. Its age status is uncertain — Precambrian to lower Palaeozoic.

The oldest formations in Mugodzhary (Fig. III-11) consist from base to top of migmatized biotite gneiss, gneiss and amphibolite (4000 m) with minor quartzites in the middle of the section and marble at the top. In addition, there is an independent alumina-rich complex (600 m) forming a syncline containing two-mica and biotite gneiss, quartz-mica and mica schist with garnet, kyanite and fibrolite. Available K–Ar dates give a wide spectrum of age values from 1020 to 380 Ma, implying that the oldest assemblages in Mugodzhary belong to the zone of Riphean and Hercynian reworking.

The Saldin complex (Fig. III-11) has the following sequence of units: the Pryanichnikov two-pyroxene gneiss, amphibolite with relict granulite and eclogite (800 m); the Brodov hornblende-biotite and magnetite-amphibole gneiss (1200 m); the Buksa amphibolites with thin marble and hornblende-biotite and hornblende gneiss (2000 m); and the Yemekh albite-epidote amphibolites (300 m). Lying above an unconformity is the Palaeozoic Istok greenschist Formation (1700 m).

Intensely migmatized amphibolites are the dominant rocks in the Saldin complex. The oldest K–Ar age date is 766 ± 53 Ma (Keilman, 1974), which appears to reflect a reworking event that affected the isotopic system in the Late Proterozoic and again during the Palaeozoic at 309–454 Ma. Saldin amphibolite pebbles are found in Riphean Medvedev Formation conglomerates. A further characteristic is the wide range of P–T metamorphic conditions, from greenschist to amphibolite facies, illustrating the polyphase nature of the metamorphism and the possibility that terrains of different ages have been structurally juxtaposed.

The oldest formations in the Polar Urals crop out in the Kharbey antiform, where they comprise the Kharbey complex, which is subdivided into three formations: the Laptayugan migmatized amphibolite and biotite and hornblende-biotite gneiss; the Khanmenkhoy amphibolite, with gneiss, quartzite, micaceous marble and eclogite also present; and the Parikvasshor gneiss and schist, containing kyanite-staurolite and staurolite-garnet varieties. Available isotopic age data do not give an unequivocal answer to the question of where in the Precambrian time scale these crystalline complexes fit. Phlogopite from ecologites has a K–Ar age in the interval 1560–615 Ma.

Summarising the available data on the Early Precambrian of the Urals, we can be fairly confident that Archaean rocks are present, initially metamorphosed at granulite facies conditions. If we take account of preliminary data on the age of the Taratash complex, 3000–3200 Ma, and the presence of medium-pressure granulite side-by-side with high-pressure granulite and ecologitic rocks which formed at the end of the Late Archaean (2700 Ma), then we have some justification for assigning these rocks to the Early Archaean, correlating them with Early Archaean formations in the basement to ancient cratons (Ivanov et al., 1982). The appearance of these low-gradient granulite and amphibolite facies metamorphic events would appear to be related to a superimposed Late Archaean tectono-metamorphic cycle.

Lower Proterozoic formations in the Urals are identified with more certainty. The Ufaley, Kharbey, Saldin, Selyankin and other complexes belong here. Their origin is related to the regional reworking (metamorphism, folding and migmatization) of older Archaean formations, which took place at 1850 Ma (Ivanov et al., 1982) under amphibolite facies conditions in variable pressure regimes.

Upper Proterozoic. The most complete upper Precambrian sedimentary successions in peripheral basins around the East European craton are found in the Urals. From geological, palaeontological and isotopic data, the Riphean is divided into four broad chronostratigraphic subdivisions: Burzyanian (Lower Riphean), Yurmatinian (Middle Riphean), Karatavian (Upper Riphean) and Kudash.

On the western slopes of the Urals, Riphean and Vendian sediments are grouped together into a single structural stage. They are best studied in the Bashkir anticlinorium, where four groups are recognised: Burzyanian (Lower Riphean), Yurmatinian (Middle Riphean), Karatavian and Kudash (Upper Riphean). The Riphean sedimentary complex rests unconformably on gneisses of the Taratash complex. Each group is a megarhythm, with conglomerate and sandstone at the base, giving way gradually to carbonate rocks at the tops of the megarhythms. Alkali basalts and tuffs are found at the base of the Burzyanian (the Aya Formation) (Kozlov, 1980). The group is over 2000 m thick.

In the eastern part of the Bashkir anticlinorium, a rhyolite-basalt unit occurs at the base of the Middle Riphean Yurmatinian Group, which wedges out westwards to be replaced by terrigenous and carbonate rocks containing identifiable stromatolites (*Baicalia baicalia* Masl., *Conophyton cylindricus* Masl., *Conophyton metula* Kir.) and microphytoliths (*Osagia tennilamellata*

Reitl., *Vesicularites flexuos*) (Kozlov, 1980). Glauconite from rocks in the upper part of the succession has a K–Ar age of 1260 Ma (Garris and Postnikov, 1970).

The Upper Riphean Karatavian consists of two macrorhythms, separated by an unconformity. Acid volcanics and diabase dykes occur at the boundary between the rhythms, which has been dated at 850 Ma (Mitrofanov et al., 1984). The lower rhythm contains mottled and red limestone, marl and claystone in a flyschoidal sequence. The group is some 5000 m thick. The microphytoliths identified in the limestones (*Osagia crispa* Z. Zhur., *Radiosus elongatus* Z. Zhur) are characteristic for the Upper Riphean (Kozlov, 1980), as are the stromatolites (*Yurusania cylindrica* Krul., *Malginella malgica* Kom. et Semikh.) (Kozlov, 1980). The age of the Karatavian is estimated to lie in the interval 1050 ± 50–680 Ma, from isotope work on glauconites (Garris, 1973).

The Kudash at the top of the Riphean, crops out on the SW limb of the Bashkir anticlinorium and rests on Karatavian sediments with a minor erosive base. The group is dominated by limestone with subordinate volumes of clastic terrigenous rocks; the limestones contain identifiable stromatolites (*Linella ukka* Kryl., *L. simica* Kryl., *Tungussia bassa* Kryl.) and microphytoliths (*Vesicularites bothrydiformis* Krasnop., *Ambliolamellatus horridus* Z. Zhur., *Osagia monolamellasa* Z. Zhur.) (Kozlov, 1980). Glauconite from rocks in the upper parts of the section has an age of 615–625 Ma (Bekker, 1975).

Overlying Kudash rocks above a sharp erosion are Vendian Asha Group rocks: sandstone, conglomerate and mottled rocks with the overall appearance of molasse, and a tillite horizon has been found at the base. Glauconite from Asha rocks gives a K–Ar age of 590–570 Ma (Garris et al., 1964).

On the whole, Riphean sediments on the western slopes formed in a miogeosynclinal basin on a down-faulted block of the East European craton (Anon., 1974).

Middle and Upper Riphean sequences are widespread in the antiforms which occupy the axial zone of the Urals. The dominant types are graphitic and siliceous schist, phyllite, porphyritoid and porphyroid. Thin limestones with stromatolites and oncolites have also been found. Due to intense dislocation, only an estimate of 8–10 km can be given for the thickness of the Riphean sequences. An estimate of 1.5–2 km can be quoted in the East Urals elevation, where structures are more flat-lying. Riphean rocks in the axial zone of the Urals are metamorphosed to greenschist and amphibolite facies grade, as distinct from those in the western zone.

Vendian sediments form part of the Vendian to Cambrian structural stage, which is exposed at the surface in the Kharbey, Lyapin-Pesovsky and Uraltau anticlinoria, also in the south of the East Urals and North Urals highs. Sediments belonging to this storey occupy rift-type graben structures. The most complete section of these rocks is found in the

Polar Urals and just to the south, where the succession is divided into two units. The lower (6–8 km) consists of volcanogenic rocks (diabase, spilite, andesite, trachyandesite, trachybasalt and trachyliparite) with thin carbonates and siliceous schist. Greywacke and arkose appear at the top of the section. The limestones contain recognisable Vendian stromatolites and archaeocyathids and Lower and Middle Cambrian trilobites. The upper unit (3–3.5 km) consists mainly of coarse-grained clastic rocks — conglomerate and sandstone with lenses of carbonaceous-siliceous shale and limestone. There are occasional occurrences of basic, acid and alkali volcanics. The upper unit has the appearance of a molasse sequence and formed during the Salair tectonic episode — it is cut by granitoids with an isotopic age of 550–450 Ma (Tseisler et al., 1984).

Low-gradient metamorphic effects constitute a highly distinctive feature of the Riphean exposed along the axial zone of the Urals. One example is the Maksyutov Formation — eclogite and glaucophane schist, found in the core of the Uraltau anticlinorium. In broad terms, Riphean eclogite-glaucophane schist complexes trace the Main Urals fault zone (Dobretsov, 1974). This does not rule out the appearance of low-gradient metamorphism in Ordovician successions, although it is of local occurrence and is mainly represented by low-temperature facies, mostly transitional to greenschist facies. A fundamental pattern in the occurrence of metamorphism in the Urals is the younging in the age of the major metamorphic episodes from west to east. For example, the effects of the Hercynian metamorphism proper are most intensely displayed only in the East Urals and Trans-Urals zones.

We conclude with a summary of the main features in the development of the Precambrian in the Urals, relying on evidence from field relations, geochronology and biostratigraphy. Archaean rocks in the Urals are broadly similar to rocks of the same age from the basement to ancient cratons. We emphasise the tendency in the evolution of these rocks towards the appearance of low-gradient regimes during the Late Archaean. Two megacycles are recognisable in post-Archaean geological history — Karelian and Riphean — with the boundary between them being around 1600 Ma and with the following internal boundaries: in the first, 2250 and 1900 Ma and in the second, 1350, 1100 and 850 Ma (Garris et al., 1979). The Karelian megacycle was accompanied by regional reworking of both Archaean and Early Proterozoic terrains. The Riphean megacycle reflects processes of taphrogenesis, which affected a mature continental crust during this period, which we could term a pre-geosynclinal phase (Ivanov, 1982). The geosynclinal process proper commenced with the formation of new oceanic crust during Vendian to Cambrian times and was completed during the Hercynian events.

2. MONGOLIA-OKHOTSK PROVINCE

The Mongolia-Okhotsk province stretches in a north-easterly direction along the edge of the Siberian craton (Fig. III-1), and is separated from the Aldan-Stanovoy Shield and the Salairides of the Baikal province by a deep fault system (Fig. III-12). The southern margin lies outside the Soviet Union. The eastern part contains the Hinggan-Burey massif, which directly borders on structural zones belonging to the Pacific belt; to the east, within the Mongolia-Okhotsk belt, lies the Variscan Dzhagda fold zone which can be traced as a narrow strip that broadens out eastwards as far as the Shantar Islands. The southern boundary of this system is marked by a fault zone which borders the Hinggan-Burey massif, and to the east it meets the approximately north–south-striking Ussuri system (Neyelov and Milkevich, 1974).

The question of the age of the fold structures in the Mongolia-Okhotsk province is debatable. Nevertheless, most authors consider this structural belt to be polycyclic with independent early Precambrian and late Proterozoic to early Palaeozoic, middle Palaeozoic and Mesozoic growth cycles to which correspond independent structural-lithological terrains.

Identifiable Precambrian formations are Archaean, early Proterozoic and late Proterozoic terrains; Vendian sediments are included with lower Palaeozoic assemblages in the same structural stage (Neyelov and Milkevich, 1974). A geosynclinal terrain proper is represented by Silurian to Lower Carboniferous successions.

Early Precambrian rocks crop out within the confines of massifs and anticlinoria such as the Priargun, Borshchovochno and Shevli-Alyan (Fig. III-12).

The oldest formations assigned to the Archaean (analogues of the Stanovoy terrain), are represented by migmatized gneiss and granite-gneiss with carbonate, quartzite, hornblende-biotite gneiss and schist with relict granulite facies rocks. Lower Proterozoic assemblages (the Undin complex) consist of biotite-epidote schist, amphibolite, two-mica gneiss and quartzite with garnet, kyanite and staurolite. The late Proterozoic terrain occupies two zones: the Acha-Borshchovocho (eugeosynclinal) and Priargun (miogeosynclinal). The first contains the Achin volcano-sedimentary assemblage, the second contains the Priargun terrigenous-carbonate assemblage. Oncolites, katagraphiids (problematic) and stromatolites have been found in rocks belonging to the Acha assemblage, allowing them to be dated as Middle to Upper Riphean (Neyelov and Milkevich, 1974).

Regional metamorphism affected the Precambrian rocks in the Mongolia-Okhotsk province several times (Milkevich, 1982). The rocks in the Stanovoy complex are characterized by amphibolite and kyanite-sillimanite facies series granulite facies metamorphism (T = 600–750°C, P = 6.0–7.5 kbar) with widespread partial melting. The Undin complex shows prograde

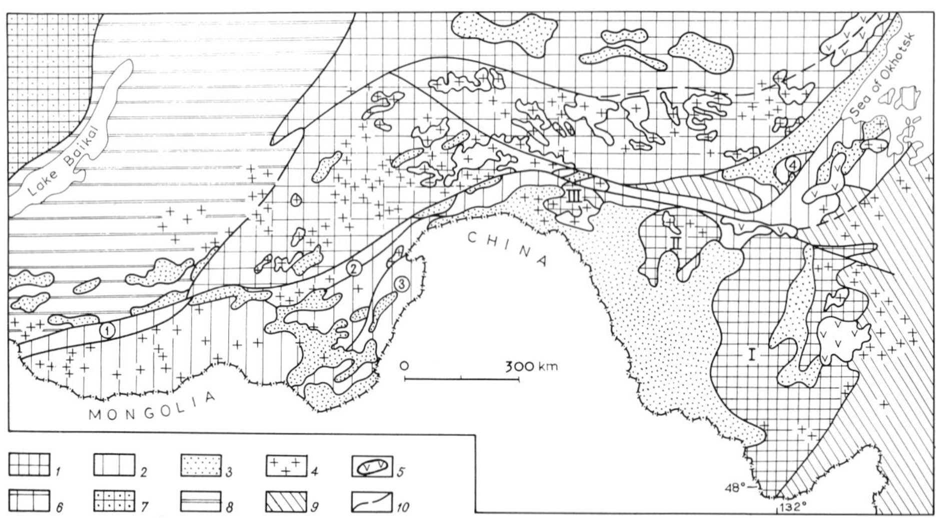

Fig. III-12. Tectonic sketch map of the Mongolia-Okhotsk fold belt (Tseisler et al., 1984, tectonic map from Anon., 1974). *1* = massifs with pre-Riphean complexes (*I* = Hingan-Burey, *II* = Mamyn, *III* = Gonzhi); *2* = fold systems of Mongolia-Okhotsk belt; *3* = Mesozoic-Cenozoic sediments in superimposed basins; *4* = granitoids; *5* = volcanics; *6* = Aldan-Stanovoy Shield; *7* = Siberian craton; *8* = Baikal fold belt; *9* = Mesozoides of Pacific fold belt; *10* = major faults. Anticlinoria (numbers in circles): *1* = Prichikoy, *2* = Borshchovochny, *3* = Priargun, *4* = Shevli-Ayan.

amphibolite facies metamorphism in the kyanite-sillimanite facies series (T = 580–600°C, P = 5.5–6.5 kbar). Differentiation of metamorphic regimes took place in Riphean times: the Acha eugeosynclinal assemblage underwent uniform glaucophane-greenschist metamorphism (T = 350–400°C, P = 4–6 kbar), and the Priargun miogeosynclinal assemblage was zonally metamorphosed under greenschist to amphibolite facies conditions in the andalusite-sillimanite facies series (in the high-temperature part, T = 500–600°C, P = 3.5–4 kbar). Vendian to Palaeozoic successions are practically unmetamorphosed — they show only local metamorphic effects up to muscovite-chlorite subfacies of the greenschist facies. Shallow zonal metamorphism from greenschist to amphibolite facies (T = 400–600°C, P = 2.5–3.5 kbar) occurred along major faults during Mesozoic activity. The oldest rocks, especially Archaean, thus experienced metamorphic effects on several occasions. The most intense episode was the late Proterozoic reworking of pre-Riphean rocks in basement inliers within antiformal structures in Palaeozoic fold belts. The most resistant areas were rigid structures in the Hinggan-Burey massif, for which Upper Proterozoic assemblages represent in most cases a weakly-metamorphosed sedimentary cover.

3. PACIFIC BELT

The Pacific belt in the USSR includes the fold belts in the Verkhoyan-Chukchi province, separated by massifs into Mesozoic fold-belt systems, the Yana-Kolyma in the west and the Anyuy-Chukchi in the east. The Late Mesozoic orogen consists of the Sikhote-Alin and Taygonoss-Koryak systems. Cenozoic fold structures include the Olyutora-Kamchatka and in part the Hokkaido-Sakhalin systems (Fig. III-1).

Several structural storeys have been erected for pre-Mesozoic formations in the Verkhoyan-Chukchi region: pre-Riphean, upper Proterozoic, lower to middle Palaeozoic and in the east upper Palaeozoic (Tseisler et al., 1984). Precambrian rocks crop out within basement massifs: Okhotsk, Omolon, Kolyma, Chukchi and others, overlain to varying degrees by a sedimentary cover. Deeply buried massifs are also thought to exist, such as the Adycha-Elgi (Fig. III-13).

Early Precambrian rocks in the Okhotsk massif are assigned to the Archaean. They form the Okhotsk complex, consisting of hornblende-biotite schist, biotite, garnet-biotite, cordierite-garnet-hypersthene and hypersthene-biotite gneiss, quartzite, calciphyre and eulysite. The rocks were migmatized at granulite facies (with the formation of charnockite and enderbite) and amphibolite facies (with the formation of alaskitic granite; Gusev and Mokshantsev, 1978). How the complex should be subdivided is still a matter of debate. Its total thickness has been estimated to be 11 km. Pb–Pb and Th–Pb isochrons for mafic schists yield ages of 4100 $\pm$ 400 Ma and 3300 $\pm$ 300 Ma; a date of 3700 $\pm$ 500 Ma is taken as the most likely age (Smirnov, 1979). Using K–Ar techniques on biotite gives ages in the interval 1240–1800 Ma and 2640 Ma (Gusev and Mokshantsev, 1978). Most workers recognise a similarity between rocks of this complex and the oldest formations in the Aldan Shield.

No Early Proterozoic rocks have been identified within the Okhotsk massif. Archaean crystalline rocks are overlain by Middle to Upper Riphean sediments above a sharp unconformity. These sediments include sandstone, quartzite, siltstone, mica-clay shale and recrystallized limestone containing stromatolites. The Middle to Upper Riphean forms part of the lower stage of the weakly dislocated sedimentary cover of the Okhotsk massif. Its thickness is 1200–1500 m.

The crystalline rocks of the Omolon massif include biotite, hornblende-biotite, pyroxene-garnet and hypersthene gneiss, two-pyroxene schist, amphibolite, calciphyre and quartzite. This is a rock complex which was initially metamorphosed at granulite facies then subsequently again at amphibolite facies conditions. There are signs that eclogitic rocks are also present besides granulites (Smirnov, 1979), suggesting that there was a low-gradient metamorphic regime beside a medium-gradient regime (medium-pressure granulite facies).

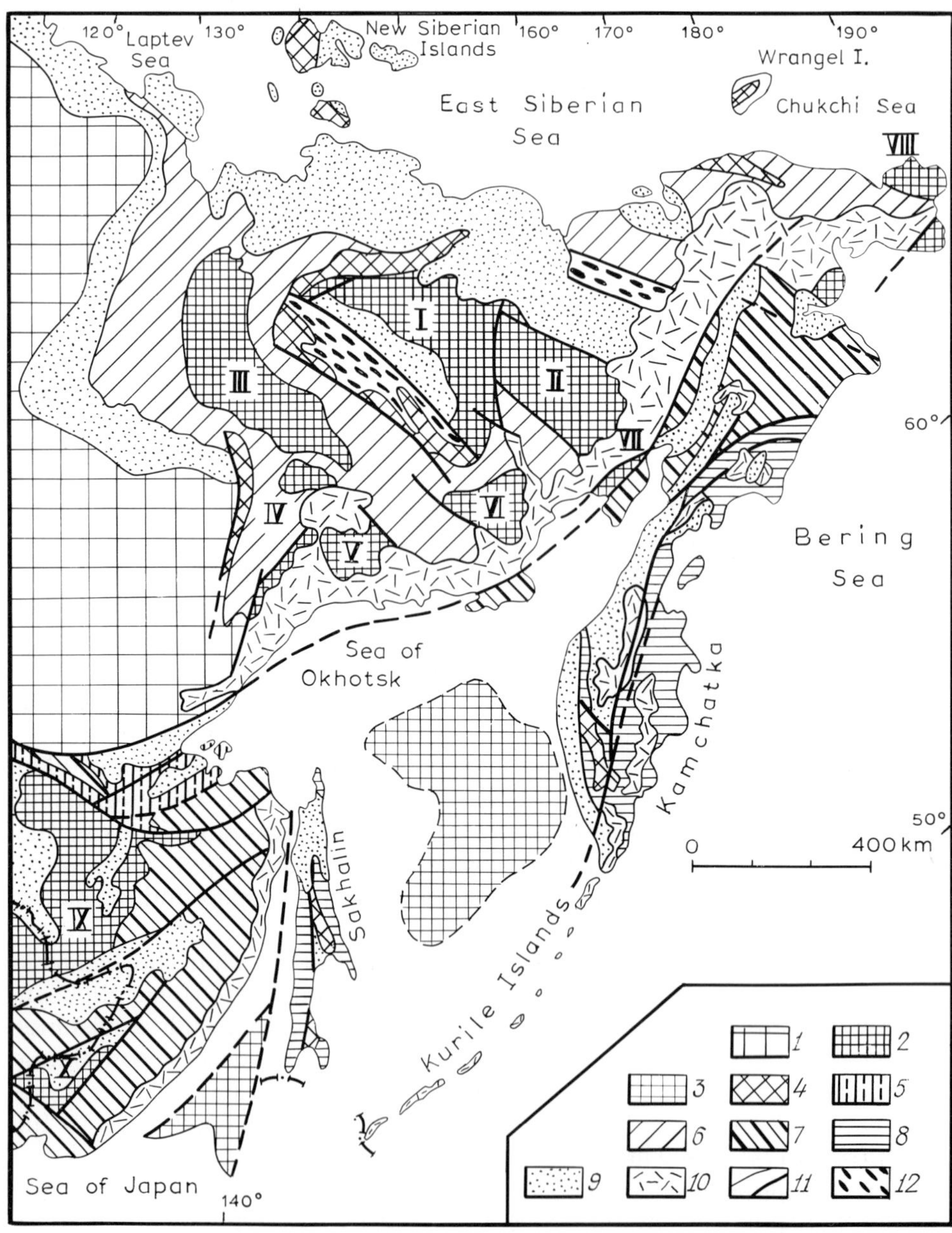

Fig. III-13. Sketch map of Precambrian in structural zones of the NW part of the Pacific fold belt (Tseisler et al., 1984; tectonic map from Anon., 1974). *1* = Siberian craton; *2* = massifs with pre-Riphean basement (*I* = Kolyma, *II* = Omolon, *III* = Adycha-Elgi, *IV* = Suntar-Labynkar, *V–VI* = Okhotsk, *VII* = Taygonos, *VIII* = Chukchi, *IX* = Khingan-Burey, *X* = Hankay); *3* = massifs beneath marginal seas; *4* = outcrops of basal assemblages in anticlinoria; *5* = Palaeozoides of Mongolia-Okhotsk fold belt; *6* = Mesozoide geosynclinal complex, Verkhoyan-Chukchi belt; *7* = late Mesozoic geosynclinal complex; *8* = Cenozoic geosynclinal complex; *9* = Mesozoic-Cenozoic sediments of orogenic complexes and covers of young platforms and massifs; *10* = volcanic belts; *11* = major faults; *12* = late Mesozoic rift-type suture zones.

Granulite-facies tonalitic gneiss yields a U–Pb isochron age on zircon of 3400 Ma (Bibikova et al., 1982), indicating that these rocks belong to the Early Archaean. This is the age of the metamorphic event, therefore the time of emplacement of the tonalites must be even older. It should be pointed out that these rocks contain a zircon generation giving an age of around 2750 Ma. Zircon from amphibolites, however, is dated at 1800 Ma. The same age is obtained by the Rb–Sr isochron method, 1800–2000 Ma.

The oldest tonalitic gneisses in the Omolon massif are similar in many respects to the "grey gneisses" in Greenland.

The crystalline rocks of the Omolon massif are overlain by upper Precambrian sediments above a sharp angular unconformity. These rocks are overlain in turn by a Palaeozoic sedimentary cover. The most complete succession with Riphean stromatolites includes a weathering crust and quartzitic sandstone; limestone, phyllitic sand-clay and graphite-clay shale and quartzose siltstone predominate in the middle and upper parts of the succession. The total thickness is 900–1000 m.

The Chukchi massif forms part of the larger Chukchi-Yukon (Eskimo) structure with the Precambrian basement, most of which is hidden under the Bering Straits and the Chukchi and Bering Seas. It is surrounded by Mesozoic fold systems, which have upper Precambrian sediments exposed in the anticlinorial zones in the region around Wrangel Island. The Precambrian formations of the Chukchi massif are divided into two complexes: the East Chukchi (Archaean) and the Wrangel (upper Precambrian). The Archaean complex includes the following formations (from the base upwards): Okatyn (<200 m) amphibolitized pyroxene schist and plagiogneiss; Litka (300 m) mafic schist, alternating with granite-gneiss; Ettelkhvyleut (650–2500 m) granite-gneiss with lenses of quartzite, amphibolite and charnockitic horizons; Puoten (700–1300 m) hornblende, pyroxene and biotite schist, granite-gneiss, high-alumina schist with kyanite, sillimanite, garnet and cordierite, quartzite and graphite lenses; Kaatap (2500 m) mafic schist with calciphyre, marble and quartzite horizons (<30 m), mica schist with kyanite, sillimanite and garnet; Runliveyem (500 m) rhythmically alternating carbonate, mica schist and hornblende schist (Gusev and Mokshantsev, 1978).

This metamorphic rock association characterises the East Chukchi complex as polymetamorphic. It is heterogeneous, since the succession contains upper Archaean and/or lower Proterozoic rocks, besides primary granulites which are provisionally assigned to the lower Archaean (Smirnov, 1979). The available K–Ar dates of 1570–1680 Ma appear to be significantly rejuvenated.

Upper Precambrian successions in the Chukchi massif constitute a greenschist complex, over 1000 m thick, consisting of epidote, chlorite-epidote and amphibole-chlorite schist, marble and porphyroid; basal conglomerates have been found at the base of the complex. As a rule, the assemblages occupy gently-dipping monoclines. The upper Precambrian complex of Wrangel Island contains hornblende, hornblende-biotite-chlorite and quartz-chlorite-biotite schist, formed from volcanics at greenschist facies metamorphism which in places reaches low grades of the amphibolite facies. Also found with them are minor amounts of sandstone, tuffaceous sandstone and marble with Middle to Late Riphean acritarchs and microphytoliths. The complex is 2000 m thick. Riphean rocks are overlain by metaconglomerate, metasandstone, actinolite-epidote-chlorite schist and quartz-albite-chlorite schist with acritarch fossils, identified as Vendian. The Vendian is 800 m thick.

Much of the Kolyma massif is hidden under a Palaeozoic and Mesozoic sedimentary cover. Rocks belonging to the pre-Riphean basement crop out in the arch of the Pre-Kolyma anticlinorium, where they constitute the Ossoli Formation, which is provisionally assigned to the lower Proterozoic. The formation consists of amphibolite and amphibolite facies gneiss in the kyanite-sillimanite facies series. Above is a greenschist facies assemblage, the Oroyeka Formation, which is assigned to the Lower Riphean (Smirnov, 1979). In the centre of the Kolyma massif, greenschist, metapelite, quartzite and quartzitic schist with marble lenses crop out on the Alazey basement high. The schists contain alkali amphiboles. Metamorphism was at glaucophane-greenschist facies. The marbles in this formation contain algae, suggesting an Early Riphean age for these rocks, while rock fragments with alkali amphiboles exist in the basal beds of a mid-Palaeozoic formation (Gusev and Mokshantsev, 1978).

Late Proterozoic sequences in the Verkhoyan-Chukchi province can be considered to be part of the sedimentary cover succession, on the grounds of their geological position and formational composition; the cover successions are similar to those on the Siberian craton, which are separated into platform and miogeosynclinal foredeep basin types (Tseisler et al., 1984). During the Riphean, the crystalline rocks which crop out within the Verkhoyan-Chukchi province formed a single cratonic structure with the pre-Riphean terrains of the Siberian craton.

The late Mesozoic fold zone of the Pacific belt within the USSR includes the Sikhote-Alin and Taygonoss-Koryak systems (Fig. III-13). The Sikhote-Alin system includes the Sikhote-Alin and Lower Amur structural zones.

The Archaean rocks in this region form the Manchurian complex, represented in the Khankay massif, where it is divided into three formations. The Matveyev Formation (3200 m) consists of biotite and garnet-cordierite-sillimanite gneiss with thin bands of garnetiferous, garnet-hypersthene, magnetite and fayalite quartzite, eulysite, graphitic schist, marble and hypersthene and diopside gneiss. The Ruzhin Formation (1000–2000 m) consists of a unit of interleaved diopside and biotite gneiss with marble, calciphyre and graphitic schist. The Turgenev Formation (2000 m) consists of biotite and hornblende-biotite gneiss and schist with thin marbles and amphibolites. These assemblages are referred to the lower Archaean and are correlated with the oldest formations in the Aldan-Stanovoy Shield (Smirnov, 1978, 1979).

The Duchun Formation (500 m), consisting of hornblendic and diopsidic amphibolite and biotite plagiogneiss is assigned to the upper Archaean. This formation has no relict granulite facies rocks and is weakly migmatized.

The Nakhimov Formation (500–3500 m thick) and the Tatyanov Formation (2200–2700 m) are considered to be lower Proterozoic in age. The former consists mainly of biotite and hornblende gneiss and schist containing lenses and thin bands of amphibolite and marble; the latter has biotite, diopside and hornblende-diopside marble with lenses of calciphyre and amphibolite (Smirnov, 1978).

The Krayevo complex (4800 m) belongs to the Lower Riphean and is exposed around the western margin of the Khankay massif. The rock assemblages of this complex can be traced into the outcrop area of the Bamyantun Formation which stretches from the northern margins of the Hinggan-Burey massif to the NE Korean Peninsula. Higher in the section, beyond the Hinggan massif, the Bamyantun succession takes on a eugeosynclinal appearance: at the base it has a spilitic association with ophiolites and higher in the section silica-clay and carbonate units make an appearance.

The 2300 m thick Spassky Formation (Smirnov, 1978, 1979) in the Khankay massif and its surroundings is Middle-Upper Riphean to Vendian in age. The Spassky Formation unconformably overlies the crystalline rocks of the lower Proterozoic Tatyanov Formation and consists of sericite, sericite-chlorite and actinolite schist, porphyroid and amphibolite. The original rocks were keratophyric — series volcanics and basic to intermediate effusive rocks. These rocks are conformably overlain by the Mitrofanov Formation (2900 m), which consists of sericite-chlorite-graphite, sericite-graphite and actinolite schist, porphyroids and phyllite. A K–Ar age date on micas from pegmatites cutting Mitrofanov Formation rocks is 635 Ma. The Rb–Sr method gives the same value.

Geological and geophysical data suggest that Precambrian complexes occur under the entire Sikhote-Alin late-Mesozoic fold system (Tseisler et al., 1984).

The oldest formations in the Taygonoss-Koryak system, which is a continuation of the Sikhote-Alin system, are found in the Taygonoss massif (Fig. III-13). They are considered to be part of the Archaean Avekov complex, divided into two formations, Purgonoss (below) and Kossov (Gusev and Mokshantsev, 1978). The Purgonoss Formation consists of biotite and pyroxene gneiss and schist containing garnet, sillimanite and kyanite. Garnet-biotite gneiss with sillimanite and graphite predominate in the bottom and top units, while the middle unit contains diopside-biotite, garnet-biotite-diopside, biotite-two-pyroxene and garnet-biotite-hypersthene gneiss, amphibolite, garnetiferous amphibolite and garnet-clinopyroxene and garnet-clinopyroxene-amphibole eclogitic rocks. The thickness has been estimated to be 3000 m. The lower member of the Kossov Formation consists mainly of amphibolite interleaved with hornblende gneiss; its composition is dominated by leucocratic plagiogneiss with thin bands of garnetiferous varieties and amphibolite lenses; relict granulite facies rocks are present. The formation is at least 6000 m thick. Overall, the formations in the Avekov complex are polymetamorphic, showing prograde metamorphism up to granulite facies level, then subjected to a second metamorphic event under amphibolite to granulite facies conditions in the kyanite-sillimanite series, with which the formation of eclogitic rocks and migmatites are probably associated. Available isotopic data give a K–Ar age for garnetiferous gneisses of 2760 Ma, and 2050 Ma for hornblende (Smirnov, 1979). A U–Pb age on zircons from amphibolites is 2880 Ma (Smirnov, 1979). These figures seem to reflect events when granulites in the Avekov complex were reworked during the Late Archaean and Early Proterozoic.

Upper Precambrian rocks in the Taygonoss massif form the Middle to Upper Riphean Zhulan assemblage (Smirnov, 1979). They overlie the crystalline basement and consist of sandstone, quartzite, phyllitic schist and stromatolitic limestone, totalling 800–950 m in thickness. These sediments are correlated with the Riphean sedimentary cover on the Omolon massif (Gusev and Mokshantsev, 1978).

The oldest rocks within the internal (eugeosynclinal) zone of the Taygonoss-Koryak system are gabbroid, amphibolite, marble and glaucophane schist, affected by greenschist facies metamorphism. They are overlain by Ordovician sediments with proven zone fossils and are therefore Precambrian to lower Palaeozoic in age.

The Olyutora-Kamchatka system is situated in the Cenozoic fold belt (Fig. III-1). It is divided into two zones, an internal one facing the Pacific Ocean and an external. In the external zone, Precambrian rocks occur in the southern part of the Central Range (Central Kamchatka horst). These are the Kolpakov gneisses with hypersthene, kyanite schist and garnetiferous amphibolite. Above is the Kamchatka Formation (400–2000 m) which is conformable or in places lies above an angular unconformity; it consists of biotite and two-mica microcline gneiss, sometimes with garnet. Thin gneisses with sillimanite, kyanite and andalusite occur among these rocks. This assemblage belongs to the Riphean or Vendian (Smirnov, 1978, 1979), but as yet there are no reliable age dates. A number of workers regard the crystalline rocks in the external zone as a sliver of the basement of a major massif, most of which is now under the Sea of Okhotsk (Fig. III-13).

Precambrian rocks in the internal zone of the Olyutora-Kamchatka system are best known from the Ganalsky Range in Kamchatka. The oldest rocks form the Ganalsky Group, which comprises hypersthene and cordierite-hypersthene plagiogneiss, quartzitic gneiss, quartzite, amphibolite and hornblende and biotite gneiss. The geological status of the granulite-facies rocks is still debatable. These rocks are considered to be either older than the amphibolite facies rocks in the Ganalsky Group, or their origin is related to contact metamorphic effects associated with the Yurchin gabbro-norite intrusion. Using Pb–Pb thermo-emission techniques on zircons from rocks in the Yurchin intrusion and from tonalites cutting granulites and Ganalsky Group amphibolite facies rocks, the maximum age is greater than 3000 Ma (up to 3290 ± 50 Ma), implying that they probably belong to the Early Archaean (Smirnov, 1978). The medium-gradient granulites originated in response to contact effects of the gabbro-norite intrusion on the prograde regional metamorphism of Ganalsky Group assemblages at amphibolite facies conditions in the kyanite-sillimanite facies series.

The next group in the succession is the essentially volcanogenic Stenovoy Group, zonally metamorphosed from greenschist facies to the andalusite-sillimanite series of the amphibolite facies. Igneous zircons from acid volcanics in the Stenovoy Group have an age of around 2600 Ma. The regional metamorphic event, which was the first to affect Stenovoy Group rocks, is superimposed on rocks belonging to the Ganalsky Group (including the granulites). Zircon ages for this metamorphism are 1800–2000 Ma. The Kizhechenok Formation is Lower Riphean in age; it consists of metasandstone and mafic metavolcanics and has been found to contain a microphytofossil assemblage comparable with Riphean assemblages (Lvov et al., 1985).

Generally speaking, the early Precambrian formations in the Mesozoides of the Pacific belt have many features similar to rocks of the same age in the Aldan-Stanovoy Shield. Late Proterozoic successions, though, are mainly sub-platform of miogeosynclinal types. Their formation was not accompanied by large-scale folding and metamorphic reworking of a pre-Riphean basement.

4. MEDITERRANEAN BELT

The Mediterranean belt separates the ancient cratons of Laurasia and Gondwana and stretches in an almost east–west direction from the Atlantic Ocean to the Himalayas. Its external north-eastern marginal part crops out within the confines of the USSR, and is bounded by the East European craton and structural zones of the Urals-Mongolian belt (Fig. III-1). Three major structural units are identified in the belt: an epi-Palaeozoic (young) platform, which originated at the end of the Palaeozoic or during the early Mesozoic, Mesozoic and Alpine fold systems (Fig. III-14).

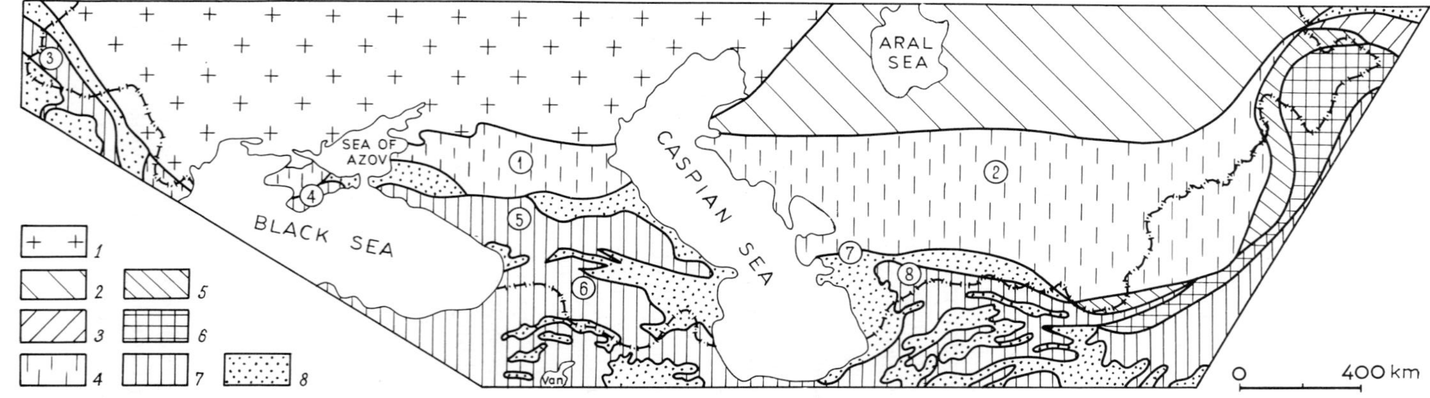

Fig. III-14. Tectonic sketch map of Mediterranean belt in the USSR and adjacent countries (Tseisler et al., 1984; tectonic map from Anon., 1974). *1* = E European craton; *2* = Urals-Mongolia belt; *3* = Kunlun Hercynian system. Mediterranean belt: *4* = epi-Palaeozoic (young) platform. Fold belts: Mesozoic: *5* = N Pamirs system, *6* = S Pamirs–Karakorum; *7* = Alpine; *8* = intermontane basins and marginal downwarps. Major structures (numbers in circles): platforms of young craton: *1* = Scythian, *2* = S Turan; fold belts: *3* = E Carpathians, *4* = Crimea, *5* = Greater Caucasus, *6* = Lesser Caucasus, *7* = Kubadag and Greater Balkhan, *8* = Kopetdag.

The epi-Palaeozoic platform

The basement (including the Precambrian) in the plates of this young platform is hidden by a sedimentary cover and has been identified in boreholes and by geophysical methods. Using such methods, it has been shown within the Scythian plate that the basement has a heterogeneous structure and is formed from pre-Palaeozoic massifs, separated by Palaeozoid fold systems. The massifs are known to contain Late Proterozoic and younger rocks. In the north Caucasus region, the oldest part of the basement belongs to the edge of the Rostov basement elevation of the East European craton. A Late Proterozoic age for the basement has also been determined for the south Caucasus region (Fig. III-15).

The basement within the South Turan plate has been down-faulted to a depth of 6–10 km. It is assumed to consist of pre-Palaeozoic massifs, separated by Hercynian and early Cimmerian fold belts. Precambrian rocks are exposed here in elevated regions of the Southwest Gissar Range and in the basement to the Afghan-Tadzhik basin in the Karatega Range.

The oldest rocks in Karatega are found in the Garm block. They may be divided into two stages. Rocks in the lower stage show prograde granulite facies metamorphism on which is superimposed a medium-pressure amphibolite facies metamorphic event; rocks in the upper stage show evidence for a low-pressure zonal metamorphic event only. These rocks are now all referred to the Karatega Group, consisting of garnet-biotite and garnet-biotite-cordierite gneiss, amphibolite, calciphyre and marble, among which pyroxene schists occur. Migmatization is evident in these rocks, and there are exposures of gneissose granite bodies. A U–Pb age date in the interval 2600–3000 Ma has been obtained for rocks with relict granulites (Anon., 1982).

The Precambrian in South-West Gissar is divided into two groups: the Baysuntau (Archaean) and the Surkhantau (lower Proterozoic). The Baysuntau Group consists of cordierite and sillimanite gneiss and schist, with subordinate amphibolite, marble and quartzite. The presence of eclogite is characteristic. The Surkhantau Group contains mainly biotite plagiogneiss with rare sillimanite-cordierite-biotite gneiss and amphibolite. The Karatega crystalline complexes in Gissar are broadly similar.

The Mesozoic fold terrain

The Mesozoic fold terrain of the Mediterranean belt in the USSR includes the Pamirs and borders on the Afghan-Tadzhik basin in the north and its north-eastern continuation. The southern border of the terrain occurs outside the USSR (Fig. III-14). Completion of the last cycle of the geosynclinal evolution of its constituent parts took place at the end of the

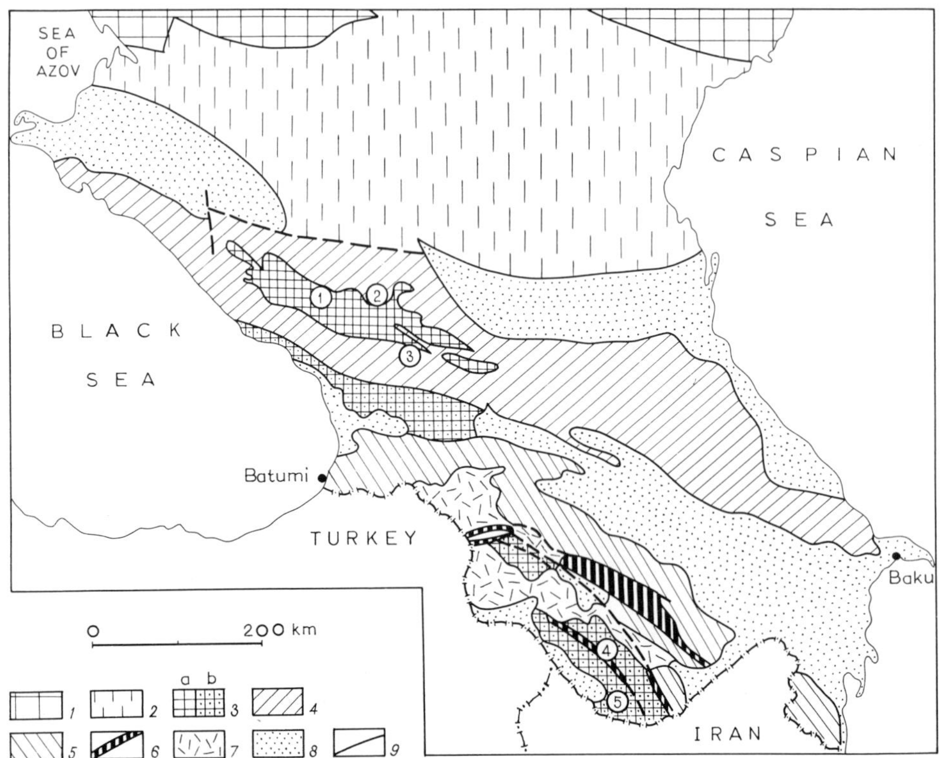

Fig. III-15. Tectonic sketch map of the Caucasus and Cis-Caucasus (modified from Tseisler et al., 1984). *1* = E European craton; *2* = cover of epi-Palaeozoic craton; *3* = crystalline complexes including Precambrian(?) (*a*: exposed, *b*: beneath Palaeozoic or Mesozoic-Cenozoic cover); *4* = sediments in Greater Caucasus mega-anticlinorium; *5* = sediments of Lesser Caucasus fold belt; *6* = ophiolite zones (Upper Jurassic–Cretaceous); *7* = volcanic plateau; *8* = intermontane basins and marginal downwarps; *9* = major faults. Structures (numbers in circles): in Greater Caucasus mega-anticlinorium: *1* = fore range zone, *2* = Begasyn anticline, *3* = Svanetia anticlinorium; in Lesser Caucasus: *4* = Miekhan-Zangazur massif, *5* = Ararat-Dzhulfa massif.

Triassic and at the Jurassic-Cretaceous boundary. The Pamirs contain the North Pamir and South Pamir–Karakorum fold systems (Fig. III-16).

The North Pamir system occupies the territory between the Darvaz-Karakorum fault and the Vanch-Tanymass thrust. The southern boundary with the South Pamir–Karakorum system is often interpreted as the line of continental collision between Laurasia and Gondwana.

Precambrian rocks form the massif of the Kurgovat zone (Fig. III-16), where they are divided into the Borshita Group (lower Proterozoic) and the Viskharv Formation (Vendian?). The base of the Borshita Group is not exposed. It consists of amphibolitic and zoisite-amphibole gneiss with thin bands of biotite gneiss and marble. Higher in the section there are high-Al

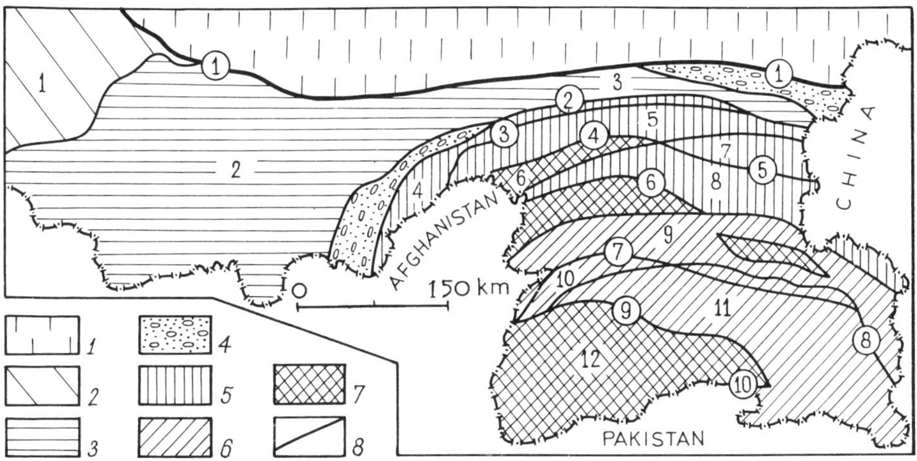

Fig. III-16. Tectonic sketch map of the Pamirs (Tseisler et al., 1984). *1* = Urals-Mongolia belt; *2* = S Turan plate; *3* = Neogene-Quaternary sediments of Afghan-Tadzhik basin; *4* = Alpine molasse (orogenic) basin; *5* = Mesozoides of N Pamir system; *6* = Mesozoides of S Pamir–Karakorum system; *7* = Precambrian blocks; *8* = boundary faults. Structures: *1* = SW Gissar high, *2* = Afghan-Tadzhik basin, *3* = Za-Alay ridge and Alay valley, *4* = Darvaz–Za-Alay zone, *5* = Kalaykhumb-Sauk zone, *6* = Kurgovat zone, *7* = Kara Kul zone, *8* = Darvaz-Sarykol zone, *9* = Central Pamir zone, *10* = Rusahn-Psharts zone, *11* = SE Pamir zone, *12* = SW Pamir Precambrian massif. Major faults (numbers in circles): *1* = Gissar (Vakhsh), *2* = Darvaz–Kara Kul, *3* = Sauksay, *4* = Viskharv-Ukbulak, *5* = Balyankiy, *6* = Vanch-Tanyman, *7* = Rushan–North Pshart, *8* = Dunkeldyk, *9* = Gunt-Alichur, *10* = S Pamir.

gneisses with garnet, staurolite, kyanite and andalusite; at the very top, more quartzite, marble and quartzose schist are encountered. The Viskharv Formation, consisting of quartzose and arkosic metasandstone with thin chlorite-sericite schist and marble, contains Vendian to Lower(?) Cambrian acritarchs at the base, and late Ordovician to Lower Silurian rugose corals at the top (Anon., 1982).

The South Pamir–Karakorum system occupies the territory between the Vanch-Tanymass thrust and the fault system in the Eastern Hindukush (Afghanistan). Precambrian metamorphic rocks constitute massifs in this region, as basement inliers amongst Palaeozoic and Mesozoic fold belts, the largest of these being the South West Pamir massif.

The massif is bordered to the north by the Gunt-Alichur fault and to the south by the South Pamir fault (Fig. III-16). The oldest rocks in the massif belong to the Goran and Shakhdar Groups and the Khorog orthocomplex separating them, which are Archaean, and the lower Proterozoic Alichur Group (Anon., 1982).

The Goran Group contains three assemblages (from base to top): marble, quartzite and gneiss; marble and gneiss; and marble and granite gneiss. Carbonate rocks amount to approximately a quarter of the volume of the Goran Group succession. Throughout the section there are bodies of eclogitic rocks and charnockite.

Three-quarters of the Khorog orthocomplex consists of igneous rocks which occur as sheets between the assemblages of the Goran and Shakhdar Groups. The majority of the rocks are amphibole gneiss (with garnet and pyroxene in places), amphibolite, eclogitic rocks and meta-ultramafics with charnockitic bodies. The massive rocks form a system of lenses, between which are amphibole orthogneiss with thin bands of aluminous gneiss containing kyanite, garnet, sillimanite and cordierite. Available age dates (U–Pb, Th–Pb and Pb–Pb methods) yield values of 2600–2750 Ma. On this basis, the rocks of the Khorog complex belong to the Late Archaean.

The Shakhdar Group is in tectonic contact with the Khorog orthocomplex. Its upper boundary is defined by the overthrust assemblages of the Alichur Group. The Shakhdar Group consists of biotite gneiss (sometimes with garnet), high-alumina gneiss with kyanite, sillimanite and cordierite, amphibolite, hornblende gneiss, garnet-two-pyroxene schist and marble. A wide spread of ages has been determined for aplitic gneissose granites, using Pb–Pb techniques: 1950–2650 Ma (Anon., 1982).

The Alichur Group (4–5 km thick) differs from Archaean groups in its formational composition and the P–T regimes displayed by internal processes. Compositionally, it is predominantly metapelitic gneiss: biotite, garnet-biotite with muscovite and andalusite varieties, dolomitic marble, calc-silicate rocks with garnet, scapolite and vesuvianite. Metamorphism of these rocks took place at amphibolite facies conditions in the andalusite-sillimanite facies series. This event is also secondary in Archaean low-gradient metamorphic rocks.

Riphean rocks have been identified in the northern margins of the Central Pamir zone (Fig. III-16), as part of the Vanch-Yazgulem Group. It consists of quartz-mica schist, quartzite, arkosic metasandstone, marble, mafic metavolcanics, conglomerate and graphitic schist. The total thickness is around 4000 m.

Vendian to lower-middle Palaeozoic sediments (predominantly carbonates) form an independent structural stage (Tseisler et al., 1984).

Alpine fold terrain

In the USSR, the Alpine fold terrain occupies the southern part of the Mediterranean belt and includes the East Carpathians, the Crimea, the Caucasus, Greater Balkhan and Kopetdag (Fig. III-14). The terrain has a complex structure, caused by the juxtaposition of Precambrian and Palaeozoic continental crustal blocks, separated by narrow zones of Mesozoic to Cenozoic geosynclinal systems.

The oldest complexes in the East Carpathians crop out only in the

Marmarosh massif, the greater part of which lies in Romania. The massif consists of schist and gneiss, the Delovetska and Belopotok Groups. The age of the Delovetska Group is defined on the basis of microphytofossils, supposedly upper Proterozoic, as well as zircon ages on lavas, which yield closely similar values in the interval 665–640 Ma (Rudakov, 1982). The gneiss and schist in the Belopotok Group are provisionally referred to an older formation, although the age of metamorphism of these rocks, from Rb–Sr and K–Ar data on biotite, has been determined at 530–550 Ma and 595–650 Ma. Galena gives an age of 610 Ma.

The nature and age of the base of the structures in the Crimean mountains is unknown. It is assumed to lie at a depth of 4–5 km and to be formed of crystalline rocks, possible Precambrian (Tseisler et al., 1984).

The mega-anticlinorium in the Greater Caucasus stretches from the Gulf of Kerch in the west to the Apsheron Peninsula in the east (Fig. III-15). The oldest metamorphic rocks in the Caucasus are divided into two complexes.

Rocks of the lower complex are exposed in anticlinoria in the Main Range, where there are outcrops of schist and gneiss with relict granulite facies rocks, marble, hornblende schist and amphibolite. Among the schists are varieties with garnet, staurolite and kyanite. Eclogites are included in associations with serpentinites in the amphibolites of the Peredovoy Ridge (Zaridze, 1982). The schists are cut by granitoids with muscovite Rb–Sr ages of 790 Ma and 850 Ma (Moralev and Perfilyev, 1978), providing an upper age boundary for these rocks.

The upper complex includes mostly metapelites in the greenschist-epidote-amphibolite facies: mica schist and amphibolite; in the upper part of the section there are mafic and intermediate metavolcanics, tuff, phyllite and conglomerate. Serpentinite bodies are encountered among the rocks in the upper complex, which are cut by granitoids with a K–Ar age of 450–470 to 500 Ma (Moralev and Perfilyev, 1978); they are provisionally referred to the Late Proterozoic.

The Lesser Caucasus are separated from the Greater Caucasus by the Trans-Caucasian massif which almost everywhere, with the exception of the Dzirul basement high, is hidden under a sedimentary cover, like the other massifs in the Lesser Caucasus (Fig. III-15). The crystalline rocks which crop out in these massifs can only provisionally be assigned to the Precambrian.

The fold systems of Kopetdag, Greater Balkhan and Kubadag in Soviet Central Asia (Fig. III-14) have a basement formed of Precambrian, Palaeozoic and Triassic rocks. The main basement outcrops are in Iran. The basement to the West Turkmen basin contains the Turkmen massif, which may be a Precambrian microcraton (Barkhatov, 1974).

5. CRUSTAL EVOLUTION

An examination of the spatial distribution of fold belts of different ages exposed at the present surface shows that the Caledonian-Hercynian belts in most cases occupy an intercratonic position and in a sense they continue the development of Precambrian belts, while Mesozoic and Cenozoic belts display clearly cross-cutting relationships with Precambrian and Palaeozoic belts, that is they are more "independent" (Condie, 1978; Kratz et al., 1982; Kozakov, 1986). Arising from this, we have more certain grounds for concluding that Palaeozoic belts have inherited certain structural characters from the Precambrian, as distinct from Mesozoic and Alpine belts.

At present we have determined the main age boundaries in the Precambrian geological evolution of younger fold belts, which enables us to note areas of similarity and difference in the Precambrian rocks found in these belts and compare them with formations in the basement to ancient cratons.

All age analogues of Archaean and Proterozoic terrains in the craton basement of North Eurasia have been identified in Palaeozoide massifs and infrastructure of the Urals-Mongolia belt.

Geological formations older than 3000 Ma in most cases are represented by metamorphic rocks formed predominantly at granulite facies conditions. As a rule, these granulites formed in stable low- or medium-pressure regimes, and the composition of these Archaean complexes is characterised by a sialic type of evolution (Mitrofanov et al., 1984; Ma and Wu, 1981), when the enderbite-granulite association formed, the primary source materials of which were tonalitic suites and their products.

Lateral differentiation of crustal regimes becomes noticeable during the Late Archaean. Regional metamorphism of Late Archaean rocks found amongst Palaeozoides frequently took place under high pressure conditions. The same conditions prevailed in structural belts in ancient cratons, which evolved simultaneously with structural belts (granite-greenstone terrains) that tended to be low-pressure high-gradient regimes (Mitrofanov et al., 1984). The formation of high-pressure complexes is typical on the whole for younger fold belts (Moskovchenko, 1982). Their origin in the Late Archaean characterizes a change from a quasi-simatic type of evolution of the belts separating the future ancient cratons. Amphibolite-gneiss terrains — the so-called "melanocratic basement" of Central Asia (Makarychev et al., 1982) — belong to this type, also the Taratash Hercynide complex in the Urals (Ivanov et al., 1982) and the many complexes surrounding the Siberian craton.

Differences in the development of internal processes in Precambrian formations of ancient cratons and in Palaeozoides of the Urals-Mongolia belt is more marked in the Early Proterozoic. During this period, epi-Archaean proto-platforms or cratons separated out within the confines of ancient cratons (Bekker, 1978; Lao et al., 1983; Ma and Wu, 1981; Salop, 1983). In individual cases they are overlain by a lower Proterozoic sedimentary cover. These proto-platforms were separated by mobile belts. Regional metamorphism typically was of variable temperature and pressure regimes: in proto-platforms, grades higher than greenschist facies are unknown, while in the mobile belts we find high-temperature subfacies of the amphibolite and even granulite facies. In pre-Riphean Palaeozoide complexes, however, the Early Proterozoic metamorphic event was ubiquitous. It affected both lower Proterozoic assemblages and Archaean terrains and was accompanied by intense retrograde metamorphism of the latter, which proceeded mainly under high-temperature regimes. Broadly speaking, we can view all the pre-Riphean Palaeozoide formations in the Urals-Mongolia belt as belonging to a single Early Proterozoic metamorphic belt.

During the Late Proterozoic, a platform regime was established within the boundaries of the present ancient cratons of Northern Eurasia. The appearance of active crust-forming processes at this time is related to the development of the Urals-Mongolia belt, in which we see a marked lateral differentiation in metamorphic regimes, in terms of both temperature and pressure. Glaucophane schist and eclogite-glaucophane schist metamorphic complexes originated at this time, while in zones parallel to these, low-pressure greenschist facies metamorphic rocks formed, as well as structural belts displaying zonal metamorphism which reached the level of the high-temperature subfacies of the amphibolite facies. Glaucophane schist complexes are known in the Caledonides of the Altay Mountains (Dobretsov et al., 1972; Dook, 1982a, b) and in Eastern Transbaikalia (Milkevich, 1982). Eclogite-greenschist rocks are represented by the Maksyutov complex in the Hercynides of the South Urals (Miller, 1977). This high-pressure metamorphism is characteristically manifest in Riphean eugeosynclinal sequences, while miogeosynclinal sediments are metamorphosed at lower pressures. The superimposition of Riphean metamorphism on Early Precambrian rocks also took place unevenly. In some cases there was intense metamorphic reworking, during which the metamorphic mineral assemblages of preceding cycles were almost completely replaced by new mineral growth; in other cases, pre-Riphean complexes are practically unaffected by Late Proterozoic retrogression. In other words, the Palaeozoides contain structural zones where metamorphic belts evolved during the Late Proterozoic, in which both Riphean and pre-Riphean complexes were reworked. These belts separated relatively rigid unreworked cratonic blocks, composed of Early Proterozoic and Archaean rocks, and overlain by a cover of volcanosedimentary Riphean rocks. In these circumstances, metamorphic processes in the mobile belts generally evolved as high- and medium-pressure regimes, while the cratonic terrains experienced low-pressure regimes. The features outlined here suggest that the metamorphic belts of this age are paired belts, indicating increasing lateral differentiation of thermodynamic regimes in the Earth's crust.

Palaeozoic regional metamorphism is typified by temperature differentiation, from greenschist to low-temperature subfacies of the amphibolite facies, against a background of predominantly high-gradient regimes.

The deformational style of structural reworking of Precambrian complexes also changed with time.

The majority of geological formations with an age greater than 3000 Ma were intensely reworked together with Late Archaean rocks and are "welded" together into a single infrastructure.

During the Early Proterozoic, structural zones emerged within the future ancient cratons which were relatively weakly affected by superimposed folding — the proto-platforms (cratons), and Early Proterozoic belts separating them, in which Archaean rocks were completely reworked (Anon., 1980; Lao et al., 1983; Ma and Wu, 1981) together with lower Proterozoic assemblages. The same style of reworking is also characteristic for all the pre-Riphean basement inliers in the Urals-Mongolia belt.

During the Late Proterozoic, reworking of the pre-Riphean basement by folding within the bounds of the ancient cratons was limited and is restricted to riftogene structures, which frequently are a continuation of the evolution of Early Proterozoic mobile zones (Milanovsky, 1983). Metamorphic belts with high- and medium-pressure regimes in the Urals-Mongolia belt contain pre-Riphean complexes deformed together with Riphean successions, forming the infrastructure of the latter. However, in blocks consisting of early Precambrian rocks with a cover of predominantly Riphean low-pressure greenschist rocks, there is no evidence of such reworking by folding, and deformational styles during the Late Proterozoic cycle in the basement here are mostly related to faulting.

Palaeozoic deformational events were responsible for refolding of rocks in the Precambrian basement on only a limited scale, both within the boundaries of the broad Palaeozoide terrains and in the ancient cratons. They are represented in the main by faulting, and in the internal structure of Precambrian metamorphic complexes, evidence for structural events belonging to the Early and Late Proterozoic tectono-metamorphic cycle is usually preserved. Reworking often affects only block margins. Riftogene structures are a reflection of Palaeozoic (and Late Proterozoic) tectonic cycles in ancient cratons (Milanovsky, 1983).

Thus, the comparative features presented above for the structural and metamorphic reworking of Precambrian complexes in the Palaeozoides of the Urals-Mongolia belt and in the basement to the ancient cratons separated by the belt illustrate the different patterns of geological processes in these major crustal structures, beginning as far back as the early Precambrian. Intense destructive processes probably occurred in the sialic proto-crust in the area occupied by Late Archaean mobile belts of the basement to ancient cratons, characterised by high-pressure regimes. In contrast to this, destruction of the crust was not so intense in the Late Archaean granite-greenstone terrains. Widespread and in places total reworking of the Archaean proto-metamorphic layer occurred during the Early Proterozoic in mobile belts which were evolving from terrains with high-pressure metamorphism. At the same time, Early Proterozoic internal processes effected crustal reworking only in zones within the future ancient cratons, leaving the epi-Archaean proto-platforms untouched. Intense reworking of early Precambrian rocks in the belts also continued in the Late Proterozoic, when the ancient cratons had already formed. In other words, in the zone of the future Urals-Mongolia belt, the most intense thermal and mass transfer between the crust and upper mantle took place over the greater part of Precambrian geological history, and this evidently predetermined the siting of Palaeozoic mobile belts here (Raaben, 1975). In general terms, if the Precambrian structure of the basement to ancient cratons is defined by the juxtaposition of epi-Archaean proto-platforms (cratons) and Early Proterozoic fold belts, then active mobile development of Early and Late Proterozoic mobile belts continued in the area occupied by Palaeozoides. These mobile belts contain no Archaean rocks that were unaffected by Early Proterozoic structural and metamorphic reworking.

By examining the composition of Precambrian rocks in the Palaeozoides, we may conclude that the destructive processes which periodically repeated themselves in the belt separating the ancient cratons did not result in the total break-up of the proto-metamorphic layer in the Precambrian. This may be related to the greater plasticity of the ancient lithosphere, due to a higher heat flow (Wynne-Edwards, 1976). On these grounds, early Precambrian mobile belts cannot be considered to be direct analogues of Phanerozoic belts. The Late Proterozoic was a transition period in geological evolution. Destructive processes were not accompanied at this time by the large-scale production of oceanic crust, but are emphasized by the emplacement of broad basic dyke swarms, while internal regimes characterize conditions of activation and greatly increased lateral differentiation of the Earth's crust. This can be viewed as a pre-geosynclinal period in crustal evolution (Ivanov et al., 1982; Ivanov, 1983). The geosynclinal process proper, with relatively large-scale new oceanic crust generation, begins at the end of the Riphean, which is fixed by ophiolite belts, dated as Upper Riphean to Vendian. The process of structural and chemical transformation of oceanic crust into continental assumes greater importance from this period, compared to processes involving the reworking of the Archaean proto-metamorphic layer and its Early Proterozoic sedimentary cover, which were characteristic for early Precambrian tectogenesis (Getsen, 1970).

The cyclical evolution of internal processes in Precambrian terrains is accompanied by repeated periods of crustal extension and contraction (Milanovsky, 1983), manifest in different ways within the present limits of

TABLE III-2

Typical ore deposits and ore shows

Deposit or show, useful mineral	Tectonic setting	Ore formational and genetic type of mineralization	Geological and mineralogical features				Geological age of host rocks mineralization
			Host or mineralized rocks	Shape of deposit, morphology of ore bodies	Ore type	Mineral composition of ores (minor phases) and hydrothermally altered rocks	
Timan-Pechora and Taymyr fold belts							
Nickel, copper, cobalt ore shows	Timan-Kanin zone	Copper-nickel sulphide; magmatic-liquation (related to gabbro-diabases)	Gabbro-diabase	Lenticular deposits	Disseminated	Pyrrhotite, pentlandite, chalcopyrite (pyrite, magnetite, etc.)	Late Riphean
Rare metal and mica (muscovite) shows	Chetlasky Kamen high	Rare metal, muscovite, associated with granites and pegmatites	Granite, pegmatite	Cavities	Disseminated, cavity infills	Cassiterite, fluorite, molybdenite, muscovite	Late Riphean
Taymyr pyrite polymetallic province; lead, zinc, copper	Tyamyr fold belt	Lead-zinc sulphide; sedimentary-volcanic	Sandstone, siltstone, volcanic and carbonate rocks	Conformable bedded and lenticular deposits	Disseminated, massive vein-type	Pyrite, sphalerite, galena (pyrrhotite, chalcopyrite, magnetite, etc.)	Middle Riphean
Sayan-Yenisey province (Yenisey Ridge)							
Lower Angara etc. (Angara-Pit basin); iron	Angara-Pit syncline	Quartz-haematite; sedimentary	Grit, sandstone, silty and calcareous shale	Bedded deposits	Densely disseminated (hematitic grit)	Hematite, hydrohematite, goethite (siderite, magnetite)	Late Riphean
Yenashim etc. (Yenashim iron ore zone); iron	Karpinsky Ridge anticlinorium	Magnetite; skarn	Skarn, granite, amphibolite, marble	Irregular deposits	Massive, disseminated	Magnetite	Early Riphean
Borisikha; titanium, iron	Borisikha intrusion	Titanomagnetite; late magmatic	Gabbro, pyroxenite	Lenticular, bed-type deposits	Disseminated, massive	Titanomagnetite, ilmenite, magnetite (hematite, martite)	Late Riphean
Tayezhnoye; manganese	Orlov (Bolshepit synclinorium)	Manganiferous graphitic-siliceous-carbonate; volcanosedimentary	Graphite-clay shale, siliceous shale	Beds, lenses	Massive	Manganocalcite, rhodochrosite (mangano-ankerite, pyrite, marcasite, pyrrhotite, gold)	Late Riphean
Porozhin; manganese	Vorogov synclinorium	Manganiferous tuffaceous-silica-carbonate; volcano-sedimentary	Dolomite, siliceous tuff	Beds, lenses	Massive	Manganite, pyrolusite, psilomelane (vernadite, pyrite, goethite, apatite)	Vendian
Gorevo; lead, zinc	Yenisey synclinorium	Galena-sphalerite; stratiform-exhalative-sedimentary	Limestone	Conformable layered and lenticular deposits	Massive, vein-type disseminations	Sphalerite, galena, pyrite, pyrrhotite (hematite, magnetite, psilomelane etc.); siderite, ankerite, quartz	Late Riphean
Lineynoye (Rassokh ore concentration); lead, zinc, copper	Orlov-Kaitbi synclinorium	Galena-sphalerite-pyrite; stratiform – volcano-sedimentary	Graphite-clay shale, graphite-silica shale acid tuffs	Bedded type deposits	Massive, disseminated	Pyrite, galena, sphalerite, (pyrrhotite, chalcopyrite, arsenopyrite, melnikovite)	Late Riphean
Uderey; antimony, gold	Tatar anticlinorium	Quartz-antimony; hydrothermal	Sericite, quartz-sericite, chlorite-sericite schists	Dendritic veins	Massive, brecciated, cavities, disseminated	Antimonite, bartierite, pyrite (arsenopyrite, siderite, etc.)	Late Riphean
Gerfed etc.: gold	Tatar anticlinorium	Gold-quartz; hydrothermal	Quartz-chlorite-sericite, quartz-sericite, graphite-clay shales	Veins	Disseminated	Pyrite, pyrrhotite, arsenopyrite, magnetite, gold (chalcopyrite, sphalerite, galena)	Riphean
Kirgitey; talc, magnesite	Angara-Pit synclinorium	Talc-magnesite; sedimentary, hydrothermal-metasomatic	Argillaceous and chloritic shales, dolomite	Bedded and lenticular deposits	Massive, brecciated	Magnesite, talc	Riphean
Kazakhstan-Tien Shan region							
Balbraun, etc. (Karsakpay group of deposits); iron	Ulutau anticlinorium	Ferruginous quartzites; metamorphic	Metamorphosed diabase porphyrites & tuffs, quartzites, marbles, schists	Conformable bedded & lenticular deposits	Densely disseminated	Hematite (magnetite, pyrite, siderite, apatite)	Riphean
Akkuduk; iron	Kendyktass high	Magnetite; skarn	Marbles, amphibolites, qz-fs-hb schists, magnesian skarns	Irregular-shaped deposits	Massive, brecciated, disseminated	Magnetite (chalcopyrite, pyrite, bornite, sphalerite)	Proterozoic
Gvardeyskoye, etc. (Zhuanto-Bin basin);	Syncline within the Zhuanto-Bin high	Ferruginous quartzites; volcano-sedimentary	Phyllitic schists, porphyroids, quartzites, marbles	Bedded, lenticular deposits	Densely disseminated	Hematite, magnetite (martite)	Riphean
Tanen, etc. (Temir ore zone)	Syncline within the Chuy high	Ferruginous quartzites; volcano-sedimentary	Phyllitic schists, quartzites, quartz-carbonate schists	Bedded, lenticular deposits	Densely disseminated	Hematite, magnetite (martite)	Riphean
Tastau; vanadium	Chuy high	Vanadium; sedimentary siliceous schists	Graphite-siliceous schists	Bedded deposits	Finely disseminated	Roskoelite (?)	Vendian
Orumbay I; lead, zinc	Karakamys block	Pyrite-polymetallic; stratiform volcano-sedimentary	Graphitic schists, marble	Bedded deposits	Massive, layer disseminations, vein-type disseminations, brecciated	Pyrite, marcasite, pyrrhotine, sphalerite, galena (chalcopyrite, arsenopyrite)	Early-Middle Proterozoic
Karakamys South-among East; tungsten	Karakamys block	Scheelite; stratiform-tactitic	Skarn marbles granitized gneisses	Bedded deposit	Finely disseminated, small cavity infills	Scheelite (bismuth, sphalerite, pyrite, marcasite)	Middle Proterozoic
Akmola; graphite	Chu-Iliy zone	Graphite, metamorphic	Gneisses, marbles, diorites	Conformable bedded and lenticular	Finely & coarsely lamellar, mostly disseminated	Graphite	Proterozoic
Muruntau; gold	Muruntau syncline	Gold-quartz-sulphide; hydrothermal	Sericite-chlorite-quartz schists, graphitic microquartzites	Stockwork	Disseminated, veins & veinlets	Pyrite, arsenopyrite, gold (pyrrhotine, scheelite, bismuthite, sphalerite, galena, etc.)	Riphean
Urals belt							
Bakalian; iron	Bashkir high	Siderite; sedimentary-hydrothermal (?)	Quartzites, calc-mica schists, dolomites	Bedded deposits	Massive	Siderite (ankerite, pyrite, etc.)	Early Riphean
Kusin; iron, titanium, vanadium	Bashkir high	Titanomagnetite; late magmatic	Amphibolites, gabbro amphibolites	Bedded and vein-type bodies	Massive, disseminated	Titanomagnetite, magnetite, ilmenite (hematite, pyrite, pyrrhotine, etc.)	Early Riphean
Komarov-Zigazin group of deposits; gold	Bashkir high	Siderite; sedimentary-hydrothermal (?)	Shales, dolomites	Bedded deposits	Massive	Siderite (ankerite, pyrite, etc.)	Late Riphean
Oreshows in the river Kharbey, Khanmey & Badya-Yugan basins; gold	Polar Urals, Kharbey anticlinorium	Ferruginous quartzites; metamorphic	Amphibole & chlorite-quartz-mica schists	Lenticular deposits	Dense disseminations	Magnetite, hematite (pyrite)	Late Riphean-Cambrian
Ore shows in river Kharmatlou, Khoy-dymyur and other basins; copper	Polar Urals, Kharbey anticlinorium	Copper pyrite; volcanogenic-hydrothermal	Andesitic porphyrites, tuffs, sericite-chlorite-quartz schists	Lenticular deposits	Disseminated, veinlets	Pyrite, chalcopyrite (sphalerite, pyrrhotine, bornite); chlorite, sericite, quartz	Late Riphean-Cambrian
Brigadnoye; aluminium	East Mugodzhary anticlinorium	Kyanite; metamorphic	Garnet-biotite & two-mica gneisses with thin qz-mica schists	Conformable bedded deposits	Massive, cavities	Kyanite (staurolite)	Riphean
Satkin; magnesite	Bashkir high	Magnesite; hydrothermal-metasomatic (?)	Dolomites	Bedded deposits	Massive	Magnesite (calcite, dolomite, ankerite, aragonite, graphite, talc, pyrite, etc.)	Early Riphean
Aktasty; graphite	East Mugodzhary anticlinorium	Graphite; metamorphic	Qz-muscovite & qz-kyanite schists	Conformable lenticular-bedded deposits	Lamellar-disseminated, massive	Graphite	Riphean
Bugetysay; anthophyllite-asbestos	East Mugodzhary anticlinorium	Anthophyllite-asbestos; metamorphic	Ultramafics	Irregular deposits	Massive	Chrysotile (bastite, antigorite)	M. Riphean | U. Riphean
Karasay etc. (East Mugodzhary zone of mica pegmatites)	East Mugodzhary anticlinorium	Muscovite; pegmatitic	High-alumina gneisses, migmatites	Pegmatite veins	Cavities-coarsely/finely tabular	Muscovite, microcline quartz, plagioclase	Late Riphean
Mongolia-Okhotsk province							
Serpukhov etc. (Khingan group of deposits); manganese	Khingan-Burey massif	Mn-Fe quartzites; sedimentary-metamorphic	Metamorphosed qz-clay shales with thin sandstones, dolomites, limestones	Bedded deposits	Dense, bedded disseminations	Braunite, hematite, gausmanite, rhodochrosite, oligonite (rhodonite, magnetite, etc.)	Riphean

ancient platforms and the Palaeozoic fold belts separating them: starting from the Late Archaean, they are more intense in the intracratonic zone, which also appear to have predetermined the siting and development of the Urals-Mongolia belt in the late Precambrian-Proterozoic.

Major Mesozoic and Cenozoic belts cut both Archaean and Proterozoic structures and Palaeozoides. The Precambrian rocks found in them bear no specific features that are characteristic of these belts themselves. For example, in the north-east USSR, the Mesozoides of the Pacific belt contain fragments of Archaean cratons which are comparatively weakly reworked by Proterozoic tectogenesis: the Okhotsk and Omolon massifs (Bibikova et al., 1982; Anon., 1980). The western part of the Pacific belt cuts Hercynian structures of the Mongolia-Okhotsk belt (Fig. III-1), whose Precambrian rocks were reworked during Early and Late Proterozoic tectogenesis, as well as in the Palaeozoides of the Urals-Mongolia belt. Farther south, Mesozoic to Cenozoic structures of the Pacific belt cut epi-Archaean proto-platforms of the North Sinian craton.

Structural zones in the Mediterranean belt occupy a similar position relative to Precambrian and Palaeozoic belts (Fig. III-1).

Thus, we encounter both fragments of epi-Archaean proto-platform massifs (cratons) and fragments of ancient crystalline rocks, reworked by Proterozoic tectogenesis among Mesozoic to Cenozoic belts. However, for the Palaeozoides of the Urals-Mongolia belt, only the latter are characteristic.

We may conclude that Precambrian prehistory confirms that it is appropriate to identify two global episodes of Phanerozoic tectogenesis: Caledonian-Hercynian and Mesozoic-Cenozoic (Kratz et al., 1982). In the first of these, the major process was evidently the break-up of Pangea continental crust, leading to the formation of individual continental masses. For the Mesozoic-Cenozoic episode, sequential destructive and constructive processes were more characteristic, with direct connections between continental and oceanic structures.

The most characteristic mineral deposits and ore shows for the Precambrian in younger fold belts and terrains are shown in Table III-2.

REFERENCES

(All references are in Russian unless otherwise stated)

Abdulin, A.A., Avdeyev, A.V., Kasymov, M.A., Matviyenko, V.N., Tokmacheva, S.G. and Yaroslavtseva, N.S., 1982. Precambrian massifs in Palaeozoic belts of Kazakhstan. In: The Precambrian in Phanerozoic fold belts, Leningrad, Nauka, pp. 113–120.

Anon., 1974a. Basement structure of cratonic regions of the USSR. Explanatory note to the tectonic map of the basement in the USSR, scale 1 : 5,000,000, Leningrad, Nauka, 400 pp.

Anon., 1974b. Tectonic map of the basement of the USSR, scale 1 : 5,000,000.

Anon., 1977. Metamorphic complexes of Asia, Novosibirsk, Nauka, 350 pp.

Anon., 1979. Tectonic map of Northern Eurasia, scale 1 : 5,000,000.

Anon., 1980. Stages and types of evolution of Precambrian crust in ancient shields, Leningrad, Nauka, 164 pp.

Anon., 1982. The Precambrian of Central Asia, Leningrad, Nauka, 263 pp.

Barkhatov, B.P., 1974. Alpine fold systems in the south and south-west of the USSR. In: Basement structure of cratonic regions of the USSR, Leningrad, Nauka, pp. 309–323.

Bekker, Yu.R., 1975. Bakeyev sediments in the Riphean stratotype succession. Izvestiya Akad. Nauk SSSR, Ser. Geologiya, No. 6.

Bekker, Yu.R., 1978. The Urals-Timan fold belt. In: Precambrian of the continents. Folded regions and younger platforms of Eastern Europe and Asia, Novosibirsk, Nauka, pp. 38–56.

Bibikova, Ye.V., Makarov, V.A., Gracheva, T.V. and Kalinkina, O.M., 1982. Ancient rocks in the Omolon massif. In: Ancient granitoids of the USSR, Leningrad, Nauka, pp. 137–146.

Chernaya, I.P., Cherny, V.G. and Raaben, M.Ye., 1979. Fundamental questions on Upper Riphean and Vendian stratigraphy of Timan. In: Stratigraphy of the upper Proterozoic in the USSR: Riphean and Vendian, Leningrad, Nauka, pp. 102–107.

Condie, K.C., 1978. Plate tectonics and crustal evolution, Pergamon Press, New York, 288 pp. [in English].

Dedeyev, V.A. and Keller, B.M. (Editors), 1986. The upper Precambrian of the European North USSR — explanatory note to the stratigraphic scheme, Syktyvkar, 41 pp.

Dobretsov, N.L., 1974. Glaucophane schist and eclogite-glaucophane schist complexes in the USSR, Novosibirsk, Nauka, 429 pp.

Dobretsov, N.L., Lepezin, G.G. and Pukinskaya, O.S., 1972. Glaucophane schists in the Altay-Sayan fold belt. Doklady Akad. Nauk SSSR, Vol. 206, No. 1, pp. 200–203.

Dook, G.G., 1982a. High-pressure greenschist belts, Leningrad, Nauka, 184 pp.

Dook, G.G., 1982b. Criteria for identifying low-temperature late Precambrian complexes in the initial stages of development of Phanerozoic fold belts. In: The Precambrian in Phanerozoic fold belts, Leningrad, Nauka, pp. 188–196.

Dunayev, V.A., Ovchinnikov, L.N. and Krasnobayev, A.A., 1969. Absolute age of geological formations in the Ilmen mountains. Doklady Akad. Nauk SSSR, Ser. Geol., Vol. 186, No. 5.

Fedorovsky, V.S., 1985. The Lower Proterozoic of the Baikal mountain region, Moscow, Nauka, 200 pp.

Garris, M.A., 1973. The late Precambrian of the Urals, Timan and Mugodzhary. In: Geochronology of the USSR, Vol. 1, Leningrad, Nedra.

Garris, M.A., Kamaletdinov, M.A. and Shvetsov, P.N., 1979. The main geochronological boundaries and tectono-magmatic episodes of Precambrian cycles in the Urals. In: Archaean and lower Proterozoic stratigraphy of the USSR, Leningrad, Nauka, pp. 192–205.

Garris, M.A., Kazakov, G.A., Keller, B.M., Polevaya, N.I. and Semikhatov, M.A., 1964. A geochronological scale for the Upper Proterozoic (Riphean and Vendian). In: The absolute age of geological formations. XXII Session, Int. Geol. Congress. Soviet Contributions, Problem 3, Moscow, Nauka.

Garris, M.A. and Postnikov, D.V., 1970. Late Precambrian geochronological boundaries. In: Proc. XV Session of the Commission for Determining the Absolute Age of Geological Formations, Moscow, Nauka.

Getsen, V.T., 1970. Stratigraphic and tectonic position of the Upper Riphean carbonate succession in Timan and the Kanin Peninsula. Byull. MOIP, Geol. Section, Vol. 45, No. 1, pp. 58–70.

Getsen, V.T., 1975. Structure of the basement in North Timan and the Kanin Peninsula, Leningrad, Nauka, 144 pp.

Gintsinger, A.B. and Vinkman, M.K., 1978. The Kuznetsky Alatau and Highland Shoriya. In: The Precambrian of the continents. Folded regions and younger platforms of Eastern Europe and Asia, Novosibirsk, Nauka, pp. 198–202.

Gintsinger, A.B., Vinkman, M.K. and Fefelov, A.D., 1979. The structure of the Precambrian succession in the Sangilen highlands, Tuva. In: Upper Precambrian of the Altay-Sayan fold belt, Novosibirsk, pp. 92–119.

Glukhovsky, M.Z. and Moralev, V.M., 1978. The Taymyr fold belt. In: The Precambrian of the continents. Folded regions and younger platforms of Eastern Europe and Asia, Novosibirsk, Nauka, pp. 147–168.

Gusev, G.S. and Mokshantsev, K.B., 1978. The Verkhoyan-Chukchi fold belt. In: The Precambrian of the continents. Folded regions and younger platforms of Eastern Europe and Asia, Novosibirsk, Nauka, pp. 123–146.

Ilyin, A.V., 1971. The Tuva-Mongolia massif. Trudy NII Zarubezhgeologiya, No. 22, pp. 67–73.

Ivanov, S.N., 1983. The prehistory of geosynclines: the Urals as a case study. Priroda, No. 6, pp. 60–68.

Ivanov, S.N., Krasnobayev, A.A. and Rusin, A.P., 1982. Precambrian of the Urals. In: The Precambrian in Phanerozoic fold belts, Leningrad, Nauka, pp. 81–94.

Keylman, G.A., 1974. Migmatite complexes in mobile belts, Moscow, Nedra, 200 pp.

Kiselev, V.V. and Koralev, V.G., 1978. Central Asia and Central Kazakhstan. In: The Precambrian of the continents. Folded regions and younger platforms of Eastern Europe and Asia, Novosibirsk, Nauka, pp. 169–193.

Kozakov, I.K., 1986a. Geotectonic implications of an analysis of the spatial setting of Precambrian and Phanerozoic mobile belts. In: Problems in the evolution of the Precambrian lithosphere, Leningrad, Nauka.

Kozakov, I.K., 1986b. Precambrian infrastructural complexes in the Palaeozoides of Mongolia, Leningrad, Nauka.

Kozlov, V.I., 1980. Riphean stratigraphy of the Southern Urals. In: The pre-Ordovician history of the Urals. 2. Stratigraphy, Sverdlovsk, Urals Scientific Centre, Acad. Sci. USSR, pp. 3–32.

Krasnobayev, A.A., 1980. Main results and problems from a geochronological study of the Urals. In: The pre-Ordovician history of the Urals. 1. General problems (preprint), Sverdlovsk, Urals Scientific Centre, Acad. Sci. USSR, pp. 28–39.

Kratz, K.O., Mitrofanov, F.P. and Khiltova, V.Ya., 1982. The Precambrian and Phanerozoic fold belts. In: The Precambrian in Phanerozoic fold belts, Leningrad, Nauka, pp. 45–52.

Kushev, V.G. and Vinogradov, D.P., 1978. Metamorphic eclogites, Novosibirsk, Nauka, 111 pp.

Lao, Q., Sno, S. and Tan, Y., 1983. Proterozoic tectonics of China. Geol. Review, Vol. 29, No. 2, pp. 111–120 [in Chinese].

Lvov, A.B., Neyelov, A.N., Bogomolov, Ye.S. and Mikhaylova, N.S., 1985. The age of metamorphic rocks in the Ganal Range, Kamchatka. Geologiya i Geofizika, No. 7, pp. 47–57.

Ma, X. and Wu, Z., 1981. Early tectonic evolution of China. Precambrian Research, Vol. 14, No. 3–4, pp. 185–202 [in English].

Ma, X., Wu, Z., Tan, Y. and Hao, C., 1979. Tectonics of the North China platform basement. Acta Geologica Sinica, Vol. 4, pp. 293–306 [in Chinese].

Makarychev, G.I., Morkovkina, V.I., Paley, I.P. and Gavrilova, S.I., 1982. Basement structure of Precambrian inliers in Central Asia. In: The Precambrian in Phanerozoic fold belts, Leningrad, Nauka, pp. 154–162.

Milanovsky, Ye.Ye., 1983. Taphrogenesis in the history of the Earth, Moscow, Nedra, 279 pp.

Milkevich, R.I., 1982. Principles of Precambrian subdivision in the Mongolia-Okhotsk fold belt, Eastern Transbaikalia. In: The Precambrian in Phanerozoic fold belts, Leningrad, Nauka, pp. 213–217.

Miller, Yu.V., 1977. The Maksyutov complex of the Southern Urals. In: Structural evolution of metamorphic complexes, Leningrad, Nauka, pp. 104–114.

Mitrofanov, F.P., Kozakov, I.K. and Paley, I.P., 1981. The Precambrian of Western Mongolia and Southern Tuva, Leningrad, Nauka, 156 pp.

Mitrofanov, F.P., Kozakov, I.K. and Zinger, T.F., 1977. Early Precambrian conglomerates with granulite boulders in the Sangilen central massif, Caledonides of Tuva. In: Problems in early Precambrian geology, Leningrad, Nauka, pp. 232–238.

Mitrofanov, F.P., Moskovchenko, N.I., Khiltova, V.Ya. and Kozakov, I.K., 1984. Problems of Archaean and Phanerozoic fold belts of Eurasia. In: Abstracts, 27th Int. Geol. Congress, Moscow, 03, pp. 344.

Moralev, V.M. and Perfilyev, Yu.S., 1978. The Alpine-Himalaya fold belt. In: The Precambrian of the continents. Folded regions and younger platforms of Eastern Europe and Asia, Novosibirsk, Nauka, pp. 224251.

Moskovchenko, N.I., 1982. Precambrian high-pressure complexes in Phanerozoic fold belts, Leningrad, Nauka, 161 pp.

Muratov, M.B., 1965. Geosynclinal fold belts of Eurasia. Geotektonika.

Neyelov, A.N. and Milkevich, R.I., 1974. Fold systems and intervening massifs in the Mongolia-Okhotsk Variscan belt. In: Basement structures in platform regions of the USSR, Leningrad, Nauka, pp. 294–303.

Plyaskin, A.M., 1979. Some problems in the Precambrian stratigraphy of Timan. In: Upper Proterozoic stratigraphy of the USSR (Riphean and Vendian), Leningrad, Nauka, pp. 109–110.

Pogrebitsky, Yu.Ye., 1972. A palaeotectonic analysis of the Taymyr fold system. Leningrad, Nauka, 248 pp.; Trudy NIIGA, Vol. 166.

Raaben, M.Ye., 1975. The Upper Riphean as a unit of the general stratigraphic scale. Moscow, Nauka, 253 pp.; Trudy GIN Akad. Nauk SSSR, Vol. 273.

Raznitsyn, V.A., 1968. Tectonics of Central Timan, Leningrad, Nauka, 221 pp.
Rudakov, S.G., 1982. The presence of Proterozoic formations in metamorphic complexes of the Marmarosh massif of the Eastern Carpathians. In: The Precambrian in Phanerozoic fold belts. Leningrad, Nauka, pp. 69–72.
Salop, L.I., 1982. Geological evolution of the Earth during the Precambrian, Leningrad, Nedra, 343 pp. [English translation 1983, Berlin and New York, Springer Verlag, 459 pp.]
Shatsky, N.S. and Bogdanov, A.A., 1961. International tectonic map of Europe, scale 1 : 2,500,000. Izvestiya Akad. Nauk SSSR, No. 4, pp. 3–25.
Smirnov, A.M., 1978. The western part of the Pacific belt. In: The Precambrian of continents. Fold belts and young platforms of Eastern Europe and Asia, Novosibirsk, Nauka, pp. 252–273.
Smirnov, A.M., 1979. Pre-Riphean Precambrian of the North-west Pacific mobile belt. In: Archaean and lower Proterozoic stratigraphy of the USSR, Leningrad, Nauka, pp. 148–163.
Tseisler, V.M., Karaulov, V.B., Uspenskaya, Ye.A., and Chernova, Ye.S., 1984. Fundamentals of the regional geology of the USSR, Moscow, Nedra, 358 pp.
Volobuyev, M.I., Zykov, S.I. and Stupnikova, N.I., 1976. Geochronology of Precambrian formations in the Sayan-Yenisey region. In: Current problems in modern geochronology, Moscow, Nauka, pp. 96–123.
Votakh, O.A., Zhabin, V.V. and Kozlov, G.V., 1978. The Yenisey ridge. In: The Precambrian of the continents. Folded regions and younger platforms of Eastern Europe and Asia, Novosibirsk, Nauka, pp. 38–56.
Wynne-Edwards, H.R., 1976. Proterozoic ensialic orogenesis: the millipede model of ductile plate tectonics. American Journal of Science, Vol. 276, No. 8, pp. 927–953 [in English].
Zaridze, G.M., 1982. Precambrian in the crystalline basement of the Caucasus. In: The Precambrian in Phanerozoic fold belts, Leningrad, Nauka, pp. 94–96.

Conclusions

CRUSTAL EVOLUTION IN THE PRECAMBRIAN

The three parts of the book have systematically described the geological structure of an extensive territory that embraces the entire northern half of the Eurasian continent. This territory includes the East European and Siberian cratons, as well as the younger fold belts surrounding and separating them. The varied geological environments which existed here and alternated with each other during the Precambrian were responsible for the broad spectrum of tectonic structures observable today, and which are described in the present book. There is no doubt that the sum total of all the material presented here provides an opportunity to unravel to a significant degree both the fundamental characteristic features of the Precambrian geological history of the continent as a whole, as well as specific features of the major types of regional tectonic structures, which differ in their geological structure and the nature of the evolution of their internal and external processes that produced the pattern seen today.

The geological description of crustal fragments assigned to the Precambrian, both those that make up the crystalline shield of ancient cratons as well as those that form part of late Precambrian to Phanerozoic mobile belts, offers an opportunity to evaluate the interrelations of these major geotectonic elements which define the fundamental features of the tectonic structure of continents. Comparative characteristics of individual tectonic provinces or domains (geoblocks) within the basement to ancient cratons allow us to identify the basic Precambrian tectonic type provinces. Finally, a comparative stage-by-stage analysis of sedimentary processes, igneous activity, metamorphism and ore formation peculiar to each period in the Precambrian evolution of the Earth's crust, enables us to uncover a directional trend in the nature of the evolution of these processes, as well as to draw conclusions about particular features of the Earth's atmosphere and hydrosphere during the early Precambrian.

We now turn our attention to some concrete conclusions that follow from the material presented in the book.

* * *

(1) The main tectonic features in the structure of the Eurasian continent are reflected in the structural position of the East European and Siberian cratons (in their relationships with younger fold belts). It is interesting

to consider this question here, since both these cratons together with the North American craton and the Greenland Shield formed the single proto-continent of Laurasia during the Precambrian.

There is much in common between the two cratons in terms of their position in the present-day structural plan. The East European craton is bounded by fold belts in the north-east, east and south-east, which show no agreement with the internal structure of the crystalline basement to the craton. In this regard, the difference between the time of formation of the basement structure and the age of the fold belts bounding the craton is here around 1.5 Ga. This time gap is not characteristic of the western boundaries of the craton. There, the structures of Proterozoic age — Svecofennian or Dalslandian — are replaced by Baikalides or Caledonides, with further "younging" of fold belts in a westerly and south-westerly direction.

The Siberian craton is also bounded by fold belts to the east and south-east, whose position does not agree with the internal structure of the basement. The difference in time between the formation of the basement structure and its folded margins is insignificant here. The change in age of fold belts is seen to be more concordant in the south-western folded margin of both the Siberian craton and the East European craton.

While turning our attention to such similarities in position, it is necessary, nevertheless, to emphasize the difference in scale between the distribution of fold belts of different ages surrounding the cratons. Thus, the Baikalides (600 ± 100 Ma), which are fairly widespread in Siberia, are weakly expressed on the margins of the East European craton. In this regard, there are zones within the Siberian Baikalides which also evolved as eugeosynclinal types.

Basic features common to the geological history of both ancient cratons can also be seen in the synchronous appearance of the most important geological time boundary, corresponding to a transition to a platform type of development proper. This boundary falls on the Early/Late Proterozoic boundary, as is the case in the North American craton.

Particular emphasis in this regard should be given to the fact that all Riphean formations, both platform-type over the craton itself and geosynclinal-type in younger belts, undoubtedly have a greater propensity towards the Phanerozoic than the lower Precambrian in terms of their tectonic aspect. In fact a new major period in the evolution of the Earth's crust begins with the Riphean–late Precambrian to Phanerozoic.

(2) The variety of early Precambrian tectonic provinces which are seen to make up the crystalline basement to the East European and Siberian cratons can be grouped into three basic types: granulite-greenstone, granite-greenstone and granite-schist. Each of these three types differs in possessing a highly distinctive evolutionary history and specific types and sequences of regional internal crust-forming processes, time of stabilization

and composition of the infrastructure thus formed. Each type of tectonic domain has its own distinctive association of more minor tectonic structures, whose evolution can be traced, if we view them as sections at different time planes. If we simplify the actual relationships, we can present an outline, which is debatable to a certain extent, in which the tectonic provinces can be viewed as geoblocks, i.e. lithospheric blocks, consolidated at different times and to varying degrees.

The possibility of identifying a single assemblage of types of tectonic domains within both ancient cratons is one definite piece of evidence in favour of assigning the Siberian craton to Laurasian-type cratons, which is often doubted. Furthermore, the different nature of the relationships between the three tectonic province types identified within the East European and Siberian cratons clearly points to the specific character of the evolution of these cratons.

The various types of tectonic domain are distributed in a roughly linear fashion in the East European craton. A north–south belt of granite-greenstone terrains can be traced over the entire territory: the Dnieper, Kursk-Voronezh, Central Russian and Karelian terrains. Blocks lying to the east of this belt are on the whole typical granulite-greenstone (granulite-gneiss) terrains. The margins of western regions in the craton are defined by granite-schist terrains.

The basement to the Siberian craton is structurally more uniform. There, most of the territory evolved according to the granulite-greenstone type. Epi-Archaean cratons of granite-greenstone terrain type are seen in the interior of the craton. An exception is the Tungusska block which is provisionally assigned to the granite-greenstone type but occupies a marginal position.

It is useful to conclude this section with a brief summary of the characteristics of the basic types of tectonic domains: Archaean granulite-greenstone and granite-greenstone terrains and Early Proterozoic granite-schist terrains.

Granulite-greenstone terrains. This type includes the Kola terrain in the Baltic Shield, the Azov terrain in the Ukrainian Shield, the Stanovoy, Batomga, Anabar and Kan–Pre-Sayan in the Siberian craton. The Early Archaean in these regions occurs as relicts among granulite-gneiss terrains. Rare linear greenstone belts formed in the Late Archaean suprastructure of these terrains, evolved along deep fault zones. The linear character of tectonic structures is more clearly expressed in superimposed Early Proterozoic volcanic (riftogene) or sedimentary troughs, which divide the tectonic provinces into higher-order blocks. Unique proto-platform sequences, such as the high-alumina sequences in the Keyvy belt in the Kola terrain, formed in the concluding evolutionary stage of the terrain (end of the Late Archaean).

Igneous activity in the granulite-greenstone terrains includes the weak development of granitic diapirism, and the wide distribution of alkaline complexes, both acid and basic in composition, the formation of which began in the final stages of crustal growth and continued right into platform conditions.

Granite-greenstone terrains. Here, Early Archaean rocks occur mostly in granite-gneiss and to a lesser extent granulite-gneiss terrains as almost completely reworked relicts. The composition of these relicts is more sialic than in granulite-greenstone terrains. High-pressure granulite facies mineral assemblages in the infrastructure to granite-greenstone terrains are usually absent. The formation of greenstone belts (volcanic palaeotroughs) is a Late Archaean event. A quasi-platform regime became established in these terrains during the Early Proterozoic, as shown by clastic terrigenous sequences with subordinate amounts of volcanics, which many workers regard as continental platform-type trap lavas.

The granite-greenstone terrains also differ markedly from other types in the nature of their magmatism. Here, granite formation was especially active at 2.6 Ga ago. Granitic diapirism is typical of magmatic processes of this age. Alkali rocks are not characteristic of this type of terrain, nor are basic intrusions, as distinct from granulite-greenstone terrains, where they are abundant.

Granite-schist terrains have been most completely studied in the Baltic Shield (the north Ladoga region) and the Ukrainian Shield (in the Kirovograd block). This type of terrain often contains fragments of granulite-greenstone and granite-greenstone terrains, from which it is possible to reconstruct an earlier, pre-Proterozoic history for the terrain. The present-day outline of granite-schist terrains was formed in the Early Proterozoic in association with the development of a network of miogeosynclinal type basins. In the orogenic period transitional to cratonization, acid volcano-plutonic igneous rock associations formed, containing gabbro, anorthosite and rapakivi granites. Cratonic conditions in these terrains begin with the Riphean.

The types of tectonic provinces found in the East European and Siberian cratons do not always have direct and complete analogues among the tectonic structures identified in ancient cratons on other continents. Even a comparison with structures in the Canadian Shield can still only be done to a first approximation. For example, one of Stockwell's provinces in the Canadian Shield, the Superior, which is considered by research workers to be a granite-greenstone terrain, has no analogues in the Baltic or Aldan Shields, whether in scale or structure. The Karelian granite-greenstone terrain is more comparable in scale and structure with one of the Superior Province greenstone belts, rather than the province as a whole. However, even in such a comparison, the terrains are significantly different in terms of the volume of basic magmatism and the range of useful mineral deposits.

Late Archaean clastic terrigenous basins, found in the eastern part of the Olyokma granite-greenstone terrain (the Olyokma block) in the Aldan Shield are comparable only in the broadest outline features with the paragneiss belts of the Superior Province, and such a comparison would be more correct if we were to compare not the individual terrigenous basins described in this book (Part II, Chapter 2), but the entire eastern part of the Olyokma terrain. In terms of size, the Olyokma basins are smaller structures than the paragneiss belts of the Superior Province. We must take into account a further fact — that the paragneiss belts in the Canadian Shield consist of rocks which are frequently considered to be relatively older in comparison to the assemblages in greenstone belts. On the other hand, a younger age has been established for the assemblages in the clastic terrigenous basins of the Aldan Shield than those in the greenstone belts of the Olondo belt in particular.

The granulite-greenstone belts identified in the East European and Siberian cratons can be compared with, for example, that part of the Canadian Shield which incorporates the Pikwitoni and Cross Lake subprovinces; however, the difference between this territory and the granulite-greenstone terrains in the USSR is that the latter have preserved more complete supracrustal assemblages, both Early and Late Archaean in age.

Analogues of Early Proterozoic granite-schist and granulite-schist terrains must be sought within the Churchill structural province.

(3) The main differences in the history of the origin of the East European and Siberian cratons result from the lack of synchroneity in the geological development of these territories. This is expressed in the basement structure mainly in the different scale of distribution of terrains in which internal processes made a final display in Late Archaean and Early Proterozoic times. Early Proterozoic processes are fairly widely manifest in the basement to the East European craton, forming granite-schist mobile regions and linear belts (the Belomoride belt, for example). Territories, occupied by tectonic structures of Early Proterozoic age, occupy almost the same area as epi-Archaean granite-greenstone terrains, where a quasi-cratonic (quasi-platform) regime was established in the Early Proterozoic. The superposition of Early Proterozoic processes added a certain mosaic pattern to the distribution of the tectonic structures of different ages in the basement to the craton. This mosaic pattern was emphasized by Early Proterozoic mobile belts, which often adhere to the junction zones between different Archaean tectonic provinces.

Over most of the area of the basement to the Siberian craton, the mobile tectonic regime ceased at the end of the Archaean. During the Early Proterozoic, it was basically narrow mobile belts that formed, such that a property of the Siberian craton was the development of mobile belts unaccompanied by sedimentation.

It has already been noted that the transition to platform conditions proper took place on both cratons generally at the same time. In addition, the initial (pre-plate) stages of platform development in the East European and Siberian cratons differ substantially in many respects. Within the East European craton a discrete system of structures — pericratonic basins and aulacogen types — was formed during the Riphean, whereas the Riphean of the Siberian craton was characterized more by basin-type structures, reminiscent in many respects of structures in the sedimentary platform cover. The platform cover itself began to form on both cratons towards the end of the Late Proterozoic.

(4) When considering the evolution of the Early Precambrian lithosphere, one of the main problems is the question of the lateral distribution of continental-type sialic crust during the earliest stages of its development (in the Early Archaean).

It has been shown that Early Precambrian terrains of the same age are present in both the basement to ancient cratons and in younger fold belts. Furthermore, early Precambrian crust-forming processes in these belts as well as in structural zones in the basement to the ancient cratons took place under different conditions — at low and high pressures. It was only during late Precambrian history that these major structures of the continent — cratons and fold belts — began to differ significantly. We may thus surmise that the oldest crust, which had formed no later than the Early Archaean, occupied the greater part of the present-day continent of Eurasia during the early Precambrian.

All the factual material presented on the tectonics of the East European and Siberian cratons, the presence of inclusions of ancient structures in Early Proterozoic (Svecofennian) belts, and the common occurrence of lower Precambrian in younger fold belts, provide evidence that the major principle in the evolution of the lithosphere lies not in the spreading outward of central cores of consolidation, but rather in the repeated reworking of crust that had formed in a previous stage of lithospheric development.

As far as the basement to the cratons is concerned, there is a particularly well-marked general tendency in the geological evolution of Precambrian structures — a change with time from regionally-distributed tectonic structures to more and more localized structures, as a reflection of a more fractional differentiation of conditions.

* * *

(1) An analysis of the material presented in the book on Precambrian geological history once more points to a regular recurrence in the Precambrian of similar geological phenomena. As before, particular interest in constructing the Precambrian geological and geochronological scale attaches to the relatively short time periods (impulses) of activity of the most intense

internal crust-forming processes. Structural and chemical reworking of the Earth's crust which occurred during such periods of high internal activity, is amenable to dating by radiometric methods, since such periods are also the time of widespread mineral and rock transformations.

Periods of intense activity, separating major relatively quiet stages in crustal evolution, appeared to a greater or lesser extent over the entire territory examined here. We merely emphasize that bursts of activity were not strictly simultaneous over the whole territory, but took place over a certain time interval, thus blurring inter-regional geological-geochronological boundaries. The asynchroneity in the appearance of the final active processes is seen not only in regard to tectonic provinces, but also in terms of the relationship to individual structures within one province. As an example we could quote the evolution of specific greenstone belts in the Karelian province.

We consider that currently, the start of geochronological dating in the USSR begins at around 3.6–3.8 Ga ago. The oldest dates belonging to this level have been obtained from basic rocks and granitoids in the Orekhov-Pavlovgrad zone in the Ukrainian Shield (3.7–3.8 Ga) and on granite-gneiss in the Aldan Shield (3.5 Ga). Age determinations in excess of 3.4 Ga have also been obtained for mafic schist in the Omolon massif in Siberia. The base level of the geochronological scale (3.7–3.8 Ga) seems to have a quite specific geological significance. The oldest tectonic and metamorphic processes belong to this time boundary, having occurred after the deposition of supracrustal assemblages. This position is confirmed by the fact that geochronologically older rocks in certain cases constitute a basement to ancient supracrustal rocks (the Aldan terrain), and in other cases they form a constituent part of the supracrustal complex.

(2) In the USSR we break down the Precambrian into two major subdivisions, lower and upper. In the geological scale there are two corresponding periods of major crustal growth, early Precambrian and late Precambrian to Phanerozoic, separated by the 1600 ± 100 Ma geochronological boundary.

Each of these major periods has its own specific complex of processes, the evolution of which led to overall consolidation and cratonization of the Earth's crust. The early Precambrian in Eurasia concluded with the formation of a thick sialic crust with a complex heterogeneous internal structure. Major fragments of this crust have been preserved in only slightly altered condition up to the present time in the territory of ancient cratons. The subsequent major period of crustal development is usually called the geosynclinal-platform period, thereby emphasising its basic geological content: the initiation and subsequent shaping of major geostructures — ancient cratons, followed by the destruction and subsequent gradual homogenization and consolidation of the Earth's crust in intracratonic domains.

Further subdivision of major periods into smaller periods has in principle a certain amount in common with John Sutton's idea that the geological history of the continents is punctuated by major craton-forming (chelogenic) cycles, which are completed by intervening stabilization of the crust. In the early Precambrian of the USSR, we can identify the Archaean and Early Proterozoic as being periods of this type, with the geological and geochronological boundary between them being put at 2600 ± 200 Ma ago.

The Archaean is characterized by intense reworking and new growth of continental crust, the area of which was approximately the same as today's by the end of the period. In many of the regions described in the book, the Late Archaean stabilization episode is clearly documented; this is related in particular to the almost final shaping of crustal structures in granite-greenstone and granulite-greenstone terrains. The greatly altered constitution of the Earth's crust led to the appearance during this period of significant volumes of late- and post-orogenic potassic granitoids, which previously had been extremely rare. The patchy distribution of a mobile regime in the subsequent period promoted the accumulation of epicratonic proto-platform sediments in the Early Proterozoic, including the oldest redbeds (for example, the Jatulian in the Karelian province).

In the late Precambrian to Phanerozoic, the geological-geochronological boundary at 1000–1100 Ma divides the entire post-early Precambrian history into two major periods. The first of these is characterized by a predominantly destructive transformation of the Earth's crust. This is expressed primarily in the widespread appearance of mobile belts without geosynclinal precursors, related to the development of regional linear zones of intense comminution and blastomylonitization of the ancient heterogeneous basement. The maximum development of such belts took place towards the end of the period. The break-up of the basement to the East European craton occurred at this particular time, accompanied by the initiation and beginning of intense development of aulacogens. Subsequently, during the last period, the mobile belts evolved mostly as geosynclinal types, with sharply distinguished mio- and eugeosynclinal zones.

(3) Data from sedimentation, magmatism and metamorphism combined with radiometric age data also allow us to detect a more fractionated periodicity in the geological development with a period of around 400 Ma. In order to explain periodicity of this scale together with the nature of internal processes, we are beginning to rely much more heavily on results emerging from an analysis of particular sedimentation features.

Broadly speaking, in each of the periods we can detect three basic characteristic episodes of sedimentation and magmatism. In each period, sedimentation commences with the formation of volcano-sedimentary assemblages. The proportion of clastic terrigenous rocks thereafter progressively increases until a transition to flyschoidal and molassic formations is reached. The accumulation of the coarsest sedimentary and volcano-sedimentary formations is restricted to the end of a period. The changing overall geodynamic constitution of the lithosphere during the period, passing through stages of extension, compression and again extension of already consolidated crust, is reflected in the predominant change in magmatism from basic to acid and back again to basic.

We can distinguish the following periods in the Precambrian of the USSR:

I. Katarchaean, 3.8(?)–3.4 Ga
II. Early Archaean, 3.4–3.0 Ga
III. Late Archaean, 3.0–2.6 Ga
IV_1. Early Proterozoic, 2.6–2.3 Ga
IV_2. Early Proterozoic, 2.2–1.8 (–1.6?) Ga

Particular difficulties arise when attempting to identify analogous periods in the Late Proterozoic. This can be explained by the fact that on cratons, although the upper Precambrian is fairly fully represented, internal processes during this period have no sharply expressed characteristics. In younger fold belts also, isotopic dating of Precambrian processes is extremely difficult due to the widespread appearance of Phanerozoic processes. As a result, the characteristics of late Precambrian periods are described below only schematically.

We should also bear in mind that the time boundaries used in considering the periods are provisional, due to the fact that boundaries change from province to province. In describing the features of the periods, we have used material from specific regions where the relevant data are available.

I. *Katarchean (>3400 Ma)*. This period is best studied in the Aldan and Anabar Shields of the Siberian craton. The crust is characteristically differentiated in terms of the regime of internal processes, proof for which is most clearly preserved in Precambrian structural belts in the Aldan Shield. Formations of this age are found in the USSR as high-grade metamorphic (granulite and high-temperature amphibolite facies) supracrustal and intrusive complexes.

The supracrustal complexes differ in the volume and composition of volcanic and sedimentary rocks present in them. Mafic schists (metavolcanics) in those instances where they have a tholeiitic composition, are frequently characterized by high Fe, Ti and P contents and are associated with primary sedimentary rocks — greywacke-type siltstones (the Kurultin Formation in the Aldan Shield and the Sharyzhalgay Formation in the Sayan region). Subalkaline basalts and andesites (the Zverev Formation, for example) are found in association with cherts and carbonate rocks, clastic rocks being absent from this association. Sedimentary rocks of this period formed by different mechanisms, and their chemical compositions form a pattern of discrete groupings.

The abundant quartzites are frequently chemogenic sediments. They form a continuous series with ferruginous quartzite, and differ from metaterrigenous rocks in a number of petrochemical features (see Part II, Chapter 2). Aluminous rocks, which are often associated with the quartzites, are chemically equivalent to siltstones and pelites (kaolin-hydromica clays with up to 30% Al_2O_3). The association of quartzites and aluminous rocks has a Na–K trend, unlike the majority of ancient rocks. A common occurrence near the top of supracrustal assemblages belonging to the oldest period is the appearance of graphitic gneiss and marble (probably volcano-sedimentary and sedimentary rocks, initially enriched in organic matter). The role of carbonate rocks increases slightly towards the top of these successions.

The existence of sulphate and chloride evaporite formations in the oldest rocks has not been proven, although scapolite-bearing and anhydrite- and barite-bearing rocks possibly belong here.

Igneous rocks of this period are characterized by geochemical heterogeneity, although petrochemically basic volcanics and their plutonic equivalents are similar.

From their REE content, metabasites (mafic schists) are divided into at least two groups (which may be of different ages). One of these is depleted in REE and has a relatively high Eu content, the second is enriched in REE and has a relatively high amount of Ce. The distribution of heavy REE in those groups is close to chondritic. The enderbite-tonalite-plagiogranite association of this period compared to metabasites is characterized by highly fractionated REE, enrichment in light REE and depletion in heavy REE. In this respect also, the Ce/Yb ratio is high. A Eu anomaly is absent from these rocks.

Metamorphic processes reflect mineral associations of both high- and medium-gradient facies series. For some regions, it has been suggested that granulite-facies metamorphism occurred at pressures of up to 9 kbar; but high-pressure metamorphic conditions have not been established in the majority of regions in this period.

II. *Early Archaean (3400–3000 Ma).* In many Precambrian regions, this period is not well studied. Tectonically active development in this period is characterized by the formation of supracrustal assemblages, preserved today in various states of reworking. Beginning with this period, geological formations of upper crustal sections are established in Precambrian regions of the USSR — supracrustal rocks, altered at greenschist facies. Practically unaltered igneous granites, reflecting the time of completion of the processes, have an age of 3110–3000 Ma in the Dnieper block in the Ukrainian Shield. This is still the single oldest granite-greenstone terrain in the USSR. Supracrustal rocks of this period are characterized on the whole by a relatively low differentiation in clastic terrigenous rock compositions.

They are represented basically by low-alumina metapelite and polymict metapsammite with greywacke dominant over plagioclase arkose. Coarsely fragmental rocks are also found among the clastic sediments. Supracrustal high-grade metamorphic complexes, including sedimentary and volcanic rocks, are characterized by Na and K–Na trends, against moderate Fe and Mg contents (the Aldan Shield).

Volcanic and plutonic activity in the initial stage is represented by significant compositional differentiation. Rocks of the basaltic suite are characterised by a sharp dominance of tholeiitic basalts, with a slight surplus of K_2O and SiO_2. A higher K content is typical for the majority of volcanics in the Aldan Shield.

Present among rocks in the basalt-picrite suite are analogues of picrites, picrite-basalts, olivine tholeiites and tholeiitic basalts. Rocks of the andesite-basalt suite are represented by analogues of basalts, andesite-basalts and andesites. Rocks of this suite sometimes have a fairly well-marked increase in alkali content.

Metamorphism associated with the development of the supracrustal assemblages outlined above, belongs to the sillimanite-andalusite series, although metamorphic processes at 3.4 and 3.1 Ga, studied in deep crustal sections (for example, in the Volyn block in the Western Ukraine), occurred under high to very high pressure conditions.

III. *Late Archaean (3000–2600 Ma).* Rocks belonging to this period have been discovered and investigated in the overwhelming majority of regions. In tectonically active regions, the supracrustal assemblages have preserved the characteristics outlined for the previous period. It should only be added that in some regions metamorphosed oligomict sandstones and high-alumina metapelites are found against a general background of weakly differentiated clastic terrigenous rocks. Carbonate assemblages on the whole have limited distribution (usually in granulite-greenstone terrains). They are usually represented by dolomites, sometimes magnesites. The dolomites often contain rather high amounts of silicate impurities, most of which is probably tuffaceous in origin.

Banded ironstones are associated with volcanics, as in formations belonging to the older period, although they do not occur directly within volcanic assemblages, but rather in sedimentary rock members, included amongst the volcanics or replacing them laterally. Ores are represented by finely banded magnetite-siderite-chert, magnetite-siderite and carbonate-sulphide rocks.

Initial volcanics are reflected in a broad spectrum of formational types — from komatiite-basalts to dacite-rhyolites. The predominant volcanic rock varieties correspond to normal tholeiitic basalts, sometimes to olivine basalts. For the basaltic series, a clear Fenner-type of differentiation has been established. For basic volcanics overall, there is a characteristic predominance of low- and medium-alumina, high-iron and high-titanium varieties. Subalkaline basalts are rare. Compared to the general volume of volcanics in the basaltic series, the volume of ultrabasic and high-magnesium basic rocks forming part of the komatiite-basaltic formation is relatively small. Dacite-andesite and dacite-rhyolite associations occupy a special position both geologically and petrochemically. The volcanics in these associations belong to the calc-alkali series, characterized by a normal differentiation trend, whereby with decreasing Mg and Fe and increasing silica and alkalis, Na maintains a predominance over K, with a lower Ti content. In a number of cases, a clear petrochemical link has been established between intermediate and acid volcanics and basaltic magma.

In broad terms, results of investigating the structural belts of the East European craton indicate that the volcanics in granite-greenstone terrains are characterized by the greatest degree of complexity and heterogeneity of volcanic series, and those in granulite-greenstone terrains show these to the least degree.

Metamorphism of this period corresponds to various facies series conditions. The metamorphic event that concludes the development of granite-greenstone terrains belongs to the sillimanite-andalusite facies series, and for granulite-greenstone terrains the sillimanite-kyanite series. In deep sections of the latter type of terrain, moderate to very high pressure granulites have been discovered (for example, in the Dzhugdzhur-Stanovoy province).

During this period, there are as well as tectonically active regions, weakly active regions (the Dnieper block) — cratons in which internal processes are reflected only in the formation of sub-platform type igneous complexes (porphyritic potassic granites and dyke swarms).

IV_1. *Early Proterozoic (2600–2200 Ma).* After the conclusion of internal crust-forming processes at the 2600 ± 100 Ma time boundary, large tracts of the present-day East European and Siberian cratons moved on to a quasi-platform evolutionary stage, preserving features of cratonic development throughout the entire succeeding geological history. Here, periodicity of geological processes becomes less obvious and is mainly seen in a change in the nature of sedimentation. The oldest sedimentary assemblages belonging to this period in such regions are characterized by large volumes of highly alumina-differentiated metamorphosed clastic terrigenous rocks — from monomict quartzose sandstones to kaolinitic pelites. The quartzites in this period frequently have SiO_2 contents exceeding 95%. Associated with oligomict and polymict metapsammites are rocks with Na clearly in excess of K. Typical of a number of formations of this period is the widespread presence of highly graphitic metaterrigenous rocks, as well as coarse clastics — conglomerates and grits.

Upwards in the succession, the degree of “maturity” of terrigenous associations is reduced and in such cases the main petrochemical feature of terrigenous rocks — metapsammites, from material concerning the Aldan and Baltic Shields — is a high total alkali content (>6%), with Na sharply dominant over K. The majority of volcano-sedimentary rocks of this period in structural belts of ancient cratons are either practically unmetamorphosed (e.g. the Sumian in Karelia) or are zonally metamorphosed at greenschist to amphibolite facies conditions (sillimanite-andalusite facies series, e.g. in the Kola region and the basement highs around the Siberian craton margins). Igneous activity in ancient cratons is to a significant extent platform-type in character. Volcanic activity is represented by contrasting rhyolite-andesite-basalt formations. Rhyolites and rhyolite-dacites typically show a rare-earth geochemical signature, while for andesite-basalts Cu and Fe contents are higher.

IV_2. *Early Proterozoic (2200–1800 Ma).* Rocks belonging to this period have been studied in various tectonic provinces on epi-Archaean cratons (the Kola, Karelia, Olyokma and other provinces) and in mobile belts.

The most typical complexes in epi-Archaean cratons (granite-greenstone terrains) are the Jatulian on the Baltic Shield and the upper part (the Kemen Group) of the Udokan complex on the Aldan Shield. Their main characteristic is the presence of terrigenous rocks, with volcanics playing a minor role. Psammitic rocks contain monomict quartzose and oligomict sandstones. They are associated with metapelites, containing high amounts of Al_2O_3 and especially K_2O. A common occurrence in the successions is the presence of horizons containing martite heavy-mineral concentrations.

Sediments belonging to this period commonly contain rock associations that show signs of alluvial origin (composition of rhythms, similar to alluvial cyclothems in Phanerozoic sediments, frequent examples of desiccation cracks in argillites, etc.). Some rock associations preserve a characteristic red staining.

Volcanic rocks on epi-Archaean cratons (granite-greenstone terrains) are characterized by exceptional uniformity and are represented by undifferentiated plateau basalts with slightly higher alkalis, titanium and alumina.

In the other type of epi-Archaean craton (Archaean granulite-greenstone terrains), where riftogene trenches formed during this part of the Early Proterozoic, volcanic events are characterized by their enormous variety — they include basalt, andesite-basalt, picrite-basalt and andesite-dacite series. Plutonic rocks are represented mostly by a gabbro-wehrlite suite, comagmatic with the volcanics. Formations belong to high-Fe type and high-Ti subalkaline type, with higher Ni, Cr and P contents. Granitoids show a sodic trend and high TiO_2 content (up to 1%).

In mobile belts of this time interval, granites in the granite-gabbro-tonalite suite are characterized by Cu, Ti, Ni, Co, Cr, V, Sr, Ba, Sc and Zr contents higher than clark values. They are oversaturated in SiO_2 and undersaturated in Al_2O_3 and have sharply lower Mo, Nb, Pb, Sn and Y contents, with W and Be almost completely absent. Rocks belonging to the anorthosite-peridotite-gabbro-norite suite (from palingenic granites to ultramafics) have the following basic petrochemical characteristic features: (1) a sharp drop in Fe, Mg and Ca oxides, with lower rock basicity and gradual increase in alkalinity (Na_2O being greatly in excess of K_2O); (2) a gradual (three-fold) increase in Fe content in the rocks, from ultrabasic to basic; and (3) an increase relative to clark values in V, Cr, Cu, Co, and Ni and a decrease in Ti, Ag, Y, Pb and Sn.

Intrusive charnockites are compositionally stable and belong to the granitoid normal series. The main varieties in the alkali granite suite have high Cr, Ta, REE, S, Zr, Nb, Pb, Zn, Sn, Ni and Ca, and lower Sr, Ba, V, Cu, Co and Cr.

In the late Precambrian, three periods have been noted in the USSR (1800–1400, 1400–1000 and 1000–600 Ma), although their main characteristic features have not yet been clearly established. The geological boundary at 1400 Ma (1350 $\pm$ 50 Ma according to other data) has as yet no clear geological significance. We should also point out that there are certain grounds for dividing the late Precambrian into two, not three, major periods of development, each some 500 Ma long (1600–1100 and 1100–600 Ma), whereby the 1800–1600 Ma time interval is regarded as belonging to the previous Early Proterozoic period.

The following remarks are pertinent to late Precambrian processes overall.

A volcano-plutonic association with rapakivi-type granites is characteristic of Early Proterozoic mobile regions which completed their evolution in the middle of the first of the periods under discussion. Upper parts of sections in this association approximates to sub-platform formations in structural terms. Metamorphic processes are either almost absent or are locally very restricted. Granite-porphyry and dolerite dyke suites play an important role among the igneous rocks, aside from rapakivi granites.

The first geosynclinal basins are seen in younger fold belts, beginning from 1500 Ma ago, and it has been noted that it is in precisely these belts that internal processes are concentrated. From 1100–1200 Ma ago, the oldest Alpine-type paired metamorphic belts began to form in younger fold belts. The most characteristic feature of late Precambrian evolution in the younger fold belts is the appearance of the first glaucophane-schist belts.

Proto-platform sedimentary cover successions continued to form during the Riphean on epi-Archaean cratons. Monomict quartzose sandstone and potassic arkose associations, often with metapelites containing high alumina (22–25%), with titanium and especially potassium playing a subordinate role, accumulated here. In many sections in Siberia, trachydacite pyroclastics have been found. The assemblages usually have textures characteristic of continental sediments. Medium-grained fractions and well-rounded grains predominate in quartzose sandstones, which may be taken as proof of the action of aeolian processes in the formation of the source material for these rocks. Stratigraphically higher in the section, red quartzose or quartz-feldspar clastic rocks usually give way to shallow-water grey terrigenous-carbonate assemblages. The volcanics associated with these sediments typically display a wide compositional spectrum (from tholeiitic basalts to rhyodacites).

Petrochemical features of basaltic vulcanism at this time are expressed in an increase of Mg, Al, K and Ti in tholeiites and a decrease in Si, Fe and Na. Differentiation of basaltic magmas is basically characterized by a Bowen trend, with an increase in Si, alkalis and P, and a decrease in Fe and Mg contents.

Most of the clastic terrigenous sediments on the East European craton formed in aulacogens. The dominant rock types are grey and mottled shallow-water sediments, clay-carbonate and rarer biogenic carbonate sediments. Continental redbeds are restricted in time to the interval 1000–700 Ma ago.

At the end of the late Precambrian, the most important specific feature of sedimentation is the widespread distribution of tillites, products of the Lapland continental glaciation. Tillites are usually represented by thick sequences of unsorted and unbedded reddish-brown or grey mixed clay-silt-sand rocks with scattered pebbles and boulders. Large boulders sometimes have longitudinal striations and scratches, reminiscent of traces left by glacial working. In many regions, Early Vendian glacial conditions gave way to intense volcanic activity. Basic volcanics and tuffs predominate among the volcanic successions.

* * *

The material on the evolution of geological processes in the Precambrian, presented by the authors of this book as well as in other publications referred to in the literature lists makes it possible to summarise existing notions on the physico-chemical conditions of mineral transformation, rock-forming and ore-forming processes.

(1) Sedimentological research in the last couple of decades has led to the construction of models for the geochemical cycle, which assume a quasi-stationary composition of reservoirs of chemically active components of the atmosphere and hydrosphere over a substantial part of geological history.

Using systems analysis methods to reconstruct the evolution of the Earth's outer layers, based on the present accumulation of analytical data, we can draw the following conclusions about particular features of the atmosphere and hydrosphere in the Precambrian, subject to the usual constraints.

(A) The build up of substantial volumes of free oxygen in the Earth's atmosphere and the changeover of the oxygen system to a quasi-stationary state occurred in the earliest stages of geological history, probably in the Early Archaean. Evidence for the antiquity of the oxygen system is as follows:

(a) high content of organic carbon in metamorphic derivatives in lower Archaean assemblages (the rate of O_2 entering atmospheric reserves equals the rate of C_{org} entering sediments);

(b) high degree of iron oxidation in lower Archaean metaterrigenous rocks compared to associated metavolcanics (in the Aldan Shield);

(c) differentiation of isotopic compositions of sulphate and sulphide sulphur (rocks in the lower Archaean Dzheltulak Group in the Aldan Shield), showing that sulphate-reduction processes were taking place at that time.

(B) The transition of the carbonate system of the hydrosphere and atmosphere to a quasi-stationary state also happened during Archaean time, particular evidence being data on the relative stability of the carbon isotopic composition in carbonates in the sedimentary layer.

(C) Side by side with directional changes in material flow rates in the geochemical cycle, other cyclical changes took place in geological history, caused by the overall cyclicity of tectonomagmatic processes. This phenomenon was responsible for cyclical fluctuations in the basic physico-chemical parameters of the atmosphere and hydrosphere: the O_2 and CO_2 contents and the Earth's surface temperatures, etc. These cyclical fluctuations had a significant masking effect on the overall trend of directional changes in the system (in its evolution).

The periodic increase in the rate of addition of reduced elements to the atmosphere and hydrosphere in an epoch of high tectonomagmatic activity led to a reduction in the oxygen content in these spheres which in turn aided the accumulation of sediments with high organic carbon contents. For example, graphitic assemblages (Sumian, Kodarian and the Onega horizon, etc.) correspond to the 2.6 Ga and 1.9 Ga diastrophisms. During periods of tectonic quiescence with low rates of addition of volatiles to the external Earth shells, the oxygen content increased, but carbon dioxide decreased, which aided the accumulation of redbeds and the development of continental glaciations.

(2) An analysis of metamorphic processes allows us to define in general terms the evolution of the thermal state of the crust.

It is now possible to trace the evolution of the thermal state more or less certainly on the basis of an analysis of metamorphic mineral assemblages and calculations on the magnitude of the geothermal gradient, obtained by V.A. Glebovitsky.

In the oldest period, metamorphic processes, which at present have been studied only for the granulite facies in the USSR, proceeded mainly at a geotherm of 27°C/km.

Beginning from 3200 Ma, we see a marked differentiation in geothermal regimes. On the one hand, broad thermal anomalies appear, for example the Iyengr block in the Aldan Shield, where the isotherm corresponded to 32°C/km, and on the other hand, high-pressure granulites formed at this time, at a geothermal gradient of 20°C/km.

The differentiation of geothermal gradients, restored from rocks of the same metamorphic facies, corresponds to their distribution in space, i.e. it reflects a horizontal series of geodynamic conditions. However, we must take into account the fact that different geothermal gradients could also constitute vertical series (from Glebovitsky's studies), due to a regular change in gradient with depth. Thus for example, if the granite melting field is reached along the geotherm of granite-greenstone terrains (26°C/km), under conditions of maximum water fugacity in the fluid flow (it will be preserved with depth), the system will turn out to be thermostationary at the amphibolite facies level and the vertical temperature gradient will be equal to zero. Granulite-facies metamorphosed sialic masses can therefore probably appear in the crust at any level.

As a result of acid melts being removed, basification occurs at lower crustal levels in granulite-gneiss terrains (in the lower sections of granulite-greenstone terrains), which further enhances the penetration of hot mafic and ultramafic mantle melts. These melts not only increased the crustal temperature at this level, but also assisted the subhorizontal movement of crustal slabs. All this created conditions for the formation of high-pressure granulites, fragments of which were elevated to the present erosion level in Early Proterozoic mafic granulite belts (the Stanovoy suture, for example).

A fundamental change in the thermal state of the lithosphere took place at the end of the Archaean. In regions of high positive anomalies in structural belts of this age, the geothermal gradient remained at 32°C/km as before, but the magnitude of the minimum gradient went down to 16°C/km. Possibly as a result of the decrease in the geothermal background in the crust, eclogite-gneiss complexes started to form, beginning only from the Late Archaean. These complexes suffered intense alteration and partial melting in the late stages of the regional metamorphic cycle. This culminated in different metamorphic belts in the formation of Na–K or Na granites. This difference is probably related to the pre-history of the terrains. Na–K granites formed in structural belts situated on a mafic granulite basement.

The tendency to significant cooling, observed at the end of the Archaean, is especially marked in the Early Proterozoic. Areas of high thermal anomalies ceased appearing at this time, and minimum anomalies appeared mainly in linear zones, in which the geothermal gradient dropped to 14°C/km. In addition to eclogite-gneiss complexes, high-pressure/medium-temperature eclogite-schist complexes also formed at this time. The latter are absent from the basement to ancient cratons and have been identified in the basement of several younger fold belts, where they reflect the specific early Precambrian prehistory of these terrains. The maximum Early Proterozoic geotherm in the most intense positive thermal anomalies does not differ from that in earlier geological epochs. However, the contraction in areas of internal processes and the lower magnitude of the minimum geothermal gradient signifies a general reduction in crustal heat flow during the Early Proterozoic.

By the Middle Riphean, the lithosphere of the East European and Siberian cratons on the whole had cooled to almost the present-day state. At this time, Alpine-type paired metamorphic belts began to form in younger fold belts. There, marginal belts in the andalusite-sillimanite facies series give way to belts in the kyanite-sillimanite series, with frequent development of eclogite-schists complexes in axial regions.

(3) Particular features of the evolution and structure of the Precambrian crust are also reflected in the general space and time distribution patterns of useful minerals. Zonal distribution of minerals, due to tectonic differentiation of the crust during the Precambrian, has been observed in a number of regions. For example, we can identify three types of metallogenic zoning in the East European craton, related to the non-uniform geological structure of the territory. The largest and youngest (pre-Vendian) zoning manifests itself in the following manner: the western zone, which is buried under a shallow cover and includes "mega-arches" (the Baltic and Ukrainian Shields and the Voronezh massif), is intensely ore-bearing. But the eastern zone, which is deeply buried and consists predominantly of granulites and partly mafic granulites, is practically barren.

Another type of zoning has been established, from a case study of the exposed basement, which is associated with the distribution on shield areas of Precambrian tectonic provinces. A uniform succession of tectonic provinces takes place from east to west within the confines of disconnected territories (the Baltic Shield, Ukrainian Shield and Voronezh massif). Particular metallogenic features correspondingly change from east to west. Eastern (NE on the Baltic Shield) parts of the shields are occupied by granulite-greenstone terrains, which are characterised by the ore content of the oldest supracrustal assemblages (iron) and linear greenstone belts with copper-nickel and copper-molybdenum mineralization and rare-metal pegmatites. In paraschist and paragneiss belts there are high-alumina schists and rare-earth pegmatites. Proterozoic riftogene basins in these terrains contain copper-nickel, iron-titanium, skarn-pyrite and copper-epidote ore mineralization. Rare earth, apatite, magnetite and several types of rare-metal mineralization are associated with alkaline intrusive complexes in these terrains.

The subsequent granite-greenstone terrains are characterized by banded ironstones, pyrite ore mineralization and a few other rare types of raw materials — gold, platinum and asbestos. In structures framing greenstone belts there are occasional occurrences of rare-metal mineralization. Disseminated copper and copper-cobalt mineralization are found in sub-cratonic basins.

Mineralization in Early Proterozoic granite-schist terrains — the extreme western terrains of shields — are characterized by skarn, greisen and cobalt rare-metal type mineral deposits. Rapakivi granites have associated pegmatites with rock crystal deposits in geodes.

The smallest scale metallogenic zoning appears in granite domes and riftogene belts.

"Crustal metallogeny" in domes has a central symmetry, with rare metals, tin and wolfram replacing each other from centre to edge. In the exocontacts of the domes there are copper, polymetallic deposits, iron and titanium.

Rift-type structures have linear symmetry, with the zoning vector cutting across the strike of the structure. Ore mineralization in these belts is predominantly mantle in nature, and changes from copper-nickel and iron-titanium at the centre to rare metal at the periphery.

The majority of Precambrian deposits evolved over a number of metallogenic epochs, but despite this, different time epochs have their own distinguishing features. Thus, the rare-metal–rare-earth pegmatites are peculiar to the 2.7 Ga epoch. Pegmatites of all types formed in the 1.8 Ga epoch, but only rock crystal pegmatites in the 1.6 Ga epoch.

The prolonged evolution of a number of Precambrian terrains sometimes leads to a telescoped structure in ore accumulation zones and causes the combination in them of mineralizations belonging to different time levels. In such terrains the basic features of metallogenic specialization are inherited from one epoch to the next. In concluding our review of the basic features in the evolution of the Precambrian crust, we may emphasize the following. Internal crust-forming processes in subsequent periods led to changes in the internal structure of continental-type crust that had basically already formed in the Early Archaean. The three-layered structure of the crust in ancient cratons appeared at the end of the Archaean, such that a granulite-mafic layer formed at the base of the crust in some regions, and in others a mafic-granitic layer with subordinate granulites.

A reduction in internal activity in the lithosphere led to the separation of the crust into active and passive tectonic structural zones, producing contrasting geodynamic regimes. A contrast in regimes was also peculiar to structurally active zones — late Precambrian fold belts, in the structure of which massifs of early Precambrian rocks played a highly significant role.

An approach to examining the evolution of the Precambrian crust from the position not of outward continental growth, but from the position of rejuvenation of continental-type crust which had already formed in the earliest stages of Earth evolution, suggests that Phanerozoic fold belts with their particular types of magmatism, metamorphism and ore mineralization were also to a large extent predetermined by Precambrian history.

The systematic study of material on the Precambrian geology of the USSR has shown how much is known about such important problems in the Precambrian as inter-regional correlation, periodicity in geological history, the nature of marker formations, geochronological and biostratigraphic age dating, and specific features of metallogeny in similar geological formations which arose at different periods in geological history, and allows us to determine the main directions for further research in Precambrian geology and geochronology.

SUBJECT INDEX

Adazh assemblage, 168
Afghan-Tadzhik basin, 483
Afghan-Tadzhik massif, 465
Aganozero block, 87
Agul basin, 381
Aim Formation, 410
Aimchan Group, 401
Aldan-Anabar-Angara block, 239, 241
Aldan crystalline massif, 265, 422
Aldan granulite-gneiss terrain, 244, 268, 269, 307, 309, 313, 315, 318, 332, 338, 346, 347, 415
Aldan-Kiliyer zone, 296
Aldan Shield, 1, 4, 7, 235, 239, 241–243, 245, 246, 265, 268, 269, 296, 298, 309, 339–343, 345, 399, 403, 410, 414, 416, 417, 419, 420, 425, 426, 428, 475, 504, 505, 507, 509, 511, 513, 516, 517
Aldan-Stanovoy Shield, 473, 479, 481
Aldan tectonic domain, 298, 308
Aldan thrust, 295, 302, 308
Aldanides, 309
Alexandrov complex, 468
Alichur Group, 485, 486
Allarechka block, 49–51
Almetyev dome, 177
Alpine fold system, 481
Altay Mountains, 489
Altay-Salair system, 460
Amga fault, 335
Ampardakh Formation, 288, 297
Anabar-Aldan granulite-gneiss terrain, 263
Anabar block, 249, 250, 256–258
Anabar dome, 410
Anabar Shield, 1, 235, 236, 238, 239, 245–247, 249–251, 253, 257, 259, 261–263, 365, 419, 420, 423, 426, 509
Angara craton, 424
Angara-Kan anticlinorium, 456
Angara-Lena step, 411
Anyuga Formation, 448
Arkhut Formation, 372, 374, 375
Arlan Formation, 182, 184, 185
Artemov Formation, 144
Arvarench Formation, 46
Arymas Formation, 408, 409
Assyntian folding, 445
Atbashi horst, 465
Atbastakh Formation, 334
Atugey-Nuyam graben, 305

Baikal deposit, 426
Baikal Group, 396
Baikal Highlands, 235, 392, 452
Baikal-Patom belt, 235, 246, 365, 389, 391, 393, 395–397, 418, 420, 422, 425
Balaganakh greenstone belt, 313, 317
Balaganakh Group, 341, 404
Baltic Shield, 1, 3, 11, 13, 19, 20, 22, 23, 25, 63, 70, 90, 127, 164, 169, 194, 197, 199, 202, 206, 209, 418, 422, 468, 503, 504, 513, 518
Bamyantun Formation, 479
Baratil Formation, 460
Barents Sea Group, 31, 32
Batomga block, 239, 242, 243, 268, 269, 287, 339, 341, 420
Batomga complex, 339, 340
Batomga Group, 339
Batomga-Timpton granulite-greenstone terrain, 245
Batomga zone, 339
Bavlin-Baltayev graben, 182, 187
Baysuntau Group, 483
Belgorod ore zone, 175
Belkin Formation, 459
Belokorovich Formation, 137
Belomorian Group, 33, 91–93, 96, 98, 100
Belomorides, 20, 91, 92, 96, 98, 100–102, 104
Belopotok Group, 487
Belorussian-Baltic block, 165, 168
Belorussian Group, 191, 193
Belorussian-Lithuanian zone, 165

Belozero Formation, 131, 143, 155
Belozero-Orekhovo zone, 155
Belyn Formation, 189
Berestovets Formation, 192, 195
Berlyk Formation, 463
Besedino complex, 171
Bezymyannaya Formation, 44
Billyakh Formation, 263
Billyakh zone, 251, 258, 259, 261–263
Biryusa block, 381, 383, 385
Biryusa complex, 381–383
Bologoye Group, 193
Bratsk synclinorium, 134
Bryansk Group, 168
Bug Group, 134, 135
Bulgunyakhtakh greenstone belt, 315, 318
Bulgunyakhtakh Group, 336
Bulun block, 372
Burakov intrusion, 208
Burovaya Formation, 406, 407
Burpala complex, 358
Burzyanian, 470
Byrylakh syncline, 333

Cape Ludovaty Formation, 449
Central Aldan block, 416
Central Aldan cratonic zone, 308
Central anticlinorium, 455, 458
Central Kamchatka horst, 480
Central Karelian zone, 76
Central Kola segment, 29, 31, 33, 35, 56
Central Pamir zone, 486
Central Russian block, 22, 23
Central subdomain, 281, 285, 286, 301, 302
Chadobets Formation, 406
Chara-Aldan lithospheric plate, 308
Chechelev Formation, 145
Chelyuskin massif, 450
Chervurt Formation, 42
Chetlass Group, 448
Chingasan Group, 458
Chistyakov Formation, 413
Chogar block, 354
Chubov complex, 178
Chudzyavr Formation, 29
Chudzyavr zone, 35
Chuga Formation, 303, 318, 321
Chulman complex, 357
Chulman depression, 348
Chumikan complex, 339
Chumikan Group, 339, 341
Chupa Formation, 101, 103
Chuya Group, 393, 398

Dalslandian epoch, 209
Daulet Formation, 464
Davangra-Khugda graben, 305
Debengda Formation, 409
Dikimda Formation, 404
Dnepropetrovsk complex, 141, 142, 146
Dnieper block, 19, 22, 127, 129, 130, 141–145, 147, 200, 204, 510, 512
Dzheltulak complex, 358, 362
Dzheltulak Group, 516
Dzheltulak suture, 348, 358
Dzhida–Upper Vitim system, 458, 459
Dzhugdzhur block, 293
Dzhugdzhur massif, 342
Dzhugdzhur-Stanovoy domain, 293, 311

Early Stanovoy cycle, 354–357, 361
Early Stanovoy complex, 356, 357, 361
Early Stanovoy tonalites, 356
East Aldan block, 239, 269
East Chukchi complex, 477
East European platform, 418
East Karelian zone, 76
Epi-Palaeozoic platform, 445, 483
Epicratonic Ripheides, 452
Eselekh Formation, 409
Estonian block, 159, 164, 165, 168, 169
Eugeosynclinal zone, 393, 452, 462, 508

Faddeyev Formation, 450
Firsov assemblage, 469
Fyodorov Formation, 295, 302, 316, 318, 415, 417, 425

Ganalsky Group, 481
Garm block, 483
Gilyuy complex, 358
Gimola Group, 106
Gimola-Kostomuksha greenstone belt, 78
Gissara-Mangyshlak fault, 445
Glavny Range intrusion, 51, 53
Glazunov Formation, 173
Golovanevsk, 137
Golovanevsk suture, 133, 134, 140
Gonam Formation, 403

Gonam-Sutam anticlinorium, 309, 319, 320
Gorod Formation, 136
Gozhan Formation, 186, 187
Granulite belt, 48, 59, 69, 201, 517
Gremyakha-Vyrmes deposit, 204
Gulyaypol Formation, 148, 149, 158
Gunt-Alichur fault, 485
Gzhatsk aulacogen, 190

Hautavaara deposit, 200
Hetolambi Formation, 96
Hoglandian, 164, 168, 170, 173

Idzhek allochthon, 293, 295, 304, 305
Idzhek fault, 292
Idzhek Formation, 292, 297
Idzhek-Nuyam fault, 295, 306
Idzhek-Sutam deep fault, 287
Idzhek-Sutam lineament, 347
Idzhek-Sutam zone, 244, 347
Igam complex, 358
Ignikan Formation, 401
Ik Group, 178
Ileir assemblage, 459
Ilikan block, 347, 348, 350, 356, 358, 361
Ilikan complex, 356, 358, 361
Ilmenogorsk Formation, 469
Ilmozero Formation, 64
Imalyk Formation, 404
Imandra-Varzuga complex, 58, 63, 65
Imandra-Varzuga structure, 43, 63, 65
Imandra-Varzuga zone, 33, 43, 56, 60, 206
Imangrakan Formation, 350, 353
Impilahti dome, 109
Inari-Tana schists, 59
Inchukalns assemblage, 167, 168
intracratonic basin, 194, 399
Inzer Formation, 189
Irgiz Formation, 189
Irkineyeva-Chadobets aulacogen, 406
Itchilyak fragment, 332
Ivyev Group, 165
Iyengr Group, 239, 279, 295, 296, 340

Jalonvaara deposit, 81
Järvi synclinorium, 107
Jatulian, 70, 87–90, 102, 106, 107, 122, 468, 508, 513
Jotnian, 89, 194

Kabyrzin Formation, 459
Kalevian, 106, 107
Kaltasy Formation, 182, 184, 185, 189
Kama-Belsk aulacogen, 181, 185, 187–189
Kama-Belsk depression, 178, 181
Kamchadal Formation, 379
Kan Group, 386
Kandalaksha Formation, 60
Kandalaksha-Kolvitsy zone, 26, 59
Karatau-Talas-Terskey fault, 464
Karatavian, 470, 471
Karatega Group, 483
Karatega horst, 466
Karatysh Formation, 149
Karelia-Kola region, 90, 91, 208, 420
Karelian complex, 27, 45, 60, 83, 89
Karelian craton, 71, 89, 90
Karelian granite-greenstone terrain, 70, 90, 200, 504
Karelian intrusion, 86
Karelian massif, 107
Karelian megacycle, 472
Karelides, 33, 60, 63, 70, 107
Karpinsky Ridge Formation, 456
Katanga saddle, 406
Kaverino graben, 188, 189
Kaverino Group, 188, 189, 193
Kazhim Formation, 188
Kemen Group, 513
Kerpyl Group, 401
Keyvy Formation, 56, 58
Keyvy structure, 56, 58
Keyvy-Sumian complex, 66
Khapchan block, 250, 256–259, 420
Khapchan Group, 239, 250, 251, 253, 257, 258
Khapchanyr Formation, 408
Kharbey complex, 469
Khashchevat-Zavelevsk, 135
Khashchevat-Zavalevsk Formation, 135
Khaypakh Formation, 408, 409
Kholbolokh assemblage, 288, 289, 291, 297
Kholodnikan complex, 357, 362
Kholodnikan greenstone belt, 423
Kholodnin Group, 391
Khonchengra massif, 317
Khorbusuonk Group, 410
Kindin assemblage, 459
Kinel complex, 177
Kirgitey Formation, 405

Kirovograd block, 127–130, 139, 144, 145, 147, 204, 504
Kirovograd-Zhitomir complex, 138
Kirovograd-Zhitomir granitoid complex, 138
Kisloguba Formation, 46
Kizhechenok Formation, 481
Kocherov Formation, 136
Kodar-Udokan copper-bearing zone, 422
Kokar graben, 182
Kokchetau Group, 464
Kokchetau massif, 462
Kola complex, 26
Kola Group, 28, 29, 33, 34, 48, 49, 56, 61, 69, 162
Kolmozero complex, 53
Kolmozero-Voronya greenstone belt, 31, 39, 53
Komi-Permyatsk arch, 188
Konka-Belozero subzone, 155
Konka Formation, 142, 143, 155
Konka-Verkhovtsev, 131
Konka-Verkhovtsev Group, 129, 138, 142, 143, 147, 155
Konzha Formation, 459
Korva-Kolvitsy zone, 56, 59, 60, 69
Koshar-Aleksandrov Formation, 135
Kossov Formation, 480
Kostomuksha deposit, 200
Kotuykan Formation, 407
Krasnoozersk Formation, 190
Krayevo complex, 479
Kresttsy-Cherepovets belt, 23
Krivoy Rog Group, 145, 146, 155, 158
Krivoy Rog–Kremenchug zone, 200, 206
Krivoy Rog structure, 146
Kukmor assemblage, 178, 179
Kulanzhi Group, 169
Kupuri block, 356
Kurbalikita Formation, 353
Kurgovat zone, 484
Kurkijoki massif, 117
Kursk-Besedino assemblage, 171
Kursk Group, 172, 174
Kursk Magnetic Anomaly (KMA), 171, 173, 174, 208
Kursk-Voronezh block, 171, 174
Kurumkan assemblage, 285, 340
Kuylyu complex, 465
Kyurikan assemblage, 298
Kyurikan Formation, 288, 290, 291, 319
Ladoga basin, 193, 194
Ladoga belt, 20, 25, 105, 122
Ladoga complex, 106, 109, 111, 119
Ladoga Group, 106–109, 111, 114, 122, 162–164, 179, 193
Ladoga-Kalevi zone, 70
Lakhanda Group, 401, 403, 408
Lamuyka zone, 261
Lapland block, 60
Lapland complex, 60
Lapland suture, 33
Larba block, 347, 355
Larba complex, 350
Late Stanovoy complex, 358
Lehti structure, 84, 206
Leonidov Formation, 187
Livvian Supergroup, 88
Loginov Group, 190
Lopian, 26, 35, 46, 48, 66, 70, 72–74, 78, 80, 81, 84, 92, 106, 206, 207
Loukhi synclinorium, 98
Lower Kara Formation, 450
Lower Khanin graben-syncline, 334
Lower Timpton dome, 309, 319, 320, 422
Lower Turkut Formation, 410
Lubosalmi graben-syncline, 87
Luchkov Formation, 192
Lukkulajsvaara intrusion, 86
Lyubim Formation, 197
Lyudenevichi Formation, 169

Magan block, 249, 250, 257–259, 261, 262, 420
Main Sayan fault, 365, 367, 378, 381
Maloarkhangelsk anomalies, 174
Maloarkhangelsk anomaly, 171
Mama-Bodaybo Group, 394, 395
Manchurian complex, 478
Mangyshlak-Gissara fault, 464
Mariupol deposit, 204
Marmarosh massif, 487
Matveyev Formation, 478
Mediterranean belt, 443, 445, 481, 483, 486, 493
Medvedev Formation, 469
Melemken and Nimnyr structural-metallogenic zone, 426
Melemken block, 285, 286, 298, 301, 302
Metamorphic facies series, 424

Microcraton, 365, 367, 375, 381, 383, 385, 386, 397, 418, 420, 422, 450, 459, 462, 487
Middle Riphean, 33, 66, 105, 184, 188, 189, 191, 193, 194, 263, 401, 403–405, 408, 448, 452, 455, 459, 464, 465, 470–172, 174
Mikulkin Group, 449
Minyara Formation, 448
Mitrofanov Formation, 479
Mizgirev Formation, 187
Mogocha block, 348
Moldoun Formation, 404
Monchegorsk intrusion, 200
Moren complex, 455
Moscow aulacogen, 190
Mukun Group, 407
Murmansk segment, 31–34, 54, 56
Muya Group, 393, 397
Muya inlier, 391

Nagorny Formation, 356
Nakhimov Formation, 479
Namsalin Formation, 334
Naryn complex, 455
Nemchan Group, 413
Nep-Botuob dome, 399, 411
Neryuyen Formation, 401
Nikel Group, 62
Nikolayev fault, 462
Nimnyr block, 301, 302, 417
Nimnyr Formation, 279, 283
Nogin Group, 190
Nokhtuy Formation, 411
North Pamir system, 484
North Turan platform, 462
Novograd-Volyn assemblage, 136, 141
Novokrivoyrog Formation, 145
Nurlat complex, 175, 177
Nyukzha belt, 361, 362
Nyukzha block, 348, 355, 359
Nyurundukan Group, 393, 397

Oboyan Group, 171
Odessa-Belotserkov metallogenic zone, 155
Ognoy Formation, 376
Okhotsk complex, 475
Okolovo Group, 167, 168, 170, 179
Olenyok arch, 408, 409
Olkhon Group, 391, 393–397, 425
Olondo Group, 326, 331, 333
Olym complex, 174
Olyokma complex, 322, 323, 341, 342
Olyokma granite-greenstone terrain, 268, 295, 303, 308, 309, 313, 314, 321, 323, 326, 329, 332–335, 341, 505
Olyokma Group, 321, 340
Olyokma infracomplex, 308, 322, 323, 329–331
Olyokma subzone, 314
Onot complex, 374, 379
Onot deposit, 424
Onot graben, 372–374, 378, 379, 424
Onot granites, 374
Orekhov-Pavlovgrad, 148, 152
Orekhov-Pavlovgrad megasuture, 127
Orekhov-Pavlovgrad suture zone, 148, 151
Orekhov-Pavlovgrad zone, 148
Orekhovo-Pavlograd belt, 206
Orekhovo-Pavlovgrad subzone, 158
Orlov Formation, 31, 32
Orsha basin, 190–192
Orsha complex, 170
Orsha Formation, 191
Osa basin, 184, 186, 187
Osipenko Formation, 131, 149, 150
Oskol Group, 172, 174
Oslyan Group, 405, 406, 458
Osnitsk complex, 136, 138
Ostrov Formation, 413
Otradnin Group, 175, 177
Ovruch graben-syncline, 139
Ovruch Group, 128, 129, 133, 139–141
Ozeryansk Formation, 137

Pachelma acritarch assemblage, 193
Pachelma aulacogen, 185, 188, 190, 194
Pachelma graben, 189
Pachelma Group, 186, 189, 190, 192, 193
Pacific belt, 443, 445, 473, 475, 478, 481, 493
Palalambi structure, 74
Pana-Babozero fault, 63
Pana-Kuolajärvi depression, 88
Parfyonov marker horizon, 411
Patom Highlands, 239, 403, 404, 411
Pavlovo-Posad, 190
Pavlovo-Posad Group, 190
Pavyuga Formation, 448
Pechenga complex, 61, 63, 64
Pechenga graben-syncline, 61

Pechenga-Imandra-Varzuga zone, 200
Pechenga structure, 50, 51, 61, 63
Penchenga Formation, 456, 458
Perm arch, 186
Perzhan metasomatites, 140, 151
Perzhan zone, 416
Petrozavodsk Formation, 89
Pioneer Formation, 399
Pitkäranta Formation, 164
Podolsk block, 129, 130, 133, 134, 136, 137, 139, 140
Pokrov Group, 171
Polmos-Poros complex, 40
Polmos-Porosozero structure, 40
Polmostundra Formation, 40
Porokhtakh Formation, 411
Poryeguba Formation, 60
Post-Bothnian cycle, 91
Potoskuy Formation, 405
Pre-Kolyma anticlinorium, 478
Pre-Stanovoy belt, 245, 246
Pre-Uralides, 466
Pre-Verkhoyan pericratonic basin, 409
Pre-Yenisey anticlinorium, 456
Pre-Yenisey belt, 386, 388
Pripyat basin, 125
Privyatsk complex, 178
Protero-Sayan anticlinorium, 452
Pudozhgora deposit, 208
Pugachev Group, 137, 139, 141
Pugachev zone, 189
Pulonga Formation, 44
Putsaari synclinorium, 107, 109, 111
Puyvi Formation, 468
Pyaloch Formation, 44

Rebolian cycle, 100, 101
Reutov zone, 314, 317, 318
Riphean assemblage, 184, 401, 403, 407, 448, 481
Riphean megacycle, 472
Ripheides, 452, 458
Rock crystal, 199, 202, 205, 209–211, 415, 417, 422, 427, 519
Rodionov Formation, 144, 145
Russian Platform, 11, 13, 14, 17, 22, 105, 175, 179, 181, 185, 188, 189, 192–195
Ryazan graben, 188
Ryazan-Saratov aulacogen, 188
Rybachy Group, 66

Saamian complex, 26, 72
Saksagan Formation, 145, 146
Salairides, 452, 458, 473
Salmi Formation, 193, 194
Saltykov complex, 172
Sandivey Formation, 450
Sangilen massif, 453
Sariolian, 84, 87
Segozero-Vodlozero belt, 76
Seletskian cycle, 102
Seligdar deposit, 415, 425
Semyonov Formation, 406
Serafimov Group, 187
Sergiyev-Abdulin aulacogen, 187, 187
Shablyk Formation, 377
Shalozero block, 87
Sharyzhalgay block, 246, 367, 368, 371, 372, 375
Sharyzhalgay Group, 391, 393, 396
Sharyzhalgay terrain, 368, 372, 378
Shchuchin Group, 165
Shincharev Formation, 460
Shoyna Formation, 449
Shtandin Formation, 186, 187
Shuntar Formation, 405–407
Siberian platform, 460
Siberian Platform, 340, 359, 365, 381, 386, 389, 396, 415, 417, 418
Siberian polymetallic belt, 418
Smolensk block, 168, 170, 171
Somovo Group, 185, 188–191
Sortavala dome, 111
Sortavala group, 109
Sortavala Group, 106, 107
Sortavala-Pitkäranta zone, 107
South Aldan deep fault, 245, 341, 361
South Aldan fault, 361
South Muya block, 459
South Pamir fault, 485
Stanovoy complex, 346, 355, 473
Stanovoy cycle, 347
Stanovoy domain, 301
Stanovoy fold belt, 268, 341, 348
Stanovoy Group, 346, 347, 356–358
Stanovoy suture, 245, 265, 268, 292, 304, 306, 308, 311, 313–315, 320, 347, 423, 517
Stanovoy zone, 342, 348
Subplatform formations, 452
Sukhopit Group, 405–407, 458

Sumian complex, 27, 54
Sumian Supergroup, 84
Superior Province, 504, 505
Sutam block, 305–308, 313–315, 320, 347, 355
Sutam Formation, 288, 290
Sutam metamorphic belt, 348
Sutam synclinorium, 309, 319
Svecofennian cycle, 91, 103, 122, 162, 166, 209
Svecofennian Domain, 20, 23, 106, 117, 119
Svisloch Formation, 192
Syrylyr fragment, 323, 330, 331, 333
Sysert-Ilmenogorsk complex, 469

Tabuyev Group, 449
Talass anticlinorium, 464
Tallinn-Loksa complex, 163
Talya Formation, 50
Tangrak block, 301
Taratash complex, 466, 468, 470
Tarkhanov Group, 449
Tarynakh ironstone deposit, 329
Taseyeva Group, 413
Tasmiyeli basin, 335
Tasmiyeli Group, 269, 335
Tasmiyeli-Udokan stage, 331
Tatar arch, 177, 187
Tatyanov Formation, 479
Tayezhnoye deposit, 415
Tectonic domain, 6, 7, 17, 23, 127, 159, 253, 261, 269, 269, 271, 287, 298, 299, 308, 426, 503
Temryuk Formation, 148, 149
Temulyakit-Tungurcha zone, 329, 332, 333
Tepsa complex, 381–383, 385
Terekta complex, 460
Terekta Formation, 460
Terekta horst, 460
Tersk-Allarechka greenstone belt, 43, 45, 48, 53, 69
Tersk Formation, 33, 66, 105
Tersk greenstone belt, 45, 63
Tersk segment, 33, 43, 54, 59
Terskey anticlinorium, 464
Teterev Group, 129, 136, 137, 140
Teya Group, 458
Timan-Pechora fold belt, 445
Timanides, 445
Timpton Group, 292
Timpton thrust, 305, 308, 309
Tomptor Formation, 410
Trans-Imandra structure, 46, 48
Tsnin Formation, 188
Tukuringra complex, 348
Tukuringra fault, 423
Tukuringra zone, 348
Tumanshet complex, 381, 383, 385
Tungurcha complex, 330
Tungurcha greenstone belt, 329
Tungurcha Group, 331–333, 341
Tungusik Group, 405–407, 458
Turgenev Formation, 478
Turgunda massif, 460
Turkmen massif, 487
Tuva-Mongolian massif, 455
Tyrnitsk Formation, 188
Tyrynakh belt, 331

Udokan basin, 417
Udokan complex, 513
Udokan Group, 326, 331, 334, 334, 338, 342
Udzha Formation, 408
Uguy graben-syncline, 334
Uguy Group, 334
Uk Formation, 186
Ukrainian metallogenic province, 152
Ukrainian Shield, 1, 3, 11, 13, 16, 19, 20, 22, 23, 125–131, 134, 136, 137, 145, 152, 153, 155, 199–202, 204–206, 208, 209, 211, 417, 503, 504, 507, 510, 518
Ulakhan-Kurunga Formation, 407
Ulkan belt, 416, 417
Uluncha Formation, 292
Umba complex, 67
Umba Formation, 64
Undin complex, 473
Ungra complex, 315, 317, 318
Ungra-Dyos-Melemken zone, 315, 318
Unguokhtakh Formation, 408
Uniy assemblage, 179
Upper Aldan Formation, 296
Upper Anabar Group, 249, 250, 252, 253, 256–258, 261
Upper Kara Formation, 450
Upper Riphean sediments, 399, 401, 404, 405, 407, 452, 459, 475
Uralides, 466
Urals fold belt, 175
Urals-Mongolia belt, 443, 445, 488–491, 493

Uraltau anticlinorium, 472
Urkan block, 348, 358, 359
Ushakov Formation, 411
Ussuri system, 473
Uy Group, 403, 410

Valaam pluton, 116
Valday Group, 181, 195
Valyukhta Group, 404
Vanch-Yazgulem Group, 486
Varzuga Group, 26, 33
Vendian, 4, 5, 27, 66, 181, 182, 185, 188–190, 192–195, 197, 265, 342, 391, 399, 409, 413, 428, 450, 452, 455, 456, 458–460, 464, 465, 468, 470–474, 478–480, 484–486, 491, 515
Vepsian Supergroup, 89
Vihanti-Outokumpu-Ladoga zone, 206
Vilyuy aulacogen, 405
Vilyuy basin, 239, 399
Vilyuy block, 239, 241
Viskharv Formation, 484, 485
Viteguba Formation, 46
Vochelambi Formation, 45, 46
Vodlozero block, 72
Vodlozero complex, 72
Volcheozero Formation, 48
Volga-Urals block, 175, 179, 202
Volyn basin, 191, 192, 195
Volyn block, 128–130, 136–140, 205, 211, 511
Volyn Group, 192, 195
Volyn-Podolsk Domain, 20
Voron Formation, 189, 190
Voronezh Formation, 173
Voronezh massif, 22, 200, 202, 204, 206, 208, 209, 211, 518
Vorontsov Group, 173
Voronyetundra Formation, 42
Vyborg rapakivi intrusion, 164
Vygozero belt, 76
Vym Group, 448

West Aldan block, 269
West Ingulets zone, 128, 130, 142, 147
West Karelian zone, 78
West Turkmen basin, 487
Western anticlinorium, 98
White Sea suture, 59

Yankan complex, 358
Yarogu graben, 337
Yarogu Group, 318, 336, 342
Yarogu structure, 336
Yarogu thrust, 283, 303, 308
Yaryshev Formation, 197
Yauriok Formation, 59, 60
Yayelakh fragment, 329–331
Yenisey fold belt, 386, 405
Yenisey ridge, 386, 413, 418, 426
Yenisey Ridge, 452, 455, 456, 458
Yoko ore show, 398
Yona synclinorium, 98
Yudin assemblage, 388
Yudomian, 399, 403, 409–411, 413, 414
Yurmatinian, 470
Yusmastakh Formation, 407

Zapolyarny Formation, 61
Zeravshan-Turkestan zone, 465
Zerendin complex, 462
Zeya assemblage, 355, 356
Zeya block, 347, 359
Zhitkovichi Group, 169, 170
Zholdybay Formation, 464
Zhulan assemblage, 480
Zhuya fault zone, 404
Zhuya Group, 404, 405
zones of tectonothermal reworking, 251
Zverev assemblage, 340
Zverev block, 299, 313–315, 317, 320, 353, 357
Zverev complex, 341, 350, 356
Zverev Formation, 353
Zverev subzone, 297, 301
Zyuratkul fault, 468